Studies in Computational Intelligence

Volume 692

Series editor

Janusz Kacprzyk, Polish Academy of Sciences, Warsaw, Poland
e-mail: kacprzyk@ibspan.waw.pl

About this Series

The series "Studies in Computational Intelligence" (SCI) publishes new developments and advances in the various areas of computational intelligence—quickly and with a high quality. The intent is to cover the theory, applications, and design methods of computational intelligence, as embedded in the fields of engineering, computer science, physics and life sciences, as well as the methodologies behind them. The series contains monographs, lecture notes and edited volumes in computational intelligence spanning the areas of neural networks, connectionist systems, genetic algorithms, evolutionary computation, artificial intelligence, cellular automata, self-organizing systems, soft computing, fuzzy systems, and hybrid intelligent systems. Of particular value to both the contributors and the readership are the short publication timeframe and the worldwide distribution, which enable both wide and rapid dissemination of research output.

More information about this series at http://www.springer.com/series/7092

Vladik Kreinovich · Songsak Sriboonchitta
Van-Nam Huynh
Editors

Robustness in Econometrics

 Springer

Editors
Vladik Kreinovich
Department of Computer Science
University of Texas at El Paso
El Paso, TX
USA

Van-Nam Huynh
Japan Advanced Institute of Science
and Technology (JAIST)
Ishikawa
Japan

Songsak Sriboonchitta
Faculty of Economics
Chiang Mai University
Chiang Mai
Thailand

ISSN 1860-949X   ISSN 1860-9503   (electronic)
Studies in Computational Intelligence
ISBN 978-3-319-84480-0   ISBN 978-3-319-50742-2   (eBook)
DOI 10.1007/978-3-319-50742-2

Printed on acid-free paper

This Springer imprint is published by Springer Nature
The registered company is Springer International Publishing AG
The registered company address is: Gewerbestrasse 11, 6330 Cham, Switzerland

Preface

Econometrics is a branch of economics that uses mathematical (especially statistical) methods to analyze economic systems, to forecast economic and financial dynamics, and to develop strategies for achieving desirable economic performance.

Most traditional statistical techniques are based on the assumption that we know the corresponding probability distributions—or at least that we know that the corresponding distribution belongs to a known finite-parametric family of distributions. In practice, such probabilistic models are only approximate. It is therefore desirable to make sure that the conclusions made based on the statistical analysis are valid not only for the corresponding (approximate) probability distributions, but also for the actual distributions—which may be somewhat different. Statistical methods which are valid not only for the approximate model but also for all the models within its neighborhood are known robust.

There is also another important aspect of robustness: in day-by-day data, we often encounter outliers that do not reflect the long-term economic tendencies, e.g., unexpected abrupt fluctuations. It is therefore important to develop and use techniques whose results are minimally affected by such outliers.

Robust statistical techniques—and their applications to real-life economic and financial situations—are the focus of this volume.

This book also contains applications of more traditional statistical techniques to econometric problems.

We hope that this volume will help practitioners to learn how to apply new robust econometric techniques, and help researchers to further improve the existing robust techniques and to come up with new ideas on how to best assure robustness in econometrics.

We want to thank all the authors for their contributions and all anonymous referees for their thorough analysis and helpful comments.

The publication of this volume is partly supported by the Chiang Mai School of Economics (CMSE), Thailand. Our thanks to Dean Pisit Leeahtam and CMSE for providing crucial support. Our special thanks to Prof. Hung T. Nguyen for his valuable advice and constant support.

We would also like to thank Prof. Janusz Kacprzyk (Series Editor) and Dr. Thomas Ditzinger (Senior Editor, Engineering/Applied Sciences) for their support and cooperation in this publication.

El Paso, TX, USA   Vladik Kreinovich
Chiang Mai, Thailand   Songsak Sriboonchitta
Ishikawa, Japan   Van-Nam Huynh
January 2017

Contents

Part III Applications

Part I
Keynote Addresses

Robust Estimation of Heckman Model

Elvezio Ronchetti

Abstract We first review the basic ideas of robust statistics and define the main tools used to formalize the problem and to construct new robust statistical procedures. In particular we focus on the influence function, the Gâteaux derivative of a functional in direction of a point mass, which can be used both to study the local stability properties of a statistical procedure and to construct new robust procedures. In the second part we show how these principles can be used to carry out a robustness analysis in [13] model and how to construct robust versions of Heckman's two-stage estimator. These are central tools for the statistical analysis of data based on non-random samples from a population.

Keywords Approximate models · Change-of-variance function · Huber function · Influence function · M-estimator · Sample selection · Two-stage estimator

1 Introduction

In science models are used as approximations to reality and statistical and econometric models are no exception. Therefore, deviations from the assumptions of classical models are typically observed on real data. They can be related to the structural form of the model (e.g. nonlinearity), the stochastic assumptions on the observations or the errors (e.g. normality), the independence assumption, the presence of heteroskedasticity etc. From a diagnostic point of view, robust statistics investigates the impact of such deviations on procedures (estimators, confidence intervals etc.) which are constructed and justified by relying on the underlying assumptions. From a more operational and preventive point of view, it derives statistical procedures which are

E. Ronchetti (✉)
Research Center for Statistics and Geneva School of Economics and Management,
University of Geneva, Geneva, Switzerland
e-mail: Elvezio.Ronchetti@unige.ch

© Springer International Publishing AG 2017
V. Kreinovich et al. (eds.), *Robustness in Econometrics*,
Studies in Computational Intelligence 692, DOI 10.1007/978-3-319-50742-2_1

3

still reliable in the presence of such deviations. More specifically, a classical (central) model is still assumed, but it is believed to be only *approximate* in the sense that the true distribution of the data lies in a (small) neighborhood of the model. Then, estimators and tests are derived which are still reliable in the full neighborhood, i.e. their statistical characteristics (such as bias, variance, confidence intervals coverage and length, level, and power) remain stable on the full neighborhood. This has the advantage to provide insurance and protection against small but harmful distributional deviations and to still benefit from the parametric structure of the model, e.g., its computational simplicity and interpretability. From a data analytic point of view, robust procedures fit the majority of the data and identifies outliers and possible substructures for further special treatment.

In situations when we are completely uncertain about the underlying distribution, the use of nonparametric methods would be in principle preferable; see the discussion in [11, p. 7]. However, nonparametric methods are not necessarily designed to be robust in the sense mentioned above. Even the arithmetic mean which is the nonparametric estimator of the expectation (if it exists) of any underlying distribution, is very sensitive to outliers and is not robust. For a detailed discussion see [17, p. 6].

Most of the classical procedures in statistics and econometrics rely on assumptions made on the structural and the stochastic parts of the model and their optimality is justified under these assumptions. However, they are typically non-robust in the presence of even small deviations from these assumptions. Standard examples are least squares estimators in linear models and their extensions, maximum likelihood estimators and the corresponding likelihood-based tests, and GMM techniques.

In the past decades, robust procedures have been developed for large classes of models both in the statistical and econometric literature; see for instance, the books by [11, 17, 18, 22] in the statistical literature and [5, 14, 26–29] in the econometric literature. Moreover, the quantile regression approach [20] has proved fruitful as a specific way to robustify classical procedures.

Here we show how these principles and tools can be applied to robustify the statistical procedures used to analyze [13] model. Introduced in his seminal paper, it is an important model in economics and econometrics and plays a central role in the analysis of data based on non-random samples from a population.

The paper is organized as follows. In Sect. 2 we introduce the basic ideas in robust statistics, provide two important robustness tools, the influence function (IF) and the change-of-variance function (CVF), and we summarize their properties. In Sect. 3 we first formalize Heckman model and analyze the classical estimators for this model from a robustness perspective. Then, we propose a robust version of Heckman's two-stage estimator. Finally, we provide numerical evidence on the performance of the new estimator. In Sect. 4 we mention some future research directions.

2 Basic Robustness Tools

2.1 Basic Approaches

Two main approaches have proved very useful for the study of the robustness properties of statistical procedures.

The first approach to formalize the robustness problem was [16] minimax theory, where statistical estimation is viewed as a game between the Nature (which chooses a distribution in the neighborhood of the model) and the statistician (who chooses a statistical procedure in a given class). The payoff of the game is the asymptotic variance of the estimator under a given distribution in the neighborhood. The statistician achieves robustness by constructing a minimax strategy which minimizes the payoff at the worst possible distribution (*least favorable distribution*) in the neighborhood.

More specifically, in the simple normal location model, for a given sample $z_1, \ldots, z_n$ the solution to this problem is the Huber estimator, an M-estimator which is the solution T_n of the estimating equation

$$\sum_{i=1}^{n} \psi_c(r_i) = 0, \tag{1}$$

where $r_i = z_i - T_n$, $\psi_c(\cdot)$ is the so-called *Huber function* shown in Fig. 1 and defined by

$$\psi_c(r) = \begin{cases} r & |r| \leq c \\ c \, \text{sign}(r) & |r| > c. \end{cases} \tag{2}$$

Fig. 1 The Huber function

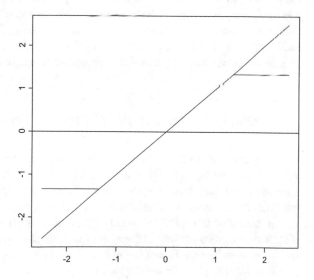

By multiplying and dividing by r_i the left-hand side of (1), the Huber estimator can be rewritten implicitly as

$$T_n = \frac{\sum_{i=1}^{n} w_c(r_i) z_i}{\sum_{i=1}^{n} w_c(r_i)},$$

where $r_i = z_i - T_n$ are the residuals and

$$w_c(r) = \psi_c(r)/r = 1 \qquad |r| \leq c$$
$$= \frac{c}{|r|} \qquad |r| > c. \qquad (3)$$

Therefore, the Huber estimator can be interpreted as an iteratively re-weighted average, where the weights are defined implicitly by (3). Outlying observations with large residuals receive smaller weights than observation belonging to the bulk of the data. Notice that the weighting scheme is not pre-imposed, but it is determined automatically by the structure of the data. The tuning constant c controls the trade-off between efficiency at the model and robustness. In particular $c = \infty$ leads to the classical efficient estimator (here the mean, which is not robust), whereas a small value of c enforces robustness at the price of some efficiency loss at the model. The Huber function is a central tool in robust statistics: "huberizing" the residuals in a given model is indeed a general way to construct new robust estimators.

The infinitesimal approach introduced by [9] in the framework of estimation, looks at the quantities of interest (for instance the bias or the variance of an estimator) as functionals of the underlying distribution and use their linear approximations to study their behavior in a neighborhood of the ideal model. A key tool is a derivative of such a functional, the influence function [10] which describes the local stability of the functional. A key point here is that this analysis can be carried out on a general functional, such as an estimator, its bias and variance, the coverage probability or the length of a confidence interval etc. In the next subsection we define this concept in a more formal way. A general review can be found in [2].

2.2 Influence Function and Change-of-Variance Function

For a given functional $T(F)$, the influence function is defined by [10] as $IF(z; T, F) = \lim_{\epsilon \to 0} [T\{(1 - \epsilon)F + \epsilon \Delta_z\} - T(F)]/\epsilon$, where Δ_z is the probability measure which puts mass 1 at the point z. The IF describes the standardized asymptotic bias on the estimator due to a small amount of contamination ϵ at the point z.

An estimator is said to be locally robust if small departures from the assumed distribution have only small effects on the estimator. Assume that we have a contaminated distribution $F_\epsilon = (1 - \epsilon)F + \epsilon G$, where G is some arbitrary distribution function. Using a [32] expansion, we can approximate the statistical functional $T(F_\epsilon)$ at the assumed distribution F as

$$T(F_\epsilon) = T(F) + \epsilon \int IF(z; T, F)dG + o(\epsilon), \tag{4}$$

and the maximum bias over the neighborhood described by F_ϵ is approximately

$$\sup_G \|T(F_\epsilon) - T(F)\| \cong \epsilon \sup_z \|IF(z; T, F)\|.$$

Therefore, a condition for local robustness is a bounded IF with respect to z, which means that if the $IF(\cdot; \cdot, \cdot)$ is unbounded then the bias of the estimator can become arbitrarily large.

The IF can also be used to compute the asymptotic variance of T, namely

$$V(T, F) = \int IF(z; T, F) \cdot IF(z; T, F)^T dF(z). \tag{5}$$

Robustness issues are not limited to the bias of an estimator, but concern also the stability of its asymptotic variance. Indeed, the latter is used to construct confidence intervals for the parameters and the influence of small deviations from the underlying distribution on their coverage probability and length should be bounded. Therefore, we investigate the behavior of the asymptotic variance of the estimator under a contaminated distribution F_ϵ and derive the CVF, which describes the influence of a small amount of contamination on the asymptotic variance of the estimator.

The CVF of an M-estimator T at a distribution F is defined by the matrix

$$CVF(z; T, F) = \left[(\partial/\partial\epsilon)V\{T, (1 - \epsilon)F + \epsilon\Delta_z\} \right]_{\epsilon=0}, \text{ for all } z \text{ where this expres-}$$

sion exists; see [7, 12]. Again a [32] expansion of $\log V(T, F_\epsilon)$ at F gives

$$V(T, F_\epsilon) \cong V(T, F) \exp\left\{ \epsilon \int \frac{CVF(z; T, F)}{V(T, F)} dG \right\}. \tag{6}$$

If the $CVF(z; T, F)$ is unbounded then the variance can become arbitrarily large or small; see [11, p. 175].

3 Application to Heckman Model

This part is a summary of [34], where all the details can be found.

3.1 Estimation of Heckman Model

Heckman model or "Tobit type-2 model" introduced by [13] plays a central role in the analysis of data based on non-random samples from a population, that is when the observations are present according to some selection rule. A simple example is the analysis of consumer expenditures, where typically the spending amount is related to the decision to spend. This type of problems arise in many research fields besides economics, including sociology, political science, finance, and many others. Let us formalize the problem by the following regression system

$$y_{1i}^* = x_{1i}^T \beta_1 + e_{1i}, \tag{7}$$
$$y_{2i}^* = x_{2i}^T \beta_2 + e_{2i}, \tag{8}$$

where the responses y_{1i}^* and y_{2i}^* are unobserved latent variables, x_{ji} is a vector of explanatory variables, β_j is a $p_j \times 1$ vector of parameters, $j = 1, 2$, and the error terms follow a bivariate normal distribution with variances $\sigma_1^2 = 1$ (to ensure identifiability), σ_2^2, and correlation ρ. Here (7) is the selection equation, defining the observability rule, and (8) is the equation of interest. The observed variables are defined by

$$y_{1i} = I(y_{1i}^* > 0), \tag{9}$$
$$y_{2i} = y_{2i}^* I(y_{1i}^* > 0), \tag{10}$$

where I is the indicator function.

Because of the presence of a selection mechanism the Ordinary Least Squares (OLS) estimator is biased and inconsistent. Heckman [13] proposed two alternative estimation procedures for this model. The first one is a Maximum Likelihood Estimator (MLE) based on the assumption of bivariate normality of the error terms. The second one is a two-stage procedure based on the equation

$$E(y_{2i}|x_{2,i}, y_{1i}^* > 0) = x_{2i}^T \beta_2 + E(e_{2i}|e_{1i} > -x_{1i}^T \beta_1),$$

which leads to the following modified regression

$$y_{2i} = x_{2i}^T \beta_2 + \beta_\lambda \lambda(x_{1i}^T \beta_1) + v_i, \tag{11}$$

where $\beta_\lambda = \rho \sigma_2$, $\lambda(x_{1i}^T \beta_1) = \phi(x_{1i}^T \beta_1)/\Phi(x_{1i}^T \beta_1)$ is the Inverse Mills Ratio (IMR), v_i is the error term with zero expectation, and $\phi(\cdot)$ denotes the density and $\Phi(\cdot)$ the cumulative distribution function of the standard normal distribution, respectively. Heckman [13] proposed then to estimate β_1 in the first stage by probit MLE and to compute estimated values of λ, and in a second stage to use OLS in (11), where the additional variable corrects for the sample selection bias.

Both estimation procedures have advantages and drawbacks, studied extensively in the literature; see e.g. the general reviews by [31, 33] and references therein. Both are sensitivite to the normality assumption of the error terms, which is often violated in practice. Another important issue is the presence of outlying observations, a well-known problem in many classical models, which is often encountered in practice. Outliers can be gross errors or legitimate extreme observations (perhaps coming from a longtailed distribution). In both cases it is of interest to identify them and this can be difficult by using only classical estimators.

Several strategies to tackle this problem can be considered. They include the development of misspecification tests for normality as in [24] to be used for diagnostic purposes, replacing the normality assumption by a more flexible class of distributions (such as the t as in [21]), the use of semiparametric [1, 23, 25], and nonparametric [6] methods. Here we show how to take a middle way between the classical strict parametric model and the fully nonparametric setup by using the ideas presented in the first part of the paper.

Although a robustification of the MLE for this model could be carried out [30], we focus here on the robustness analysis of Heckman's two-stage procedure for the model specified by (7)–(10). It is structurally simpler, has a straightforward interpretation, and leads to a robust estimator, which is computationally simple.

Let us consider a parametric sample selection model $\{F_\theta\}$, where $\theta = (\beta_1, \beta_2, \sigma_2, \rho)$ lies in Θ, a compact subset of $\mathbb{R}^{p_1} \times \mathbb{R}^{p_2} \times \mathbb{R}^+ \times [-1, 1]$. Let F_N be the empirical distribution function putting mass $1/N$ at each observation $z_i = (z_{1i}, z_{2i})$, where $z_{ji} = (x_{ji}, y_{ji})$, $j = 1, 2, i = 1, \ldots, N$, and let F be the distribution function of z_i. The Heckman's estimator can be represented as a two-stage M-estimator, with probit MLE in the first stage and OLS in the second stage. Define two statistical functionals S and T corresponding to the estimators of the first and second stage, respectively. Then, the two-stage estimator can be expressed as a solution of the empirical counterpart of the system:

$$\int \Psi_1\{(x_1, y_1); S(F)\}dF = 0, \tag{12}$$

$$\int \Psi_2[(x_2, y_2); \lambda\{(x_1, y_1); S(F)\}, T(F)]dF = 0, \tag{13}$$

where $\Psi_1(\cdot; \cdot)$ and $\Psi_2(\cdot; \cdot, \cdot)$ are the score functions of the first and second stage estimators, respectively. In the classical case $\Psi_1(\cdot; \cdot)$ and $\Psi_2(\cdot; \cdot, \cdot)$ are given by

$$\Psi_1\{(x_1, y_1); S(F)\} = \{y_1 - \Phi(x_1^T \beta_1)\} \frac{\phi(x_1^T \beta_1)}{\Phi(x_1^T \beta_1)\{1 - \Phi(x_1^T \beta_1)\}} x_1, \tag{14}$$

$$\Psi_2[(x_2, y_2); \lambda\{(x_1, y_1); S(F)\}, T(F)] = (y_2 - x_2^T \beta_2 - \lambda\beta_\lambda) \begin{pmatrix} x_2 \\ \lambda \end{pmatrix} y_1. \tag{15}$$

3.2 Influence Function and Change-of-Variance Function of Heckman's Two-Stage Estimator

For the model (7)–(10), the IF of the Heckman's two-stage estimator is

$$
IF(z; T, F) = \left\{ \int \begin{pmatrix} x_2 x_2^T & \lambda x_2 \\ \lambda x_2^T & \lambda^2 \end{pmatrix} y_1 dF \right\}^{-1} \left\{ (y_2 - x_2^T \beta_2 - \lambda \beta_\lambda) \begin{pmatrix} x_2 \\ \lambda \end{pmatrix} y_1 \right.
$$

$$
\left. + \int \begin{pmatrix} x_2 \beta_\lambda \\ \lambda \beta_\lambda \end{pmatrix} y_1 \lambda' dF \cdot IF(z; S, F) \right\},
\tag{16}
$$

where

$$
IF(z; S, F) = \left(\int \left[\frac{\phi(x_1^T \beta_1)^2 x_1 x_1^T}{\Phi(x_1^T \beta_1)\{1 - \Phi(x_1^T \beta_1)\}} \right] dF \right)^{-1} \{y_1 - \Phi(x_1^T \beta_1)\} \cdot
$$

$$
\frac{\phi(x_1^T \beta_1) x_1}{\Phi(x_1^T \beta_1)\{1 - \Phi(x_1^T \beta_1)\}}.
\tag{17}
$$

The first term of (16) is the score function of the second stage and it corresponds to the IF of a standard OLS regression. The second term contains the IF of the first stage estimator. Clearly, the first term is unbounded with respect to y_2, x_2 and λ. Notice that the function λ is unbounded from the left, and tends to zero from the right. From (17) we can see that the second term is also unbounded, which means that there is a second source of unboundedness arising from the selection stage. Therefore, the estimator fails to be locally robust. A small amount of contamination is enough for the estimator to become arbitrarily biased.

The expression of the asymptotic variance for the two-stage estimator has been derived by [13], and later corrected by [8]. It can be obtained by means of (5). Specifically, denote the components of the IF as follows:

$$
a(z) = (y_2 - x_2^T \beta_2 - \lambda \beta_\lambda) \begin{pmatrix} x_2 \\ \lambda \end{pmatrix} y_1,
$$

$$
b(z) = \left\{ \int \begin{pmatrix} x_2 \beta_\lambda \\ \lambda \beta_\lambda \end{pmatrix} y_1 \lambda' dF \right\} \cdot IF(z; S, F),
$$

$$
M(\Psi_2) = \int \begin{pmatrix} x_2 x_2^T & \lambda x_2 \\ \lambda x_2^T & \lambda^2 \end{pmatrix} y_1 dF.
\tag{18}
$$

Then the asymptotic variance of Heckman's two-stage estimator is

$$V(T, F) =$$
$$M(\Psi_2)^{-1} \int \left\{ a(z)a(z)^T + a(z)b(z)^T + b(z)a(z)^T + b(z)b(z)^T \right\} dF(z) \, M(\Psi_2)^{-1}.$$

and its CVF

$$CVF(z; S, T, F) = V - M(\Psi_2)^{-1} \left\{ \int D_H dF + \begin{pmatrix} x_2 x_2^T & \lambda x_2 \\ \lambda x_2^T & \lambda^2 \end{pmatrix} y_1 \right\} V$$

$$+ M(\Psi_2)^{-1} \int \left\{ A_H a^T + A_H b^T + B_H b^T \right\} dF M(\Psi_2)^{-1}$$

$$+ M(\Psi_2)^{-1} \int \left\{ a A_H^T + b A_H^T + b B_H^T \right\} dF M(\Psi_2)^{-1}$$

$$+ M(\Psi_2)^{-1} \left\{ a(z)a(z)^T + a(z)b(z)^T + b(z)a(z)^T + b(z)b(z)^T \right\} M(\Psi_2)^{-1}$$

$$- V \left\{ \int D_H dF + \begin{pmatrix} x_2 x_2^T & \lambda x_2 \\ \lambda x_2^T & \lambda^2 \end{pmatrix} y_1 \right\} M(\Psi_2)^{-1} \qquad (19)$$

The CVF has several sources of unboundedness. The first term of (19) contains the derivative of the score function $\Psi_2(\cdot; \cdot, \cdot)$ with respect to the parameter which is unbounded. The same holds for the last term. Finally, in the fourth term there are two components depending on the score functions of two estimators which are unbounded. Clearly, the CVF is unbounded, which means that the variance can become arbitrarily large. Taking into account that the two-stage estimator by definition is not efficient, we can observe a combined effect of inefficiency with non-robustness of the variance estimator. These problems can lead to incorrect confidence intervals.

A similar analysis can be done on the test for selection bias, i.e. the t-test of the coefficient β_λ; see [34]. Not surprisingly, it turns out that the IF of the test statistic is also unbounded and this implies non-robustness of the level and of the power of the test.

3.3 A Robust Two-Stage Estimator

From the expression of the IF in (16), it is natural to construct a robust two-stage estimator by robustifying the estimators in both stages. The idea is to obtain an estimator with bounded bias in the first stage, then compute λ, which will transfer potential leverage effects from the first stage to the second, and use the robust estimator in the second stage, which will correct for the remaining outliers.

Consider the two-stage M-estimation framework given by (12) and (13). We can obtain a robust estimator by bounding ("huberizing") both score functions. In the first stage, we construct a robust probit estimator. We use a general class of M-estimators of Mallows' type, where the influence of deviations on y_1 and x_1 are bounded separately; see [4]. The estimator is defined by the following score function:

$$\Psi_1^R\{z_1; S(F)\} = \nu(z_1; \mu)\omega_1(x_1)\mu' - \alpha(\beta_1),\tag{20}$$

where $\alpha(\beta_1) = \frac{1}{n}\sum_{i=1}^n E\{\nu(z_{1i}; \mu_i)\}\omega_1(x_{1i})\mu_i'$ is a term to ensure the unbiasedness of the estimating function with the expectation taken with respect to the conditional distribution of $y|x$, $\nu(\cdot|\cdot)$, $\omega_1(x_1)$ are weight functions defined below, $\mu_i = \mu_i(z_{1i}, \beta_1) = \Phi(x_{1i}^T\beta_1)$, and $\mu_i' = \frac{\partial}{\partial\beta_1}\mu_i$.

The weight functions are defined by

$$\nu(z_{1i}; \mu_i) = \psi_{c_1}(r_i)\frac{1}{V^{1/2}(\mu_i)},$$

where $r_i = \frac{y_{1i}-\mu_i}{V^{1/2}(\mu_i)}$ are Pearson residuals and ψ_{c_1} is the Huber function defined by (2). The tuning constant c_1 is chosen to ensure a given level of asymptotic efficiency at the normal model. A typical value is 1.345, as advocated by [4] in the GLM setting. A simple choice of the weight function $\omega_1(\cdot)$ is $\omega_{1i} = \sqrt{1 - H_{ii}}$, where H_{ii} is the ith diagonal element of the hat matrix $H = X(X^TX)^{-1}X^T$. More sophisticated choices for ω_1 are available, e.g. the inverse of the robust Mahalanobis distance based on high breakdown robust estimators of location and scatter of the x_{1i}. For the probit case we have that $\mu_i = \Phi(x_{1i}^T\beta_1)$, $V(\mu_i) = \Phi(x_{1i}^T\beta_1)\{1 - \Phi(x_{1i}^T\beta_1)\}$ and hence the quasi-likelihood estimating equations are

$$\sum_{i=1}^n \left\{\psi_{c_1}(r_i)\omega_1(x_{1i})\frac{1}{[\Phi(x_{1i}^T\beta_1)\{1 - \Phi(x_{1i}^T\beta - 1)\}]^{1/2}}\phi(x_{1i}^T\beta_1)x_{1i} - \alpha(\beta_1)\right\} = 0,$$

and $E\{\psi_{c_1}(r_i)\}$ in the $\alpha(\beta_1)$ term is equal to

$$E\left[\psi_{c_1}\left\{\frac{y_{1i} - \mu_i}{V^{1/2}(\mu_i)}\right\}\right] = \psi_{c_1}\left\{\frac{-\mu_i}{V^{1/2}(\mu_i)}\right\}\{1 - \Phi(x_{1i}^T\beta_1)\}$$

$$+\psi_{c_1}\left\{\frac{1 - \mu_i}{V^{1/2}(\mu_i)}\right\}\Phi(x_{1i}^T\beta_1).$$

This estimator has a bounded IF and ensures robustness of the first estimation stage.

To obtain a robust estimator for the equation of interest (second stage), we propose to use an M-estimator of Mallows-type with the following Ψ-function:

$$\Psi_2^R(z_2; \lambda, T) = \psi_{c_2}(y_2 - x_2^T \beta_2 - \lambda \beta_\lambda)\omega(x_2, \lambda)y_1, \tag{21}$$

where $\psi_{c_2}(\cdot)$ is the Huber function defined by (2), but with possibly a different tuning constant c_2, $\omega(\cdot)$ is a weight function on the x's, which can also be based on the robust Mahalanobis distance $d(x_2, \lambda)$, e.g.

$$\omega(x_2, \lambda) = \begin{cases} x_2 & \text{if } d(x_2, \lambda) < c_m \\ \frac{x_2 c_m}{d(x_2, \lambda)} & \text{if } d(x_2, \lambda) \geq c_m, \end{cases} \tag{22}$$

where c_m is chosen according to the level of tolerance, given that the squared Mahalanobis distance follows a χ^2-distribution. The choices of c_2, $\omega(\cdot)$, and c_m come from the results in the theory of robust linear regression; see [11]. In our numerical applications, we use $c_2 = 1.345$ and c_m corresponding to 5% critical level.

The robust estimator derived above assumes implicitly the presence of exclusion restrictions, i.e. $x_1 \neq x_2$, but often in practice the sets of explanatory variables are the same for both selection and outcome equations, i.e. $x_1 = x_2$. This issue can lead to multicollinearity because of quasi-linearity of the inverse Mills ratio in a substantial range of its support. In practice it is recommended that there should be a predictor which explains y_1 and is not significant for y_2, although it might not be easy to find such a variable. From a robustness perspective, we would like our estimator to be still reliable also when the exclusion restriction is not available. Therefore, a slight modification of the robust estimator developed above is necessary to cover this situation.

In the presence of a high degree of correlation between the explanatory variables, the Mahalanobis distance can become inflated. This leads to an increase in the number of zero weights in (22) and, hence, to an additional loss of efficiency. Given that the source of the multicollinearity is known, i.e. λ can be (approximately) expressed as a linear combination of x_2's, a simple solution is to split the design space (x_2, λ) while computing the robustness weights $\omega(x_2, \lambda)$. We split the (approximately) linearly dependent components $\left(x_2^{(1)}, \ldots, x_2^{(p_2)}, \lambda\right)$ into two independent components $\left(x_2^{(1)}, \ldots, x_2^{(q)}\right)$ and $\left(x_2^{(q+1)}, \ldots, x_2^{(p_2)}, \lambda\right)$ and compute the robustness weights $\omega\left(x_2^{(1)}, \ldots, x_2^{(q)}\right)$ and $\omega\left(x_2^{(q+1)}, \ldots, x_2^{(p_2)}, \lambda\right)$. Then, we combine these weights as $\omega(x_2, \lambda) = \omega\left(x_2^{(1)}, \ldots, x_2^{(q)}\right)\omega\left(x_2^{(q+1)}, \ldots, x_2^{(p_2)}, \lambda\right)$, which guarantee robustness in this case. As a general rule, we suggest to group λ with variable(s) having the smallest correlation with it.

Remark In the likelihood framework, the Huber function defines the most efficient estimator, subject to a bounded influence function. Therefore, in addition to its computational simplicity, it seems natural to use this function in our case. Of course, in principle other bounded score functions could be used, such as that defining the maximum likelihood estimator under a t_ν distribution; see in a more restricted setting [21].

3.4 A Simulation Study

We present some simulation results to illustrate the robustness issues and to compare different estimators. In our experiment we generate $y_{1i}^* = x_{11i} + x_{12i} + 0.75x_{13i} + e_{1i}$, where $x_{11i} \sim N(0, 1)$, $x_{12i} \sim N(-1, 0.5)$, and $x_{13i} \sim N(1, 1)$. For the equation of interest when the exclusion restriction is not available, we use the same explanatory variables $x_2 = x_1$. When it is available, the variable x_{23i} is generated independently from x_{13i}, and follows the same distribution. The errors e_1 and e_2 are from a bivariate normal distribution with expectation 0, $\sigma_1 = \sigma_2 = 1$, and $\rho = 0.7$, which gives $\beta_\lambda = 0.7$. The degree of censoring is controlled by the intercept in the selection equation, denoted by β_{10} and set to 0, which corresponds to approximately 45% of censoring. In the equation of interest the intercept is $\beta_{20} = 0$ and the slope coefficients are $\beta_2 = (1.5, 1, 0.5)^T$. We find the estimates of β_1 and β_2 without contamination and with two types of contamination. In the first scenario we contaminate x_1 when the

Table 1 Bias, Variance and MSE of the classical, robust probit, and semiparametric binary regression estimator at the model and under two types of contamination

$N = 1000$	Not contaminated			x_1 is contaminated, $y_1 = 1$			x_1 is contaminated, $y_1 = 0$		
	Bias	Var	MSE	Bias	Var	MSE	Bias	Var	MSE
Classical									
β_{10}	−0.013	0.015	0.015	−0.074	0.011	0.016	−0.194	0.012	0.050
β_{11}	0.007	0.005	0.005	−0.323	0.005	0.110	−0.291	0.005	0.089
β_{12}	0.002	0.011	0.011	−0.456	0.013	0.221	−0.419	0.013	0.189
β_{13}	0.008	0.004	0.004	−0.274	0.004	0.079	−0.247	0.004	0.065
Robust									
β_{10}	−0.011	0.016	0.016	−0.011	0.016	0.016	−0.013	0.016	0.016
β_{11}	0.008	0.006	0.006	0.006	0.006	0.006	0.003	0.006	0.006
β_{12}	0.004	0.013	0.013	0.002	0.013	0.013	−0.003	0.013	0.013
β_{13}	0.009	0.004	0.004	0.007	0.004	0.004	0.004	0.005	0.005
Klein-Spady + Probit									
β_{10}	−0.014	0.020	0.020	0.024	0.010	0.011	−0.102	0.011	0.021
β_{11}	0.005	0.005	0.005	−0.364	0.006	0.138	−0.330	0.005	0.114
β_{12}	−0.001	0.015	0.015	−0.352	0.011	0.135	−0.320	0.011	0.113
β_{13}	0.007	0.005	0.005	−0.267	0.004	0.075	−0.240	0.004	0.062

corresponding $y_1 = 0$. We generate observations from the model described above and replace them with probability $\epsilon = 0.01$ by a point mass at $(x_{11}, x_{12}, x_{13}, y_1, y_2) = (2, 0, 3, 0, 1)$. In this case we study the effect of leverage outliers when they are not transferred to the main equation. In the second scenario we contaminate x_1 when the corresponding $y_1 = 1$. We use the same type of contamination as in the first scenario, but the point mass is at $(-2, -2, -1, 1, 0)$. Notice that the contaminating point deviates by two standard deviations from the centers of distributions of the explanatory variables, which is very hard to identify using standard exploratory analysis. The sample size is $N = 1000$ and we repeat the experiment 500 times.

We compare Heckman's estimator with two robust versions derived in Sect. 3.3, i.e. the robust probit with OLS and the robust two-stage (2S) estimator. Moreover, when an exclusion restriction is available, we add the quantile regression estimator (QRE). This is the estimator proposed by [3] and extended by [15], which is a combination of a semiparametric binary regression as in [19] in the first stage and quantile regression in the second stage. It is computed using the code kindly provided by M. Huber.

In Table 1 we first consider only the first stage. We notice that the three estimators perform well at the model (with very small efficiency losses for our robust proposal

Table 2 Bias, Variance and MSE of the classical and robust two-stage estimators at the model and under two types of contamination, when the exclusion restriction is not available

$N = 1000$	Not contaminated			x_1 is contaminated, $y_1 = 1$			x_1 is contaminated, $y_1 = 0$		
	Bias	Var	MSE	Bias	Var	MSE	Bias	Var	MSE
Classical									
β_{20}	0.000	0.064	0.064	−1.872	0.445	3.947	−0.695	0.339	0.822
β_{21}	−0.004	0.016	0.016	0.615	0.044	0.422	0.197	0.046	0.085
β_{22}	0.000	0.023	0.023	0.406	0.040	0.205	0.111	0.041	0.053
β_{23}	0.001	0.011	0.011	0.411	0.022	0.191	0.129	0.025	0.041
β_λ	−0.003	0.073	0.073	2.237	0.491	5.497	0.682	0.350	0.815
Robust probit + OLS									
β_{20}	0.001	0.064	0.064	−0.520	0.051	0.322	−0.004	0.065	0.065
β_{21}	−0.005	0.016	0.016	0.229	0.012	0.064	−0.003	0.016	0.016
β_{22}	−0.001	0.024	0.024	0.217	0.021	0.068	0.001	0.024	0.024
β_{23}	−0.001	0.011	0.011	0.172	0.008	0.038	0.002	0.011	0.011
β_λ	−0.005	0.073	0.073	0.653	0.040	0.466	0.001	0.074	0.074
Robust 2S									
β_{20}	−0.027	0.080	0.081	−0.072	0.075	0.080	−0.030	0.081	0.082
β_{21}	−0.005	0.020	0.020	0.025	0.018	0.019	0.006	0.020	0.020
β_{22}	0.009	0.027	0.027	0.028	0.026	0.027	0.008	0.028	0.028
β_{23}	0.008	0.013	0.013	0.022	0.012	0.013	0.008	0.013	0.013
β_λ	0.019	0.099	0.099	0.078	0.088	0.094	0.021	0.100	0.100

and for semiparametric binary regression with respect to the classical one). However, under contamination only the robust proposal remains nearly unbiased. In Tables 2, 3 and Figs. 2, 3 and 4 we consider classical and robust two-stage estimators. The QRE is included in the table only when an exclusion restriction is available.

Again all estimators perform well without contamination. As expected, under contamination Heckman's estimator breaks down. This effect can be seen in Fig. 3 (with exclusion restriction) and Fig. 4 (without exclusion restriction). When the exclusion restriction is not available the magnitude of the bias of the classical estimator is considerably higher than that when the exclusion restriction is available. The estimators of the slope coefficients by QRE are robust. However, the estimators of the intercept and β_λ become severely biased. While it is true that often one is mostly interested

Table 3 Bias, Variance and MSE of the classical, robust two-stage estimators, and QRE at the model and under two types of contamination, when the exclusion restriction is available

$N = 1000$	Not contaminated			x_1 is contaminated, $y_1 = 1$			x_1 is contaminated, $y_1 = 0$		
	Bias	Var	MSE	Bias	Var	MSE	Bias	Var	MSE
Classical									
β_{20}	0.006	0.015	0.015	−0.638	0.062	0.469	−0.249	0.032	0.094
β_{21}	−0.000	0.005	0.005	0.153	0.009	0.032	0.036	0.006	0.008
β_{22}	0.002	0.011	0.011	0.002	0.015	0.015	−0.024	0.012	0.012
β_{23}	−0.004	0.002	0.002	−0.041	0.002	0.004	−0.004	0.002	0.002
β_λ	−0.003	0.016	0.016	1.018	0.090	1.127	0.225	0.038	0.089
Robust probit + OLS									
β_{20}	0.007	0.015	0.015	−0.150	0.017	0.039	0.005	0.015	0.015
β_{21}	−0.000	0.005	0.005	0.078	0.005	0.011	0.000	0.005	0.005
β_{22}	0.002	0.011	0.011	0.056	0.012	0.016	0.002	0.011	0.011
β_{23}	−0.004	0.002	0.002	−0.028	0.002	0.003	−0.004	0.002	0.002
β_λ	−0.003	0.017	0.016	0.368	0.012	0.148	−0.001	0.017	0.017
Robust 2S									
β_{20}	−0.001	0.017	0.017	−0.011	0.017	0.017	−0.003	0.017	0.017
β_{21}	−0.001	0.005	0.005	0.003	0.005	0.005	−0.001	0.005	0.005
β_{22}	0.004	0.012	0.012	0.008	0.012	0.012	0.004	0.012	0.012
β_{23}	−0.004	0.002	0.002	−0.005	0.002	0.002	−0.004	0.002	0.002
β_λ	−0.001	0.021	0.021	0.023	0.021	0.021	0.001	0.022	0.022
QRE									
β_{20}	0.020	0.168	0.168	−0.029	5.998	5.999	0.099	3.299	3.310
β_{21}	−0.009	0.010	0.010	0.011	0.009	0.009	−0.011	0.009	0.009
β_{22}	−0.001	0.021	0.021	0.005	0.021	0.021	0.002	0.021	0.021
β_{23}	−0.005	0.003	0.003	−0.005	0.003	0.003	−0.006	0.003	0.003
β_λ	−0.234	11.370	11.370	−0.575	338.820	339.150	−1.306	174.335	176.040

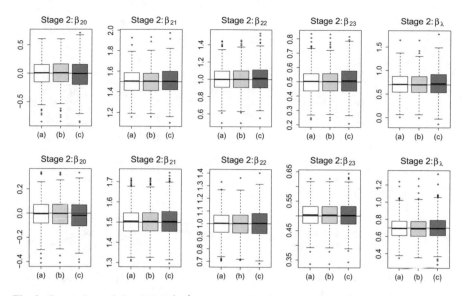

Fig. 2 Comparison of classical and robust two-stage estimators without contamination, when the exclusion restriction is not available (*top panel*), and when it is available (*bottom panel*). Case **a** corresponds to the classical estimator, **b** corresponds to robust probit with OLS in the second stage, and **c** to robust two-stage. *Horizontal lines* mark the true values of the parameters

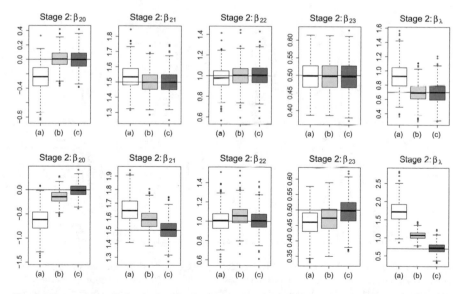

Fig. 3 Comparison of classical and robust two-stage estimators with contamination, when the exclusion restriction is available. *Top panel* corresponds to $y_1 = 0$, and *bottom panel* corresponds to $y_1 = 1$. Case **a** corresponds to the classical estimator, **b** corresponds to robust probit with OLS in the second stage, and **c** to robust two-stage. *Horizontal lines* mark the true values of the parameters

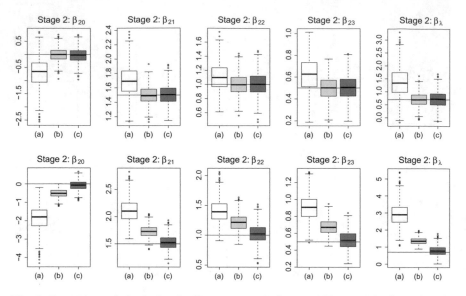

Fig. 4 Comparison of classical and robust two-stage estimators with contamination, when the exclusion restriction is not available. *Top panel* corresponds to $y_1 = 0$, and *bottom panel* corresponds to $y_1 = 1$. Case **a** corresponds to the classical estimator, **b** corresponds to robust probit with OLS in the second stage, and **c** to robust two-stage. *Horizontal lines* mark the true values of the parameters

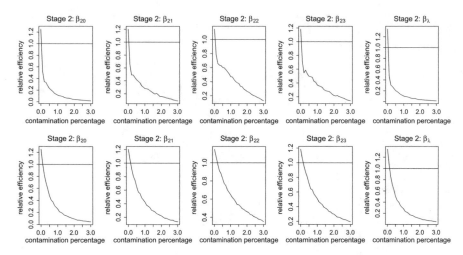

Fig. 5 Relative efficiency of the robust two-stage estimator to the classical two-stage estimator. The *top* and *bottom panels* correspond to the cases when the exclusion restriction is not available and when it is available, respectively. The x axis corresponds to the proportion of contamination ϵ, where x_1 is contaminated and the corresponding $y_1 = 1$

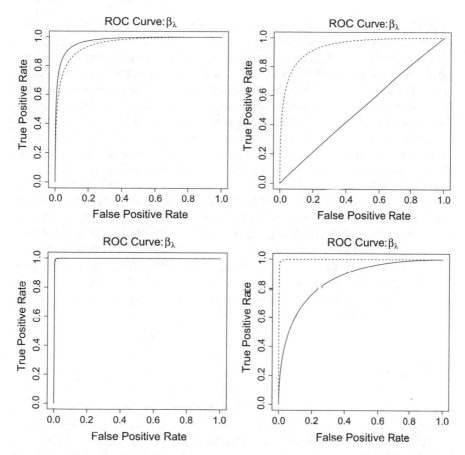

Fig. 6 ROC curves for the sample selection bias test. *Top left* plot corresponds to the case without exclusion restriction without contamination, *top right* plot corresponds to the case without exclusion restriction with contamination, *bottom left* plot corresponds to the case with exclusion restriction without contamination, and *bottom right* plot corresponds to the case with exclusion restriction with contamination. The *solid line* corresponds to the classical test and the *dashed line* corresponds to the robust test

only in the slopes, the non-robustness with respect to β_λ affects the subsequent test for selectivity. Finally, notice that the QRE of the slopes have larger MSE than those of the robust two-stage.

In the case when the outlier is not transferred to the equation of interest (Figs. 3 and 4 top panels) it is enough to use a robust probit, but when the outlier emerges in the equation of interest (Figs. 3 and 4 bottom panels), a robust estimation of the second stage is necessary. In this case the outliers influence not only both estimation stages directly, but the effect of contamination is amplified by the influence through λ. The behavior of the variances of the robust estimators remains stable, while the variance of the classical estimator is seriously affected by the contamination.

In Fig. 5 we study the efficiency of the estimators. We present the plots of the relative efficiency of the robust two-stage estimator versus the classical estimator, depending on the amount of contamination ϵ, which varies from zero to 3%. We show the figures for the case when the contaminated observations emerge at both stages ($y_{1i} = 1$). As it is expected from the theory, the robust estimator is less efficient than the classical one, when the distributional assumptions hold exactly. However, when a small amount of contamination is introduced, the situation changes completely. For instance, when the exclusion restriction is not available and the contaminated observations emerge in the second stage, the classical estimator of β_λ becomes less efficient than the robust one with only 0.1% contamination (top panel of Fig. 5). The efficiency loss of the classical estimator concerns not only the IMR parameter, but also the other explanatory variables. Note that the behavior of the variance of the robust estimator remains stable under contamination (Tables 1 and 2). Finally, in Fig. 6 we plot the Receiver Operating Characteristic (ROC) curves of the test for selectivity. The data generating process (DGP) is as discussed above except for $\rho = -0.7$, and the contamination is very mild ($\epsilon = 0.001$). We study the case when $y_1 = 1$ and put the contaminating point mass at $(-1.5, -1.75, -0.5, 1, 0)$. Without contamination the curves are close, however when the data are slightly contaminated the classical test loses its power.

4 Conclusion

We showed how to use basic ideas and tools from robust statistics to develop a framework for robust estimation and testing for sample selection models. These methods allow to deal with data deviating from the assumed model and to carry out reliable inference even in the presence of small deviations from the assumed normality model. Although we focused on the basic sample selection model, our methodology can be easily extended to more general frameworks beyond simple regression, such as the Switching Regression Model and the Simultaneous Equations Model with Selectivity.

References

1. Ahn H, Powell JL (1993) Semiparametric estimation of censored selection models with a nonparametric selection mechanism. J Econometrics 58:3–29
2. Avella Medina M, Ronchetti E (2015) Robust statistics: a selective overview and new directions. WIREs Comput Stat 7:372–393
3. Buchinsky M (1998) The dynamics of changes in the female wage distribution in the USA: a quantile regression approach. J Appl Econometrics 13:1–30
4. Cantoni E, Ronchetti E (2001) Robust inference for generalized linear models. J Am Stat Assoc 96:1022–1030
5. Cowell F, Victoria-Feser M (1996) Robustness properties of inequality measures. Econometrica 64:77–101

6. Das M, Newey WK, Vella F (2003) Nonparametric estimation of sample selection models. Rev Econ Stud 70:33–58
7. Genton MG, Rousseeuw PJ (1995) The change-of-variance function of M-estimators of scale under general contamination. J Comput Appl Math 64:69–80
8. Greene WH (1981) Sample selection bias as a specification error: comment. Econometrica 49:795–798
9. Hampel F (1968) Contribution to the theory of robust estimation. Ph.D thesis, University of California, Berkeley
10. Hampel F (1974) The influence curve and its role in robust estimation. J Am Stat Assoc 69:383–393
11. Hampel F, Ronchetti EM, Rousseeuw PJ, Stahel WA (1986) Robust statistics: the approach based on influence functions. Wiley, New York
12. Hampel F, Rousseeuw PJ, Ronchetti E (1981) The change-of-variance curve and optimal redescending M-estimators. J Am Stat Assoc 76:643–648
13. Heckman J (1979) Sample selection bias as a specification error. Econometrica 47:153–161
14. Hill JB, Prokhorov A (2016) GEL estimation for heavy-tailed GARCH models with robust empirical likelihood inference. J Econometrics 190:18–45
15. Huber M, Melly B (2015) A test of the conditional independence assumption in sample selection models. J Appl Econometrics 30:1144–1168
16. Huber PJ (1964) Robust estimation of a location parameter. Ann Math Stat 35:73–101
17. Huber PJ (1981) Robust statistics. Wiley, New York
18. Huber PJ, Ronchetti E (2009) Robust statistics, 2nd edn. Wiley, New York
19. Klein RW, Spady RH (1993) Efficient semiparametric estimator for binary response models. Econometrica 61:387–421
20. Koenker R (2005) Quantile regression. Cambridge University Press, New York
21. Marchenko YV, Genton MG (2012) A Heckman selection-t model. J Am Stat Assoc 107:304–317
22. Maronna RA, Martin GR, Yohai VJ (2006) Robust statistics: theory and methods. Wiley, Chichester
23. Marra G, Radice R (2013) Estimation of a regression spline sample selection model. Comput Stat Data Anal 61:158–173
24. Montes-Rojas GV (2011) Robust misspecification tests for the Heckman's two-step estimator. Econometric Rev 30:154–172
25. Newey WK (2009) Two-step series estimation of sample selection models. Econometrics J 12:S217–S229
26. Peracchi F (1990) Bounded-influence estimators for the Tobit model. J Econometrics 44:107–126
27. Peracchi F (1991) Robust M-tests. Econometric Theory 7:69–84
28. Ronchetti E, Trojani F (2001) Robust inference with GMM estimators. J Econometrics 101:37–69
29. Sakata S, White H (1998) High breakdown point conditional dispersion estimation with application to S & P 500 daily returns volatility. Econometrica 66:529–567
30. Salazar L (2008) A robustness study of Heckman's model. Master's thesis, University of Geneva, Switzerland
31. Vella F (1998) Estimating models with sample selection bias: a survey. J Hum Res 33:127–169
32. von Mises R (1947) On the asymptotic distribution of differentiable statistical functions. Ann Math Stat 18:309–348
33. Winship C, Mare RD (1992) Models for sample selection bias. Ann Rev Soc 18:327–350
34. Zhelonkin M, Genton MG, Ronchetti E (2016) Robust inference in sample selection models. J R Stat Soc Ser B 78:805–827

Part II
Fundamental Theory

Sequential Monte Carlo Sampling for State Space Models

Mario V. Wüthrich

Abstract The aim of these notes is to revisit sequential Monte Carlo (SMC) sampling. SMC sampling is a powerful simulation tool for solving non-linear and/or non-Gaussian state space models. We illustrate this with several examples.

1 Introduction

In these notes we revisit sequential Monte Carlo (SMC) sampling for (non-linear and non-Gaussian) state space models in discrete time. SMC sampling and non-linear particle filters were introduced in the 1990s by Gordon et al. [7] and Del Moral [3]. Meanwhile there is a vast literature on SMC sampling and there are excellent (overview) contributions such as Del Moral et al. [4, 5], Johansen and Evers [9], Doucet and Johansen [6] and Creal [2]. In fact, we have learned SMC methods from these references, in particular, from Doucet and Johansen [6]. The reason for writing these notes is that we had to prepare for a tutorial lecture on SMC sampling. For this purpose it is always advantageous to develop and implement own examples to understand and back-test the algorithms. These own examples and their implementation are probably our only real contributions here, but nevertheless they might be helpful to a wider audience who wants to get familiar with SMC sampling.

Organization. We start by giving three explicit examples of state space models in Sects. 2 and 3, the sampling algorithms are only presented later in Sect. 4. In Sect. 2 we give two examples of *linear* state space models: (1) a *Gaussian linear* state space model and (2) a *non-Gaussian linear* state space model. These models can be solved with the Kalman filter technique that is exact in the former case and that is an approximation in the latter case. In Sect. 3 we consider (3) a *non-Gaussian and non-linear* state space model. Moreover, we present the corresponding densities of all three models. Section 4 is devoted to the sampling algorithms. We start with importance sampling, then discuss sequential importance sampling (SIS) and the last

M.V. Wüthrich (✉)
RiskLab, Department of Mathematics, ETH Zurich, 8092 Zurich, Switzerland
e-mail: mario.wuethrich@math.ethz.ch

© Springer International Publishing AG 2017
V. Kreinovich et al. (eds.), *Robustness in Econometrics*,
Studies in Computational Intelligence 692, DOI 10.1007/978-3-319-50742-2_2

algorithm presented is a SMC sampling one. These algorithms are useful to solve the three models introduced above. This is demonstrated in the examples Sect. 5. In this section we also provide another practical example that corresponds to a stochastic volatility model (that is inspired by the Heston [8] model) and we describe backward smoothing of the resulting estimates.

2 Linear State Space Models

In this section we present two explicit examples of *linear* state space models: (1) a *Gaussian* one in Sect. 2.1 and (2) a *non-Gaussian* one in Sect. 2.2. Moreover, we present the Kalman filter technique that solves these models.

2.1 Gaussian Linear State Space Models and the Kalman Filter

In many situations Gaussian linear state space models are studied. These are either exact or used as an approximation to the full problem. Therefore, we start by describing Gaussian linear state space models. Such models typically consist of two processes: (i) a *transition system*, which describes the latent risk factor process, and (ii) a *measurement system*, that describes the observable process.

The following structure gives a (one-dimensional) Gaussian linear state space model:

(i) The *transition system* is described by a process $(\Theta_t)_{t \in \mathbb{N}_0}$ with $\Theta_0 = \theta_0$ and for $t \geq 1$

$$\Theta_t = a + b\Theta_{t-1} + \tau \eta_t, \tag{1}$$

for $a, b \in \mathbb{R}$, $\tau > 0$ and $(\eta_t)_{t \geq 1}$ being i.i.d. standard Gaussian distributed.

(ii) The *measurement system* is described by a process $(X_t)_{t \in \mathbb{N}}$ with for $t \geq 1$

$$X_t = \Theta_t + \sigma \varepsilon_t, \tag{2}$$

for $\sigma > 0$ and $(\varepsilon_t)_{t \geq 1}$ being i.i.d. standard Gaussian distributed and being independent of the process $(\eta_t)_{t \geq 1}$.

For given parameters $\theta_0, a, b, \tau, \sigma$ one aims at inferring the (unobservable, latent) vector $\Theta_{1:t} = (\Theta_1, \ldots, \Theta_t)$ from given observations $X_{1:t} = (X_1, \ldots, X_t)$. The technique usually used is the so-called Kalman filter [10] which can be interpreted as an exact linear credibility estimator, see Sect. 9.5 in Bühlmann and Gisler [1]. The Kalman filter provides the following algorithm:

Step 1 (anchoring). Initialize

$$\theta_{1|0} = \mathbb{E}\left[\Theta_1 | X_{1:0}\right] = a + b\theta_0 \quad \text{and} \quad \tau_{1|0}^2 = \mathrm{Var}\left(\Theta_1 \mid X_{1:0}\right) = \tau^2,$$

where the empty vector $X_{1:0}$ is assumed to generate the trivial σ-field, leading to $\mathbb{E}[\Theta_1 | X_{1:0}] = \mathbb{E}[\Theta_1]$ and $\mathrm{Var}(\Theta_1 | X_{1:0}) = \mathrm{Var}(\Theta_1)$.

Step 2 (forecasting the measurement system). At time $t \geq 1$ we obtain forecast

$$x_{t|t-1} = \mathbb{E}\left[X_t | X_{1:t-1}\right] = \mathbb{E}\left[\Theta_t | X_{1:t-1}\right] = \theta_{t|t-1},$$

and prediction variance

$$s_{t|t-1}^2 = \mathrm{Var}\left(X_t | X_{1:t-1}\right) = \mathrm{Var}\left(\Theta_t | X_{1:t-1}\right) + \sigma^2 = \tau_{t|t-1}^2 + \sigma^2.$$

One period later, for given observation X_t, we receive (observable) prediction error

$$\zeta_t = X_t - \mathbb{E}\left[X_t | X_{1:t-1}\right] = X_t - x_{t|t-1}.$$

Step 3 (Bayesian inference of the transition system). This prediction error ζ_t is used to update the transition system at time t. Using inference we obtain Bayesian estimate

$$\theta_{t|t} = \mathbb{E}\left[\Theta_t | X_{1:t}\right] = \mathbb{E}\left[\Theta_t | X_{1:t-1}\right] + K_t \zeta_t = \theta_{t|t-1} + K_t \zeta_t,$$

with the so-called Kalman gain matrix (credibility weight)

$$K_t = \mathrm{Var}\left(\Theta_t | X_{1:t-1}\right) \mathrm{Var}\left(X_t | X_{1:t-1}\right)^{-1} = \tau_{t|t-1}^2 / s_{t|t-1}^2,$$

and the variance $\tau_{t|t-1}^2$ is updated by

$$\tau_{t|t}^2 = \mathrm{Var}\left(\Theta_t | X_{1:t}\right) = (1 - K_t) \mathrm{Var}\left(\Theta_t | X_{1:t-1}\right) = (1 - K_t) \tau_{t|t-1}^2.$$

Step 4 (forecasting the transition system). For the latent risk factor we obtain forecast

$$\theta_{t+1|t} = \mathbb{E}\left[\Theta_{t+1} | X_{1:t}\right] = a + b\mathbb{E}\left[\Theta_t | X_{1:t}\right] = a + b\theta_{t|t},$$

and prediction variance

$$\tau_{t+1|t}^2 = \mathrm{Var}\left(\Theta_{t+1} | X_{1:t}\right) = b^2 \mathrm{Var}\left(\Theta_t | X_{1:t}\right) + \tau^2 = b^2 \tau_{t|t}^2 + \tau^2.$$

Remark We emphasize the distinguished meanings of $\theta_{t|t-1}$, $x_{t|t-1}$ and $\theta_{t|t}$. The former two $\theta_{t|t-1}$ and $x_{t|t-1}$ are *predictors* to forecast Θ_t and X_t based on the information $X_{1:t-1}$; the latter $\theta_{t|t}$ is an *estimator* for the latent Θ_t based on the information $X_{1:t}$. These predictors and estimators are exact and optimal (in a Bayesian way) for Gaussian innovations in linear state space models. In fact, we obtain the following exact credibility formula in Step 3 (weighted average between observation X_t and (prior) forecast $\theta_{t|t-1}$):

$$\theta_{t|t} = \mathbb{E}\left[\Theta_t | X_{1:t}\right] = \theta_{t|t-1} + K_t \zeta_t = K_t X_t + (1 - K_t)\theta_{t|t-1},$$

with credibility weight (Kalman gain matrix)

$$K_t = \frac{\tau_{t|t-1}^2}{\tau_{t|t-1}^2 + \sigma^2} = \frac{1}{1 + \sigma^2/\tau_{t|t-1}^2} \in (0, 1).$$

This credibility estimator $\theta_{t|t}$ is exact in the Gaussian linear state space model and it can be used as best linear approximation (for the quadratic loss function) for other state space models, see Chap. 9 in Bühlmann and Gisler [1].

The main question we would like to address here is: how can we optimally infer Θ_t in non-Gaussian and non-linear state space models? Before addressing this question we briefly consider a non-Gaussian linear state space model.

2.2 Non-Gaussian Linear State Space Models

We present for illustration one example of a non-Gaussian linear state space model. Therefore, we replace (1)–(2) by the following structure:

(i) The *transition system* is described by a process $(\Theta_t)_{t \in \mathbb{N}_0}$ with $\Theta_0 = \theta_0$ and for $t \geq 1$

$$\Theta_t = b\Theta_{t-1} + \eta_t, \tag{3}$$

for $b \in \mathbb{R}$ and $(\eta_t)_{t \geq 1}$ being i.i.d. gamma distributed with $\mathbb{E}[\eta_t] = a$ and $\mathrm{Var}(\eta_t) = \tau^2$.

(ii) The *measurement system* is described by a process $(X_t)_{t \in \mathbb{N}}$ with for $t \geq 1$

$$X_t = \Theta_t + \sigma \varepsilon_t, \tag{4}$$

for $\sigma > 0$ and $(\varepsilon_t)_{t \geq 1}$ being i.i.d. standard Gaussian distributed and being independent of process $(\eta_t)_{t \geq 1}$.

Observe that the measurement systems (2) and (4) are identical, conditionally given $(\Theta_t)_{t \in \mathbb{N}_0}$. The transition systems (1) and (3) differ, but not in the first two moments, that is,

$$\mathbb{E}\left[\Theta_t | \Theta_{t-1}\right] = a + b\Theta_{t-1} \quad \text{and} \quad \mathrm{Var}\left(\Theta_t | \Theta_{t-1}\right) = \tau^2. \tag{5}$$

This implies that linear credibility filtering provides the same Kalman filter results in both models, see Chap. 9 in Bühlmann and Gisler [1].

3 Non-Gaussian and Non-linear State Space Models

3.1 *Illustrative Example*

We consider the following non-Gaussian and non-linear state space model:

(i) The *transition system* is described by a process $(\Theta_t)_{t\in\mathbb{N}_0}$ with $\Theta_0 = \theta_0 = 1$ and for $t \geq 1$

$$\Theta_t = b\Theta_{t-1} + \sqrt{\Theta_{t-1}}\eta_t, \tag{6}$$

for $b > 0$ and $(\eta_t)_{t\geq 1}$ being i.i.d. gamma distributed with $\mathbb{E}[\eta_t] = a$ and $\mathrm{Var}(\eta_t) = \tau^2$.

(ii) The *measurement system* is described by a process $(X_t)_{t\in\mathbb{N}}$ with for $t \geq 1$

$$X_t = \Theta_t + \sigma\varepsilon_t, \tag{7}$$

for $\sigma > 0$ and $(\varepsilon_t)_{t\geq 1}$ being i.i.d. standard Gaussian distributed and being independent of process $(\eta_t)_{t\geq 1}$.

Note that in all three examples we consider the same measurement system (2), (4) and (7), but the three transition systems (1), (3) and (6) differ. Below, we will choose $\theta_0 = 1$ and $b = 1 - a \in (0, 1)$, see (21). These parameter choices imply for the first two *linear* state space models (1) and (3), see (5),

$$\mathbb{E}[\Theta_t] = 1 \quad\text{and}\quad \mathrm{Var}(\Theta_t) = \tau^2 \frac{1 - b^{2t}}{1 - b^2} \leq \tau^2 \frac{1}{1 - b^2}.$$

The *non-linear* model (6) is mean reverting in the following sense, assume $b = 1 - a \in (0, 1)$,

$$\mathbb{E}[\Theta_t \mid \Theta_{t-1}] = (1 - a)\Theta_{t-1} + \sqrt{\Theta_{t-1}}\,a \begin{cases} < \Theta_{t-1} & \text{if } \Theta_{t-1} > 1, \\ = \Theta_{t-1} & \text{if } \Theta_{t-1} = 1, \\ > \Theta_{t-1} & \text{if } \Theta_{t-1} < 1. \end{cases}$$

For $t = 1$ and $\Theta_0 = \theta_0 = 1$ we obtain

$$\mathbb{E}[\Theta_1 \mid \Theta_0] = (1 - a)\Theta_0 + a\sqrt{\Theta_0} = 1.$$

By induction, using the Markov property of $(\Theta_t)_{t\in\mathbb{N}_0}$, the tower property of conditional expectations and applying Jensen's inequality, we obtain for $t \geq 2$

$$\mathbb{E}\left[\Theta_t \mid \Theta_0\right] = \mathbb{E}\left[\mathbb{E}\left[\Theta_t \mid \Theta_{t-1}\right] \mid \Theta_0\right] = \mathbb{E}\left[(1-a)\,\Theta_{t-1} + a\sqrt{\Theta_{t-1}} \mid \Theta_0\right]$$

$$< (1-a)\,\mathbb{E}\left[\Theta_{t-1} \mid \Theta_0\right] + a\mathbb{E}\left[\Theta_{t-1} \mid \Theta_0\right]^{1/2} \leq 1.$$

3.2 Bayesian Inference of the Transition System

The state space models introduced above can be interpreted as Bayesian models. This is highlighted next. In general, we will use letter π to denote (conditional) densities that belong to the transition system and letter f for (conditional) densities that belong to the measurement system.

We start with the non-Gaussian and non-linear state space model (6)–(7). Choose parameters $\gamma, c > 0$ such that $\mathbb{E}[\eta_t] = \gamma/c = a$ and $\mathrm{Var}(\eta_t) = \gamma/c^2 = \tau^2$. Given the transition system $\Theta_{1:t}$, the observations $X_{1:t}$ have the following joint (product) density

$$X_{1:t}\big|_{\{\Theta_{1:t}=\theta_{1:t}\}} \sim f\left(x_{1:t} \mid \theta_{1:t}\right) = \prod_{s=1}^{t} f\left(x_s \mid \theta_s\right)$$

$$= \prod_{s=1}^{t} \frac{1}{\sqrt{2\pi}\sigma} \exp\left\{-\frac{(x_s - \theta_s)^2}{2\sigma^2}\right\}.$$

The joint (prior) density of the vector $\Theta_{1:t}$ is given by (for later purposes we indicate θ_0 in the notation)

$$\Theta_{1:t}\big|_{\theta_0} \sim \pi\left(\theta_{1:t} \mid \theta_0\right) = \prod_{s=1}^{t} \pi\left(\theta_s \mid \theta_{s-1}\right)$$

$$= \prod_{s=1}^{t} \frac{1}{\sqrt{\theta_{s-1}}} \frac{c^\gamma}{\Gamma(\gamma)} \left(\frac{\theta_s - b\theta_{s-1}}{\sqrt{\theta_{s-1}}}\right)^{\gamma-1} \exp\left\{-c\left(\frac{\theta_s - b\theta_{s-1}}{\sqrt{\theta_{s-1}}}\right)\right\}$$

$$\times 1_{\{\theta_s \geq b\theta_{s-1}\}}.$$

This implies that the posterior density of $\Theta_{1:t}$, conditionally given $(X_{1:t}, \theta_0)$, satisfies

$$\pi\left(\theta_{1:t} \mid X_{1:t}, \theta_0\right) \propto f\left(X_{1:t} \mid \theta_{1:t}\right) \pi\left(\theta_{1:t} \mid \theta_0\right) = \prod_{s=1}^{t} f\left(X_s \mid \theta_s\right) \pi\left(\theta_s \mid \theta_{s-1}\right)$$

$$\propto \prod_{s=1}^{t} \frac{1}{\sqrt{\theta_{s-1}}} \left(\frac{\theta_s - b\theta_{s-1}}{\sqrt{\theta_{s-1}}}\right)^{\gamma-1} \exp\left\{-\frac{(X_s - \theta_s)^2}{2\sigma^2} - c\left(\frac{\theta_s - b\theta_{s-1}}{\sqrt{\theta_{s-1}}}\right)\right\}$$

$$\times 1_{\{\theta_s \geq b\theta_{s-1}\}}. \tag{8}$$

Thus, we can determine the posterior density $\pi(\theta_{1:t}|X_{1:t}, \theta_0)$ in model (6)–(7) up to the normalizing constant, but we do not immediately recognize that it comes from a well-understood (multivariate) distribution function. Therefore, we determine the posterior distribution numerically.

For completeness we also provide the posterior distribution in the linear state space models of Sect. 2. In model (1)–(2) it is derived as follows. Given the vector $\Theta_{1:t}$ the observations $X_{1:t}$, have the following joint (product) density

$$X_{1:t}|_{\{\Theta_{1:t}=\theta_{1:t}\}} \sim f(x_{1:t}|\theta_{1:t}) = \prod_{s=1}^{t} \frac{1}{\sqrt{2\pi}\sigma} \exp\left\{-\frac{(x_s - \theta_s)^2}{2\sigma^2}\right\}.$$

The joint (prior) density of the vector $\Theta_{1:t}$ is given by

$$\Theta_{1:t}|_{\theta_0} \sim \pi(\theta_{1:t}|\theta_0) = \prod_{s=1}^{t} \frac{1}{\sqrt{2\pi}\tau} \exp\left\{-\frac{(\theta_s - a - b\theta_{s-1})^2}{2\tau^2}\right\}.$$

This implies that the posterior density of $\Theta_{1:t}$, conditionally given $(X_{1:t}, \theta_0)$, satisfies

$$\pi(\theta_{1:t}|X_{1:t}, \theta_0) \propto f(X_{1:t}|\theta_{1:t})\pi(\theta_{1:t}|\theta_0) = \prod_{s=1}^{t} f(X_s|\theta_s)\pi(\theta_s|\theta_{s-1})$$

$$\propto \exp\left\{-\sum_{s=1}^{t} \frac{(X_s - \theta_s)^2}{2\sigma^2} + \frac{(\theta_s - a - b\theta_{s-1})^2}{2\tau^2}\right\}. \tag{9}$$

From this we see that the posterior of $\Theta_{1:t}$, given $(X_{1:t}, \theta_0)$, is a multivariate Gaussian distribution (with known parameters) and any problem can directly be solved from this knowledge. Observe that this slightly differs from the Kalman filter of Sect. 2. In the Kalman filter we were estimating (the next) Θ_t based on observations $X_{1:t}$, which provided Bayesian estimate $\theta_{t|t}$. The full posterior $\pi(\theta_{1:t}|X_{1:t}, \theta_0)$ now also allows us for backward smoothing, that is, we can study the Bayesian estimator of Θ_s for any earlier time point $s = 1, \ldots, t$ given by

$$\theta_{s|t} = \mathbb{E}[\Theta_s|X_{1:t}].$$

The posterior distribution in model (3)–(4) is given by

$$\pi(\theta_{1:t}|X_{1:t}, \theta_0) \propto f(X_{1:t}|\theta_{1:t})\pi(\theta_{1:t}|\theta_0) \tag{10}$$

$$\propto \prod_{s=1}^{t} (\theta_s - b\theta_{s-1})^{\gamma-1} \exp\left\{-\frac{(X_s - \theta_s)^2}{2\sigma^2} - c(\theta_s - b\theta_{s-1})\right\} 1_{\{\theta_s \geq b\theta_{s-1}\}}.$$

The aim in the next section is to describe algorithms that allow us to simulate directly from the posterior densities (8)–(10), respectively.

4 Sequential Monte Carlo Sampling

In this section we follow Sect. 3 of Doucet and Johansen [6]. Throughout we assume that all terms considered are well-defined, for instance, concerning integrability. The aim is to sample from the (posterior) densities $\pi(\theta_{1:t}|X_{1:t}, \theta_0)$ that are known up to the normalizing constants, that is,

$$\pi\left(\theta_{1:t}|X_{1:t}, \theta_0\right) \propto f\left(X_{1:t}|\theta_{1:t}\right)\pi\left(\theta_{1:t}|\theta_0\right) =: \gamma_t\left(\theta_{1:t}\right),$$

where the last identity defines γ_t which describes the functional form of the (posterior) density up to a normalizing constant Z_t that is given by

$$Z_t = \int \gamma_t\left(\theta_{1:t}\right) d\theta_{1:t}.$$

In particular, this implies that we have (posterior) density

$$\pi\left(\theta_{1:t}|X_{1:t}, \theta_0\right) = Z_t^{-1}\gamma_t(\theta_{1:t}).$$

We remark that the following algorithms are quite general. The careful reader will notice that they require much less structure than the three models introduced above possess.

4.1 Importance Sampling

A general way to obtain samples from a density $\pi(\theta_{1:t}|X_{1:t}, \theta_0)$ that is only known up to a normalizing constant is to apply importance sampling. Assume h is a well-behaved measurable function and we aim at calculating the (posterior) mean

$$\mathbb{E}\left[h(\Theta_{1:t})|X_{1:t}\right] = \int h(\theta_{1:t})\pi\left(\theta_{1:t}|X_{1:t}, \theta_0\right) d\theta_{1:t} = \frac{\int h(\theta_{1:t})\gamma_t\left(\theta_{1:t}\right) d\theta_{1:t}}{\int \gamma_t\left(\theta_{1:t}\right) d\theta_{1:t}}.$$

For importance sampling we choose an importance density q_t that has at least the same support as γ_t and from which we can (easily) sample. The latter is important because otherwise the problem will not be solved. Using this importance density q_t and assuming that $\widetilde{\Theta}_{1:t} \sim q_t$ we can rewrite the above (posterior) mean as follows

$$\mathbb{E}\left[h(\Theta_{1:t})|X_{1:t}\right] = \frac{\int h(\theta_{1:t})\frac{\gamma_t(\theta_{1:t})}{q_t(\theta_{1:t})}q_t(\theta_{1:t})d\theta_{1:t}}{\int \frac{\gamma_t(\theta_{1:t})}{q_t(\theta_{1:t})}q_t(\theta_{1:t})d\theta_{1:t}} = \frac{\mathbb{E}\left[h(\widetilde{\Theta}_{1:t})w_t(\widetilde{\Theta}_{1:t})|X_{1:t}\right]}{\mathbb{E}\left[w_t(\widetilde{\Theta}_{1:t})|X_{1:t}\right]},$$

$$(11)$$

where we have defined the unnormalized importance weights

$$w_t(\theta_{1:t}) = \frac{\gamma_t(\theta_{1:t})}{q_t(\theta_{1:t})}.$$

Remarks

- Identity (11) says that we can sample from a tractable density $\widetilde{\Theta}_{1:t} \sim q_t$. To obtain samples from γ_t we simply need to re-weight these samples using the importance weights w_t. Note that this requires that $\mathrm{supp}(\gamma_t) \subseteq \mathrm{supp}(q_t)$.
- Efficient algorithms to evaluate (11) numerically for arbitrary functions h will consider importance densities q_t such that w_t has a small variance. This leads to a fast convergence in the normalizing constant Z_t (which is the denominator of (11)). Ideally, one also wants to have fast convergence in the numerator of (11). However, since this is not possible for arbitrary function h, one only focuses on the importance weights for the normalizing constant.
- Note that we condition on $\sigma\{X_{1:t}, \theta_0\}$ in (11) because (strictly speaking) the importance weights w_t depend on these observations (if we are aiming at calculating the posterior distributions). Moreover, also the choice of $q_t(\cdot) = q_t(\cdot | X_{1:t}, \theta_0)$ may depend on these observations.

We now evaluate (11) empirically. Choose $I \in \mathbb{N}$ independent samples $\widetilde{\Theta}_{1:t}^{(i)} \sim q_t$, $i = 1, \ldots, I$. We obtain empirical estimate

$$\widehat{\mathbb{E}}^{(I)}[h(\Theta_{1:t}) | X_{1:t}] = \frac{\frac{1}{I}\sum_{i=1}^{I} h(\widetilde{\Theta}_{1:t}^{(i)}) w_t(\widetilde{\Theta}_{1:t}^{(i)})}{\frac{1}{I}\sum_{i=1}^{I} w_t(\widetilde{\Theta}_{1:t}^{(i)})} \qquad (12)$$

$$= \sum_{i=1}^{I} h(\widetilde{\Theta}_{1:t}^{(i)}) \frac{w_t(\widetilde{\Theta}_{1:t}^{(i)})}{\sum_{j=1}^{I} w_t(\widetilde{\Theta}_{1:t}^{(j)})}.$$

This importance sampling algorithm proposes to evaluate the function h under the empirical (discrete) distribution

$$\widehat{\pi}^{(I)}(\theta_{1:t} | X_{1:t}, \theta_0) = \sum_{i=1}^{I} W_t(\widetilde{\Theta}_{1:t}^{(i)}) \, \delta_{\widetilde{\Theta}_{1:t}^{(i)}}(\theta_{1:t}),$$

with normalized importance weights

$$W_t(\widetilde{\Theta}_{1:t}^{(i)}) = \frac{w_t(\widetilde{\Theta}_{1:t}^{(i)})}{\sum_{j=1}^{I} w_t(\widetilde{\Theta}_{1:t}^{(j)})} = \frac{\gamma_t\left(\widetilde{\Theta}_{1:t}^{(i)}\right)/q_t(\widetilde{\Theta}_{1:t}^{(i)})}{\sum_{j=1}^{I} \gamma_t\left(\widetilde{\Theta}_{1:t}^{(j)}\right)/q_t(\widetilde{\Theta}_{1:t}^{(j)})}.$$

Estimate (12) is consistent satisfying the central limit theorem with asymptotic variance as $I \to \infty$, see (27) in Doucet and Johansen [6],

$$\frac{1}{I} \int \frac{\pi(\theta_{1:t} | X_{1:t}, \theta_0)^2}{q_t(\theta_{1:t})} (h(\theta_{1:t}) - \mathbb{E}[h(\Theta_{1:t}) | X_{1:t}])^2 \, d\theta_{1:t}.$$

Moreover, as mentioned in Doucet and Johansen [6], the asymptotic bias of this empirical estimate is of order $\mathcal{O}(1/I)$, the asymptotic variance of order $\mathcal{O}(1/I)$, and the mean squared error is asymptotically dominated by the variance term, see also Theorem 2.2 in Johansen and Evers [9].

Note that so far we have not used the sequential product structure (8)–(10) of our problems. This structure will help to control the computational complexity, this we are going to explore next.

4.2 Sequential Importance Sampling

Observe that the evaluation of the importance weights w_t can be complex if we do not benefit from the additional Markovian structure of problems (8)–(10). This can be achieved by considering a product structure for the importance density q_t, i.e. we choose (by a slight abuse of notation)

$$q_t\,(\theta_{1:t}) = q_t\,(\theta_{1:t}|\,X_{1:t},\theta_0) = \prod_{s=1}^{t} q_s\,(\theta_s|\,X_{1:s},\theta_{0:s-1})\,. \tag{13}$$

These (conditional) importance densities $q_s\,(\theta_s|X_{1:s},\theta_{0:s-1})$ may also depend on $X_{1:s}$, often this is not highlighted in the notation. In the sequel we drop "conditional" in the terminology because notation already indicates this. Using (8) we calculate the importance weights recursively

$$\begin{aligned} w_t(\theta_{1:t}) &= \prod_{s=1}^{t} \frac{f\,(X_s|\theta_s)\,\pi\,(\theta_s|\theta_{s-1})}{q_s\,(\theta_s|X_{1:s},\theta_{0:s-1})} \\ &= w_{t-1}(\theta_{1:t-1})\,\frac{f\,(X_t|\theta_t)\,\pi\,(\theta_t|\theta_{t-1})}{q_t\,(\theta_t|X_{1:t},\theta_{0:t-1})}, \end{aligned}$$

with initialization $w_0(\theta_{1:0}) = 1$. This allows us to define the incremental importance weights

$$\alpha_t(\theta_{1:t}) = \frac{f\,(X_t|\,\theta_t)\,\pi\,(\theta_t|\,\theta_{t-1})}{q_t\,(\theta_t|\,X_{1:t},\theta_{0:t-1})},$$

and then the unnormalized importance weights under (8)–(10) and (13) are written as

$$w_t(\theta_{1:t}) = w_{t-1}(\theta_{1:t-1})\,\alpha_t(\theta_{1:t}) = \prod_{s=1}^{t} \alpha_s(\theta_{1:s}).$$

Here, we see the sequential nature of the algorithm!

In view of the Markovian structure in (8)–(10) it makes sense to also choose a Markovian structure in (13) because the numerator of the incremental importance weights $\alpha_t(\theta_{1:t})$ only depends on X_t, θ_t and θ_{t-1}, thus we choose importance density

$$q_t\ (\theta_t|\,X_t,\theta_{t-1}) \equiv q_t\ (\theta_t|\,X_{1:t},\theta_{0:t-1})\,. \tag{14}$$

This provides incremental importance weights

$$\alpha_t(\theta_{t-1:t}) = \alpha_t(\theta_{1:t}) = \frac{f\ (X_t|\,\theta_t)\,\pi\ (\theta_t|\,\theta_{t-1})}{q_t\ (\theta_t|\,X_t,\theta_{t-1})}\,. \tag{15}$$

We arrive at the following algorithm under (8)–(10), (13) and (14).

Sequential importance sampling (SIS) algorithm.
▷ Set $\widetilde{\Theta}_0^{(i)} = \theta_0$ and $w_0(\widetilde{\Theta}_{1:0}^{(i)}) = 1$ for $i = 1,\dots,I$.
▷ Repeat for $s = 1,\dots,t$:

- repeat for $i = 1,\dots,I$:
  - sample $\widetilde{\Theta}_s^{(i)} \sim q_s\left(\cdot \,\middle|\, X_s, \widetilde{\Theta}_{s-1}^{(i)}\right)$;
  - calculate the importance weights

$$\alpha_s(\widetilde{\Theta}_{s-1:s}^{(i)}) = \frac{f\left(X_s \,\middle|\, \widetilde{\Theta}_s^{(i)}\right)\,\pi\left(\widetilde{\Theta}_s^{(i)} \,\middle|\, \widetilde{\Theta}_{s-1}^{(i)}\right)}{q_s\left(\widetilde{\Theta}_s^{(i)} \,\middle|\, X_s, \widetilde{\Theta}_{s-1}^{(i)}\right)},$$

$$w_s(\widetilde{\Theta}_{1:s}^{(i)}) = w_{s-1}(\widetilde{\Theta}_{1:s-1}^{(i)})\,\alpha_s(\widetilde{\Theta}_{s-1:s}^{(i)});$$

- calculate for $i = 1,\dots,I$ the normalized importance weights

$$W_s(\widetilde{\Theta}_{1:s}^{(i)}) = \frac{w_s(\widetilde{\Theta}_{1:s}^{(i)})}{\sum_{j=1}^{I} w_s(\widetilde{\Theta}_{1:s}^{(j)})} \propto w_s(\widetilde{\Theta}_{1:s}^{(i)}).$$

This SIS algorithm provides empirical distributions for any $s = 1,\dots,t$

$$\widehat{\pi}^{(I)}\ (\theta_{1:s}|\,X_{1:s},\theta_0) = \sum_{i=1}^{I} W_s(\widetilde{\Theta}_{1:s}^{(i)})\,\delta_{\widetilde{\Theta}_{1:s}^{(i)}}\ (\theta_{1:s})\,. \tag{16}$$

If we are only interested in $s = t$ we would not need to calculate $W_s(\widetilde{\Theta}_{1:s}^{(i)})$ for $s < t$ in the SIS algorithm, however this is going to be important in the refinement of the SIS algorithm. Note that any marginal $\widehat{\pi}^{(I)}(\theta_s|X_{1:t},\theta_0)$ can easily be obtained, for $s = t$ this refers to Step 3 in the Kalman filter, for $s < t$ this refers to backward smoothing.

A main deficiency of the SIS algorithm is that the variance increases rapidly in the number of periods t considered, and thus a large number I of simulations is needed in order to get accurate results, Doucet and Johansen [6] provide an example in Sect. 3.3. Therefore, variance reduction techniques should be applied and the SIS algorithm needs to be refined.

4.3 Sequential Monte Carlo with Adaptive Resampling

The SIS algorithm provides empirical distributions $\widehat{\pi}^{(I)}(\theta_{1:s}|X_{1:s}, \theta_0)$ for $s = 1$, $\ldots, t$, see (16). These empirical distributions are estimates for the true distributions $\pi(\theta_{1:s}|X_{1:s}, \theta_0)$. Resampling the particle system means that we sample from these empirical distributions $\widehat{\pi}^{(I)}(\theta_{1:s}|X_{1:s}, \theta_0)$, that is, we may sample $\Theta_{1:s}^{(j)} \sim \widehat{\pi}^{(I)}(\theta_{1:s}|X_{1:s}, \theta_0)$ i.i.d. for $j = 1, \ldots, I$. Denote for $i = 1, \ldots, I$

$$N_s^{(i)} = \sum_{j=1}^{I} \delta_{\widetilde{\Theta}_{1:s}^{(i)}} \left(\Theta_{1:s}^{(j)} \right),$$

the number of times that $\widetilde{\Theta}_{1:s}^{(i)}$ was re-chosen among the I trials $\Theta_{1:s}^{(1)}, \ldots, \Theta_{1:s}^{(I)}$. This provides a second (resampled) empirical distribution

$$\overline{\pi}^{(I)} (\theta_{1:s}| X_{1:s}, \theta_0) = \sum_{i=1}^{I} \frac{N_s^{(i)}}{I} \delta_{\widetilde{\Theta}_{1:s}^{(i)}} (\theta_{1:s}). \tag{17}$$

This resampled empirical distribution $\overline{\pi}^{(I)} (\theta_{1:s}| X_{1:s}, \theta_0)$ serves as an approximation to the empirical distribution $\widehat{\pi}^{(I)}(\theta_{1:s}|X_{1:s}, \theta_0)$ and henceforth to $\pi(\theta_{1:s}|X_{1:s}, \theta_0)$.

The important remark here is that this resampling does not necessarily reduce the variance, but it may remove particles $\widetilde{\Theta}_{1:s}^{(i)}$ that have low weights $W_s(\widetilde{\Theta}_{1:s}^{(i)})$ (are in an unlikely region of the probability space) and we only work in the part of the probability space that has a sufficiently high probability mass. There are the following important remarks:

- There are more efficient resampling techniques than the i.i.d. resampling one proposed above (which in fact provides a multinomial distribution). Doucet and Johansen [6] support the systematic resampling technique. It samples $U_1 \sim$ Uniform[0, 1] and then defines $U_{i+1} = U_1 + i/I$ for $i = 1, \ldots, I - 1$. An unbiased resampled distribution is obtained by setting

$$N_s^{(i)} = \sum_{j=1}^{I} 1_{\left\{ \sum_{k=1}^{i-1} W_s(\widetilde{\Theta}_{1:s}^{(k)}) \leq U_j \leq \sum_{k=1}^{i} W_s(\widetilde{\Theta}_{1:s}^{(k)}) \right\}}. \tag{18}$$

- For convergence results we refer to the literature mentioned in Doucet and Johansen [6].

- The resampling step may lead to degeneracy of $\bar{\pi}^{(I)}(\theta_{1:s}|X_{1:s}, \theta_0)$ with positive probability. Therefore, one should always back-test whether the resulting empirical distribution is sufficiently rich for the indexes $s = 1, \ldots, t$ under consideration.
- In many cases one applies adaptive resampling, i.e. the resampling step is only applied if the weights are too disperse. One way to measure dispersion is the effective sample size (ESS) defined by

$$\text{ESS}_s = \left(\sum_{i=1}^{I} \left(W_s(\tilde{\Theta}_{1:s}^{(i)}) \right)^2 \right)^{-1} \in [1, I]. \tag{19}$$

The resampling is then only applied if the ESS is too small. Note that if all particles have the same weight $1/I$, then ESS_s is equal to I, if one particle concentrates the entire probability mass, then ESS_s is equal to 1.

This provides the following algorithm under (8)–(10), (13) and (14) and given resampling threshold $\chi \in [1, I]$.

Sequential Monte Carlo (SMC) with adaptive resampling algorithm.
▷ Set $\tilde{\Theta}_0^{(i)} = \theta_0$ and $w_0(\tilde{\Theta}_{1:0}^{(i)}) = 1$ for $i = 1, \ldots, I$.
▷ Repeat for $s = 1, \ldots, t$:

- repeat for $i = 1, \ldots, I$:
  - sample $\tilde{\Theta}_s^{(i)} \sim q_s\left(\cdot \Big| X_s, \tilde{\Theta}_{s-1}^{(i)}\right)$;
  - calculate the importance weights

$$\alpha_s(\tilde{\Theta}_{s-1:s}^{(i)}) = \frac{f\left(X_s \Big| \tilde{\Theta}_s^{(i)}\right) \pi\left(\tilde{\Theta}_s^{(i)} \Big| \tilde{\Theta}_{s-1}^{(i)}\right)}{q_s\left(\tilde{\Theta}_s^{(i)} \Big| X_s, \tilde{\Theta}_{s-1}^{(i)}\right)},$$

$$w_s(\tilde{\Theta}_{1:s}^{(i)}) = w_{s-1}(\tilde{\Theta}_{1:s-1}^{(i)}) \, \alpha_s(\tilde{\Theta}_{s-1:s}^{(i)});$$

- calculate for $i = 1, \ldots, I$ the normalized importance weights

$$W_s(\tilde{\Theta}_{1:s}^{(i)}) = \frac{w_s(\tilde{\Theta}_{1:s}^{(i)})}{\sum_{j=1}^{I} w_s(\tilde{\Theta}_{1:s}^{(j)})} \propto w_{s-1}(\tilde{\Theta}_{1:s-1}^{(i)}) \, \alpha_s(\tilde{\Theta}_{s-1:s}^{(i)}),$$

and the corresponding ESS_s according to (19);
- if $\text{ESS}_s \leq \chi$ resample $\Theta_{1:s}^{(1)}, \ldots, \Theta_{1:s}^{(I)}$ from (16) and set for $i = 1, \ldots, I$

$$w_s(\tilde{\Theta}_{1:s}^{(i)}) = 1, \qquad W_s(\tilde{\Theta}_{1:s}^{(i)}) = \frac{1}{I} \qquad \text{and} \qquad \tilde{\Theta}_{1:s}^{(i)} \leftarrow \Theta_{1:s}^{(i)}.$$

Operation $\tilde{\Theta}_{1:s}^{(i)} \leftarrow \Theta_{1:s}^{(i)}$ is an assignment in the R coding language sense.

This SMC with adaptive resampling algorithm provides empirical distributions for any $s = 1, \ldots, t$

$$\widehat{\pi}^{(I)} \left(\theta_{1:s} \mid X_{1:s}, \theta_0\right) = \sum_{i=1}^{I} W_s(\widetilde{\Theta}_{1:s}^{(i)}) \, \delta_{\widetilde{\Theta}_{1:s}^{(i)}}(\theta_{1:s}) . \tag{20}$$

Note that the weights $W_s(\widetilde{\Theta}_{1:s}^{(i)})$ and particles $\widetilde{\Theta}_{1:s}^{(i)}$ may now differ from the ones of the SIS algorithm (16) due to the potential resampling step that is performed whenever $\mathrm{ESS}_u \leq \chi$ for $u \leq s$. Often one chooses resampling threshold $\chi = I/2$.

5 Examples and Backward Smoothing

In Sect. 5.1 we study the two linear state space models introduced in Sect. 2, in Sect. 5.2 we explore the non-Gaussian and non-linear state space model of Sect. 3, and in Sect. 5.3 we consider a (new) model that may serve as a stochastic volatility model for asset prices.

5.1 Linear State Space Models

We start by considering the linear state space models (1)–(2) and (3)–(4) in the Gaussian and the gamma case, respectively. As parameters we choose

$$\theta_0 = 1, \quad a = 1/10, \quad b = 9/10, \quad \tau = \sqrt{1/10}, \quad \sigma = 1/2. \tag{21}$$

The transition systems (1) and (3) have the same first two moments, but different distributional properties, in particular, the gamma one is bounded from below by zero, whereas the Gaussian one is unbounded from below. In Fig. 1 (lhs) we plot 10 sample paths $\Theta_{1:t}$, $t = 100$, in the transition system for each of the two linear models. This figure is complemented by the empirical means and the confidence bounds of 2 empirical standard deviations (of 1'000 simulations). We see that these measures coincide for the Gaussian and gamma cases, however, the sample paths look rather different in the two models. Based on two selected samples $\Theta_{1:t}$ (darker trajectories) in the transition system we draw a sample $X_{1:t}$ (for each model) in the measurement system according to (2) and (4), respectively. The ones given in Fig. 1 (rhs) are used in the further analysis in order to infer Θ_t from $X_{1:t}$, i.e. we aim at calculating the Bayesian estimate

$$\theta_{t|t} = \mathbb{E}\left[\Theta_t \mid X_{1:t}\right].$$

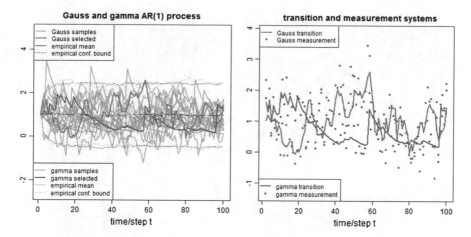

Fig. 1 (*lhs*) Simulated sample paths $\Theta_{1:100}$ of the Gaussian linear transition system (1) and the gamma linear transition system (3) for parameters (21) complemented by the empirical means and the confidence bounds of 2 empirical standard deviations; the darker sample paths were selected for the subsequent state space model analysis; (*rhs*) empirical samples $X_{1:100}$ and $\Theta_{1:100}$ in the two linear state space models (1)–(2) and (3)–(4) for parameters (21) and the selected sample paths of the (*lhs*)

In the Gaussian linear state space model we can calculate $\theta_{t|t}$ exactly, using the Kalman filter; in the gamma linear state space model the Kalman filter provides the best linear (credibility) approximation to the conditional mean of Θ_t, given $X_{1:t}$, see Chap. 9 in Bühlmann and Gisler [1]. The Kalman filter results are presented in Fig. 2. We observe that the Kalman filter achieves to estimate the true $\Theta_{1:t}$ quite accurately, however the noise in $X_{1:t}$ slightly distorts these estimates. Of course, the bigger the parameter σ the harder it becomes to infer the transition system $\Theta_{1:t}$ from the observations $X_{1:t}$.

Next we explore the SIS and the SMC algorithms and compare the results to the Kalman filter ones. We therefore need to choose importance densities q_t. A simple (non-optimal) way is to choose $q_t(\cdot|X_t, \theta_{t-1}) = \pi(\cdot|\theta_{t-1})$, see also (14). This choice provides incremental importance weights under model assumptions (2) and (4) given by

$$\alpha_t(\theta_{t-1:t}) = f(X_t|\theta_t) = \exp\left\{-\frac{(X_t - \theta_t)^2}{2\sigma^2}\right\}. \tag{22}$$

Note that these incremental importance weights are uniformly bounded. In the case of the Gaussian innovations (1) the choice of the importance density q_t could be improved because we can directly simulate from the posterior distribution (which is

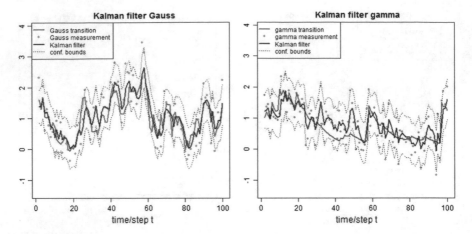

Fig. 2 (Estimated) posterior means $\theta_{t|t} = \mathbb{E}[\Theta_t | X_{1:t}]$ using the Kalman filter complemented by the (estimated) confidence bounds of 2 posterior standard deviations $\text{Var}(\Theta_t | X_{1:t})^{1/2}$: (*lhs*) (exact) Gaussian linear state space model posterior means and (*rhs*) (estimated) gamma linear state space model posterior means

a multivariate normal one). However, we refrain from doing so because our choice works in all three models introduced above and leads to identical incremental importance weights (22).

Finally, we choose $I = 10'000$ independent samples and for the SMC with adaptive resampling algorithm we choose resampling threshold $\chi = I/2$. All parameters are now determined and we can sample from to the SIS and SMC algorithms. In Fig. 3

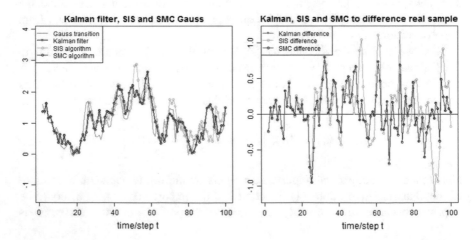

Fig. 3 Comparison between the true sample $\Theta_{1:t}$, the Kalman filter estimate $\theta_{t|t}$, the SIS estimate and the SMC estimate in the Gaussian linear state space model (1)–(2): (*lhs*) estimates and (*rhs*) resulting differences to the true sample $\Theta_{1:t}$

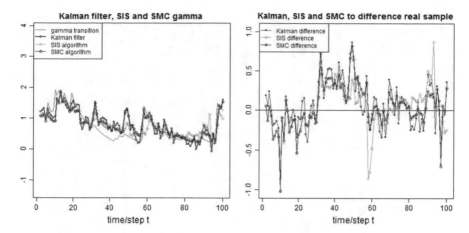

Fig. 4 Comparison between the true sample $\Theta_{1:t}$, the Kalman filter estimate $\theta_{t|t}$, the SIS estimate and the SMC estimate in the gamma linear state space model (3)–(4): (*lhs*) estimates and (*rhs*) resulting differences to the true sample $\Theta_{1:t}$

(lhs) we provide the results for the Gaussian linear state space model (1)–(2). We observe that the SMC estimates coincide almost perfectly with the (exact) Kalman filter estimates $\theta_{t|t}$. On the other hand the SIS estimates start to deviate from $\theta_{t|t}$ after roughly $t = 20$ time steps because the normalized importance weights $W_t(\widetilde{\Theta}_{1:t}^{(i)})$ start to be too disperse and a resampling step should be applied. Figure 3 (rhs) shows the differences between the estimates $\theta_{t|t}$ and the true factors Θ_t. Also here we see that the SIS estimates start to have difficulties with increasing t.

Next we analyze the same plots for the gamma linear state space model (3)–(4). In this model the Kalman filter gives a best linear credibility approximation to the true posterior mean $\mathbb{E}[\Theta_t|X_{1:t}]$, and the SIS and SMC estimates should be exact up to simulation error. Also here we see that the SIS algorithm has a poor behavior for bigger t and one should prefer the SMC estimate. Interestingly, the SMC estimate clearly differs from the Kalman filter estimate because of the different distributional properties from the Gaussian ones. In particular, the Kalman filter has difficulties to cope with the tails which leads to too extreme estimates. We conclude that we should choose the SMC estimates, as soon as the number of samples I is sufficiently large (Fig. 4).

In Fig. 5 we plot the posterior standard deviations $\mathrm{Var}(\Theta_t|X_{1:t})^{1/2}$. In the Gaussian model the SMC and the Kalman filter estimates coincide, whereas the SIS estimate has a poor volatile behavior. In the gamma model the SMC estimate is also volatile and of smaller size than the Kalman filter estimate (which is also not exact in the

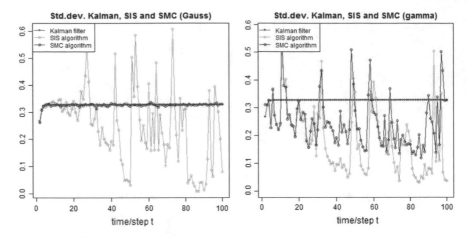

Fig. 5 (Estimated) posterior standard deviation $\text{Var}(\Theta_t | X_{1:t})^{1/2}$ from the Kalman filter, the SIS and SMC algorithms for the (*lhs*) Gaussian and the (*rhs*) the gamma linear state space models

non-Gaussian case). Here one should probably use a smoothed version of the SMC estimate because the Kalman filter over-estimates the posterior standard deviation because it cannot cope with the tail of the gamma distribution.

5.2 Non-Gaussian and Non-linear State Space Models

In this section we explore the non-linear state space model (6)–(7). The posterior density is given by (8) and under the choice $q_t(\cdot | X_t, \theta_{t-1}) = \pi(\cdot | \theta_{t-1})$ we obtain incremental importance weights (22). These are, of course, again bounded and we can apply the algorithm from before, the only change lies in the choice of the importance distribution which now has a non-linear scaling, see (8).

In Fig. 6 (rhs) we provide an explicit sample $\Theta_{1:t}$ for the transition system and a corresponding sample $X_{1:t}$ for the measurement system. Note that we provide exactly the same random samples, but with scaling (6) instead of scaling (3). Figure 7 then shows the resulting Kalman filter approximations, where for the non-linear model (6) we use first order Taylor approximation $\sqrt{\Theta_{t-1}} \approx 1$ in the Kalman filter application. Note that this could be refined by a second order Taylor approximation $\sqrt{\Theta_{t-1}} \approx 1 + (\Theta_{t-1} - 1)/2$.

In Fig. 8 we then compare the Kalman filter, SIS and SMC estimates of the posterior means $\mathbb{E}[\Theta_t | X_{1:t}]$. We observe that the Kalman filter receives too high peaks and should not be used for the gamma non-linear state space model. The SIS estimate

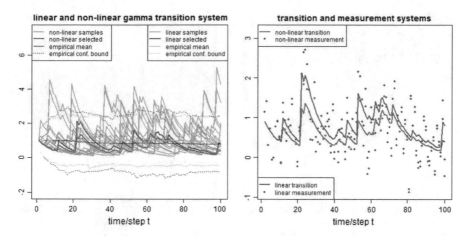

Fig. 6 (*lhs*) Simulated sample paths $\Theta_{1:100}$ of the gamma linear transition system (3) and the gamma non-linear transition system (6) for parameters (21) complemented by the empirical means and the confidence bounds of 2 empirical standard deviations; the darker sample paths were selected for the state space model analysis; (*rhs*) empirical samples $X_{1:100}$ and $\Theta_{1:100}$ in the two models (note that we give exactly the same random samples, only scaling in the transition system differs)

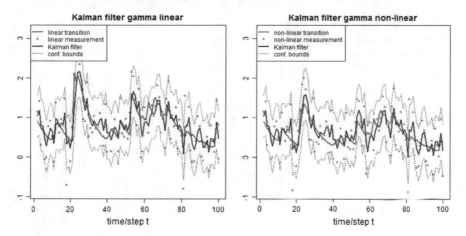

Fig. 7 Estimated posterior means $\theta_{t|t} = \mathbb{E}[\Theta_t | X_{1:t}]$ using the Kalman filter complemented by the estimated confidence bounds of 2 posterior standard deviations: (*lhs*) gamma linear and (*rhs*) gamma non-linear state space models; in the latter we approximate $\sqrt{\Theta_{t-1}} \approx 1$

becomes poorer for bigger t, hence the SMC estimate, that looks reasonable in Fig. 8 (rhs), should be preferred. In Fig. 9 we also see that the Kalman filter over-estimates posterior variance because it cannot cope with the gamma distribution and it cannot interpret scaling $\sqrt{\Theta_{t-1}}$ in (6). For this reason a smoothed version of the SMC posterior standard deviation estimate should be used.

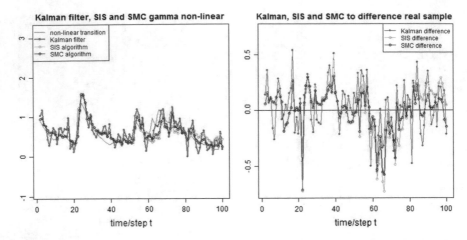

Fig. 8 Comparison between the true sample $\Theta_{1:t}$, the Kalman filter estimate $\theta_{t|t}$, the SIS estimate and the SMC estimate in the gamma non-linear state space model (6)–(7): (*lhs*) estimates and (*rhs*) resulting differences to the true sample $\Theta_{1:t}$

Fig. 9 Estimated posterior standard deviation $\mathrm{Var}(\Theta_t | X_{1:t})^{1/2}$ from the Kalman filter, the SIS and SMC algorithms for the gamma non-linear state space model

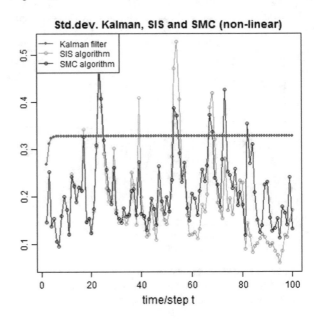

5.3 Stochastic Volatility Model for Asset Prices

We close this section with an example that considers stochastic volatility modeling in the transition system and (de-trended) logarithmic asset prices in the measurement system. Inspired by the Heston [8] model we consider a gamma non-linear transition system for the stochastic volatility process and a log-normal model for the de-trended asset prices. Note that de-trended asset prices means that the log-normal distribution has to have a negative mean parameter being equal to minus one half of the variance parameter. This motivates the following model:

(i) The *transition system* is described by a process $(\Theta_t)_{t \in \mathbb{N}_0}$ with $\Theta_0 = \theta_0 = 1$ and for $t \geq 1$

$$\Theta_t = b\Theta_{t-1} + \sqrt{\Theta_{t-1}}\eta_t, \tag{23}$$

for $b \in \mathbb{R}$ and $(\eta_t)_{t \geq 1}$ being i.i.d. gamma distributed with $\mathbb{E}[\eta_t] = a$ and $\text{Var}(\eta_t) = \tau^2$.

(ii) The *measurement system* is described by a process $(X_t)_{t \in \mathbb{N}}$ with for $t \geq 1$

$$X_t = -\sigma^2 \Theta_t / 2 + \sigma \sqrt{\Theta_t} \varepsilon_t, \tag{24}$$

for $\sigma > 0$ and $(\varepsilon_t)_{t \geq 1}$ being i.i.d. standard Gaussian distributed and being independent of process $(\eta_t)_{t \geq 1}$.

The posterior density of $\Theta_{1:t}$ for given observations $X_{1:t}$ is given by

$$\pi(\theta_{1:t} | X_{1:t}, \theta_0) \propto f(X_{1:t} | \theta_{1:t}) \pi(\theta_{1:t} | \theta_0) \tag{25}$$

$$\propto \prod_{s=1}^{t} \frac{1}{\sqrt{\theta_s \theta_{s-1}}} \left(\frac{\theta_s - b\theta_{s-1}}{\sqrt{\theta_{s-1}}} \right)^{\gamma - 1}$$

$$\times \exp\left\{ -\frac{(X_s + \sigma^2 \theta_s / 2)^2}{2\sigma^2 \theta_s} - c\left(\frac{\theta_s - b\theta_{s-1}}{\sqrt{\theta_{s-1}}} \right) \right\} 1_{\{\theta_s \geq b\theta_{s-1}\}}.$$

The transition system of the stochastic volatility process $(\Theta_t)_{t \in \mathbb{N}_0}$ given in (23) is exactly the same as (6). Therefore, Fig. 6 (*lhs*) provides typical trajectories for the parameters (21). In contrast to the previous models we now also have a non-linearity in the measurement system (24). We would like to indicate two different extreme cases: (i) for $\sigma \gg 1$ very large we obtain

$$X_t = -\sigma^2 \Theta_t / 2 + \sigma \sqrt{\Theta_t} \varepsilon_t \approx -\sigma^2 \Theta_t / 2. \tag{26}$$

For this reason we expect to detect the transition system rather accurately in this case (in fact the filter becomes almost superfluous); (ii) for $\sigma \ll 1$ very small we obtain

$$X_t = -\sigma^2 \Theta_t / 2 + \sigma \sqrt{\Theta_t} \varepsilon_t \approx \sigma \sqrt{\Theta_t} \varepsilon_t. \tag{27}$$

In this case we expect the de-trending term $-\sigma^2 \Theta_t / 2$ to be almost useless in supporting the filtering algorithm.

For the transition system we use the parameters (21) and the importance density $q_t(\cdot | X_t, \theta_{t-1})$ is chosen as $\pi(\cdot | \theta_{t-1})$. In Fig. 10 we present the SIS and SMC algorithm results for $\sigma = 0.25$. These results are compared to the SMC algorithm results provided by model (27) (since $\sigma = 0.25$ is comparably small). The first observation

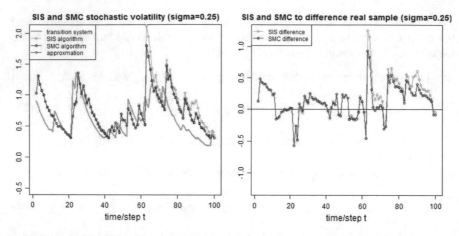

Fig. 10 Comparison between the true sample $\Theta_{1:t}$, the SIS estimate and the SMC estimate in the stochastic volatility model (23)–(24) for $\sigma = 0.25$: (*lhs*) estimates and (*rhs*) resulting differences to the true sample $\Theta_{1:t}$; the approximation on the (*lhs*) uses (27) with SMC and it is almost identically equal to the original SMC estimate (and therefore not visible in the plot)

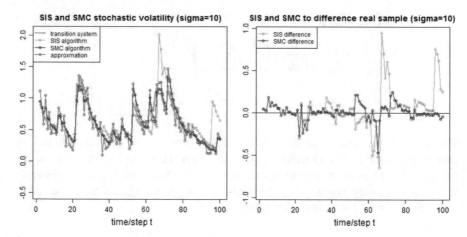

Fig. 11 Comparison between the true sample $\Theta_{1:t}$, the SIS estimate and the SMC estimate in the stochastic volatility model (23)–(24) for $\sigma = 10$: (*lhs*) estimates and (*rhs*) resulting differences to the true sample $\Theta_{1:t}$; the approximation on the (*lhs*) uses (26)

is that we cannot distinguish the SMC results from models (24) and (27), thus, our de-trending term is too small to be helpful to improve inference of the transition system. Secondly, we observe based on scaling (27) that the transition innovation η_t and the measurement innovation ε_t live on a competing scale which makes it difficult to infer Θ_t from $X_{1:t}$. In fact, as can be seen from Fig. 10 (lhs), this leads to a visible delay in the filtered estimation of Θ_t. This is quite a typical phenomenon in filtering and it comes from the fact that "we cannot look into the future for smoothing".

In our second example we choose a large $\sigma = 10$ and compare the solution of model (24) to the (deterministic) one of model (26), see Fig. 11 (lhs). We see that

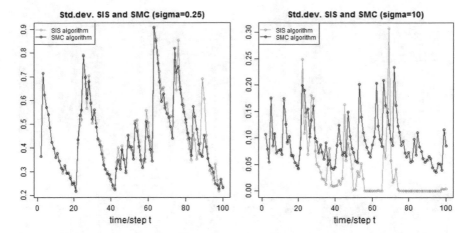

Fig. 12 Estimated posterior standard deviation $\mathrm{Var}(\Theta_t|X_{1:t})^{1/2}$ from the SIS and SMC algorithms for stochastic volatility model (23)–(24) for (*lhs*) $\sigma = 0.25$ and (*rhs*) $\sigma = 10$

approximation (26) clearly fluctuates around the true Θ_t, but fluctuation is still a bit too large to directly extract Θ_t from the observations X_t. This means that σ is not sufficiently large and we should use SMC filtering. The SMC filter provides very good results, in fact much better results than in the previous example $\sigma = 0.25$, because the predictive power of the de-trending term is already quite large in this situation. This can also be seen by comparing Figs. 10 (rhs) and 11 (rhs). Finally, in Fig. 12 we present the posterior standard deviations which (in a smoothed version) allow us to construct confidence bounds for the prediction. For $\sigma = 0.25$ they are in the range of 0.4, for $\sigma = 10$ they are of size 0.1 which is clearly smaller (due to the higher predictive power of the de-trending term).

5.4 Backward Smoothing and Resample-Moves

In Fig. 10 (lhs) we have seen that the filter estimates $\theta_{t|t}$ always come with a delay in reaction. Backward smoothing means that we use later information to re-assess the value of Θ_t. We have already met this idea in Sect. 3.2 and after the description of the SIS algorithm. Basically this means that we infer Θ_t by considering

$$\theta_{t|T} = \mathbb{E}\left[\Theta_t | X_{1:T}\right] \qquad \text{for later time points } T \geq t.$$

Using the simulated samples we calculate empirically at time $T \geq t$, see also (12),

$$\widehat{\mathbb{E}}^{(I)}\left[\Theta_t | X_{1:T}\right] = \sum_{i=1}^{I} \widetilde{\Theta}_t^{(i)} \frac{w_T(\widetilde{\Theta}_{1:T}^{(i)})}{\sum_{j=1}^{I} w_T(\widetilde{\Theta}_{1:T}^{(j)})}.$$

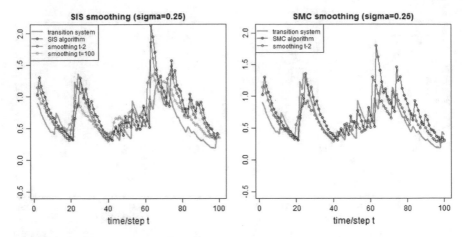

Fig. 13 Comparison between the true sample $\Theta_{1:t}$, the SIS estimate and the SMC estimate in the stochastic volatility model (23)–(24) for $\sigma = 0.25$: (*lhs*) SIS estimates of $\theta_{t|t}$ and of $\theta_{t|t+2}$ and $\theta_{t|100}$ (*backward smoothed*) and (*lhs*) SMC estimates of $\theta_{t|t}$ and of $\theta_{t|t+2}$ (*backward smoothed*)

This seems straightforward, however, this empirical method needs some care! It can be done in a direct manner for the SIS algorithm. We provide the results in Fig. 13 (lhs) for the stochastic volatility model with $\sigma = 0.25$ for $T = t + 2$ and $T = 100$. We see a clear left shift of the estimates $\theta_{t|T}$, that is, for more information $X_{1:T}$ we can better distinguish the competing innovations ε_t and η_t. If T is too large the model looks like it is over-smoothing, which means that an appropriate time lag $T - t$ needs to be determined for smoothing.

For the SMC algorithm backward smoothing is much more delicate due to the resampling step. Observe that the resampling step (17) at time s tends to select only the particles $\widetilde{\Theta}_{1:s}^{(i)}$ that have a sufficiently large importance weight $W_s(\widetilde{\Theta}_{1:s}^{(i)})$. From these selected particles I new trajectories are simulated into the future after time s. But, this selection also implies a thinning of the past trajectories (because part of the particles are dropped in the resampling step and, thus, also their history). Applying the resampling step several times therefore leads to very poor properties at the beginning of the trajectories because of the successive selection of the fittest particles. For this reason in SMC sampling it is preferable to do backward smoothing for time lags that are smaller than the time lags between adaptive resampling steps. An example for $T = t + 2$ is presented in Fig. 13 (rhs). We see that also here backward smoothing leads to better inference compared to the true value Θ_t (which can be seen by the left shift of $\theta_{t|t+2}$ versus $\theta_{t|t}$).

There are ways to deal with the deficiency of the SMC algorithm that it cannot be used for arbitrary backward smoothing because of potential degeneracy of trajectories for big time lags. Ways to fix these problems are, for instance, a resample-move or a block sample step. Such methods mainly aim at spreading the degenerate part of the trajectories by a Markov chain Monte Carlo (MCMC) step using the Metropolis–

Hastings algorithm, the Gibbs sampler or related techniques. We briefly explain the resample-move, for more details we refer to Doucet and Johansen [6].

Consider the posterior density $\pi(\theta_{1:t}|X_{1:t},\theta_0)$ as the invariant (stationary limit) distribution of a Markov process $(\Theta_{1:t}^{(s)})_{s\in\mathbb{N}}$ having transition kernel $K(\theta_{1:t}^{(s+1)}|\theta_{1:t}^{(s)})$. As a consequence we obtain identity

$$\int \pi\left(\theta_{1:t}|\,X_{1:t},\theta_0\right) K\left(\theta'_{1:t}\,\middle|\,\theta_{1:t}\right) d\theta_{1:t} = \pi\left(\theta'_{1:t}\,\middle|\,X_{1:t},\theta_0\right).$$

This immediately implies that for given $\Theta_{1:t}\sim\pi(\cdot|X_{1:t},\theta_0)$ we can resample from the transition kernel $\Theta'_{1:t}\sim K(\cdot|\Theta_{1:t})$ and the resulting sample is still distributed according to $\pi(\cdot|X_{1:t},\theta_0)$. As a consequence if we obtain in the adaptive resampling step of the SMC algorithm $N_t^{(i)}>1$ particles that have the same past history $\widetilde{\Theta}_{1:t}^{(i)}$, see (17), we can spread these particles by applying an independent resample-move to each particle using transition kernel $K(\cdot|\widetilde{\Theta}_{1:t}^{(i)})$. In addition, MCMC sampling theory provides explicit constructions of transition kernels K for given invariant distributions $\pi(\cdot|X_{1:t},\theta_0)$ as soon as the latter are known up to the normalizing constants (which is the case in our situation, see for instance (8)). In practice, only a fixed time lag is resampled by this MCMC step, firstly because then one does not need to deal with different lengths of resample-moves as t increases, and secondly because filtering is also only applied to limited time lags (to preserve stationarity in real world time series).

6 Conclusions and Outlook

Gaussian linear state space models can be solved with Kalman filter techniques. Non-Gaussian or/and non-linear state space models can only be solved numerically. A powerful simulation method is sequential Monte Carlo (SMC) sampling. The resulting sampler is a version of importance sampling that benefits from the underlying Markovian structure of state space models. We have presented the SMC sampler and we have illustrated it in terms of several examples.

This outline and the examples presented were always based on known densities (up to the normalizing constants). Unknown model parameters add an additional complexity to the problem. Solving the latter problem may take advantage of Markov chain Monte Carlo (MCMC) methods, in particular, the particle marginal Metropolis-Hastings (PMMH) algorithm may be useful.

References

1. Bühlmann H, Gisler A (2005) A course in credibility theory and its applications. Springer, Berlin
2. Creal D (2012) A survey of sequential Monte Carlo methods for economics and finance. Econ Rev 31(3):245–296
3. Del Moral P (1996) Non linear filtering: interacting particle solution. Markov Process Relat Fields 2(4):555–580
4. Del Moral P, Doucet A, Jasra A (2006) Sequential Monte Carlo samplers. J R Stat Soc Ser B 68(3):411–436
5. Del Moral P, Peters GW, Vergé C (2012) An introduction to stochastic particle integration methods: with applications to risk and insurance. In: Dick J, Kuo FY, Peters GW, Sloan IH (eds) Monte Carlo and Quasi-Monte Carlo Methods. Springer Proceedings in Mathematics & Statistics. Springer, Berlin
6. Doucet A, Johansen AM (2011) A tutorial on particle filtering and smoothing: fifteen years later. In: Crisan D, Rozovsky B (eds) Handbook of nonlinear filtering. Oxford University Press, Oxford
7. Gordon NJ, Salmond DJ, Smith AFM (1993) Novel approach to nonlinear/non-Gaussian Bayesian state estimation. IEE Proc F Radar Sig Process 140(2):107–113
8. Heston SL (1993) A closed-form solution for options with stochastic volatility with applications to bond and currency options. Rev Financ Stud 6(2):327–343
9. Johansen AM, Evers L (2007) Monte Carlo Methods. Lecture Notes. University of Bristol, Bristol
10. Kalman RE (1960) A new approach to linear filtering and prediction problems. J Basic Eng 82:35–45

Robustness as a Criterion for Selecting a Probability Distribution Under Uncertainty

Songsak Sriboonchitta, Hung T. Nguyen, Vladik Kreinovich and Olga Kosheleva

Abstract Often, we only have partial knowledge about a probability distribution, and we would like to select a single probability distribution $\rho(x)$ out of all probability distributions which are consistent with the available knowledge. One way to make this selection is to take into account that usually, the values x of the corresponding quantity are also known only with some accuracy. It is therefore desirable to select a distribution which is the most robust—in the sense the x-inaccuracy leads to the smallest possible inaccuracy in the resulting probabilities. In this paper, we describe the corresponding most robust probability distributions, and we show that the use of resulting probability distributions has an additional advantage: it makes related computations easier and faster.

1 Formulation of the Problem

Need to make decisions under uncertainty. One of the main objectives of science is to understand the world, to predict the future state of the world under different

S. Sriboonchitta (✉) · H.T. Nguyen
Faculty of Economics, Chiang Mai University, Chiang Mai, Thailand
e-mail: songsakecon@gmail.com

H.T. Nguyen
Department of Mathematical Sciences, New Mexico State University,
Las Cruces, NM 88003, USA
e-mail: hunguyen@nmsu.edu

V. Kreinovich · O. Kosheleva
University of Texas at El Paso, 500 W. University, El Paso, TX 79968, USA
e-mail: vladik@utep.edu

O. Kosheleva
e-mail: olgak@utep.edu

© Springer International Publishing AG 2017
V. Kreinovich et al. (eds.), *Robustness in Econometrics*,
Studies in Computational Intelligence 692, DOI 10.1007/978-3-319-50742-2_3

possible decisions—and then, to use these predictions to select the decision for which the corresponding prediction is the most preferable.

When we have the full knowledge of the situation, the problem of selecting the best decision becomes a straightforward optimization problem. In practice, however, we rarely have the full knowledge. Usually, we have some uncertainty about the future situations. It is therefore important to make decisions under uncertainty.

Traditional decision making assumes that we know the probabilities. There exist many techniques for decision making under uncertainty. Most of these techniques assume that we know the probabilities of different outcomes—i.e., in precise terms, that we know the probability distribution on the set of all possible outcomes; see, e.g., [2, 6, 7, 14].

In practice, we often have only partial knowledge about the probabilities. In many real-life random phenomena, we only have partial knowledge about the corresponding probability distributions. In such situations, several different probability distributions are consistent with the available data.

The resulting need to select a single probability distribution. As we have mentioned, most decision making techniques use a single probability distribution. So, to be able to apply these techniques to the practical situations, when *several* different probability distributions are consistent with our knowledge, we need to be able to select a *single* probability distribution—and use it in decision making.

What we do in this paper. To select a probability distribution, we can take into account that, in addition to imprecise knowledge about *probabilities* of different values of the corresponding quantity x (or quantities), we also have imprecise knowledge about the actual *values* of these quantities.

Indeed, the knowledge about these values comes from measurements, and measurements are never absolutely accurate: there is always a difference between the measurement result and the actual value, the difference known as the *measurement error*; see, e.g., [13]. In other words, when the measurement result is $\tilde{x}$, the actual value x can be (and usually is) slightly different. It is therefore reasonable to select a probability distribution which is the most *robust*, i.e., for which the change from $\tilde{x}$ to x has the smallest possible effect on the resulting probabilities.

In this paper, we show that this robustness idea indeed enables us to select a single distribution.

2 Robustness: From an Informal General Idea to a Precise Description

1-D case: analysis of the problem. Let us start with a 1-D case, when we have a single quantity x. In this case, we are interested in the probability of different events related to this quantity, i.e., in mathematical terms, in the probabilities of different subsets of the real line.

In many cases, it makes sense to limit ourselves to connected sets. In the 1-D case, the only connected sets are intervals $[\underline{x}, \overline{x}]$ (finite or infinite).

This make practical sense: e.g., it corresponds to checking whether x is larger than or equal to a certain lower threshold $\underline{x}$ and/or checking whether x is smaller than or equal to a certain upper threshold $\overline{x}$, or to checking whether x belongs to the given tolerance interval $[\underline{x}, \overline{x}]$.

From this viewpoint, all we need is for different intervals $[\underline{x}, \overline{x}]$, to find the probability that the value x belongs to this interval.

A 1-D probability distribution can be naturally described in terms of the corresponding probability density function $\rho(x)$. In terms of this function, the desired probability is equal to the integral $P = \int_{\underline{x}}^{\overline{x}} \rho(x) \, dx$.

Local robustness. As we have mentioned earlier, all the values of the quantity—in particular, the threshold values—and known with uncertainty. Let us consider, for example, the effect of uncertainty in $\underline{x}$ on the resulting probability. If we replace the value $\underline{x}$ with a slightly different value $\underline{x}' = \underline{x} + \Delta x$, then the original probability P changes to the slightly different probability

$$
\begin{aligned}
P' &= \int_{\underline{x}+\Delta x}^{\overline{x}} \rho(x) \, dx \\
&= \int_{\underline{x}}^{\overline{x}} \rho(x) \, dx - \int_{\underline{x}}^{\underline{x}+\Delta x} \rho(x) \, dx = P - \int_{\underline{x}}^{\underline{x}+\Delta x} \rho(x) \, dx.
\end{aligned}
\tag{1}
$$

When the value Δx is small, we can, in the first approximation, ignore the changes of the function $\rho(x)$ on the narrow interval $[\underline{x}, \underline{x} + \Delta x]$ and thus, get $\int_{\underline{x}}^{\underline{x}+\Delta x} \rho(x) \, dx \approx \rho(\underline{x}) \cdot \Delta x$. Then, the resulting change in probability $\Delta P \stackrel{\text{def}}{=} P' - P$ can be described as $\Delta P \approx -\rho(\underline{x}) \cdot \Delta x$, so $|\Delta P| \approx \rho(\underline{x}) \cdot |\Delta x|$.

Thus, the effect of the uncertainty Δx (with which we know $\underline{x}$) on the change in probability P is determined by the value $\rho(\underline{x})$. Similarly, the effect of the uncertainty Δx with which we know $\overline{x}$ on the change in probability P is determined by the value $\rho(\overline{x})$.

We can summarize both cases by saying that for any point x, the effect of the uncertainty Δx (with which we know x) on the change in probability P is determined by the value $\rho(x)$. This value $\rho(x)$ thus serves as a measure of local robustness at the point x.

From local robustness to global robustness. For different values x, the local robustness degree is, in general, different. To select a distribution, we need to combine these values into a single criterion.

Local robustness values are proportional to approximation errors caused by uncertainty Δx. There are two natural ways to combine different approximation errors:

- we can consider the worst-case error, or
- we can consider the mean squared error.

The worst-case error corresponds to selecting the largest possible value of the approximation error, i.e., in our terms, the largest possible value $\max_x \rho(x)$.

The mean squared error means considering the mean value of the squared error, i.e., equivalently, of the squared coefficient $\rho^2(x)$. In contrast to the worst-case approach, where the global criterion is uniquely determined, here, we have two possible choices:

- we can interpret mean as the average over all possible x, i.e., as a quantity proportional to the integral $\int (\rho(x))^2 \, dx$;
- alternatively, we can interpret mean as averaging over the probability distribution characterized by the probability density $\rho(x)$; in this case, as a criterion of global robustness, we get the quantity

$$\int \rho(x) \cdot (\rho(x))^2 \, dx = \int (\rho(x))^3 \, dx. \qquad (2)$$

Thus, we arrive at the following conclusion.

Resulting criteria of global robustness. We have three possible choices of selecting the most robust probability distribution:

- we can select a probability distribution $\rho(x)$ for which the maximum $\max_x \rho(x)$ attains the smallest possible value;
- we can select a probability distribution $\rho(x)$ for which the integral $\int (\rho(x))^2 \, dx$ attains the smallest possible value; and
- we can select a probability distribution $\rho(x)$ for which the integral $\int (\rho(x))^3 \, dx$ attains the smallest possible value.

Relation to maximum entropy approach. Traditionally in probability theory, when we only have partial knowledge about the probability distribution, we select a distribution for which the entropy $-\int \rho(x) \cdot \ln(\rho(x)) \, dx$ attains the largest possible value (see, e.g., [3]), or, equivalently, for which the integral $\int \rho(x) \cdot \ln(\rho(x)) \, dx$ attains the smallest possible value.

It is worth mentioning that, in general, if we assume that the criterion for selecting a probability distribution is scale-invariant (in some reasonable sense), then this criterion is equivalent to optimizing either entropy, or generalized entropy $\int \ln(\rho(x)) \, dx$ or $\int \rho^\alpha(x) \, dx$, for some $\alpha > 0$; see, e.g., [5]. Our analysis shows that the generalized entropy corresponding to $\alpha = 2$ and $\alpha = 3$ describes mean-squared robustness.

The worst-case criterion can also be thus interpreted. Indeed, it is known that for non-negative values $v_1, \ldots, v_n$, we have

$$\max(v_1, \ldots, v_n) = \lim_{p \to \infty} ((v_1)^p + \ldots + (v_n)^p)^{1/p} \qquad (3)$$

and similarly,

$$\max_x \rho(x) = \lim_{p \to \infty} \left(\int (\rho(x))^p \, dx \right)^{1/p}. \qquad (4)$$

Thus, minimizing $\max\limits_{x} \rho(x)$ is, for large enough p, equivalent to minimizing the expression $\left(\int (\rho(x))^p \, dx\right)^{1/p}$ and hence, equivalent to minimizing the corresponding generalized entropy $\int (\rho(x))^p \, dx$.

Multi-D case. In the multi-D case, when the probability density function $\rho(x)$ depends on several variables $x = (x_1, \ldots, x_m)$, we can also consider general connected sets S. Similarly to the 1-D case, if we add, to the set S, a small neighborhood of a point x, of volume ΔV, then the resulting change in probability is equal to $\Delta P = \rho(x) \cdot \Delta V$. Vice versa, if the set S contained the point x with some neighborhood, and we delete an x-neighborhood of volume ΔV from the set S, then we get $\Delta P = -\rho(x) \cdot \Delta V$.

In both cases, we have $|\Delta P| = \rho(x) \cdot \Delta V$. Thus, in the multi-D case too, the value $\rho(x)$ serves as a measure of local robustness at a point x. So, when we apply the usual techniques for combining local robustness measures into a single global one, we get one of three criteria described above.

What we do in the following sections. Now that we know that we have three possible ways of selecting the most robust probability distribution, let us consider these three ways one by one. For each way, on several simple examples, we explain what exactly probability distribution will be thus selected.

Comment. It is worth mentioning that a similar idea of selecting the most robust description is actively used in fuzzy logic [4, 11, 16]; namely, in [8–11], it is shown how we can select the most robust membership functions and the most robust "and"- and "or"-operations.

While our problem is different, several related formulas are similar—and this similarity helped us with our results.

3 Selecting a Probability Distribution that Minimizes $\int (\rho(x))^2 \, dx$

General idea. In this section, we will describe, for several reasonable types of partial knowledge, which probability distribution corresponds to the smallest possible values of the global robustness criterion $\int (\rho(x))^2 \, dx$.

Types of partial knowledge about the probability distribution. What type of partial knowledge do we have about a random variable? For example, about a random measurement error?

First, we can have lower and upper bounds on the measurement error (and, more generally, on the possible values of the random variable).

Second, we may know:

- the mean value, i.e., the first moment of the corresponding random variable,
- the variance (i.e., equivalently, the second moment),

- sometimes the skewness (i.e., equivalently, the third moment) that characterizes the distribution's asymmetry, and
- the excess (i.e., equivalently, the fourth moment) that describes how heavy are the distribution's tails.

In general, we will therefore consider the cases when we know the bounds and some moments (maybe none).

Simplest case, when we only know the bounds. Let us start with the simplest case, when we only know the bounds $\underline{a}$ and $\overline{a}$ on the values of the corresponding random variable x, i.e., we know that always $\underline{a} \le x \le \overline{a}$ and thus, that $\rho(x) = 0$ for values x outside the interval $[\underline{a}, \overline{a}]$.

In this case, the problem of selecting the most robust distribution takes the following form: minimize $\int_{\underline{a}}^{\overline{a}} (\rho(x))^2 \, dx$ under the constraints that $\int_{\underline{a}}^{\overline{a}} \rho(x) \, dx = 1$ and $\rho(x) \ge 0$ for all x. To solve this constrained optimization problem, we can apply the Lagrange multiplier methods to reduce it to an easier-to-solve unconstrained optimization problem

$$\int_{\underline{a}}^{\overline{a}} (\rho(x))^2 \, dx + \lambda \cdot \left(\int_{\underline{a}}^{\overline{a}} \rho(x) \, dx - 1 \right) \to \min_{\rho(x)}, \tag{5}$$

under the condition that $\rho(x) \ge 0$ for all x.

According to calculus, for every x, when the value $\rho(x)$ corresponding to the optimum is inside the corresponding range $(0, \infty)$, the derivative of the above objective function with respect to $\rho(x)$ should be equal to 0. Differentiating the above expression and equating its derivative to 0, we get $2\rho(x) + \lambda = 0$, hence $\rho(x) = c$ for some constant c (equal to $-\lambda/2$; strictly speaking, we should be talking here about *variational* derivative, not a regular derivative).

So, for every x from the interval $[\underline{a}, \overline{a}]$, $\rho(x) > 0$ implies that $\rho(x) = c$. In other words, for every $x \in [\underline{a}, \overline{a}]$, we have either $\rho(x) = 0$ or $\rho(x) = c$.

Let S denote the set of all the points $x \in [\underline{a}, \overline{a}]$ for which $\rho(x) > 0$. Let L denote the total length (1-D Lebesgue measure) of this set. Then, the condition $\int_{\underline{a}}^{\overline{a}} \rho(x) \, dx = \int_S \rho(x) \, dx = 1$ implies that $c \cdot L = 1$, hence $c = \dfrac{1}{L}$. Thus, the value of the desired objective function takes the form

$$\int_{\underline{a}}^{\overline{a}} (\rho(x))^2 \, dx = L \cdot \left(\frac{1}{L} \right)^2 = \frac{1}{L}. \tag{6}$$

One can easily see that this value is the smallest if and only if the length L is the largest.

The largest possible length of a set $S \subseteq [\underline{a}, \overline{a}]$ is attained when this subset coincide with the interval—and is equal to the length $\overline{a} - \underline{a}$ of this interval. In this case, $\rho(x) = \text{const}$ for all points $x \in [\underline{a}, \overline{a}]$.

Thus, in this case, the most robust distribution is the uniform distribution on the interval $[\underline{a}, \overline{a}]$.

Comment. It is worth mentioning that in this case, when we only know the bounds $\underline{a}$ and $\overline{a}$ on the values of the corresponding random variable x, maximum entropy method leads to the exact same uniform distribution.

What if we also know the mean? Let us now consider the next case, when, in addition to the bounds $\underline{a}$ and $\overline{a}$ on the values of the corresponding random variable x, we also know its mean μ.

In this case, we need to minimize the functional $\int (\rho(x))^2 \, dx$ under the constraints $\int \rho(x) \, dx = 1$, $\int x \cdot \rho(x) \, dx = \mu$, and $\rho(x) \geq 0$. By using the Lagrange multiplier method, we can reduce this constraint optimization problem to the following unconstrained optimization problem:

$$\int_{\underline{a}}^{\overline{a}} (\rho(x))^2 \, dx + \lambda \cdot \left(\int_{\underline{a}}^{\overline{a}} \rho(x) \, dx - 1 \right) + \lambda_1 \cdot \left(\int_{\underline{a}}^{\overline{a}} x \cdot \rho(x) \, dx - \mu \right) \to \min$$

(7)

under the constraint that $\rho(x) \geq 0$ for all $x \in [\underline{a}, \overline{a}]$.

Similarly to the previous case, for the points x for which $\rho(x) > 0$, the derivative of the above expression relative to $\rho(x)$ should be equal to 0, so we conclude that for some x, we have $\rho(x) = p_0 + q \cdot x$ for appropriate constants $p_0 = -\lambda/2$ and $q = -\lambda_1/2$. In other words, the probability density $\rho(x)$ is either determined by a linear expression or it is equal to 0.

One can check that, in general, the desired minimum is attained when

$$\rho(x) = \max(0, p_0 + q \cdot x). \tag{8}$$

In particular, in the case when $\rho(x) > 0$ for all $x \in [\underline{a}, \overline{a}]$, the probability density function $\rho(x)$ is linear for all $x \in [\underline{a}, \overline{a}]$: $\rho(x) = p_0 + q \cdot x$. We can get explicit expressions for p_0 and q if we reformulate this linear function in an equivalent form $\rho(x) = p_0 + q \cdot (x - \widetilde{a})$, where $\widetilde{a} \stackrel{\text{def}}{=} \dfrac{\underline{a} + \overline{a}}{2}$ is the interval's midpoint. In this case, the condition $\int_{\underline{a}}^{\overline{a}} \rho(x) \, dx = 1$ takes the form $\int_{-\Delta}^{\Delta} (p_0 + q \cdot t) \, dt = 1$, where $t \stackrel{\text{def}}{=} x - \widetilde{a}$ and $\Delta \stackrel{\text{def}}{=} \dfrac{\overline{a} - \underline{a}}{2}$ is the half-width (radius) of the interval $[\underline{a}, \overline{a}]$. The integral of an odd function t over a symmetric interval $[-\Delta, \Delta]$ is equal to 0, so we have $2\Delta \cdot p_0 = 1$ and thus,

$$p_0 = \frac{1}{2\Delta} = \frac{1}{\overline{a} - \underline{a}}, \tag{9}$$

exactly the value corresponding to the uniform distribution on the interval $[\underline{a}, \overline{a}]$.

The value q can be determined by the condition $\int_{\underline{a}}^{\overline{a}} x \cdot \rho(x)\,dx = \mu$. Since $\int_{\underline{a}}^{\overline{a}} \rho(x)\,dx = 1$, this condition is equivalent to $\int_{\underline{a}}^{\overline{a}} (x - \widetilde{a}) \cdot \rho(x)\,dx = \mu - \widetilde{a}$ and thus, to

$$\int_{-\Delta}^{\Delta} t \cdot (\mu_0 + q \cdot t)\,dt = \int_{-\Delta}^{\Delta} (t \cdot \mu_0 + q \cdot t^2)\,dt = \mu - \widetilde{a}. \qquad (10)$$

Here similarly, the integral of t is equal to 0, and the integral of t^2 is equal to

$$\int_{-\Delta}^{\Delta} t^2\,dt = \left.\frac{t^3}{3}\right|_{-\Delta}^{\Delta} = \frac{2\Delta^3}{3}, \qquad (11)$$

thus the above condition leads to $q \cdot \dfrac{2\Delta^3}{3} = \mu - \widetilde{a}$ and to

$$q = \frac{3(\mu - \widetilde{a})}{2\Delta^3}. \qquad (12)$$

Substituting, into this formulas, the definition of half-width in terms of the bounds $\underline{a}$ and $\overline{a}$, we get an equivalent formula

$$q = \frac{12 \cdot (\mu - \widetilde{a})}{(\overline{a} - \underline{a})^3}. \qquad (12a)$$

The resulting linear formula $\rho(x) = \rho_0 + q \cdot (x - \widetilde{a})$ only works when the resulting expression is non-negative for all x, i.e., when $\rho_0 + q \cdot t \geq 0$ for all $t \in [-\Delta, \Delta]$. This, in its turn, it equivalent to $\rho_0 \geq |q| \cdot \Delta$, i.e., to $\dfrac{1}{2\Delta} \geq \dfrac{3 \cdot |\mu - \widetilde{a}|}{2\Delta^2}$, and, equivalently, to $|\mu - \widetilde{a}| \leq \dfrac{1}{3} \cdot \Delta$.

When $|\mu - \widetilde{a}| > \dfrac{1}{3} \cdot \Delta$, we have to consider probability density functions $\rho(x)$ which are equal to 0 on some subinterval of the interval $[\underline{a}, \overline{a}]$. For a random variable $x \in [\underline{a}, \overline{a}]$, its means value μ also has to be within the same interval, so we must have $\mu \in [\underline{a}, \overline{a}]$ and $\mu - \widetilde{a} \in [-\Delta, \Delta]$.

- When $\mu - \widetilde{a} \to \Delta$, i.e., when $\mu \to \overline{a}$, the corresponding probability distribution get concentrated on a narrower and narrower interval containing the point $x = \overline{a}$.
- Similarly, when $\mu - \widetilde{a} \to -\Delta$, i.e., when $\mu \to \underline{a}$, the corresponding probability distribution get concentrated on a narrower and narrower interval containing the point $x = \underline{a}$.

Comment. If instead of our robustness criterion, we would look for the probability distribution with the largest entropy, then the corresponding derivative would take a form $-\ln(\rho(x)) - 1 + \lambda + \lambda_1 \cdot x = 0$, so $\ln(\rho(x)) = a + b \cdot x$, where $a = \lambda - 1$ and $b = \lambda_1$, and we would get an exponential distribution $\rho(x) = \exp(a + b \cdot x)$.

What if we also know the first two moments? Let us now consider the next case, when, in addition to the bounds $\underline{a}$ and $\overline{a}$ on the values of the corresponding random variable x, and the mean μ, we also know the second moment M_2—or, equivalently, the variance $V = \sigma^2 = M_2 - \mu^2$.

In this case, we need to minimize the functional $\int_{\underline{a}}^{\overline{a}} (\rho(x))^2 \, dx$ under the constraints $\int_{\underline{a}}^{\overline{a}} \rho(x) \, dx = 1$, $\int_{\underline{a}}^{\overline{a}} x \cdot \rho(x) \, dx = \mu$, $\int_{\underline{a}}^{\overline{a}} x^2 \cdot \rho(x) \, dx = M_2$, and $\rho(x) \geq 0$. By using the Lagrange multiplier method, we can reduce this constraint optimization problem to the following unconstrained optimization problem:

$$
\int_{\underline{a}}^{\overline{a}} (\rho(x))^2 \, dx + \lambda \cdot \left(\int_{\underline{a}}^{\overline{a}} \rho(x) \, dx - 1 \right) +
$$

$$
\lambda_1 \cdot \left(\int_{\underline{a}}^{\overline{a}} x \cdot \rho(x) \, dx - \mu \right) + \lambda_2 \cdot \left(\int_{\underline{a}}^{\overline{a}} x^2 \cdot \rho(x) \, dx - M_2 \right) \to \min \qquad (13)
$$

under the constraint that $\rho(x) \geq 0$ for all $x \in [\underline{a}, \overline{a}]$.

Similarly to the previous case, for the points x for which $\rho(x) > 0$, the derivative of the above expression relative to $\rho(x)$ should be equal to 0, so we conclude that for some x, we have $\rho(x) = p_0 + q \cdot x + r \cdot x^2$, where $p_0 = -\lambda/2$, $q = -\lambda_1/2$, and $r = -\lambda_2/2$. In other words, the probability density $\rho(x)$ is either determined by a quadratic expression or it is equal to 0. One can check that, in general, the desired minimum is attained when

$$
\rho(x) = \max \left(0, \, p_0 + q \cdot x + r \cdot x^2 \right). \qquad (14)
$$

It should be mentioned that for the maximum entropy case, similar arguments lead to the Gaussian distribution $\rho_G(x) = \text{const} \cdot \exp \left(-\dfrac{(x - \mu_0)^2}{2\sigma_0^2} \right)$ truncated to some interval $[\underline{b}, \overline{b}] \subseteq [\underline{a}, \overline{a}]$ of the given interval $[\underline{a}, \overline{a}]$: $\rho(x) = \rho_G(x)$ for $x \in [\underline{b}, \overline{b}]$ and $\rho(x) = 0$ for all other x.

Let us consider particular cases. When $r < 0$, we get a bell-shaped distribution— i.e., somewhat similar in shape to the Gaussian distribution. However, the new distribution has several advantages over the Gaussian distribution:

- first, the new distribution is more robust—it is actually the most robust of all the distributions on the given interval with the given two moments (this is how we selected it);
- second, the new probability distribution function $\rho(x)$ is continuous on the entire real line—while, due to the fact that the probability density of a Gaussian distribution is always positive, the pdf of the truncated Gaussian distribution is discontinuous at the endpoints $\underline{b}$ and $\overline{b}$ of the corresponding interval;
- third, the new distribution is computationally easier, since computation with polynomials (e.g., computing probability over different intervals or different moments) is much easier than computation with the Gaussian pdf.

When the variance is sufficiently high, we get $r > 0$, which corresponds to a *bimodal distribution*. Bimodal distributions are common in measuring instruments (see, e.g., [12]). There are two main reasons for the bimodal distribution. The first is the effect of the sinusoid signal in the electric grid. Electric grids are ubiquitous, and the electromagnetic field created by the electric plugs affects all electromagnetic devices. The resulting noise is proportional to $\sin(\omega \cdot t)$ at a random time t—and the resulting random variable indeed has a bimodal distribution.

The second reason is related to the very process of manufacturing the corresponding measuring instrument. Indeed, usually, we have a desired upper bound on the measurement error. At first, the measurement error of the newly manufactured measuring instrument is normally distributed. This can be explained by the fact that there are many different independent factors that contribute to this original measurement error and thus, due to the Central Limit Theorem, we expect the overall effect of these factors to be approximately normally distributed; see, e.g., [13, 15]. However, the range of the corresponding errors Δx is usually much wider than the desired tolerance bounds. Thus, the manufacturers start tuning the instrument until it fits into the bounds; this tuning stops as soon as we get into the desired intervals $[-\Delta, \Delta]$. As a result:

- all the cases when originally, we had $\Delta x \leq -\Delta$ are converted to $-\Delta$ and
- all the cases when originally, we had $\Delta x \geq \Delta$ are converted to Δ.

Hence, the vicinities of the two extreme values $-\Delta$ and Δ get a high probability—and thus, very high values of probability density $\rho(x)$. So, we get a distribution which is either bimodal or even tri-modal (with a smaller original peak).

In our robust approach, we cover bimodal distributions by using the same easy-to-process quadratic formulas as the more usual unimodal ones—a clear advantage over the more traditional approach, when bimodal distributions are modeled by using much more computationally complex expressions.

What if we also know higher moments? In many cases, we also know higher moments. For example, often, we know third and/or fourth moments, i.e., equivalently, skewness and excess. For such situations, traditionally, there are no easy-to-use expression. However, in our case, we do get such an expression.

Namely, let us now consider the case, when, in addition to the bounds $\underline{a}$ and $\overline{a}$ on the values of the corresponding random variable x, we also know the values of the first m moments $\int_{\underline{a}}^{\overline{a}} x^k \cdot \rho(x)\, dx = M_k$, $k = 1, 2, \ldots, m$.

In this case, we need to minimize the functional $\int_{\underline{a}}^{\overline{a}} (\rho(x))^2\, dx$ under the constraints $\int_{\underline{a}}^{\overline{a}} \rho(x)\, dx = 1$, $\int_{\underline{a}}^{\overline{a}} x^k \cdot \rho(x)\, dx = M_k$ for $k = 1, \ldots, m$, and $\rho(x) \geq 0$ for all x. By using the Lagrange multiplier method, we can reduce this constraint optimization problem to the following unconstrained optimization problem:

$$\int_{\underline{a}}^{\overline{a}} (\rho(x))^2 \, dx + \lambda \cdot \left(\int_{\underline{a}}^{\overline{a}} \rho(x) \, dx - 1 \right) +$$

$$\sum_{k=1}^{m} \lambda_k \cdot \left(\int_{\underline{a}}^{\overline{a}} x^k \cdot \rho(x) \, dx - M_k \right) \to \min \tag{15}$$

under the constraint that $\rho(x) \geq 0$ for all $x \in [\underline{a}, \overline{a}]$.

For the points x for which $\rho(x) > 0$, the derivative of the above expression relative to $\rho(x)$ should be equal to 0, so we conclude that for some x, we have $\rho(x) = p_0 + \sum_{k=1}^{m} q_k \cdot x^k$, where $p_0 = -\lambda/2$ and $q_k = -\lambda_k/2$. In other words, the probability density $\rho(x)$ is either determined by a polynomial expression or it is equal to 0. One can check that, in general, the desired minimum is attained when

$$\rho(x) = \max \left(0, \, p_0 + \sum_{k=1}^{m} q_k \cdot x^k \right). \tag{16}$$

This polynomial expression is easy to process, so we have a distribution whose processing is computationally easy—as opposed to the usual not-so-computationally easy approaches of dealing with, e.g., skew-normal distributions [1].

Multi-D case: good news. What is we want to analyze a joint distribution of several variables? Similarly to the 1-D case, if we know several moments, then the most robust pdf $\rho(x_1, \ldots, x_d)$ on a given box $[\underline{a}_1, \overline{a}_1] \times \ldots \times [\underline{a}_d, \overline{a}_d]$ is described by a polynomial, or, to be more precise, by an expression

$$\rho(x_1, \ldots, x_d) = \max(0, P(x_1, \ldots, x_d)) \tag{17}$$

for some polynomial $P(x_1, \ldots, x_d)$.

The degree of this polynomial depends on what moments we know:

- if we do not know any moments, then $P(x_1, \ldots, x_d)$ is a constant, and thus, we get a uniform distribution—similarly to what we get if we use the maximum entropy approach;
- if we only know the means $E[x_i]$, then $P(x_1, \ldots, x_d)$ is a linear function;
- if we also know second moments $E[(x_i)^2]$ and $E[x_i \cdot x_j]$—i.e., equivalently, the covariance matrix—then $P(x_1, \ldots, x_d)$ is a quadratic function;
- if we also know third (and fourth) order moments, then $P(x_1, \ldots, x_d)$ is a cubic (quartic) polynomial, etc.

These polynomial pdf's are not only more robust, but they are also much easier to process than Gaussian or other usually used pdf's.

But maybe we are missing something? Not really, since, as it is well known, polynomials are *universal approximators*—in the sense that any arbitrary continuous function on a given box can be, with any desired accuracy, approximated by a polynomial.

Multi-D case: remaining challenges. While, as we have mentioned on several example, the robust approach to selecting a probability distribution has many advantages
over the maximum entropy approach, there are situations in which the use of the
robust approach faces some challenges.

One such situation is when we know the marginal distributions $\rho_1(x_1)$ and $\rho_2(x_2)$,
and we need to reconstruct the original 2-D distribution $\rho(x_1, x_2)$. In the usual
maximum entropy approach, the corresponding optimization problem leads to the
independence-related formula $\rho(x_1, x_2) = \rho_1(x_1) \cdot \rho_2(x_2)$; see, e.g., [3]. This makes
perfect sense: if we know nothing about the relation between two random variables,
it is reasonable to assume that they are independent.

For our robust approach, however, the situation is less intuitive. Specifically, we
want to find a distribution $\rho(x_1, x_2) \geq 0$ on the box $A = [\underline{a}_1, \overline{a}_1] \times [\underline{a}_2, \overline{a}_2]$ for which
the following conditions are satisfied:

- $\int_A \rho(x_1, x_2) \, dx_1 \, dx_2 = 1$,
- $\int_{\underline{a}_2}^{\overline{a}_2} \rho(x_1, x_2) \, dx_2 = \rho_1(x_1)$ for all x_2, and
- $\int_{\underline{a}_1}^{\overline{a}_1} \rho(x_1, x_2) \, dx_1 = \rho_2(x_2)$ for all x_2.

(Strictly speaking, we do not need the first condition, since it automatically follows
from, e.g., the second one if we integrate both sides over x_1.)

For this constraint optimization problem, the Lagrange multiplier technique means
minimizing the functional

$$\int_A (\rho(x_1, x_2))^2 \, dx_1 \, dx_2 + \int_{\underline{a}_1}^{\overline{a}_1} dx_1 \, \lambda_1(x_1) \cdot \left(\int_{\underline{a}_2}^{\overline{a}_2} \rho(x_1, x_2) \, dx_2 \right) +$$

$$\int_{\underline{a}_2}^{\overline{a}_2} dx_2 \, \lambda_2(x_2) \cdot \left(\int_{\underline{a}_1}^{\overline{a}_1} \rho(x_1, x_2) \, dx_1 \right)$$

for appropriate values $\lambda_i(x_i)$. When $\rho(x_1, x_2) > 0$, differentiating this objective
function with respect to $\rho(x_1, x_2)$ leads to $\rho(x_1, x_2) = a_1(x_1) + a_2(x_2)$, where
$a_1(x_1) = -\lambda_1(x_1)/2$ and $a_2(x_2) = -\lambda_2(x_2)/2$.

In general, we get

$$\rho(x_1, x_2) = \max(0, a_1(x_1) + a_2(x_2)). \tag{18}$$

In particular, when $\rho(x_1, x_2) > 0$ for all $x_1 \in [\underline{a}_1, \overline{a}_1]$ and $x_2 \in [\underline{a}_2, \overline{a}_2]$, then we
get $\rho(x_1, x_2) = a_1(x_1) + a_2(x_2)$. Integrating over x_2, we conclude that

$$\rho_1(x_1) = (\overline{a}_2 - \underline{a}_2) \cdot a_1(x_1) + C_1,$$

where $C_1 \overset{\text{def}}{=} \int_{\underline{a}_1}^{\overline{a}_1} a_2(x_2)$. Thus, $a_1(x_1) = \dfrac{1}{\overline{a}_2 - \underline{a}_2} \cdot \rho_1(x_1) + C_1$, for some
constant C_1.

Similarly, we get $a_2(x_2) = \dfrac{1}{\bar{a}_1 - \underline{a}_1} \cdot p_2(x_2) + C_2$, for some constant C_1. So,

$$\rho(x_1, x_2) = \frac{1}{\bar{a}_2 - \underline{a}_2} \cdot p_1(x_1) + \frac{1}{\bar{a}_1 - \underline{a}_1} \cdot p_2(x_2) + C, \qquad (19)$$

where $C \stackrel{\text{def}}{=} C_1 + C_2$. We can find the constant C if we integrate both sides of this equality over all $x_1 \in [\underline{a}_1, \bar{a}_1]$ and all $x_2 \in [\underline{a}_2, \bar{a}_2]$; we then get

$$1 = 1 + 1 + C \cdot (\bar{a}_1 - \underline{a}_1) \cdot (\bar{a}_2 - \underline{a}_2). \qquad (20)$$

Thus, $C = -\dfrac{1}{(\bar{a}_1 - \underline{a}_1) \cdot (\bar{a}_2 - \underline{a}_2)}$ and so,

$$\rho(x_1, x_2) = \frac{1}{\bar{a}_2 - \underline{a}_2} \cdot p_1(x_1) + \frac{1}{\bar{a}_1 - \underline{a}_1} \cdot p_2(x_2) - \frac{1}{(\bar{a}_1 - \underline{a}_1) \cdot (\bar{a}_2 - \underline{a}_2)}. \qquad (21)$$

When both x_1 and x_2 are uniformly distributed, the result is the uniform distribution on the box in which the random variables x_1 and x_2 are independent—similarly to the maximum entropy approach. However, in general, this is *not* independence, it is a *mixture* of the two distributions—and it is not very clear what is the intuitive meaning of this mixture.

4 Selecting a Probability Distribution that Minimizes $\int (\rho(x))^3 \, dx$

General idea. As we have mentioned earlier, one of the possible ways to describe robustness is to select the probability distribution corresponds to the smallest possible values of the global robustness criterion $\int (\rho(x))^3 \, dx$.

Simplest case, when we only know the bounds. Let us start with the simplest case, when we only know the bounds $\underline{a}$ and $\bar{a}$ on the values of the corresponding random variable x, i.e., we know that always $\underline{a} \le x \le \bar{a}$ and thus, that $\rho(x) = 0$ for values x outside the interval $[\underline{a}, \bar{a}]$.

In this case, the problem of selecting the most robust distribution takes the following form: minimize $\int_{\underline{a}}^{\bar{a}} (\rho(x))^3 \, dx$ under the constraints that $\int_{\underline{a}}^{\bar{a}} \rho(x) \, dx = 1$ and $\rho(x) \ge 0$ for all x. To solve this constrained optimization problem, we can apply the Lagrange multiplier methods to reduce it to an easier-to-solve unconstrained optimization problem

$$\int_{\underline{a}}^{\bar{a}} (\rho(x))^3 \, dx + \lambda \cdot \left(\int_{\underline{a}}^{\bar{a}} \rho(x) \, dx - 1 \right) \to \min_{\rho(x)}. \qquad (22)$$

When $\rho(x) > 0$, then differentiation over $\rho(x)$ leads to $3(\rho(x))^2 + \lambda = 0$, i.e., to $\rho(x) = c$, where $c = \sqrt{-\lambda}$.

Similarly to the case of the criterion $\int (\rho(x))^2 \, dx$, we can conclude that the smallest value of the robustness criterion is attained when $\rho(x) > 0$ for all $x \in [\underline{a}, \overline{a}]$, i.e., when we have a uniform distribution on the given interval. In other words, in this simplest case, we have the same distribution as when we use the first robustness criterion or when we use the maximum entropy approach.

What if we also know several moments? Let us now consider the case, when, in addition to the bounds $\underline{a}$ and $\overline{a}$ on the values of the corresponding random variable x, we also know the values of the first m moments

$$\int_{\underline{a}}^{\overline{a}} x^k \cdot \rho(x) \, dx = M_k, \quad k = 1, 2, \ldots, m. \tag{23}$$

In this case, we need to minimize the functional $\int_{\underline{a}}^{\overline{a}} (\rho(x))^3 \, dx$ under the constraints $\int_{\underline{a}}^{\overline{a}} \rho(x) \, dx = 1$, and $\int_{\underline{a}}^{\overline{a}} x^k \cdot \rho(x) \, dx = M_k$ for $k = 1, \ldots, m$. By using the Lagrange multiplier method, we can reduce this constraint optimization problem to the following unconstrained optimization problem:

$$\int_{\underline{a}}^{\overline{a}} (\rho(x))^3 \, dx + \lambda \cdot \left(\int_{\underline{a}}^{\overline{a}} \rho(x) \, dx - 1 \right) +$$

$$\sum_{k=1}^{m} \lambda_k \cdot \left(\int_{\underline{a}}^{\overline{a}} x^k \cdot \rho(x) \, dx - M_k \right) \to \min \tag{24}$$

under the constraint that $\rho(x) \geq 0$ for all $x \in [\underline{a}, \overline{a}]$.

For the points x for which $\rho(x) > 0$, the derivative of the above expression relative to $\rho(x)$ should be equal to 0, so we conclude that for some x, we have $(\rho(x))^2 = p_0 + \sum_{k=1}^{m} q_k \cdot x^k$, where $p_0 = -\lambda/3$ and $q_k = -\lambda_k/3$. In other words, the probability density $\rho(x)$ is either determined by a square root of a polynomial expression or it is equal to 0. One can check that, in general, the desired minimum is attained when

$$\rho(x) = \sqrt{\max\left(0, p_0 + \sum_{k=1}^{m} q_k \cdot x^k \right)}. \tag{25}$$

Similarly, in the multi-D case, we get

$$\rho(x_1, \ldots, x_d) = \sqrt{\max(0, P(x_1, \ldots, x_d))}, \tag{26}$$

for an appropriate polynomial $P(x_1, \ldots, x_d)$.

The results of this approach are less desirable than the results of using the first robustness criterion. From the computational viewpoint, integrating polynomials is easy, but integrating square roots of polynomials is not easy. From this viewpoint, the first robustness criterion—that was analyzed in the previous section—is much more computationally advantageous that the second robustness criterion that we analyze in this section.

It turns out that square roots also lead to less accurate approximations. Let us illustrate it on the example of approximating a Gaussian distribution by a quadratic polynomial versus by a square root of a quadratic polynomial. Without losing generality, we can restrict ourselves to a standard normal distribution with 0 mean and standard deviation 1, for which the probability density is proportional to

$$f(x) \overset{\text{def}}{=} \exp\left(-\frac{x^2}{2}\right) = 1 - \frac{x^2}{2} + \frac{x^4}{8} + \ldots \tag{27}$$

If we approximate this expression in the vicinity of 0, then the best quadratic approximation corresponds to taking the first two terms in the above Taylor expansion $f_1(x) = 1 - \frac{x^2}{2}$, and the accuracy $\delta_1 = |f_1(x) - f(x)|$ of this approximation is largely determined by the first ignored terms: $\delta_1 \approx \frac{x^4}{8}$.

On the other hand, if we use square roots of quadratic polynomials, then, due to the symmetry of the problem with respect to the transformation $x \rightarrow -x$, we have to use symmetric quadratic polynomials $a \cdot (1 + b \cdot x^2)$. For this polynomial, we have $\sqrt{a \cdot (1 + b \cdot x^2)} = \sqrt{a} \cdot \left(1 + \frac{b \cdot x^2}{2}\right) + \ldots$ For these terms to coincide with the first two terms in the Taylor expansion of the function $f(x)$, we must therefore take $a = 1$ and $b = -1$. For the resulting approximating function $f_2(x) = \sqrt{1 - x^2}$, we have

$$f_2(x) = \sqrt{1 - x^2} = 1 - \frac{x^2}{2} - \frac{x^4}{8} + \ldots \tag{28}$$

Here, the approximation accuracy is equal to $f_2(x) - f(x) = -\frac{x^4}{8} + \ldots$ So asymptotically, the approximation error has the form $\delta_2 = |f_2(x) - f(x)| \approx \frac{x^4}{4}$—which is twice larger than when we use the first robustness criterion

$$\int (\rho(x))^2 \, dx \rightarrow \min. \tag{29}$$

5 Selecting a Probability Distribution that Minimizes $\max\limits_{x} \rho(x)$

Reminder. If we use the worst-case description of robustness, then we should select a distribution $\rho(x)$ for which the value $\max\limits_{x} \rho(x)$ is the smallest possible.

Analysis of the problem. In this case, whatever moments conditions we impose, if there is a point x_0 in the vicinity of which $0 < \rho(x_0) < \max\limits_{x} \rho(x)$, then we can decrease all the value $\rho(x)$ for which $\rho(x) = \max$ by some small amount—compensating it with an appropriate increase in the vicinity of x_0, and satisfy the same criteria while decreasing the value $\max \rho(x)$.

Thus, when the desired criterion $\max\limits_{x} \rho(x)$ is the smallest possible, then for every x, we either have $\rho(x) = 0$ or $\rho(x)$ is equal to this maximum.

This somewhat informal argument can be formally confirmed if we take into account that, that we have mentioned earlier, the worst-case criterion can be viewed as a limit, when $p \to \infty$, of the criteria $\int_{\underline{a}}^{\overline{a}} (\rho(x))^p \, dx \to \min$.

Let us thus consider the case, when, in addition to the bounds $\underline{a}$ and $\overline{a}$ on the values of the corresponding random variable x, we also know the values of the first m moments $\int_{\underline{a}}^{\overline{a}} x^k \cdot \rho(x) \, dx = M_k$, $k = 1, 2, \ldots, m$. In this case, we need to minimize the functional $\int_{\underline{a}}^{\overline{a}} (\rho(x))^p \, dx$ under the constraints $\int_{\underline{a}}^{\overline{a}} \rho(x) \, dx = 1$, and $\int_{\underline{a}}^{\overline{a}} x^k \cdot \rho(x) \, dx = M_k$ for $k = 1, \ldots, m$. By using the Lagrange multiplier method, we can reduce this constraint optimization problem to the following unconstrained optimization problem:

$$\int_{\underline{a}}^{\overline{a}} (\rho(x))^p \, dx + \lambda \cdot \left(\int_{\underline{a}}^{\overline{a}} \rho(x) \, dx - 1 \right) +$$

$$\sum_{k=1}^{m} \lambda_k \cdot \left(\int_{\underline{a}}^{\overline{a}} x^k \cdot \rho(x) \, dx - M_k \right) \to \min \tag{30}$$

under the constraint that $\rho(x) \geq 0$ for all $x \in [\underline{a}, \overline{a}]$.

For the points x for which $\rho(x) > 0$, the derivative of the above expression relative to $\rho(x)$ should be equal to 0, so we conclude that for some x, we have $(\rho(x))^{p-1} = p_0 + \sum_{k=1}^{m} q_k \cdot x^k$, where $p_0 = -\lambda/p$ and $q_k = -\lambda_k/p$. Thus, for such points x, we have $\rho(x) = \text{const} \cdot (P(x))^{1/(p-1)}$ for some polynomial $P(x)$. When $p \to \infty$, we have $1/(p-1) \to 0$, and the 0-th power of a positive number is always 1. Thus, we indeed have $\rho(x) = \text{const}$ whenever $\rho(x) > 0$. A similar conclusion can be made in the multi-D case. So, we arrive at the following conclusion.

Resulting formulas. For the worst-case robustness criterion, for the optimal distribution $\rho(x)$, the probability density is either equal to 0, or to some constant.

To be more precise, both in the 1-D case and in the multi-D case, the zone at which $\rho(x) > 0$ is determined by some polynomial $P(x)$, i.e., we have:

- $\rho(x) = 0$ when $P(x) \leq 0$ and
- $\rho(x) = c$ when $P(x) > 0$.

The value c is determined by the condition that the total probability should be equal to 1: $\int \rho(x)\, dx = 1$, hence $c = \dfrac{1}{A}$, where A is the Lebesque measure (length, areas, volume, etc., depending on the dimension d) of the set of all the points $x = (x_1, \ldots, x_d)$ for which $P(x_1, \ldots, x_d) > 0$.

Shall we recommend this approach? It depends on what we want:

- If the goal is to get a good approximation to the original cdf, then clearly no: in contrast to polynomials, these functions do not have a universal approximation property.
- On the other hand, in critical situations, when we want to minimize worst-case dependence on the input's uncertainty, these are the distributions that we should use.

Acknowledgements We acknowledge the partial support of the Center of Excellence in Econometrics, Faculty of Economics, Chiang Mai University, Thailand. This work was also supported in part by the National Science Foundation grants HRD-0734825 and HRD-1242122 (Cyber-ShARE Center of Excellence) and DUE-0926721, and by an award "UTEP and Prudential Actuarial Science Academy and Pipeline Initiative" from Prudential Foundation.

References

1. Azzalini A, Capitanio A (2013) The Skew-Normal and Related Families. Cambridge University Press, Cambridge, Massachusetts
2. Fishburn PC (1969) Utility Theory for Decision Making. Wiley, New York
3. Jaynes ET, Bretthorst GL (2003) Probability theory: the logic of science. Cambridge University Press, Cambridge
4. Klir G, Yuan B (1995) Fuzzy sets and fuzzy logic. Prentice Hall, Upper Saddle River
5. Kreinovich V, Kosheleva O, Nguyen HT, Sriboonchitta S (2016) Why some families of probability distributions are practically efficient: a symmetry-based explanation. In: Huynh VN, Kreinovich V, Sriboonchitta S (eds) Causal inference in econometrics. Springer, Cham, pp 133–152
6. Luce RD, Raiffa R (1989) Games and decisions: introduction and critical survey. Dover, New York
7. Nguyen HT, Kosheleva O, Kreinovich V (2009) Decision making beyond Arrow's 'impossibility theorem', with the analysis of effects of collusion and mutual attraction. Int J Intell Syst 24(1):27–47
8. Nguyen HT, Kreinovich V, Lea B, Tolbert D (1992) How to control if even experts are not sure: robust fuzzy control. In: Proceedings of the second international workshop on industrial applications of fuzzy control and intelligent systems, College Station, Texas, 2–4 December 1992, pp 153–162

9. Nguyen HT, Kreinovich V, Tolbert D (1993) On robustness of fuzzy logics. In: Proceedings of the 1993 IEEE international conference on fuzzy systems FUZZ-IEEE'93, San Francisco, California, March 1993, vol 1, pp 543–547
10. Nguyen HT, Kreinovich V, Tolbert D (1994) A measure of average sensitivity for fuzzy logics. Int J Uncertainty Fuzziness Knowl Based Syst 2(4):361–375
11. Nguyen HT, Walker EA (2006) A first course in fuzzy logic. Chapman and Hall/CRC, Boca Raton, Florida
12. Novitskii PV, Zograph IA (1991) Estimating the measurement errors. Energoatomizdat, Leningrad (in Russian)
13. Rabinovich SG (2005) Measurement errors and uncertainty. Theory and practice. Springer, Berlin
14. Raiffa H (1970) Decision analysis. Addison-Wesley, Reading
15. Sheskin DJ (2011) Handbook of parametric and nonparametric statistical procedures. Chapman & Hall/CRC, Boca Raton, Florida
16. Zadeh LA (1965) Fuzzy sets. Inf Control 8:338–353

Why Cannot We Have a Strongly Consistent Family of Skew Normal (and Higher Order) Distributions

Thongchai Dumrongpokaphan and Vladik Kreinovich

Abstract In many practical situations, the only information that we have about the probability distribution is its first few moments. Since many statistical techniques require us to select a single distribution, it is therefore desirable to select, out of all possible distributions with these moments, a single "most representative" one. When we know the first two moments, a natural idea is to select a normal distribution. This selection is *strongly consistent* in the sense that if a random variable is a sum of several independent ones, then selecting normal distribution for all of the terms in the sum leads to a similar normal distribution for the sum. In situations when we know three moments, there is also a widely used selection—of the so-called skew-normal distribution. However, this selection is not strongly consistent in the above sense. In this paper, we show that this absence of strong consistency is not a fault of a specific selection but a general feature of the problem: for third and higher order moments, no strongly consistent selection is possible.

1 Formulation of the Problem

Need to select a distribution based on the first few moments. In many practical situations, we only have a partial information about the probability distribution. For example, often, all we know is the values of the first few moments.

Most probabilistic and statistical techniques assume that we know the exact form of a probability distribution; see, e.g., [5]. In situations when we only have partial

---

T. Dumrongpokaphan (✉)
Department of Mathematics, Faculty of Science, Chiang Mai University, Chiang Mai, Thailand
e-mail: tcd43@hotmail.com

V. Kreinovich
Department of Computer Science, University of Texas at El Paso,
500 W. University, El Paso, TX 79968, USA
e-mail: vladik@utep.edu

© Springer International Publishing AG 2017
V. Kreinovich et al. (eds.), *Robustness in Econometrics*,
Studies in Computational Intelligence 692, DOI 10.1007/978-3-319-50742-2_4

information about the probability distribution, there are many probability distributions which are consistent with our knowledge. To use the usual techniques in such a situation, we therefore need to select, from all possible distributions, a single one.

In situations when we know the first two moments, there is a strongly consistent way of selecting a single distribution. In situations when all we know is the first two moments $\mu = \int x \cdot \rho(x)\, dx$ and $M_2 = \int x^2 \cdot \rho(x)\, dx$, a natural idea is to select a distribution for which the entropy (uncertainty) $S = -\int \rho(x) \cdot \ln(\rho(x))\, dx$ is the largest possible; see, e.g., [3].

By applying the Lagrange multiplier techniques, one can easily check that maximizing entropy under the constraints $\int \rho(x)\, dx = 1$, $\mu = \int x \cdot \rho(x)\, dx$ and $M_2 = \int x^2 \cdot \rho(x)\, dx$ leads to the Gaussian (normal) distribution, with probability density

$$\rho(x) = \frac{1}{\sqrt{2\pi}\sigma} \cdot \exp\left(-\frac{(x-\mu)^2}{2\sigma^2}\right),$$

where $\sigma^2 = M_2 - \mu^2$.

This selection is *strongly consistent* in the following sense. Often, the random variable of interest has several components. For example, an overall income consists of salaries, pensions, unemployment benefits, interest on bank deposits, etc. Each of these categories, in its turn, can be subdivided into more subcategories. If for each of these categories, we only know the first two moments, then, in principle, we can apply the selection:

- either to the overall sum,
- or separately to each term,
- or we can go down to the next level of granularity and apply the selection to each term on this granularity level, etc.

It seems reasonable to require that whichever granularity level we select, the resulting distribution for the overall sum should be the same. This is indeed true for normal distributions. Indeed, knowing μ and M_2 is equivalent to knowing μ and the variance σ^2, and it is known that for the sum $X = X_1 + X_2$ of two independent random variables, its mean and variance are equal to the sum of the means and variances of the two components: $\mu = \mu_1 + \mu_2$ and $\sigma^2 = \sigma_1^2 + \sigma_2^2$. So:

- If we apply the selection to the sum itself, we then get a normal distribution with mean μ and standard deviation σ.
- Alternatively, if we first apply the selection to each component, we conclude that X_1 is normally distributed with mean μ_1 and standard deviation σ_1 and X_2 is normally distributed with mean μ_2 and standard deviation σ_2.

It is known that the sum of two independent normally distributed random variables is also normally distributed, with the mean $\mu = \mu_1 + \mu_2$ and the variance $\sigma^2 = \sigma_1^2 + \sigma_2^2$. Thus, in both cases, we indeed get the same probability distribution. This *strong consistency* is one of the reasons why selecting a normal distribution is indeed ubiquitous in practical applications.

Selection of a skew normal distribution is not strongly consistent. A natural next case is when, in addition to the first two moments μ and M_2, we also know the third moment M_3. Alternatively, this can be described as knowing the mean μ, the variance $V = \sigma^2$, and the third central moment

$$m_3 \overset{\text{def}}{=} E[(X - \mu)^3].$$

In this case, we can no longer use the Maximum Entropy approach to select a single distribution. Indeed, if we try to formally maximize the entropy under the four constrains corresponding to the condition $\int \rho(x)\, dx = 1$ and to the three moments, then the Lagrange multiplier method leads to the function

$$\rho(x) = \exp(a_0 + a_1 \cdot x + a_2 \cdot x^2 + a_3 \cdot x^3)$$

which does not satisfy the requirement $\int \rho(x)\, dx = 1$:

- when $a_3 > 0$, then $\rho(x) \to \infty$ as $x \to +\infty$, so $\int_{-\infty}^{\infty} \rho(x)\, dx = \infty$; and
- when $a_3 < 0$, then $\rho(x) \to \infty$ as $x \to -\infty$, so also $\int_{-\infty}^{\infty} \rho(x)\, dx = \infty$.

There is a widely used selection, called a *skew normal* distribution (see, e.g., [2, 4, 6]), when we choose a distribution with the probability density function

$$\rho(x) = \frac{1}{2\omega} \cdot \phi\left(\frac{x - \eta}{\omega}\right) \cdot \Phi\left(\alpha \cdot \frac{x - \eta}{\omega}\right),$$

where:

- $\phi(x) \overset{\text{def}}{=} \dfrac{1}{\sqrt{2\pi}} \cdot \exp\left(-\dfrac{x^2}{2}\right)$ is the pdf of the standard Gaussian distribution, with mean 0 and standard deviation 1, and
- $\Phi(x)$ is the corresponding cumulative distribution function

$$\Phi(x) = \int_{-\infty}^{x} \phi(t)\, dt.$$

For this distribution,

- $\mu = \eta + \omega \cdot \delta \cdot \sqrt{\dfrac{2}{\pi}}$, where $\delta \overset{\text{def}}{=} \dfrac{\alpha}{\sqrt{1 + \alpha^2}}$,
- $\sigma^2 = \omega^2 \cdot \left(1 - \dfrac{2\delta^2}{\pi}\right)$, and
- $m_3 = \dfrac{4 - \pi}{2} \cdot \sigma^3 \cdot \dfrac{(\delta \cdot \sqrt{2/\pi})^3}{(1 - 2\delta^2/\pi)^{3/2}}.$

The skew normal distribution has many applications, but it is *not* strongly consistent in the above sense: in general, the sum of two independent skew normal variables is *not* skew normal, and thus, the result of applying the selection depends on the level of granularity to which this selection is applied.

Natural question. Since the usual selection corresponding to three moments is not strongly consistent, a natural question is:

- is this a fault of this particular selection—and an alternative strongly consistent selection *is* possible,
- or is this a feature of the problem—and for the case of three moments, a strongly consistent selection is not possible?

In this paper, we show that under the reasonable assumption of scale-invariance, for three and more moments, a strongly consistent selection is not possible—and thus, the absence of string consistency is a feature of the problem and not a limitation of the current selection of skew normal distributions.

2   Analysis of the Problem and the Main Result

Let us formulate the selection problem in precise terms. We want to assign, to each triple (μ, V, m_3) consisting of the mean, the variance, and the central third moment m_3, a probability distribution $\rho(x, \mu, V, m_3)$.

Let us list the natural properties of this assignment.

First property: continuity. Moments are rarely known exactly, we usually know them with some accuracy. It is thus reasonable to require that if the moments change slightly, then the corresponding distribution should not change much. In other words, it is reasonable to require that the function $\rho(x, \mu, V, m_3)$ is a continuous function of μ, V, and m_3.

Comment. As we can see from our proof, to prove the impossibility, it is sufficient to impose an even weaker requirement: that the dependence of $\rho(x, \mu, V, m_3)$ on μ, V, and m_3 is *measurable*.

Second property: strong consistency. We require that if X_1 and X_2 are independent random variables for which:

- X_1 is distributed according to the distribution $\rho(x, \mu_1, V_1, m_{31})$, and
- X_2 is distributed according to the distribution $\rho(x, \mu_2, V_2, m_{32})$,

then the sum $X = X_1 + X_2$ is distributed according to the distribution

$$\rho(x, \mu_1 + \mu_2, V_1 + V_2, m_{31} + m_{32}).$$

Final property: scale-invariance. Numerical values of different quantities depend on the choice of a measuring unit. For example, an income can be described in Baht or—to make comparison with people from other countries easier—in dollars. If we change the measuring unit to a new one which is λ times smaller, then the actual incomes will not change, but the numerical values will change: all numerical values will be multiplied by λ: $x \to x' = \lambda \cdot x$.

If we perform the selection in the original units, then we select a distribution with the probability density function $\rho(x, \mu, V, m_3)$. If we simply re-scale x to $x' = \lambda \cdot x$, then for x', we get a new distribution $\rho'(x') = \dfrac{1}{\lambda} \cdot \rho\left(\dfrac{x'}{\lambda}, \mu, V, m_3\right)$.

We should get the exact same distribution if we make a selection *after* the re-scaling to the new units. After this re-scaling:

- the first moment is multiplied by λ: $\mu \to \lambda \cdot \lambda$,
- the variance is multiplied by λ^2: $V \to \lambda^2 \cdot V$, and
- the central third moment is multiplied by λ^3: $m_3 \to \lambda^3 \cdot m_3$.

So, in the new units, we get a probability distribution $\rho(x', \lambda \cdot \mu, \lambda^2 \cdot V, \lambda^3 \cdot m_3)$. A natural requirement is that the resulting selection should be the same, i.e., that we should have

$$\frac{1}{\lambda} \cdot \rho\left(\frac{x'}{\lambda}, \mu, V, m_3\right) = \rho(x', \lambda \cdot \mu, \lambda^2 \cdot V, \lambda^3 \cdot m_3)$$

for all x, λ, μ, V, and m_3.

Comment. One can easily check that the both the above selection corresponding to skew normal distributions is scale-invariant (and that for the case of two moments, the standard normal-distribution selection is also scale-invariant).

Now, we can formulate the problem in precise terms.

Definition 1

- We say that a tuple (μ, V, m_3) is possible *if there exists a probability distribution with mean μ, variance V, and central third moment m_3.*
- By a 3-selection, *we mean a measurable mapping that maps each possible tuple (μ, V, m_3) into a probability distribution $\rho(x, \mu, V, m_3)$.*
- We say that a 3-selection is strongly consistent *if for every two possible tuples, if $X_1 \sim \rho(x, \mu_1, V_1, m_{31})$, $X_2 \sim \rho(x, \mu_2, V_2, m_{32})$, and X_1 and X_2 are independent, then $X_1 + X_2 \sim \rho(x, \mu_1 + \mu_2, V_1 + V_2, m_{31} + m_{32})$.*
- We say that a 3-selection is scale-invariant *if for every possible tuple (μ, V, m_3), for every $\lambda > 0$, and for all x', we have*

$$\frac{1}{\lambda} \cdot \rho\left(\frac{x'}{\lambda}, \mu, V, m_3\right) = \rho(x', \lambda \cdot \mu, \lambda^2 \cdot V, \lambda^3 \cdot m_3).$$

Proposition 1 *No 3-selection is strongly consistent and scale-invariant.*

Comment. A similar result can be formulated—and similarly proven—for the case when we also know higher order moments. In this case, instead of the original moments, we can consider *cumulants* κ_n: terms at $\dfrac{i^n \cdot t^n}{n!}$ in the Taylor expansion of the corresponding generating function $\ln(E[\exp(i \cdot t \cdot X)])$. For $n = 1$, $n = 2$, and $n = 3$, we get exactly the mean, the variance, and the central third moment. In general, cumulants are additive: if $X = X_1 + X_2$ and X_1 and X_2 are independent, then $\kappa_n(X) = \kappa_n(X_1) + \kappa_n(X_2)$.

Discussion. Since we cannot make a strongly consistent selection, what should we do? One possible idea is to use the fact that, in addition to the sum, min and max are also natural operations in many applications. For example, in econometrics, if there are several ways to invest money with the same level of risk, then an investor selects the one that leads to the largest interest rate.

From this viewpoint, once we have normally distributed random variables, it is also reasonable to consider minima and maxima of normal variables. Interestingly, in some cases, these minima and maxima are distributed according to the skew normal distribution. This may be an additional argument in favor of using these distributions.

Proof It is known that when we deal with the sum of independent random variables $X = X_1 + X_2$, then, instead of the original probability density functions, it is more convenient to consider *characteristic functions*

$$\chi_{X_1}(\omega) \stackrel{\text{def}}{=} E[\exp(i \cdot \omega \cdot X_1)], \quad \chi_{X_2}(\omega) \stackrel{\text{def}}{=} E[\exp(i \cdot \omega \cdot X_2)],$$

and

$$\chi_X(\omega) \stackrel{\text{def}}{=} E[\exp(i \cdot \omega \cdot X)].$$

Indeed, for these characteristic functions, we have

$$\chi_X(\omega) = \chi_{X_1}(\omega) \cdot \chi_{X_2}(\omega).$$

Comment. To avoid possible confusion, it is worth noticing that this frequency ω is unrelated to the parameter ω of the skew normal distribution.

Thus, instead of considering the original probability density functions

$$\rho(x, \mu, V, m_3),$$

let us consider the corresponding characteristic functions

$$\chi(\omega, \mu, V, m_3) \stackrel{\text{def}}{=} \int \exp(i \cdot \omega \cdot x) \cdot \rho(x, \mu, V, m_3) \, dx.$$

Since the original dependence $\rho(x, \mu, V, m_3)$ is measurable, its Fourier transform $\chi(\omega, \mu, V, m_3)$ is measurable as well.

In terms of the characteristic functions, the strong consistency requirement takes a simpler form

$$\chi(\omega, \mu_1 + \mu_2, V_1 + V_2, m_{31} + m_{32}) = \chi(\omega, \mu_1, V_1, m_{31}) \cdot \chi(\omega_2, \mu_2, V, m_{32}).$$

This requirement becomes even simpler if we take logarithm of both sides. Then, for the auxiliary functions $\ell(\omega, \mu, V, m_3) \stackrel{\text{def}}{=} \ln(\chi(\omega, \mu, V, m_3))$, we get an even simpler form of the requirement:

$$\ell(\omega, \mu_1 + \mu_2, V_1 + V_2, m_{31} + m_{32}) = \ell(\omega, \mu_1, V_1, m_{31}) + \ell(\omega_2, \mu_2, V, m_{32}).$$

It is known (see, e.g., [1]) that the only measurable functions with this additivity property are linear functions, so

$$\ell(\omega, \mu, V, m_3) = \mu \cdot \ell_1(\omega) + V \cdot \ell_2(\omega) + m_3 \cdot \ell_3(\omega)$$

for some functions $\ell_1(\omega)$, $\ell_2(\omega)$, and $\ell_3(\omega)$.

Let us now use the scale invariance requirement. When we re-scale a random variable X, i.e., replace its numerical values x to new numerical values $x' = \lambda \cdot x$, then, for the new random variable $X' = \omega \cdot X$, we have

$$\chi_{X'}(\omega) = E[\exp(i \cdot \omega \cdot X')] = E[\exp(i \cdot \omega \cdot (\lambda \cdot X))] = E[\exp(i \cdot (\omega \cdot \lambda) \cdot X)] = \chi_X(\lambda \cdot \omega).$$

Thus re-scaled characteristic function $\chi(\lambda \cdot \omega, \mu, V, m_3)$ should be equal to the characteristic function obtained when we use re-scaled values of the moments $\chi(\omega, \lambda \cdot \mu, \lambda^2 \cdot V, \lambda^3 \cdot m_3)$:

$$\chi(\lambda \cdot \omega, \mu, V, m_3) = \chi(\omega, \lambda \cdot \mu, \lambda^2 \cdot V, \lambda^3 \cdot m_3).$$

Their logarithms should also be equal, so:

$$\ell(\lambda \cdot \omega, \mu, V, m_3) = \ell(\omega, \lambda \cdot \mu, \lambda^2 \cdot V, \lambda^3 \cdot m_3).$$

Substituting the above linear expression for the function $\ell(\omega, \mu, V, m_3)$ into this equality, we conclude that

$$\mu \cdot \ell_1(\lambda \cdot \omega) + V \cdot \ell_2(\omega \cdot \omega) + m_3 \cdot \ell_3(\lambda \cdot \omega) =$$
$$\lambda \cdot \mu \cdot \ell_1(\omega) + \lambda^2 \cdot V \cdot \ell_2(\omega) + \lambda^3 \cdot m_3 \cdot \ell_3(\omega).$$

This equality must hold for all possible triples (μ, V, m_3). Thus, the coefficient at μ, V, and m_3 on both sides must coincide.

- By equating coefficients at μ, we conclude that $\ell_1(\lambda \cdot \omega) = \lambda \cdot \ell_1(\omega)$. In particular, for $\omega = 1$, we conclude that $\ell_1(\lambda) = \lambda \cdot \ell_1(1)$, i.e., that $\ell_1(\omega) = c_1 \cdot \omega$ for some constant c_1.

- By equating coefficients at V, we conclude that $\ell_2(\lambda \cdot \omega) = \lambda^2 \cdot \ell_1(\omega)$. In particular, for $\omega = 1$, we conclude that $\ell_2(\lambda) = \lambda^2 \cdot \ell_1(1)$, i.e., that $\ell_2(\omega) = c_2 \cdot \omega^2$ for some constant c_2.
- By equating coefficients at m_3, we conclude that $\ell_3(\lambda \cdot \omega) = \lambda^3 \cdot \ell_3(\omega)$. In particular, for $\omega = 1$, we conclude that $\ell_3(\lambda) = \lambda \cdot \ell_3(1)$, i.e., that $\ell_3(\omega) = c_3 \cdot \omega^3$ for some constant c_3.

Thus, we get $\ell(\omega, \mu, V, m_3) = c_1 \cdot \mu \cdot \omega + c_2 \cdot V \cdot \omega^2 + c_3 \cdot m_3 \cdot \omega^3$. Since, by definition, $\ell(\omega, \mu, V, m_3)$ is the logarithm of the characteristic function, we thus conclude that the characteristic function has the form

$$\chi(\omega, u, V, m_3) = \exp(c_1 \cdot \mu \cdot \omega + c_2 \cdot V \cdot \omega^2 + c_3 \cdot m_3 \cdot \omega^3).$$

In principle, once we know the characteristic function, we can reconstruct the probability density function by applying the inverse Fourier transform. The problem here is that, as one can easily check by numerical computations, the Fourier transform of the above expression is, in general, *not* an everywhere non-negative function—and thus, cannot serve as a probability density function.

This proves that a strongly consistent selection of a probability distribution is indeed impossible.

Comment. If we only consider two moments, then the above proof leads to the characteristic function $\chi(\omega, \mu, V) = \exp(c_1 \cdot \mu \cdot \omega + c_2 \cdot V \cdot \omega^2)$ that describes the Gaussian distribution. This, we have, in effect proven the following auxiliary result:

Definition 2

- *We say that a tuple (μ, V) is* possible *if there exists a probability distribution with mean μ and variance V.*
- *By a 2-selection, we mean a measurable mapping that maps each possible tuple (μ, V) into a probability distribution $\rho(x, \mu, V)$.*
- *We say that a 2-selection is* strongly consistent *if for every two possible tuples, if $X_1 \sim \rho(x, \mu_1, V_1)$, $X_2 \sim \rho(x, \mu_2, V_2)$, and X_1 and X_2 are independent, then $X_1 + X_2 \sim \rho(x, \mu_1 + \mu_2, V_1 + V_2)$.*
- *We say that a 2-selection is* scale-invariant *it if for every possible tuple (μ, V), for every $\lambda > 0$, and for all x', we have*

$$\frac{1}{\lambda} \cdot \rho\left(\frac{x'}{\lambda}, \mu, V\right) = \rho(x', \lambda \cdot \mu, \lambda^2 \cdot V).$$

Proposition 2 *Every strongly consistent and scale-invariant 2-selection assigns, to each possible tuple (μ, V), a Gaussian distribution with mean μ and variance V.*

Acknowledgements This work was supported by Chiang Mai University, Thailand. This work was also supported in part by the National Science Foundation grants HRD-0734825 and HRD-1242122 (Cyber-ShARE Center of Excellence) and DUE-0926721, and by an award "UTEP and Prudential Actuarial Science Academy and Pipeline Initiative" from Prudential Foundation.

References

1. Aczél J, Dhombres J (2008) Functional equations in several variables. Camridge University Press, Cambridge
2. Azzalini A, Capitanio A (2013) The skew-normal and related families. Cambridge University Press, Cambridge
3. Jaynes ET, Bretthorst GL (2003) Probability theory: the logic of science. Cambridge University Press, Cambridge
4. Li B, Shi D, Wang T (2009) Some applications of one-sided skew distributions. Int J Intell Technol Appl Stat 2(1):13–27
5. Sheskin DJ (2011) Handbook of parametric and nonparametric statistical procedures. Chapman and Hall/CRC, Boca Raton
6. Wang T, Li B, Gupta AK (2009) Distribution of quadratic forms under skew normal settings. J Multivar Anal 100:533–545

Econometric Models of Probabilistic Choice: Beyond McFadden's Formulas

Olga Kosheleva, Vladik Kreinovich and Songsak Sriboonchitta

Abstract Traditional decision theory assumes that for every two alternatives, people always make the same (deterministic) choice. In practice, people's choices are often probabilistic, especially for similar alternatives: the same decision maker can sometimes select one of them and sometimes the other one. In many practical situations, an adequate description of this probabilistic choice can be provided by a logit model proposed by 2001 Nobelist D. McFadden. In this model, the probability of selecting an alternative a is proportional to $\exp(\beta \cdot u(a))$, where $u(a)$ is the alternative's utility. Recently, however, empirical evidence appeared that shows that in some situations, we need to go beyond McFadden's formulas. In this paper, we use natural symmetries to come up with an appropriate generalization of McFadden's formulas.

1 Need to Go Beyond McFadden's Probabilistic Choice Models: Formulation of the Problem

Traditional (deterministic choice) approach to decision making. In the traditional (deterministic) approach to decision making (see, e.g., [2, 4, 5, 8, 9]), we assume that for every two alternatives a and b:

- either the decision maker always prefers the alternative a,
- or the decision maker always prefers the alternative b,

O. Kosheleva
University of Texas at El Paso, 500 W. University, El Paso, TX 79968, USA
e-mail: olgak@utep.edu

V. Kreinovich (✉)
Department of Computer Science, University of Texas at El Paso,
500 W. University, El Paso, TX 79968, USA
e-mail: vladik@utep.edu

S. Sriboonchitta
Faculty of Economics, Chiang Mai University, Chiang Mai, Thailand
e-mail: songsakecon@gmail.com

© Springer International Publishing AG 2017
V. Kreinovich et al. (eds.), *Robustness in Econometrics*,
Studies in Computational Intelligence 692, DOI 10.1007/978-3-319-50742-2_5

- or the decision maker always states that the alternatives a and b are absolutely equivalent to him/her.

Under this assumption, preferences of a decision maker can be described by a utility function which can be defined as follows. We select two alternatives which are not present in the original choices:

- a very bad alternative a_0, and
- a very good alternative a_1.

Then, each actual alternative a is better than the very bad alternative a_0 and worse that the very good alternative a_1: $a_0 < a < a_1$. To gauge the quality of the alternative a to the decision maker, we can consider lotteries $L(p)$ in which we get a_1 with probability p and a_0 with the remaining probability $1 - p$.

In accordance with our assumption, for every p, we either have $L(p) < a$ or $a < L(p)$, or we have equivalence $L(p) \sim a$.

When $p = 1$, the lottery $L(1)$ coincides with the very good alternative a_1 and is, thus, better than a: $a < L(1)$. When $p = 0$, the lottery $L(0)$ coincides with the very bad alternative a_0 and is, thus, worse than a: $L(0) < a$. Clearly, the larger the probability p of the very good outcome, the better the lottery; thus, if $p < p'$, then:

- $a < L(p)$ implies $a < L(p')$, and
- $L(p') < a$ implies $L(p) < a$.

Therefore, we can conclude that $\sup\{p : L(p) < a\} = \inf\{p : a < L(p)\}$. This joint value

$$u(a) \overset{\text{def}}{=} \sup\{p : L(p) < a\} = \inf\{p : a < L(p)\}$$

has the following properties:

- if $p < u(a)$, then $L(p) < a$; and
- if $p > u(a)$, then $a < L(p)$.

In particular, for every small $\varepsilon > 0$, we have $L(u(a) - \varepsilon) < a < L(u(a) + \varepsilon)$. In other words, modulo arbitrary small changes in probabilities, the alternative a is equivalent to the lottery $L(p)$ in which a_1 is selected with the probability $p = u(a)$:

$$a \equiv L(u(a)).$$

This probability $u(a)$ is what is known as *utility*.

Once we know all the utility values, we can decide which alternative the decision maker will choose: the one with the largest utility. Indeed, as we have mentioned, $p < p'$ implies that $L(p) < L(p')$, so when $u(a) < u(b)$, we have

$$a \equiv L(u(a)) < L(u(b)) \equiv b$$

and thus, $a < b$.

The above definition of utility depends on the choice of two alternatives a_0 and a_1. If we select two different benchmarks a_0' and a_1', then, as one can show, the new values of the utility are linearly related to the previous ones: $u'(a) = k \cdot u(a) + \ell$, for some real numbers $k > 0$ and ℓ. Thus, *utility is defined modulo linear transformation.*

Actual choices are often probabilistic. In practice, people sometimes make different choices when repeatedly presented with the same pair of alternatives a and b. This is especially true when the compared alternatives a and b are close in value. In such situations, we cannot predict which of the alternatives will be chosen.

The best we can do is try to predict the frequency (probability) $P(a, b)$ with which the decision maker will select a over b. More generally, we would like to predict the probability $P(a, A)$ of selecting an alternative a from a given set of alternatives A that contains a.

In the probabilistic situation, we can still talk about utilities. In the probabilistic case, we can still have a deterministic distinction:

- for some pairs (a, b), the decision maker selects a more frequently than b: $P(a, b) > 0.5$; in such situations, we can say that a is preferable to b ($b < a$);
- for some other pairs (a, b), the decision maker selects b more frequently than a: $P(a, b) < 0.5$; in such situations, we can say that b is preferable to a ($a < b$);
- finally, for some pairs (a, b), the decision maker selects a exactly as many times as b ($a \sim b$): $P(a, b) = 0.5$; in such situations, we can say that to this decision maker, a and b are equivalent.

Usually, the corresponding preference relations are transitive. For example, if $a < b$ and $b < c$, i.e., if in most situations, the decision maker selects b rather than a and c rather than b, then we should expect $a < c$, i.e., we should expect that in most cases, the decision maker will prefer c to a.

Because of this, we can still perform the comparison with lotteries, and thus, come up with the utility $u(a)$ of each alternative—just like we did in the deterministic case. The main difference from the deterministic case is that:

- in the deterministic case, once we know all the utilities, we can uniquely predict which decision the decision maker will make;
- in contrast, in the probabilistic case, after we know the utility values $u(a)$ and $u(b)$, we can predict which of the two alternatives will be selected more frequently, but we still need to find out the probability $P(a, b)$.

Natural assumption. Since the alternatives can be described by their utility values, it is reasonable to assume that the desired probability $P(a, A)$ of selecting an alternative a from the set $A = \{a, \ldots, b\}$ of alternatives depends only on the utilities $u(a)$, ..., $u(b)$ of these alternatives.

McFadden's formulas for probabilistic selection. The 2001 Nobelist D. McFadden proposed the following formula for the desired probability $P(a, A)$:

$$P(a, A) = \frac{\exp(\beta \cdot u(a))}{\sum_{b \in A} \exp(\beta \cdot u(b))};$$

see, e.g., [6, 7, 10]. In many practical situations, this formula indeed describes people's choices really well.

Need to go beyond McFadden's formulas. While McFadden's formula works in many practical situations, in some case, alternative formulas provide a better explanation of the empirical choices; see, e.g., [3] and references therein.

In this paper, we use natural symmetries to come up with an appropriate generalization of McFadden's formulas.

2   Analysis of the Problem

A usual important assumption and its consequences (see, e.g., [4]). In principle, we may have many different alternatives $a, b, \ldots$ In some cases, we prefer a, in other cases, we prefer b. It is reasonable to require that once we have decided on selecting either a or b, then the relative frequency of selecting a should be the same as when we simply select between a and b, with no other alternatives present:

$$\frac{P(a, A)}{P(b, A)} = \frac{P(a, b)}{P(b, a)} = \frac{P(a, b)}{1 - P(a, b)}.$$

Once we make this assumption, we can then describe the general probabilities $P(a, A)$ in terms of function of one variable. Indeed, let us add a new alternative a_n to our list of alternatives. Then, because of our assumption, we have:

$$\frac{P(a, A)}{P(a_n, A)} = \frac{P(a, a_n)}{1 - P(a, a_n)},$$

hence

$$P(a, A) = P(a_n, A) \cdot f(a),$$

where we denoted $f(a) \stackrel{\text{def}}{=} \dfrac{P(a, a_n)}{1 - P(a, a_n)}$. In other words, for every alternative a, we have $P(a, A) = c \cdot f(a)$, where $c \stackrel{\text{def}}{=} P(a_n, A)$ does not depend on a. This constant c can then be found from the condition that one of the alternatives $b \in A$ will be selected, i.e., that $\sum_{b \in A} P(b, A) = 1$. Substituting $p(b, A) = c \cdot f(b)$ into this formula, we conclude that $c \cdot \sum_{b \in A} f(b) = 1$, hence

$$c = \frac{1}{\sum\limits_{b \in A} f(b)},$$

and thus,

$$P(a, A) = \frac{f(a)}{\sum_{b \in A} f(b)}.$$

So, the probabilities are uniquely determined by some values $f(a)$ corresponding to different alternatives. Since we assumed that the probabilities depend only on the utilities $u(a)$, we thus conclude that $f(a)$ must depend only on the utilities, i.e., that we have $f(a) = F(u(a))$ for some function $F(u)$. In terms of this function, the above formula for the probabilistic choice takes the form

$$P(a, A) = \frac{F(u(a))}{\sum_{b \in A} F(u(b))}. \tag{1}$$

From this viewpoint, all we need to do is to find an appropriate function $F(u)$.

The function $F(u)$ must be monotonic. The better the alternative a, i.e., the larger its utility $u(a)$, the higher should be the probability that we select this alternative. Thus, it is reasonable to require that the function $F(u)$ is an increasing function of the utility u.

The function $F(u)$ is defined modulo a constant factor. The above formula does not uniquely define the function $F(u)$: indeed, if we multiply all the values of $F(u)$ by a constant, i.e., consider the new function $F'(u) = C \cdot F(u)$, then in the Formula (1), constants C in the numerator and the denominator will cancel each other, and thus, we will get the exact same probabilities.

Vice versa, if two functions $F(u)$ and $F'(u)$ always lead to the same probabilities, this means, in particular, that for every two utility values u_1 and u_2, we have

$$\frac{F(u_2)}{F(u_1) + F(u_2)} = \frac{F'(u_2)}{F'(u_1) + F'(u_2)}.$$

Reversing both sides of this equality and subtracting 1 from both sides, we conclude that

$$\frac{F(u_1)}{F(u_2)} = \frac{F'(u_1)}{F'(u_2)},$$

i.e., equivalently, that

$$\frac{F'(u_1)}{F(u_1)} = \frac{F'(u_2)}{F(u_2)}.$$

In other words, the ratio $\dfrac{F'(u)}{F(u)}$ is the same for all utility values u, and is, therefore, a constant C. Thus, in this case, $F'(u) = C \cdot F(u)$.

So, two functions $F(u)$ and $F'(u)$ always lead to the same probabilities if and only if their differ by a constant factor.

In these terms, how can we explain the original McFadden's formulas. As we have mentioned earlier, utilities are defined modulo a general linear transformation. In particular, it is possible to add a constant to all the utility values

$$u(a) \rightarrow u'(a) = u(a) + c$$

and still get the description of exactly the same preferences. Since this shift does not change the preferences, it is therefore reasonable to require that after such a shift, we get the exact same probabilities.

Using new utility values $u'(a) = u(a) + c$ means that we replace the values $F(u(a))$ with the values $F(u'(a)) = F(u(a) + c)$. This is equivalent to using the *original* utility values but with a *new function* $F'(u) \stackrel{\text{def}}{=} F(u + c)$.

As we have mentioned earlier, the requirement that the two functions $F(u)$ and $F'(u)$ describe the same probabilities is equivalent to requiring that $F'(u) = C \cdot F(u)$ for some constant C, so $F(u + c) = C \cdot F(u)$. The factor C is, in general, different for different shifts c: $C = C(c)$. Thus, we conclude that

$$F(u + c) = C(c) \cdot F(u). \tag{2}$$

It is known (see, e.g., [1]) that every monotonic solution to this function equation has the form $F(u) = C_0 \cdot \exp(\beta \cdot u)$. This is exactly McFadden's formula.

Discussion. For a general monotonic function, the proof of the function-equation result may be somewhat complicated. However, under a natural assumption that the function $F(u)$ is differentiable, this result can be proven rather easily.

First, we take into account that $C(c)$ is a ratio of two differentiable functions $F(u + c)$ and $F(u)$, and is, thus, differentiable itself. Since both functions $F(u)$ and $C(c)$ are differentiable, we can differentiate both sides of the equality (2) by c and take $c = 0$. As a result, we get the following equality:

$$\frac{dF}{du} = \beta \cdot F,$$

where we denoted $\beta \stackrel{\text{def}}{=} \dfrac{dC}{dc}_{|c=0}$. By moving all the terms containing the unknown F to one side and all other terms to the other side, we conclude that:

$$\frac{dF}{F} = \beta \cdot du.$$

Integrating both sides of this equality, we get $\ln(F) = \beta \cdot u + C_1$, where C_1 is the integration constant. Thus, $F(u) = \exp(\ln(F)) = C_0 \cdot \exp(\beta \cdot u)$, where we denoted $C_0 \stackrel{\text{def}}{=} \exp(C_1)$.

Our main idea and the resulting formulas. Please note that while adding a constant to all the utility values does not change the probabilities computed by using

McFadden's formula, multiplying all the utility values by a constant—which is also a legitimate transformation for utilities—*does* change the probabilities.

Therefore, we cannot require that the probability Formula (1) not change for all possible linear transformations of utility: once we require shift-invariance, we get McFadden's formula which is not scale-invariant.

Since we cannot require invariance with respect to *all* possible re-scalings of utility, we should require invariance with respect to *some* family of re-scalings.

If a formula does not change when we apply each transformation, it will also not change if we apply them one after another, i.e., if we consider a composition of transformations. Each shift can be represented as a superposition of many small (infinitesimal) shifts, i.e., shifts of the type $u \to u + B \cdot dt$ for some B. Similarly, each scaling can be represented as a superposition of many small (infinitesimal) scalings, i.e., scalings of the type $u \to (1 + A \cdot dt) \cdot u$. Thus, it is sufficient to consider invariance with respect to an infinitesimal transformation, i.e., a linear transformation of the type

$$u \to u' = (1 + A \cdot dt) \cdot u + B \cdot dt.$$

Invariance means that the values $F(u')$ lead to the same probabilities as the original values $F(u)$, i.e., that $F(u')$ is obtained from $F(u)$ by an appropriate (infinitesimal) re-scaling $F(u) \to (1 + C \cdot dt) \cdot F(u)$. In other words, we require that

$$F((1 + A \cdot dt) \cdot u + B \cdot dt) = (1 + C \cdot dt) \cdot F(u),$$

i.e., that

$$F(u + (A \cdot u + B) \cdot dt) = F(u) + C \cdot F(u) \cdot dt. \tag{3}$$

Here, by definition of the derivative, $F(u + q \cdot dt) = F(u) + \dfrac{dF}{du} \cdot q \cdot dt$. Thus, from (3), we conclude that

$$F(u) + (A \cdot u + B) \cdot \frac{dF}{du} \cdot dt = F(u) + C \cdot F(u) \cdot dt.$$

Subtracting $F(u)$ from both sides and dividing the resulting equality by dt, we conclude that

$$(A \cdot u + B) \cdot \frac{dF}{du} = C \cdot F(u).$$

We can separate the variables by moving all the terms related to F to one side and all the terms related to u to another side. As a result, we get

$$\frac{dF}{F} = C \cdot \frac{du}{A \cdot u + b}.$$

We have already shown that the case $A = 0$ leads to McFadden's formulas. So, to get a full description of all possible probabilistic choice formulas, we need to consider the cases when $A \neq 0$. In these cases, for $x \stackrel{\text{def}}{=} u + k$, where $k \stackrel{\text{def}}{=} \dfrac{B}{A}$, we have

$$\frac{dF}{F} = c \cdot \frac{dx}{x},$$

where $c \stackrel{\text{def}}{=} \dfrac{C}{A}$. Integration leads to $\ln(F) = c \cdot \ln(x) + C_0$ for some constant C_0, thus $F = C_1 \cdot x^c$ for $C_1 \stackrel{\text{def}}{=} \exp(C_0)$, i.e., to

$$F(u) = C_1 \cdot (u + k)^c. \tag{4}$$

Conclusions and discussion. In addition to the original McFadden's formula for the probabilistic choice, we can also have the case when $F(u)$ is described by the formula (4) and where, therefore, the probabilistic choice is described by the formula

$$P(a, A) = \frac{(u(a) + k)^c}{\sum_{b \in A} (u(b) + k)^c}.$$

This expression is in good accordance with the empirical dependencies described in [3] that also contain power-law terms.

It is worth mentioning that while we derived the new formula as an *alternative* to McFadden's formula, this new formula can be viewed as a *generalization* of McFadden's formula. Indeed, it is known that $\exp(u) = \lim\limits_{n \to \infty} \left(1 + \dfrac{u}{n}\right)^n$ and thus, $\exp(\beta \cdot u) = \lim\limits_{n \to \infty} \left(1 + \dfrac{\beta \cdot u}{n}\right)^n$. Thus, when n is large, the use of McFadden's expression $F(u) = \exp(\beta \cdot u)$ is practically indistinguishable from the use of the power-law expression $F_{\approx}(u) = \left(1 + \dfrac{\beta \cdot u}{n}\right)^n$. This power-law expression, in its turn, can be represented in the form (4), with $c = n$, $k = \dfrac{n}{\beta}$, and $C_1 = \left(\dfrac{\beta}{n}\right)^n$.

So, instead of a 1-parametric McFadden's formula, we now have a 2-parametric formula. We can use this additional parameter to get an even more accurate description of the actual probabilistic choice.

Acknowledgements This work was supported by Chiang Mai University, Thailand. This work was also supported in part by the National Science Foundation grants HRD-0734825 and HRD-1242122 (Cyber-ShARE Center of Excellence) and DUE-0926721, and by an award "UTEP and Prudential Actuarial Science Academy and Pipeline Initiative" from Prudential Foundation.

References

1. Aczél J, Dhombres J (2008) Functional equations in several variables. Camridge University Press, Cambridge, UK
2. Fishburn PC (1969) Utility theory for decision making. Wiley Inc., New York
3. Jakubczyk M (2016) Estimating the membership function of the fuzzy willingness-to-pay/accept for health via Bayesin modeling. In: Proceedings of the 6th world conference on soft computing, Berkeley, California, 22–25 May 2016
4. Luce D (2005) Inividual choice behavior: a theoretical analysis. Dover, New York
5. Luce RD, Raiffa R (1989) Games and decisions: introduction and critical survey. Dover, New York
6. McFadden D (1974) Conditional logit analysis of qualitative choice behavior. In: Zarembka P (ed) Frontiers in econometrics. Academic Press, New York, pp 105–142
7. McFadden D (2001) Economic choices. Am Econ Rev 91:351–378
8. Nguyen HT, Kosheleva O, Kreinovich V (2009) Decision making beyond Arrow's 'impossibility theorem', with the analysis of effects of collusion and mutual attraction. Int J Intell Syst 24(1):27–47
9. Raiffa H (1970) Decision analysis. Addison-Wesley, Reading
10. Train K (2003) Discrete choice methods with simulation. Cambridge University Press, Cambridge

How to Explain Ubiquity of Constant Elasticity of Substitution (CES) Production and Utility Functions Without Explicitly Postulating CES

Olga Kosheleva, Vladik Kreinovich and Thongchai Dumrongpokaphan

Abstract In many situations, the dependence of the production or utility on the corresponding factors is described by the CES (Constant Elasticity of Substitution) functions. These functions are usually explained by postulating two requirements: an economically reasonable postulate of homogeneity (that the formulas should not change if we change a measuring unit) and a less convincing CSE requirement. In this paper, we show that the CES requirement can be replaced by a more convincing requirement—that the combined effect of all the factors should not depend on the order in which we combine these factors.

1 Formulation of the Problem

CES production functions and CES utility function are ubiquitous. Most observed data about production y is well described by the *CES production function*

$$y = \left(\sum_{i=1}^{n} a_i \cdot x_i^r \right)^{1/r}, \tag{1}$$

O. Kosheleva
University of Texas at El Paso, 500 W. University, El Paso, TX 79968, USA
e-mail: olgak@utep.edu

V. Kreinovich (✉)
Department of Computer Science, University of Texas at El Paso, 500 W. University, El Paso, TX 79968, USA
e-mail: vladik@utep.edu

T. Dumrongpokaphan
Department of Mathematics, Faculty of Science, Chiang Mai University, Chiang Mai, Thailand
e-mail: tcd43@hotmail.com

© Springer International Publishing AG 2017
V. Kreinovich et al. (eds.), *Robustness in Econometrics*,
Studies in Computational Intelligence 692, DOI 10.1007/978-3-319-50742-2_6

where x_i are the numerical measures of the factors that influence production, such as amount of capital, amount of labor, etc.; see, e.g., [6, 17, 18, 23].

A similar Formula (1) describes how the person's utility y depends on different factors x_i such as amounts of different types of consumer goods, utilities of other people, etc.; see, e.g., [7, 12, 13, 28].

How this ubiquity is explained now. The current explanation for the empirical success of CES function is based on the following two requirements.

The first requirement is that the corresponding function $y = f(x_1, \ldots, x_n)$ is *homogeneous*, i.e., that:

$$f(\lambda \cdot x_1, \ldots, \lambda \cdot x_n) = \lambda \cdot f(x_1, \ldots, x_n). \tag{2}$$

This requirement makes perfect economic sense: e.g., we can describe different factors by using different monetary units, and the results should not change if we replace the original unit by a one which is λ times smaller. After this replacement, the numerical value of each factor changes from x_i to $\lambda \cdot x_i$ and y is replace by $\lambda \cdot y$. The value $f(\lambda \cdot x_1, \ldots, \lambda \cdot x_n)$ that we obtain by using the new units should thus be exactly λ times larger than the value $f(x_1, \ldots, x_n)$ obtained in the original units—and this is exactly the requirement (2).

The second requirement is that the corresponding function $f(x_1, \ldots, x_n)$ should provide *constant elasticity of substitution* (CES). The requirement is easier to explain for the case of two factors $n = 2$. In this case, this requirement deals with "substitution" situations in which we change x_1 and then change the original value x_2 to the new value $x_2(x_1)$ so that the overall production or utility remain the same.

The corresponding substitution rate can then be calculated as $s \overset{\text{def}}{=} \dfrac{dx_2}{dx_1}$. The substitution function $x_2(x_1)$ is explicitly defined by the equation $f(x_1, x_2(x_1)) = \text{const}$. By using the formula for the derivative of the implicit function, we can conclude that the substitution rate has the form

$$s = -\frac{f_{,1}(x_1, x_2)}{f_{,2}(x_1, x_2)},$$

where we denoted

$$f_{,1}(x_1, x_2) \overset{\text{def}}{=} \frac{\partial f}{\partial x_1}(x_1, x_2) \text{ and } f_{,2}(x_1, x_2) \overset{\text{def}}{=} \frac{\partial f}{\partial x_2}(x_1, x_2).$$

The requirement is that for each percent of the change in ratio $\dfrac{x_2}{x_1}$, we get the same constant number of percents change in s:

$$\frac{ds}{d\left(\dfrac{x_2}{x_1}\right)} = \text{const}.$$

This explanation needs strengthening. While homogeneity is a reasonable require-ment, the above CES condition sounds somewhat too mathematical to be fully con-vincing for economists.

To explain the ubiquity of CSE production and utility functions, it is therefore desirable to come up with additional—hopefully, more convincing—arguments in favor of these functions. This is what we intend to do in this paper.

2 Main Idea Behind a New Explanation

Main idea. In our explanation, we will use the fact that in most practical situations, we combine several factors. We can combine these factors in different order:

- For example, we can first combine the effects of capital and labor into a single characteristic that describes the joint even of both factors, and then combine it with other factors.
- Alternatively, we can first combine capital with other factors, and only then com-bine the resulting combined factor with labor, etc.

The result should not depend on the order in which we perform these combinations.

What we do in this paper. In this paper, we show that this idea implies the CES functions. Thus, we indeed get a new explanation for the ubiquity of CES production and utility functions.

3 Derivation of the CES Functions from the Above Idea

Towards formalizing our idea. Let us denote a function that combines factors i and j into a single quantity x_{ij} by $f_{i,j}(x_i, x_j)$. Similarly, let us denote a function that combines the values x_{ij} and $x_{k\ell}$ into a single quantity $x_{ijk\ell}$ by $f_{ij,k\ell}(x_{ij}, x_{k\ell})$. In these terms, the requirement that the resulting values do not depend on the order implies, e.g., that we always have

$$f_{12,34}(f_{1,2}(x_1, x_2), \ f_{3,4}(x_3, x_4)) = f_{13,24}(f_{1,3}(x_1, x_3), \ f_{2,4}(x_2, x_4)). \tag{3}$$

Additional requirement. In both production and utility situations, for each i and j, the combination function $f_{i,j}(x_i, x_j)$ is an increasing function of both variables x_i and x_j. It is reasonable to require that it is continuous, and then when one of the factors tends to infinity, the result also tends to infinity. Under these reasonable assumptions, the combination functions tends out to be *invertible* in the following sense:

Definition 1 *A function* $f : A \times B \to C$ *is called* invertible *if the following two conditions are satisfied:*

- *for every* $a \in A$ *and for every* $c \in C$, *there exists a unique value* $b \in B$ *for which* $c = f(a, b)$;
- *for every* $b \in B$ *and for every* $c \in C$, *there exists a unique value* $a \in A$ *for which* $f(a, b) = c$.

Comment. In mathematics, functions invertible in the sense of Definition 1 are called *generalized quasigroups*; see, e.g., [5].

Let us now formalize the above requirement.

Definition 2 *Let* X_i, X_{ij}, *and* X *be sets, where* $i = 1, 2, 3, 4$. *We say that invertible operations* $f_{i,j} : X_i \times X_j \to X_{ij}$ *and* $f_{ij,k\ell} : X_{ij} \times X_{k\ell} \to X$ *(for different* i, j, k, *and* ℓ) *satisfy the generalized associativity requirement if for all* $x_i \in X_i$, *we have*

$$f_{12,34}(f_{1,2}(x_1, x_2), \ f_{3,4}(x_3, x_4)) = f_{13,24}(f_{1,3}(x_1, x_3), \ f_{2,4}(x_2, x_4)). \quad (4)$$

Comment. In mathematical terms, this requirement is known as *generalized mediality* [5].

Groups and Abelian groups: reminder. To describe operations that satisfy the generalized associativity requirement, we need to recall that a set G with an associative operation $g(a, b)$ and a unit element e (for which $g(a, e) = g(e, a) = a$) is called a *group* if every element is invertible, i.e., if for every a, there exists an a' for which $g(a, a') = e$. A group in which the operation $g(a, b)$ is commutative is known as *Abelian*.

Proposition [3–5, 25–27] *For every set of invertible operations that satisfy the generalized associativity requirement, there exists an Abelian group* G *and 1–1 mappings* $r_i : X_i \to G$, $r_{ij} : X_{ij} \to G$ *and* $r_X : X \to G$ *for which, for all* $x_i \in X_i$ *and* $x_{ij} \in X_{ij}$, *we have*

$$f_{ij}(x_i, x_j) = r_{ij}^{-1}(g(r_i(x_i), r_j(x_j))) \ \ and$$

$$f_{ij,kl}(x_{ij}, x_{k\ell}) = r_X^{-1}(g(r_{ij}(x_{ij}), r_{k\ell}(x_{k\ell}))).$$

Discussion. All continuous 1-D Abelian groups with order-preserving operations are isomorphic to the additive group of real numbers, with $g(a, b) = a + b$. Thus, we can conclude that all combining operations have the form

$$f_{ij}(x_i, x_j) = r_{ij}^{-1}(r_i(x_i) + r_j(x_j)), \quad (5)$$

i.e., equivalently, $f_{ij}(x_i, x_j) = y$ means that

$$r_{ij}(y) = r_i(x_i) + r_j(x_j). \quad (6)$$

Let us use homogeneity: result. We will now prove that homogeneity leads exactly to the desired CES combinations. This will give us the desired new explanation of the ubiquity of the CES operations.

Homogeneity leads to CES operations: proof. Homogeneity means that if the relation (6) holds for some values x_i, x_j, and y, then, for every λ, a similar relation holds for re-scaled values $\lambda \cdot x_i$, $\lambda \cdot x_j$, and $\lambda \cdot y$, i.e.:

$$r_{ij}(\lambda \cdot y) = r_i(\lambda \cdot x_i) + r_j(\lambda \cdot x_j).$$

To utilize this requirement, let us use the idea of substitution: for each possible value $x_i' = x_i + \Delta x_i$, let us find the corresponding value $x_j' = x_j + \Delta x_j$ for which the right-hand side of the Formula (6) remains the same—and thus, the combined value y remains the same:

$$r_i(x_i') + r_j(x_j') = r_i(x_i + \Delta x_i) + r_j(x_j + \Delta x_j) = r_i(x_i) + r_j(x_j). \qquad (7)$$

In general, the substitute value x_j' is a function of x_i'. $x_j' = x_j'(x_i')$. When $\Delta x_i = 0$, i.e., when $x_i' = x_i$, we clearly have $x_j' = x_j$, so $\Delta x_j = 0$. For small Δx_i, we get $y_j' = y_j + k \cdot \Delta x_i + o(\Delta x_i)$, where $k \stackrel{\text{def}}{=} \dfrac{dx_j'}{dx_i'}$, so $\Delta x_j = k \cdot \Delta x_i + o(\Delta x_i)$ for some real number k.

Here, $r_i(x_i + \Delta x_i) = r_i(x_i) + r_i'(x_i) \cdot \Delta x_i + o(\Delta x_i)$, where, as usual, f' denotes the derivative. Similarly,

$$r_j(x_j + \Delta x_j) = r_j(x_j + k \cdot \Delta x_i + o(\Delta x_i)) = r_j(x_j) + k \cdot r_j'(x_j) \cdot \Delta x_i + o(\Delta x_i).$$

Thus, the condition (7) takes the form

$$r_i(x_i) + r_j(x_j) + (r_i'(x_i) + k \cdot r_j'(x_j)) \cdot \Delta x_i + o(\Delta x_i) = r_i(x_i) + r_j(x_j).$$

Subtracting the right-hand side from the both sides, dividing both sides of the resulting equation by Δx_i, and tending Δx_i to 0, we conclude that

$$r_i'(x_i) + k \cdot r_j'(x_j) = 0,$$

i.e., that

$$k = -\frac{r_i'(x_i)}{r_j'(x_j)}. \qquad (8)$$

Homogeneity means, in particular, that if now apply the combination function r_{ij} to the values

$$\lambda \cdot x_i' = \lambda \cdot x_i + \lambda \cdot \Delta x_i$$

and
$$\lambda \cdot x'_j = \lambda \cdot x_j + \lambda \cdot k \cdot \Delta x_i + o(\Delta x_i),$$

then we should get the value $\lambda \cdot y$. So:

$$r_i(\lambda \cdot x_i + \lambda \cdot \Delta x_i) + r_j(\lambda \cdot x_j + \lambda \cdot k \cdot \Delta x_i + o(\Delta x_i)) = \\ r_i(\lambda \cdot x_i) + r_j(\lambda \cdot x_j). \tag{9}$$

For small Δx_i, we have

$$r_i(\lambda \cdot x_i + \lambda \cdot \Delta x_i) = r(\lambda \cdot x_i) + \lambda \cdot \Delta x_i \cdot r'_i(\lambda \cdot x_i) + o(\Delta x_i),$$

where f' denote a derivative, and similarly,

$$r_j(\lambda \cdot x_j + \lambda \cdot k \cdot \Delta x_i + o(\Delta x_1)) = r(\lambda \cdot x_2) + \lambda \cdot k \cdot \Delta x_i \cdot r'_j(\lambda \cdot x_j) + o(\Delta x_i).$$

Substituting these expressions into the Formula (9), we conclude that

$$r_i(\lambda \cdot x_i) + \lambda \cdot \Delta x_i \cdot r'(\lambda \cdot x_i) + r_j(\lambda \cdot x_j) + \lambda \cdot k \cdot r'_j(\lambda \cdot x_j) \cdot \Delta x_i + o(\Delta x_i) = \\ r_i(\lambda \cdot x_i) + r_j(\lambda \cdot x_j).$$

Subtracting the right-hand side from the left-hand side, dividing the result by Δx_i and tending Δx_i to 0, we conclude that

$$r'(\lambda \cdot x_i) + k \cdot r'_j(\lambda \cdot x_j) = 0,$$

i.e., in view of the Formula (8), that

$$r'(\lambda \cdot x_i) - \frac{r'_i(x_i)}{r'_j(x_j)} \cdot r'_j(\lambda \cdot x_j) = 0.$$

Moving the second term to the right-hand side and dividing both sides by $r'_i(x_i)$, we conclude that

$$\frac{r'_i(\lambda \cdot x_i)}{r'_i(x_i)} = \frac{r'_j(\lambda \cdot x_j)}{r'_j(x_j)}.$$

The right-hand side of this formula does not depend on x_i at all, thus, the left-hand side also does not depend on x_i, it only depends on λ:

$$\frac{r'_i(\lambda \cdot x_i)}{r'_i(x_i)} = c(\lambda).$$

for some function $c(\lambda)$. Thus, the derivative $R_i(x_i) \overset{\text{def}}{=} r_i'(x_i)$ satisfies the functional equation

$$R_i(\lambda \cdot x_i) = R_i(x_i) \cdot c(\lambda)$$

for all λ and x_i.

It is know that every continuous solution to this equation has the form $r_i'(x_i) = R_i(x_i) = A_i \cdot x_i^{\alpha_i}$ for some A_i and α_i; see, e.g., [5]. For differentiable functions, this can be easily proven if we differentiate both sides of this equation by c and take $c = 1$. Then, we get $x_i \cdot \dfrac{dR_i}{dc_i} = c \cdot R_i$. Separating variables, we get

$$\frac{dR_i}{R_i} = c \cdot \frac{dx_i}{x_i}.$$

Integration leads to $\ln(R_i) = c \cdot \ln(x_i) + C_1$ and thus, to the desired formula.

Integrating the above expression for $r_i'(x_i)$, we get $r_i(x_i) = a_i \cdot x_i^{\beta_i} + C_i$ and similarly, $r_j(x_j) = a_j \cdot x_j^{\beta_j} + C_j$. One can easily check that homogeneity implies that $\beta_i = \beta_j$ and $C_i + C_j = 0$, so the sum $r_i(x_i) + r_j(x_j)$ takes the form $a_i \cdot x_i^r + a_j \cdot x_j^r$.

By considering a similar substitution between x_i and y (in which x_j remains intact), we conclude that $r_{ij}(y) = \text{const} \cdot y^r$, so we indeed get the desired formula $r_{ij}(x_i, x_j) = (a_i \cdot x_i^r + a_j \cdot x_j^r)^{1/r}$. By using similar formulas to combine x_{ij} with x_k, etc., we get the desired CES combination function.

4 Possible Application to Copulas

What is a copula: a brief reminder. Specifically, a 1-D probability distribution of a random variable X can be described by its *cumulative distribution function* (cdf) $F_X(x) \overset{\text{def}}{=} \text{Prob}(X \le x)$. A 2-D distribution of a random vector (X, Y) can be similarly described by its 2-D cdf $F_{XY}(x, y) = \text{Prob}(X \le x \,\&\, Y \le y)$.

It turns out that we can always describe $F(x, y)$ as

$$F_{XY}(x, y) = C_{XY}(F_X(x), F_Y(y))$$

for an appropriate function $C_{XY} : [0, 1] \times [0, 1] \to [0, 1]$ known as a *copula*; see, e.g., [20, 22].

For a joint distribution of several random variables $X, Y, \ldots, Z$, we can similarly write

$$F_{XY\ldots Z}(x, y, \ldots, z) \overset{\text{def}}{=} \text{Prob}(X \le x \,\&\, Y \le y \,\&\, \ldots \,\&\, Z \le z) =$$

$$C_{XY\ldots Z}(F_X(x), F_Y(y), \ldots, F_Z(z))$$

for an appropriate multi-D copula $C_{XY\ldots Z}$.

Vine copulas. When we have many ($n \gg 1$) random variables, then to exactly describe their joint distribution, we need to describe a general function of n variables. Even if we use two values for each variable, we get 2^n combinations, which for large n can be astronomically large. Thus, a reasonable idea is to approximate the multi-D distribution.

A reasonable way to approximate is to use 2-D copulas. For example, to describe a joint distribution of three variables X, Y, and Z, we first describe the joint distribution of X and Y as $F_{XY}(x, y) = C_{XY}(F_X(x), F_Y(y))$, and then use an appropriate copula $C_{XY,Z}$ to combine it with $F_Z(z)$:

$$F_{XYZ}(x, y, z) \approx C_{XY,Z}(F_{XY}(x, y), F_Z(z)) = C_{XY,Z}(C_{XY}(F_X(x), F_Y(y), F_Z(z)).$$

Such an approximation, when copulas are applied to one another like a vine, are known as *vine copulas*; see, e.g., [1, 2, 8, 9, 11, 14–16, 19, 21, 24].

Natural analogue of associativity. It is reasonable to require that the result of the vine copula approximation should not depend on the order in which we combine the variables. In particular, for four random variables X, Y, Z, and T, we should get the same result in the following two situations:

- if we first combine X with Y, Z and T, and then combine the two results; or
- if we first combine X with Z, Y with T, and then combine the two results.

Thus, we require that for all possible real numbers x, y, z, and t, we get

$$C_{XY,ZT}(C_{XY}(F_X(x), F_Y(y)), C_{ZT}(F_Z(z), F_T(t))) =$$

$$C_{XZ,YT}(C_{XZ}(F_X(x), F_Z(z)), C_{YT}(F_Y(y), F_T(t))).$$

If we denote $a = F_X(x)$, $b = F_Y(y)$, $c = F_Z(z)$, and $d = F_T(t)$, we conclude that for every a, b, c, and d, we have

$$C_{XY,ZT}(C_{XY}(a, b), C_{ZT}(c, d)) = C_{XZ,YT}(C_{XZ}(a, c), C_{YT}(b, d)).$$

This is exactly the generalized associativity requirement. Thus, if we extend copulas to invertible operations, then we can conclude that copulas can be *re-scaled to associative operations*—in the sense of the above Proposition.

Acknowledgements This work was supported in part by the National Science Foundation grants HRD-0734825 and HRD-1242122 (Cyber-ShARE Center of Excellence) and DUE-0926721, and by an award "UTEP and Prudential Actuarial Science Academy and Pipeline Initiative" from Prudential Foundation.

References

1. Kurowicka D, Joe H (eds) (2010) Dependence modeling: Vine copula handbook. World Scientific, Singapore
2. Aas K, Czado C, Frigessi A, Bakken H (2009) Pair-copula constructions of multiple dependence. Insurance Math Econ 44:82–198
3. Aczeél J (1965) Quasigroup-net-nomograms. Adv Math 1:383–450
4. Aczél J, Belousov VD, Hosszú M (1960) Generalized associativity and bisymmetry on quasigroups. Acta Math Acad Sci Hungar 11:127–136
5. Aczél J, Dhombres J (2008) Functional equations in several variables. Camridge University Press, Cambridge
6. Arrow KJ, Chenery HB, Minhas BS, Solow RM (1961) Capital-labor substitution and economic efficiency. Rev Econ Stat 43(3):225–250
7. Baltas G (2001) Utility-consistent brand demand systems with endogeneous category consumption: principles and marketing applications. Decis Sci 32(3):399–421
8. Bedford T, Cooke RM (2001) Monte Carlo simulation of vine dependent random variables for applications in uncertainty analysis. In: Proceedings of European safety and reliability conference ESREL'2001, Turin, Italy
9. Bedford T, Cooke RM (2002) Vines—a new graphical model for dependent random variables. Ann Stat 30(4):1031–1068
10. Buchanan BG, Shortliffe EH (1984) Rule based expert systems: the MYCIN experiments of the stanford heuristic programming project. Addison-Wesley, Reading
11. Czado C (2010) Pair-copula constructions of multivariate copulas. In: Jaworski P (ed) Copula theory and its applications. Lecture notes in statistics, vol 198. Springer, Berlin, pp 93–109
12. Dixit A, Steiglitz J (1977) Monopolistic competition and optimum product diversity. Am Econ Rev 67(3):297–308
13. Fisman R, Jakiela P, Kariv S, Markovits D (2015) The distributional preferences of an elite. Science 349(6254):1300. Artile aab096
14. Joe H (2010) Dependence comparisons of Vine copulae with four or more variables. In: Kurowicka D, Joe H (eds) Dependence modeling: Vine copula handbook. World Scientific, Singapore
15. Joe H, Hu T (1996) Multivariate distributions from mixtures of max-infinitely divisible distributions. J Multivar Anal 57(2):240–265
16. Joe H, Li H, Nikoloulopoulos AK (2010) Tail dependence functions and Vine copulas. J Multivar Anal 101:252–270
17. Jorgensen DW (2000) Econometrics. Economic modelling of producer behavior, vol 1. MIT Press, Cambridge
18. Klump R, McAadam P, Willaims A (2007) Factor substitution and factor augmenting technical progress in the US: a normalized supply-side system approach. Rev Econ Statisticsm 89(1):183–192
19. Kurowicka D, Cooke RM (2006) Uncertainty analysis with high dimensional dependence modelling. Wiley, New York
20. Nelsen RB (1999) An introduction to copulas. Springer, New York
21. Nikoloulopoulos AK (2012) Vine copulas with asymmetric tail dependence and applications to financial return data. Comput Stat Data Anal 56:3659–3673
22. Sklar A (1959) Fonctions de répartition á n dimensions et leurs marges. Publ Inst Statist Univ Paris 8:229–231
23. Solow RM (1956) A contribution to the theory of economic growth. Q J Econ 70:65–94
24. Sriboonchitta S, Kosheleva O, Nguyen HT (2015) Why are Vine copulas so successful in econometrics? Int J Uncertain Fuzziness Knowl Based Syst (IJUFKS) 23(Suppl 1):133–142

25. Taylor MA (1972) The generalized equations of bisymmetry, associativity and transitivity on quasigroups. Can Math Bull 15:119–124
26. Taylor MA (1973) Certian functional equations on groupoids weaker than quasigroups. Aequationes Mathematicae 9:23–29
27. Taylor MA (1978) On the generalized equations of associativity and bisymmetry. Aequationes Mathematicae 17:154–163
28. Varian H (1992) Microeconometric analysis. Norton, New York

How to Make Plausibility-Based Forecasting More Accurate

Kongliang Zhu, Nantiworn Thianpaen and Vladik Kreinovich

Abstract In recent papers, a new plausibility-based forecasting method was proposed. While this method has been empirically successful, one of its steps—selecting a uniform probability distribution for the plausibility level—is heuristic. It is therefore desirable to check whether this selection is optimal or whether a modified selection would like to a more accurate forecast. In this paper, we show that the uniform distribution does not always lead to (asymptotically) optimal estimates, and we show how to modify the uniform-distribution step so that the resulting estimates become asymptotically optimal.

1 Plausbility-Based Forecasting: Description, Successes, and Formulation of the Problem

Need for prediction. One of the main objectives of science is, given the available data $x_1, \ldots, x_n$, to predict future values of different quantities y.

The usual approach to solving this problem consists of two stages:

- first, we find a *model* that describes the observed data; and
- then, we use this model to predict the future value of each of the quantities y.

In some cases, it is sufficient to have a *deterministic model*, that describes the dependence of each observed value on the known values describing the i-th observation

---

K. Zhu · N. Thianpaen
Faculty of Economics, Chiang Mai University, Chiang Mai, Thailand
e-mail: 258zkl@gmail.com

N. Thianpaen
Faculty of Management Sciences, Suratthani Rajabhat University, Surat Thani, Thailand
e-mail: nantiworn@outlook.com

V. Kreinovich (✉)
Department of Computer Science, University of Texas at El Paso,
500 W. University, El Paso, TX 79968, USA
e-mail: vladik@utep.edu

© Springer International Publishing AG 2017   99
V. Kreinovich et al. (eds.), *Robustness in Econometrics*,
Studies in Computational Intelligence 692, DOI 10.1007/978-3-319-50742-2_7

and on the (unknown) parameters p of the model: $x_i = f_i(p)$. In this case, we can predict the value y as $y = f(p)$ for an appropriate function $f(p)$.

For example, in Newtonian's model of the Solar system, once we know the initial locations, initial velocities, and masses of all the celestial bodies (which, in this case, are the parameters p), we can predict the position and velocity of each body at any future moment of time.

In this deterministic case, we can use the known observed values to estimate the parameters p of the corresponding probabilistic model, and then we can use these parameters to predict the desired future values. This is how, e.g., solar eclipses can be predicted for centuries ahead.

Need for statistical prediction. In most practical problems, however, a fully deterministic prediction is not possible, since, in addition to the parameters p, both the observed values x_i and the future value y are affected by other parameters beyond our control, parameters that can be viewed as *random*. Thus, instead of a deterministic model, we have a general *probabilistic model* $x_i = f(p, z_1, \ldots, z_m)$ and $y = f(p, z_1, \ldots, z_m)$, where z_j are random variables.

Usually, we do not know the exact probability distribution for the variables z_i, but we know a finite-parametric family of distributions that contains the actual (unknown) distribution. For example, we may know that the distribution is Gaussian, or that it is uniform. Let q denote the parameter(s) that describe this distribution.

In this case, both x_i and y are random variables whose distribution depends on all the parameters $\theta = (p, q)$: $x_i \sim f_{i,\theta}$ and $y \sim f_\theta$.

In this case, to identify the model:

- we first estimate the parameters θ based on the observations $x_1, \ldots, x_n$, and then
- we use the distribution f_θ corresponding to these parameter values to predict the values y—or, to be more precise, to predict the probability of different values of y.

Need for a confidence interval. Since in the statistical case, we cannot predict the *exact* value of y, it is desirable to predict the *range* of possible values of y.

For many distributions—e.g., for a (ubiquitous) normal distribution—it is, in principle, possible to have arbitrarily small and arbitrarily large values, just the probability of these values is very small. In such situations, there is no *guaranteed* range of values of y.

However, we can still try to estimate a *confidence interval*, i.e., for a given small value $\alpha > 0$, an interval $[\underline{y}_\alpha, \overline{y}_\alpha]$ that contains the actual value y with confidence $1 - \alpha$. In other words, we would like to find an interval for which $\text{Prob}(y \in [\underline{y}_\alpha, \overline{y}_\alpha]) \geq \alpha$.

In the idealized situation, when we know the probabilities of different values of y—i.e., in precise terms, when we know the corresponding cumulative distribution function (cdf) $F(y) \stackrel{\text{def}}{=} \text{Prob}(Y \leq y)$—then we know that $Y \leq F^{-1}\left(\dfrac{\alpha}{2}\right)$ with probability $\alpha/2$ and that $Y > F^{-1}\left(1 - \dfrac{\alpha}{2}\right)$ with probability $\alpha/2$. Thus, with probability $1 - \alpha$, we have $y \in [\underline{y}_\alpha, \overline{y}_\alpha]$, where $\underline{y}_\alpha = F^{-1}\left(\dfrac{\alpha}{2}\right)$ and

$$\overline{y}_\alpha = F^{-1}\left(1 - \frac{\alpha}{2}\right).$$

In general, a statistical estimate based on a finite sample is only approximate. Thus, based on a finite sample, we can predict the value of the parameters θ only approximately—and therefore, we only have an approximate estimate of the probabilities of different values of y. So, instead of the actual cdf $F(y)$, we only know the bounds on the cdf: $\underline{F}(y) \le F(y) \le \overline{F}(y)$. We want to select the interval $[\underline{y}_\alpha, \overline{y}_\alpha]$ in such a way that the probability of being outside this interval is guaranteed not to exceed α.

For the lower bound $\underline{y}_\alpha$, all we know about the probability $F(\underline{y}_\alpha)$ of being smaller than this bound is that this probability is bounded, from above, by the known value $\overline{F}(\underline{y}_\alpha)$: $F(\underline{y}_\alpha) \le \overline{F}(\underline{y}_\alpha)$. Thus, to guarantee that this probability does not exceed $\frac{\alpha}{2}$, we must select a bound $\underline{y}_\alpha$ for which $\overline{F}(\underline{y}_\alpha) = \frac{\alpha}{2}$. In other words, we should take

$$\underline{y}_\alpha = (\overline{F})^{-1}\left(\frac{\alpha}{2}\right).$$

Similarly, the probability $1 - F(\overline{y}_\alpha)$ of being larger than the upper bound $\overline{y}_\alpha$ is bounded, from above, by the known value $1 - \underline{F}(\overline{y}_\alpha)$: $1 - F(\overline{y}_\alpha) \le \underline{F}(\overline{y}_\alpha)$. Thus, to guarantee that this probability does not exceed $\frac{\alpha}{2}$, we must select a bound $\overline{y}_\alpha$ for which $1 - \underline{F}(\overline{y}_\alpha) = \frac{\alpha}{2}$. In other words, we should take

$$\overline{y}_\alpha = (\underline{F})^{-1}\left(1 - \frac{\alpha}{2}\right).$$

Plausibility-based forecasting: a brief reminder. In [1, 3, 4], a new forecasting method was proposed. In this method, we start by forming a *likelihood function*, i.e., a function that describes, for each possible value θ, the probability (density) of observing the values $x = (x_1, \dots, x_n)$. If we assume that the probability density function corresponding to each observation x_i has the form $f_{i,\theta}(x_i)$, then, under the natural assumption that the observations $x_1, \dots, x_n$ are independent, we conclude that:

$$L_x(\theta) = \prod_{i=1}^{n} f_{\theta_i}(x_i).$$

The likelihood function is normally used to find the *maximum likelihood* estimate for the parameters θ, i.e., the estimate $\hat{\theta}$ for which $L_x\left(\hat{\theta}\right) = \max_\theta L_x(\theta)$.

In the plausibility-based approach to forecasting, instead of simply computing this value $\hat{\theta}$, we use the likelihood function to define the *plausibility function* as

$$\mathrm{pl}_x(\theta) = \frac{L_x(\theta)}{\sup\limits_{\theta'} L_x(\theta')} = \frac{L_x(\theta)}{L_x\left(\hat{\theta}\right)}.$$

Based on this plausibility function, we define, for each real number $\omega \in [0, 1]$, a *plausibility region*

$$\Gamma_x(\omega) = \{\theta : \mathrm{pl}_x(\theta) \geq \omega\}.$$

We then represent a probability distribution for y as $y = g(\theta, z)$ for an auxiliary variable z whose distribution does not depend on θ. Usually, as z, we select a random variable which is uniformly distributed on the interval $[0, 1]$. Such a representation is possible for each random variable with a probability density function $f_\theta(y)$ and corresponding cumulative distribution function $F_\theta(y)$: namely, we can simply take $g(\theta, z) = F_\theta^{-1}(z)$, where F^{-1} denotes an inverse function, i.e., a function for which $F_\theta^{-1}(F_\theta(x)) = x$ for all x.

Based on the plausibility regions, we then compute the belief and plausibility of each set A of possible values of θ as follows:

$$\mathrm{Bel}(A) = \mathrm{Prob}(g(\Gamma_x(\omega), z) \subseteq A)$$

and

$$\mathrm{Pl}(A) = \mathrm{Prob}(g(\Gamma_x(\omega), z) \cap A \neq \emptyset),$$

where both ω and z are uniformly distributed on the interval $[0, 1]$. After that, we compute the lower and upper bounds on the cdf $F(y)$ for y as

$$\underline{F}(y) = \mathrm{Bel}((-\infty, y])$$

and

$$\overline{F}(y) = \mathrm{Pl}((-\infty, y]).$$

Then, for any given small value $\alpha > 0$, we predict that y is, with confidence $1 - \alpha > 0$, contained in the interval $[\underline{y}_\alpha, \overline{y}_\alpha]$, where $\underline{y}_\alpha = (\overline{F})^{-1}\left(\frac{\alpha}{2}\right)$ and $\overline{y}_\alpha = (\underline{F})^{-1}\left(1 - \frac{\alpha}{2}\right)$.

Remaining problem. While the new approach has led to interesting applications, the motivations for this approach are not very clear. To be more precise:

- it is clear why, to simulate z, we use a uniform distribution on the interval $[0, 1]$— because we represent the corresponding probabilistic model for y as $y = g(\theta, z)$ for exactly this distribution for z;
- what is less clear is why we select a uniform distribution for ω.

Yes, this ω-distribution sounds like a reasonable idea: we know that ω is located on the interval $[0, 1]$, we do not know which values ω are more probable and which are less probable, so we select a uniform distribution. However, since we are not just making reasonable estimates, we are making predictions with confidence, it is desirable to come up with a more convincing justification for selecting the probability distribution for ω: namely, a justification that would explain why we believe that the predicted value y belongs to the above-constructed confidence interval.

Maybe we can get a justification, or maybe we can conclude that the above interval is only an approximation—and by selecting a different probability distribution for ω, we can make the resulting forecasting more accurate.

This is the problem that we will be analyzing in this paper.

2   On a Simple Example, Let Us Compare Plausibility-Based Forecasting with Known Methods

Let us consider the simplest possible situation. To analyze this problem, let us consider the simplest case:

- when we have only one parameter $\theta = \theta_1$,
- when the predicted value y simply coincides with the value of this parameter, i.e., when the probabilistic model $y = g(\theta, z)$ has the form $g(\theta, z) = \theta$, and
- when the likelihood $L_x(\theta)$ is continuous and strictly decreasing as we move away from the maximum likelihood estimate $\hat{\theta}$; in other words, we assume that:

   - for $\theta \le \hat{\theta}$ the likelihood function strictly increases, while
   - for $\theta \ge \hat{\theta}$ the likelihood function strictly decreases.

Comment. While the first two conditions are really restrictive, the third condition— monotonicity—is not very restrictive, it is true in the overwhelming majority of practical situations.

Let us analyze this simplest possible situation. Since in our case, $y = \theta$, the desired bounds on the predicted value y are simply bounds on the value θ of the corresponding parameter, bounds that contain θ with a given confidence α. In other words, what we want is a traditional confidence interval for θ.

In the above simplest possible situation, we can explicitly express the resulting confidence interval in terms of the likelihood function. According to the plausibility-based forecasting method, we select

$$\underline{F}(y) = \mathrm{Bel}((-\infty, y]) = \mathrm{Prob}(\{\theta : \mathrm{pl}_x(\theta) \ge \omega\} \subseteq (-\infty, y]).$$

Since we assumed that the likelihood function $L_x(\theta)$ is increasing for $\theta \le \hat{\theta}$ and decreasing for $\theta \ge \hat{\theta}$, the plausibility function $\mathrm{pl}_x(\theta)$ – which is obtained by $L_x(\theta)$ by dividing by a constant—also has the same property:

- the function $\mathrm{pl}_x(\theta)$ is increasing for $\theta \le \hat{\theta}$, and
- the function $\mathrm{pl}_x(\theta)$ is decreasing for $\theta \ge \hat{\theta}$.

In this case, the set $\{\theta : \mathrm{pl}_x(\theta) \geq \omega\}$ is simply an interval $[\theta^-, \theta^+]$, whose endpoints can be described as follows:

- the lower endpoint θ^- is the value to the left of $\hat{\theta}$ for which $\mathrm{pl}_x(\theta) = \omega$, and
- the upper endpoint θ^+ is the value to the right of $\hat{\theta}$ for which $\mathrm{pl}_x(\theta) = \omega$.

In these terms, the condition that the set $\{\theta : \mathrm{pl}_x(\theta) \geq \omega\} = [\theta^-, \theta^+]$ is contained in $(-\infty, y]$ simply means that $\theta^+ \leq y$.

Since $\theta^+ \geq \hat{\theta}$, we thus have $y \geq \hat{\theta}$ as well. The plausibility function is strictly decreasing for $\theta \geq \hat{\theta}$, the inequality $\theta^+ \leq y$ is equivalent to $\mathrm{pl}(\theta^+) \geq \mathrm{pl}(y)$. By the construction of the value θ^+, we know that $\mathrm{pl}(\theta^+) = \omega$. Thus, the condition $\{\theta : \mathrm{pl}_x(\theta) \geq \omega\} \subseteq (-\infty, y]$ is simply equivalent to $\omega \geq \mathrm{pl}_x(y)$. Hence,

$$\underline{F}(y) = \mathrm{Prob}(\omega \geq \mathrm{pl}_x(y)).$$

When ω is uniformly distributed on the interval $[0, 1]$, then, for all z, the probability $\mathrm{Prob}(\omega \geq z)$ that ω is in the interval $[z, 1]$, is simply equal to the width of bthis interval, i.e., to $1 - z$. In particular, for $z = \mathrm{pl}_x(y)$, we have $\underline{F}(y) = 1 - \mathrm{pl}_x(y)$. In these terms, in the plausibility-based forecasting method, as the upper bound $\overline{\theta}_\alpha$ of the confidence interval, we select the value $\overline{\theta}_\alpha$ for which $1 - \mathrm{pl}_x(\overline{\theta}_\alpha) = 1 - \dfrac{\alpha}{2}$, i.e., for which

$$\mathrm{pl}_x(\overline{\theta}_\alpha) = \frac{\alpha}{2}.$$

Similarly, the condition that the set $\{\theta : \mathrm{pl}_x(\theta) \geq \omega\} = [\theta^-, \theta^+]$ has a non-empty intersection with $(-\infty, y]$ simply means that $\theta^- \leq y$.

Since $\theta^- \leq \hat{\theta}$, this inequality is always true for $y \geq \hat{\theta}$. So, for $y \geq \hat{\theta}$, we have $\overline{F}(y) = 1$. For $y \leq \hat{\theta}$, the inequality $\theta^- \leq y$ is equivalent to $\mathrm{pl}(\theta^-) \leq \mathrm{pl}(y)$. By the construction of the value θ^-, we know that $\mathrm{pl}(\theta^-) = \omega$. Thus, the condition $\{\theta : \mathrm{pl}_x(\theta) \geq \omega\} \cap (-\infty, y] \neq \emptyset$ is simply equivalent to $\omega \leq \mathrm{pl}_x(y)$. Hence,

$$\overline{F}(y) = \mathrm{Prob}(\omega \leq \mathrm{pl}_x(y)).$$

When ω is uniformly distributed on the interval $[0, 1]$, then, for all z, the probability $\mathrm{Prob}(\omega \leq z)$ is simply equal to z. In particular, for $z = \mathrm{pl}_x(y)$, we have $\overline{F}(y) = \mathrm{pl}_x(y)$. In these terms, in the plausibility-based forecasting method, as the lower bound $\underline{\theta}_\alpha$ of the confidence interval, we select the value $\underline{\theta}_\alpha$ for which

$$\mathrm{pl}_x(\underline{\theta}_\alpha) = \frac{\alpha}{2}.$$

Thus, the confidence interval obtained by using the plausibility method is the interval $[\underline{\theta}_\alpha, \overline{\theta}_\alpha]$ between the two values $\underline{\theta}_\alpha < \hat{\theta} < \overline{\theta}_\alpha$ for which

$$\mathrm{pl}_x(\underline{\theta}_\alpha) = \mathrm{pl}_x(\overline{\theta}_\alpha) = \frac{\alpha}{2}.$$

The confidence interval $[\underline{\theta}_\alpha, \overline{\theta}_\alpha]$ consists of all the values θ for which

$$\mathrm{pl}_x(\theta) \geq \frac{\alpha}{2}.$$

In terms of the likelihood function $L_x(\theta)$, this means that, as the confident interval, we select the set of all the values θ for which

$$\frac{L_x(\theta)}{L_x\left(\hat{\theta}\right)} \geq \frac{\alpha}{2},$$

i.e., equivalently, for which

$$\ln\left(L_x(\theta)\right) \geq \ln\left(L_x\left(\hat{\theta}\right)\right) - (\ln(2) + |\ln(\alpha)|). \tag{1}$$

Let us compare the resulting confident interval with the traditional likelihood-based confidence interval. In traditional statistics, one of the methods to estimate the confidence interval based on the likelihood function—based on Wilks's theorem—is to select the set of all possible values θ for which

$$\ln\left(L_x(\theta)\right) \geq \ln\left(L_x\left(\hat{\theta}\right)\right) - \frac{1}{2} \cdot \chi^2_{1,1-\alpha}; \tag{2}$$

see, e.g., [2], where $\chi^2_{1,1-\alpha}$ is the threshold for which, for the χ^2_1-distribution—i.e., for the square of the standard normally distributed random variable, with 0 means and standard deviation 1, we have $\mathrm{Prob}(\chi^2 \leq \chi^2_{1,1-\alpha}) = 1 - \alpha$.

The corresponding confidence interval (2) is somewhat different from the interval (1) obtained by using plausibility-based forecasting. It is known that Wilks's theorem provides an *asymptotically accurate* description of the confidence region when the number of observations n increases.

It is desirable to modify the plausibility-based forecasting method to make it asymptotically optimal. It is desirable to modify the plausibility-based forecasting method to make its results asymptotically optimal.

3 How to Best Modify the Current Plausibility-Based Forecasting Method: Analysis of the Problem

Problem: reminder. In the previous section, we have shown that the use of a (heuristically selected) uniform distribution for the variable ω, while empirically efficient, does not always lead us to asymptotically optimal estimates. Let us therefore try to find an alternative distribution for ω for which, in the above case, the resulting confidence interval will be asymptotically optimal.

Which distribution for ω we should select: analysis of the problem. In the general case, we still have $\overline{F}(y) = \text{Prob}(\omega \le p_x(y))$. We want to make sure that for the Wilks's bound, this probability is equal to $\dfrac{\alpha}{2}$.

For the Wilks's bound, by exponentiating both sides of the Formula (2), we conclude that

$$p_x(y) = \frac{L_x(\theta)}{L_x(\hat{\theta})} = \exp\left(-\frac{1}{2} \cdot \chi^2_{1,1-\alpha}\right);$$

thus, we conclude that

$$\text{Prob}\left(\omega \le \exp\left(-\frac{1}{2} \cdot \chi^2_{1,1-\alpha}\right)\right) = \frac{\alpha}{2}. \tag{3}$$

By definition of $\chi^2_{1,1-\alpha}$, if we take a variable n which is normally distributed with 0 mean and standard deviation 1, then we have:

$$\text{Prob}\left(n^2 \le \chi^2_{1,1-\alpha}\right) = 1 - \alpha.$$

Thus, for the opposite event, we have

$$\text{Prob}\left(n^2 \ge \chi^2_{1,1-\alpha}\right) = (1 - (1 - \alpha)) = \alpha.$$

The inequality $n^2 \ge \chi^2_{1,1-\alpha}$ occurs in two equally probable situations:

- when n is positive and $n \ge \sqrt{\chi^2_{1,1-\alpha}}$ and
- when n is negative and $n \le -\sqrt{\chi^2_{1,1-\alpha}}$.

Thus, the probability of each of these two situations is equal to $\dfrac{\alpha}{2}$; in particular, we have:

$$\text{Prob}\left(n \le -\sqrt{\chi^2_{1,1-\alpha}}\right) = \frac{\alpha}{2}. \tag{4}$$

Let us transform the desired inequality (3) to this form. The inequality

$$\omega \le \exp\left(-\frac{1}{2} \cdot \chi^2_{1,1-\alpha}\right)$$

is equivalent to

$$\ln(\omega) \le -\frac{1}{2} \cdot \chi^2_{1,1-\alpha},$$

hence to

$$-2 \ln(\omega) \geq \chi^2_{1,1-\alpha},$$

$$\sqrt{-2 \ln(\omega)} \geq \sqrt{\chi^2_{1,1-\alpha}},$$

and

$$-\sqrt{-2 \ln(\omega)} \leq -\sqrt{\chi^2_{1,1-\alpha}}.$$

Thus, the desired inequality (3) is equivalent to

$$\text{Prob}\left(-\sqrt{-2 \ln(\omega)} \leq -\sqrt{\chi^2_{1,1-\alpha}}\right) = \frac{\alpha}{2}.$$

In view of the Formula (4), this equality is attained if we have $n = -\sqrt{-2 \ln(\omega)}$. In this case, $-2 \ln(\omega) = n^2$, hence

$$\ln(\omega) = -\frac{n^2}{2},$$

and thus,

$$\omega = \exp\left(-\frac{n^2}{2}\right). \tag{5}$$

So, we arrive at the following conclusion.

Conclusion. Instead of a uniformly distributed random variable ω, we need to use a variable (5), where n is a random variable distributed according to the standard normal distribution with 0 means and standard deviation 1.

What is the probability density function of this distribution? In general, if we have a random variable with a probability density function $\rho_X(x)$, then for any function $f(x)$, for the random variable $Y = f(X)$, we can determine its probability density function $\rho_Y(y)$ from the condition that for $y = f(x)$, we have

$$\text{Prob}(f(x) \leq Y \leq f(x+dx)) = \text{Prob}(x \leq X \leq x+dx) = \rho_X(x) \cdot dx.$$

Here, $f(x) = y$ band

$$f(x+dx) = f(x) + f'(x) \cdot dx = y + f'(x) \cdot dx,$$

hence

$$\text{Prob}(f(x) \leq Y \leq f(x+dx)) = \text{Prob}(y \leq Y \leq y + f'(x) \cdot dx) = \rho_Y(y) \cdot |f'(x)| \cdot dx.$$

Equating these two expressions, we conclude that for $y = f(x)$, we have

$$\rho_Y(y) = \frac{\rho_X(x)}{|f'(x)|}.$$

In our case,

$$\rho_X(x) = \frac{1}{\sqrt{2\pi}} \cdot \exp\left(-\frac{x^2}{2}\right)$$

and

$$f(x) = \exp\left(-\frac{x^2}{2}\right),$$

hence

$$f'(x) = -\exp\left(-\frac{x^2}{2}\right) \cdot x.$$

Thus,

$$\rho_Y(y) = \frac{1}{\sqrt{2\pi} \cdot |x|}.$$

From $y = \exp\left(-\dfrac{x^2}{2}\right)$, we conclude that $\dfrac{x^2}{2} = -\ln(y)$, thus, $|x| = \sqrt{2 \cdot |\ln(y)|}$. So, the probability distribution function for $y = \omega$ has the form

$$\rho(\omega) = \frac{1}{\sqrt{2\pi} \cdot \sqrt{2 \cdot |\ln(\omega)|}} = \frac{1}{2 \cdot \sqrt{\pi} \cdot \sqrt{|\ln(\omega)|}}.$$

This distribution is indeed close to uniform. The value $\ln(\omega)$ is changing very slowly, so, in effect, the resulting probability density function is close to a constant, and thus, the corresponding probability distribution is close to the uniform one.

4   Resulting Recommendations

As a result of the above analysis, we arrive at the following modification of the plausibility-based forecasting algorithm.

In this modification, first, we define the likelihood function $L_x(\theta)$ and then find its largest possible value $L_x\left(\hat{\theta}\right) = \max_\theta L_x(\theta)$.

Then, we define the plausibility function as

$$\mathrm{pl}_x(\theta) = \frac{L_x(\theta)}{\sup_{\theta'} L_x(\theta')} = \frac{L_x(\theta)}{L_x\left(\hat{\theta}\right)}.$$

Based on this plausibility function, we define, for each real number $\omega \in [0, 1]$, a plausibility region

$$\Gamma_x(\omega) = \{\theta : \text{pl}_x(\theta) \geq \omega\}.$$

We then represent a probability distribution for y as $y = g(\theta, z)$ for an auxiliary variable z which is uniformly distributed on the interval $[0, 1]$.

Based on the plausibility regions, we then compute the belief and plausibility of each set A of possible values of θ as follows:

$$\text{Bel}(A) = \text{Prob}(g(\Gamma_x(\omega), z) \subseteq A)$$

and

$$\text{Pl}(A) = \text{Prob}(g(\Gamma_x(\omega), z) \cap A \neq \emptyset),$$

where both z is uniformly distributed on the interval $[0, 1]$, and ω is distributed in accordance with the probability density

$$\rho(\omega) = \frac{1}{2 \cdot \sqrt{\pi} \cdot \sqrt{|\ln(\omega)|}}.$$

The corresponding random variable can be simulated as

$$\omega = \exp\left(-\frac{n^2}{2}\right),$$

where n is a standard normally distributed random variable, with 0 mean and standard deviation 1.

After that, we compute the lower and upper bounds on the cdf $F(y)$ for y as

$$\underline{F}(y) = \text{Bel}((-\infty, y])$$

and

$$\overline{F}(y) = \text{Pl}((-\infty, y]).$$

Then, for any given small value $\alpha > 0$, we predict that y is, with confidence $1 - \alpha > 0$, contained in the interval $[\underline{y}_\alpha, \overline{y}_\alpha]$, where $\underline{y}_\alpha = (\overline{F})^{-1}\left(\frac{\alpha}{2}\right)$ and $\overline{y}_\alpha = (\underline{F})^{-1}\left(1 - \frac{\alpha}{2}\right)$.

Acknowledgements We acknowledge the partial support of the Center of Excellence in Econometrics, Faculty of Economics, Chiang Mai University, Thailand. This work was also supported in part by the National Science Foundation grants HRD-0734825 and HRD-1242122 (Cyber-ShARE Center of Excellence) and DUE-0926721, and by an award "UTEP and Prudential Actuarial Science Academy and Pipeline Initiative" from Prudential Foundation.

The authors are greatly thankful to Hung T. Nguyen for valuable suggestions.

References

1. Abdallah NB, Voyeneau NM, Denoeux T (2012) Combining statistical and expert evidence within the D-S framework: application to hydrological return level estimation. In: Denoeux T, Masson M-H (eds) Belief functions: theory and applications: proceedings of the 2nd international conference on belief functions, Compiègne, France, May 9–11, 2012. Springer, Berlin, pp 393–400
2. Abramovich F, Ritov Y (2013) Statistical theory: a concise introduction. CRC Press, Boca Raton
3. Kanjanatarakul O, Sriboonchitta S, Denoeux T (2014) Forecasting using belief functions: an application to marketing econometrics. Int J Approx Reason 55:1113–1128
4. Thianpaen N, Liu J, Sriboonchitta S Time series using AR-belief approach. Thai J Math 14(3):527–541
5. Martin, R (2015) Plausibility functions and exact frequentist inference. Journal of the American Statistical Association 110:1552–1561

Structural Breaks of CAPM-type Market Model with Heteroskedasticity and Quantile Regression

Cathy W.S. Chen, Khemmanant Khamthong and Sangyeol Lee

Abstract In this study we analyze the market beta coefficients of two large capitalization stocks, American International Group (AIG) and Citigroup, from 2005 to 2016 based on a capital asset pricing model (CAPM) Since the daily returns of stock prices experience structural changes in their underlying CAPM-type models, we detect the number and locations of change employing the residual-based cumulative sum (CUSUM) of squares test and then estimate the parameters for each sub-period to evaluate market risk. Moreover, using the quantile regression method, we explore the different behaviors of the market beta and lagged autoregressive effects for different sub-periods and quantile levels. Our final result pertains to the relationship between time-varying betas and structural breaks.

Keywords CAPM · Change point test · CUSUM test · Beta coefficient · Asymmetric effect · GARCH model · Quantile regression

1 Introduction

American International Group, Inc. (AIG) is an insurance company that provides property casualty insurance, life insurance, retirement products, mortgage insurance, and other financial services to customers in over 100 countries and jurisdictions. Citigroup Inc. is a financial services holding company whose businesses provide consumers, corporations, governments, and institutions with a wide range of financial products and services, including consumer banking and credit, corporate and investment banking, securities brokerage, trade and securities services, and wealth

C.W.S. Chen (✉) · K. Khamthong
Department of Statistics, Feng Chia University, Taichung, Taiwan
e-mail: chenws@mail.fcu.edu.tw

S. Lee
Department of Statistics, Seoul National University, Seoul, Korea

© Springer International Publishing AG 2017   111
V. Kreinovich et al. (eds.), *Robustness in Econometrics*,
Studies in Computational Intelligence 692, DOI 10.1007/978-3-319-50742-2_8

management. Both AIG and Citigroup are two large market capitalization (large-cap) stocks in the U.S. A large-cap (or big-cap) stock refers to a company with a market capitalization value of more than \$5 billion. This study aims to explore the performances of these two stock's market beta coefficients during 2005–2016.

The capital asset pricing model (CAPM) is one of the most commonly applied models in finance and investment. The portfolio selection model, introduced by Markowitz [21], offers a suitable portfolio based only on the mean and variance for the returns of assets contained in the portfolio. The original CAPM of Sharp [23] and Lintner [19] proposes efficient and intuitive predictions to measure risk and the relation between expected return and market risk premiums. CAPM is still widely used in many applications such as in estimating the cost of capital for firms and in evaluating the performance of managed portfolios. Here, we consider CAPM as follows:

$$E(R_i) = R_f + \beta_i(E(R_m) - R_f),$$

where $E(R_i)$ and $E(R_m)$ are the expected returns of the capital asset and the market portfolio, respectively; R_f is the risk-free rate; $E(R_m) - R_f$ is called the market risk premium; and β_i stands for the sensitivity of the expected asset returns to the expected market returns and can be estimated as:

$$\frac{\text{Cov}(R_t - R_f, R_m - R_f)}{\text{Var}(R_m - R_f)}.$$

The CAPM obtains historic achievements in asset pricing and portfolio selection, but still has some empirical defects in that the constant beta coefficient and time invariant variance in the original market model make it less convincing to capture the dynamics of the real financial market (see Engle and Rodrigues [9]). The autoregressive conditionally heteroskedastic (ARCH) and generalized ARCH (GARCH) models of Engle [8] and Bollerslev [2] provide an efficient technique to measure price volatility. Bollerslev et al. [4] are the first to model a dynamic market beta in terms of time-varying variances and covariance, via a multivariate GARCH model. Harvey [12] uses the generalized method of moments, whereas Schwert and Seguin [22] employ the weighted least-squares estimation approach that is robust to heteroskedasticity for portfolio returns. Both studies provide evidence that does not favor the Sharpe-Lintner CAPM. The existing market model is usually set up like a CAPM-type model, which takes the form of a GARCH specification [6, 18] or exhibits a nonlinear form [7]. Chen et al. [6] propose an asymmetric market model with heteroskedasticity and use a quantile regression to show that each market beta varies with different quantile levels. This set-up helps capture different states of market conditions and asymmetric risk through market beta and negative news. Although the asymmetric market model of Chen et al. [6] is favorable and successfully obtains model adequacy for many stock's returns traded in the Dow Jones Industrial Average, it is not directly applicable to the stocks of AIG and Citigroup, because their daily returns might have structural changes in their underlying models. Like most financial corporations in 2008,

Citigroup's massive derivative portfolios were primarily made up of Collateralized Debt Obligations (CDOs) and Mortgage Backed Securities (MBSs), with the majority of its CDOs and MBSs of subprime quality. When the mortgage bubble burst, Citigroup, Lehman Brothers, and AIG among many others were hit incredibly hard for their bets on subprime lending. The authorities let Lehman Brothers go bankrupt, while Citigroup was teetering on the edge of bankruptcy. Leading up to the crisis, AIG had transformed itself from an insurance company into one of the leading players in the new credit default swap market. AIG was in the business of insuring leveraged debt just at the time when the financial system was on the brink of collapse. As a result of its large default swap positions, the U.S. government took over AIG under an $85 billion bailout package in 2008. In this study we focus on the market risk of AIG and Citigroup stocks. For this task, we check the existence of structural changes in the CAPM-type model with heteroskedasticity, using the residual-based cumulative sum (CUSUM) of squares test for return volatility. The CUSUM of squares test is a device that detects any changes of model parameters, because it has the ability to note the change point and to allocate its location accurately. In order to test for a variance change, Inclan and Tiao [11] propose the CUSUM of squares test for independent and identically distributed (i.i.d.) normal random variables, based on the earlier work of Brown et al. [5] who consider it to test for the constancy of the regression coefficients. The CUSUM of squares test can also detect multiple change points in the unconditional variance of possibly heteroskedastic time series [14]. Lee et al. [17] and Lee and Lee [16] later use the residual-based CUSUM of squares test to detect a parameter change in GARCH(1,1) models. This test's merits help overcome certain drawbacks, such as size distortions, that the estimates-based CUSUM test [15] might have and improves the power of the CUSUM of squares test of Kim et al. [14]. The stability of the test is due to the removal of trend and correlations. Under the assumption of no change points, the test is simplified into testing for the variance change of the i.i.d. error terms. Our goal is to explore the implications of long-run changes in market betas for CAPM-based models. This research analyzes AIG and Citigroup stock returns over 12 years (January 3, 2005–March 31, 2016) with daily data, particularly applying the CUSUM of squares test to detect the change points in a market model with GARCH effect. Our first set of results renders the piecewise CAPM-based market model with GARCH effect, wherein we determine the number of change points in order to divide the whole time series into several sub-series of homogeneity, estimate the model parameters for each sub-period, and evaluate the market risk. Next, our second set of results concerns a piecewise quantile regression, so as to explore the different behaviors in the market beta and lagged autoregressive effect for different sub-periods and quantile levels. Some researchers expect the market beta to be time-varying. Our last set of results pertains to the relationship between time-varying betas and structural breaks. The intuition behind the empirical success of our piecewise CAPM-based model is as follows. In our approach based on the CUSUM test, given the structural break points, the proposed model captures asymmetric risk by allowing the market beta to change discretely over time when driven by market information. Our empirical results show that the CUSUM of squares test is very powerful to detect changes in the model parameters. This paper is organized as follows. Section 2 presents the CAPM-based market model with structural breaks and employs the CUSUM of squares test. Section 3 uses the piecewise CAPM-based model to amend the CAPM-type model with heteroskedasticity for the daily excess returns of the AIG and Citigroup stocks and adopts the quantile regression method to explore the market beta coefficients of these two stocks for different sub-periods and quantile levels. We further model time-varying betas using the CAPM-based market model with GARCH(1,1) innovation. Section 4 offers a comparison of time-varying betas and structural breaks from the CUSUM of squares test. Section 5 provides concluding remarks.

2 Structural Change Market Models

The market model in finance relates the excess return of an individual stock to that of a market index (see [24]), displayed as follows:

$$r_t = \beta_0 + \beta_1 r_{m,t} + a_t, \ t = 1, \ldots, n, \qquad (1)$$
$$r_t = (\ln P_t - \ln P_{t-1}) \times 100 - r_{f,t},$$
$$r_{m,t} = (\ln P_{m,t} - \ln P_{m,t-1}) \times 100 - r_{f,t},$$

where P_t and $P_{m,t}$ are the individual stock and market price; $r_{f,t}$ are the risk-free return; $r_t, r_{m,t}$ are the individual stock and market excess return, and a_t is the error term. The parameters β_0 and β_1 denote the excess return, with respect to the market. Here, β_1 is a coefficient that measures the expected change in r_t given a change in $r_{m,t}$. Numerous works have demonstrated that employing the GARCH process for a_t in a market model can achieve better performance.

Chen et al. [6] propose a GARCH-type market model that is an asymmetric market model with heteroscedasticity by adding one more parameter in the regression to investigate the magnitude of negative and positive news from the previous day to the response of the stock market return. This is a measure of the leverage effect of $r_{m,t-1}$. In general, the GARCH(1,1) model is sufficient to capture volatility clustering in most financial applications [3]. To incorporate conditional variance into the system, Chen et al. [6] consider the asymmetric market model with GARCH errors as follows:

$$r_t = \beta_0 + \phi_1 r_{t-1} + \beta_1 r_{m,t} + \beta_2 I_{(r_{m,t-1}<0)} r_{m,t-1} + a_t, \qquad (2)$$
$$a_t = \sigma_t \varepsilon_t, \ \varepsilon_t \ i.i.d. \sim t^*(v)$$
$$\sigma_t^2 = \alpha_0 + \alpha_1 a_{t-1}^2 + \gamma_1 \sigma_{t-1}^2,$$

where $t^*(v)$ represents a standardized Student-t distribution with zero mean and unit variance; σ_t^2 denotes the conditional variance at time t, $\alpha_0 > 0$, $0 \leq \alpha_1$, $\gamma_1 < 1$, $\alpha_1 + \gamma_1 < 1$; and $I_{(r_{m,t-1}<0)}$ is an indicator function:

$$I_{(r_{m,t-1}<0)} = \begin{cases} 1 & \text{if } r_{m,t-1} < 0 \\ 0 & \text{if } r_{m,t-1} \geq 0. \end{cases}$$

Therefore, when $r_{m,t-1} < 0$, the lag effect from the previous day is measured by β_2 in the model; otherwise, there is no lag effect in the model. This model contains the lagged value of the excess return, which helps identify whether the return series shows mean reversion or market efficiency.

Although the asymmetric market model of Chen et al. [6] is quite promising and appropriate in describing many stocks traded in the Dow Jones Industrial Average, it cannot be directly applied to AIG and Citigroup, because their daily stock returns might have structural changes during the period of investigation. In this section we provide a method to check for the structural changes in model (1), related to the excess returns of both AIG and Citigroup to a market index. Our procedure is as follows.

Step 1: Identify structural changes by fitting the asymmetric market model in (1) to each returns series and apply the CUSUM of squares test to their residuals to detect the change in volatility of each series.

Step 2: Detect the change in volatility of each series based on the CUSUM of squares test based on the residuals from Step 1:

$$T_n = \frac{1}{\sqrt{n}\hat{\tau}} \max_{1 \leq k \leq n} \left| \sum_{t=1}^{k} \hat{\varepsilon}_t^2 - \frac{k}{n} \sum_{t=1}^{n} \hat{\varepsilon}_t^2 \right|,$$

where $\hat{\tau}^2 = \widehat{\text{Var}}(\varepsilon_1^2)$ and $\hat{\varepsilon}_t = \dfrac{\hat{a}_t}{\hat{\sigma}_t}$,

with $\hat{a}_t$ are the estimates of a_t from the asymmetric market model in (1). Those estimators play an important role to detect changes in the parameters in the presence of any such changes: the i.i.d. property of the true errors still remains therein when no changes exist.

Step 3: If change points exist, divide the returns series into sub-sample periods based on their volatility.

For the computational task, we use the "changepoint" package in R. We then examine the relationship between the excess return of an individual stock and that of a market index for each sub-period. Next, we estimate model parameters to evaluate market risk.

We further inspect the beta coefficients of the two firms' financial investments under extreme market conditions rather than for normal market conditions. Here, an asymmetric market model with quantile regression is applied as well to explore the different behaviors in the market beta and lagged autoregressive effect for different sub-periods and quantile levels. The model by Chen et al. [6] is written as follows:

$$Q^{(\tau)}(r_t|F_{t-1}) = \beta_0^{(\tau)} + \phi_1^{(\tau)} r_{t-1} + \beta_1^{(\tau)} r_{m,t} + \beta_2^{(\tau)} I_{(r_{m,t-1}<0)} r_{m,t-1}, \qquad (3)$$

where $Q^{(\tau)}(r_t)$ represents the conditional τth quantile estimates of the stock excess return at time t, and F_{t-1} is information set up to time $t-1$. The quantile regression minimizes an asymmetrically weighted sum of absolute errors as follows:

$$\min_{\beta^{(\tau)}} \sum_{t=1}^{n} \rho_\tau \left(r_t - Q^{(\tau)}(r_t | F_{t-1})\right),$$

where $\rho_\tau(u) = u \times [\tau - I_{(u<0)}]$ and $\beta^{(\tau)} = (\beta_0^{(\tau)}, \phi_1^{(\tau)}, \beta_1^{(\tau)}, \beta_2^{(\tau)})^T$. When $\tau = .05$ and $u < 0$, the weight is $-.95$, while when $u \geq 0$, the weight is 0.05. In other words, when $u < 0$, the absolute error's weight is 0.95, while for $u \geq 0$, the absolute error's weight is 0.05.

3 Analytics Results

In our study we use the S&P500 Index as the market portfolio proxy for analyzing AIG and Citigroup share prices. The datasets are downloaded from Yahoo Finance, containing the daily three-month U.S Treasury bill rate and the level of the S&P500 Index. The sample period begins from January 3, 2005 to March 31, 2016 and consists of 2,830 observations at maximum. For our analysis, we follow the steps below.

Step 1: Transform the daily three-month U.S Treasury bill rate i_t into the risk-free rate $r_{f,t}$:

$$r_{f,t} = \left(\left(1 + \frac{i_t}{100}\right)^{\frac{1}{365}} - 1 \right) \times 100.$$

Step 2: Calculate the excess returns of the individual stock and the market portfolio based on stock price P_t and the value of the market portfolio $P_{m,t}$ on day t, via Eq. (1).

Figure 1 presents the time series plots of daily excess returns for S&P500, T-bill, AIG, and Citigroup from January 3, 2005 to March 31, 2016, showing that the two series of returns for AIG and Citigroup are much more volatile, especially their severe downturn during the global financial crisis from 2008 to 2009 versus other periods. Since a direct usage of model (1) is not appropriate owing to the different pattern as seen in Fig. 1, we consider model (1) with structural breaks.

We now employ Steps 1 and 2 to detect the change points in the market model fitted to AIG and Citigroup returns. Figure 2 provides time series plots with multiple change points from market model (1), detected from the residual-based CUSUM of squares test. More precisely, AIG excess returns have three change points at 123, 697 and 1338, which are the respective observations at June 28, 2005, October 9, 2007, and April 27, 2010; whereas Citigroups excess returns have two change points of detection at 696 and 1773, which are the respective observations at October 8,

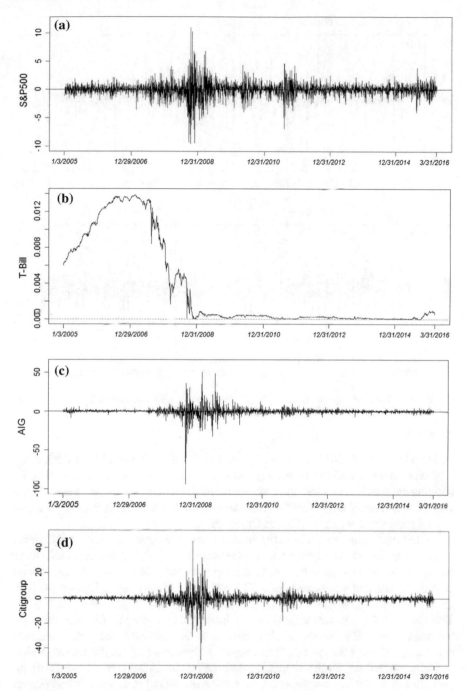

Fig. 1 Time series plots: **a** daily excess returns of S&P500, **b** risk-free rate, **c** daily excess returns of AIG, and **d** daily excess returns of Citigroup from January 3, 2005 to March 31, 2016

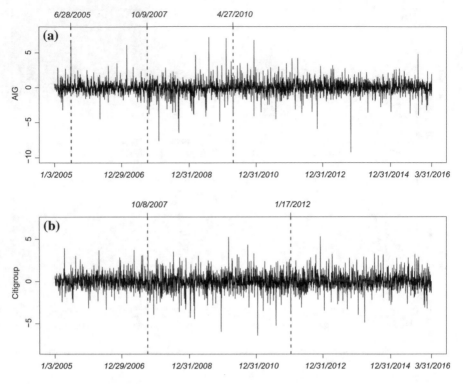

Fig. 2 Detection of change points from CAPM-based model residuals using CUSUM of squares test: AIG and Citigroup from January 3, 2005 to March 31, 2016

2007 and January 1, 2012. Here, we ignore the change point of 123 for AIG, since otherwise there would not be enough observations in the first sub-series, thereby dividing each time series into three sub-periods. The two stock returns naturally are very volatile during the global financial crisis from 2008 to 2009, contained in the second sub-period for both AIG and Citigroup.

Table 1 lists summary statistics for the three sub-series of each stock's daily return, including the market (S&P500) and the Treasury bill (T-bill). The excess returns of the three sub-periods for AIG and Citigroup have the widest range in the second period, i.e. the 2nd sub-periods of AIG (-93.63 to 50.68) and Citigroup (49.5 to 45.6), respectively, whereas the market return (S&P500) ranges from -9.5 to 11. This shows that the excess returns on stocks are much more volatile than those on the market portfolio. All the stock returns have fat tails with the excess kurtosis exceeding 3 (except for the T-bill) and negative skewness (except the 1st sub-period of Citigroup and the T-bill), indicating that the left tail is heavier. The last column presents the correlation coefficients between the individual stock returns and market returns in each period, all of which are positive around 0.4791 to 0.7470, meaning that an increase in the market value is associated with that of the value of the stock returns

Table 1 Summary statistics: two daily excess returns and three subseries of each stock from January 3, 2005 to March 31, 2016

Stock	Mean	SD	Q1	Median	Q3	Min	Max	Skewness	Excess kurtosis	r
T-bill	0.0035	0.005	0.0001	0.0004	0.0076	0	0.0139	1.0857	−0.5236	–
S&P500	0.0153	1.2612	−0.459	0.0662	0.5575	−9.4704	10.9565	−0.3212	10.7681	–
AIG	−0.1067	4.7317	−1.036	−0.0005	0.9604	−93.6299	50.6812	−2.9752	83.0946	0.4946
AIG 1	0.0003	1.1069	−0.5166	0.0153	0.5756	−8.3991	5.8181	−0.3486	7.9537	0.5844
AIG 2	−0.5354	9.3612	−3.369	−0.3685	2.5705	−93.6299	50.6812	−1.5556	21.4598	0.4791
AIG 3	0.0275	2.0467	−0.9492	0.0646	1.0357	−17.484	12.3709	−0.333	6.7937	0.7189
Citigroup	−0.0798	3.6574	−1.0218	−0.0211	0.9641	−49.4656	45.6316	−0.4913	35.0599	0.6627
Citigroup 1	0.0162	1.0284	−0.4733	0	0.5171	−5.3753	5.593	0.0354	4.686	0.7189
Citigroup 2	−0.2492	5.6097	−2.2273	−0.2122	1.8275	−49.4695	45.6316	−0.2692	14.9821	0.6638
Citigroup 3	0.0296	1.7394	−0.8905	0.0331	0.9506	−8.5534	7.0631	−0.1625	1.94	0.747

Date of three subseries for each stock:

AIG 1: 1/3/2005–10/8/2007, AIG 2: 10/9/2007–4/26/2010, AIG 3: 4/27/2010–3/31/2016

Citigroup 1: 1/3/2005–10/5/2007, Citigroup 2: 10/8/2007–1/13/2012, Citigroup 3: 1/17/2012–3/31/2016

in each period. It shows that the highest and lowest correlations are respectively the third sub-period of Citigroup and the second sub-period of AIG, which represents their strong and weak linear dependence with the market.

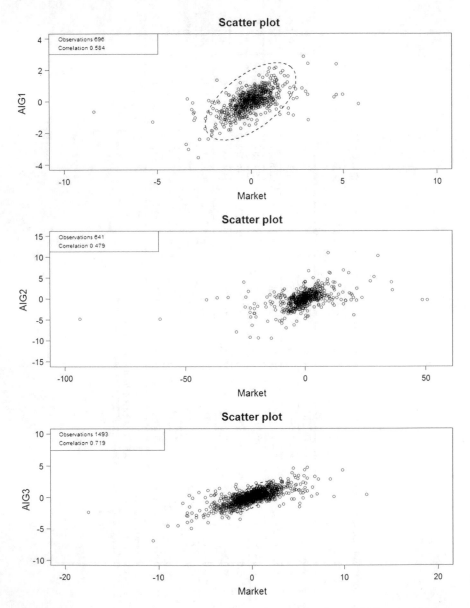

Fig. 3 Scatter plot of daily excess returns between AIG stock (Y-axis) and S&P composite index from January 2005 to March 2016. AIG 1: 1/3/2005–10/8/2007, AIG 2: 10/9/2007–4/26/2010, AIG 3: 4/27/2010–3/31/2016

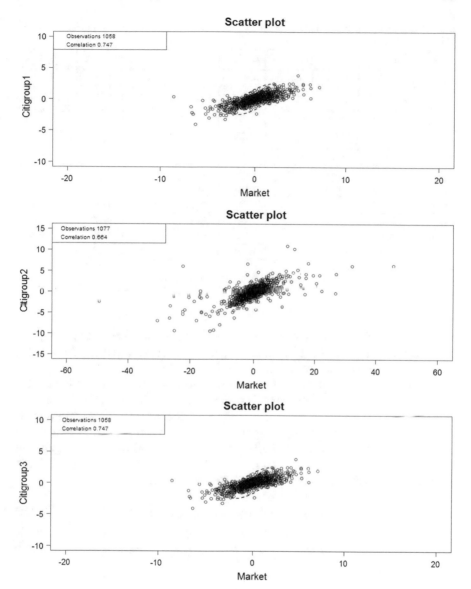

Fig. 4 Citigroup 1: 1/3/2005–10/5/2007, Citigroup 2: 10/9/2007–4/26/2010, Citigroup 3: 1/17/2012–3/31/2016

Figures 3 and 4 show scatterplots of the market-to-stock excess returns of the stocks for each sub-period. These plots display a positive correlation between each excess stock return and the market for each sub-period. We see that 95% of observations for the first sub-series lie inside the ellipse, but the other subseries show that a lot of observations are spread outside the ellipse, especially in the second series of

Table 2 Parameter estimates and unconditional variance of three subseries for the market model with GARCH effect

Stock	β_0	ϕ_1	β_1	α_0	α_1	γ_1	ν	$\dfrac{\alpha_0}{1-\alpha_1-\gamma_1}$
AIG 1	−0.030	0.149	0.783	0.060	0.228	0.705	3.805	0.896
s.e.	0.025	0.040	0.030	0.029	0.084	0.097	0.547	
AIG 2	−0.281	−0.008	1.769	2.581	0.332	0.667	2.847	2580.950
s.e.	0.114	0.040	0.084	0.990	0.077	0.332	0.225	
AIG 3	−0.007	−0.035	1.244	0.004	0.018	0.979	4.021	1.173
s.e.	0.023	0.024	0.030	0.002	0.002	0.001	0.402	
Citigroup 1	−0.012	0.039	0.916	0.080	0.114	0.733	4.404	0.526
s.e.	0.023	0.039	0.036	0.041	0.050	0.110	0.725	
Citigroup 2	−0.213	0.047	1.694	0.337	0.231	0.761	4.482	40.951
s.e.	0.057	0.032	0.053	0.107	0.047	0.039	0.609	
Citigroup 3	−0.057	0.076	1.387	0.006	0.023	0.974	4.672	1.446
s.e.	0.031	0.029	0.037	0.003	0.004	0.003	0.672	

Date of time period for each stock:
AIG 1: 1/3/2005–10/8/2007, AIG 2: 10/9/2007–4/26/2010, AIG 3: 4/27/2010–3/31/2016
Citigroup 1: 1/3/2005–10/5/2007, Citigroup 2: 10/8/2007–1/13/2012, Citigroup 3: 1/17/2012–3/31/2016

each stock. In other words, the scatter plots provide empirical evidence to support our quantile regression setting for examining the behavior and variation of the market beta, especially under extreme market conditions.

To estimate model parameters for each sub-period and also evaluate the market risk of model (1), we use the "rugrach" package in R, available on CRAN (https://cran.r-project.org/web/packages/changepoint/changepoint). The initial results show that there is no need to fit an asymmetric effect on each sub-period any longer: the proposed market model is given in (4). Indeed, the market beta has no significant asymmetric impact on each sub-period for both AIG and Citigroup. Table 2 shows the estimation results on market model (4) for each sub-period:

$$r_t = \beta_{0,i} + \phi_{1,i} r_{t-1} + \beta_{1,i} r_{m,t} + a_t, \tag{4}$$
$$a_t = \sigma_t \varepsilon_t, \ \varepsilon_t \ \text{i.i.d.} \ \sim t^*(\nu)$$
$$\sigma^2 = \alpha_{0,i} + \alpha_{1,i} a_{t-1}^2 + \gamma_{1,i} \sigma_{t-1}^2,$$
$$i = \begin{cases} 1 \ \text{if } t < c_1 \\ 2 \ \text{if } c_1 \leq t < c_2, \\ 3 \ \text{if } t \geq c_2 \end{cases}$$

where c_1 and c_2 are pre-fixed change points obtained based on the CUSUM test. To ensure stationarity and positive volatilities, we impose the following constraints on the volatility parameters:

$$\alpha_{0,i} > 0, \ 0 \leq \alpha_{1,i}, \ \gamma_{1,i} < 1, \ \alpha_{1,i} + \gamma_{1,i} < 1, \ i = 1, 2, 3.$$

We note the significant AR(1) coefficient at the 5% level for the first sub-period of AIG and the third sub-period of Citigroup. The estimates of β_1 appear to be crucial for all sub-periods, having the relationship of $\beta_{1,2} > \beta_{1,3} > 1 > \beta_{1,1}$, where $\beta_{1,i}$ denotes the market beta of sub-period i. The last column in Table 2 shows the average unconditional variance, $\alpha_{0,i}/1 - \alpha_{1,i} - \gamma_{1,i}$. The average unconditional variances of the second sub-period exhibit more extreme values than other sub-periods of each stock: they are respectively 2581 and 41 for AIG and Citigroup during the global

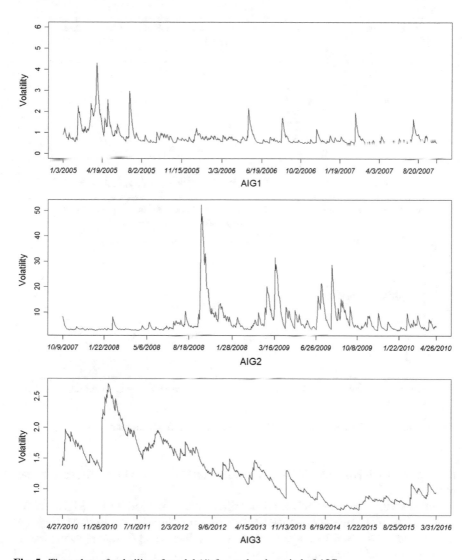

Fig. 5 Time plots of volatility of model (4) for each sub-period of AIG

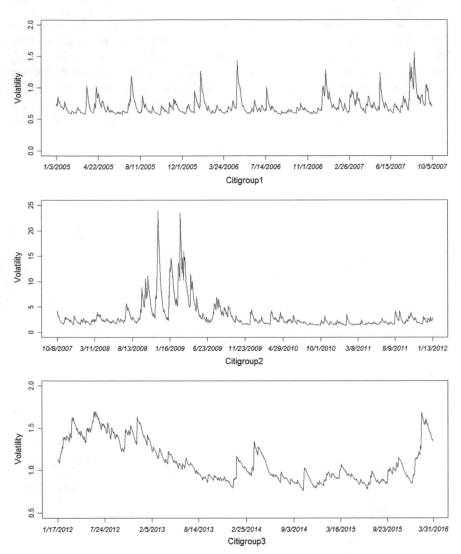

Fig. 6 Time plots of volatility of model (4) for each sub-period of Citigroup

financial crisis. It is obvious that one should separate these extreme periods in model building.

Figures 5 and 6 provide estimated volatilities of all sub-periods for AIG and Citigroup. These plots demonstrate that the estimated volatilities of the second sub-period have tremendous volatility magnitude. Note that Figs. 5 and 6 do not have the same scale in the y-axis; the first and third periods have the same scale, while the second period of the two excess stock returns looks much more volatile, especially from 2008 to 2009.

Table 3 Ljung-Box statistics from the standardized residuals, $\tilde{a}_{t-1}$, and squared standardized residuals, $\tilde{a}_{t-1}^2$, based on Model in (4)

Stock	$Q(1)$	p-value	$Q^2(1)$	p-value	$Q(5)$	p-value	$Q^2(5)$	p-value
AIG1	0.63	0.4274	0.0195	0.8889	3.084	0.4105	0.5751	0.9455
AIG2	3.317	0.0686	0.0117	0.9138	5.044	0.1027	0.1131	0.9976
AIG3	0.9492	0.3299	1.854	0.1733	1.1051	0.9299	2.382	0.5314
Citigroup1	1.376	0.2408	0.3561	0.5507	3.877	0.2456	0.989	0.8622
Citigroup2	2.251	0.1336	0	0.9993	3.062	0.4158	0.4716	0.962
Citigroup3	0.1731	0.6773	1.996	0.1577	1.3912	0.8723	2.464	0.5135

Date of three subseries for each stock:
AIG1: 1/3/2005–10/8/2007, AIC2: 10/9/2007–4/26/2010, AIG3: 4/27/2010–3/31/2016
Citigroup1: 1/3/2005–10/5/2007, Citigroup2: 10/8/2007–1/13/2012, Citigroup3: 1/17/2012–3/31/2016

Diagnostic checking is an important task in model fitting. Here, we use the Ljung-Box test [20] to examine model adequacy based on the autocorrelation functions (ACFs) of standardized residuals $\tilde{a}_{t-1} = (a_{t-1}/\sigma_{t-1})$ to check the adequacy of the mean equation as well as those of the squared standardized residuals $\tilde{a}_{t-1}^2$ to check the validity of the volatility equation as shown in Table 3. The Ljung-Box statistics show that market model (4) is adequate for all sup-periods: all p-values appear to be greater than 0.05. Figures 7 and 8 present the ACF plots of squared standardized residuals for all three subseries of AIG and Citigroup, respectively. To save space, we do not list ACF plots of standardized residuals for all three subseries for the two stocks. No significant autocorrelations are shown in the residuals for all subseries of the two, thus indicating that the market model with structural breaks fits quite well to AIG and Citigroup stocks in each sub-period.

We finally consider the quantile regression on the asymmetric market mode and estimate market model (5) as follows:

$$Q^{(\tau)}(r_t | F_{t-1}) = \beta_{o,i}^{(\tau)} + \phi_{1,i}^{(\tau)} r_{t-1} + \beta_{1,i}^{(\tau)} r_{m,t} + \beta_{2,i}^{(\tau)} I_{(r_{m,t-1}<0)} r_{m,t-1}, \tag{5}$$

$$i = \begin{cases} 1 \text{ if } t < c_1 \\ 2 \text{ if } c_1 \le t < c_2, \\ 3 \text{ if } t \ge c_2 \end{cases}$$

where $Q^{(\tau)}(r_t)$ represents the conditional τth quantile estimates of stock excess return at time t, and F_{t-1} is information set up to time $t-1$. Again, c_1 and c_2 are pre-fixed change points based on the CUSUM test. In Table 4 the estimate results of the numbers represent the quantile estimates and the corresponding least square estimates (LSEs) for the asymmetric quantile model (5). It shows that the intercept $\beta_{0,i}^{(\tau)}$ increases in τ as expected; the intercepts are negative at low quantile levels and positive at high quantile levels for each sub-period. The quantile estimates of $\beta_{1,i}^{(\tau)}$ behave very differently across the quantile levels and sub-periods. For example, the relation $\beta_{1,1}^{(\tau)} < \beta_{1,3}^{(\tau)} < \beta_{1,2}^{(\tau)}$ is true for all quantile levels for AIG and Citigroup.

Table 4 Parameter estimates and LSE for the asymmetric quantile market model in (5) over various quantile levels for AIG and Citigroup returns

τ	0.1	0.2	0.3	0.4	0.5	0.6	0.7	0.8	0.9	LSE
AIG										
$\beta_{0,1}$	-0.8100	-0.5196	-0.3429	-0.2163	-0.0760	0.0463	0.2125	0.3678	0.7156	-0.0574
$\phi_{1,1}$	0.0596	0.0664	0.0517	0.0283	0.0530	0.0435	0.0113	0.0350	0.0232	0.0328
$\beta_{1,1}$	0.9572	0.8564	0.8686	0.8619	0.8608	0.8247	0.8452	0.8577	0.8889	0.8876
$\beta_{2,1}$	0.0320	-0.0560	-0.0530	-0.0846	-0.1119	-0.1050	-0.1505	-0.3125	-0.3511	-0.1348
$\beta_{0,2}$	-6.9067	-3.1166	-1.8552	-1.1365	-0.3518	0.3348	0.9834	2.0290	5.3616	-0.6543
$\phi_{1,2}$	0.2165	0.1010	0.1092	0.0817	0.0769	0.0916	0.1270	0.1230	0.1627	0.2390
$\beta_{1,2}$	2.2540	1.9374	1.9470	1.8164	1.8877	1.9241	2.0186	1.9242	1.9812	2.2719
$\beta_{2,2}$	-0.0626	0.3042	0.0882	-0.1707	-0.1770	-0.3031	-0.5587	-0.5750	-0.6385	-0.4936
$\beta_{0,3}$	-1.2747	-0.8013	-0.5271	-0.3002	-0.0432	0.2218	0.4639	0.7792	1.3247	0.0231
$\phi_{1,3}$	-0.0483	-0.0424	-0.0382	-0.0366	-0.0432	-0.0528	-0.0346	-0.0109	-0.0218	-0.0628
$\beta_{1,3}$	1.4214	1.4477	1.3738	1.3686	1.3695	1.3781	1.4008	1.4255	1.4760	1.4245
$\beta_{2,3}$	0.1901	0.1743	0.0325	0.0006	0.0357	0.1002	-0.0473	-0.1053	-0.013	0.1023
Citigroup										
$\beta_{0,1}$	-0.7992	-0.4946	-0.2892	-0.1561	0.0007	0.1220	0.2285	0.4422	0.6988	-0.0366
$\phi_{1,1}$	-0.0144	-0.0241	-0.0293	-0.0609	-0.0309	-0.0064	0.0162	0.0306	0.0278	0.0196
$\beta_{1,1}$	1.1134	1.0091	0.9816	0.9258	0.8982	0.9277	0.9546	0.9651	1.0158	1.0165
$\beta_{2,1}$	1.1134	0.0852	0.0798	0.0612	0.0064	-0.0574	-0.1349	-0.0559	-0.2458	-0.1018
$\beta_{0,2}$	-2.8739	-1.7736	-1.1866	-0.7102	-0.2817	0.1761	0.6418	1.3524	2.4551	-0.1248
$\phi_{1,2}$	0.0811	0.0588	0.0543	0.081	0.0677	0.0668	0.1099	0.1309	0.1655	0.0949
$\beta_{1,2}$	2.0444	1.9178	1.8885	1.8436	1.8629	1.8855	1.9108	1.8760	1.9561	2.1164
$\beta_{2,2}$	0.7604	0.4654	0.2158	0.132	-0.0229	-0.2838	-0.509	-0.6269	-1.1790	0.1000
$\beta_{0,3}$	-1.2321	-0.8074	-0.4616	-0.2587	-0.0907	0.1332	0.3895	0.6780	1.2471	-0.0285
$\phi_{1,3}$	-0.0149	0.0069	0.0160	0.0395	0.0499	0.0651	0.0679	0.0751	0.0343	0.0186
$\beta_{1,3}$	1.6111	1.4575	1.4099	1.4353	1.4521	1.4432	1.4800	1.5814	1.6983	1.5663
$\beta_{2,3}$	0.1844	0.0745	0.1315	0.0839	0.0469	0.0377	-0.0570	-0.2858	-0.3846	0.0103

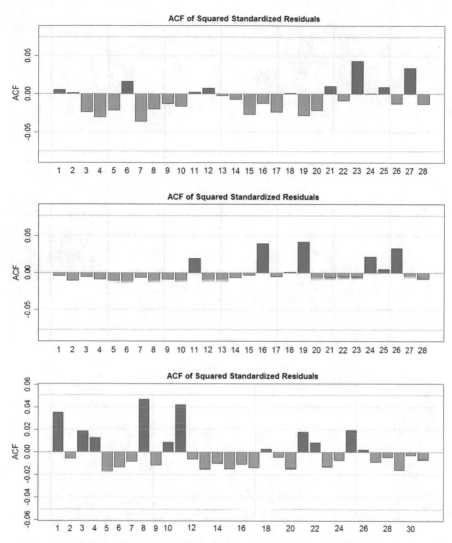

Fig. 7 ACF plots of squared standardized for AIG1, AIG2, and AIG3 based on the market model with GARCH effects in (4), respectively

Moreover, the estimates of $\beta_{1,2}^{(0.1)}$ are the highest among all quantile levels for both stocks, i.e. $\beta_{1,2}^{(0.1)} = 2.2540$ and $\beta_{1,2}^{(0.1)} = 2.0444$ for AIG and Citigroup, respectively.

There are some asymmetric effects during the second sub-period for AIG and Citigroup for some τ levels. Table 4 shows that $\beta_{2,2}^{(\tau)}$ is significantly positive at lower quantile levels when $\tau = 0.1$ and 0.2 for Citigroup, while $\beta_{2,2}^{(\tau)}$ is significantly negative at higher quantile levels for AIG ($\tau = 0.7, 0.8$) and Citigroup ($\tau = 0.7, 0.8, 0.9$).

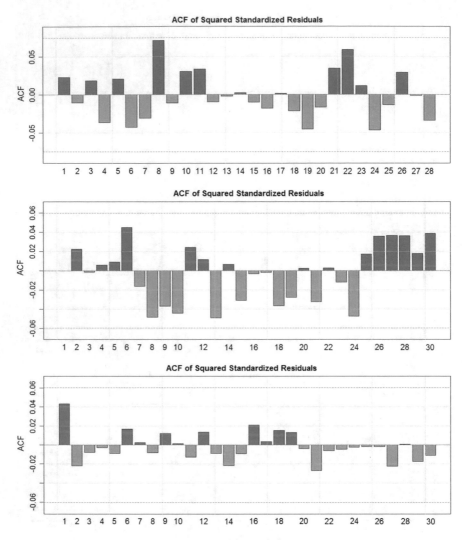

Fig. 8 ACF plots of squared standardized for Citigroup1, Citigroup2, and Citigroup3 based on the market model with GARCH effects in (4), respectively

The AR(1) coefficients, $\phi_{1,1}^{(\tau)}$ and $\phi_{1,3}^{(\tau)}$, are insignificant for all τ. We observe that some of the coefficients of $\phi_{1,2}^{(\tau)}$ are significant in the lower regime and/or the upper regime in extreme quantile levels. This implies that investors react differently and have different preferences when facing varied market news, especially under extreme quantile levels. Figures 9 and 10 provide all quantile levels versus the corresponding parameter estimates, including 95% confidence intervals for model (5) in each sub-period for AIG and Citigroup, respectively. The LSE and 95% confidence intervals

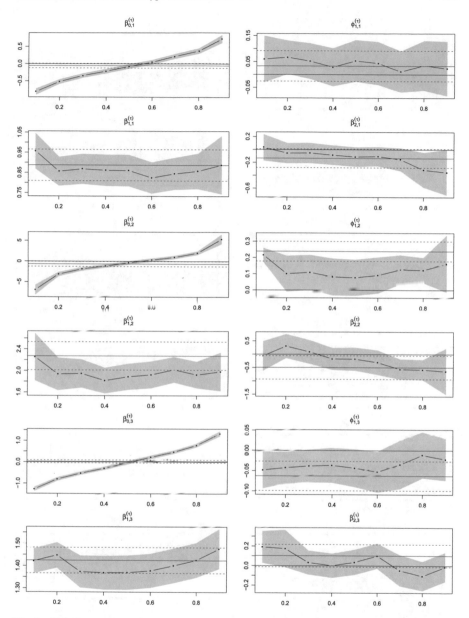

Fig. 9 AIG: Parameter estimates and the corresponding 95% confidence intervals for model (5) under all quantile levels. The first two rows denote the quantile plots of the first sub-period. The 3rd and 4th rows denote the quantile plots of the second sub-period. The 5th and 6th rows denote the quantile plots of the third sub-period

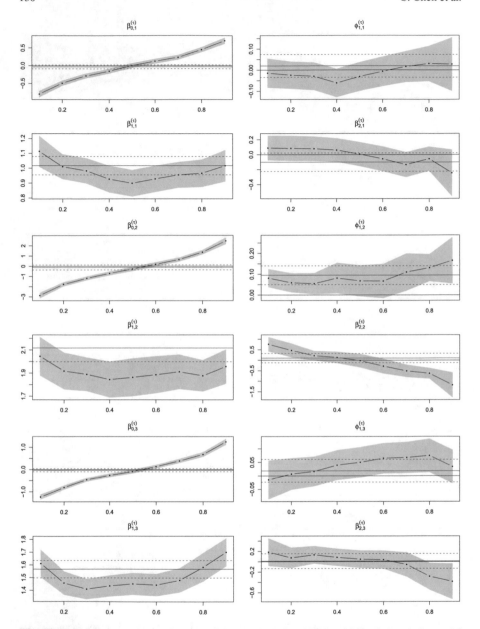

Fig. 10 Citigroup: Parameter estimates and the corresponding 95% confidence intervals for model (5) under all quantile levels. The first two rows denote the quantile plots of the first sub-period. The 3rd and 4th rows denote the quantile plots of the second sub-period. The 5th and 6th rows denote the quantile plots of the third sub-period

are denoted by solid and dash lines, respectively. We find that the estimates of $\beta_{2,2}^{(\tau)}$ behave very differently from LSE, which is lower than that of LSE in most cases.

4 Times-Varying Betas

For the CAPM-type market model in Eq. (1), we have:

$$\beta = \frac{\text{Cov}(r_t, r_{m,t})}{\text{Var}(r_{m,t})}.$$

Ferson and Harvey [10] propose a conditional CAPM that describes the time-varying dynamics of the market beta. Adrian and Franzoni [1] introduce 'learning' into standard conditional CAPM models, by estimating betas with the Kalman filter. Tsay [25] also demonstrates time-varying betas based on the CAPM-type model with GARCH (1,1) innovations. In this section we compare structural changes of beta coefficients and time-varying betas by using the CAPM-based model in (1) with GARCH (1,1) errors, which model the time-varying β. Following Tsay's [25] suggestion, we estimate $\text{Cov}(r_t, r_{m,t})$ as follows.

$$\text{Var}(r_i + r_{m,t}) = \text{Var}(r_t) + 2\text{Cov}(r_t, r_{m,t}) + \text{Var}(r_{m,t}),$$
$$\text{Var}(r_i - r_{m,t}) = \text{Var}(r_t) - 2\text{Cov}(r_t, r_{m,t}) + \text{Var}(r_{m,t}).$$

We thus obtain:

$$\text{Cov}(r_i, r_{m,t}) = \frac{\text{Var}(r_i + r_{m,t}) - \text{Var}(r_i - r_{m,t})}{4} \tag{6}$$

This indicates that time-varying covariances between excess returns $r_{m,t}$ and r_t can be obtained by the volatilities of $r_t + r_{m,t}$ and $r_t - r_{m,t}$. Since the estimate of $\text{Var}(r_{m,t})$ can be obtained by the volatility of the GARCH(1,1) model with Gaussian errors, β_t can be estimated via the ratio of Eq. 6 and the estimate of $\text{Var}(r_{m,t})$. Figures 11 and 12 illustrate the time-varying β for the daily excess returns of AIG and Citigroup stocks, indicating that the betas fluctuate substantially in the period between the suggested two break points and is high in the latter part of 2008. The magnitude of variations of the time-varying betas is much more severe for AIG within the range of (-7.7, 7.3). This result highlights that it is reasonable to discretize time-varying betas over time into three sub-periods based on the CUSUM of squares test.

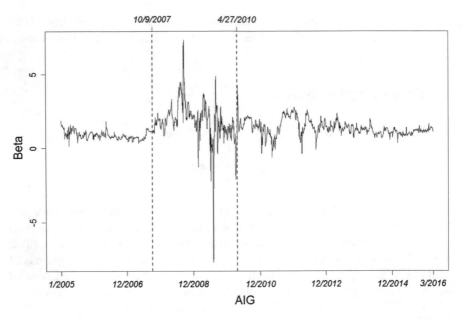

Fig. 11 Time-varying betas for the daily excess returns of AIG stock from January 3, 2005 to December 31, 2016. The daily excess returns of the S&P 500 index are used as the market returns. The vertical lines denote the two structure breaks detected by the CUSUM test

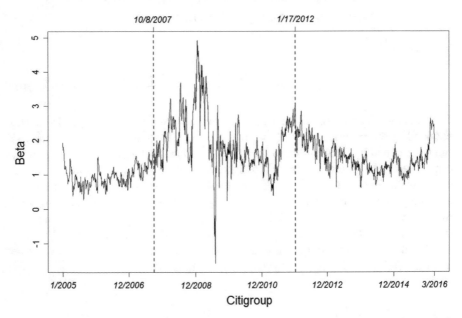

Fig. 12 Time-varying betas for the daily excess returns of Citigroup stock from January 3, 2005 to December 31, 2016. The daily excess returns of the S&P 500 index are used as the market returns. The vertical lines denote the two structure breaks detected by the CUSUM test

5 Conclusion

The world economy faced one of its most considerable systemic risk crises in 2008. In fact, the contagion began in 2007 after home prices in the U.S. had just peaked, leading to the values of subprime mortgage portfolios plunging downward first for the entire U.S. financial sector and then to financial markets overseas. On September 15, 2008, the government decided to not bail out Lehman Brothers, allowing one of the largest investment banks in the U.S. to go bankrupt. One day later, the government instead rescued AIG, because it was deemed too big to fail (see [13]), bailing out the insurance company to the amount of $85 billion.

In this work we use the CAPM-type model with heteroskedasticity for AIG and Citigroup stock returns during 2005 to 2016 with two structural breaks to discriminate the period influenced by the global financial crisis, subsequently observing a dramatic change in the beta coefficient and volatility in the second sub-period for both stocks. We confirm the time-varying nature of market risk in response to changes in the market, and that this variation can change over different time periods. Next, the quantile regression method reveals a different behavior in the market beta and lagged autoregressive effect for different sub-periods and quantile levels. Our findings are consistent with the fact that the beta coefficients of AIG and Citigroup stocks in the second sub-period show a more extreme trend, and that the third sub-period has the second highest beta coefficient, indicating high risk for AIG and Citigroup even for the last sub-period relative to other listed stocks in the market. All these results confirm the validity of our method in examining the behavior of the beta coefficient and volatility of big-cap stocks such as AIG and Citigroup.

Acknowledgements We thank the editor and anonymous referee for their helpful comments. Cathy W.S. Chen's research is funded by the Ministry of Science and Technology, Taiwan (MOST 105-2118-M-035-003-MY2 and MOST 103-2118-M-035-002-MY2). Sangyeol Lee's research is supported by the Basic Science Research Program through the National Research Foundation of Korea (NRF) funded by the Ministry of Science, ICT and Future Planning (No. 2015R1A2A2A010003894).

References

1. Adrian T, Franzoni F (2009) Learning about beta: Time-varying factor loadings, expected returns, and the conditional CAPM. J Emp Finan 16:537–556
2. Bollerslev T (1986) Generalized autoregressive conditional heteroskedasticity. J Econometr 31:307–327
3. Bollerslev T, Chou RY, Kroner KP (1992) ARCH modeling in finance: A review of the theory and empirical evidence. J Econometr 52:5–59
4. Bollerslev T, Engle RF, Wooldridge JM (1988) A capital assert pricing model with times-varying covariances. J Polit Econ 96:116–131
5. Brown RL, Durbin J, Evans JM (1975) Techniques for testing the constancy of regression relationships over time. J R Stat Soc Ser B 35:149–192
6. Chen CWS, Li M, Nguyen NTH, Sriboonchitta S (2015) On asymmetric market model with heteroskedasticty and quantile regression. Comput Econ. doi:10.1007/s10614-015-9550-3

7. Chen CWS, Gerlach R, Lin AMH (2011) Multi-regime nonlinear capital asset pricing model. Quant Finan 11:1421–1438
8. Engle RF (1982) Autoregressive conditional heteroscedasticity with estimates of the variance of United Kingdom inflation. Econometrica 50:987–1008
9. Engle C, Rodrigues AP (1989) Test of international CAPM with time-varying covariances. J Appl Econometr 4:119–138
10. Ferson W, Harvey C (1999) Conditioning variables and the cross-section of stock returns. J Finan 54:1325–1360
11. Inclan C, Tiao GC (1994) Use of the cumulative sums of squares for retrospective detection of changes of variance. J Am Stat Assoc 89:913–923
12. Harvey AC (1989) Forecasting, structural time series models and the Kalman filter. Cambridge University Press, Cambridge
13. Karnitschnig M, Solomon D, Pleven L, Hilsenrath JE (2008) U.S. to take over AIG in $85 billion bailout; central banks inject cash as credit dries up. The Wall Street J. http://www.wsj.com/articles/SB122156561931242905
14. Kim S, Cho S, Lee S (2000) On the CUSUM test for parameter changes in GARCH(1,1) models. Commun Stat Theor Methods 29:445–462
15. Lee S, Ha J, Na O, Na S (2003) The CUSUM test for parameter change in time series models. Scand J Stat 30:781–796
16. Lee S, Lee J (2014) Residual based cusum test for parameter change in AR-GARCH models. In: Huynh VN, Kreinovich V, Sriboonchitta S (eds) Modeling dependence in econometrics. Advances in intelligent systems and computing, vol 251, Springer, Switzerland, pp 101–111
17. Lee S, Tokutsu Y, Maekawa K (2004) The CUSUM test for parameter change in regression models with ARCH errors. J Jpn Stat Soc 34:173–188
18. Lilian Ng (1991) Tests of the CAPM with time-varying covariances: a multivariate GARCH approach. J Finan 46:1507–1521
19. Lintner J (1965) The valuation of risk assets and selection of risky investments in stock portfolios and capital budgets. Rev Econ Stat 47:13–37
20. Ljung G, Box GEP (1978) On a measure of lack of fit in time series models. Biometrika 66:67–72
21. Markowitz HM (1952) Portfolio selection. J Finan 7:77–91
22. Schwert GW, Seguin PJ (1990) Heteroskedasticty in stock returns. J Financ 4:1929–1965
23. Sharpe WF (1964) Capital asset price: a theory market equilibrium under conditions of risks. J Financ 19:425–442
24. Tsay Ruey S (2010) Analysis of financial time series, 3rd edn. John Wiley & Sons Inc, New Jersey
25. Tsay Ruey S (2012) An introduction to analysis of financial data with R. John Wiley & Sons Inc, New Jersey

Weighted Least Squares and Adaptive Least Squares: Further Empirical Evidence

Martin Sterchi and Michael Wolf

Abstract This paper compares ordinary least squares (OLS), weighted least squares (WLS), and adaptive least squares (ALS) by means of a Monte Carlo study and an application to two empirical data sets. Overall, ALS emerges as the winner. It achieves most or even all of the efficiency gains of WLS over OLS when WLS outperforms OLS, but it only has very limited downside risk compared to OLS when OLS outperforms WLS.

1 Introduction

The linear regression model is still a cornerstone of empirical work in the social sciences. The standard textbook treatment assumes conditional homoskedasticity of the error terms. When this assumption is violated—that is, when conditional heteroskedasticity is present—standard inference is no longer valid. The current practice in such a setting is to estimate the model by ordinary least squares (OLS) and use heteroskedasticity-consistent (HC) standard errors; this approach dates back to [14].

[13] propose to 'resurrect' the previous practice of using weighted least squares (WLS), which weights the data before applying OLS. The weighting scheme is based on an estimate of the skedastic function, that is, of the function that determines the conditional variance of the error term given the values of the regressors. In practice, the model for estimating the skedastic function may be misspecified. If this is the case, using standard inference based on the weighted data will not be valid. Therefore,

M. Sterchi (✉)
University of Applied Sciences and Arts Northwestern Switzerland, Olten, Switzerland
e-mail: martin.sterchi@fhnw.ch

M. Wolf
Department of Economics, University of Zurich, Zurich, Switzerland
e-mail: michael.wolf@econ.uzh.ch

© Springer International Publishing AG 2017
V. Kreinovich et al. (eds.), *Robustness in Econometrics*,
Studies in Computational Intelligence 692, DOI 10.1007/978-3-319-50742-2_9

[13] propose to also use HC standard errors for weighted data (as would be done for the original data) and prove asymptotic validity of the resulting inference under suitable regularity conditions.

[13] also propose adaptive least squares (ALS) where a pretest for conditional heteroskedasticity decides whether the applied researcher should use OLS (with HC standard errors) or WLS (with HC standard errors). Asymptotic validity of the resulting inference is established as well.

In addition to providing asymptotic theory, [13] examine finite-sample performance of WLS and ALS compared to OLS via Monte Carlo simulations. But these simulations are restricted to univariate regressions (that is, regressions where there is only one regressor in addition to the constant). In applied work, though, multivariate regressions are more common.

The purpose of this paper is two-fold. On the one hand, we provide extensive Monte Carlo simulations comparing WLS and ALS to OLS in multivariate regressions, covering both estimation and inference. On the other hand, we compare the results of WLS and ALS to OLS for two empirical data sets.

The remainder of the paper is organized as follows. Section 2 gives a brief description of the methodology for completeness. Section 3 examines finite-sample performance via a Monte Carlo study. Section 4 provides an application to two empirical data sets. Section 5 concludes.

2 Brief Description of the Methodology

For completeness, we give a brief description of the methodology for WLS and ALS here. More details can be found in [13].

2.1 The Model

We maintain the following set of assumptions throughout the paper.

(A1) The linear model is of the form

$$y_i = x_i'\beta + \varepsilon_i \quad (i = 1, \ldots, n), \tag{1}$$

where $x_i \in \mathbb{R}^K$ is a vector of explanatory variables (regressors), $\beta \in \mathbb{R}^K$ is a coefficient vector, and ε_i is the unobservable error term with certain properties to be specified below.

(A2) The sample $\{(y_i, x_i')\}_{i=1}^n$ is independent and identically distributed (i.i.d.).

(A3) All the regressors are predetermined in the sense that they are orthogonal to the contemporaneous error term:

$$\mathbb{E}(\varepsilon_i | x_i) = 0. \tag{2}$$

(A4) The $K \times K$ matrix $\Sigma_{xx} := \mathbb{E}(x_i x_i')$ is nonsingular (and hence finite). Furthermore, $\sum_{i=1}^{n} x_i x_i'$ is invertible with probability one.

(A5) The $K \times K$ matrix $\Omega := \mathbb{E}(\varepsilon_i^2 x_i x_i')$ is nonsingular (and hence) finite.

(A6) There exists a nonrandom function $v : \mathbb{R}^K \to \mathbb{R}_+$ such that

$$\mathbb{E}(\varepsilon_i^2 | x_i) = v(x_i). \tag{3}$$

Therefore, the *skedastic function* $v(\cdot)$ determines the functional form of the conditional heteroskedasticity. Note that under (A6),

$$\Omega = \mathbb{E}\big[v(x_i) \cdot x_i x_i'\big].$$

It is useful to introduce the customary vector-matrix notations

$$y := \begin{bmatrix} y_1 \\ \vdots \\ y_n \end{bmatrix}, \quad \varepsilon := \begin{bmatrix} \varepsilon_1 \\ \vdots \\ \varepsilon_n \end{bmatrix}, \quad X := \begin{bmatrix} x_1' \\ \vdots \\ x_n' \end{bmatrix} = \begin{bmatrix} x_{11} \cdots x_{1K} \\ \vdots \cdots \vdots \\ x_{n1} \cdots x_{nK} \end{bmatrix},$$

so that Eq. (1) can be written more compactly as

$$y = X\beta + \varepsilon. \tag{4}$$

Furthermore, assumptions (A2), (A3), and (A5) imply that

$$\text{Var}(\varepsilon | X) = \begin{bmatrix} v(x_1) \\ & \ddots \\ & & v(x_n) \end{bmatrix}.$$

2.2 Estimators: OLS, WLS, and ALS

The well-known ordinary least squares (OLS) estimator of β is given by

$$\hat{\beta}_{\text{OLS}} := (X'X)^{-1}X'y.$$

Under the maintained assumptions, the OLS estimator is unbiased and consistent. This is the good news.

A more efficient estimator can be obtained by reweighting the data (y_i, x_i') and then applying OLS in the transformed model

$$\frac{y_i}{\sqrt{v(x_i)}} = \frac{x_i'}{\sqrt{v(x_i)}}\beta + \frac{\varepsilon_i}{\sqrt{v(x_i)}}. \tag{5}$$

Letting

$$V := \begin{bmatrix} v(x_1) & & \\ & \ddots & \\ & & v(x_n) \end{bmatrix},$$

the resulting estimator can be written as

$$\hat{\beta}_{\text{BLUE}} := (X'V^{-1}X)^{-1}X'V^{-1}y. \tag{6}$$

It is the best linear unbiased estimator (BLUE) and is consistent; in particular, it is more efficient than the OLS estimator. However, it is generally not a feasible estimator, since the skedastic function $v(\cdot)$ is generally unknown.

A feasible approach is to estimate the skedastic function $v(\cdot)$ from the data in some way and to then apply OLS in the model

$$\frac{y_i}{\sqrt{\hat{v}(x_i)}} = \frac{x_i'}{\sqrt{\hat{v}(x_i)}}\beta + \frac{\varepsilon_i}{\sqrt{\hat{v}(x_i)}}, \tag{7}$$

where $\hat{v}(\cdot)$ denotes the estimator of $v(\cdot)$. The resulting estimator is the weighted least squares (WLS) estimator. Letting

$$\hat{V} := \begin{bmatrix} \hat{v}(x_1) & & \\ & \ddots & \\ & & \hat{v}(x_n) \end{bmatrix},$$

the WLS estimator can be written as

$$\hat{\beta}_{\text{WLS}} := (X'\hat{V}^{-1}X)^{-1}X'\hat{V}^{-1}y.$$

It is not necessarily unbiased. If $\hat{v}(\cdot)$ is a consistent estimator of $v(\cdot)$, than WLS is asymptotically more efficient than OLS. But even if $\hat{v}(\cdot)$ is an inconsistent estimator of $v(\cdot)$, WLS can result in large efficiency gains over OLS in the presence of noticeable conditional heteroskedasticity; see Sect. 3.

The idea of adaptive least squares (ALS) is that we let the data 'decide' whether to use OLS or WLS for the estimation. Intuitively, we only want to use WLS if there is 'noticeable' conditional heteroskedasticity present in the data. Here, 'noticeable' is with respect to the model used for estimating the skedastic function in practice.

[13] suggest applying a test for conditional heteroskedasticity. Several such tests exists, the most popular ones being the tests of [2, 14]; also see [9, 10]. If the null hypothesis of conditional homoskedasticity it not rejected by such a test, use the OLS estimator; otherwise, use the WLS estimator. The resulting estimator is nothing else than the ALS estimator.

2.3 Parametric Model for Estimating the Skedastic Function

In order to estimate the skedastic function $v(\cdot)$, [13] suggest the use of the following parametric model:

$$v_\theta(x_i) := \exp\big(\nu + \gamma_2 \log |x_{i,2}| + \ldots + \gamma_K \log |x_{i,K}|\big), \tag{8}$$

with $\theta := (\nu, \gamma_2, \ldots, \gamma_K)'$, assuming that $x_{i,1} \equiv 1$ (that is, the original regression contains a constant). Otherwise, the model should be

$$v_\theta(x_i) := \exp\big(\nu + \gamma_1 \log |x_{i,1}| + \gamma_2 \log |x_{i,2}| + \ldots + \gamma_K \log |x_{i,K}|\big),$$

with $\theta := (\nu, \gamma_1, \ldots, \gamma_K)'$. Such a model is a special case of the form of multiplicative conditional heteroskedasticity previously proposed by [5] and Sect. 9.3 of [8], among others.

Assuming model (8), the test for conditional heteroskedasticity specifies

$$H_0 : \gamma_2 = \ldots = \gamma_K = 0 \quad \text{versus} \quad H_1 : \text{at least one } \gamma_k \neq 0 \ (k = 2, \ldots, K).$$

To carry out the test, fix a small constant $\delta > 0$, estimate the following regression by OLS:

$$\log\big[\max(\delta^2, \hat{\varepsilon}_i^2)\big] = \nu + \gamma_2 \log |x_{i,2}| + \ldots + \gamma_K \log |x_{i,K}| + u_i, \tag{9}$$

with $\hat{\varepsilon}_i := y_i - x_i'\hat{\beta}_{\text{OLS}}$, and denote the resulting R^2-statistic by R^2.[1] Furthermore, denote by $\chi^2_{K-1,1-\alpha}$ the $1 - \alpha$ quantile of the chi-squared distribution with $K - 1$ degrees of freedom. Then the test rejects conditional homoskedasticity at nominal level α if $n \cdot R^2 > \chi^2_{K-1,1-\alpha}$.

Last but not least, the estimate of the skedastic function is given by

$$\hat{v}(\cdot) := v_{\hat{\theta}}(\cdot),$$

where $\hat{\theta}$ is an estimator of θ obtained by the OLS regression (9).

---

[1] The reason for introducing a small constant $\delta > 0$ on the left-hand side of (9) is that, because one is taking logs, one needs to avoid a residual of zero, or even very near zero. The choice $\delta = 0.1$ seems to work well in practice.

2.4 Inference: OLS, WLS, and ALS

2.4.1 Confidence Intervals

A nominal $1 - \alpha$ confidence interval for β_k based on OLS is given by

$$\hat{\beta}_{k,\text{OLS}} \pm t_{n-K,1-\alpha/2} \cdot \text{SE}_{\text{HC}}(\hat{\beta}_{k,\text{OLS}}), \tag{10}$$

where $t_{n-K,1-\alpha/2}$ denotes the $1 - \alpha/2$ quantile of the t distribution with $n - K$ degrees of freedom. Here $\text{SE}_{\text{HC}}(\cdot)$ denotes a HC standard error. Specifically [13] suggest to use the HC3 standard error introduced by [12].

A nominal $1 - \alpha$ confidence interval for β_k based on WLS is given by

$$\hat{\beta}_{k,\text{WLS}} \pm t_{n-K,1-\alpha/2} \cdot \text{SE}_{\text{HC}}(\hat{\beta}_{k,\text{WLS}}), \tag{11}$$

where again [13] suggest to use the HC3 standard error.

A nominal $1 - \alpha$ confidence interval for β_k based on ALS is given by either (10) or (11), depending on whether the ALS estimator is equal to the OLS estimator or to the WLS estimator.

2.4.2 Testing a Set of Linear Restrictions

Consider testing a set of linear restrictions on β of the form

$$H_0 : R\beta = r,$$

where $R \in \mathbb{R}^{p \times K}$ is matrix of full row rank specifying $p \leq K$ linear combinations of interest and $r \in \mathbb{R}^p$ is a vector specifying their respective values under the null.

A HC Wald statistic based on the OLS estimator is given by

$$W_{\text{HC}}(\hat{\beta}_{\text{OLS}}) := \frac{n}{p} \cdot (R\hat{\beta}_{\text{OLS}} - r)' \left[R\, \widehat{\text{Avar}}_{\text{HC}}(\hat{\beta}_{\text{OLS}}) R' \right]^{-1} (R\hat{\beta}_{\text{OLS}} - r).$$

Here $\widehat{\text{Avar}}_{\text{HC}}(\hat{\beta}_{\text{OLS}})$ denotes a HC estimator of the asymptotic variance of $\hat{\beta}_{\text{OLS}}$, that is, of the variance of the limiting multivariate normal distribution of $\hat{\beta}_{\text{OLS}}$. More specifically, if

$$\sqrt{n}(\hat{\beta}_{\text{OLS}} - \beta) \xrightarrow{d} N(0, \Sigma),$$

where the symbol $\xrightarrow{d}$ denotes convergence in distribution, then $\widehat{\text{Avar}}_{\text{HC}}(\hat{\beta}_{\text{OLS}})$ is an estimator of Σ. Related details can be found in Sect. 4 of [13]; in particular, it is again recommended to use a HC3 estimator.

A HC Wald statistic based on the WLS estimator is given by

$$W_{\mathrm{HC}}(\hat{\beta}_{\mathrm{WLS}}) := \frac{n}{p} \cdot (R\hat{\beta}_{\mathrm{WLS}} - r)'\left[R\,\widehat{\mathrm{Avar}}_{\mathrm{HC}}(\hat{\beta}_{\mathrm{WLS}})R'\right]^{-1}(R\hat{\beta}_{\mathrm{WLS}} - r).$$

For a generic Wald statistic W, the corresponding p-value is obtained as

$$PV(W) := \mathrm{Prob}\{F \geq \tilde{W}\}, \quad \text{where } F \sim F_{p,n}.$$

Here, $F_{p,n}$ denotes the F distribution with p and n degrees of freedom.

HC inference based on the OLS estimator reports $PV(W_{\mathrm{HC}}(\hat{\beta}_{\mathrm{OLS}}))$ while HC inference based on the WLS estimator reports $PV(W_{\mathrm{HC}}(\hat{\beta}_{\mathrm{WLS}}))$. Depending on the outcome of the test for conditional heteroskedasticity, ALS inference either coincides with OLS inference (namely, if the test does not reject conditional homoskedasticity) or coincides with WLS inference (namely, if the test rejects conditional homoskedasticity).

3 Monte Carlo Evidence

3.1 Configuration

We consider the following multivariate linear regression model

$$y_i = \beta_0 + \beta_1 x_{i,1} + \beta_2 x_{i,2} + \beta_3 x_{i,3} + \varepsilon_i. \tag{12}$$

The regressors are first generated according to a uniform distribution between 1 and 4, denoted by $U[1, 4]$. The simulation study is then repeated with the regressors generated according to a Beta distribution with the parameters $\alpha = 2$ and $\beta = 5$, denoted by Beta(2,5). In order to guarantee a range of values comparable to the one for the uniformly distributed regressors, the Beta distributed regressors have been multiplied by five. [11] chooses a standard lognormal distribution for the regressors and points out that, as a result, HC inference becomes particularly difficult because of a few extreme observations for the regressors. Since both the standard lognormal distribution and the Beta(2,5) distribution are right-skewed, the second part of the simulation study is in the spirit of the one in [11].

The error term model in (12) is given by

$$\varepsilon_i := \sqrt{v(x_i)}z_i \tag{13}$$

where $z_i \sim N(0, 1)$ and z_i is independent of all explanatory variables x_i. Here, $v(\cdot)$ corresponds to the skedastic function and will be specified below. Alternatively, a setting with error terms following a t-distribution with five degrees of freedom (scaled

Table 1 Parametric specifications of the skedastic function

S.1	$v(x_i) = z(\gamma) \cdot |x_{i,1}|^{\gamma} \cdot |x_{i,2}|^{\gamma} \cdot |x_{i,3}|^{\gamma}$	with $\gamma \in \{0, 1, 2, 4\}$
S.2	$v(x_i) = z(\gamma)\big(\gamma|x_{i,1}| + \gamma|x_{i,2}| + \gamma|x_{i,3}|\big)$	with $\gamma \in \{1, 2, 3\}$
S.3	$v(x_i) = z(\gamma) \exp\big(\gamma|x_{i,1}| + \gamma|x_{i,2}| + \gamma|x_{i,3}|\big)$	with $\gamma \in \{0.5, 1\}$
S.4	$v(x_i) = z(\gamma)\big(|x_{i,1}| + |x_{i,2}| + |x_{i,3}|\big)^{\gamma}$	with $\gamma \in \{2, 4\}$

to have variance one) will be tested. Without loss of generality, the parameters in (12) are all set to zero, that is, $(\beta_0, \beta_1, \beta_2, \beta_3) = (0, 0, 0, 0)$.

We consider four parametric specifications of the skedastic function as shown in Table 1. For the sake of simplicity, all specifications use only one parameter γ. (For example, Specification S.1 uses a common power γ on the absolute values of $x_{i,1}$, $x_{i,2}$, and $x_{i,3}$.) It would in principle be possible to use more than one parameter in a given specification, but then the number of scenarios in our Monte Carlo study would become too large. [11] proposes the use of a scaling factor for the specifications in order to make sure that the conditional variance of ε_i is on average one, while the degree of heteroskedasticity remains the same. For that reason, all the specifications in Table 1 contain a scaling factor $z(\gamma)$. [4] suggest measuring the aforementioned degree of heteroskedasticity by the ratio of the maximal value of $v(x)$ to the minimal value of $v(x)$. Consequently, in the case of conditional homoskedasticity, the degree of heteroskedasticity is one. The full set of results is presented in Table 4; note that in specification S.2, the degree of heteroskedasticity does not depend on the value of γ.

3.2 Estimation of the Skedastic Function

The following parametric model is used to estimate the skedastic function:

$$v_\theta(x_i) = \exp(\upsilon + \gamma_1 \log|x_{i,1}| + \gamma_2 \log|x_{i,2}| + \gamma_3 \log|x_{i,3}|). \tag{14}$$

It can be reformulated as

$$v_\theta(x_i) = \exp(\upsilon) \cdot |x_{i,1}|^{\gamma_1} \cdot |x_{i,2}|^{\gamma_2} \cdot |x_{i,3}|^{\gamma_3}. \tag{15}$$

Formulation (15) is equivalent to specification S.1 with $\exp(\upsilon) = z(\gamma_i)$. Hence, in the case of specification S.1, we assume the correct functional form of the skedastic function when estimating it. For all other specifications mentioned in the previous section—namely S.2–S.4—model (14) is misspecified.

The parameters of model (14) will be estimated by the following OLS regression:

$$\log[\max(\delta^2, \hat{\varepsilon}_i^2)] = \upsilon + \gamma_1 \log |x_{i,1}| + \gamma_2 \log |x_{i,2}| + \gamma_3 \log |x_{i,3}| + u_i, \qquad (16)$$

where the $\hat{\varepsilon}_i^2$ are the squared OLS residuals from regression (12). [13] suggest using a small constant $\delta > 0$ on the left-hand side of (16) in order to avoid taking the logarithm of squared OLS residuals near zero; as they do, we use $\delta = 0.1$.

Denote the fitted values of the regression (16) by $\hat{g}_i$. Then weights of the data for the application of WLS are simply given by $\hat{v}_i := \exp(\hat{g}_i)$, for $i = 1, \ldots, n$.

3.3 Estimation, Inference, and Performance Measures

The parameters in the regression model (12) are estimated using OLS and WLS. In addition, we include the ALS estimator. As suggested in Remark 3.1 of [13], a Breusch-Pagan test will be applied in order to determine the ALS estimator. Conditional homoskedasticity is rejected if $nR^2 > \chi^2_{3,0.9}$, where the R^2 statistic in this test is taken from the OLS regression (16). If conditional homoskedasticity is rejected, ALS coincides with WLS; otherwise ALS coincides with OLS.

To measure the performance of the different estimators, we use the empirical mean squared error (eMSE) given by

$$\text{eMSE}(\tilde{\beta}_k) := \frac{1}{B} \sum_{b=1}^{B} (\tilde{\beta}_{k,b} - \beta_k)^2, \qquad (17)$$

where $\tilde{\beta}_k$ denotes a generic estimator (OLS, WLS, or ALS) of the true parameter β_k. As is well known, the population mean squared error (MSE) can be broken down into two components as follows:

$$\text{MSE}(\tilde{\beta}_k) = \text{Var}(\tilde{\beta}_k) + \text{Bias}^2(\tilde{\beta}_k). \qquad (18)$$

Thus, the MSE corresponds to the sum of the variance of an estimator $\tilde{\beta}_k$ and its squared bias. While OLS is unbiased even in the case of conditional heteroskedasticity, WLS and ALS can be biased. Therefore, using the eMSE makes sure that OLS, WLS, and ALS are compared on equal footing.

We also assess the finite-sample performance of confidence intervals of the type

$$\tilde{\beta}_k \pm t_{n-4,1-\alpha/2} \cdot \text{SE}(\tilde{\beta}_k), \qquad (19)$$

where SE is either the HC standard error or the maximal (Max) standard error[2] of the corresponding estimator $\tilde{\beta}_k$ and $t_{n-K,1-\alpha/2}$ denotes the $1 - \alpha/2$ quantile of the t distribution with $n - K$ degrees of freedom.

First, we compute the empirical coverage probability of nominal 95% confidence intervals. Second, for OLS-Max, WLS-HC, WLS-Max, ALS-HC and ALS-Max, we compute the ratio of the average length of the confidence interval to the average length of the OLS-HC confidence interval, which thus serves as the benchmark. All the performance measures are chosen as in [13] to facilitate comparability of the results.

3.4 Results

We discuss separately the results for estimation and inference. For compactness of the exposition, we only report results for β_1. (The results for β_2 and β_3 are very similar and are available from the authors upon request.)

3.4.1 Estimation

Tables 5 and 6 in the appendix present the basic set of results when the regressors are generated according to a uniform distribution while the error terms are normally distributed. If the specification used to estimate the weights corresponds to the true specification of the skedastic function (Table 5), WLS is generally more efficient than OLS, except for the case of conditional homoskedasticity ($\gamma = 0$). For $\gamma = 0$, OLS is more efficient than WLS, which is reflected by ratios of the eMSE's (WLS/OLS) that are higher than one for all of the sample sizes. As n increases the ratios get closer to one, indicating a smaller efficiency loss of WLS compared to OLS. On the other hand, for positive values of γ, WLS is always more efficient than OLS and the efficiency gains can be dramatic for moderate and large sample sizes ($n = 50, 100$) and for noticeable conditional heteroskedasticity ($\gamma = 2, 4$). ALS offers an attractive compromise between OLS and WLS. Under conditional homoskedasticity ($\gamma = 0$), the efficiency loss compared to OLS is negligible, as all the eMSE ratios are no larger than 1.03. Under conditional heteroskedasticity, the efficiency gains over OLS are not as large as for WLS for small sample sizes ($n = 20$) but they are almost as large as for WLS for moderate sample sizes ($n = 50$) and equally as large as for WLS for large sample sizes ($n = 100$) (Tables 2 and 3).

---

[2]See Sect. 4.1 of [13] for a detailed description of the Max standard error. In a nutshell, the Max standard error is the maximum of the HC standard error and the 'textbook' standard error from an OLS regression, which assumes conditional homoskedasticity.

Table 2 OLS and WLS results for the CEO salaries data set. WLS/OLS denotes the ratio of the WLS-HC standard error to the OLS-HC standard error. For this data set, ALS coincides with WLS

Response variable: log(*salary*)

OLS

Coefficient	Estimate	SE-HC	t-stat	
constant	4.504	0.290	15.54	
log(*sales*)	0.163	0.039	4.15	
log(*mktval*)	0.109	0.052	2.11	
ceoten	0.012	0.008	1.54	
$R^2 = 0.32$	$\bar{R}^2 = 0.31$	$s = 0.50$	$F = 26.91$	

WLS

Coefficient	Estimate	SE-HC	t-stat	WLS/OLS
constant	4.421	0.240	18.45	0.83
log(*sales*)	0.152	0.037	4.13	0.94
log(*mktval*)	0.126	0.044	2.91	0.84
ceoten	0.015	0.007	2.31	0.88
$R^2 = 0.33$	$\bar{R}^2 = 0.32$	$s = 1.73$	$F = 29.04$	

Table 3 OLS and WLS results for the housing prices data set. WLS/OLS denotes the ratio of the WLS-HC standard error to the OLS-HC standard error. For this data set, ALS coincides with WLS

Response variable: log(*price*)

OLS

Coefficient	Estimate	SE (HC)	t-stat	
constant	11.084	0.383	28.98	
log(*nox*)	−0.954	0.128	−7.44	
log(*dist*)	−0.134	0.054	−2.48	
rooms	0.255	0.025	10.10	
stratio	−0.052	0.005	−11.26	
$R^2 = 0.58$	$\bar{R}^2 = 0.58$	$s = 0.27$	$F = 175.90$	

WLS

Coefficient	Estimate	SE (HC)	t-stat	WLS/OLS
constant	10.195	0.272	37.43	0.71
log(*nox*)	−0.793	0.097	−8.17	0.76
log(*dist*)	−0.127	0.035	−3.62	0.65
rooms	0.307	0.016	19.23	0.63
stratio	−0.037	0.004	−8.78	0.90
$R^2 = 0.68$	$\bar{R}^2 = 0.68$	$s = 1.33$	$F = 267.8$	

The higher the degree of heteroskedasticity, the higher the efficiency gain is of WLS over OLS. For instance, $\gamma = 4$ results in very strong conditional heteroskedasticity, as can be seen in Table 4. As a result, the ratio of the eMSE of WLS to the eMSE of OLS is below 0.05 for large sample sizes ($n = 100$). However, in the case of conditional homoskedasticity ($\gamma = 0$), OLS is more efficient than WLS, which is reflected by ratios of the eMSE's (WLS/OLS) that are higher than one for all of the sample sizes (though getting closer to one as n increases).

Figure 1 displays density plots of the three estimators of β_1 in the case of the four different parameter values of specification S.1 and for $n = 100$. The four plots visualize the potential efficiency gains of WLS and ALS over OLS as presented in Table 5 numerically. In the cases of $\gamma = 2$ and $\gamma = 4$, the density of ALS is virtually equal to the density of WLS, as there is no visible difference. It can be clearly seen how the variances of WLS and ALS get smaller relative to OLS when the degree of conditional heteroskedasticity increases.

What changes if the specification used to estimate the skedastic function does not correspond to the true specification thereof? The results for this case are presented in Table 6. First of all, the linear specification S.2 results in WLS being less efficient than OLS. Although the linear specification represents a form of conditional heteroskedasticity, it is of a different form than our parametric model used to estimate the skedastic function (that is, misspecified model). Due to the linearity of specification S.2, any choice of γ will result in the same degree of heteroskedasticity, given the sample size n. Therefore, the results of the simulation study were the same for different values of γ. Next, in specification S.3, WLS is more efficient than OLS for both choices of γ and all sample sizes. Finally, specification S.4 results in WLS being less efficient than OLS for small and moderate sample sizes ($n = 20$ and $n = 50$) and $\gamma = 2$, whereas WLS is clearly more efficient when $\gamma = 4$. Unsurprisingly, $\gamma = 4$ corresponds to a considerably higher degree of heteroskedasticity than $\gamma = 2$. Again, ALS offers an attractive compromise. It is never noticeably less efficient than OLS (that is, eMSE ratios never larger than 1.03) but is nearly as efficient ($n = 50$) or as efficient ($n = 100$) as WLS when WLS outperforms OLS.

Do the results differ if the regressors are not uniformly distributed or if the error terms are not normally distributed? In order to answer this question, the simulation study has been repeated with two different settings.

First, the regressors were chosen to follow a Beta(2,5) distribution, as specified in Sect. 3.1. As a consequence, the degree of heteroskedasticity is higher in most cases (except for specification S.3). compared to when the regressors follow a uniform distribution; see Table 4. A comparison of the two results reveals that, once again, the main factor relevant for the efficiency of WLS compared to OLS seems to be the degree of heteroskedasticity. Interestingly though, these results do not seem to apply to any degree of heteroskedasticity. Consider for example the first specification S.1. In the case of conditional homoskedasticity, the ratios of the eMSE's are similar, whereas introducing conditional heteroskedasticity ($\gamma = 1$ and $\gamma = 2$) leads to considerably stronger efficiency gains of WLS compared to OLS in the case of the Beta-distributed regressors. Unsurprisingly, the degree of heteroskedasticity for these two specifications is substantially higher in the case of Beta-distributed regressors. However, for $\gamma = 4$, WLS is more efficient in the case of uniformly distributed regressors, although the degree of heteroskedasticity is considerably lower than with

Beta-distributed regressors. The results for the other specifications (S.2–S.4) generally support the findings described in this paragraph.

Second, the basic setting has been changed by letting z_i follow a t-distribution with five degrees of freedom (scaled to have variance one). For small and moderate sample sizes ($n = 20, 50$), the efficiency gains of WLS over OLS are more pronounced compared to normally distributed z_i, whereas the efficiency gains are similar for $n = 100$.

As before, ALS offers an attractive compromise: (i) it is never noticeably less efficient than OLS and (ii) it enjoys most ($n = 50$) or practically all ($n = 100$) of the efficiency gains of WLS in case WLS outperforms OLS.

Remark 1 (Graphical Comparison) We find it useful to 'condense' the information on the ratios of the eMSE's contained in Tables 5, 6, 7, 8, 9 and 10 into a single Fig. 2. For each sample size ($n = 20, 50, 100$) and each method (WLS and ALS) there are 27 eMSE ratios compared to OLS. Here the number 27, corresponds to all combinations of specification of the skedastic function, corresponding parameter, distribution of the regressors, and distribution of the error term. For each sample size ($n = 20, 50, 100$), two boxplots are juxtaposed: one for the 27 eMSE ratios of WLS and one for the 27 eMSE ratios of ALS. In each case, a dashed horizontal line indicates the value of 1.0 (that is, same efficiency as OLS).

It can be seen that for each sample size, ALS has smaller risk of efficiency loss (with respect to OLS) than WLS: the numbers above the horizontal 1.0-line do not extend as far up. On the other hand, ALS also has a smaller chance of efficiency gain (with respect to OLS) than WLS: the numbers below the horizontal 1.0-line do not extend as far down. But the corresponding differences diminish with the sample size: There is a marked difference for $n = 20$, a moderate difference for $n = 50$, and practically no difference for $n = 100$.

Therefore, it can also be seen graphically that ALS offers an attractive compromise: (i) it is never noticeably less efficient than OLS and (ii) it enjoys most ($n = 50$) or practically all ($n = 100$) of the efficiency gains of WLS in case WLS outperforms OLS.   □

3.4.2   Inference

As described in Sect. 3.3, we use two performance measures to evaluate confidence intervals: the empirical coverage probability of a nominal 95% confidence interval and the ratio of the average length of a confidence interval to the average length of the OLS-HC confidence interval.[3]

The results for the basic setting, in which the regressors are uniformly distributed and the error terms are normally distributed, are presented in Tables 11 and 12.

---

[3] The second performance measure does not depend on the nominal confidence level, since by definition (19), it is equivalent to the ratio of the average standard error of a given method to the average OLS-HC standard error.

In general, confidence intervals based on WLS-HC standard errors tend to under-cover for small and moderate sample sizes ($n = 20, 50$). The empirical coverage probabilities for the OLS-HC confidence intervals, on the other hand, are generally satisfactory. Based on the theory, we would expect that all the HC confidence inter-vals tend to undercover in small samples due to the bias and increased variance of HC standard error estimates. Yet, the results here indicate that the HC confidence inter-vals for the WLS estimator are more prone to liberal inference. [6, p. 137] points out that the large-sample approximations for WLS are often unsatisfactory because WLS requires the estimation of more parameters (the parameters of the skedastic function) than OLS. Increasing the sample size improves the adequacy of the WLS-HC con-fidence intervals and the empirical coverage probabilities are always above 94% for $n = 100$. ALS-HC confidence intervals exhibit better coverage than WLS-HC confi-dence intervals: Already for $n = 50$, the empirical coverage probabilities are always over 94%.

When the degree of heteroskedasticity is high, then the average length of WLS-HC confidence intervals can be substantially shorter than the average length of OLS-HC confidence intervals. For instance, for specification S.1 with $\gamma = 4$ and $n = 100$, the average length of the WLS-HC confidence interval amounts to only 18% of the average length of the OLS-HC confidence interval, while the empirical coverage probability is more than satisfactory (95.8%). It is important to note that on aver-age short confidence intervals are only desirable if, at the same time, the empirical coverage probability is satisfactory. These findings have important implications for empirical research. It is crucial to only apply WLS in combination with HC standard errors when the sample size is large enough, that is, $n \geq 100$. For smaller sample sizes, the results of the simulation study have shown that the empirical coverage probabilities can be too low. On the other hand, the ALS-HC confidence interval appears trustworthy for moderate sample sizes already, that is, for $n \geq 50$. Further-more, the efficiency gains of the ALS-HC confidence interval over the OLS-HC (in terms of average length) are generally also substantial in the presence of notice-able conditional heteroskedasticity. For instance, for specification S.1 with $\gamma = 4$ and $n = 100$, the average length of the ALS-HC confidence interval also amounts to only 18% of the average length of the OLS-HC confidence interval, while the empirical coverage probability is more than satisfactory (95.8%).

As before, we want to analyze what happens when the regressors follow a Beta distribution as specified in Sect. 3.1, instead of a uniform distribution. As can be seen in Tables 13 and 14, for most of the specifications, the WLS-HC confidence intervals do not have a satisfactory empirical coverage probability, especially for small sample sizes. In the case of S.1 with $\gamma = 2$ or $\gamma = 4$, however, the empirical coverage probability is surprisingly high even for small sample sizes. [3] note that in the case of severe heteroskedasticity, the HC standard errors might be upward biased. In fact, the degree of heteroskedasticity is quite extreme for these two specifications and it is much higher than in the case of uniformly distributed regressors; see Table 4.

In contrast to the WLS-HC confidence intervals, the ALS-HC confidence intervals exhibit satisfactory coverage for moderate and large sample sizes ($n = 50, 100$) with all empirical coverage probabilities exceeding 94%.

The main result shown in [3] is that the bias of HC standard errors not only depends on the sample size, but also on whether or not a sample contains high leverage points. In empirical work, an observation is usually considered as a high leverage point if its diagonal element of the hat matrix is larger than $2p/n$, where p is the rank of the design matrix X.[4] A comparison of the diagonal elements of the hat matrix for both distributional assumptions of the regressors reveals that the samples created by Beta-distributed regressors generally contain more high leverage points. For instance, when $n = 100$, the sample with Beta distributed regressors contains six high leverage points, while the sample with uniformly distributed regressors only contains two high leverage points. Interestingly, for $n = 100$, the empirical coverage probability, for both OLS-HC and WLS-HC, is always larger for uniformly distributed regressors, that is, samples with fewer high leverage points, except for S.1 with $\gamma = 2, 4$ (which was discussed above).

Remark 2 (Maximal Standard Errors) The problem of undercoverage for small and moderate sample sizes (=20, 50) can be mitigated by using maximal standard errors, that is, by the use of WLS-Max and ALS-Max. Using maximal standard errors is proposed in Sect. 8.1 of [1], for example. However, these intervals can overcover by a lot for large sample sizes ($n = 100$), exhibiting empirical coverage probabilities sometimes near 100%. (This is also true for OLS-Max, although to a lesser extent.) Therefore, using maximal standard errors to mitigate undercoverage for small and moderate sample sizes seems a rather crude approach. A more promising approach, not leading to sizeable overcoverage for large sample sizes, would be the use of bootstrap methods. This topic is currently under study.   □

Remark 3 (Graphical Comparison) We find it useful to 'condense' the information on the ratios of the average lengths of confidence intervals contained in Tables 11, 12, 13, 14, 15 and 16 into a single Fig. 3. We only do this for the sample size $n = 100$ to ensure a fair comparison. Comparisons for $n = 20, 50$ would not be really fair to OLS, given that WLS confidence intervals tend to undercover for $n = 20, 50$ and that ALS confidence intervals tend to undercover for $n = 20$.

It can be seen that both WLS and ALS are always weakly more efficient than OLS in the sense that none of the average-length ratios are above 1.0. It can also be seen that, for all practical purposes, ALS is as efficient as OLS.   □

4 Empirical Applications

This section examines the application of OLS, WLS, and ALS to two empirical data sets. As will be seen the use of WLS and ALS can lead to much smaller standard

---

[4]It can be shown [7, e.g.] that p/n corresponds to the average element of the hat matrix.

errors (and thus much shorter confidence intervals) in the presence of noticeable conditional heteroskedasticity.

The two data sets are taken from [15].[5] In the first example, we model CEO salaries while in the second example, we model housing prices.

4.1 CEO Salaries

This cross-sectional data set from 1990 contains the salaries of 177 CEOs as well as further variables describing attributes of the CEOs and the corresponding companies. The model considered in this section tries to explain the log of the CEO salaries.[6] The variables (one response and three explanatory) used in the regression model under consideration are as follows:

$$
\begin{aligned}
&\log(salary): \quad \text{log of CEO's salary (in US\$1,000)}\\
&\log(sales): \quad \text{log of firm sales (in million US\$)}\\
&\log(mktval): \quad \text{log of market value (in million US\$)}\\
&ceoten: \quad \text{years as the CEO of the company}
\end{aligned}
$$

The sample size is $n = 177$ and the number of regressors (including the constant) is $K = 4$. Based on the results of the Monte Carlo study in Sect. 3, the sample size is large enough so that WLS and ALS inference can both be trusted.

The model is specified as in [15, p. 213] and is first estimated using OLS. The results are shown in the upper part of Table 2. The estimated coefficients are all positive, which intuitively makes sense. Examining the t-statistics (based on HC standard errors) shows that all estimated coefficients are significant at the 5% level except for the estimated coefficient on *ceoten*, which is insignificant.

The lower part of Table 2 shows the WLS results. The WLS estimates do not substantially differ from the OLS estimates. However, the HC standard errors are always smaller for WLS compared to OLS and generally noticeably so, with the ratios ranging from 0.93 to 0.84. In particular, now all estimated coefficients are individually significant at the 5% level, including the estimated coefficient on *ceoten*.

To determine the nature of ALS, we run a Breusch-Pagan test as described in Sect. 2.3.[7] The critical value of the test is $\chi^2_{3,0.90} = 6.25$ and the value of the test statistic is 8.25. Hence, the test detects conditional heteroskedasticity and ALS coincides with WLS.

---

[5]The two data sets are available under the names CEOSAL2 and HPRICE2, respectively at http://fmwww.bc.edu/ec-p/data/wooldridge/datasets.list.html.

[6]The log always corresponds to the natural logarithm.

[7]This regression results in taking the log of $\log(sales)$ and $\log(mktval)$ on the right-hand side; taking absolute values is not necessary, since $\log(sales)$ and $\log(mktval)$ are always positive. Furthermore, some observations have a value of zero for *ceoten*; we replace those values by 0.01 before taking logs.

4.2   Housing Prices

This cross-sectional data set from 1970 contains 506 observations from communities in the Boston area. The aim is to explain the median housing price in a community by means of the level of air pollution, the average number of rooms per house and other community characteristics. The variables (one response and four explanatory) used in the regression model under consideration are as follows:

log(*price*): log of median housing price (in US$)
log(*nox*):   log of nitrogen oxide in the air (in parts per million)
log(*dist*):   log of weighted distance from 5 employment centers (in miles)
rooms: average number of rooms per house
stratio: average student-teacher ratio

   The sample size is $n = 506$ and the number of regressors (including the constant) is $K = 5$. Based on the results of the Monte Carlo study in Sect. 3, the sample size is large enough so that WLS and ALS inference can both be trusted.

   The model follows an example in [15, p. 132]. The results from the OLS estimation are reported in the upper part of Table 3. All the estimated coefficients have the expected sign and are significant at the 1% level.

   The lower part of Table 3 shows the WLS results. The WLS estimates do not substantially differ from the OLS estimates. However, the HC standard errors are always smaller for WLS compared to OLS and generally noticeably so, with the ratios ranging from 0.90 to 0.63. As for OLS, all estimated coefficients are significant at the 1% level. But the corresponding confidence intervals based on WLS are shorter compared to OLS due to the smaller standard errors, which results in more informative inference. For example, a 95% confidence interval for the coefficient on *rooms* is given by [0.276, 0.338] based on WLS and by [0.258, 0.356] based on OLS. Needless to say, the smaller standard errors for WLS compared to OLS would also result in more powerful hypothesis tests concerning the various regression coefficients.

   To determine the nature of ALS, we run a Breusch-Pagan test as described in Sect. 2.3. The critical value of the test is $\chi^2_{4,0.90} = 7.78$ and the value of the test statistic is 92.08. Hence, the test detects conditional heteroskedasticity and ALS coincides with WLS.

5   Conclusion

The linear regression model remains a cornerstone of applied research in the social sciences. Many real-life data sets exhibit conditional heteroskedasticity which makes text-book inference based on ordinary least squares (OLS) invalid. The current practice in analyzing such data sets—going back to [14]—is to use OLS in conjunction with heteroskedasticity consistent (HC) standard errors.

In a recent paper, [13] suggest to return to the previous practice of using weighted least squares (WLS), also in conjunction with HC standard errors. Doing so ensures validity of the resulting inference even if the model for estimating the skedastic function is misspecified. In addition, they make the new proposal of adaptive least squares (ALS), where it is 'decided' from the data whether the applied researcher should use either OLS or WLS, in conjunction with HC standard errors.

This paper makes two contributions. On the one hand, we have compared finite-sample performance of OLS, WLS, and ALS for multivariate regressions via a Monte Carlo study. On the other hand, we have compared OLS, WLS, and ALS when applied to two empirical data sets.[8]

The results of the Monte Carlo study point towards ALS as the overall winner. When WLS outperforms OLS, then ALS achieves most (for moderate sample sizes) or even all (for large sample sizes) of the gains of WLS; and these gains can be dramatic. When OLS outperforms WLS, then it also outperforms ALS but by a much smaller margin. Consequently, when comparing ALS to OLS, there is large upside potential and only very limited downside risk.

The application to two empirical data sets have shown that WLS and ALS can achieve large efficiency gains over OLS in the presence of noticeable conditional heteroskedasticity. Namely, smaller standard errors result in shorter (and thus more informative) confidence intervals and in more powerful hypothesis tests.

A Figures and Tables

---

[8][13] only use univariate regressions in their Monte Carlo study and do not provide any applications to empirical data sets.

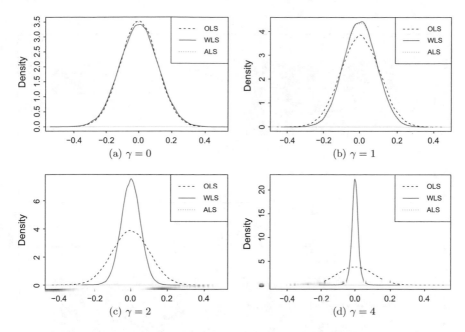

Fig. 1 Density plots for the estimators of β_1 for Specification S.1 and its four parameter values. The sample size is 100, the regressors are $U[1, 4]$-distributed and the error terms follow a standard normal distribution

Table 4 Degree of heteroskedasticity for the different specifications of the scedastic function. The degree of heteroskedasticity is measured as $\max(v(x))/\min(v(x))$

	S.1		S.2		S.3		S.4	
	Uniform	Beta	Uniform	Beta	Uniform	Beta	Uniform	Beta
	$\gamma = 0$		$\gamma = 1$		$\gamma = 0.5$		$\gamma = 2$	
$n = 20$	1.0	1.0	1.9	3.0	14.4	10.0	3.8	8.7
$n = 50$	1.0	1.0	2.0	5.3	15.2	24.0	4.1	28.0
$n = 100$	1.0	1.0	2.8	6.4	34.0	25.2	7.9	41.1
	$\gamma = 1$		$\gamma = 2$		$\gamma = 1$		$\gamma = 4$	
$n = 20$	9.7	174.1	1.9	3.0	206.2	99.5	14.3	76.0
$n = 50$	10.4	439.3	2.0	5.3	231.8	576.9	16.5	781.9
$n = 100$	24.3	682.5	2.8	6.4	1,157.5	633.8	62.5	1,689.3
	$\gamma = 2$		$\gamma = 3$					
$n = 20$	93.3	30,323.5	1.9	3.0				
$n = 50$	108.3	193,011.0	2.0	5.3				
$n = 100$	590.5	465,764.5	2.8	6.4				
	$\gamma = 4$							
$n = 20$	8,699.6	0.92×10^9						
$n = 50$	11,737.4	37×10^9						
$n = 100$	348,646.3	217×10^9						

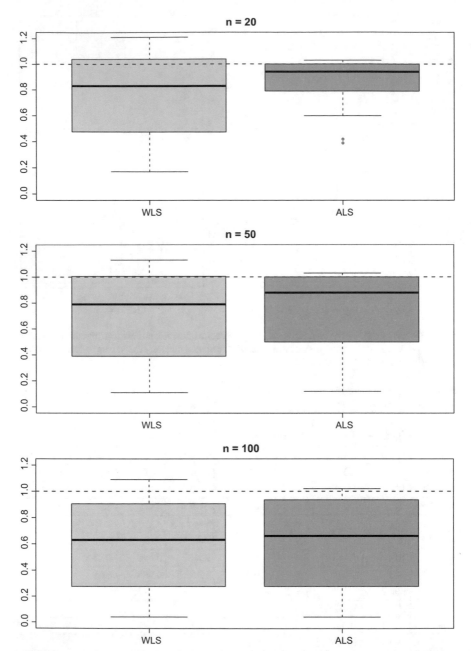

Fig. 2 Boxplots of the ratios of the eMSE of WLS (*left*) and ALS (*right*) to the eMSE of OLS. For a given sample size $n = 20, 50, 100$, the boxplots are over all 27 combinations of specification of the skedastic function, parameter value, distribution of the regressors, and distribution of the error terms

Fig. 3 Boxplots of the ratios of the average length of WLS confidence intervals for β_1 (*left*) and ALS confidence intervals for β_1 (*right*) to the average length of OLS confidence intervals for β_1. For the given sample size $n = 100$, the boxplots are over all 27 combinations of specification of the skedastic function, parameter value, distribution of the regressors, and distribution of the error terms

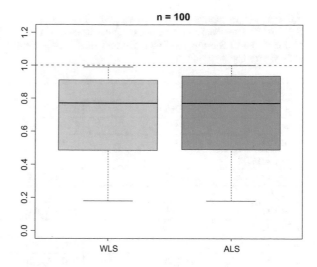

Table 5 Empirical mean squared errors (eMSEs) of estimators of β_1 in the case of Specification S.1. The numbers in parentheses express the ratios of the eMSE of a given estimator to the eMSE of the OLS estimator. The regressors are $U[1, 4]$-distributed and the error terms follow a standard normal distribution.

	OLS	WLS	ALS
S.1 ($\gamma = 0$)			
$n = 20$	0.064	0.077 (1.19)	0.066 (1.03)
$n = 50$	0.029	0.032 (1.13)	0.029 (1.03)
$n = 100$	0.013	0.014 (1.08)	0.013 (1.02)
S.1 ($\gamma = 1$)			
$n = 20$	0.071	0.065 (0.92)	0.070 (0.98)
$n = 50$	0.026	0.022 (0.85)	0.025 (0.93)
$n = 100$	0.011	0.008 (0.72)	0.008 (0.73)
S.1 ($\gamma = 2$)			
$n = 20$	0.084	0.042 (0.50)	0.062 (0.73)
$n = 50$	0.028	0.012 (0.42)	0.014 (0.49)
$n = 100$	0.010	0.003 (0.27)	0.003 (0.27)
S.1 ($\gamma = 4$)			
$n = 20$	0.097	0.019 (0.20)	0.041 (0.42)
$n = 50$	0.034	0.004 (0.10)	0.004 (0.12)
$n = 100$	0.010	0.000 (0.04)	0.000 (0.04)

Table 6 Empirical mean squared errors (eMSEs) of estimators of β_1 in the case of Specifications S.2–S.4. The numbers in parentheses express the ratios of the eMSE of a given estimator to the eMSE of the OLS estimator. The regressors are $U[1, 4]$-distributed and the error terms follow a standard normal distribution

	OLS	WLS	ALS
S.2 ($\gamma > 0$)			
$n = 20$	0.066	0.077 (1.17)	0.068 (1.03)
$n = 50$	0.028	0.030 (1.10)	0.028 (1.03)
$n = 100$	0.012	0.013 (1.04)	0.012 (1.02)
S.3 ($\gamma = 0.5$)			
$n = 20$	0.077	0.064 (0.83)	0.073 (0.94)
$n = 50$	0.028	0.022 (0.79)	0.024 (0.88)
$n = 100$	0.011	0.007 (0.65)	0.008 (0.67)
S.3 ($\gamma = 1$)			
$n = 20$	0.092	0.036 (0.39)	0.058 (0.63)
$n = 50$	0.030	0.010 (0.33)	0.012 (0.39)
$n = 100$	0.011	0.002 (0.20)	0.002 (0.20)
S.4 ($\gamma = 2$)			
$n = 20$	0.069	0.074 (1.08)	0.070 (1.02)
$n = 50$	0.027	0.028 (1.03)	0.027 (1.01)
$n = 100$	0.012	0.011 (0.92)	0.011 (0.93)
S.4 ($\gamma = 4$)			
$n = 20$	0.076	0.063 (0.83)	0.072 (0.94)
$n = 50$	0.027	0.021 (0.79)	0.024 (0.88)
$n = 100$	0.011	0.007 (0.61)	0.007 (0.62)

Table 7 Empirical mean squared errors (eMSEs) of estimators of β_1 in the case of Specification S.1. The numbers in parentheses express the ratios of the eMSE of a given estimator to the eMSE of the OLS estimator. The regressors are Beta(2,5)-distributed and the error terms follow a standard normal distribution

	OLS	WLS	ALS
S.1 ($\gamma = 0$)			
$n = 20$	0.142	0.172 (1.21)	0.147 (1.03)
$n = 50$	0.032	0.037 (1.13)	0.033 (1.03)
$n = 100$	0.013	0.014 (1.09)	0.013 (1.02)
S.1 ($\gamma = 1$)			
$n = 20$	0.122	0.081 (0.66)	0.106 (0.87)
$n = 50$	0.034	0.016 (0.46)	0.020 (0.58)
$n = 100$	0.016	0.006 (0.36)	0.006 (0.36)
S.1 ($\gamma = 2$)			
$n = 20$	0.129	0.049 (0.38)	0.095 (0.74)
$n = 50$	0.033	0.006 (0.18)	0.010 (0.31)
$n = 100$	0.017	0.002 (0.13)	0.002 (0.13)
S.1 ($\gamma = 4$)			
$n = 20$	0.136	0.038 (0.28)	0.115 (0.84)
$n = 50$	0.025	0.003 (0.13)	0.013 (0.52)
$n = 100$	0.014	0.003 (0.18)	0.003 (0.19)

Table 8 Empirical mean squared errors (eMSEs) of estimators of β_1 in the case of Specifications S.2–S.4. The numbers in parentheses express the ratios of the eMSE of a given estimator to the eMSE of the OLS estimator. The regressors are Beta(2,5)-distributed and the error terms follow a standard normal distribution

	OLS	WLS	ALS
S.2 ($\gamma > 0$)			
$n = 20$	0.131	0.152 (1.16)	0.134 (1.02)
$n = 50$	0.033	0.035 (1.04)	0.034 (1.01)
$n = 100$	0.014	0.014 (0.97)	0.014 (0.99)
S.3 ($\gamma = 0.5$)			
$n = 20$	0.123	0.121 (0.99)	0.122 (0.99)
$n = 50$	0.035	0.029 (0.81)	0.032 (0.91)
$n = 100$	0.018	0.013 (0.70)	0.014 (0.72)
S.3 ($\gamma = 1$)			
$n = 20$	0.111	0.070 (0.63)	0.098 (0.88)
$n = 50$	0.036	0.013 (0.37)	0.018 (0.50)
$n = 100$	0.025	0.007 (0.28)	0.007 (0.28)
S.4 ($\gamma = 2$)			
$n = 20$	0.123	0.124 (1.01)	0.123 (1.00)
$n = 50$	0.035	0.029 (0.82)	0.032 (0.92)
$n = 100$	0.016	0.012 (0.72)	0.012 (0.74)
S.4 ($\gamma = 4$)			
$n = 20$	0.115	0.079 (0.69)	0.103 (0.89)
$n = 50$	0.037	0.016 (0.44)	0.020 (0.54)
$n = 100$	0.021	0.007 (0.33)	0.007 (0.33)

Table 9 Empirical mean squared errors (eMSEs) of estimators of β_1 in the case of Specification S.1. The numbers in parentheses express the ratios of the eMSE of a given estimator to the eMSE of the OLS estimator. The regressors are $U[1, 4]$-distributed but the error terms follow a t-distribution with five degrees of freedom

	OLS	WLS	ALS
S.1 ($\gamma = 0$)			
$n = 20$	0.064	0.070 (1.10)	0.064 (1.01)
$n = 50$	0.028	0.030 (1.08)	0.029 (1.01)
$n = 100$	0.013	0.013 (1.04)	0.013 (1.01)
S.1 ($\gamma = 1$)			
$n = 20$	0.071	0.060 (0.84)	0.067 (0.94)
$n = 50$	0.026	0.022 (0.82)	0.024 (0.91)
$n = 100$	0.011	0.008 (0.70)	0.008 (0.72)
S.1 ($\gamma = 2$)			
$n = 20$	0.084	0.038 (0.45)	0.058 (0.70)
$n = 50$	0.028	0.011 (0.41)	0.014 (0.50)
$n = 100$	0.011	0.003 (0.28)	0.003 (0.28)
S.1 ($\gamma = 4$)			
$n = 20$	0.096	0.016 (0.17)	0.038 (0.39)
$n = 50$	0.034	0.004 (0.11)	0.004 (0.13)
$n = 100$	0.011	0.001 (0.05)	0.001 (0.05)

Table 10 Empirical mean squared errors (eMSEs) of estimators of β_1 in the case of Specification S.2–S.4. The numbers in parentheses express the ratios of the eMSE of a given estimator to the eMSE of the OLS estimator. The regressors are $U[1, 4]$-distributed but the error terms follow a t-distribution with five degrees of freedom

	OLS	WLS	ALS
S.2 ($\gamma > 0$)			
$n = 20$	0.065	0.070 (1.07)	0.066 (1.00)
$n = 50$	0.027	0.029 (1.05)	0.028 (1.01)
$n = 100$	0.012	0.012 (1.00)	0.012 (1.00)
S.3 ($\gamma = 0.5$)			
$n = 20$	0.077	0.058 (0.76)	0.069 (0.90)
$n = 50$	0.027	0.021 (0.76)	0.024 (0.87)
$n = 100$	0.012	0.007 (0.63)	0.008 (0.66)
S.3 ($\gamma = 1$)			
$n = 20$	0.091	0.031 (0.35)	0.055 (0.60)
$n = 50$	0.030	0.010 (0.32)	0.012 (0.40)
$n = 100$	0.011	0.002 (0.21)	0.002 (0.21)
S.4 ($\gamma = 2$)			
$n = 20$	0.068	0.068 (1.00)	0.067 (0.99)
$n = 50$	0.027	0.026 (0.98)	0.027 (0.99)
$n = 100$	0.012	0.011 (0.89)	0.011 (0.94)
S.4 ($\gamma = 4$)			
$n = 20$	0.076	0.058 (0.76)	0.068 (0.90)
$n = 50$	0.027	0.020 (0.76)	0.023 (0.87)
$n = 100$	0.012	0.007 (0.60)	0.007 (0.62)

Table 11 Empirical coverage probabilities of nominal 95% confidence intervals for β_1 in the case of Specification S.1 (in percent). The numbers in parentheses express the ratios of the average length of a given confidence interval to the average length of OLS-HC. The regressors are $U[1, 4]$-distributed and the error terms follow a standard normal distribution

	OLS-HC	OLS-Max	WLS-HC	WLS-Max	ALS-HC	ALS-Max
S.1 ($\gamma = 0$)						
$n = 20$	96.4	97.1 (1.02)	92.7 (0.91)	93.6 (0.93)	95.4 (0.98)	96.1 (1.00)
$n = 50$	95.5	96.2 (1.02)	93.3 (0.97)	94.1 (0.99)	94.9 (0.99)	95.7 (1.01)
$n = 100$	95.4	95.9 (1.02)	94.1 (0.99)	94.7 (1.01)	95.1 (1.00)	95.6 (1.01)
S.1 ($\gamma = 1$)						
$n = 20$	96.6	97.1 (1.01)	93.0 (0.81)	93.9 (0.82)	95.0 (0.91)	95.6 (0.93)
$n = 50$	95.7	96.7 (1.04)	93.9 (0.85)	94.7 (0.88)	94.2 (0.91)	95.1 (0.93)
$n = 100$	95.5	96.7 (1.06)	94.3 (0.81)	95.2 (0.84)	94.1 (0.82)	95.1 (0.85)
S.1 ($\gamma = 2$)						
$n = 20$	96.3	96.6 (1.00)	92.9 (0.58)	93.9 (0.59)	94.4 (0.70)	94.4 (0.71)
$n = 50$	95.4	96.3 (1.03)	94.1 (0.60)	95.1 (0.62)	94.3 (0.62)	94.8 (0.64)
$n = 100$	95.4	97.2 (1.09)	94.3 (0.50)	96.7 (0.56)	94.3 (0.50)	96.7 (0.56)
S.1 ($\gamma = 4$)						
$n = 20$	96.1	96.2 (1.00)	94.2 (0.31)	94.9 (0.32)	94.2 (0.40)	94.8 (0.41)
$n = 50$	94.8	95.2 (1.01)	94.6 (0.27)	97.1 (0.31)	94.5 (0.27)	97.0 (0.31)
$n = 100$	95.7	97.6 (1.11)	95.8 (0.18)	99.9 (0.32)	95.8 (0.18)	99.9 (0.32)

Table 12 Empirical coverage probabilities of nominal 95% confidence intervals for β_1 in the case of Specification S.2–S.4 (in percent). The numbers in parentheses express the ratios of the average length of a given confidence interval to the average length of OLS-HC. The regressors are $U[1, 4]$-distributed and the error terms follow a standard normal distribution

	OLS-HC	OLS-Max	WLS-HC	WLS-Max	ALS-HC	ALS-Max
S.2 ($\gamma > 0$)						
$n = 20$	96.5	97.1 (1.02)	92.8 (0.90)	93.6 (0.92)	95.4 (0.97)	96.1 (0.99)
$n = 50$	95.7	96.4 (1.03)	93.5 (0.96)	94.2 (0.98)	95.0 (0.99)	95.7 (1.01)
$n = 100$	95.5	96.0 (1.03)	94.2 (0.97)	94.7 (0.99)	94.9 (0.99)	95.4 (1.01)
S.3 ($\gamma = 0.5$)						
$n = 20$	96.5	96.9 (1.01)	92.7 (0.76)	93.5 (0.78)	94.6 (0.88)	95.1 (0.89)
$n = 50$	95.7	96.5 (1.03)	93.9 (0.82)	94.5 (0.84)	94.1 (0.88)	94.7 (0.89)
$n = 100$	95.5	96.5 (1.05)	94.2 (0.77)	94.9 (0.79)	94.1 (0.78)	94.8 (0.80)
S.3 ($\gamma = 1$)						
$n = 20$	96.5	96.7 (1.00)	92.4 (0.49)	93.1 (0.50)	93.1 (0.61)	93.7 (0.62)
$n = 50$	95.3	95.8 (1.02)	93.9 (0.53)	94.7 (0.54)	94.1 (0.54)	94.4 (0.55)
$n = 100$	95.5	96.9 (1.07)	94.1 (0.43)	96.6 (0.47)	94.1 (0.43)	96.6 (0.47)
S.4 ($\gamma = 2$)						
$n = 20$	96.5	97.1 (1.01)	92.9 (0.87)	93.7 (0.89)	95.3 (0.96)	95.9 (0.97)
$n = 50$	95.7	96.5 (1.03)	93.6 (0.93)	94.4 (0.95)	94.7 (0.97)	95.6 (0.99)
$n = 100$	95.4	96.2 (1.03)	94.2 (0.92)	94.8 (0.93)	94.4 (0.94)	95.0 (0.96)
S.4 ($\gamma = 4$)						
$n = 20$	96.6	97.0 (1.01)	92.9 (0.76)	93.7 (0.78)	94.7 (0.88)	95.3 (0.89)
$n = 50$	95.7	96.6 (1.03)	94.0 (0.82)	94.7 (0.84)	94.1 (0.87)	94.9 (0.90)
$n = 100$	95.5	96.5 (1.05)	94.1 (0.75)	94.9 (0.77)	94.0 (0.75)	94.9 (0.77)

Table 13 Empirical coverage probabilities of nominal 95% confidence intervals for β_1 in the case of Specification S.1 (in percent). The numbers in parentheses express the ratios of the average length of a given confidence interval to the average length of OLS-HC. The regressors are Beta(2,5)-distributed and the error terms follow a standard normal distribution

	OLS-HC	OLS-Max	WLS-HC	WLS-Max	ALS-HC	ALS-Max
S.1 ($\gamma = 0$)						
$n = 20$	96.0	97.0 (1.03)	91.2 (0.89)	92.7 (0.92)	94.7 (0.97)	95.8 (1.00)
$n = 50$	95.3	95.9 (1.02)	93.1 (0.97)	94.1 (0.99)	94.7 (0.99)	95.4 (1.01)
$n = 100$	95.2	96.0 (1.03)	93.7 (0.99)	94.7 (1.01)	94.8 (1.00)	95.6 (1.02)
S.1 ($\gamma = 1$)						
$n = 20$	97.7	98.6 (1.05)	93.9 (0.67)	95.6 (0.71)	95.6 (0.83)	96.9 (0.88)
$n = 50$	95.4	95.9 (1.02)	94.1 (0.62)	95.6 (0.65)	94.5 (0.65)	95.0 (0.69)
$n = 100$	94.7	95.3 (1.02)	94.0 (0.55)	95.7 (0.60)	94.0 (0.55)	95.7 (0.60)
S.1 ($\gamma = 2$)						
$n = 20$	98.2	98.8 (1.04)	96.8 (0.50)	98.1 (0.53)	97.4 (0.73)	98.4 (0.76)
$n = 50$	95.4	95.7 (1.01)	96.3 (0.38)	98.2 (0.46)	95.9 (0.42)	97.6 (0.49)
$n = 100$	94.7	95.3 (1.02)	95.7 (0.31)	90.2 (0.39)	95.7 (0.31)	98.2 (0.39)
S.1 ($\gamma = 4$)						
$n = 20$	98.7	99.1 (1.01)	99.0 (0.45)	99.5 (0.48)	98.7 (0.84)	99.1 (0.85)
$n = 50$	97.2	97.3 (1.01)	98.6 (0.38)	99.6 (0.53)	98.5 (0.57)	99.1 (0.65)
$n = 100$	95.3	96.4 (1.05)	96.6 (0.35)	98.4 (0.46)	96.5 (0.36)	98.3 (0.47)

Table 14 Empirical coverage probabilities of nominal 95% confidence intervals for β_1 in the case of Specification S.2–S.4 (in percent). The numbers in parentheses express the ratios of the average length of a given confidence interval to the average length of OLS-HC. The regressors are Beta(2,5)-distributed and the error terms follow a standard normal distribution

	OLS-HC	OLS-Max	WLS-HC	WLS-Max	ALS-HC	ALS-Max
S.2 ($\gamma > 0$)						
$n = 20$	96.5	97.5 (1.04)	91.8 (0.87)	93.2 (0.90)	95.0 (0.96)	96.1 (1.00)
$n = 50$	95.2	95.8 (1.02)	93.0 (0.92)	93.9 (0.95)	94.2 (0.97)	95.0 (0.99)
$n = 100$	94.9	95.4 (1.02)	93.5 (0.93)	94.2 (0.94)	94.0 (0.96)	94.5 (0.97)
S.3 ($\gamma = 0.5$)						
$n = 20$	97.0	97.9 (1.04)	92.2 (0.81)	93.8 (0.84)	95.1 (0.93)	96.2 (0.97)
$n = 50$	95.4	95.8 (1.01)	93.1 (0.82)	93.8 (0.83)	94.0 (0.89)	94.3 (0.89)
$n = 100$	94.7	94.9 (1.01)	93.2 (0.78)	93.5 (0.79)	93.7 (0.82)	93.9 (0.82)
S.3 ($\gamma = 1$)						
$n = 20$	97.7	98.4 (1.04)	94.5 (0.64)	95.9 (0.67)	95.9 (0.84)	97.0 (0.87)
$n = 50$	95.9	96.0 (1.01)	94.0 (0.56)	95.3 (0.59)	94.1 (0.62)	94.9 (0.63)
$n = 100$	94.6	94.8 (1.00)	93.2 (0.47)	93.9 (0.48)	93.6 (0.48)	94.1 (0.49)
S.4 ($\gamma = 2$)						
$n = 20$	97.0	97.9 (1.04)	92.5 (0.82)	94.1 (0.85)	95.3 (0.94)	96.4 (0.98)
$n = 50$	95.3	95.8 (1.01)	93.1 (0.83)	94.0 (0.85)	93.8 (0.89)	94.3 (0.90)
$n = 100$	94.7	95.1 (1.01)	93.4 (0.79)	93.9 (0.80)	93.8 (0.82)	94.2 (0.83)
S.4 ($\gamma = 4$)						
$n = 20$	97.8	98.5 (1.05)	94.8 (0.68)	96.1 (0.71)	96.2 (0.85)	97.2 (0.89)
$n = 50$	95.6	95.8 (1.01)	93.8 (0.60)	95.0 (0.63)	93.9 (0.65)	94.5 (0.65)
$n = 100$	94.5	94.7 (1.00)	93.3 (0.52)	94.2 (0.53)	93.8 (0.53)	94.4 (0.54)

Table 15 Empirical coverage probabilities of nominal 95% confidence intervals for β_1 in the case of Specification S.1 (in percent). The numbers in parentheses express the ratios of the average length of a given confidence interval to the average length of OLS-HC. The regressors are $U[1, 4]$-distributed and the error terms follow a t-distribution with five degrees of freedom

	OLS-HC	OLS-Max	WLS-HC	WLS-Max	ALS-HC	ALS-Max
S.1 ($\gamma = 0$)						
$n = 20$	97.1	97.8 (1.03)	93.3 (0.88)	94.3 (0.90)	96.1 (0.97)	96.8 (0.99)
$n = 50$	95.8	96.6 (1.04)	93.5 (0.95)	94.6 (0.98)	95.1 (0.98)	96.1 (1.02)
$n = 100$	95.4	96.0 (1.03)	94.3 (0.97)	95.0 (1.00)	95.1 (0.99)	95.7 (1.02)
S.1 ($\gamma = 1$)						
$n = 20$	97.2	97.7 (1.02)	93.7 (0.79)	94.6 (0.81)	95.7 (0.91)	96.4 (0.92)
$n = 50$	96.0	97.0 (1.05)	93.9 (0.84)	95.1 (0.87)	94.5 (0.90)	95.7 (0.94)
$n = 100$	95.5	96.8 (1.07)	94.2 (0.81)	95.4 (0.85)	94.2 (0.82)	95.4 (0.86)
S.1 ($\gamma = 2$)						
$n = 20$	97.0	97.2 (1.01)	93.7 (0.58)	94.6 (0.59)	94.4 (0.71)	95.2 (0.72)
$n = 50$	95.8	96.8 (1.04)	94.3 (0.60)	95.7 (0.64)	94.0 (0.63)	95.4 (0.67)
$n = 100$	95.4	97.2 (1.11)	94.5 (0.51)	97.1 (0.58)	94.5 (0.51)	97.1 (0.58)
S.1 ($\gamma = 4$)						
$n = 20$	96.6	96.7 (1.00)	95.0 (0.32)	95.6 (0.33)	94.9 (0.41)	95.5 (0.42)
$n = 50$	95.4	96.0 (1.02)	95.5 (0.28)	97.7 (0.33)	95.4 (0.29)	97.6 (0.33)
$n = 100$	95.5	97.6 (1.12)	96.0 (0.20)	99.9 (0.35)	96.0 (0.20)	99.9 (0.35)

Table 16 Empirical coverage probabilities of nominal 95% confidence intervals for β_1 in the case of Specification S.2–S.4 (in percent). The numbers in parentheses express the ratios of the average length of a given confidence interval to the average length of OLS-HC. The regressors are $U[1, 4]$-distributed and the error terms follow a t-distribution with five degrees of freedom

	OLS-HC	OLS-Max	WLS-HC	WLS-Max	ALS-HC	ALS-Max
S.2 ($\gamma > 0$)						
$n = 20$	97.1	97.7 (1.03)	93.3 (0.87)	94.2 (0.89)	96.0 (0.96)	96.7 (0.99)
$n = 50$	95.9	96.8 (1.04)	93.5 (0.94)	94.6 (0.97)	95.1 (0.98)	96.0 (1.02)
$n = 100$	95.4	96.2 (1.04)	94.2 (0.95)	95.0 (0.98)	94.9 (0.98)	95.6 (1.02)
S.3 ($\gamma = 0.5$)						
$n = 20$	97.1	97.5 (1.01)	93.4 (0.75)	94.2 (0.76)	95.3 (0.88)	95.9 (0.89)
$n = 50$	95.9	96.8 (1.04)	93.9 (0.81)	94.9 (0.83)	94.2 (0.87)	95.2 (0.90)
$n = 100$	95.5	96.6 (1.06)	94.2 (0.77)	95.2 (0.80)	94.1 (0.77)	95.1 (0.81)
S.3 ($\gamma = 1$)						
$n = 20$	97.0	97.1 (1.00)	93.2 (0.49)	94.0 (0.50)	93.9 (0.62)	94.6 (0.63)
$n = 50$	95.7	96.4 (1.03)	94.2 (0.53)	95.3 (0.55)	94.0 (0.55)	95.0 (0.57)
$n = 100$	95.4	96.9 (1.08)	94.3 (0.44)	97.0 (0.50)	94.3 (0.44)	97.0 (0.50)
S.4 ($\gamma = 2$)						
$n = 20$	97.2	97.7 (1.02)	93.4 (0.85)	94.4 (0.87)	95.9 (0.95)	96.5 (0.97)
$n = 50$	95.9	96.9 (1.05)	93.7 (0.91)	94.8 (0.94)	94.9 (0.96)	95.9 (1.00)
$n = 100$	95.4	96.4 (1.05)	94.2 (0.90)	95.0 (0.93)	94.4 (0.93)	95.3 (0.97)
S.4 ($\gamma = 4$)						
$n = 20$	97.2	97.6 (1.01)	93.6 (0.75)	94.4 (0.77)	95.5 (0.88)	96.1 (0.89)
$n = 50$	96.0	96.9 (1.05)	94.0 (0.81)	95.1 (0.84)	94.3 (0.87)	95.4 (0.91)
$n = 100$	95.5	96.6 (1.06)	94.2 (0.74)	95.3 (0.78)	94.1 (0.75)	95.2 (0.78)

References

1. Angrist JD, Pischke J-S (2009) Mostly harmless econometrics. Princeton University Press, Princeton
2. Breusch T, Pagan A (1979) A simple test for heteroscedasticity and random coefficient variation. Econometrica 47:1287–1294
3. Chesher A, Jewitt I (1987) The bias of a heteroskedasticity consistent covariance matrix estimator. Econometrica 55(5):1217–1222
4. Cribari-Neto F (2004) Asymptotic inference under heterskedasticty of unknown form. Comput Stat Data Anal 45:215–233
5. Harvey AC (1976) Estimating regression models with multiplicative heteroscedasticity. Econometrica 44:461–465
6. Hayashi F (2000) Econometrics. Princeton University Press, Princeton
7. Hoaglin DC, Welsch RE (1978) The hat matrix in regression and ANOVA. Am Stat 32(1):17–22
8. Judge GG, Hill RC, Griffiths WE, Lütkepohl H, Lee T-C (1988) Introduction to the theory and practice of econometrics, 2nd edn. Wiley, New York
9. Koenker R (1981) A note on studentizing a test for heteroscedasticity. J Econometrics 17:107–112
10. Koenker R, Bassett G (1982) Robust tests for heteroscedasticity based on regression quantiles Econometrica 50:43–61
11. MacKinnon JG (2012) Thirty years of heteroskedasticity-robust inference. In: Chen X, Swanson N (eds) Recent advances and future directions in causality, prediction, and specification analysis. Springer, New York, pp 437–461
12. MacKinnon JG, White HL (1985) Some heteroskedasticity-consistent covariance matrix estimators with improved finite-sample properties. J Econometrics 29:53–57
13. Romano JP, Wolf M (2016) Resurrecting weighted least squares. Working paper ECON 172, Department of Economics, University of Zurich
14. White HL (1980) A heteroskedasticity-consistent covariance matrix estimator and a direct test of heteroskedasticity. Econometrica 48:817–838
15. Wooldridge JM (2012) Introductory econometrics, 5th edn. South-Western, Mason

Prior-Free Probabilistic Inference for Econometricians

Ryan Martin

Abstract The econometrics literature is dominated by the frequentist school which places primary emphasis on the specification of methods that have certain long-run frequency properties, mostly disavowing any notion of inference based on the given data. This preference for frequentism is at least partially based on the belief that probabilistic inference is possible only through a Bayesian approach, the success of which generally depends on the unrealistic assumption that the prior distribution is meaningful in some way. This paper is intended to inform econometricians that an alternative *inferential model* (IM) approach exists that can achieve probabilistic inference without a prior and while enjoying certain calibration properties essential for reproducibility, etc. Details about the IM construction and its properties are presented, along with some intuition and examples.

Keywords Belief · Calibration · Inferential model · Marginal inference · Plausibility · Random set · Validity

1 Introduction

There is a general impression among statisticians, econometricians, and others that "probabilistic inference" can be achieved only through a Bayesian approach, i.e., where a prior distribution and likelihood function is converted into a posterior distribution, via Bayes's theorem, from which inference can be drawn. But the difficulties involved in eliciting or otherwise constructing/justifying a suitable prior distribution explains why, despite the benefits of working with probabilities, researchers tend to shy away from the Bayesian approach. The more familiar alternative to the Bayesian

R. Martin (✉)
Department of Statistics, North Carolina State University, 2311 Stinson Drive,
Campus Box 8203, Raliegh, NC 27695, USA
e-mail: rgmarti3@ncsu.edu

© Springer International Publishing AG 2017   169
V. Kreinovich et al. (eds.), *Robustness in Econometrics*,
Studies in Computational Intelligence 692, DOI 10.1007/978-3-319-50742-2_10

approach, often called "frequentist," abandons the notion of probabilistic inference in exchange for certain error rate control based on repeated sampling. The question to be considered here is whether it is necessary for one to choose between the two approaches. In other words, is it possible to carry out probabilistic inference, like a Bayesian, but without a prior, and while retaining the benefits of the frequentist's repeated sampling guarantees? My answer to this question is YES, and this paper intends to justify this claim.

Before introducing the proposal to be discussed herein, it is worthwhile to comment briefly on some previous attempts at this goal of prior-free probabilistic inference.

- *Objective Bayes*. Here the idea is to introduce a prior distribution which is "objective" or "non-informative" in some way, and then carry out the usual Bayesian prior-to-posterior update; see, for example, Ghosh et al. [17, Chap. 5], Berger [1], and Ghosh [18]. To me, a more accurate name for these priors is "default," since the idea is that the user will not have to make their own choices for their own specific applications. From a foundational point of view, perhaps the most important concern is that a prior distribution cannot be non-informative, i.e., the prior will always have some kind of a influence, and this may be harmful unless it is based on genuine prior information or the sample size is large. From a practical point of view, one of the main issues is that the default-prior Bayes approach does not provide any error rate guarantees, except perhaps in the large-sample limit. Another concern is that one of the selling points of the Bayesian approach, namely, the ability to obtain marginal inference via integration, is lost because a "good" default prior for the full parameter may not be a good default prior for a function thereof.
- *Confidence distributions*. Frequentists can define a probability distribution on the parameter space for inference, called a "confidence distribution" (e.g., [38, 39, 45]). To me, however, confidence distributions are only a summary of the output of a particular method and, therefore, do not provide any new methods or understanding. For example, one can use a likelihood ratio test or the bootstrap to construct a confidence distribution, but, e.g., the interval estimates it produces are exactly the same as that from the likelihood ratio test or the bootstrap behind the scenes. With this understanding, it is clear that, if the method use to generate the confidence distribution have certain desirable properties, then so will the confidence distribution. However, like objective Bayes, integrating a joint confidence distribution may not give a confidence distribution on the specified margin.
- *Fiducial and Dempster–Shafer*. The goal of prior-free probabilistic inference dates back to Fisher's fiducial ideas (e.g., [15, 47]) and to the developments of Dempster [4–8] and Shafer [40, 42]. Most would agree that there is something intriguing about these two approaches, but they have not gained popularity in the statistics or econometrics literature—indeed, Efron [10, p. 105] refers to fiducial inference as "Fisher's biggest blunder." There are a number of explanations for why these fiducial-like methods are not fully satisfactory, and these focus primarily on Fisher's over-confidence in the idea, but perhaps a more important concern is that,

in general, these methods do not have any frequency calibration. Given the importance of reproducibility in all areas of science, this kind of calibration is essential.
Recent extensions of the fiducial ideas, e.g., Hannig [20] and Hannig et al. [21],
focus on an asymptotic calibration property. However, like the objective Bayes
and confidence distributions, these also have difficulties with marginalization.

This somewhat critical assessment is not intended to say that these other methods
have no value. In my opinion, all of these methods provide a sort of approximation
to whatever is the "best possible" mode of inference. That these methods fall short
of a completely satisfactory solution is not a serious criticism, given the importance
and difficulty of the problem at hand; as (e.g., [11], p. 41) writes

> ...perhaps the most important unresolved problem in statistical inference is the use of Bayes
> theorem in the absence of prior information.

Efron later[1] called this the "Holy grail" of statistics. From our efforts to understand
these existing methods, especially fiducial and Dempster–Shafer, Chuanhai Liu and
I have developed what we believe to be a solution to this fundamental problem.

This new approach is called the *inferential model* (IM) approach, and was introduced in Martin and Liu [27]; a thorough introduction can be found in Martin and
Liu [30]. The reasoning behind the name is as follows: like we use sampling models
to describe how data are generated, we use inferential models to describe how data
are processed for the purpose of inference. The goal here is to present both some
intuition and some details behind the IM approach to an econometrics audience.

Why should econometricians be interested in IMs? The ability to provide probabilistic inference with a prior or a Bayesian approach, as discussed above, is one
benefit, but there are others. First, a feature of the IM construction is its ability to
produce inferential output which is *valid* or, in other words, calibrated with respect to
the posited sampling model; see Sect. 2. This validity property is automatic from the
construction, and does not require any large-sample approximations, etc. Therefore,
procedures derived from the IM's output have frequentist error rate control guarantees, e.g., the IM's 95% interval estimates have guaranteed coverage probability at
least 0.95. Second, the starting point of the IM construction is what we have termed
an "association," a known relationship between the observable data, the unknown
parameters, and some unobservable auxiliary variables. Econometricians often call
these "functional" or "structural" models and they are very comfortable with this
formulation, perhaps more so than statisticians. The result is that econometricians
do not need to rethink how they formulate their problems before they can reconsider
how they carry out their analysis.

Naturally, the benefits of the IM approach do not come without a cost. From
a technical point of view, in order to follow the general development and theory,
the user would need some basic understanding of random sets; however, in specific
applications, the necessary probability calculations for these random sets can often

---

[1] At his keynote address for a workshop on foundations at Rutgers University in April 2016, http://
statistics.rutgers.edu/bff2016.

be streamlined, re-expressed in terms of ordinary probability calculations for random variables. From an interpretation point of view, the user would need to rethink the role that probability plays in statistical inference. Indeed, inference requires a summary of the evidence in the observed data (and in any other input) supporting the truthfulness and falsity of various assertions or hypotheses of interest. The natural way to encode such a summary of evidence is via the belief and plausibility functions, and this is form of the IM's output. Details about the random sets and the IM output are provided in Sect. 2 below. More generally, since the goal of the IM approach is to provide the "best possible" inference, care is needed at every step. In particular, the dimension of the auxiliary variables mentioned above is important in terms of the IM's efficiency, so reducing the dimension as much as possible is critical; see Sect. 3.

In Sect. 4, I will put the IM machinery into action in several different variations on the linear regression model, commonly in econometrics applications. Finally, some concluding remarks are given in Sect. 5.

2 Basic IM Framework

2.1 Objectives

Before describing the details of the IM approach, it will help to give some motivation by describing what, in my opinion, is the goal of statistical inference. According to Martin and Liu [32], probabilistic inference obtains from

a function $b_y : 2^\Theta \to [0, 1]$ such that, for any assertion or hypothesis A about the parameter of interest, $b_y(A)$ represents the data analyst's degree of belief in the truthfulness of the claim "$\theta \in A$" based on the observed data $Y = y$.

In other words, the inferential output b_y encodes the evidence in the observed data (and any other relevant input) concerning the parameter θ. Of course, b_y could be a probability measure, e.g., Bayesian posterior distribution, but that is not necessary from the definition. Indeed, given the "evidence" interpretation, there is reason to think that the mathematical tools used to describe evidence, such as belief and plausibility functions [40] might be preferred. It turns out that Shafer's generalization of probability will appear quite naturally in the IM construction; see Sect. 2.3.

An important question is the following: why should one data analyst's degrees of belief be meaningful to anyone else? It is essential that these beliefs, which necessarily have some personal or subjective element, be calibrated to a common scale so that others can interpret their values. While various scales could be considered, a natural choice is a scale based on frequencies relative to the sampling model $\mathsf{P}_{Y|\theta}$. Along this line, the probabilistic inference based on the function b_y is called *valid* if

$$\sup_{\theta \in A^c} \mathsf{P}_{Y|\theta}\{b_Y(A) \geq 1 - \alpha\} \leq \alpha \quad \forall \, \alpha \in (0, 1), \quad \forall \, A \subseteq \Theta. \tag{1}$$

In words, the validity condition (1) says that, if the assertion A is false, then $b_Y(A)$, as a function of $Y \sim P_{Y|\theta}$, with $\theta \notin A$, is stochastically no larger than $\mathsf{Unif}(0, 1)$. In addition to calibrating the degrees of belief, validity has some important consequences; see, e.g., (10). Remarkably, it turns out to be relatively straightforward to construct an IM such that the corresponding degrees of belief satisfy (1).

2.2 Intuition

To build up some intuition for how the IM approach can meet the goals described above, consider the following basic representation of the sampling model, $Y \sim P_{Y|\theta}$, for the observable data Y, i.e.,

$$Y = a(\theta, U), \quad U \sim P_U, \tag{2}$$

where $\theta \in \Theta$ is the unknown parameter of interest, and U is an unobservable auxiliary variable, taking values in a space $\mathbb{U}$, with a known distribution P_U. A classical example of this is the case of the simple normal mean model, $Y \sim N(\theta, 1)$, which can be rewritten in association form, $Y = \theta + U$, where $U \sim N(0, 1)$. Models written in the association form (2) are standard in econometrics; see Sect. 4. Suppose, for the moment, that, for a given (y, u), there is a unique solution $\theta = \theta(y, u)$ to the equation in (2). Since $Y = y$ is observable, if I also happened to know the value u^* of U that corresponds to the observed y and the true $\theta = \theta^*$, then I can easily and exactly solve the inference problem, since $\theta^* = \theta(y, u^*)$. Of course, I can never see u^* so what I just described cannot be put into practice. However, it does provide some important intuition, namely, that the inference problem can— and arguably should—be cast in terms of U and u^*, instead of θ. This observation is important because there is information in the specified P_U concerning the likely values of U, whereas typically nothing really concrete is known about θ. Therefore, the starting point for the IM approach is the idea that it suffices to take u^* as the inferential target, instead of θ, but the question remains of how to harness the information in P_U to make valid inference on θ through U.

The careful reader will recognize that my argument above relies heavily on the assumption that the Eq. (2) has a unique solution, and this rarely holds even for very simple models. I will discuss this issue in detail in Sect. 3 but, for now, suffice it to say that most problems will admit an association of the form (2) for which a unique solution to the equation exists. So, working from the assumption that there is a unique solution is essentially without loss of generality.

2.3 Construction

Based on the shift of focus from θ to U, Martin and Liu [27] put forth a concise three-step construction of an IM. The previous subsection essentially described the first step, the A-step. The second and third steps are where things get more interesting. In particular, the P-step suggests the use of a random set to predict[2] the unobserved value of U with a random set, and the C-step combines the results of the A- and P-steps to produce a new random set whose distribution produces the relevant inferential output.

A-step Define an association $Y = a(\theta, U)$, with $U \sim \mathsf{P}_U$, of the form (2) that specifies the relationship between the observable data Y, the unknown parameter θ, and the unobservable auxiliary variable U. Depending on the application and the goal, some preprocessing may be recommended before writing the association; see Sect. 3. The result is a set-valued function $u \mapsto \Theta_y(u)$, indexed by the observed $Y = y$, given by

$$\Theta_y(u) = \{\theta \in \Theta : y = a(\theta, u)\}, \quad u \in \mathbb{U}. \tag{3}$$

In the case discussed above where there is a unique solution to $y = a(\theta, u)$, then $\Theta_y(u)$ is a singleton, but there are examples, e.g., discrete-data problems, where $\Theta_y(u)$ will have even uncountably many elements.

P-step Define a *predictive random set* $\mathcal{S} \sim \mathsf{P}_\mathcal{S}$ supported on a collection $\mathbb{S}$ of subsets of $\mathbb{U}$, i.e., the realizations of $\mathcal{S}$ are subsets of $\mathbb{U}$, to predict the unobserved value, say, $u^\star$ of U. Thorough introductions of the theory of random sets are presented in, e.g., Molchanov [34] and Nguyen [36], but only some very basic notions are required here. Certain conditions on $\mathsf{P}_\mathcal{S}$ will be required for the inferential output to be meaningful but, roughly, $\mathsf{P}_\mathcal{S}$ must be specified such that the "net" $\mathcal{S}$ cast to hit $u^\star$ of U should hit its target with high probability for all likely values of $u^\star$; see (7) below. This condition seems rather complicated but, fortunately, it turns out to be relatively easy to arrange.

C-step Combine the results of the A- and P-steps by constructing new random set, defined on subsets of the parameter space Θ, that corresponds to those θ values consistent with both the association (3) and the random set $\mathcal{S}$. That is, write

$$\Theta_y(\mathcal{S}) = \bigcup_{u \in \mathcal{S}} \Theta_y(u), \quad \mathcal{S} \sim \mathsf{P}_\mathcal{S}. \tag{4}$$

This new random set represents the set of all θ values consistent with y and a set $\mathcal{S}$ of (presumably) "reasonable" set of u values. The key property is this: $\Theta_y(\mathcal{S})$

---

[2]This is not an ideal choice of word because "predict" has a particular meaning in statistics and econometrics, but the meaning here is different. "Guess" or "impute" are other potential words to describe the operation in consideration, but both still miss the mark slightly. A more accurate description of what I have in mind is "to cast a net" at the target.

contains the true θ if and only if S contains the value u* corresponding to the observed y and the true θ. Therefore, as hinted at by the intuition above, the quality of inference about θ is tied directly to how well S predicts the unobserved value of U. This inference about θ is summarized by the function[3]

$$b_y(A) = b_y(A; P_S) = P_S\{\Theta_y(S) \subseteq A\}, \quad A \subseteq \Theta. \tag{5}$$

That is, the degree of belief assigned to the claim "$\theta \in A$" based on data $Y = y$ is the P_S-probability that $\Theta_y(S)$ is consistent with the claim. This function is a *belief function* in the sense of, e.g., Shafer [40, 41, 43] and Nguyen [35], a generalization of a probability measure, and some consequences of this will be discussed below.

To summarize, the IM construction proceeds by specifying an association and a predictive random set that acts like a net cast to catch the unobserved value of the auxiliary variable. These two pieces are combined in the C-step to generate a new random set whose distribution is to be used for inference on the parameter. This distribution, the IM's output, takes the form of a belief function defined on the parameter space. This generalizes the familiar Bayesian approach in the sense that, if a genuine prior for θ is available, then it can be incorporated and the corresponding IM output would be the Bayesian posterior ([29], Remark 4). However, the IM construction can be carried out without a prior, and apparently the only cost is that the output is a belief function instead of a probability measure—but the output is still "probabilistic" in the sense of Sect. 2.1. There is nothing particularly special about probability measures beyond that they are familiar to statisticians and econometricians. In fact, one could argue that probability may not be appropriate for summarizing evidence or degrees of belief, which is essentially what inference is about. For example, if evidence is described via a probability P, then the complementation formula, $P(A^c) = 1 - P(A)$, implies that evidence in support of the truthfulness of A is evidence against the truthfulness of A^c. However, in a statistical inference problem, it is possible that the data is not particularly informative for certain assertions. In such cases, it would be reasonable to say that there is little evidence to support the truthfulness of either A or A^c. Probabilities cannot facilitate this conclusion, but belief functions can, so perhaps the latter are more suitable; see, e.g., Shafer [40] and Martin and Liu [32] for more on this.

2.4 Properties

The above construction shows how that shift of focus from the parameter θ to the auxiliary variable U can lead to probabilistic inference, and the logic behind this

---

[3]The formula (5) silently assumes that $\Theta_y(S)$ is non-empty with P_S-probability 1 for all y. This holds automatically in many cases, often because of the preprocessing discussed in Sect. 3, but not in all. The ideal remedy seems to be choosing a predictive random set which is *elastic* in a certain sense; see Ermini Leaf and Liu [12].

construction appears to be sound. However, the belief function b_y in (5) is largely determined by the predictive random set $S \sim \mathsf{P}_S$, which is specified by the data analyst. What makes one data analyst's degrees of belief meaningful to another? What kind of properties of the degrees of belief are desirable, and what conditions are required to achieve these? This section will address these questions.

As a preview for the IM approach and its output, I presented the validity condition, Eq. (1), in Sect. 2.1. In particular, validity implies that the event "$b_Y(A) \geq 0.9$" has probability no more than 0.1 when $Y \sim \mathsf{P}_{Y|\theta}$ with $\theta \notin A$; this makes sense because $b_y(A)$ should tend to be small when A is false. Therefore, the validity condition calibrates the belief function values, telling users what are "small" and "large" belief values. Bayesian posterior probabilities, for example, are not guaranteed to be valid, as the example in Martin and Liu [32, Sect. 5.2] clearly demonstrates.

How can validity be achieved? Since the belief function values are determined by the predictive random set, it is reasonable to suspect that validity can be achieved through a judicious choice of $S \sim \mathsf{P}_S$. Indeed, Martin and Liu [27] give sufficient conditions on S such that the corresponding belief function is provably valid. They provide general conditions for constructing a suitable predictive random set, but a simpler condition, which is easy to check but not helpful for construction, is as follows. Let

$$\gamma(u) = \gamma(u; \mathsf{P}_S) = \mathsf{P}_S\{S \ni u\}, \quad u \in \mathbb{U}, \tag{6}$$

denote the contour function of S, i.e., the probability that S catches a specified target u. Intuitively, the contour function should be large for all but the "extreme values" of u relative to P_U. Formally, Martin and Liu [27] say that S is *valid* if

$$\mathsf{P}_U\{\gamma(U) \leq \alpha\} \leq \alpha, \quad \forall \alpha \in (0, 1), \tag{7}$$

or, in other words, $\gamma(U)$ is stochastically no smaller than $\mathsf{Unif}(0, 1)$ as a function of $U \sim \mathsf{P}_U$. Theorem 2 in Martin and Liu [27] says that if S satisfies (7) [and is such that $\Theta_y(S)$ is non-empty with P_S-probability 1 for all y], then the IM output satisfies the validity condition (1). The most common scenario, and one of the simplest, is when P_U is $\mathsf{Unif}(0, 1)$ and, for this case, there is a "default" predictive random set S of the form

$$S = \{u \in [0, 1] : |u - 0.5| \leq |U' - 0.5|\}, \quad U' \sim \mathsf{Unif}(0, 1), \tag{8}$$

i.e., S is a symmetric interval about 0.5 with a random width, and this satisfies the condition (7). This can be verified directly by first writing down the contour function,

$$\gamma(u) = \mathsf{P}\{|\mathsf{Unif}(0, 1) - 0.5| \geq |u - 0.5|\} = 1 - |2u - 1|,$$

and then recognizing that $1 - |2U - 1|$ is $\mathsf{Unif}(0, 1)$ when U is. I will use this default predictive random set in the examples in Sect. 4, but this is mainly for simplicity—in some cases, there are better choices of S in terms of IM efficiency; see Sect. 5.

There are many ways to calibrate the numerical values of the belief function output, but there are some additional advantages to doing so relative to the sampling model. This is related to the question about repeated sampling properties of the inferential output, and is particularly relevant given the recent concerns about reproducibility in scientific research; see, e.g., Nuzzo [37] and the collection of reports published by *Nature*[4] on the "Challenges in Irreproducible Research." Simply put, the validity result implies that decision procedures constructed based on the IM output are guaranteed to control frequentist error rates at the specified level. Here I will focus only on the case of set/interval estimation. As mentioned above, the belief function output differs from an ordinary probability. One difference is that it may not be additive, i.e.,

$$b_y(A) + b_y(A^c) \geq 1 \quad \forall \, A \subseteq \Theta.$$

Define the corresponding *plausibility function* p_y as

$$p_y(A) = 1 - b_y(A^c), \tag{9}$$

and observe that non-additivity implies that $b_y(A) \leq p_y(A)$ for all A. This inequality explains why b_y and p_y are sometimes referred to as lower and upper probabilities. Anyway, the plausibility function is a useful tool for designing decision procedures. In particular, the $100(1 - \alpha)\%$ *plausibility region* for θ, defined as

$$\Pi_\alpha(Y) = \{\vartheta \in \Theta : p_y(\vartheta) \geq \alpha\}, \quad \text{where} \quad p_y(\vartheta) = p_y(\{\vartheta\}), \tag{10}$$

is the IM counterpart of a frequentist confidence region or a Bayesian credible region. It is the set made up of parameter values which are, pointwise, sufficiently plausible based on the observed data $Y = y$. This is a nice interpretation, arguably easier to understand and more meaningful than the explanation of confidence intervals. Moreover, the IM's validity property guarantees that the plausibility region in (10) has the nominal coverage probability, i.e.,

$$\mathsf{P}_{Y|\theta}\{\Pi_\alpha(Y) \ni \theta\} \geq 1 - \alpha, \quad \forall \, \theta \in \Theta.$$

This is a direct consequence of the validity property, and is not a large-sample limit approximation. Compared to existing methods which rely almost entirely on asymptotic approximations for justification, the IM approach needs no asymptotics: it stands on its own from a logic point of view, and even provides stronger results in terms of sampling distribution properties than the traditional "frequentist" approaches.

As a last point about the IM output, it is worth mentioning that the plausibility function in (9) is closely related to the familiar p-value introduced in standard introductory textbooks for the purpose of testing hypotheses. P-values have always been quite controversial and, recently, the journal *Basic and Applied Social Psychology* has banned

---

[4]http://www.nature.com/nature/focus/reproducibility/index.html.

their use [44]; see also, the recent statement[5] issued by American Statistical Association. Much of this controversy exists because there is confusion about what a p-value actually is; a quick look at any standard introductory text will reveal a list of incorrect interpretations of p-values, but none address the question directly. Martin and Liu [28] demonstrate that, for any p-value, there exists a valid IM whose plausibility function, at the specified assertion, is numerically equal to the given p-value. Therefore, p-values are plausibilities and this provides an easy interpretation which is directly in line with how they are used in practice. Moreover, their demonstration makes clear that p-values should not be taken as some "magic number" that contains any especially powerful information—it is only one feature of a proper assessment of the information in the available data concerning inference on θ. Of course, if the goal of defining a p-value is simply to construct a testing rule with suitable frequentist error rate control, then there is no trouble with the mathematics; in fact, the IM's validity property can be used to prove such claims.

3 Dimension Reduction Techniques

3.1 What's the Issue?

In the basic IM construction, the intuition was designed around cases where the association (2) had a unique solution $\theta = \theta(y, u)$ for any given (y, u) pair. While the construction and validity properties discussed above do not require this uniqueness, except in how it affects the (non) emptiness of $\Theta_y(\mathcal{S})$, it turns out that having a unique solution is beneficial in terms of the efficiency[6] of the IM. Unfortunately, the uniqueness condition rarely holds when considering an association in terms of the full sampling model. To see this, suppose that $Y = (Y_1, \ldots, Y_n)$ is a vector of iid $N(\theta, 1)$ random variables. Then the naive association, $Y = \theta 1_n + U$, where 1_n is an n-vector of unity and $U = (U_1, \ldots, U_n)$ is a vector of n iid $N(0, 1)$ random variables, does not admit a unique solution for θ unless y and u differ only by a constant shift. The problem in this case is differing dimensions, i.e., the auxiliary variable U is n-dimensional while the parameter is a scalar. Fortunately, there often is a transformation which will allow for the dimension of the auxiliary variable to be reduced to that of the parameter, so that a unique solution is available.

Here I will comment on two cases where the dimension reduction steps have been studied. The first is a general reduction step—conditioning—that can be carried out

---

[5]http://amstat.tandfonline.com/doi/abs/10.1080/00031305.2016.1154108.

[6]Efficiency is essentially the rival to validity, and the goal is to balance between the two. For example, a belief function that always takes value 0 for all $A \neq \Theta$, which is achieved by a choosing an extreme predictive random set $\mathcal{S} \equiv \mathbb{U}$, is sure to be valid, but then the corresponding plausibility function would always be 1; consequently, the plausibility regions would be unbounded and practically useless. So, roughly, the idea behind efficiency is to have the smallest predictive random set such that the IM is still efficient. More on this in Sect. 5.

in most problems, while the second step—marginalization—is specific to the case that only some feature of the full parameter is of interest. This discussion focuses only on the A-step portion of the IM construction. Once the dimension of the auxiliary variable has been reduced as far as possible, one can proceed with the P- and C-steps to get valid inference on the interest parameter.

3.2 Conditioning

Consider the naive baseline association $Y = a(\theta, U)$ that describe the sampling model, and may or may not admit a unique solution for θ. Suppose that there exists two pairs of functions (τ, η) and (T, H) such that both $u \mapsto (\tau(u), \eta(u))$ and $y \mapsto (T(y), H(y))$ are one-to-one and the association equation maps as follows:

$$T(Y) = b(\theta, \tau(U)) \quad \text{and} \quad H(Y) = \eta(U).$$

Since the functions are one-to one, nothing is lost in going from the original naive association to this one. What is gained, however, is that the parameter is gone from the second equation. This means that the feature of $\eta(U)$ is actually observed, so there is no need to predict its value with a random set, etc. Moreover, the information in its observed value can potentially be used to help improve the prediction of the complementary feature $\tau(U)$. This suggests working with the reduced auxiliary variable $V_1 = \tau(U)$ and the association

$$T(Y) = b(\theta, V_1), \quad V_1 \sim \mathsf{P}_{V_1|V_2=H(y)},$$

where $V_2 = \eta(U)$ and the latter is the conditional distribution of V_1, given that V_2 equals the observed value $H(y)$. Often, the transformations can be chosen such that the reduced association admits a unique solution for θ, in terms of $T(Y)$ and V_1, which is desirable.

One kind of reduction that can always be carried out is a reduction to a minimal sufficient statistic. Indeed, if $T(Y)$ denotes a minimal sufficient statistic and $H(Y)$ is complement, making $y \mapsto (T(y), H(y))$ one-to-one, then the sampling model can be characterized through the marginal distribution of $T(Y)$ together with the conditional distribution of $H(Y)$, given $T(Y)$. The latter conditional distribution, by the definition of sufficiency, does not depend on the parameter, has no effect on the inference problem and, therefore, can be dropped. However, it is interesting to note that, outside these "regular" problems where the minimal sufficient statistic has the same dimension as the parameter, there are opportunities to reduce the dimension beyond that provided by sufficiency alone. Some examples of that are given in Martin and Liu [29], Cheng et al. [2], and Martin and Liu [25]. This is consistent with the claim in Fraser [16] that conditioning is often more useful than sufficiency in applications.

The tool that has been used to reduce the dimension of the auxiliary variable beyond that provide by sufficiency is an approach based on solving a suitable differential equation. Intuitively, a function $\eta(U)$ of the auxiliary variable can be fully observed from the sample data if $\eta(u_{y,\theta})$ is not sensitive to changes in θ, where $u_{y,\theta}$ is the solution to the baseline association for u based on a given (y, θ). This insensitivity can be described by requiring that η satisfy the following differential equation:

$$\frac{\partial \eta(u_{y,\theta})}{\partial \theta} = 0.$$

I will not go into any details here but this approach is both interesting and has proved successful in the applications mentioned above. Further investigation is needed.

3.3 Marginalization

While the dimension-reduction strategy based on conditioning is general, the next approach is for the case where only some feature of the full parameter θ is of interest. Towards this, suppose that θ can be expressed in terms of a lower-dimensional interest parameter ψ and a nuisance parameter λ. Since the dimension of ψ is smaller than that of θ, even after a conditioning argument, there is an opportunity to further reduce the auxiliary variable dimension.

Suppose that the baseline association, $Y = a(\psi, \lambda; U)$, which could be the result of the conditioning argument above, can be rewritten as

$$G(Y, \psi) = b(\psi, V_1) \quad \text{and} \quad K(Y, \psi, \lambda, V_2) = 0,$$

where $V = (V_1, V_2)$ is a new pair of auxiliary variables; note that the first equation does not depend on the nuisance parameter λ so V_1 should be of dimension smaller than U. When the function K is such that, for any (y, ψ, v_2), there exists λ such that the above equality holds, Martin and Liu [31] call the association "regular," and they prove that a suitable marginal association for ψ obtains by simply ignoring the second equation that involves λ, working with only

$$G(Y, \psi) = b(\psi, V_1), \quad V_1 \sim \mathsf{P}_{V_1},$$

where P_{V_1} is the marginal distribution of V_1 derived from the joint distribution of (V_1, V_2). This characterization of "regular" marginal inference problems, and the corresponding strategy for carrying out marginalization, has already proved to be useful in other contexts; see Theorem 2.3 in Hannig et al. [21]. In addition to the examples below, Martin and Liu [31] present marginal IM solutions to Stein's many-normal-means problem, the Behrens–Fisher problem, and the gamma mean problem; also, Martin and Lingham [26] present an IM approach to prediction, which can be viewed as an extreme marginalization problem in which θ itself is a nuisance parameter.

The examples in Sect. 4 below should shed some light on the kind of calculations involved to make valid marginal inference. To conclude this section, I would like to make some comments about the difficulty of valid marginal inference in general. The key point is that marginalization needs to be considered as from the outset, i.e., one cannot construct a method for inference on θ and then expect that its desirable properties will carry over to marginal inference on a function $\psi = \psi(\theta)$. The considerable work (e.g., [46]) on constructing default priors on the full parameter θ for marginal inference on various ψ is a testament to this difficulty, as are the striking results of Gleser and Hwang [19] that raise serious concerns about the use of Wald-style plug-in confidence intervals in certain marginal inference problems. These comments are intended to make clear that there is good reason for putting so much care into marginal inference.

4 Examples

4.1 Linear Regression

As a first illustrative example, consider the standard linear regression model. That is,

$$Y = X\beta + \sigma Z,$$

where Y is a n-dimensional vector of response variables, X is a fixed $n \times p$ matrix of predictor variables, β is a p-dimensional vector of unknown slope coefficients, σ is an unknown scale parameter, and Z is a n-dimensional vector of iid $N(0, 1)$ random variables. Assume that $p < n$ and that X has rank p. This is a classic model in both statistics and econometrics texts. Below we will consider several different interest parameters.

First, since Z is n-dimensional but there are $p + 1 \leq n$ parameters, there is an immediate opportunity to reduce the auxiliary variable dimension via conditioning. It turns out that, in this case, the first dimension reduction step corresponds to working with an association based on the minimal sufficient statistics. That is, a reduced-form (conditional) association is

$$\hat{\beta} = \beta + \sigma V_1 \quad \text{and} \quad \hat{\sigma} = \sigma V_2, \tag{11}$$

where $(\hat{\beta}, \hat{\sigma}^2)$ are the usual least-squares estimators, and (V_1, V_2) are independent, with $V_1 \sim N_p(0, (X^\top X)^{-1})$ and $(n - p - 1)V_2^2 \sim \text{ChiSq}(n - p - 1)$. As an aside, this conditional association can also be derived by using the differential equation-based technique described in Sect. 3; take $\eta(z) = (1_p^\top Mz)^{-1}(Mz)$, where 1_p is the p-vector of unity, and $M = I_n - X(X^\top X)^{-1}X^\top$ projects onto the space orthogonal to that spanned by the columns of X. The new auxiliary variable, $V = (V_1, V_2)$, is of the same dimension as the parameter θ but, for scalar parameters of interest, further reduction is possible. This is the baseline association from which various different IMs can be constructed. Interest here is in features of the β vector, but it is possible to construct a marginal IM for σ ([30], Sect. 8.2.4) and even a marginal IM for predicting a future observation Y_{n+1} [26].

A straightforward manipulation of (11) gives

$$\hat{\beta} = \beta + \hat{\sigma} V_1/V_2 \quad \text{and} \quad \hat{\sigma} = \sigma V_2.$$

Since σ only appears in the second equation, and that a solution exists for all data and all auxiliary variables, a marginal association for the vector β obtains by simply ignoring the second equation. Moreover, the ratio V_1/V_2 becomes a single p-dimensional auxiliary variable having a scaled multivariate t-distribution, i.e., $U = V_1/V_2 \sim \mathsf{t}_p(0, (X^\top X)^{-1}; n - p - 1)$. Therefore, we have the following marginal association for β:

$$\hat{\beta} = \beta + \hat{\sigma} U, \quad U \sim \mathsf{t}_p(0, (X^\top X)^{-1}; n - p - 1). \tag{12}$$

From here, one can construct an IM for simultaneous inference on the β vector, or marginal IMs for various features of β. Here are two marginalization problems.

- A first and relatively simple feature to consider is, say, $\psi = \beta_1$. It follows immediately from (12) that a marginal association for β_1 is

$$\hat{\beta}_1 = \beta_1 + \hat{\sigma} U_1,$$

and this completes the A-step of the IM construction. For the P-step, the optimal predictive random set is a symmetric interval around zero. That is, make a change-of-auxiliary variable and write $W = F_n(U_1)$, where F_n is the distribution function of the appropriate Student-t distribution. Then the optimal predictive random set $\mathcal{S}$ for W is the default (8). This completes the P-step, and the C-step is straightforward. Indeed, it is easy to see that the corresponding plausibility intervals (10) for β_1 are exactly the usual Student-t intervals based on the distribution theory of the least-squares estimator. In particular, the guaranteed validity of the marginal IM provides an alternative indirect prove of the well-known result that these standard Student-t confidence intervals have exact coverage probability.

- Next, consider a ratio of regression coefficients, say, $\psi = \beta_2/\beta_1$, as the interest parameter. This is a difficult marginal inference problem, similar to that considered by Fieller [14] and Creasy [3] related to the *instrumental variable* models widely used in econometrics, a special case of the general situation studied by Gleser and Hwang [19] and Dufour [9]. From the discussion above, the following reduced-form association can be reached immediately:

$$\hat{\beta}_1 = \beta_1 + \hat{\sigma} U_1 \quad \text{and} \quad \hat{\beta}_2 = \beta_2 + \hat{\sigma} U_2.$$

Substituting $\beta_2 = \psi\beta_1$ into the second equation, it follows that

$$\hat{\beta}_2 - \psi\hat{\beta}_2 = \hat{\sigma}(U_2 - \psi U_1) \quad \text{and} \quad \hat{\beta}_1 = \beta_1 + \hat{\sigma}U_1.$$

Based on the general ideas in Sect. 3.3, the latter equation can be ignored, leading to the one-dimensional association

$$\hat{\beta}_2 - \psi\hat{\beta}_1 = \hat{\sigma}(U_2 - \psi U_1).$$

To complete the A-step, it is desirable to rewrite the equation in terms of a single scalar auxiliary variable, instead of the pair (U_1, U_2). The simplest strategy is to let F_ψ denote the distribution function of $U_2 - \psi U_1$, and set $W = F_\psi(U_2 - \psi U_1)$. Then the above association can be written as

$$\hat{\beta}_2 - \psi\hat{\beta}_1 = \hat{\sigma}F_\psi^{-1}(W), \quad W \sim \mathsf{Unif}(0, 1). \tag{13}$$

The only difficulty with this approach is that, to my knowledge, the distribution of $U_2 - \psi U_1$ is not a standard one with a closed-form distribution function. But this is only a minor obstacle, since it is straightforward to evaluate F_ψ via Monte Carlo. Having completed the A-step in (13), the P-step requires a predictive random set for the unobserved value of $W \sim \mathsf{Unif}(0, 1)$. Here I will suggest—for simplicity and not necessarily for optimality as in the previous example—to use the default predictive random set in (8). Then the belief and plausibility function for ψ obtain from the C-step. In particular, for constructing a plausibility region for ψ as in (10), the pointwise plausibility function is

$$p_y(\psi) = \mathsf{P}_\mathcal{S}\left\{\mathcal{S} \ni F_\psi\left(\hat{\sigma}^{-1}(\hat{\beta}_2 - \psi\hat{\beta})\right)\right\}$$
$$= 1 - \left|2F_\psi\left(\hat{\sigma}^{-1}(\hat{\beta}_2 - \psi\hat{\beta}_1)\right) - 1\right|.$$

It follows from the general IM validity theory that the $100(1 - \alpha)\%$ plausibility region (10) based on the plausibility function above has the nominal coverage probability for all n, not just asymptotically. To put this in perspective, recall the unexpected result in Gleser and Hwang [19] that says the usual Wald-style plug-in intervals based on the delta theorem are not valid, moreover, their coverage probability is zero. It is possible to construct genuine confidence intervals using techniques other than the IM approach described here, e.g., using the deviance-based technique advocated in Schweder and Hjort [39], but the IM approach seems to handle the Gleser–Hwang challenge automatically, whereas other approaches require some special care and/or modifications.

4.2 Panel Data

For the sake of space, I will not go into details here about the IM approach to the problem of panel or longitudinal data. However, the models used for analyzing these data are ones that have attracted some attention from IM users. In particular, Cheng et al. [2] studied the linear mixed effects model with special interest in marginal inference on the heritability coefficient, i.e., the proportion of total variance that can be attributed to the individual effect. There, in fact, they employ the differential equation-based approach mentioned briefly in Sect. 3.2 to achieve maximal reduction of the auxiliary variable dimension and exact marginal inference. Ongoing work of Chuanhai Liu and myself will further investigate both the fixed and random effects models like those often used in panel data applications.

5 Discussion

This paper was intended to give econometricians an introduction to the logic behind and the mechanics involved in implementing the new IM approach. I don't know much about econometrics, but I think that IMs have a lot of potential because the structural/functional models are so common and can be immediately recast into the IM's "association" language, with auxiliary variables, etc. Some feedback I have received from statisticians is that they are uncomfortable with the association setup— likelihoods are more familiar—in part because that representation is not unique. Non-uniqueness is not a real concern to me, nor is it for the econometricians who are already working with these associations, so the IM approach seems like a good fit for them. Beyond being a good fit, I think the IM's emphasis on a deeper level of thinking about inference would be beneficial to econometrics in general which, based on my minimal experience, seems to be focused primarily on finding estimators with desirable asymptotic properties and, therefore, not really addressing the issue of uncertainty.

  Though the story presented here may appear to be nearly complete, there are still lots of open questions to consider. First, I did not really address here any "real" econometrics applications and, to my knowledge, these have yet to be touched by anyone who has any experience with IMs. This is important not only for the development of IMs but also for econometrics: showing that the new approach makes sense and actually works well (probably better than existing methods) in real and challenging problems, such as instrumental variable models, could have a huge impact in the econometrics field. Second, the question about how to choose the "best" predictive random set for a given problem is almost completely open. It may be that there is no realistic notion of optimality here, but there should be some guidelines for users to help ensure a certain level of efficiency. Third, there is the fundamental question about model assessment, i.e., how can the IM approach, which is designed for inference on the model-specific parameters, provide a valid assessment of the uncertainty in the model itself? Some first steps towards this have been made in Martin et al. [33] but more work is needed.

Finally, I want to mention some alternative ways to view and/or present the IM approach. As an attempt to bridge the gap between the foundational aspects of the IM approach and the actual methodology, I wanted to develop an alternative way to sell the IM-based methodology to users without them having to get through all the technicalities of random sets, etc. Two papers [22, 23] show that it is possible to construct an IM based on some default or generalized associations, e.g., using likelihood ratios, apparently without loss of efficiency; in a third paper [24], I propose an intermediate-level statistics theory course that teaches students some of the IM-based reasoning but using the more familiar language of p-values.

Acknowledgements The author thanks the organizers of the 10th International Conference of the Thailand Econometric Society, in particular, Professor Hung Nguyen, for the invitation to present at the conference and to submit a paper for the proceedings.

References

1. Berger J (2006) The case for objective Bayesian analysis. Bayesian Anal 1(3):385–402
2. Cheng Q, Gao X, Martin R (2014) Exact prior-free probabilistic inference on the heritability coefficient in a linear mixed model. Electron J Stat 8(2):3062–3076
3. Creasy MA (1954) Symposium on interval estimation: limits for the ratio of means. J R Stat Soc Ser B 16:186–194
4. Dempster AP (1966) New methods for reasoning towards posterior distributions based on sample data. Ann Math Stat 37:355–374
5. Dempster AP (1967) Upper and lower probabilities induced by a multivalued mapping. Ann Math Stat 38:325–339
6. Dempster AP (1968) A generalization of Bayesian inference (With discussion). J R Stat Soc Ser B 30:205–247
7. Dempster AP (2008) The Dempster-Shafer calculus for statisticians. Int J Approx Reason 48(2):365–377
8. Dempster AP (2014) Statistical inference from a Dempster–Shafer perspective. In: Lin X, Genest C, Banks DL, Molenberghs G, Scott DW, Wang J-L (eds) Past, present, and future of statistical science. Chapman & Hall/CRC Press. Chap 24
9. Dufour J-M (1997) Some impossibility theorems in econometrics with applications to structural and dynamic models. Econometrica 65(6):1365–1387
10. Efron B (1998) R. A. Fisher in the 21st century. Stat Sci 13(2):95–122
11. Efron B (2013) Discussion: "Confidence distribution, the frequentist distribution estimator of a parameter: a review" [mr3047496]. Int Stat Rev 81(1):41–42
12. Ermini Leaf D, Liu C (2012) Inference about constrained parameters using the elastic belief method. Int J Approx Reason 53(5):709–727
13. Fidler F, Thomason N, Cummings G, Fineh S, Leeman J (2004) Editors can lead researchers to confidence intervals, but can't make them think. Psychol Sci 15:119–126
14. Fieller EC (1954) Symposium on interval estimation: some problems in interval estimation. J R Stat Soc Ser B 16:175–185
15. Fisher RA (1973) Statistical methods and scientific inference, 3rd edn. Hafner Press, New York
16. Fraser DAS (2004) Ancillaries and conditional inference. Stat Sci 19(2):333–369 With comments and a rejoinder by the author

17. Ghosh JK, Delampady M, Samanta T (2006) An introduction to Bayesian analysis. Springer, New York
18. Ghosh M (2011) Objective priors: an introduction for frequentists. Stat Sci 26(2):187–202
19. Gleser LJ, Hwang JT (1987) The nonexistence of $100(1 - \alpha)\%$ confidence sets of finite expected diameter in errors-in-variables and related models. Ann Stat 15(4):1351–1362
20. Hannig J (2009) On generalized fiducial inference. Stat Sinica 19(2):491–544
21. Hannig J, Iyer H, Lai RCS, Lee TCM (2016) Generalized fiducial inference: a review and new results. J Am Stat Assoc 111(515):1346–1361
22. Martin R (2015) Plausibility functions and exact frequentist inference. J Am Stat Assoc 110:1552–1561
23. Martin R (2017) On an inferential model construction using generalized associations. J Stat Plann Infer. doi:10.1016/j.jspi.2016.11.006 (to appear)
24. Martin R (2017) A statistical inference course based on p-values. Am Stat. doi:10.1080/00031305.2016.1208629 (to appear)
25. Martin R, Lin Y (2016) Exact prior-free probabilistic inference in a class of non-regular models. Statistics 5(1):312–321
26. Martin R, Lingham R (2016) Prior-free probabilistic prediction of future observations. Technometrics 58(2):225–235
27. Martin R, Liu C (2013) Inferential models: a framework for prior-free posterior probabilistic inference. J Am Stat Assoc 108(501):301–313
28. Martin R, Liu C (2014) A note on p-values interpreted as plausibilities. Stat Sinica 24:1703–1716
29. Martin R, Liu C (2015) Conditional inferential models: combining information for prior-free probabilistic inference. J R Stat Soc Ser B 77(1):195–217
30. Martin R, Liu C (2015) Inferential models: reasoning with uncertainty. Monographs in statistics and applied probability series. Chapman & Hall/CRC Press, Boca Raton
31. Martin R, Liu C (2015) Marginal inferential models: prior-free probabilistic inference on interest parameters. J Am Stat Assoc 110:1621–1631
32. Martin R, Liu C (2016) Validity and the foundations of statistical inference. arXiv:1607.05051 (Unpublished manuscript)
33. Martin R, Xu H, Zhang Z, Liu C (2016) Valid uncertainty quantification about the model in a linear regression setting. arXiv:1412.5139 (Unpublished manuscript)
34. Molchanov I (2005) Theory of random sets. Probability and its applications (New York). Springer, London
35. Nguyen HT (1978) On random sets and belief functions. J Math Anal Appl 65(3):531–542
36. Nguyen HT (2006) An introduction to random sets. Chapman & Hall/CRC, Boca Raton
37. Nuzzo R (2014) Scientific method: statistical errors. Nature 506:150–152
38. Schweder T, Hjort NL (2002) Confidence and likelihood. Scand J Stat 29(2):309–332
39. Schweder T, Hjort NL (2016) Confidence, likelihood, probability: statistical inference with confidence distributions. Cambridge University Press, Cambridge
40. Shafer G (1976) A mathematical theory of evidence. Princeton University Press, Princeton
41. Shafer G (1979) Allocations of probability. Ann Probab 7(5):827–839
42. Shafer G (1982) Belief functions and parametric models. J Roy Stat Soc Ser B 44(3):322–352 With discussion
43. Shafer G (1987) Belief functions and possibility measures. In: Bezdek JC (ed) The analysis of fuzzy information. Mathematics and logic, vol 1. CRC, pp 51–84
44. Trafimowa D, Marks M (2015) Editorial. Basic Appl Soc Psychol 37(1):1–2
45. Xie M, Singh K (2013) Confidence distribution, the frequentist distribution of a parameter—a review. Int Stat Rev 81(1):3–39
46. Yang R, Berger JO (1998) A catalog of non-informative priors. http://www.stats.org.uk/priors/noninformative/YangBerger1998.pdf (Unpublished manuscript)
47. Zabell SL (1992) R. A. Fisher and the fiducial argument. Stat Sci 7(3):369–387

Robustness in Forecasting Future Liabilities in Insurance

W.Y. Jessica Leung and S.T. Boris Choy

Abstract The Gaussian distribution has been widely used in statistical modelling. Being susceptible to outliers, the distribution hampers the robustness of statistical inference. In this paper, we propose two heavy-tailed distributions in the normal location-scale family and show that they are superior to the Gaussian distribution in the modelling of claim amount data from multiple lines of insurance business. Moreover, they also enable better forecasts of future liabilities and risk assessment and management. Implications on risk management practices are also discussed.

Keywords Heavy-tailed distribution · Bayesian inference · Markov chain Monte Carlo · Loss reserve · Risk diversification

1 Introduction

As a specialised and unique part of the actuarial endeavour, loss reserving has been one of the most classical, yet challenging, problems in the insurance industry. The fundamental purpose of loss reserving is to ensure sufficient capital is set aside to fund outstanding claim payments, avoiding the risk of insolvency. In the insurance industry, outstanding claims can be viewed as losses that are incurred, but yet to be developed. These claims remain to be the greatest source of liability that the insurance company is legally obliged to cover. From the perspective of enterprise risk management, loss incurred from a particular line or multiple lines of business is regarded as financial risk. Intuitively, a tailor-made model for the loss data in multiple lines is of great interest to the insurer to understand the stochastic nature of the data from different lines of business. With the use of statistical and econometrics tools, the uncertainty

W.Y. Jessica Leung (✉) · S.T. Boris Choy
The University of Sydney, Sydney, Australia
e-mail: jessica.leung@sydney.edu.au

S.T. Boris Choy
e-mail: boris.choy@sydney.edu.au

in relation to the variability of the point estimates can be quantified, enabling a more accurate forecast of future liability and better risk assessments and management.

From an insurer's perspective, the motivation for accurately forecasting loss reserves is twofold. Externally, regulators worldwide are moving towards a risk-based capital framework, mandating disclosure of uncertainty in loss reserve estimates and calculation of various risk measures. For example, the second-generation of China's Risk-Oriented Solvency System has been implemented in early 2016 [11]. Likewise, Singapore has adopted a one year Value-at-Risk (VaR) approach as a solvency requirement which is consistent with the Solvency II requirement in the European Union. Similarly, the Swiss Solvency Test and the Australian Prudential Regulation Authority prudential standard GPS320 are also analogues of risk-based capital requirements in Switzerland and Australia respectively [2]. All of the above risk measures require information regarding the variability of the loss reserve in addition to point estimates. Typically, this refers to certain percentile of the estimated future claim amount distribution or a point estimate of the loss reserve, on top of which a margin to be held for prudence can be added [5]. As such, it is crucial that insurance companies, especially those that operate across various countries, align their loss reserving methodologies to comply with the legal requirements in different jurisdictions by adopting a risk-based approach and embracing changes in the regulatory environment.

Internally, inaccurate loss reserve estimates will either prejudice the profitability of an insurance company or increase the risk of financial distress. Overestimating the required loss reserve contravenes the principle of efficient capital use, inducing opportunity costs. Underestimating the required loss reserve and failure in reserving sufficient funds for future claims may result in insolvency and bankruptcy. Therefore, it is important that insurance companies accurately forecast their future liabilities. This enables them to have greater insight into their cost base, which facilitates the optimisation in capital allocation decisions and improvement of premium pricing strategies.

However, in reality, the inherent time delay involved in the process of reporting claims and settlements adds a layer of complexity in forecasting loss reserve. Moreover, loss reserve data always follow a heavy-tailed distribution and may contain some extreme values that distort statistical inference. In this paper, we consider two robustifying bivariate distributions, namely Student-t and variance gamma (VG) distributions, to compete with the Gaussian distribution in fitting loss reserve data from two lines of business: the personal and commercial automobile. These distributions are extremely good in capturing tail behaviour of the data and hence provide a robust analysis and better forecasts.

The aim of this paper is to provide a risk management perspective in the context of the loss reserving problem. More specifically, we highlight the importance of incorporating heavy-tailed distributions in actuarial models in order to enhance model robustness. It is shown that robustness is critical in determining an insurance companys reserve level and risk capital. It is vital for the company to come up with robust and informative solutions in actuarial modelling in order to fulfil the risk-based capital requirement and mitigate the risk of insolvency or bankruptcy. In the empirical study of loss reserve data from two lines of automobile business, we adopt a two-way analysis of variance (AVOVA) model for the loss reserve data and use data augmentation technique to handle the Student-t and VG distributions in a Bayesian framework. The Bayesian Markov chain Monte Carlo techniques produce realisations from the intractable joint posterior distribution for Bayesian inference and predictive modelling. The entire predictive distribution of the loss reserve data can be obtained, enabling analyst to quantify the financial risks in the risk capital analysis. To demonstrate, risk measures including VaR and Tail value-at-risk (TVaR) are calculated for the insurers personal and commercial automobile businesses individually, as well as the combined portfolio.

This paper is structured as follows. Section 1 briefly introduces the loss reserving in risk management and actuarial context. Section 2 explains the run-off triangle for loss reserve data and proposes robust models for fitting observed loss data and forecasting unpaid losses. Section 3 presents statistical inference of an empirical study on two lines of automobile business and models are evaluated and compared. Forecasts of future losses and risk measures are provided in Sect. 4. Finally, this paper is concluded in Sect. 5.

2 Methodology

2.1 Run-Off Triangle

Let Y_{ij} where $i = 1, ..., I$ and $j = 1, ..., J$ be the incremental claim amount paid by an insurer in accident year i (the origin year that claims are filed) and settled in development year j (the number of years before claims are settled). The observed annual claim amounts are typically presented in a run-off triangle, or loss triangle, as shown in Table 1.

Table 1 Run-off triangle

Accident Year i	Development Year j					
	1	2	...	j	...	J
1						
2			Observations of Y_{ij}			
⋮						
i						
⋮				Forecasts of Y_{ij}		
I						

The shaded area in the upper triangle contains the observed claim data that are known as at year I, while the unshaded area in the lower triangle contains unobserved future claim amounts which will be forecasted. Claim amounts in a particular calendar year are the sum of the values along the off-diagonal. In this paper, we consider run-off triangles that have the same number of development years and accident years, i.e. $I = J$.

In real world, insurance companies have multiple lines of business and each line has its own run-off triangle. The understanding of the dependence amongst the claim amount data in the run-off triangles is crucial in the management of actuarial risk. This paper primarily focuses on two lines of business. Claim amount data in two run-off triangles are assumed to be dependent and are modelled by a bivariate distribution.

2.2 Two-Way ANOVA Model

Inspired by the multiplicative model in De Vylder [4], Kremer [9] suggested a fixed effects two-way analysis of variance (ANOVA) model for the logarithmic transformed claim amounts.

Let $z_{ij} = (z_{ij1}, z_{ij2}) = (\ln y_{ij1}, \ln y_{ij2})$ contains the log-claim amounts from two lines of business. The two-way ANOVA model is specified by:

$$z_{ij} = \mu + \alpha_i + \beta_j + \epsilon_{ij}$$

where $\mu = (\mu_1, \mu_2)'$ is the vector of overall means, $\alpha_i = (\alpha_{i,1}, \alpha_{i,2})', i = 1, ..., I$, is the vector of accident year effects, $\beta_j = (\beta_{j,1}, \beta_{j,2})', j = 1, ..., J$ is the vector of development year effects, and $\epsilon_{ij} = (\epsilon_{i,j,1}, \epsilon_{i,j,2})', i = 1, ..., I, j = 1, ..., J$ is the vector of random errors which is assumed follow a bivariate distribution D with a zero mean vector and a covariance matrix given by:

$$\Sigma = \begin{bmatrix} \sigma_1^2 & \rho\sigma_1\sigma_2 \\ \rho\sigma_1\sigma_2 & \sigma_2^2 \end{bmatrix}$$

As the ANOVA is a fixed effects model, we impose either one of the following two constraints on the parameter values.

Constraint 1: $\alpha_{1,1} = \alpha_{1,2} = \beta_{1,1} = \beta_{1,2} = 0$ and all other α and β are free parameters.

Constraint 2: $\sum_{i=1}^{I} \alpha_{i,k} = \sum_{j=1}^{J} \beta_{j,k} = 0$ for $k = 1, 2$.
In this paper, Constraint 2 is adopted.

2.3 Error Distributions

The choice of an appropriate error distribution plays an important role in statistical modelling. Inappropriate choices may lead to biased and non-robust analysis. For symmetric error distributions, the Gaussian distribution has been the most popular error distribution in many applications but it fails to provide a robust analysis should outliers exist. Heavy-tailed distributions are insensitive to outliers and therefore they provide a robust analysis. Amongst many heavy-tailed distributions, the Student-t distribution is most widely used. In this paper, the random errors are assumed to follow either a bivariate Student-t distribution or a bivariate VG distribution.

Both univariate and multivariate t and VG distributions are members of the scale mixture of normal (SMN) family [1, 3]. If the random errors in the bivariate two-way ANOVA model follow a bivariate t distribution, then its joint probability density function (PDF) can be given in the following integral form.

$$f(\epsilon_{ij}|\boldsymbol{\Sigma}, \nu) = \int_0^\infty N_2(\epsilon_{ij}|\mathbf{0}, \lambda_{ij}\boldsymbol{\Sigma})IG\left(\lambda_{ij}\Big|\frac{\nu}{2}, \frac{\nu}{2}\right)d\lambda$$

where $N_2(\cdot|\cdot, \cdot)$ is the bivariate Gaussian PDF, $IG(\cdot|\cdot, \cdot)$ is the inverse gamma PDF, ν is the degrees of freedom and λ_{ij} are the auxiliary variables, known as the scale mixture variables, which can be used to identify potential outliers. For bivariate VG error distribution, the PDF can also be expressed as

$$f(\epsilon_{ij}|\boldsymbol{\Sigma}, \nu) = \int_0^\infty N_2(\epsilon_{ij}|\mathbf{0}, \lambda_{ij}\boldsymbol{\Sigma})Ga\left(\lambda_{ij}\Big|\frac{\nu}{2}, \frac{\nu}{2}\right)d\lambda$$

where $Ga(\cdot|\cdot, \cdot)$ is the gamma PDF and ν shape parameter of the VG distribution.

The SMN representation for the PDF of a heavy-tailed distribution facilitates the data augmentation technique that enables efficient Bayesian computation. The introduction of the scale mixture variables in Bayesian Markov chain Monte Carlo (MCMC) results in a simpler Gibbs sampling scheme for simulating posterior samples for statistical analysis.

2.4 Markov Chain Monte Carlo

Since higher-order integrals are often encountered in Bayesian analysis, the joint posterior distribution and the predictive distribution are always intractable. The MCMC algorithms [7] have become a popular way to obtain Bayesian solutions via stochastic simulation. Amongst the MCMC techniques, Gibbs sampling scheme [6, 8] remains the mostly used algorithm. It generates posterior realisations from the successively updated full conditional distributions and these realisations mimic a random sample from the intractable joint posterior distribution. Realisations of the predictive distributions can be simulated in a similar way. In this paper, we implement a Gibbs sampler in OpenBUGS, an open source software which conducts Bayesian inference using Gibbs sampling scheme. Vague and Noninformative priors are adopted for all model parameters to allow for objectivity in the analysis. The only exception is the degrees of freedom of the Student-t distribution has a truncated exponential prior distribution. In model implementation, we run a Gibbs sampler for 10,000 iterations a burn-in period of 1,000 iterations.

3 Empirical Study

This section presents an empirical study of claim amount data from two lines of automobile business of a major US insurer. The data are modelled by a bivariate two-way ANOVA model as described in Sect. 2.2 with different error distributions. Models are then compared and evaluated with Goodness-of-fit measurements and model selection criteria. Parameter estimates are also shown.

3.1 Data Set

In this empirical study, 10 years of incremental claim history (1988–1997) for 2 different lines of automobile business, namely the personal automobile and commercial automobile, are analysed. The run-off triangles are sourced from the Schedule P of the National Association of Insurance Commissioners' (NAIC) database. Subject to the willingness of insurers to disclose information, the availability of run-off triangles of this size from the same company across multiple lines of business is scarce. This dataset is also considered by other researchers in a different context (Shi and Frees [10]; Avanzi et al. [2]).

For simplicity, the claim amount data are log-transformed, i.e. $z_{ij} = \ln(y_{ij})$. Tables 2 and 3 display the data for the personal automobile and commercial automobile, respectively.

3.1.1 Parameter Estimation

The log-claim amount data in two run-off triangles are modelled by a bivariate two-way ANOVA model with fixed effects. Data in the run-off triangles are assumed to be dependent and independent, respectively under different error distributions to generate different results for comparison. Table 4 presents the parameter estimates and several conclusions can be made. Firstly, it is observed that the overall mean estimate for the personal automobile line is relatively consistent across the different models. The overall mean estimate for the commercial automobile from the heavy-tailed distributions tends to be larger than that of the Gaussian distribution.

Table 2 Log-transformed incremental paid losses for personal automobile

		1	2	3	4	5	6	7	8	9	10
		\multicolumn{10}{c}{Development Year}									
Accident Year	1	14.13	14.01	13.19	12.66	12.03	11.29	10.58	9.62	9.29	8.32
	2	14.27	14.18	13.39	12.74	12.15	11.25	10.46	9.75	9.43	
	3	14.38	14.25	13.43	12.81	12.08	11.27	10.77	10.14		
	4	14.39	14.22	13.40	12.68	12.00	11.35	10.89			
	5	14.43	14.27	13.33	12.61	12.08	11.57				
	6	14.49	14.23	13.27	12.76	12.38					
	7	14.53	14.17	13.36	12.98						
	8	14.54	14.19	13.50							
	9	14.61	14.23								
	10	14.61									

Table 3 Log-transformed incremental paid losses for commercial automobile

		1	2	3	4	5	6	7	8	9	10
		\multicolumn{10}{c}{Development Year}									
Accident Year	1	10.43	10.72	10.75	10.47	10.06	9.43	8.80	8.12	7.77	6.66
	2	10.54	10.85	10.62	10.29	9.45	9.32	8.66	8.35	7.55	
	3	10.61	10.94	10.94	10.39	9.67	9.04	8.87	7.03		
	4	10.61	10.81	10.58	10.09	9.48	9.44	7.98			
	5	10.52	10.84	10.44	10.14	9.87	8.65				
	6	10.62	10.88	10.60	10.59	8.80					
	7	10.96	11.13	11.36	9.80						
	8	11.03	11.79	9.95							
	9	11.63	10.41								
	10	10.53									

Secondly, the correlation parameter ρ are modelled as negative across the models, indicating a potential negative correlation between the random error terms of the personal and commercial automobile lines. Thirdly, the degrees of freedom parameter is estimated to be approximately 4 in the dependence case and approximately 3 in independence case, indicating the impact of modelling correlations. Moreover, it is possible that the assumption of the same degrees of freedom across different lines of business is restricting the model from capturing the actual degrees of freedom. In fact, a substantial difference in degrees of freedom exists between two lines of business when the triangles are modelled individually in a univariate setting.

Table 4 Posterior means of model parameters under six different two-way ANOVA models

Parameter	Dependent			Independent		
	N	t	VG	N	t	VG
μ_1	11.720	11.700	11.710	11.720	11.690	11.710
$\alpha_{1,1}$	−0.209	−0.189	−0.193	−0.209	−0.184	−0.191
$\alpha_{2,1}$	−0.118	−0.089	−0.092	−0.118	−0.081	−0.088
$\alpha_{3,1}$	−0.005	−0.009	−0.008	−0.005	−0.009	−0.011
$\alpha_{4,1}$	−0.042	−0.040	−0.041	−0.040	−0.041	−0.042
$\alpha_{5,1}$	−0.027	−0.026	−0.026	−0.027	−0.029	−0.028
$\alpha_{6,1}$	0.023	0.007	0.008	0.023	0.000	0.004
$\alpha_{7,1}$	0.050	0.029	0.035	0.049	0.025	0.029
$\alpha_{8,1}$	0.064	0.060	0.062	0.064	0.062	0.065
$\alpha_{9,1}$	0.096	0.090	0.089	0.096	0.092	0.091
$\alpha_{10,1}$	0.169	0.166	0.166	0.168	0.165	0.171
$\beta_{1,1}$	2.717	2.731	2.727	2.717	2.738	2.730
$\beta_{2,1}$	2.493	2.524	2.519	2.493	2.535	2.528
$\beta_{3,1}$	1.670	1.689	1.684	1.670	1.697	1.691
$\beta_{4,1}$	1.073	1.085	1.084	1.074	1.090	1.088
$\beta_{5,1}$	0.464	0.453	0.454	0.463	0.453	0.452
$\beta_{6,1}$	−0.296	−0.298	−0.301	−0.296	−0.301	−0.305
$\beta_{7,1}$	−0.955	−0.945	−0.947	−0.954	−0.945	−0.947
$\beta_{8,1}$	−1.778	−1.837	−1.821	−1.777	−1.849	−1.834
$\beta_{9,1}$	−2.193	−2.200	−2.201	−2.195	−2.199	−2.204
$\beta_{10,1}$	−3.195	−3.203	−3.198	−3.196	−3.219	−3.199
μ_2	9.186	9.290	9.263	9.185	9.321	9.287
$\alpha_{1,2}$	0.133	0.034	0.059	0.137	0.000	0.033
$\alpha_{2,2}$	0.031	−0.077	−0.060	0.034	−0.105	−0.080
$\alpha_{3,2}$	−0.034	−0.036	−0.027	−0.037	−0.040	−0.031
$\alpha_{4,2}$	−0.140	−0.178	−0.159	−0.141	−0.181	−0.161
$\alpha_{5,2}$	−0.152	−0.188	−0.180	−0.156	−0.192	−0.183
$\alpha_{6,2}$	−0.137	−0.093	−0.088	−0.133	−0.078	−0.083
$\alpha_{7,2}$	0.166	0.213	0.203	0.169	0.237	0.212
$\alpha_{8,2}$	0.157	0.289	0.240	0.152	0.307	0.241
$\alpha_{9,2}$	0.190	0.187	0.195	0.189	0.233	0.224
$\alpha_{10,2}$	−0.215	−0.150	−0.183	−0.213	−0.182	−0.172
$\beta_{1,2}$	1.563	1.417	1.454	1.562	1.385	1.427
$\beta_{2,2}$	1.720	1.656	1.666	1.724	1.623	1.643
$\beta_{3,2}$	1.466	1.453	1.450	1.470	1.435	1.440
$\beta_{4,2}$	1.087	1.078	1.082	1.087	1.068	1.076
$\beta_{5,2}$	0.420	0.430	0.426	0.422	0.435	0.428
$\beta_{6,2}$	0.028	0.036	0.046	0.024	0.040	0.047
$\beta_{7,2}$	−0.607	−0.590	−0.589	−0.610	−0.573	−0.574
$\beta_{8,2}$	−1.394	−1.205	−1.274	−1.397	−1.165	−1.235
$\beta_{9,2}$	−1.608	−1.604	−1.602	−1.609	−1.604	−1.604
$\beta_{10,2}$	−2.677	−2.671	−2.660	−2.674	−2.643	−2.648
df	–	4.770	4.806	–	3.625	3.584
ρ	−0.341	−0.228	−0.262	–	–	–

3.2 Model Comparison

For model evaluation purposes, two error measurements and one model selection criteria are used to choose the best model for the log-claim amount data. Error measurements compare different models by accounting the deviation of the estimation from the actual observations. In this paper, in-sample estimation errors (in the upper triangle) are measured by
(1) Mean absolute deviation (MAD):

$$MAD = \frac{1}{2n} \sum_{i=1, j=1}^{n,2} |z_{i,j} - \hat{z}_{i,j}|$$

and

(2) Mean absolute percentage error (MAPE):

$$MAPE = \frac{1}{2n} \sum_{i=1, j=1}^{n,2} \frac{|z_{i,j} - \hat{z}_{i,j}|}{z_{i,j}} \times 100\%$$

For model selection criteria, a measure of model adequacy and a measure of model complexity are traded off in order to avoid the problem of overfitting. In this paper, models are selected based on Bayesian information criteria (BIC)

$$BIC = -2 \ln(\hat{L}) + k \ln(n)$$

where $\hat{L}$ is the maximum value of the likelihood function, k is the number of model parameters and n is the number of observations in the model. The most competitive model is the one which has the smallest BIC value, MAD and MAPE. The results of model comparisons are given in Table 5.

From Table 5, it is observed that both the Student-t and VG distributions outperformed the Gaussian distribution. Under the dependence assumption, the Student-t distribution is slightly better than the VG distribution in all three model comparison criteria. It has the smallest $MAD = 0.1422$, $MAPE = 2.030\%$ and BIC (127.40) among the three models with dependent random errors. However, under

Table 5 Model comparisons of the six models using MAD, MAPE and BIC

	Dependence			Independence		
	N	t	VG	N	t	VG
MAD	0.1510	0.1422	0.1424	0.1507	0.1423	0.1414
MAPE (%)	2.161	2.030	2.033	2.157	2.032	2.019
BIC	160.01	127.40	130.36	161.62	112.75	119.59

the independence assumption, the Student-t distribution defects the VG distribution only in the BIC while the VG distribution has the smallest $MAD = 0.1414$ and $MAPE = 2.019\%$.

It is observed that the VG and Student-t distributions yield similar results. Some may argue that they are indifferent from each other. However, the performance of the distribution largely depends on the characteristics of the data set. In fact, in order to investigate the nature of the individual triangles, the models are run individually on a univariate basis. It is found that the VG distribution has a better fit in the personal automobile line while the Student-t distribution has a better fit in the commercial automobile line. Therefore, applying bivariate distributions that restrict the same marginal distribution in all lines of business may compromise in model accuracy as it is possible that the characteristic of individual lines of business is substantially different from each other. Nevertheless, the focus of this study is on the robustness of the models. It is evident that the Gaussian error distribution is not robust to outliers whereas heavy-tailed distributions are better at capturing the tail behaviour.

Another observation from the goodness-of-fit measurement is that the independent models performed relatively better than the dependent models. Yet, the independent Student-t model which is the second best independent model, did not outperformed other dependent models. This may suggest moderate correlations which is discussed in the next section.

4 Estimation of Future Liabilities

4.1 Reserve Forecasts

The ultimate objective of loss reserving analysis is to forecast future liabilities so that sufficient fund can be allocated to the reserve to settle future payments. For the personal automobile and commercial automobile lines, outstanding claim amounts are the exponent of the unobserved values in the low triangle. Therefore, the loss reserve forecast of a line of business is the sum of all outstanding claim amounts in the lower triangle. From an accounting or risk management point of view, apart from the total loss reserve, it is also the interest of reserve actuaries to forecasts loss reserve for next calendar year, which is the sum of claim amounts in the first off-diagonal.

A major advantage of using the Bayesian approach for statistical analysis and predictive analysis is that the Gibbs sampling scheme can easily simulate the outstanding claim amounts and the predictive distributions become accessible. The posterior mean, median, standard deviation, percentiles, credible interval and other risk measures of the next year loss reserve and total reserve can be obtained from the Gibbs output.

Tables 6 and 7 presents the posterior mean and standard deviation of the total reserve and next calendar year reserve (in million dollars), respectively. Since the data were recorded from 1988 to 1997, the next calendar year is 1998.

Table 6 Total reserve forecasts for the six models

Reserve	N		t		VG	
	Mean	Std Dev	Mean	Std Dev	Mean	Std Dev
Dependent						
Personal auto	6.4800	0.4634	6.4230	0.5023	6.4410	0.4966
Commercial auto	0.4437	0.0996	0.5348	0.1696	0.5011	0.1304
Total reserve	6.9240	0.4431	6.9580	0.5146	6.9420	0.5054
Independent						
Personal auto	6.4770	0.4578	6.4020	0.5211	6.4610	0.4774
Commercial auto	0.4408	0.0927	0.5584	0.2234	0.5190	0.1344
Total reserve	6.9170	0.4685	6.9600	0.5696	6.9800	0.5023

Table 7 Next calendar year reserves for the six models

Reserve	N		t		VG	
	Mean	Std Dev	Mean	Std Dev	Mean	Std Dev
Dependent						
Personal auto	3.2700	0.2388	3.2710	0.2589	3.2750	0.2568
Commercial auto	0.1766	0.0356	0.2090	0.0633	0.1964	0.0473
Total NY reserve	3.4470	0.2309	3.4800	0.2619	3.4720	0.2588
Independent						
Personal auto	3.2680	0.2360	3.2690	0.2695	3.2940	0.2477
Commercial auto	0.1758	0.0333	0.2163	0.0810	0.2018	0.0478
Total NY reserve	3.4440	0.2388	3.4850	0.2822	3.4960	0.2550

From Tables 6 and 7, we have the following observations. Firstly, the posterior means of the reserves are close to one another in all models. This is in consistent with the close estimates of overall mean across different models as shown in Table 4. Secondly, reserve predictions from heavy-tailed models are larger than those from the Gaussian distribution. This means that models with heavy-tailed error distributions are more conservative in loss reserving, protecting the insurance company against insolvency. On the contrary, models with Gaussian error distribution may lead to the underestimation of the loss reserves. Thirdly, the standard deviations of the entire portfolio are smaller than that of the individual lines of business. It is a direct result of the diversification of risks, suggesting a negative correlation coefficient for the random errors in the two-way ANOVA models for the personal automobile and commercial automobile claim amount data. This is also in consistent with the parameter estimates shown in Table 4.

Table 8 Value-at-Risks of loss reserves under different models

Personal automobile

	Dependent			Independent		
	N	t	VG	N	t	VG
Percentile	Total reserve					
70th	6.6950	6.6300	6.6523	6.6950	6.5970	6.6680
90th	7.0840	7.0330	7.0351	7.0800	6.9630	7.0630
95th	7.2791	7.2730	7.2821	7.2610	7.1780	7.2731
99th	7.6890	7.8550	7.8620	7.6340	7.8041	7.7690
Percentile	Next calendar year reserve					
70th	0.4784	0.5829	0.5413	0.4754	0.6056	0.5634
90th	0.5145	0.6353	0.5828	0.5089	0.6553	0.6097
95th	0.5371	0.6702	0.6103	0.5295	0.6893	0.6404
99th	0.6629	0.8645	0.7756	0.6454	0.8704	0.8099

Commercial automobile

Percentile	Total reserve					
70th	3.3810	3.3750	3.3810	3.3770	3.3660	3.4010
90th	3.5820	3.5870	3.5830	3.5750	3.5580	3.6050
95th	3.6791	3.7111	3.7100	3.6830	3.6720	3.7140
99th	3.8970	3.9990	4.0110	3.8670	4.0090	3.9800
Percentile	Next calendar year reserve					
70th	0.1894	0.2268	0.2115	0.1888	0.2344	0.2179
90th	0.2219	0.2757	0.2499	0.2187	0.2804	0.2609
95th	0.2393	0.3063	0.2761	0.2357	0.3062	0.2882
99th	0.2870	0.3968	0.3431	0.2758	0.3918	0.3483

4.2   Implications in Risk Management Practice

Information regarding the variability of the loss reserve forecast has important implications in decision making of capital allocation. The predictive variability enables
reserve actuaries to have greater insight into the reasonable range of capital to be
set aside for outstanding claims and the risk margin. As such, considering the variety of risk management tools that can be implemented, a predictive distribution of
loss reserve from a stochastic model is more preferred by reserve actuaries compared to merely a point estimate obtained form a deterministic model. This section
computes the risk measures, Value-at-Risk (VaR) and Tail Conditional Expectation
(TCE), of loss reserves under different two-way ANOVA models for the personal
and commercial automobile lines.

Risk margin is the amount of capital to be held in addition to a point estimate for
prudential risk management purposes. VaR is a measure of risk that can be interpreted
as the maximum loss that can occur with $(1 - \alpha)$ confidence over a certain period.

Table 9 Tail conditional expectation of loss reserves under different models

Total reserve

TVaR	Dependence			Independence		
	N	t	VG	N	t	VG
Personal automobile						
90%	7.3559	7.3814	7.3951	7.3339	7.3695	7.3677
95%	7.5363	7.6252	7.6461	7.5057	7.6800	7.5757
99%	7.9285	8.1975	8.2607	7.9021	8.7643	8.0724
Commercial automobile						
90%	0.6492	0.8733	0.7685	0.6310	0.9341	0.7970
95%	0.7075	0.9960	0.8579	0.6824	1.1084	0.8818
99%	0.8619	1.4613	1.1504	0.8065	2.0421	1.1444

Next calendar year reserve

Personal automobile						
90%	3.7210	3.7686	3.7682	3.7119	3.7705	3.7672
95%	3.8126	3.8958	3.8994	3.8010	3.9322	3.8795
99%	4.0138	4.2040	4.2151	4.0034	4.5036	4.1450
Commercial automobile						
90%	0.2497	0.3345	0.2937	0.2437	0.3536	0.3013
95%	0.2695	0.3796	0.3257	0.2611	0.4155	0.3297
99%	0.3218	0.5501	0.4297	0.3040	0.7468	0.4143

In fact, the VaR(α) is simply the $100(1 - \alpha)$th percentile of the predictive distribution. In addition to the commonly used 90th, 95th and 99th percentile, the 70th percentile is also considered as a risk margin in the insurance industry. Table 8 displays the Value-at-Risks (in million US dollars) at various levels for the total reserves and next calendar year reserves in the two lines of automobile business.

From Table 8, both Student-t distribution and the VG distribution provide more conservative estimates, whereas the Gaussian distribution reserves substantially less than those of the heavy-tailed distributions. The difference between the Gaussian distribution and heavy-tailed distributions is amplified with higher VaRs in the next calendar year reserve.

Another risk measure that is commonly used in the industry and often demanded in regulatory requirements is the TCE which is the expected loss given the occurrence of an extreme event beyond a certain probability $(1 - \alpha)$. In the context, the $100(1 - \alpha)\%$ TCE is the expected loss reserve given that this reserve is greater than the $100(1 - \alpha)\%$ VaR. Table 9 exhibits the 90th, 95th and 99th TCE values (in million US dollars) of the reserves under different two-way ANOVA models.

From Table 9, it is observed that the dependence assumption in the models tends to produce larger TCEs. Similar to VaRs, the Student-t and VG distributions produce more conservative risk measures than the Gaussian distribution. In this case, the difference in the TCE is more obvious between the Gaussian and heavy-tailed distributions.

5 Conclusion

In this paper, we provide a robust inference in the analysis of loss reserve data in two lines of insurance business. The proposed methodology can be straightforwardly applied to multiple lines of business. Although the Gaussian distribution is the most widely used error distribution in many statistical modelling, being sensitive to outliers and unable to provide a robust analysis are its main defects. In the forecast of future liabilities, our empirical study shows that the Gaussian distribution provides a less conservative forecast than its two rivals and this may result in a high risk of insolvency. Moreover, the Gaussian distribution may also overestimate the linear correlation between two lines of business, falling into the trap of correlation illusion. On the contrary, the heavy-tailed Student-t and VG distributions suggest no linear correlation between the two lines of business. In fact, these two distributions are shown to be superior to the Gaussian distribution in terms of estimation errors and model selection criteria. Risk measures and risk margins can be evaluated from the full predictive distribution of loss reserve accordingly with the use of Bayesian MCMC methods. Finally, we make no claim that we have found the best model for the claim reserve data in this paper. Analysis on the individual lines of business suggests the VG distribution fits the personal automobile loss data better while the Student-t distribution fits the commercial automobile loss data better. Our work triggers the further study of more advanced models for loss reserving.

References

1. Andrews DF, Mallows CL (1974) Scale mixtures of normal distributions. J Roy Stat Soc Ser B (Methodol) 99–102
2. Avanzi B, Taylor G, Wong B (2016) Correlations between insurance lines of business: an illusion or a real phenomenon? Some methodological considerations. Astin Bull 46(02):225–263
3. Choy STB, Chan JS (2008) Scale mixtures distributions in statistical modelling. Aust New Zealand J Stat 50(2):135–146
4. De Vylder F (1978) Estimation of IBNR claims by least squares. Bull Assoc. Swiss Actuaries 78:247–254
5. England PD, Verrall RJ (2002) Stochastic claims reserving in general insurance. Br Actuarial J 8(03):443–518
6. Gelfand AE, Smith AF (1990) Sampling-based approaches to calculating marginal densities. J Am Statis Assoc 85(410):398–409
7. Gilks WR, Richardson S, Spiegelhalter D (1995) Markov Chain Monte Carlo in practice. Chapman & Hall/CRC Interdisciplinary Statistics
8. Jacquier E, Jarrow R (2000) Bayesian analysis of contingent claim model error. J Econometrics 94(1):145–180
9. Kremer E (1982) IBNR-claims and the two-way model of ANOVA. Scand Actuarial J 1982(1):47–55
10. Shi P, Frees EW (2011) Dependent loss reserving using copulas. Astin Bull 41(02):449–486
11. Xie Z (2013) Risk and Regulation-a broader view on their consistency. Ann Actuarial Sci 7(2):169

On Conditioning in Multidimensional Probabilistic Models

Radim Jiroušek

Abstract Graphical Markov models, and above all Bayesian networks have become a very popular tool for multidimensional probability distribution representation and processing. The technique making computation with several hundred dimensional probability distribution possible was suggested by Lauritzen and Spiegelhalter. However, to employ it one has to transform a Bayesian network into a *decomposable model*. This is because decomposable models (or more precisely their building blocks, i.e., their low-dimensional marginals) can be reordered in many ways, so that each variable can be placed at the beginning of the model. It is not difficult to show that there is a much wider class of models possessing this property. In compositional models theory we call these models flexible. It is the widest class of models for which one can always restructure the model in the way that any variable can appear at the beginning of the model. But until recently it had been an open problem whether this class of models is closed under conditioning; i.e., whether a conditional of a flexible model is again flexible. In the paper we will show that this property holds true, which proves the importance of flexible models for practical applications.

1 Introduction

Graphical Markov models [12] have become a very popular way for the representation of multidimensional probability distributions. This is not only because these models efficiently represent multidimensional models with a reasonable number of parameters, but mainly because one can get a ready-made software that makes the application of graphical models to practical problems easier [14, 22]. The first algorithms for knowledge propagation (conditioning) in singly connected Markov and Bayesian networks were proposed by Judea Pearl [17], who also suggested

R. Jiroušek (✉)
Faculty of Management, University of Economics, Prague,
Jarošovská 1117/II, 377 01 Jindřich Hradec, Czech Republic
e-mail: radim@utia.cas.cz

© Springer International Publishing AG 2017
V. Kreinovich et al. (eds.), *Robustness in Econometrics*,
Studies in Computational Intelligence 692, DOI 10.1007/978-3-319-50742-2_12

collapsing variables into a single node to cope with "loops" (for historical notes see [18]). In a way, this idea appeared later also in the Lauritzen Spiegelhalter method (discussed below in more details).

In our best knowledge the only computational processes performing the computations directly with Bayesian networks, which are not based on simulations, are those by Ross Shachter [19, 20]. His procedures, based on two rules called *node deletion* and *edge reversal*, were designed to realize both basic procedures, of which all computations with probabilistic models consists: *marginalization* and *conditioning*. Regarding marginalization, Shachter's node deletion rule is based on the idea that some variables may easily be deleted because of their special position in the model. In fact, this property holds also for other graphical models. For Bayesian networks it is about variables having no children in the respective acyclic digraph. For decomposable models it is about variables whose nodes are *simplicial* in the respective chordal graph. Similarly, when considering non-graphical approaches it is about variables appearing only in one of the building blocks of the model. We do not go into details here, because when speaking about computational procedures in this paper we will primarily have in mind the other computational procedures, those realizing computation of conditional distributions.[1]

Naturally, not all graphical models are equally convenient for computation of conditionals. Undoubtedly, from the mentioned point of view the most advantageous are decomposable models, for which Lauritzen and Spiegelhalter [13] suggested a computational process realizing computation of conditionals locally. This process takes advantage of the following two facts. First, the building blocks of decomposable models, i.e., their low-dimensional marginals, can be ordered to meet so called Running Intersection Property (RIP). This is the very property that makes the process of computation of conditionals local (the process does not need any auxiliary space, the computations can be performed in the space required for the model representation). Second advantageous property is that there are many such RIP orderings; for each variable one can always find an ordering, in which the selected variable can be placed at the beginning of the model. And this second property makes computations of all conditionals possible. And this is the property that will be studied in this paper in detail.

To make our considerations more general, we will not deal with a specific class of graphical models. Instead, we will consider a general class of models that factorizes into a system of its marginals (or generally, into a system of general low-dimensional distributions), i.e., so called *compositional models*. Therefore, the next section introduces notation and basic concepts necessary to describe compositional models and the rest of this paper is organized as follows. Section 3 is devoted to the presentation of principles making the local computations with decomposable models possible. Introduction of flexible models, as well as new results enabling us to prove the main assertion of this paper are presented in Sect. 4. The last section

---

[1] The reader interested in marginalization procedures is referred to algorithms designed for computations with compositional models either in Malvestuto's papers [15, 16] and/or in [2].

concludes the paper referring to the relation of the presented results with other research areas and presenting an open problem.

2 Basic Notions and Notation

In this text we use almost the same notation as in [9] presented at the 6th International Conference of the Thailand Econometric Society in 2013. We deal with a finite system of finite-valued random variables N. For $u \in N$, $\mathbb{X}_u$ denote the respective finite (nonempty) set of values of variable u. The set of all combinations of the considered values will be denoted $\mathbb{X}_N = \times_{u \in N} \mathbb{X}_u$. Analogously, for $K \subset N$, $\mathbb{X}_K = \times_{u \in K} \mathbb{X}_u$.

Distributions of the considered variables will be denoted by Greek letters (π, κ, λ, μ, δ with possible indices); thus $\pi(K)$ denote a $|K|$-dimensional distribution and $\pi(x)$ a value of probability distribution π for point $x \in \mathbb{X}_K$. Its *marginal distribution* for $J \subseteq K$ will be denoted $\pi^{\downarrow J}$. Analogously, $x^{\downarrow J}$ denote the *projection* of $x \in \mathbb{X}_K$ into $\mathbb{X}_J$. When computing marginal distributions we do not exclude situations when $J = \emptyset$. In this case, naturally, $\pi^{\downarrow \emptyset} = 1$.

Consider a distribution $\pi(N)$, and three disjoint subset $K, L, M \subset N$, $K \neq \emptyset$, $L \neq \emptyset$. We say that for distribution π variables K and L are *conditionally independent given variables M*, if for all $x \in \mathbb{X}_{K \cup L \cup M}$

$$\pi^{\downarrow K \cup L \cup M}(x) \cdot \pi^{\downarrow M}(x^{\downarrow M}) = \pi^{\downarrow K \cup M}(x^{\downarrow K \cup M}) \cdot \pi^{\downarrow L \cup M}(x^{\downarrow L \cup M}).$$

This independence will be denoted $K \perp\!\!\!\perp L | M[\pi]$.

Two distributions $\kappa(K)$ and $\lambda(L)$ are said to be *consistent* if their joint marginals coincide: $\kappa^{\downarrow K \cap L} = \lambda^{\downarrow K \cap L}$.

Having two distributions defined for the same set of variables $\pi(K)$ and $\kappa(K)$, we say that κ *dominates* π (in symbol $\pi \ll \kappa$) if for all $x \in \mathbb{X}_K$

$$\kappa(x) = 0 \implies \pi(x) = 0.$$

2.1 Operator of Composition

Prior to introducing compositional models we have to present a formal definition of the operator of composition.

Definition 1 For arbitrary two distributions $\kappa(K)$ and $\lambda(L)$, for which $\kappa^{\downarrow K \cap L} \ll \lambda^{\downarrow K \cap L}$ their *composition* is for each $x \in \mathbb{X}_{(L \cup K)}$ given by the following formula

$$(\kappa \triangleright \lambda)(x) = \frac{\kappa(x^{\downarrow K})\lambda(x^{\downarrow L})}{\lambda^{\downarrow K \cap L}(x^{\downarrow K \cap L})}.$$

In case that $\kappa^{\downarrow K \cap L} \not\ll \lambda^{\downarrow K \cap L}$ the composition remains undefined.

The following assertion summarizes the basic properties of this operator. The respective proofs can be found in [6, 7].

Theorem 1 *Suppose* $\kappa(K)$, $\lambda(L)$ *and* $\mu(M)$ *are probability distributions. The following statements hold under the assumption that the respective compositions are defined:*

1. (Domain): $\kappa \triangleright \lambda$ *is a probability distribution for* $K \cup L$.
2. (Composition preserves first marginal): $(\kappa \triangleright \lambda)^{\downarrow K} = \kappa$.
3. (Reduction): *If* $L \subseteq K$ *then,* $\kappa \triangleright \lambda = \kappa$.
4. (Extension): *If* $M \subseteq K$ *then,* $\kappa^{\downarrow M} \triangleright \kappa = \kappa$.
5. (Perfectization): $\kappa \triangleright \lambda = \kappa \triangleright (\kappa \triangleright \lambda)^{\downarrow L}$.
6. (Commutativity under consistency): *In general,* $\kappa \triangleright \lambda \neq \lambda \triangleright \kappa$, *however,* κ *and* λ *are consistent if and only if* $\kappa \triangleright \lambda = \lambda \triangleright \kappa$.
7. (Associativity under RIP): *In general,* $(\kappa \triangleright \lambda) \triangleright \mu \neq \kappa \triangleright (\lambda \triangleright \mu)$. *However, if* $K \supset (L \cap M)$, *or* $L \supset (K \cap M)$ *then,* $(\kappa \triangleright \lambda) \triangleright \mu = \kappa \triangleright (\lambda \triangleright \mu)$.
8. (Exchangeability): *If* $K \supset (L \cap M)$ *then,* $(\kappa \triangleright \lambda) \triangleright \mu = (\kappa \triangleright \mu) \triangleright \lambda$.
9. (Simple marginalization): *If* $(K \cap L) \subseteq M \subseteq K \cup L$ *then,* $(\kappa \triangleright \lambda)^{\downarrow M} = \kappa^{\downarrow K \cap M} \triangleright \lambda^{\downarrow L \cap M}$.
10. (Conditional independence): $(K \setminus L) \perp\!\!\!\perp (L \setminus K)|(K \cap L)[\kappa \triangleright \lambda]$.
11. (Factorization): *Let* $M \supseteq K \cup L$. $(K \setminus L) \perp\!\!\!\perp (L \setminus K)|(K \cap L)[\mu]$ *if and only if* $\mu^{\downarrow K \cup L} = \mu^{\downarrow K} \triangleright \mu^{\downarrow L}$.

When computing conditionals we will need a degenerate one-dimensional distribution expressing certainty. Consider variable u and its value $\mathbf{a} \in \mathbb{X}_u$. The distribution $\delta_{\mathbf{a}}(u)$ expressing for certain that variable $u = \mathbf{a}$ is defined for each $x \in \mathbb{X}_u$ as

$$\delta_{\mathbf{a}}(x) = \begin{cases} 1, & \text{if } x = \mathbf{a}; \\ 0, & \text{otherwise}. \end{cases}$$

For the proof of the following assertion showing how to compute conditional distributions see Theorem 2.3 in [3].

Theorem 2 *Consider a distribution* $\kappa(K)$, *variable* $u \in K$, *its value* $\mathbf{a} \in \mathbb{X}_u$, *and* $L \subseteq K \setminus \{u\}$. *If* $\kappa^{\downarrow \{u\}}(\mathbf{a}) > 0$, *then the corresponding conditional distribution* $\kappa(L|u = \mathbf{a})$ *can be computed*

$$\kappa(L|u = \mathbf{a}) = (\delta_{\mathbf{a}}(u) \triangleright \kappa)^{\downarrow L}.$$

2.2 Compositional Models

To enable the reader understanding of the following text without necessity to look up the respective parts from the previous papers, we briefly define a compositional model and present two basic properties, proofs of which can be found in [7]. To simplify the exposition let us first make the following conventions.

In the rest of the paper we will consider a system of n oligodimensional distributions $\kappa_1(K_1), \kappa_2(K_2), \ldots, \kappa_n(K_n)$. Therefore, whenever we speak about distribution κ_k, we will assume it is a distribution of variables K_k, i.e., $\kappa_k(K_k)$. Moreover, whenever we use the operator of composition, we will assume that the composed distributions meet the assumption from Definition 1, and therefore we assume that the corresponding composition is defined. Thus, formulas $\kappa_1 \triangleright \kappa_2 \triangleright \ldots \triangleright \kappa_n$ will always be defined and determine the distributions of variables $K_1 \cup K_2 \cup \ldots \cup K_n$.

To avoid necessity to write too many parentheses in the formulas, let us accept a convention that we will apply the operators from left to right. Thus

$$\kappa_1 \triangleright \kappa_2 \triangleright \kappa_3 \triangleright \ldots \triangleright \kappa_n = (\ldots((\kappa_1 \triangleright \kappa_2) \triangleright \kappa_3) \triangleright \ldots \triangleright \kappa_n),$$

and the parentheses will be used only when we will want to change this default ordering.

Definition 2 A compositional model $\kappa_1 \triangleright \kappa_2 \triangleright \ldots \triangleright \kappa_n$ is said to be *perfect* if

$$\kappa_1 \triangleright \kappa_2 = \kappa_2 \triangleright \kappa_1,$$
$$\kappa_1 \triangleright \kappa_2 \triangleright \kappa_3 = \kappa_3 \triangleright (\kappa_1 \triangleright \kappa_2),$$
$$\vdots$$
$$\kappa_1 \triangleright \kappa_2 \triangleright \ldots \triangleright \kappa_n = \kappa_n \triangleright (\kappa_1 \triangleright \ldots \triangleright \kappa_{n-1}).$$

Theorem 3 *A compositional model $\kappa_1 \triangleright \kappa_2 \triangleright \ldots \triangleright \kappa_n$ is perfect if and only if all distributions κ_k ($k = 1, \ldots, n$) are the marginals of the distribution $(\kappa_1 \triangleright \kappa_2 \triangleright \ldots \triangleright \kappa_n)$.*

The following assertion, which is in fact a generalization of Property 5 of Theorem 1, shows that each compositional model can be transformed into a perfect one.

Theorem 4 *For a compositional model $\kappa_1 \triangleright \kappa_2 \triangleright \ldots \triangleright \kappa_n$ the sequence $\mu_1, \mu_2, \ldots, \mu_n$ computed by the following process*

$$\mu_1 = \kappa_1,$$
$$\mu_2 = \mu_1^{\downarrow K_2 \cap K_1} \triangleright \kappa_2,$$

$$\mu_3 = (\mu_1 \triangleright \mu_2)^{\downarrow K_3 \cap (K_1 \cup K_2)} \triangleright \kappa_3,$$

$$\vdots$$

$$\mu_n = (\mu_1 \triangleright \ldots \triangleright \mu_{n-1})^{\downarrow K_n \cap (K_1 \cup \ldots \cup K_{n-1})} \triangleright \kappa_n$$

defines a perfect model, for which

$$\kappa_1 \triangleright \ldots \triangleright \kappa_n = \mu_1 \triangleright \ldots \triangleright \mu_n.$$

3 Decomposable Models and Local Computations

The purpose of this short section is to explain which properties of decomposable model make local computations possible [5], and why we are interested in a class of flexible models that will be introduced in the next section.

Definition 3 We call a compositional model $\kappa_1 \triangleright \kappa_2 \triangleright \ldots \triangleright \kappa_n$ *decomposable* if the corresponding sequence of variable sets $K_1, K_2, \ldots, K_n$ meets the *running intersection property* (RIP), i.e., if

$$\forall i = 2, \ldots, n \ \exists j (1 \leq j < i) \left(K_i \cap \left(\bigcup_{k=1}^{i-1} K_k \right) \subseteq K_j \right).$$

The important properties of decomposable models are expressed by the following two lemmas. The first one follows from the existence of a join tree [1], the proof of the latter one can be found in [7],

Lemma 1 *If $K_1, \ldots, K_n$ meets RIP, then for each $\ell \in \{1, \ldots, n\}$ there exists a permutation $i_1, \ldots, i_n$ such that $K_{i_1} = K_\ell$, and $K_{i_1}, K_{i_2}, \ldots, K_{i_n}$ meets RIP.*

Lemma 2 *If $\kappa_1, \ldots, \kappa_n$ is a sequence of pairwise consistent probability distributions such that $K_1, \ldots, K_n$ meets RIP, then the compositional model $\kappa_1 \triangleright \kappa_2 \triangleright \ldots \triangleright \kappa_n$ is perfect.*

Let us explain general principles why the mentioned properties guarantee the possibility to compute conditionals locally (for more details see [13], or e.g. [8], where the reasoning is based on compositional models). Due to Theorem 2, the computation of conditionals from a model $\kappa_1 \triangleright \ldots \triangleright \kappa_n$ means to compute

$$\delta_{\mathbf{a}}(u) \triangleright (\kappa_1 \triangleright \kappa_2 \triangleright \ldots \triangleright \kappa_n).$$

This is an easy task in case that $u \in K_1$, because, in this case,

$$\delta_{\mathbf{a}}(u) \triangleright (\kappa_1 \triangleright \kappa_2 \triangleright \ldots \triangleright \kappa_n) = (\delta_{\mathbf{a}}(u) \triangleright \kappa_1) \triangleright \kappa_2 \triangleright \ldots \triangleright \kappa_n. \tag{1}$$

This equality can be proven by the multiple application of Property 7 of Theorem 1 (in more details it will be shown in the proof of Theorem 7). Now, consider the application of the perfectization procedure from Theorem 4 to the right hand side of Eq. (1). The reader can notice that if the sequence $K_1, K_2, \ldots, K_n$ meets RIP, then $K_3 \cap (K_1 \cup K_2)$ equals either $K_3 \cap K_1$ or $K_3 \cap K_2$. Similarly, $K_4 \cap (K_1 \cup K_2 \cup K_3)$ equals $K_4 \cap K_j$ for some $j \leq 3$. Formally, thanks to RIP we can define a function

$$f : \{3, 4, \ldots, n\} \longrightarrow \{1, 2, \ldots, n-1\},$$

such that for all $i = 3, \ldots, n$

$$f(i) < i, \text{ and } K_{f(i)} \supseteq K_i \cap (K_1 \cup K_2 \cup \ldots \cup K_{i-1}).$$

It means that for all $i = 3, 4, \ldots, n$ the marginal distributions necessary in the perfectization process

$$
\begin{aligned}
\mu_1 &= \delta_\mathbf{a}(u) \triangleright \kappa_1, \\
\mu_2 &= \mu_1^{\downarrow K_2 \cap K_1} \triangleright \kappa_2, \\
\mu_3 &= (\mu_1 \triangleright \mu_2)^{\downarrow K_3 \cap (K_1 \cup K_2)} \triangleright \kappa_3 = \mu_{f(3)}^{\downarrow K_3 \cap K_{f(3)}} \triangleright \kappa_3, \\
&\vdots \\
\mu_n &= (\mu_1 \triangleright \ldots \triangleright \mu_{n-1})^{\downarrow K_n \cap (K_1 \cup \ldots \cup K_{n-1})} \triangleright \kappa_n = \mu_{f(n)}^{\downarrow K_n \cap K_{f(n)}} \triangleright \kappa_n
\end{aligned}
\tag{2}
$$

can be computed from $\mu_{f(i)}$ as $\mu_{f(i)}^{\downarrow K_i \cap K_{f(i)}}$, because $\mu_1 \triangleright \ldots \triangleright \mu_{i-1}$ is a perfect model and therefore $\mu_{f(i)}$ is marginal to $\mu_1 \triangleright \ldots \triangleright \mu_{i-1}$. All this means that for decomposable models the process of perfectization can be performed locally, and that the resulting perfect model $\mu_1 \triangleright \ldots \triangleright \mu_n$ is decomposable. The last statement follows from Lemma 2 because $\mu_1, \ldots, \mu_n$ are defined for $K_1, \ldots, K_n$, which meet RIP, and all μ_i are pairwise consistent because they are marginals of $\mu_1 \triangleright \ldots \triangleright \mu_n$ (Theorem 3).

From the point of view of the following section it is important to realize that the process described by Expressions (2) can always be performed, though for general (non-decomposable) models this process may be computationally more space-demanding. The question to be answered is whether one gets a possibility to consider a wider class of probability distributions when giving up the requirement that the computations must be local.

4 Flexible Models

Regardless the fact whether the model is decomposable or not, Equality (1) holds in case that $u \in K_1$. It means that computations of conditionals is algorithmically simple for conditioning by variables from K_1. Therefore, in this section we are interested in models that can be equivalently expressed in many ways, so that each variable may appear among the arguments of the first distribution.

Definition 4 A model $\kappa_1 \rhd \kappa_2 \rhd \ldots \rhd \kappa_n$ is called *flexible* if for each $u \in K_1 \cup \ldots \cup K_n$ there exists a permutation $i_1, i_2, \ldots, i_n$ such that $u \in K_{i_1}$ and

$$\kappa_{i_1} \rhd \kappa_{i_2} \rhd \ldots \rhd \kappa_{i_n} = \kappa_1 \rhd \kappa_2 \rhd \ldots \rhd \kappa_n.$$

Lemmas 1 and 2 say that a decomposable model consisting of pairwise consistent distribution is flexible. Therefore, any decomposable model can be transformed into a flexible one by the application of the perfectization procedure from Theorem 4. Nevertheless, it is important to realize that flexibility, in contrast to decomposability, is not a structural property [11]. Decomposability is a property of a sequence of sets $K_1, K_2, \ldots, K_n$, and any compositional model with a sequence of variable sets $K_1, K_2, \ldots, K_n$ meeting RIP is decomposable. Contrarily, for any sequence of variable sets $K_1, K_2, \ldots, K_n$ one can find a compositional model that is flexible. A trivial example confirming this assertion is a model $\kappa_1 \rhd \kappa_2 \rhd \ldots \rhd \kappa_n$, where all distributions κ_i are uniform. In this case $\kappa_1 \rhd \kappa_2 \rhd \ldots \rhd \kappa_n$ is a uniform multidimensional distribution regardless the ordering of the distributions in the model.

To illustrate their properties, let us present two nontrivial examples of flexible models.

4.1 Examples of Flexible Models

Circle. Consider a compositional model composed of two-dimensional distributions

$$\kappa_1(u_1, u_2) \rhd \kappa_2(u_2, u_3) \rhd \ldots \rhd \kappa_{n-1}(u_{n-1}, u_n) \rhd \kappa_n(u_n, u_1).$$

If the considered distributions are pairwise consistent, then this model is flexible, because of Property 3 of Theorem 1,

$$\kappa_1(u_1, u_2) \rhd \kappa_2(u_2, u_3) \rhd \ldots \rhd \kappa_{n-1}(u_{n-1}, u_n) \rhd \kappa_n(u_n, u_1)$$
$$= \kappa_1(u_1, u_2) \rhd \kappa_2(u_2, u_3) \rhd \ldots \rhd \kappa_{n-1}(u_{n-1}, u_n),$$

which shows that the model in the right hand side of this expression is in fact decomposable. Realize, however, that from the given set of pairwise consistent two-dimensional distributions we can set up a number of other (generally different) flexible models. For example, cyclic re-orderings

$$\kappa_i(u_i, u_{i+1}) \rhd \ldots \rhd \kappa_n(u_n, u_1) \rhd \kappa_1(u_1, u_2) \rhd \ldots \rhd \kappa_{i-1}(u_{i-1}, u_i)$$

yield other $n - 1$ decomposable, and thus flexible models. In fact any permutation of these distributions bring forth a flexible model. When considering $n = 8$ as in Fig. 1, taking all their permutations into account, one gets 46 of possibly different models, and all of them are flexible (because all of them are decomposable).

Two channel transmission. The flexible models from the preceding example were decomposable. To see that not all flexible models are, or can be easily transformed into decomposable ones, consider the following compositional model consisting of four distributions (see Fig. 2)

$$\kappa_1(u_1, u_2, u_3) \triangleright \kappa_2(u_2, v_2) \triangleright \kappa_3(u_3, v_3) \triangleright \kappa_4(v_1, v_2, v_3).$$

Further assume that $\mathbb{X}_{u_2} = \mathbb{X}_{v_2}$, and $\mathbb{X}_{u_3} = \mathbb{X}_{v_3}$, and that the distributions κ_2 and κ_3 realize noiseless duplex transmission:

$$\kappa_2(x, y) > 0 \text{ if and only if } x = y,$$
$$\kappa_3(x, y) > 0 \text{ if and only if } x = y.$$

Fig. 1 Eight two-dimensional distributions forming a cycle

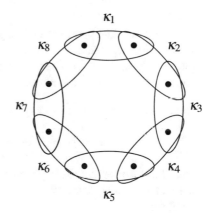

Fig. 2 Six-dimensional flexible model

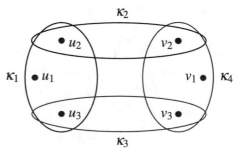

In case that distributions $\kappa_1, \kappa_2, \kappa_3$ and κ_4 are pairwise consistent, it is not difficult to show that

$$\kappa_1(u_1, u_2, u_3) \triangleright \kappa_2(u_2, v_2) \triangleright \kappa_3(u_3, v_3) \triangleright \kappa_4(v_1, v_2, v_3)$$
$$= \kappa_4(v_1, v_2, v_3) \triangleright \kappa_2(u_2, v_2) \triangleright \kappa_3(u_3, v_3) \triangleright \kappa_1(u_1, u_2, u_3), \quad (3)$$

which means that the model is flexible. To show that Equality (3) holds it is enough to use Properties 6 and 8 of Theorem 1:

$$\kappa_1(u_1, u_2, u_3) \triangleright \kappa_2(u_2, v_2) \triangleright \kappa_3(u_3, v_3) \triangleright \kappa_4(v_1, v_2, v_3)$$
$$= \kappa_2(u_2, v_2) \triangleright \kappa_1(u_1, u_2, u_3) \triangleright \kappa_3(u_3, v_3) \triangleright \kappa_4(v_1, v_2, v_3)$$
$$= \kappa_2(u_2, v_2) \triangleright \kappa_3(u_3, v_3) \triangleright \kappa_1(u_1, u_2, u_3) \triangleright \kappa_4(v_1, v_2, v_3)$$
$$= \kappa_2(u_2, v_2) \triangleright \kappa_3(u_3, v_3) \triangleright \kappa_4(v_1, v_2, v_3) \triangleright \kappa_1(u_1, u_2, u_3)$$
$$= \kappa_2(u_2, v_2) \triangleright \kappa_4(v_1, v_2, v_3) \triangleright \kappa_3(u_3, v_3) \triangleright \kappa_1(u_1, u_2, u_3)$$
$$= \kappa_4(v_1, v_2, v_3) \triangleright \kappa_2(u_2, v_2) \triangleright \kappa_3(u_3, v_3) \triangleright \kappa_1(u_1, u_2, u_3).$$

4.2   Conditioning in Flexible Models

Let us turn our attention to the main problem of this paper that had remained open for more than five years: *Are conditionals of a flexible model flexible?* The answer to this question, given by Theorem 7 below, is based on the following, surprisingly simple assertion and its corollary.

Theorem 5 *Let $\pi(K \cup L) = \kappa(K) \triangleright \lambda(L)$ be defined, and either $M \subseteq K$, or $M \subseteq L$. Then*

$$\mu(M) \triangleright \pi(K \cup L) = (\mu \triangleright \pi)^{\downarrow K} \triangleright (\mu \triangleright \pi)^{\downarrow L}. \quad (4)$$

Proof First, notice that the right hand side of Equality (4) is a composition of marginals of the same distribution. Therefore it is undefined if and only if $\mu \triangleright \pi$ is undefined.

Assuming $K \supseteq M$ we can compute

$$\mu \triangleright \pi = (\mu \triangleright \pi)^{\downarrow K} \triangleright (\mu \triangleright \pi) \qquad \text{Property 4, Theorem 1}$$
$$= (\mu \triangleright \pi)^{\downarrow K} \triangleright \mu \triangleright \pi \qquad \text{Property 7, Theorem 1}$$
$$= (\mu \triangleright \pi)^{\downarrow K} \triangleright \mu \triangleright \left(\pi^{\downarrow K} \triangleright \pi^{\downarrow L}\right) \qquad \text{Properties 10 and 11, Theorem 1}$$
$$= (\mu \triangleright \pi)^{\downarrow K} \triangleright \mu \triangleright \pi^{\downarrow K} \triangleright \pi^{\downarrow L} \qquad \text{Property 7, Theorem 1}$$
$$= (\mu \triangleright \pi)^{\downarrow K} \triangleright \mu \triangleright \pi^{\downarrow L} \qquad \text{Property 3, Theorem 1}$$
$$= (\mu \triangleright \pi)^{\downarrow K} \triangleright \left(\mu \triangleright \pi^{\downarrow L}\right) \qquad \text{Property 7, Theorem 1}$$
$$= (\mu \triangleright \pi)^{\downarrow K} \triangleright (\mu \triangleright \pi)^{\downarrow L}. \qquad \text{Property 9, Theorem 1}$$

In case that $M \subseteq L$, using the same computations as above we get

$$\mu \triangleright \pi = (\mu \triangleright \pi)^{\downarrow L} \triangleright (\mu \triangleright \pi)^{\downarrow K} = (\mu \triangleright \pi)^{\downarrow K} \triangleright (\mu \triangleright \pi)^{\downarrow L},$$

where the last equality holds because of Property 6 of Theorem 1. □

Corollary 1 *Let* $\pi(K \cup L) = \kappa(K) \triangleright \lambda(L)$ *be defined. Then for any* $u \in K \cup L$ *and* $\mathbf{a} \in \mathbb{X}_u$ *such that* $\pi^{\downarrow\{u\}}(\mathbf{a}) > 0$, *the conditional* $\vartheta(K \cup L) = \delta_{\mathbf{a}}(u) \triangleright \pi(K \cup L)$ *can be expressed in the form of a composition*

$$\vartheta = \vartheta^{\downarrow K} \triangleright \vartheta^{\downarrow L}.$$

Repeat, flexible sequences are those, which can be reordered in many ways so that each variable can appear among the arguments of the first distribution. As showed in the example *two channel transmission* it does not mean that each distribution appears at the beginning of the generating sequence. Therefore, the property of flexibility is not a structural property [11] (it is not a property of a sequence $K_1, K_2, \ldots, K_n$ like decomposability), and therefore we have to prove that the perfectization procedure from Theorem 3 preserves flexibility.

Theorem 6 *If a model* $\kappa_1 \triangleright \kappa_2 \triangleright \ldots \triangleright \kappa_n$ *is flexible then its perfectized form* $\mu_1 \triangleright \mu_2 \triangleright \ldots \triangleright \mu_n$ *defined by the procedure*

$$\mu_1 = \kappa_1,$$
$$\mu_2 = \mu_1^{\downarrow K_2 \cap K_1} \triangleright \kappa_2,$$
$$\mu_3 = (\mu_1 \triangleright \mu_2)^{\downarrow K_3 \cap (K_1 \cup K_2)} \triangleright \kappa_3,$$
$$\vdots$$
$$\mu_n = (\mu_1 \triangleright \ldots \triangleright \mu_{n-1})^{\downarrow K_n \cap (K_1 \cup \ldots \cup K_{n-1})} \triangleright \kappa_n$$

is also flexible.

Proof To prove this assertion it is enough to show that for each permutation $i_1, i_2, \ldots, i_n$ for which

$$\kappa_{i_1} \triangleright \kappa_{i_2} \triangleright \ldots \triangleright \kappa_{i_n} = \kappa_1 \triangleright \kappa_2 \triangleright \ldots \triangleright \kappa_n, \tag{5}$$

and for each $j = 1, \ldots, n$, distribution $\mu_{i_1} \triangleright \mu_{i_2} \triangleright \ldots \triangleright \mu_{i_j}$ is marginal to $\kappa_1 \triangleright \ldots \triangleright \kappa_n$, which means that

$$\mu_{i_1} \triangleright \mu_{i_2} \triangleright \ldots \triangleright \mu_{i_j} = (\kappa_1 \triangleright \kappa_2 \triangleright \ldots \triangleright \kappa_n)^{\downarrow K_{i_1} \cup \ldots \cup K_{i_j}}. \tag{6}$$

Consider a permutation $i_1, i_2, \ldots, i_n$ for which Eq. (5) holds true. For $j = 1$ Equality (6) holds trivially because all distributions $\mu_1, \mu_2, \ldots, \mu_n$ are marginals of $\mu_1 \rhd \ldots \rhd \mu_n = \kappa_1 \rhd \ldots \rhd \kappa_n$ (Theorem 3). So, to finish the proof by an induction, let us show that if $\mu_{i_1} \rhd \ldots \rhd \mu_{i_{j-1}}$ is the marginal of $\kappa_1 \rhd \ldots \rhd \kappa_n$, the same must hold also for

$$\mu_{i_1} \rhd \ldots \rhd \mu_{i_j} = (\mu_{i_1} \rhd \ldots \rhd \mu_{i_{j-1}}) \rhd \mu_{i_j}.$$

So, we assume that

$$\mu_{i_1} \rhd \ldots \rhd \mu_{i_{j-1}} = (\kappa_1 \rhd \ldots \rhd \kappa_n)^{\downarrow K_{i_1} \cup \ldots \cup K_{i_{j-1}}}, \tag{7}$$

which means (because of Eq. (5) and Property 2 of Theorem 1) that

$$\mu_{i_1} \rhd \ldots \rhd \mu_{i_{j-1}} = \kappa_{i_1} \rhd \ldots \rhd \kappa_{i_{j-1}},$$

and therefore also

$$(\kappa_1 \rhd \kappa_2 \rhd \ldots \rhd \kappa_n)^{\downarrow K_{i_1} \cup \ldots \cup K_{i_j}} = \kappa_{i_1} \rhd \kappa_{i_2} \rhd \ldots \rhd \kappa_{i_j} = \mu_{i_1} \rhd \ldots \rhd \mu_{i_{j-1}} \rhd \kappa_{i_j}$$
$$= \mu_{i_1} \rhd \ldots \rhd \mu_{i_{j-1}} \rhd \mu_{i_j},$$

where the last equality holds due to Property 5 of Theorem 1, because

$$\mu_{i_j} = (\kappa_1 \rhd \kappa_2 \rhd \ldots \rhd \kappa_n)^{\downarrow K_{i_j}}.$$

$\square$

The following theorem, which is the main result of this paper, states that the computation of a conditional from a flexible model does not spoil its flexibility.

Theorem 7 *Consider a flexible model* $\pi = \kappa_1 \rhd \kappa_2 \rhd \ldots \rhd \kappa_n$, *variable* $u \in K_1$, *its value* $\mathbf{a} \in \mathbb{X}_u$ *such that* $\pi^{\downarrow \{X\}}(\mathbf{a}) > 0$, *and the corresponding conditional distribution* $\vartheta = \delta_{\mathbf{a}}(u) \rhd (\kappa_1 \rhd \kappa_2 \rhd \ldots \rhd \kappa_n)$. *Then*

$$\vartheta = \vartheta^{\downarrow K_1} \rhd \vartheta^{\downarrow K_2} \rhd \ldots \rhd \vartheta^{\downarrow K_n},$$

is a flexible model.

Proof First, let us start with showing that

$$\vartheta = \delta_{\mathbf{a}}(u) \rhd (\kappa_1 \rhd \kappa_2 \rhd \ldots \rhd \kappa_n) = \delta_{\mathbf{a}}(u) \rhd \kappa_1 \rhd \kappa_2 \rhd \ldots \rhd \kappa_n. \tag{8}$$

For this, it is enough to apply Property 7 of Theorem 1 and the fact that $u \in K_1$, and therefore $u \in K_1 \cup \ldots \cup K_i$ for all $i = 1, \ldots, n-1$. In this way we get

$$\delta_{\mathbf{a}}(u) \rhd (\kappa_1 \rhd \kappa_2 \rhd \ldots \rhd \kappa_n) = \delta_{\mathbf{a}}(u) \rhd \big((\kappa_1 \rhd \kappa_2 \rhd \ldots \rhd \kappa_{n-1}) \rhd \kappa_n\big)$$
$$= \delta_{\mathbf{a}}(u) \rhd (\kappa_1 \rhd \kappa_2 \rhd \ldots \rhd \kappa_{n-1}) \rhd \kappa_n$$
$$= \delta_{\mathbf{a}}(u) \rhd (\kappa_1 \rhd \kappa_2 \rhd \ldots \rhd \kappa_{n-2}) \rhd \kappa_{n-1} \rhd \kappa_n$$
$$= \ldots = \delta_{\mathbf{a}}(u) \rhd \kappa_1 \rhd \kappa_2 \rhd \ldots \rhd \kappa_n.$$

The rest of the proof follows the idea from the proof of Theorem 6. The only difference is, as the reader will see, that we have to employ Corollary 1.

To get a perfect model for conditional distribution ϑ, apply the perfectization procedure to $(\delta_{\mathbf{a}}(u) \triangleright \kappa_1) \triangleright \kappa_2 \triangleright \ldots \triangleright \kappa_n$:

$$\mu_1 = \delta_{\mathbf{a}}(u) \triangleright \kappa_1,$$
$$\mu_2 = \mu_1^{\downarrow K_2 \cap K_1} \triangleright \kappa_2,$$

$$\vdots$$

$$\mu_n = (\mu_1 \triangleright \ldots \triangleright \mu_{n-1})^{\downarrow K_n \cap (K_1 \cup \ldots \cup K_{n-1})} \triangleright \kappa_n.$$

We will again show that for each permutation $i_1, i_2, \ldots, i_n$ for which

$$\kappa_{i_1} \triangleright \kappa_{i_2} \triangleright \ldots \triangleright \kappa_{i_n} = \kappa_1 \triangleright \kappa_2 \triangleright \ldots \triangleright \kappa_n,$$

and for each $j = 1, \ldots, n$, distribution $\mu_{i_1} \triangleright \mu_{i_2} \triangleright \ldots \triangleright \mu_{i_j}$ is marginal to ϑ, which means that

$$\mu_{i_1} \triangleright \mu_{i_2} \triangleright \ldots \triangleright \mu_{i_j} = (\vartheta)^{\downarrow K_{i_1} \cup \ldots \cup K_{i_j}}. \tag{9}$$

For $j = 1$ Equality (9) holds because Theorem 3 guarantees that all the distributions from a perfect sequence are marginal to the resulting distribution. To conclude the induction we will show that if $\mu_{i_1} \triangleright \ldots \triangleright \mu_{i_{j-1}}$ is the marginal of ϑ, the same must hold also for

$$\mu_{i_1} \triangleright \ldots \triangleright \mu_{i_j} = (\mu_{i_1} \triangleright \ldots \triangleright \mu_{i_{j-1}}) \triangleright \mu_{i_j}.$$

Assume

$$\mu_{i_1} \triangleright \ldots \triangleright \mu_{i_{j-1}} = \vartheta^{\downarrow K_{i_1} \cup \ldots \cup K_{i_{j-1}}}. \tag{10}$$

The considered permutation is selected in the way that $\left(\kappa_{i_1} \triangleright \ldots \triangleright \kappa_{i_{j-1}} \right) \triangleright \kappa_{i_j}$ is marginal to $\kappa_1 \triangleright \ldots \triangleright \kappa_n$, and therefore, due to Property 10 of Theorem 1,

$$(K_{i_1} \cup \ldots \cup K_{i_{j-1}}) \setminus K_{i_j} \perp\!\!\!\perp K_{i_j} \setminus (K_{i_1} \cup \ldots \cup K_{i_{j-1}})[\kappa_1 \triangleright \ldots \triangleright \kappa_n].$$

Due to Corollary 1, the same conditional independence relation holds also for distribution ϑ, and therefore, due to Property 11 of Theorem 1,

$$\vartheta^{\downarrow K_{i_1} \cup \ldots \cup K_{i_j}} = \vartheta^{\downarrow K_{i_1} \cup \ldots \cup K_{i_{j-1}}} \triangleright \vartheta^{\downarrow K_{i_j}}.$$

Since $\vartheta = \mu_1 \triangleright \ldots \triangleright \mu_n$ is a perfect model, $\mu_{i_j} = \vartheta^{\downarrow K_{i_j}}$, and Eq. (9) holds because we assume that Eq. (10) holds true. $\qquad\square$

5 Concluding Remarks and Open Problem

In this paper we have proven that when computing conditionals from a flexible model, the resulting model, after perfectization, is again flexible. Though one cannot expect that the respective computational processes will be "local", the revealed property speaks out in favor of the efforts aiming to the design of new computational procedures similar to "message passing" propagation algorithms [4] for computations in join trees, or Shenoy-Shafer architecture algorithms [21]. This is a good piece of news, because it means that flexible models, more general models than decomposable ones, can be used for knowledge representation in computer aided decision systems.

The class of the considered models would increase even more, if instead of general flexible models introduced in Definition 4 one considered M-flexible models, where set M contains those variables that are required to appear at the beginning of the model. It means that M-flexible models are suitable for computations of conditionals with conditioning variables from M. It is obvious that, in some applications (such as computer-aided diagnosis making), the user can specify in advance a set of variables, conditioning by which has a sense. When trying to specify a proper diagnosis we can use a model, in which conditioning by the diagnosis and/or other hidden variables is not allowed. It means models, in which conditioning is allowed only for features that are measurable (observable).

Another remark concerns causal models. Since the compositional models are also used for modeling causality [10], it is important to realize that the results presented in this paper are not applicable to causal models. This is because the reordering of distributions of a causal model is not allowed (except for very special cases). On the other hand side, these results are unnecessary for the computation of impact of intervention, because the application of the Pearl's *do-operator* is in compositional models always easy to compute (for details see [10]).

The last remark concerns the relation of decomposable and flexible models. As said above, the class of flexible models covers the class of decomposable models. However, in [7], decomposable models were defined in a different, much broader, sense. Let us call them *weakly decomposable* here.

Definition 5 A model $\kappa_1 \triangleright \kappa_2 \triangleright \ldots \triangleright \kappa_n$ is called *weakly decomposable* if there exists a decomposable model $\lambda_1(L_1) \triangleright \lambda_2(L_2) \triangleright \ldots \triangleright \lambda_m(L_m)$ such that

- $\lambda_1 \triangleright \lambda_2 \triangleright \ldots \triangleright \lambda_m = \kappa_1 \triangleright \kappa_2 \triangleright \ldots \triangleright \kappa_n$, and
- for each $j = 1, 2, \ldots, m$ there is $i \in \{1, 2, \ldots, n\}$, for which $\lambda_j(L_j) = \kappa_i^{\downarrow L_j}$.

It means that weakly decomposable models can be transformed into decomposable ones without increase in space requirements. Example 13.3 in [7] presents a weakly decomposable model that is neither flexible nor perfect. It is a five-dimensional model whose structure is depicted in the left hand side of Fig. 3 (the structure of the equivalent decomposable model is in the right hand side of this figure).

Above introduced *two channel transmission* example is not weakly decomposable, but still there is a way (rather nontrivial) how to transform it into a decomposable

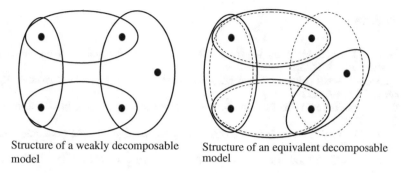

Structure of a weakly decomposable model

Structure of an equivalent decomposable model

Fig. 3 Structure of a weakly decomposable model and its decomposable equivalent

Fig. 4 Six-dimensional decomposable model

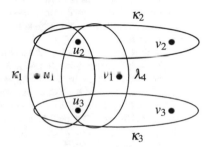

model without increase in space requirements. The reader can easily verify that it is the model (Fig. 4)

$$\kappa_1(u_1, u_2, u_3) \rhd \lambda_4(u_1, u_2, v_3) \rhd \kappa_2(u_2, v_2) \rhd \kappa_3(u_3, v_3),$$

where

$$\lambda_4(u_1, u_2, v_3) = (\kappa_1(u_1, u_2, u_3) \rhd \kappa_2(u_2, v_2) \rhd \kappa_3(u_3, v_3) \rhd \kappa_4(v_1, v_2, v_3))^{\downarrow \{u_1, u_2, v_3\}}.$$

In fact, all flexible models we dealt with were models that could be in a way transformed into decomposable ones. So, we can conclude the paper by presenting an open question: *Does there exist a flexible models for which all equivalent decomposable models are defined with distributions defined by a larger number of probabilities than the original flexible model?*

Acknowledgements This research was supported in part by the National Science Foundation of the Czech Republic by grant no. GACR 15-00215S. The paper is an extended version of the contribution presented at the *19th Czech-Japan Seminar on Data Analysis and Decision Making under Uncertainty* held in Matsumoto, Japan, September 5-7, 2016.

References

1. Beeri C, Fagin R, Maier D, Yannakakis M (1983) On the desirability of acyclic database schemes. J ACM 30(3):479–513
2. Bína V, Jiroušek R (2006) Marginalization in multidimensional compositional models. Kybernetika 42(4):405–422
3. Bína V, Jiroušek R (2015) On computations with causal compositional models. Kybernetika 51(3):525–539
4. Dawid AP (1992) Applications of a general propagation algorithm for probabilistic expert systems. Stat Comput 2(1):25–26
5. Jensen FV (2001) Bayesian networks and decision graphs. IEEE Computer Society Press, New York
6. Jiroušek R (2002) Decomposition of multidimensional distributions represented by perfect sequences. Ann Math Artif Intell 35(1–4):215–226
7. Jiroušek R (2011) Foundations of compositional model theory. Int J Gen Syst 40(6):623–678
8. Jiroušek R (2012) Local computations in dempster-shafer theory of evidence. Int J Approx Reason 53(8):1155–1167
9. Jiroušek R, (2013) Brief introduction to probabilistic compositional models. In: Huynh VN, Kreinovich V, Sriboonchita S, Suriya K (eds) Uncertainty analysis in econometrics with applications AISC 200. Springer, Berlin, pp 49–60
10. Jiroušek R, (2016) Brief introduction to causal compositional models. In: Huynh VN, Kreinovich V, Sriboonchita S (eds) Causal inference in econometrics. Studies in computational intelligence (SCI), vol 622. Springer, Cham, pp 199–211
11. Jiroušek R, Kratochvíl V (2015) Foundations of compositional models: structural properties. Int J Gen Syst 44(1):2–25
12. Lauritzen SL (1996) Graphical models. Oxford University Press, Oxford
13. Lauritzen SL, Spiegelhalter D (1988) Local computation with probabilities on graphical structures and their application to expert systems. J R Stat Soc B 50:157–224
14. Madsen AL, Jensen F, Kjærulff UB, Lang M (2005) HUGIN-the tool for bayesian networks and influence diagrams. Int J Artif Intell Tools 14(3):507–543
15. Malvestuto FM (2014) Equivalence of compositional expressions and independence relations in compositional models. Kybernetika 50(3):322–362
16. Malvestuto FM (2015) Marginalization in models generated by compositional expressions. Kybernetika 51(4):541–570
17. Pearl J (1982) Reverend Bayes on inference engines: a distributed hierarchical approach. In: Proceedings of the national conference on artificial intelligence, Pittsburgh, pp 133–136
18. Pearl J (2014) Probabilistic reasoning in intelligent systems: networks of plausible inference. Morgan Kaufmann, Burlington (Revised Second Printing)
19. Shachter RD (1986) Evaluating influence diagrams. Oper Res 34(6):871–882
20. Shachter RD (1988) Probabilistic inference and influence diagrams. Oper Res 36(4):589–604
21. Shenoy PP, Shafer G (2008) Axioms for probability and belief-function propagation. In: Classic works of the dempster-shafer theory of belief functions. Springer, Berlin, pp 499–528
22. Netica NORSYS (2008) Bayesian networks graphical application. https://www.norsys.com/

New Estimation Method for Mixture of Normal Distributions

Qianfang Hu, Zheng Wei, Baokun Li and Tonghui Wang

Abstract Normal mixture models are widely used for statistical modeling of data, including classification and cluster analysis. However the popular EM algorithms for normal mixtures may give imprecise estimates due to singularities or degeneracies. To avoid this, we propose a new two-step estimation method: first truncate the whole data set to tail data sets that contain points belonging to one component normal distribution with very high probability, and obtain initial estimates of parameters; then upgrade the estimates to better estimates recursively. The initial estimates are simply Method of Moments Estimates in this paper. Empirical results show that parameter estimates are more accurate than that with traditional EM and SEM algorithms.

1 Introduction

The mixture of normal distributions is used in many areas, such as biology, genetics, medicinal science, economics and so on. In the research literature of mixture of normal distributions, the most popular algorithm to estimate the parameters is Expectation Maximization (EM) algorithm proposed by Arthur Dempster, Nan Laird and Donald Rubin in 1977. EM is an iterative method for finding maximum likelihood estimates or maximum a posteriori (MAP) estimates of parameters in statistical models that contain unobserved latent variables. However, EM typically converges to a local optimum–not necessarily the global optimum, there is no bound on the

Q. Hu · B. Li (✉)
School of Statistics, Southwestern University of Finance and Economics, Chengdu, China
e-mail: bali@swufe.edu.cn

Z. Wei
Department of Mathematical Sciences, University of Massachusetts-Amherst, Amherst, USA

T. Wang (✉)
Department of Mathematical Sciences, New Mexico State University, Las Cruces, USA
e-mail: twang@math.nmsu.edu

© Springer International Publishing AG 2017
V. Kreinovich et al. (eds.), *Robustness in Econometrics*,
Studies in Computational Intelligence 692, DOI 10.1007/978-3-319-50742-2_13

convergence rate in general, and it is sensitive to initial values. For these weak points there are some improvements, such as [1–5], etc. All these improvements do not give much more accurate estimates for mixture distribution, especially for high dimensional mixture data. Hence in this paper we propose a new estimation method and show that it dominates previous estimations in our empirical study.

For a mixture of normal distributions, suppose the samples of the component normal distributions could be separated completely, there should be no difficulty in estimating all the parameters using the Maximum Likelihood Estimation (MLE) or the Method of Moments (MM). While we can not separate the samples of the component distributions from the whole sample data completely, we are able to get such samples respectively with very high probability, because it is much more possible for points in a truncated sample far from the center of the data to follow one distribution than to follow another distribution. Then we could use the truncated samples to estimate the parameters of the component normal distributions.

It is not uncommon that truncated samples are used to estimate the parameters of original distribution ([6–9], etc.). After applying usual estimation methods for truncated normal distribution, for example MM, we propose an iterative upgrading process which is used to approach the real parameters step by step starting from MM. We proved that under some conditions the upgraded estimates are better than initial estimates of the original normal distribution.

To estimate the parameters of one-dimensional mixture of normal distributions, we propose a new method which is called Truncation and Upgrading (TU) algorithm. In the truncation step, we set a reasonable cut point which ensures that points in the truncated sample belong to one component distribution with high probability, and the sample size is as large as possible. In the upgrading step, we apply the approaching process illustrated in Sect. 2 to the truncated sample to obtain more accurate estimates than initial estimates from an usual estimation method, here we use MME with this algorithm estimation of parameters in one component distribution is followed by estimation of parameters in another component distribution. From the simulation comparisons in Sect. 4, TU algorithm does give more accurate results than other methods, such as EM, SEM.

Due to hardship for parameter estimation of multivariate mixture of normal distributions using usual methods, simulation analysis on this topic is rarely seen. However, with the TU algorithm, parameter estimation for multivariate mixture of normal distributions is almost as easy as that for one dimensional mixture of normal distributions. It is well known that the marginal distributions of a multivariate normal distribution are also normal distributions. Similar statement holds true for a multivariate mixed normal distributions too. Because of this property, the estimation problem of a multivariate mixture of normal model is reduced to the estimation the parameters of one dimensional mixture of normal distributions.

This paper is organized as follows. After the introduction, Sect. 2 introduces the details of the parameter estimates upgrade process for truncated normal distribution; Sect. 3 describes the steps of TU algorithm for different mixture of normal distributions; Sect. 4 shows comparison among estimates from several methods with the same simulation plans. And Sect. 5 is the conclusion of the study.

2 Estimate Upgrade for Truncated Normal Distribution Parameters

2.1 Truncated Normal Distribution

Suppose $X \sim N(\mu, \sigma^2)$ is a normal distribution. When $X \in (a, b), -\infty \leq a < b \leq \infty$, then X conditional on $a < X < b$ is a truncated normal distribution. Let Y be the truncated normal distribution, the density function $f(y)$ is given by:

$$f(y) = \frac{\frac{1}{\sigma}\phi(\frac{y-\mu}{\sigma})}{\Phi(\frac{b-\mu}{\sigma}) - \Phi(\frac{a-\mu}{\sigma})} I(a < y < b).$$

where $\phi(.)$ and $\Phi(.)$ are the density and cumulative functions of the standard normal distribution, and $I(.)$ is an indicator function. When $a = -\infty$, the truncated normal distribution is a right truncated normal distribution. When $b = \infty$, the truncated normal distribution is a left truncated distribution.

The corresponding mean and variance are

$$E(Y) = \mu - \sigma \frac{\phi(a') - \phi(b')}{\Phi(b') - \Phi(a')},$$

$$Var(Y) = \sigma^2 \left[1 + \frac{a'\phi(a') - b'\phi(b')}{\Phi(b') - \Phi(a')} - \left(\frac{\phi(a') - \phi(b')}{\Phi(b') - \Phi(a')} \right)^2 \right],$$

where $a' = \frac{a-\mu}{\sigma}$, and $b' = \frac{b-\mu}{\sigma}$.

2.2 Estimate Upgrade for Truncated Normal Distribution Parameters

Intuitively, the estimation based on the truncated sample will be inferior to the estimation based on complete sample data. Therefore if we could make up the missing part of the normal distribution properly, we should have got better estimators of the normal distribution than the estimators dependent only on the truncated normal distribution samples. In the following we will introduce the estimate upgrade with initial estimates coming from the usual method of moments. The initial estimates could come from other estimation methods. Here for simplicity, we get them from method of moments. Also we show that for given initial estimates of the truncated normal distribution, under some conditions the procedure does improve the estimation accuracy.

Theorem 2.1 *Given a sample $x_1, x_2, \ldots, x_{m_1}$ from a normal distribution $N(\mu, \sigma^2)$ with $a < x_i < b$, and $i = 1, 2, \ldots, m_1,$. Denote the parameter estimates from MM are $\hat{\mu}_0, \hat{\sigma}_0^2, p_0 = \int_a^b f(x|\hat{\mu}_0, \hat{\sigma}_0^2)dx, p = \int_a^b f(x|\mu, \sigma^2)dx,$ and $p \geq 0.2$. Then*

I: assume $|p_0 - p| < 0.3$ and $\hat{\sigma}_0^2 = \sigma^2$, define $\hat{\mu}_t = I\hat{\mu}_{t-1,x} + I\hat{\mu}_{t-1,missing}$, $t > 0$, where $I\hat{\mu}_{t-1,x} = p_{t-1} \times \frac{1}{m_1} \sum_{i=1}^{m_1} x_i,$

$$I\hat{\mu}_{t-1,missing} = \int_{-\infty}^a x f(x|\hat{\mu}_{t-1}, \hat{\sigma}_{t-1}^2)dx + \int_b^{\infty} x f(x|\hat{\mu}_{t-1}, \hat{\sigma}_{t-1}^2)dx,$$

$p_{t-1} = \int_a^b f(x|\hat{\mu}_{t-1}, \hat{\sigma}_{t-1}^2)dx$. *Repeat the upgrading process, the upgraded estimate sequence converges.*

II: assume $|p_0 - p| < 0.5$ and $\hat{\mu}_0 = \mu$, define $\hat{\sigma}_t^2 = I\hat{\sigma}^2_{t-1,x} + I\hat{\sigma}^2_{t-1,missing}$, $t > 0$, where $I\hat{\sigma}^2_{t-1,x} = p_{t-1} \times \frac{1}{m_1} \sum_{i=1}^{m_1}(x_i - \hat{\mu}_{t-1})^2,$

$$I\hat{\sigma}^2_{t-1,missing} = \int_{-\infty}^a (x - \hat{\mu}_{t-1})^2 f(x|\hat{\mu}_{t-1}, \hat{\sigma}_{t-1}^2)dx + \int_b^{\infty}(x - \hat{\mu}_{t-1})^2 f(x|\hat{\mu}_{t-1}, \hat{\sigma}_{t-1}^2)dx,$$

$p_{t-1} = \int_a^b f(x|\hat{\mu}_{t-1}, \hat{\sigma}_{t-1}^2)dx$. *Repeat the upgrading process, the upgraded estimate sequence converges.*

This theorems proof is in the appendix.

3 TU for the Mixture of Normal Models

3.1 Mixture of Normal Distributions

Given an observed sample $X = (X_1, \ldots, X_n)$, where $X_j \in R^d$ follows some mixture of normal distributions. The density function of the mixture is given as follows.,

$$p(X, \Theta) = \sum_{j=1}^k w_j G(x, m_j, \Sigma_j), \qquad w_j \geq 0 \text{ and } \sum_{j=1}^k w_j = 1,$$

where $G(x, m_j, \Sigma_j) = \frac{1}{(2\pi)^{d/2}|\Sigma_j|^{1/2}} \exp\left(-\frac{1}{2}(x - m_j)^T \Sigma_j(x - m_j)\right), j = 1, 2, \ldots, k$. In estimation of mixture of normal models, the parameters to be estimated are w_j, m_j, and Σ_j.

3.2 TU Algorithm

To obtain better parameter estimates of mixture of normal models, we propose a new method, Truncation and Upgrade (TU) algorithm. In the truncation step, we try to find a reasonable truncated point which ensures that points in the truncated sample belong to one component normal distribution with high probability and the sample size is as large as possible. In the upgrade step, we apply the estimate upgrade method in Sect. 2 to upgrade initial estimates from MME based on the truncated sample. With TU algorithm the parameters of one component distribution are estimated followed by estimation of the parameters of another distribution. This method not only works well in the univariate mixture cases, but also in the multivariate mixture cases.

For simplicity, we mainly consider the mixture of only two normal distributions, since mixture with several normal distributions could be generalized from this. In this section, we will describe the TU algorithm for univariate and multivariate mixed normal models respectively.

For an univariate mixture random variable X with $w_1 N(\mu_1, \sigma_1^2) + w_2 N(\mu_2, \sigma_2^2)$, its mean and variance are as follows.

$$\mu = w_1\mu_1 + w_2\mu_2, \tag{1}$$

$$\sigma^2 = w_1\sigma_1^2 + w_2\sigma_2^2 + w_1w_2(\mu_1 - \mu_2)^2. \tag{2}$$

Given a random sample distributed as X, there are two frequently seen patterns about its fitted density curve. In one pattern the curve has one peak and is symmetric like Fig. 1 left, which implies that the means of the two distributions are very close. Thus we assume the two means are the same. In the other pattern the curve has two peaks

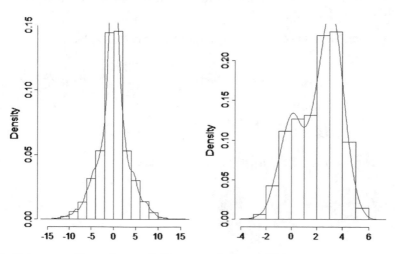

Fig. 1 Histograms with equal means (*left*) or unequal means (*right*)

clearly or is asymmetric like Fig. 1 right. For this pattern we assume the two means are different. We will describe TU algorithm steps for each pattern. The thumb rule to decide the same mean case is this: if the highest peak of the total sample is in the 5% confidence interval of the total sample mean, the means are all equal to the total sample mean. Otherwise the two distributions have different means.

3.2.1 The TU Algorithm for the Equal Mean Case

The TU algorithm is inspired with the following observation. For a random sample with the mixture of normal distributions described as above, there is about 95.45% of probability that a point in distribution N_2 lies in the interval $(\mu_2 - 1.96\sigma_2, \mu_2 + 1.96\sigma_2)$ when $\mu_1 = \mu_2$ and $\sigma_1^2 > \sigma_2^2$. If a sample point is not in this interval, then this point belongs to N_1 with a large probability in general. By this observation, we could ensure some truncated sample from one normal distribution approximately.

In practice, for the same mean normal mixture model to be estimable, it has to satisfy two requirements. Firstly, the difference between w_1 and w_2 is required to be small, say $|w_1 - w_2| < 0.4$. Secondly the variances σ_1^2 and σ_2^2 are required not be very close. Without loss of generality, we assume $\sigma_1^2 > \sigma_2^2$. Because the variance of the mixture is between the two single variances σ_1^2 and σ_2^2 from Eq. (2), the sample points beyond the interval $[\hat{\mu} - 2\hat{\sigma}, \hat{\mu} + 2\hat{\sigma}]$ belong to the distribution $N_1(\mu_1, \sigma_1^2)$ with very high possibility.

T-Step Compute the sample mean and variance as the parameter estimates $\hat{\mu}$ and $\hat{\sigma}^2$ of the mixture. Extract the samples beyond the interval $[\hat{\mu} - 2\hat{\sigma}, \hat{\mu} + 2\hat{\sigma}]$ and denote as S_1, and the sample size is m_1. From the above discussion, almost all points in S_1 come from $N_1(\mu_1, \sigma_1^2)$, therefore we assume it is a truncated normal sample. With S_1 we can get the initial estimates for distribution $N_1(\mu_1, \sigma_1^2)$, namely $\hat{\mu}_{1,0}$ and $\hat{\sigma}_{1,0}$ from the usual MME.

U-Step Using the initial estimates $\hat{\mu}_{1,0}$, $\hat{\sigma}_{1,0}$ of μ_1 and σ_1, calculate

$$p_{t-1} = P(x \notin [\hat{\mu} - 2\hat{\sigma}, \hat{\mu} + 2\hat{\sigma}]) = 1 - \int_{\hat{\mu}-2\hat{\sigma}}^{\hat{\mu}+2\hat{\sigma}} f(x|\hat{\mu}_{1,0}, \hat{\sigma}_{1,0}^2)dx,$$

$$I\hat{\mu}_{2\sigma} = \int_{\hat{\mu}-2\hat{\sigma}}^{\hat{\mu}+2\hat{\sigma}} xf(x|\hat{\mu}_{1,0}, \hat{\sigma}_{1,0}^2)dx,$$

$$I\hat{\sigma}_{2\sigma}^2 = \int_{\hat{\mu}-2\hat{\sigma}}^{\hat{\mu}+2\hat{\sigma}} (x - \hat{\mu}_{1,0})^2 f(x|\hat{\mu}_{t-1}, \hat{\sigma}_{1,0}^2)dx,$$

$$I\hat{\sigma}_x^2 = p_{t-1}\frac{1}{m_1}\sum_{i=1}^{m_1}(x_i - \hat{\mu}_{1,0})^2.$$

Let $\hat{\mu}_{1,1} = I\hat{\mu}_{x_1} + I\hat{\mu}_{2\sigma}, I\hat{\sigma}_{1,1}^2 = I\hat{\sigma}_x^2 + I\hat{\sigma}_{2\sigma}^2$. Then we have the new estimates $\hat{\mu}_{1,1}$, $\hat{\sigma}_{1,1}$ of μ_1 and σ_1, Repeat this process to get $\hat{\mu}_{1,t}, \hat{\sigma}_{1,t}^2$ $t = 2, 3, \ldots$, till the updated estimates converge. Denote $\hat{\mu}_1 = \lim_{t \to \infty} \hat{\mu}_{1,t}$, and $\hat{\sigma}_1^2 = \lim_{t \to \infty} \hat{\sigma}_{1,t}^2$.

Calculate the estimator of sample size n_1, i.e., $\hat{n}_1 = \frac{m_1}{p_{t-1}}$, we get the estimate of the distribution weight $\hat{w}_1 = \hat{n}_1/n$, then we could obtain all the other parameter estimates from Eqs. (1) and (2).

3.2.2 TU Algorithm for Different Means Case

For the mixture of normal distributions with two different means, without loss of generality, we assume that $\mu_1 < \mu_2, \sigma_1^2 \geq \sigma_2^2$. So the left distribution is $N_1(\mu_1, \sigma_1^2)$ and the right distribution is $N_2(\mu_2, \sigma_2^2)$. The TU steps are as follows.

T-Step In Fig. 2 (right), a, b, c, d represent four points in a two-mode density curve: the 2.5% quantile, the first peak, the second peak, the 97.5% quantile. In case there is only one mode observed, then we have $b = c$ as a special case of two-mode density curve. Let X_1 be the sample of points falling in the interval $(-\infty, 2c - d]$, and its size is m_1. Taking X_1 as a truncated normal sample, initial estimates of the parameters in $N_1(\mu_1, \sigma_1^2)$ are obtained with MME. And the estimates are denoted as $\hat{\mu}_{1,0}, \hat{\sigma}_{1,0}^2$.

The Truncation step is based on the fact that almost all points in X_1 come from the distribution $N_1(\mu_1, \sigma_1^2)$. Since c is very close to the μ_2, in the histogram the points in the interval $[c, d]$ account for approximately $48-50\%$ of all the sample points from $N_2(\mu_2, \sigma_2^2)$. By symmetry of the normal distribution, there is same percentage of points from N_2 in the interval $[2c - d, c)$. Thus interval $(2c - d, +\infty)$ contains

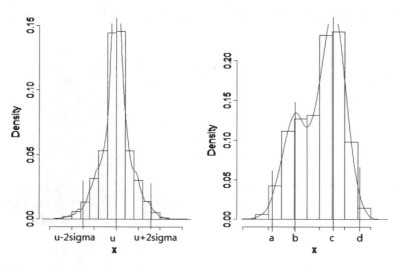

Fig. 2 Truncated samples for equal means or unequal means

more than 96% of points from N_2. So we conclude that the sample X_1 containing points in the interval $(-\infty, 2c - d]$ mostly come from the distribution N_1.

U-Step Using the initial estimates $\hat{\mu}_{1,0}$, $\hat{\sigma}_{1,0}$ of μ_1 and σ_1, calculate

$$p_{t-1} = \int_{-\infty}^{2c-d} f(x|\hat{\mu}_{1,0}, \hat{\sigma}_{1,0}^2)dx,$$

$$I\hat{\mu}_{right} = \int_{2c-d}^{+\infty} xf(x|\hat{\mu}_{1,0}, \hat{\sigma}_{1,0}^2)dx,$$

$$I\hat{\sigma}_{right}^2 = \int_{2c-d}^{+\infty} (x - \hat{\mu}_{1,0})^2 f(x|\hat{\mu}_{1,0}, \hat{\sigma}_{1,0}^2)dx,$$

$$I\hat{\sigma}_x^2 = p_{t-1}\frac{1}{m_1}\sum_{i=1}^{m_1}(x_i - \hat{\mu}_{1,0})^2,$$

Let $\hat{\mu}_{1,1} = I\hat{\mu}_{x_1} + I\hat{\mu}_{right}$, $I\hat{\sigma}_{1,1}^2 = I\hat{\sigma}_x^2 + I\hat{\sigma}_{right}^2$. Then we have the new estimates $\hat{\mu}_{1,1}$, $\hat{\sigma}_{1,1}$ of μ_1 and σ_1. Repeat this process to get $\hat{\mu}_{1,t}$, $\hat{\sigma}_{1,t}^2$ $t = 2, 3, \ldots$, until the upgraded estimates converge. Denote $\hat{\mu}_1 = \lim_{t\to\infty} \hat{\mu}_{1,t}$, and $\hat{\sigma}_1^2 = \lim_{t\to\infty} \hat{\sigma}_{1,t}^2$.

The T-Step is slightly different for the parameter estimation of $N_2(\mu_2, \sigma_2^2)$ depending on the size of $|\mu_1 - \mu_2|$. When the difference of the two means is significantly different from 0, the T-Step is the same as above. Otherwise we take out m_2 points randomly from the truncated sample containing all points greater than c. Here $m_2 = m_1 p_2/p_1$, $p_2 = P(X > c)$, and X follows $N_1(\hat{\mu}_1, \hat{\sigma}_1^2)$. With the truncated sample repeat the U-Step till we get $\hat{\mu}_2$ and $\hat{\sigma}_2^2$. The distribution weights are estimated as $\hat{w}_1 = \frac{n_1}{n_1+n_2}$, $\hat{w}_2 = 1 - \hat{w}_1$. Where $n_1 = m_1/p_1$, it is the estimated number of sample points following distribution $N_1(\mu_1, \sigma_1^2)$, and n_2 is computed similarly.

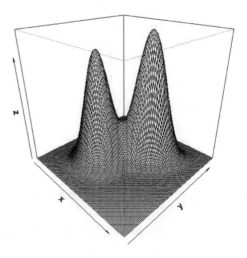

Fig. 3 Density contour of a bivariate normal mixture distribution

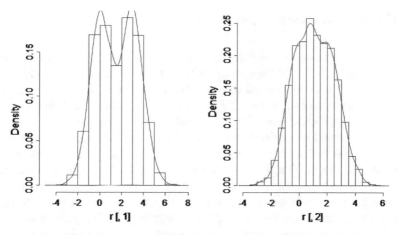

Fig. 4 Histograms of projections of a mixture of bivariate normal distributions

The TU algorithm could also be used to estimate the parameters of multivariate normal mixture distribution. Figure 3 is the density contour of a mixture of bivariate normal distributions, the density function is $f(x, y) = w_1 f(x_1, y_1) + w_2 f(x_2, y_2)$. Figure 4 contains the projections of Fig. 3 on x and y axis, both projections are mixtures of univariate normal distributions. The left of Fig. 4 is the projection on x-axis and the right is on y-axis. The parameter estimation steps are as follows.

(1) Project all the points on each axis.
(2) Use the above TU algorithm to obtain upgraded parameter estimates in mixture of univariate normal distributions.
(3) Estimate the correlation coefficient using conditional distribution of one variable on the other variable.

For the correlation coefficient between two random normal variables X and Y, it is estimated using the fact

$$(Y|X = x) \sim N(\mu_1 + \frac{\sigma_2}{\sigma_1}\rho(x - \mu_1), \sigma_2^2(1 - \rho^2)).$$

Notice that the variance of the condition distribution has no relation with x. Suppose we have sufficiently large size of data, the estimation could be carried out by computing the sample variance of a small neighborhood of x.

For instance, if $a > \mu_1$, then the neighborhood could be set as $[\mu_1 - \epsilon, \mu_1 + \epsilon]$, where ϵ is a small positive number. Denote the sample in the neighborhood as Y_1, the corresponding sample variance is $\hat{\sigma}_{Y_1}^2$, approximately we have $\hat{\sigma}_{Y_1}^2 = \hat{\sigma}_2^2(1 - \rho^2)$. Then we get the estimate of correlation coefficient $\hat{\rho}$. If $a < \mu_1$, then the neighborhood could be set as $[a - \epsilon, a]$, similarly we get $\hat{\rho}$.

4   Comparison Among Estimation Methods

For comparison purpose, we will test our new algorithm on the same data as EM and
SEM were tested in the 1995 paper of Celeux et al. Just like the 1995 paper, each
result table in this article is based on a simulation consisting of n iterations, in each
iteration is generated a sample of size N of a mixture of normal distributions.

In every table, one column is for the real values of the parameters, the subsequent
columns give the estimates for the 3 selected algorithms. The results of EM are just
the result of EM in the 1995 paper. The values in the first row of the intersections
of parameter and algorithm are mean values of estimates. The values in presences in
the second row are the standard deviations computed over the n simulations.

In this paper the main concern is the estimate accuracy because the convergence
times are very short, even for the most time consuming TU algorithm it only takes
several seconds to converge with a usual laptop. So in the following tables, we
just provide the mean value and standard deviation of parameter estimates for each
parameter under each algorithm.

4.1   Estimation of Normal Mixture Distribution with Same Means

For the mixture of normal distributions $\frac{1}{3}N(0, 1) + \frac{2}{3}N(0, 16)$, a random sample of
600 points is generated in iteration. After 200 iterations means and variances of the
estimates with EM, SEM, and TU algorithms are summarized in Table 1.

From Table 1, we can conclude that except the mean of estimates of p_1 by TU is
slightly worse than that by SEM, TU dominates both EM and SEM algorithms in the
sense of sample mean and variance of estimates.

Table 1 Means and variances of the estimates for $1/3N(0, 1) + 2/3N(0, 16)$

Parameter	Real value	EM	SEM	TU
p_1	0.333	0.38 (0.28)	0.33 (0.16)	0.32 (0.01)
μ_1	0.000	0.05 (3.67)	0.04 (1.67)	0.01 (0.01)
σ_1^2	1.000	2.03 (2.15)	1.17 (1.31)	1.11 (0.08)
μ_2	0.000	0.40 (2.56)	0.01 (1.49)	−0.01 (0.02)
σ_2^2	16.000	12.69 (5.72)	14.24 (3.97)	15.53 (1.91)

4.2 Mixture of Different Mean Normal Distributions

For the normal mixture distribution $\frac{1}{3}N(0, 1) + \frac{2}{3}N(0.8, 2.25)$, a random sample of 600 points is generated in each iteration, after 200 iterations Table 2 gives the comparison of the results with EM, SEM, and TU algorithms.

For the mixture of normal distributions $\frac{1}{3}N(0, 1) + \frac{2}{3}N(0.8, 2.25)$, a random sample of 600 points is generated in iteration. After 200 iterations means and variances of the estimates with EM, SEM, and TU algorithms are summarized in Table 2.

Table 2 continuously shows that TU dominates EM and SEM in almost all parameter estimates. Except this, when the estimates of μ_1 and σ_2^2 are relatively bad in EM and SEM, TU still give much better estimates.

4.3 Mixture of Four Normal Distributions

The mixture of normal of four distributions $\frac{1}{4}N(0, 1) + \frac{1}{4}N(2, 2.25) + \frac{1}{4}N(9, 2.25) + \frac{1}{4}N(15, 2.25)$ is used by the same paper as above.

Figure 5 is the histogram of a sample of the mixture distribution; it is easy to distinguish the right two distributions. So first we estimate the parameters in the right most distribution, after that the sample points from this distribution are taken

Table 2 Means and variances of the estimates for $1/3N(0, 1) + 2/3N(0.8, 2.25)$

Parameter	Real value	EM	SEM	TU
p_1	0.333	0.27 (0.21)	0.17 (0.25)	0.38 (0.04)
μ_1	0.000	0.58 (1.70)	0.33 (1.54)	0.07 (0.10)
σ_1^2	1.000	0.86 (0.99)	0.32 (0.58)	1.14 (0.04)
μ_2	0.800	0.83 (0.57)	0.78 (0.54)	0.77 (0.10)
σ_2^2	2.250	1.57 (0.54)	1.68 (0.61)	2.13 (0.14)

Fig. 5 A mixture of four normal distributions

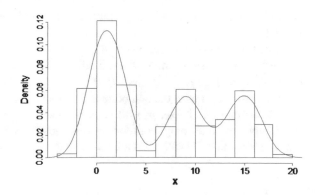

Table 3 Means and variances of the estimates for the mixture of 4 normal distributions

Parameter	Real value	EM	SEM	TU
p_1	0.25	0.30 (0.14)	0.33 (0.14)	0.26 (0.05)
μ_1	0.00	1.07 (2.79)	1.02 (2.39)	0.04 (0.09)
σ_1^2	1.00	1.07 (0.77)	1.28 (0.86)	1.00 (0.10)
p_2	0.25	0.21 (0.14)	0.19 (0.14)	0.24 (0.05)
μ_2	2.00	3.31 (3.34)	4.31 (4.23)	1.84 (0.13)
σ_2^2	2.25	1.43 (1.40)	1.46 (1.42)	2.35 (0.22)
p_3	0.25	0.22 (0.15)	0.23 (0.08)	0.25 (0.01)
μ_3	9.00	7.84 (4.16)	9.28 (0.99)	8.98 (0.06)
σ_3^2	2.25	3.46 (4.99)	2.47 (2.13)	2.44 (0.21)
μ_4	15.00	12.86 (4.41)	14.41 (2.72)	14.98 (0.06)
σ_4^2	2.25	4.06 (4.38)	2.43 (1.02)	2.27 (0.15)

out from the whole sample. The other three component distributions are estimated consecutively.

The simulation contains 100 iterations, in each iteration a random sample of 600 points is generated. Table 3 shows that the TU performs much better than other estimation methods.

4.4 Mixture of Multi-dimensional Normal Distributions

TU algorithm is also useful in estimation of multi-dimensional mixture of normal distributions. We use the same simulation plan as in Table 4 for the distribution $\frac{1}{3}N(\mu_1, \Sigma_1) + \frac{2}{3}N(\mu_2, \Sigma_2)$, where $\mu_1 = (0, 0)^T$, $\mu_2 = (0, 2)^T$, $\Sigma_1 = \begin{pmatrix} 1 & 0.5 \\ 0.5 & 1 \end{pmatrix}$ and $\Sigma_2 = \begin{pmatrix} 16 & 3 \\ 3 & 2.25 \end{pmatrix}$. Figure 6 left and right are the projections of the distribution on axis x and axis y respectively. Table 4 is the result of TU algorithm after 100 iterations.

From the mixture distribution of Fig. 6 (left), we estimate the parameters p_1, μ_{x_1}, μ_{x_2}, $\sigma_{x_1}^2$, and $\sigma_{x_2}^2$. From the mixture distribution of Fig. 6 (right), we estimate the parameters p_1, μ_{y_1}, μ_{y_2}, $\sigma_{y_1}^2$, and $\sigma_{y_2}^2$. For the two estimate of p_1, we choose one from the projection with the component distributions could be better distinguished. In this example we choose p_1 estimated from axis x. In estimating the correlation coefficient, the small neighborhood of x is chosen from the projection on x axis. Table 4 shows that TU estimates are all close to the real parameters.

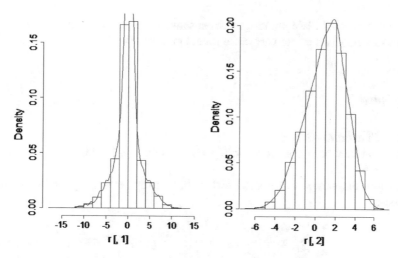

Fig. 6 Histograms of projections of a mixture of bivariate normal distributions

Table 4 Means and variances of the estimates for the two dimensional normal mixture with TU

Parameter	Real value	TU
p_1	0.67	0.69(0.11)
μ_1	$(0, 0)^T$	$(0.11, 0.03)^T$ $(0.05, 0.02)^T$
Σ_1	$\begin{pmatrix} 1 & 0.5 \\ 0.5 & 1 \end{pmatrix}$	$\begin{pmatrix} 1.19 & 0.48 \\ 0.48 & 0.93 \end{pmatrix}$ $\begin{pmatrix} 0.43 & 0.03 \\ 0.03 & 0.12 \end{pmatrix}$
μ_2	$(0, 2)^T$	$(-0.08, 2.11)^T$ $(0.002, 0.21)^T$
Σ_2	$\begin{pmatrix} 16 & 3 \\ 3 & 2.25 \end{pmatrix}$	$\begin{pmatrix} 15.11 & 3.23 \\ 3.23 & 2.40 \end{pmatrix}$ $\begin{pmatrix} 2.11 & 0.26 \\ 0.26 & 1.08 \end{pmatrix}$

5 Conclusion

To overcome the weakness of low precision by EM algorithms in parameter estimation of mixture of normal distributions, we propose a new estimation method TU to get more accurate estimates. Though the convergence of the upgrade process is proved with strict conditions, the conditions could be loosed a lot in real applications, and the convergence of the upgraded estimates could still be accomplished. In this paper empirical results with simulation data show that TU algorithm not only dominates EM and SEM algorithms when sample size is more than 100, it always results in estimates that are very close to real parameters.

Appendix

Proof of Theorem 2.1

For convenience, we set $a = -\infty$. These results also hold true for $b = \infty$ and $a < X < b$.

I: Since the variance of normal distribution always exists, by the law of large numbers:

$$\frac{1}{m_1} \sum_{i=1}^{m_1} x_i \xrightarrow{p} \frac{1}{p} I\mu_x = \frac{1}{p} \int_{-\infty}^{b} xf(x|\mu, \sigma^2)dx.$$

Then

$$\hat{\mu}_t \approx p_{t-1} \frac{I\mu_x}{p} + I\hat{\mu}_{t-1,missing},$$

where

$$I\mu_x = \int_{-\infty}^{b} xf(x|\mu, \sigma^2)dx$$

$$= \int_{-\infty}^{b} (x - \mu + \mu)f(x|\mu, \sigma^2)dx$$

$$= \mu F(b|\mu, \sigma^2) - \sigma^2 f(b|\mu, \sigma^2).$$

Similarly,

$$I\hat{\mu}_{t-1,x} = \int_{-\infty}^{b} xf(x|\hat{\mu}_{t-1}, \hat{\sigma}^2_{t-1})dx$$

$$= \hat{\mu}_{t-1} F(b|\hat{\mu}_{t-1}, \hat{\sigma}^2_{t-1}) - \hat{\sigma}^2_{t-1} f(b|\hat{\mu}_{t-1}, \hat{\sigma}^2_{t-1}).$$

Because the following equation is always true:

$$\hat{\mu}_{t-1} = \int_{-\infty}^{b} xf(x|\hat{\mu}_{t-1}, \hat{\sigma}^2_{t-1})dx + \int_{b}^{+\infty} xf(x|\hat{\mu}_{t-1}, \hat{\sigma}^2_{t-1})dx$$

$$= I\hat{\mu}_{t-1,x} + I\hat{\mu}_{t-1,missing}$$

$$= p_{t-1} \frac{I\hat{\mu}_{t-1,x}}{p_{t-1}} + I\hat{\mu}_{t-1,missing},$$

$$\hat{\mu}_t - \hat{\mu}_{t-1} = \left(p_{t-1} \frac{\hat{I\mu}_x}{p} + \hat{I\mu}_{t-1,missing} \right) - \left(p_{t-1} \frac{\hat{I\mu}_{t-1,x}}{p_{t-1}} + \hat{I\mu}_{t-1,missing} \right)$$

$$= p_{t-1} \left(\left(\mu - \sigma^2 \frac{f(b|\mu,\sigma^2)}{F(b|\mu,\sigma^2)} \right) - \left(\hat{\mu}_{t-1} - \hat{\sigma}_{t-1}^2 \frac{f(b|\hat{\mu}_{t-1},\hat{\sigma}_{t-1}^2)}{F(b|\hat{\mu}_{t-1},\hat{\sigma}_{t-1}^2)} \right) \right)$$

$$= p_{t-1} \left(\mu - \hat{\mu}_{t-1} - \sigma^2 \frac{f(b|\mu,\sigma^2)}{F(b|\mu,\sigma^2)} + \hat{\sigma}_{t-1}^2 \frac{f(b|\hat{\mu}_{t-1},\hat{\sigma}_{t-1}^2)}{F(b|\hat{\mu}_{t-1},\hat{\sigma}_{t-1}^2)} \right).$$

Suppose $\hat{\sigma}_0^2 = \sigma^2$, let $g(\mu) = \sigma^2 \frac{f(b|\mu,\sigma^2)}{F(b|\mu,\sigma^2)}$, then

$$g'(\mu) = \sigma^2 \left(\frac{f(b|\mu,\sigma^2)}{F(b|\mu,\sigma^2)} \right)' = \frac{sf(s)F(s) - f^2(s)}{F^2(s)},$$

where $s = \frac{b-\mu}{\sigma}$, and $f(s)$, $F(s)$ are the density function and cumulative function of the standard normal distribution. Then

$$\hat{\mu}_t - \hat{\mu}_{t-1} = p_{t-1} \left(\mu - \hat{\mu}_{t-1} - g(\mu) + g(\hat{\mu}_{t-1}) \right). \tag{3}$$

By the Lagrange's mean value theorem we have

$$g(\mu) - g(\hat{\mu}_{t-1}) = \left(\mu - \hat{\mu}_{t-1} \right) g'(c),$$

where c is between μ and $\hat{\mu}_{t-1}$. So the above Eq. (3) is

$$\hat{\mu}_t - \hat{\mu}_{t-1} = p_{t-1}(1 - g'(c)) \left(\mu - \hat{\mu}_{t-1} \right).$$

Denote $s_1 = \frac{b-c}{\sigma}$. Then $1 - g'(c) > 0$ and $0 < F(s_1)(1 - g'(c)) < 1$ are always true. And when $0.2 < F(s_1) < 0.8$, $0 < (F(s_1) + 0.3)(1 - g'(c)) < 2$.

If $\mu < \hat{\mu}_{t-1}$, then $p_{t-1} < F(s_1) < p$, therefor $0 < p_{t-1}(1 - g'(c)) < F(s_1)(1 - g'(c)) < 1$ is always true. So $\mu < \hat{\mu}_t < \hat{\mu}_{t-1}$. And then the upgraded estimate sequence converges.

If $\mu > \hat{\mu}_{t-1}$, then $\hat{\mu}_t > \hat{\mu}_{t-1}$, and $p < F(s_1) < p_{t-1}$. When $p > 0.05$, $0 < p_{t-1}(1 - g'(c)) < 1$. Then $\hat{\mu}_{t-1} < \hat{\mu}_1 < \mu$. When $0.2 < p < 0.5$, so if $|p_{t-1} - p| < 0.3$, we have $0 < p_{t-1}(1 - g'(c)) < (F(s_1) + 0.3)(1 - g'(c)) < c$. This implies $\hat{\mu}_{t-1} < \hat{\mu}_t < \mu + (\mu - \hat{\mu}_{t-1})$.

If $\hat{\mu}_t > \mu$, that is to say $\mu < \hat{\mu}_t < \mu + (\mu - \hat{\mu}_{t-1})$, then from the above paragraph the sequence of following upgraded estimate converges. If $\hat{\mu}_t < \mu$, that is to say $\hat{\mu}_{t-1} < \hat{\mu}_t < \mu$. Then we could also have the conclusion that the upgraded estimate sequence converges.

So from the above we can conclude that when $\hat{\sigma}_0^2 = \sigma^2$, the upgraded estimate sequence converges. The results hold true for left truncated and both sides truncated normal distributions.

II: Since the variance of normal distribution always exists, by the law of large numbers:

$$\frac{\hat{I\sigma}_x^2}{p_{t-1}} \xrightarrow{P} \frac{1}{p} I\sigma_x^2 = \frac{1}{p}\int_{-\infty}^b (x-\mu)^2 f(x|\mu,\sigma^2)dx$$

And

$$I\sigma_x^2 = \int_{-\infty}^b (x-\mu)^2 f(x|\mu,\sigma^2)dx$$

$$= \sigma^2\int_{-\infty}^b \frac{x-\mu}{\sigma}\frac{1}{\sqrt{2\pi}}\exp\{-\frac{(x-\mu)^2}{2\sigma^2}\}d\frac{(x-\mu)^2}{2\sigma^2}$$

$$= \sigma^2[F(t)-tf(t)],$$

$$\frac{I\sigma_x^2}{p} = \sigma^2\left(1-\frac{sf(s)}{F(s)}\right),$$

where $s = \frac{b-\mu}{\sigma}$, and $f(s)$, $F(s)$ are the density function and cumulative function of the standard normal distribution.

Similarly:

$$\frac{\hat{I\sigma}_{t-1,x}^2}{p_{t-1}} = \hat{\sigma}_{t-1}^2\left(1-\frac{\hat{t}f(\hat{t})}{F(\hat{t})}\right),$$

where $\hat{s} = \frac{b-\mu}{\hat{\sigma}}$, and $f()$, $F()$ are the density function and cumulative function of the standard normal distribution.

Assume $\hat{\mu}_0 = \mu$, Let $g(\sigma^2) = \sigma^2\left(1-\frac{sf(s)}{F(s)}\right)$, then

$$g'(\sigma^2) = 1-\frac{sf(s)}{F(s)} - \frac{1/2s^3f(s)+1/2tf(s)}{F(s)} - \frac{1/2s^2f^2(s)}{F^2(s)}.$$

By the Lagrange's mean value theorem we have

$$g(\sigma^2) - g(\hat{\sigma}_{t-1}^2) = (\sigma^2-\hat{\sigma}_{t-1}^2)g'(d^2),$$

where d^2 is between σ^2 and $\hat{\sigma}_{t-1}^2$.

And we also have this equation

$$\hat{\sigma}_{t-1}^2 = \hat{I\sigma}_{t-1,x}^2 + \hat{I\sigma}_{t-1,missing}^2,$$

So

$$\hat{\sigma}_t^2 - \hat{\sigma}_{t-1}^2 = (\hat{I\sigma}_x^2 + \hat{I\sigma}_{t-1,missing}^2) - (\hat{I\sigma}_{t-1,x}^2 + \hat{I\sigma}_{t-1,missing}^2)$$

$$= p_{t-1}\left(\frac{I\sigma_x^2}{p} - \frac{\hat{I\sigma}_{t-1,x}^2}{p_{t-1}}\right)$$

$$= p_{t-1}g'(d)(\sigma^2 - \hat{\sigma}_{t-1}^2)$$

Denote $s_2 = \frac{b-\mu}{d}$. Then from R software $g'(d) > 0$ and $0 < F(s_2)g'(d) < 1$ are always true. And when $0 < F(s_2) < 0.5$, $0 < (F(s_2) + 0.5)g'(d) < 1$.

If $\hat{\sigma}_{t-1}^2 < \sigma^2$, then $p_{t-1} < F(s_2) < p$, then $0 < p_{t-1}g'(d) < F(s_2)g'(d) < 1$ is always true. So $\hat{\sigma}_{t-1}^2 < \hat{\sigma}_t^2 < \sigma^2$ is always true. And then the upgrading process of estimators converges.

If $\hat{\sigma}_{t-1}^2 > \sigma^2$, then $\hat{\sigma}_t^2 < \hat{\sigma}_{t-1}^2$, $p < F(s_2) < p_{t-1}$. As $|p_{t-1} - p| < 0.5$, when $F(s_2) > 0.5$, $0 < p_{t-1}g'(d) < 1$, when $F(s_2) < 0.5$, $0 < p_{t-1}g'(d) < (F(s_2) + 0.5)g'(d) < 1$. That is to say $\sigma^2 < \hat{\sigma}_t^2 < \hat{\sigma}_{t-1}^2$. Then we could also have the conclusion that the upgrading process of estimator converges.

So from the above we can conclude that when $\hat{\mu}_{t-1} = \mu$, the upgrading process of estimators converges to σ^2.

References

1. Dias JG, Wedel M (2004) An empirical comparison of EM, SEM and MCMC performance for problematic Gaussian mixture likelihoods. Stat Comput 14:323–332
2. Celeux G, Chauveau D, Diebolt J (1995) On stochastic versions of the EM algorithm. Institute National de Recherche en Informatique et en Automatique, Mars, pp 1–22
3. Karlis D, Xekalaki E (2003) Choosing initial values for the EM algorithm for finite mixtures. Comput Stat Data Anal 41:577–590
4. Yao W (2013) A note on EM algorithm for mixture models. Stat Probab Lett 83:519–526
5. Chen LS, Prentice RL, Wang P (2014) A penalized EM algorithm incorporating missing data mechanism for gaussian parameter estimation. Biometrics 70:312–322
6. Horrace WC (2005) Notes: some results on the multivariate truncated normal distribution. J Multivariate Anal 94:209–221
7. Horrace WC (2015) Moments of the truncated normal distribution. J Prod Anal 43:133–138
8. del Castillo J, Daoudi J (2009) The mixture of left–right truncated normal distributions. J Stat Plann Infer 139:3543–3551
9. Emura T, Konno Y (2014) Erratum to: multivariate normal distribution approaches for dependently truncated data. Stat Papers 55:1233–1236

EM Estimation for Multivariate Skew Slash Distribution

Weizhong Tian, Guodong Han, Tonghui Wang and
Varith Pipitpojanakarn

Abstract In this paper, the class of multivariate skew slash distributions under different type of setting is introduced and its density function is discussed. A procedure to obtain the Maximum Likelihood estimators for this family is studied. In addition, the Maximum Likelihood estimators for the mixture model based on this family are discussed. For illustration of the main results, we use the actual data coming from the Inner Mongolia Academy of Agriculture and Animal Husbandry Research Station to show the performance of the proposed algorithm.

1 Introduction

Despite the central role played by the magic bell-shaped normal distribution in statistics, there has been a sustained interest among statisticians in constructing more challenging distributions for their procedures. A first family of scenarios can be represented by a finite mixture of normal distributions [15]. Another family of scenarios is described by the standard slash distribution was introduced by Rogers and

W. Tian
Department of Mathematical Sciences, Eastern New Mexico University, Portales, USA
e-mail: weizhong.tian@enmu.edu

G. Han
College of Ecological and Environmental Science,
Inner Mongolia Agricultural University, Hohhot, China
e-mail: nmghanguodong@163.com

T. Wang (✉)
Department Mathematical Sciences, New Mexico State University, Las Cruces, USA
e-mail: twang@nmsu.edu

V. Pipitpojanakarn
Faculty of Economics, Chiang Mai University, Chiang Mai, Thailand
e-mail: Varith-p@cmu.ac.th

© Springer International Publishing AG 2017
V. Kreinovich et al. (eds.), *Robustness in Econometrics*,
Studies in Computational Intelligence 692, DOI 10.1007/978-3-319-50742-2_14

Tukey [18], representing the distribution of the ratio $X = ZU^{-\frac{1}{q}}$, where Z is a standard normal variate independent of $U \sim U(0, 1)$, a standard uniform variate and $q > 0$. We obtain the canonical slash when $q = 1$, whereas $q \to \infty$ yields the normal distribution. The probability density function of the univariate slash distribution is symmetric about the origin and has heavier tails than those of the normal density, with, for the canonical slash, the same tail heaviness as the Cauchy.

The slash distribution has been mainly used in simulation studies because it represents an extreme situation, see Andrews et al. [1], Gross [12], and Morgenthaler and Tukey [16]. Recently, Gomez et al. [11] introduced extension of univariate and multivariate slash distributions as a scale mixture of elliptically contour distributions. Arslan and Genc [6] introduced a generalization of multivariate slash distribution by using Kotz-type distribution, and Reyes et al. [17] introduced the modified slash distribution in univariate and multivariate settings as a scale mixture of normal distribution and exponential distribution with the scale parameter 2. They gave some distributional properties and parameter estimations.

The multivariate skew slash distributions were introduced by Wang and Genton [21], under the multivariate skew normal setting, see [3, 8, 13, 20], which is an extension of the skew normal distribution [7, 9]. Later on, Arslan [4, 5] discussed about two alternative types of skew slash distribution in multivariate setting. In this paper, I will discuss another type of multivariate skew slash distribution, under the multivariate skew normal distribution introduced by Sahu et al. [19], which is given in the following.

A random vector $\mathbf{Y}$ is said to follow a p-dimensional skew normal distribution with a location vector $\boldsymbol{\xi} \in \Re^p$, a positive definite scale covariance matrix $\Sigma \in M_{p \times p}$, and a skewness vector $\boldsymbol{\Lambda} \in \Re^p$, if its density function is

$$f(\mathbf{y}; \boldsymbol{\xi}, \Sigma, \boldsymbol{\Lambda}) = 2\phi_p(\mathbf{y}; \boldsymbol{\xi}, \Omega)\Phi_p(\boldsymbol{\Lambda}'\Omega^{-1}(\mathbf{y} - \boldsymbol{\xi}); \Delta), \tag{1}$$

where $\Omega = \Sigma + \boldsymbol{\Lambda}\boldsymbol{\Lambda}'$ and $\Delta = (I_p + \boldsymbol{\Lambda}'\Sigma^{-1}\boldsymbol{\Lambda})^{-1}$, with I_p is a $p \times p$ identity matrix, $\phi_p(\cdot; \boldsymbol{\mu}; \Sigma)$ is the p-dimensional normal density function with mean vector $\boldsymbol{\mu}$ and covariance matrix Σ, and $\Phi_p(\cdot; \Sigma)$ is the p-dimensional distribution function with mean vector $\mathbf{0}$ and covariance matrix Σ. Let us denote this distribution by $\mathbf{Y} \sim SN_p(\boldsymbol{\xi}, \Sigma, \boldsymbol{\Lambda})$. By Propositions of Arellano-Valle et al. [2], Azzalini and Dalla [9], it turns out that (1) has a stochastic representation

$$\mathbf{Y} = \boldsymbol{\xi} + \boldsymbol{\Lambda}T + \mathbf{Z}, \tag{2}$$

where T is a standard half normal variate independent of $\mathbf{Z}$, p-dimensional normal variate with mean vector $\mathbf{0}$ and covariance matrix Σ.

The remainder of this paper is organized as follows. The definition of multivariate skew slash distribution and its density function are discussed in Sect. 2. A procedure to obtain the Maximum Likelihood (ML) estimators for the parameters of the proposed distribution is provided in Sect. 3. The ML estimators for the mixture model based on multivariate skew slash distributions are discussed in Sect. 4. A application is studied to show the performance of the proposed algorithm in Sect. 5.

2 Multivariate Skew Slash Distribution

In this section, the definition of this new type of multivariate slash distribution, its stochastic representation and density function are studied.

Definition 1 A random vector $\mathbf{X} \in \Re^p$ has a *p-dimensional skew slash distribution* with location parameter $\boldsymbol{\mu}$, positive definite scale matrix parameter Σ, and tail parameter $q > 0$, denoted by $\mathbf{X} \sim SSL_p(\boldsymbol{\mu}, \Sigma, \Lambda, q)$, if

$$\mathbf{X} = \boldsymbol{\mu} + \mathbf{Y}U^{-\frac{1}{q}} \tag{3}$$

where $\mathbf{Y} \sim SN_p(\mathbf{0}, \Sigma, \Lambda)$ is independent of $U \sim U(0, 1)$

Remark 1
(i) $\mathbf{X}$ in (3) has a standard multivariate skew slash distribution when $\boldsymbol{\mu} = 0$ and $\Sigma = I_p$, and the probability density function is

$$f_p(\mathbf{x}; \boldsymbol{\mu}, \Sigma, \Lambda) = 2q \int_0^1 u^{q+p-1} \phi_p(u\mathbf{x}; \mathbf{0}, \Omega) \Phi(u\Lambda'\Omega^{-1}\mathbf{x}; \Lambda) du. \tag{4}$$

(ii) $\mathbf{X}$ in (3) reduces to the multivariate slash distribution $SL_p(\boldsymbol{\mu}, \Sigma, q)$ when $\Lambda = 0$ and has the density

$$f_p(\mathbf{x}; \boldsymbol{\mu}, \Sigma) = q \int_0^1 u^{q+p-1} \phi_p(u\mathbf{x}; u\boldsymbol{\mu}, \Omega) du. \tag{5}$$

(iii) $\mathbf{X}$ in (3) reduces to the skew normal distribution $SN_p(\boldsymbol{\mu}, \Sigma, \Lambda)$ when $q \to \infty$.
(iv) $\mathbf{X}$ in (3) reduces to the normal distribution $N_p(\boldsymbol{\mu}, \Sigma)$ when both $\Lambda = 0$ and $q \to \infty$.

The multivariate skew slash distribution includes a wide variety of contour shapes. To illustrate the skewness and tail behavior of the skew slash, we draw the density of the univariate skew slash distribution $SSL_1(0, 1, 1, 1)$ together with the densities of the standard normal distribution $N(0, 1)$ and slash distribution $SL_1(0, 1, 1, 1)$ (Fig. 1).

Next we consider the linear transformation $\mathbf{Y} = \mathbf{b} + A\mathbf{X}$, where $\mathbf{X} \sim SL_p$ $(\boldsymbol{\mu}, \Sigma, \Lambda)$, $\mathbf{b} \in \Re^k$, and A is a nonsingular matrix.

Proposition 1 *If* $\mathbf{X} \sim SSL_p(\boldsymbol{\mu}, \Sigma, \Lambda)$, *then its linear transformation* $\mathbf{Y} = \mathbf{b} + A\mathbf{X} \sim SSL_p(\mathbf{b} + A\boldsymbol{\mu}, A\Sigma A', A\Lambda, q)$.

The above result implies that the class of skew slash distributions is invariant under linear transformations.

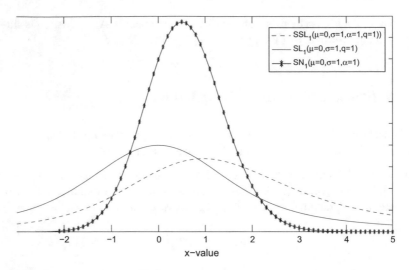

Fig. 1 Density curves of $SSL_1 (0, 1, 1, 1)$ (*black-**), $SL_1 (0, 1, 1)$ (*blue-*) and $SN_1 (0, 1, 1)$ (*red-*)

3 EM Estimation for Multivariate Skew Slash Distribution

The EM algorithm is a popular iterative algorithm for ML estimation in models with incomplete data, which was introduced by Dempster et al. [10]. Recently, Lin [14] proposed EM algorithm to compute maximum likelihood estimates of model parameters for skew normal mixture models. Arslan [6] provided ML estimators for the parameters of the proposed distribution based on the EM algorithm.

Assume that we have independent and identically distributed data $\mathbf{X}_1, \mathbf{X}_2, \cdots,$ $\mathbf{X}_n \in \Re^p$ and wish to fit a multivariate skew slash $SSL_p(\boldsymbol{\mu}, \boldsymbol{\Sigma}, \boldsymbol{\Lambda}, q)$ distribution with the unknown parameters $\boldsymbol{\mu}, \boldsymbol{\Sigma}$ and $\boldsymbol{\Lambda}$, and we assume that $q > 0$ is known. The log-likelihood function that we need to maximize is

$$\ell(\boldsymbol{\mu}, \boldsymbol{\Sigma}, \boldsymbol{\Lambda}) = \sum_{i=1}^{n} \log f_p(\mathbf{x}_i; \boldsymbol{\mu}, \boldsymbol{\Sigma}, \boldsymbol{\Lambda}). \tag{6}$$

Since the maximization of this function is not very tractable, the maximum likelihood estimators of the parameters cannot be easily obtained. However, because of the advantage of the scale mixture representation of this distribution, EM algorithm procedure can be applied to find the maximum likelihood estimators for the parameters of the multivariate skew slash distribution.

From (3), a hierarchical representation of the multivariate skew slash distribution is given by

$$\mathbf{X}|(T = t, U = u) \sim N_p(\boldsymbol{\mu} + u^{-\frac{1}{q}}\boldsymbol{\Lambda}t, u^{-\frac{2}{q}}\boldsymbol{\Sigma}), \tag{7}$$

$$\mathbf{X}|(U = u) \sim SN_p(\boldsymbol{\mu}, u^{-\frac{2}{q}}\boldsymbol{\Sigma}, u^{-\frac{1}{q}}\boldsymbol{\Lambda}), \tag{8}$$

$$\mathbf{X}|(T = t) \sim SL_p(\boldsymbol{\mu} + \boldsymbol{\Lambda}t, \boldsymbol{\Sigma}). \tag{9}$$

Suppose that the latent mixing variables $(t_1, u_1), (t_2, u_2), \cdots, (t_n, u_n)$ from the mixture representation in (7) are also observable and we define $(\mathbf{X}_i, t_i, u_i)$ for $i = 1, 2, \cdots, n$ as the complete data, where $\mathbf{X}_i$ and (t_i, u_i) are called observed and missing data, respectively.

By (7)–(9), we can obtain the joint density of $(\mathbf{X}, t, u)$ is

$$f(\mathbf{X}, t, u) = 2(2\pi)^{-\frac{p}{2}} \left|u^{-\frac{2}{q}}\boldsymbol{\Sigma}\right|^{-\frac{1}{2}} \exp\left\{-\frac{t^2}{2}\right\} \tag{10}$$

$$\times \exp\left\{-\frac{1}{2}tr\left[\left(u^{-\frac{2}{q}}\boldsymbol{\Sigma}\right)^{-1}\left(\mathbf{x} - \boldsymbol{\mu} - u^{-\frac{1}{q}}\boldsymbol{\Lambda}t\right)\left(\mathbf{x} - \boldsymbol{\mu} - u^{-\frac{1}{q}}\boldsymbol{\Lambda}t\right)'\right]\right\},$$

where $tr(A)$ is the trace of a square matrix A. The log-likelihood function for the complete data $(\mathbf{X}_i, t_i, u_i)$ for $i = 1, 2, \cdots, n$ is

$$\ell(\boldsymbol{\mu}, \boldsymbol{\Sigma}, \boldsymbol{\Lambda}) = n\log 2 - \frac{pn}{2}\log(2\pi) - \frac{n}{2}\log|\boldsymbol{\Sigma}| + \frac{1}{q}\sum_{i=1}^{n}\log(u_i) - \frac{1}{2}\sum_{i=1}^{n}t_i^2 \tag{11}$$

$$-\frac{1}{2}\sum_{i=1}^{n}tr\left[\left(u^{-\frac{2}{q}}\boldsymbol{\Sigma}\right)^{-1}\left(\mathbf{x} - \boldsymbol{\mu} - u^{-\frac{1}{q}}\boldsymbol{\Lambda}t\right)\left(\mathbf{x} - \boldsymbol{\mu} - u^{-\frac{1}{q}}\boldsymbol{\Lambda}t\right)'\right].$$

Note that since $\sum_{i=1}^{n}\log(u_i)$ and $\sum_{i=1}^{n}t_i^2$ do not contain any unknown parameters, they can be ignored. We knew only the observed data $\mathbf{X}_i$, while the missing data (t_i, u_i) are unknown, for $i = 1, \cdots, n$. To overcome this problem, the conditional expectation of $L(\boldsymbol{\mu}, \boldsymbol{\Sigma}, \boldsymbol{\Lambda})$ given the observed data $\mathbf{X}_i$ and the current estimates $\hat{\boldsymbol{\mu}}$, $\hat{\boldsymbol{\Sigma}}$ and $\hat{\boldsymbol{\Lambda}}$ are obtained. After taking the conditional expectation of $L(\boldsymbol{\mu}, \boldsymbol{\Sigma}, \boldsymbol{\Lambda})$, we get

$$Q(\boldsymbol{\mu}, \boldsymbol{\Sigma}, \boldsymbol{\Lambda}) = E\left[\ell(\boldsymbol{\mu}, \boldsymbol{\Sigma}, \boldsymbol{\Lambda})|\mathbf{X}_i, \hat{\boldsymbol{\mu}}, \hat{\boldsymbol{\Sigma}}, \hat{\boldsymbol{\Lambda}}\right] \tag{12}$$

$$= -\frac{pn}{2}\log(2\pi) - \frac{n}{2}\log|\boldsymbol{\Sigma}|$$

$$-\frac{1}{2}\sum_{i=1}^{n}E\left[tr\left[\left(u_i^{-\frac{2}{q}}\boldsymbol{\Sigma}\right)^{-1}\left(\mathbf{x}_i - \boldsymbol{\mu} - u_i^{-\frac{1}{q}}\boldsymbol{\Lambda}t_i\right)\right.\right.$$

$$\left.\left.\left(\mathbf{x}_i - \boldsymbol{\mu} - u_i^{-\frac{1}{q}}\boldsymbol{\Lambda}t_i\right)'\right]|\mathbf{X}_i, \hat{\boldsymbol{\mu}}, \hat{\boldsymbol{\Sigma}}, \hat{\boldsymbol{\Lambda}}\right],$$

after simplifying, the last term in (12) is

$$tr\left[\Sigma^{-1}\left(u_i^{-\frac{2}{q}}(\mathbf{x}_i-\boldsymbol{\mu})(\mathbf{x}_i-\boldsymbol{\mu})'-u_i^{\frac{1}{q}}t_i\left[\Lambda(\mathbf{x}_i-\boldsymbol{\mu})'+(\mathbf{x}_i-\boldsymbol{\mu})\Lambda'\right]+\Lambda\Lambda't_i^2\right)\right].$$

To compute this conditional expectation, we have to find the conditional distribution of U, T and UT given $\mathbf{X}$. After some straightforward algebra, the probability density function of this conditional distributions can be obtained as follows

$$f_{U|\mathbf{x}}(u)=\frac{etr\left[-\frac{1}{2}u^{\frac{2}{q}}\Omega^{-1}(\mathbf{x}-\boldsymbol{\mu})(\mathbf{x}-\boldsymbol{\mu})'\right]\Phi\left(u^{-\frac{1}{q}}\Lambda'\Omega^{-1}(\mathbf{x}-\boldsymbol{\mu});\Delta\right)}{q\displaystyle\int_0^1 u^{p+q-1}etr\left[-\frac{u^2}{2}\Omega^{-1}(\mathbf{x}-\boldsymbol{\mu})(\mathbf{x}-\boldsymbol{\mu})'\right]\Phi\left(u\Lambda'\Omega^{-1}(\mathbf{x}-\boldsymbol{\mu});\Delta\right)du},$$

$$f_{T|\mathbf{x}}(t)=\frac{\exp\left\{-\frac{t^2}{2}\right\}\displaystyle\int_0^1 u^{p+q-1}etr\left[-\frac{u^2}{2}\Sigma^{-1}(\mathbf{x}-\boldsymbol{\mu}-\Lambda t)(\mathbf{x}-\boldsymbol{\mu}-\Lambda t)'\right]du}{\displaystyle\int_0^1 u^{p+q-1}etr\left[-\frac{u^2}{2}\Omega^{-1}(\mathbf{x}-\boldsymbol{\mu})(\mathbf{x}-\boldsymbol{\mu})'\right]\Phi\left(u\Lambda'\Omega^{-1}(\mathbf{x}-\boldsymbol{\mu});\Delta\right)du},$$

where $etr(A)=\exp\{tr(A)\}$ and tr denotes trace.

Thus, using this conditional distribution above, we calculate and denote

$$a_i=E\left[u_i^{-\frac{2}{q}}|\mathbf{X}_i,\hat{\boldsymbol{\mu}},\hat{\Sigma},\hat{\Lambda}\right],$$

$$b_i=E\left[u_i^{\frac{1}{q}}t_i|\mathbf{X}_i,\hat{\boldsymbol{\mu}},\hat{\Sigma},\hat{\Lambda}\right],$$

$$c_i=E\left[t_i^2|\mathbf{X}_i,\hat{\boldsymbol{\mu}},\hat{\Sigma},\hat{\Lambda}\right].$$

Replace $E[u_i^{\frac{2}{q}}|\mathbf{X}_i,\hat{\boldsymbol{\mu}},\hat{\Sigma},\hat{\Lambda}]$, $E[u_i^{\frac{1}{q}}t_i|\mathbf{X}_i,\hat{\boldsymbol{\mu}},\hat{\Sigma},\hat{\Lambda}]$ and $E[t_i^2|\mathbf{X}_i,\hat{\boldsymbol{\mu}},\hat{\Sigma},\hat{\Lambda}]$ by a_i, b_i and c_i in Eq. (12), we obtain the following objective function which will be maximized with respect to $\boldsymbol{\mu}$, Σ, Λ,

$$Q(\boldsymbol{\mu},\Sigma,\Lambda)=E\left[\ell(\boldsymbol{\mu},\Sigma,\Lambda)|\mathbf{X}_i,\hat{\boldsymbol{\mu}},\hat{\Sigma},\hat{\Lambda}\right] \qquad (13)$$

$$=-\frac{pn}{2}\log(2\pi)-\frac{n}{2}\log|\Sigma|-\frac{1}{2}\sum_{i=1}^n tr\left[\Sigma^{-1}\left[a_i(\mathbf{x}_i-\boldsymbol{\mu})(\mathbf{x}_i-\boldsymbol{\mu})'\right.\right.$$
$$\left.\left.-b_i\left[\Lambda(\mathbf{x}_i-\boldsymbol{\mu})'+(\mathbf{x}_i-\boldsymbol{\mu})\Lambda'\right]+\Lambda\Lambda'c_i\right]\right].$$

By the Eq. (13), we can obtain the following estimators

$$\hat{\mu} = \frac{\sum_{i=1}^{n} \left(a_i \mathbf{x}_i - b_i \hat{\Lambda} \right)}{\sum_{i=1}^{n} a_i},$$

$$\hat{\Lambda} = \frac{\sum_{i=1}^{n} b_i \left(\mathbf{x}_i - \hat{\mu} \right)}{\sum_{i=1}^{n} c_i},$$

$$\hat{\Sigma} = \frac{1}{n} \sum_{i=1}^{n} \left[a_i \left(\mathbf{x}_i - \hat{\mu} \right) \left(\mathbf{x}_i - \hat{\mu} \right)' - b_i \left[\hat{\Lambda} (\mathbf{x}_i - \hat{\mu})' + (\mathbf{x}_i - \hat{\mu}) \hat{\Lambda}' \right] + c_i \hat{\Lambda} \hat{\Lambda}' \right].$$

Using the steps of EM algorithm, we can formulate the following simple iteratively algorithm to calculate the ML estimates of the parameters. The algorithm is iterated until a reasonable convergence criterion is reached. This algorithm can be easily implemented and the convergence is guaranteed since it is an EM algorithm. The iteratively algorithm is as following,

(1) Set iteration number $k - 1$ and select initial estimates for the parameters μ, Σ, Λ.
(2) Use $\hat{\mu}^{(k)}$, $\hat{\Sigma}^{(k)}$, $\hat{\Lambda}^{(k)}$ and $\mathbf{X}_i$ to calculate $a_i^{(k)}$, $b_i^{(k)}$ and $c_i^{(k)}$ for $i = 1, \cdots, n$.
(3) Use the following updating equations to calculate the new estimates,

$$\hat{\mu}^{(k+1)} = \frac{\sum_{i=1}^{n} \left(a_i^{(k)} \mathbf{x}_i - b_i^{(k)} \hat{\Lambda}^{(k)} \right)}{\sum_{i=1}^{n} a_i^{(k)}},$$

$$\hat{\Lambda}^{(k+1)} = \frac{\sum_{i=1}^{n} b_i^{(k)} \left(\mathbf{x}_i - \hat{\mu}^{(k+1)} \right)}{\sum_{i=1}^{n} c_i^{(k)}},$$

$$\hat{\Sigma}^{(k+1)} = \frac{1}{n} \sum_{i=1}^{n} \left[a_i^{(k)} \left(\mathbf{x}_i - \hat{\mu}^{(k+1)} \right) \left(\mathbf{x}_i - \hat{\mu}^{(k+1)} \right)' - b_i^{(k)} \left[\hat{\Lambda}^{(k+1)} \left(\mathbf{x}_i - \hat{\mu}^{(k+1)} \right)' \right. \right.$$
$$\left. \left. + \left(\mathbf{x}_i - \hat{\mu}^{(k+1)} \right) \hat{\Lambda}^{(k+1)'} \right] + c_i^{(k)} \hat{\Lambda}^{(k+1)} \hat{\Lambda}^{(k+1)'} \right].$$

(4) Repeat these steps until convergence.

4 EM Estimation for Multivariate Skew Slash Mixture Models

In this section the EM estimation for a k-component mixture model in which a set of random sample $\mathbf{X}_1, \cdots, \mathbf{X}_n \in \Re^p$ follows a mixture of multivariate skew slash distributions will be considered. Its probability density function of the mixture model can be written as

$$\mathbf{X}_j \sim \sum_{i=1}^{k} \omega_i f_p(\mathbf{x}_i; \boldsymbol{\mu}_i, \Sigma_i, \Lambda_i), \quad \omega_i \leq 0, \quad \sum_{i=1}^{k} \omega_i = 1, \tag{14}$$

where $\boldsymbol{\Theta} = (\boldsymbol{\Theta}_1, \cdots, \boldsymbol{\Theta}_k)$ with $\boldsymbol{\Theta}_i = (\omega_i, \boldsymbol{\mu}_i, \Sigma_i, \Lambda_i)$ being the unknown parameters of component i, and ω_i's being the mixing probabilities for $i = 1 \cdots, k$.

The EM estimates $\hat{\boldsymbol{\Theta}}$ based on a set of independent observations $\mathbf{X} = (\mathbf{X}_i', \cdots, \mathbf{X}_n')'$, is

$$\hat{\boldsymbol{\Theta}} = \underset{\boldsymbol{\Theta}}{\operatorname{argmax}} \, \ell(\boldsymbol{\Theta}|\mathbf{X}),$$

where

$$\ell(\boldsymbol{\Theta}|\mathbf{X}) = \sum_{j=1}^{n} \left(\sum_{i=1}^{k} \omega_i f_p(\mathbf{x}_i; \boldsymbol{\mu}_i, \Sigma_i, \Lambda_i) \right), \tag{15}$$

is called the observed log-likelihood function. Generally, there is no explicit analytical solution of $\hat{\boldsymbol{\Theta}}$, but it can be achieved iteratively by using the EM algorithm under the complete data framework discussed later.

In the context of hierarchical mixture modeling, for each $\mathbf{X}_j$, it is convenient to introduce a set of zero-one indicator variables $\mathbf{Z}_j = (Z_{1j}, \cdots, Z_{kj})'$ for $j = 1, \cdots, n$, which is a multinomial random vector with 1 trial and cell probabilities $\omega_1, \cdots, \omega_k$, denoted as $\mathbf{Z}_j \sim \mathcal{M}(1; \omega_1, \cdots, \omega_k)$. Note that the rth element $Z_{rj} = 1$ if $\mathbf{X}_j$ arises from component r. From (3), with the inclusion of indicator variables $\mathbf{Z}_j$'s, a hierarchical representation of (14) with $q > 0$ is given by

$$\mathbf{X}_j|(T = t_j, U = u_j, Z_{ij} = 1) \sim N_p \left(\boldsymbol{\mu}_i + u_j^{-\frac{1}{q}} \Lambda t_j, u^{-\frac{2}{q}} \Sigma_i \right), \tag{16}$$

$$\mathbf{X}_j|(U = u_j, Z_{ij} = 1) \sim SN_p \left(\boldsymbol{\mu}_i, u_j^{-\frac{2}{q}} \Sigma_i, u_j^{-\frac{1}{q}} \Lambda_i \right), \tag{17}$$

$$\mathbf{X}_j|(T = t_j, Z_{ij} = 1) \sim SL_p \left(\boldsymbol{\mu}_i + \Lambda_i t_j, \Sigma_i \right). \tag{18}$$

Denoted $\mathbf{Z} = (\mathbf{Z}_1', \cdots, \mathbf{Z}_n')$, $\mathbf{t} = (t_1, \cdots, t_n)'$, $\mathbf{u} = (u_1, \cdots, u_n)'$, $\boldsymbol{\omega} = (\omega_1, \cdots, \omega_k)'$, $\boldsymbol{\mu} = (\boldsymbol{\mu}_1', \cdots, \boldsymbol{\mu}_k')'$, $\Lambda = (\Lambda_1', \cdots, \Lambda_k')'$ and $\Sigma = (\Sigma_1, \cdots, \Sigma_k)$. The complete data log-likelihood function is

$$\ell(\boldsymbol{\Theta}|\mathbf{X}, \mathbf{t}, \mathbf{u}, \mathbf{Z}) \tag{19}$$

$$= \sum_{j=1}^{n} \sum_{i=1}^{k} Z_{ij} \left\{ -\frac{p}{2} \log(2\pi) - \log(\omega_i) - \frac{1}{2} \log|\Sigma_i| + \frac{1}{q} \log(u_j) - \frac{1}{2} t_j^2 \right\}$$

$$- \frac{1}{2} \sum_{j=1}^{n} \sum_{i=1}^{k} Z_{ij} \left\{ tr \left[\left(u_j^{-\frac{2}{q}} \Sigma_i \right)^{-1} \left(\mathbf{x}_j - \boldsymbol{\mu}_i - u_j^{-\frac{1}{q}} \Lambda_i t_j \right) \left(\mathbf{x}_j - \boldsymbol{\mu}_i - u_j^{-\frac{1}{q}} \Lambda_i t_j \right)' \right] \right\}.$$

Note that $\log(2\pi)Z_{ij}$, $Z_{ij} \log(u_i)$ and $Z_{ij} t_i^2$ does not contain any unknown parameters $\boldsymbol{\Theta}$, thus they can be ignored, and we knew only the observed data $\mathbf{X}_j$, while the missing data (t_j, u_j, Z_{ij}) are unknown, for $j = 1, \cdots, n$ and $i = 1, \cdots, k$. Similarly, we will take the conditional expectation of $\ell(\boldsymbol{\Theta})$ given the observed data $\mathbf{X}$ and the current estimates $\hat{\omega}$, $\hat{\boldsymbol{\mu}}$, $\hat{\Sigma}$ and $\hat{\boldsymbol{\Lambda}}$.

After taking the conditional expectation of $\ell(\boldsymbol{\Theta})$, we get

$$
Q(\boldsymbol{\Theta}) = E\left[\ell(\boldsymbol{\Theta})|\mathbf{X}, \hat{\boldsymbol{\Theta}}\right] \tag{20}
$$

$$
= \sum_{j=1}^{n}\sum_{i=1}^{k} Z_{ij}\left\{ -\log(\omega_i) - \frac{1}{2}\log|\Sigma_i| - tr\left[\Sigma^{-1}\left(u_j^{-\frac{2}{q}}(\mathbf{x}_j - \boldsymbol{\mu}_i)(\mathbf{x}_j - \boldsymbol{\mu}_i)'\right.\right.\right.
$$
$$
\left.\left.\left. - 2u_j^{\frac{1}{q}}t_j\left[\boldsymbol{\Lambda}(\mathbf{x}_j - \boldsymbol{\mu}_i)' + (\mathbf{x}_j - \boldsymbol{\mu}_i)\boldsymbol{\Lambda}'\right] + \boldsymbol{\Lambda}_i\boldsymbol{\Lambda}_i't_j^2\right)|\mathbf{X}, \hat{\boldsymbol{\Theta}}\right]\right\}.
$$

Define

$$
E\left[Z_{ij}|\mathbf{X}_j, \hat{\boldsymbol{\Theta}}\right] = \frac{\hat{\omega}_i f_p(\mathbf{x}_j; \hat{\boldsymbol{\mu}}_i, \hat{\Sigma}_i, \hat{\boldsymbol{\Lambda}}_i)}{\sum_{i=1}^{k}\hat{\omega}_i f_p(\mathbf{x}_j; \hat{\boldsymbol{\mu}}_i, \hat{\Sigma}_i, \hat{\boldsymbol{\Lambda}}_i)} = \hat{z}_{ij},
$$

$$
E\left[Z_{ij}u_j^{-\frac{2}{q}}|\mathbf{X}_j, \hat{\boldsymbol{\Theta}}\right] = \hat{z}_{ij}a_{ij},
$$

$$
E\left[Z_{ij}u_j^{\frac{1}{q}}t_j|\mathbf{X}_j, \hat{\boldsymbol{\Theta}}\right] = \hat{z}_{ij}b_{ij}
$$

$$
E\left[Z_{ij}t_j^2|\mathbf{X}_j, \hat{\boldsymbol{\Theta}}\right] = \hat{z}_{ij}c_{ij},
$$

where $a_{ij} = E[u_j^{\frac{2}{q}}|Z_{ij} = 1, \mathbf{X}_j, \hat{\boldsymbol{\Theta}}]$, $b_{ij} = E[u_j^{\frac{1}{q}}t_j|Z_{ij} = 1, \mathbf{X}_j, \hat{\boldsymbol{\Theta}}]$ and $c_{ij} = E[t_j^2|Z_{ij} = 1, \mathbf{X}_j, \hat{\boldsymbol{\Theta}}]$ can be obtained by (16), (17) and (18). The function (20) can be written as

$$
Q(\boldsymbol{\Theta}) = E\left[\ell(\boldsymbol{\Theta})|\mathbf{X}, \hat{\boldsymbol{\Theta}}\right] \tag{21}
$$

$$
= \sum_{j=1}^{n}\sum_{i=1}^{k} \hat{z}_{ij}\left\{ -\log(\omega_i) - \frac{1}{2}\log|\Sigma_i| - tr\left[\Sigma^{-1}\left(a_{ij}(\mathbf{x}_j - \boldsymbol{\mu}_i)(\mathbf{x}_j - \boldsymbol{\mu}_i)'\right.\right.\right.
$$
$$
\left.\left.\left. - b_{ij}\left[\boldsymbol{\Lambda}(\mathbf{x}_j - \boldsymbol{\mu}_i)' + (\mathbf{x}_j - \boldsymbol{\mu}_i)\boldsymbol{\Lambda}'\right] + \boldsymbol{\Lambda}_i\boldsymbol{\Lambda}_i'c_{ij}\right)|\mathbf{X}, \hat{\boldsymbol{\Theta}}\right]\right\}.
$$

By the Eq. (21), we can obtain the following estimators,

$$
\hat{\omega}_i = \frac{\sum_{j=1}^{n}\hat{z}_{ij}}{n},
$$

$$
\hat{\boldsymbol{\mu}}_i = \frac{\sum_{j=1}^{n}\left(\hat{z}_{ij}a_{ij}\mathbf{x}_j - \hat{z}_{ij}b_{ij}\hat{\boldsymbol{\Lambda}}_i\right)}{\sum_{j=1}^{n}\hat{z}_{ij}a_{ij}},
$$

$$\hat{\Lambda}_i = \frac{\sum_{j=1}^{n} \hat{z}_{ij} b_{ij} (\mathbf{x}_j - \hat{\boldsymbol{\mu}}_i)}{\sum_{j=1}^{n} \hat{z}_{ij} c_{ij}},$$

$$\hat{\Sigma}_i = \frac{1}{n} \sum_{j=1}^{n} \hat{z}_{ij} \left[a_{ij} (\mathbf{x}_j - \hat{\boldsymbol{\mu}}_i)(\mathbf{x}_j - \hat{\boldsymbol{\mu}}_i)' - b_{ij} \left[\hat{\Lambda} (\mathbf{x}_j - \hat{\boldsymbol{\mu}}_i)' + (\mathbf{x}_j - \hat{\boldsymbol{\mu}}_i)\hat{\Lambda}' \right] \right.$$
$$\left. + c_{ij} \hat{\Lambda}_i \hat{\Lambda}_i' \right].$$

Similarly, the iteratively algorithm is,

(1) Set iteration number $m = 1$ and select initial estimates for the parameters ω_i, $\boldsymbol{\mu}_i$, Σ_i, Λ_i, for $i = 1, \cdots, k$.
(2) Using $\hat{\omega}_i^{(m)}$, $\hat{\boldsymbol{\mu}}_i^{(m)}$, $\hat{\Sigma}_i^{(m)}$, $\hat{\Lambda}_i^{(m)}$ and $\mathbf{X}_j$ to calculate $\hat{z}_{ij}^{(m)}$, $a_{ij}^{(m)}$, $b_{ij}^{(m)}$ and $c_{ij}^{(m)}$ for $i = 1, \cdots, k$ and $j = 1, \cdots, n$.
(3) Using the following updating equations to calculate the new estimates, for $i = 1, \cdots, k$,

$$\hat{\omega}_i^{(m+1)} = \frac{\sum_{j=1}^{n} \hat{z}_{ij}^{(m)}}{n},$$

$$\hat{\boldsymbol{\mu}}_i^{(m+1)} = \frac{\sum_{j=1}^{n} \left(\hat{z}_{ij}^{(m+1)} a_{ij}^{(m+1)} \mathbf{x}_j - \hat{z}_{ij}^{(m+1)} b_{ij}^{(m+1)} \hat{\Lambda}_i^{(m+1)} \right)}{\sum_{j=1}^{n} \hat{z}_{ij}^{(m)} a_{ij}^{(m)}},$$

$$\hat{\Lambda}_i^{(m+1)} = \frac{\sum_{j=1}^{n} \hat{z}_{ij}^{(m)} b_{ij}^{(m)} \left(\mathbf{x}_j - \hat{\boldsymbol{\mu}}_i^{(m)} \right)}{\sum_{j=1}^{n} \hat{z}_{ij}^{(m)} c_{ij}^{(m)}},$$

$$\hat{\Sigma}_i^{(m+1)} = \frac{1}{n} \sum_{j=1}^{n} \hat{z}_{ij}^{(m)} \left[a_{ij}^{(m)} \left(\mathbf{x}_j - \hat{\boldsymbol{\mu}}_i^{(m)} \right) \left(\mathbf{x}_j - \hat{\boldsymbol{\mu}}_i^{(m)} \right)' - b_{ij} \left[\hat{\Lambda}_i^{(m)} \left(\mathbf{x}_j - \hat{\boldsymbol{\mu}}_i^{(m)} \right)' \right. \right.$$
$$\left. \left. + \left(\mathbf{x}_j - \hat{\boldsymbol{\mu}}_i^{(m)} \right) \hat{\Lambda}_i'^{(m)} \right] + c_{ij}^{(m)} \hat{\Lambda}_i^{(m)} \hat{\Lambda}_i'^{(m)} \right].$$

(4) Repeat these steps until convergence.

5 Simulation and Application

In this section, we give a small simulation study on bivariate skew slash distribution to show that the iteratively algorithm is working as claimed, and then, we present a practical application of the bivariate skew slash distribution.

We generate two different sample data using scale mixture representation given in (2) and Definition 1 with $q = 2$. Data I coming from $\boldsymbol{\mu}_1 = (1, 2)'$, $\boldsymbol{\Lambda}_2 = (1, -1)'$ and $\Sigma_1 = \begin{pmatrix} 1 & 0.5 \\ 0.5 & 1 \end{pmatrix}$. Data II coming from $\boldsymbol{\mu}_2 = (1, 1, 1)'$, $\boldsymbol{\Lambda}_2 = (1, 0, -1)'$ and $\Sigma_2 = \begin{pmatrix} 1 & 0.3 & 0.7 \\ 0.3 & 1 & 0.4 \\ 0.7 & 0.4 & 1 \end{pmatrix}$.

The data are generated as follows, for each $i = 1, \cdots, n$, we generate T_i from univariate standard normal distribution, Z_i from the bivariate normal distribution with location parameter $\mathbf{0}$ and scale parameter Σ_j with $j = 1, 2$, and U_i from the uniform distribution on $(0, 1)$ with power of $-1/q$.

In the simulation study, we take the sample size $n = 50$, 100, 200. To see the performance of the estimators, we used the Euclidean norm between the estimates and the true values of the parameters. Here, the Euclidean norm is defined as $\|\hat{\boldsymbol{\mu}} - \boldsymbol{\mu}\| = \sqrt{(\hat{\boldsymbol{\mu}} - \boldsymbol{\mu})'(\hat{\boldsymbol{\mu}} - \boldsymbol{\mu})}$, $\|\hat{\boldsymbol{\Lambda}} - \boldsymbol{\Lambda}\| = \sqrt{(\hat{\boldsymbol{\Lambda}} - \boldsymbol{\Lambda})'(\hat{\boldsymbol{\Lambda}} - \boldsymbol{\Lambda})}$ and $\|\hat{\Sigma} - \Sigma\| = \sqrt{(\hat{\Sigma} - \Sigma)'(\hat{\Sigma} - \Sigma)}$. We stop the algorithm when the Euclidean norms are less than 10^{-3}. For each simulation scheme, we repeat the simulation 100 times.

Simulation results are given in Tables 1 and 2. In these tables, we give the estimator for μ, Λ and Σ, we also provided the standard error for the 100 times run.

From these tables, we can see that when the sample size increases, the standard error for all the parameters decrease. Also, the sample size affects the estimator of parameters. When the sample size is getting bigger, the estimators are getting better. In summary, our simulation study works for some fixed situation, but here I just consider the bivariate case. As a result of this limited simulation study, we can say that the overall performance of the algorithm is satisfactory to compute the estimators for the parameters of the multivariate skew slash distribution.

In the next, we discuss a practical application on the actual data. The study was conducted at an experimental site of the Inner Mongolia Academy of Agriculture and Animal Husbandry Research Station. The site has an elevation of 1450 m and is in a temperate continental climate, characterized by a short growing season and long cold winter with a frost-free period of 175 days. The data was collected here about

Table 1 Estimator and standard error ($\pm$) for μ_1, Λ_1 and Σ_1 with different size of data

n	$\hat{\mu}_1$	S.E of $\hat{\mu}_1$	$\hat{\Lambda}_1$	S.E of $\hat{\Lambda}_1$	$\hat{\Sigma}_1$	S.E of $\hat{\Sigma}_1$
50	$\begin{pmatrix} 1.101 \\ 1.868 \end{pmatrix}$	0.237	$\begin{pmatrix} 1.112 \\ -1.097 \end{pmatrix}$	0.271	$\begin{pmatrix} 0.907 & 0.401 \\ 0.401 & 0.916 \end{pmatrix}$	0.310
100	$\begin{pmatrix} 1.036 \\ 1.986 \end{pmatrix}$	0.073	$\begin{pmatrix} 1.014 \\ -0.964 \end{pmatrix}$	0.032	$\begin{pmatrix} 0.973 & 0.472 \\ 0.472 & 0.976 \end{pmatrix}$	0.101
200	$\begin{pmatrix} 1.005 \\ 2.002 \end{pmatrix}$	0.010	$\begin{pmatrix} 0.998 \\ -1.002 \end{pmatrix}$	0.011	$\begin{pmatrix} 0.989 & 0.496 \\ 0.496 & 1.012 \end{pmatrix}$	0.039

Table 2 Estimator and standard error ($\pm$) for μ_2, Λ_2 and Σ_2 with different size of data

n	$\hat{\mu}_2$	S.E of $\hat{\mu}_1$	$\hat{\Lambda}_2$	S.E of $\hat{\Lambda}_2$	$\hat{\Sigma}_2$	S.E of $\hat{\Sigma}_2$
50	$\begin{pmatrix} 1.101 \\ 1.103 \\ 1.101 \end{pmatrix}$	0.197	$\begin{pmatrix} 1.101 \\ -0.107 \\ -0.892 \end{pmatrix}$	0.217	$\begin{pmatrix} 0.906\ 0.410\ 0.809 \\ 0.410\ 0.910\ 0.302 \\ 0.809\ 0.302\ 0.913 \end{pmatrix}$	0.341
100	$\begin{pmatrix} 1.041 \\ 1.039 \\ 1.029 \end{pmatrix}$	0.092	$\begin{pmatrix} 1.101 \\ -0.048 \\ -0.953 \end{pmatrix}$	0.104	$\begin{pmatrix} 0.977\ 0.351\ 0.760 \\ 0.351\ 0.979\ 0.453 \\ 0.760\ 0.453\ 0.969 \end{pmatrix}$	0.187
200	$\begin{pmatrix} 1.003 \\ 1.004 \\ 1.002 \end{pmatrix}$	0.011	$\begin{pmatrix} 1.002 \\ 0.004 \\ -0.993 \end{pmatrix}$	0.015	$\begin{pmatrix} 0.992\ 0.310\ 0.691 \\ 0.310\ 1.007\ 0.404 \\ 0.691\ 0.404\ 1.004 \end{pmatrix}$	0.032

the growth nutrient of root for different plants in 2013. The root were buried in the different districts under different treatments in May, took out in August, and dried them to a constant weight. We chose 617 roots and focus on two different variables of the nutrient, Carbon and Nitrogen. The content of total Carbon and total Nitrogen in root were calculated in percentage (Fig. 2).

We fit a bivariate skew slash distribution to these data with $q = 2$. The fitted parameters, obtained by EM estimation method, are $\hat{\mu} = (14.753, 0.836)'$, $\hat{\Lambda} = (-2.042, 1.276)'$ and $\hat{\Sigma} = \begin{pmatrix} 3.241 & 0.291 \\ 0.291 & 0.927 \end{pmatrix}$. We draw the fitted bivariate skew slash distribution density function in Fig. 3.

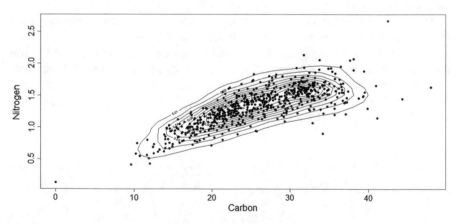

Fig. 2 Scatter plot and fitted contours of Carbon and Nitrogen

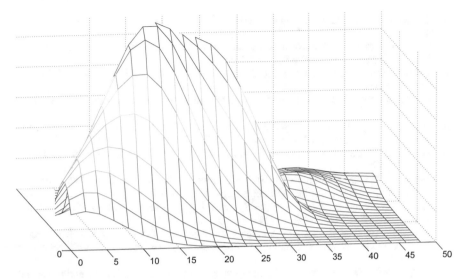

Fig. 3 Bivariate skew slash distribution with $\hat{\mu}$, $\hat{\Lambda}$ and $\hat{\Sigma}$

6 Discussion

We have introduced a multivariate skew slash distribution, which based on the a different type of skew normal distributed variable. A simulation and practical application are given based on the EM estimation algorithm. But, our study is limited and we only consider the case when q is known. As a result of this limited study, we can say that the overall performance of the algorithm is satisfactory to compute the estimations for the parameters of the multivariate variate skew slash distribution. In the future, we will analyze the case when q is unknown and the matrix variate skew slash distribution.

Acknowledgements This material is based upon work funded by National Natural Science Foundation of China (Grant No. IRT1259).

We gratefully acknowledge referees for their valuable comments and suggestions which greatly improve this paper.

References

1. Andrews DF, Bickel PJ, Hampel FR, Huber PJ, Rogers WH, Tukey JW (1972) Robust estimates of location: survey and advances. Princeton University Press, Princeton
2. Arellano-Valle RB, Bolfarine H, Lachos VH (1907) Bayesian inference for skew normal linear mixed models. J Appl Stat 34(6):663–682
3. Arellano-Valle R, Ozan S, Bolfarine H, Lachos V (2005) Skew normal measurement error models. J Multivar Anal 96(2):265–281

4. Arslan O (2008) An alternative multivariate skew slash distribution. Stat Prob Lett 78(16):2756–2761
5. Arslan O (2009) Maximum likelihood parameter estimation for the multivariate skew slash distribution. Stat Prob Lett 79(20):2158–2165
6. Arslan O, Genc AI (2009) A generalization of the multivariate slash distribution. J Stat Plan Inf 139(3):1164–1170
7. Azzalini A (1985) A class of distributions which includes the normal ones. Scandinavian J Stat 12(2):171–178
8. Azzalini A, Capitanio A (1999) Statistical applications of the multivariate skew normal distribution. J R Stat Soc Ser B (Stat Methodol) 61(3):579–602
9. Azzalini A, Dalla Valle A (1996) The multivariate skew normal distribution. Biometrika 83(4):715–726
10. Dempster AP, Laird NM, Rubin DB (1977) Maximum likelihood from incomplete data via the EM algorithm. J R Stat Soc Ser B (Methodological) 39(1):1–38
11. Gomez HW, Quintana FA, Torres FJ (2007) A new family of slash-distributions with elliptical contours. Stat Prob Lett 77(7):717–725
12. Gross AM (1973) A Monte Carlo swindle for estimators of location. Appl Stat 22:347–353
13. Gupta A, Huang W (2002) Quadratic forms in skew normal variates. J Math Anal Appl 273(2):558–564
14. Lin TI (2009) Maximum likelihood estimation for multivariate skew normal mixture models. J Multivar Anal 100(2):257–265
15. MacKenzie G, Peng D (2014) Introduction. Springer International Publishing, Switzerland, pp 1–6
16. Morgenthaler S, Tukey JW (1991) Configural polysampling: a route to practical robustness. Wiley, New York
17. Reyes J, Gmez HW, Bolfarine H (2013) Modified slash distribution. Statistics 47(5):929–941
18. Rogers WH, Tukey JW (1972) Understanding some long-tailed symmetrical distributions. Statistica Neerlandica 26(3):211–226
19. Sahu SK, Dey DK, Branco MD (2003) A new class of multivariate skew distributions with applications to Bayesian regression models. Can J Stat 31(2):129–150
20. Sekely G, Rizzo M (2005) A new test for multivariate normality. J Multivar Anal 93(1):58–80
21. Wang J, Genton MG (2006) The multivariate skew slash distribution. J Stat Plan Inf 136(1):209–220

Constructions of Multivariate Copulas

Xiaonan Zhu, Tonghui Wang and Varith Pipitpojanakarn

Abstract In this chapter, several general methods of constructions of multivariate copulas are presented, which are generalizations of some existing constructions in bivariate copulas. Dependence properties of new families are explored and examples are given for illustration of our results.

1 Introduction

In recent years, copulas are hot topics in probability and statistics. By *Sklar theorem* [16], the importance of copulas comes from two aspects, (1) describing dependence properties of random variables, such as Joe [6], Nelsen [11], Siburg [15], Tasena [17], Shan [14], Wei [20]; and (2) constructing the joint distributions of random variables. In the second direction, there are many papers devoting to the constructions of bivariate copulas, such as Rodríguez-Lallena [12], Kim [7], Durante [4], Mesiar [9], Aguilo [1], Mesiar [10], but few of constructions of multivariate copulas, such as Liebscher [8], Durante [3].

In this paper, we discussed several general methods of constructing multivariate copulas, which are generalizations of some bivariate results. The paper is organized as follows: In Sect. 2, we introduce some necessary definitions and existing results. Several general methods for constructing multivariate copulas are provided in Sect. 3 and their dependence properties are discussed in Sect. 4. Finally, two examples are given in Sect. 5.

X. Zhu · T. Wang (✉)
Department of Mathematical Sciences, New Mexico State University, Las Cruces, USA
e-mail: xzhu@nmsu.edu
e-mail: twang@nmsu.edu

V. Pipitpojanakarn
Faculty of Economics, Chiang Mai University, Chiang Mai, Thailand
e-mail: Varithp@cmu.ac.th

© Springer International Publishing AG 2017
V. Kreinovich et al. (eds.), *Robustness in Econometrics*,
Studies in Computational Intelligence 692, DOI 10.1007/978-3-319-50742-2_15

249

2 Definitions and Existing Results

A function $C : I^n \to I$ is called an *n-copula* [11], where $I = [0, 1]$, if C satisfies the following properties:

(i) C is *grounded*, i.e., for any $\mathbf{u} = (u_1, \cdots, u_n)' \in I^n$, if at least one $u_i = 0$, then $C(\mathbf{u}) = 0$,

(ii) One-dimensional marginals of C are uniformly distributed, i.e., for any $u_i \in I$, $i = 1, \cdots, n$,
$$C(1, \cdots, 1, u_i, 1, \cdots, 1) = u_i,$$

(iii) C is *n-increasing*, i.e., for any $\mathbf{u}, \mathbf{v} \in I^n$ such that $\mathbf{u} \le \mathbf{v}$, we have
$$V_C([\mathbf{u}, \mathbf{v}]) = \sum sgn(\mathbf{a}) C(\mathbf{a}) \ge 0,$$

where the sum is taken over all vertices $\mathbf{a}$ of the *n*-box $[\mathbf{u}, \mathbf{v}] = [u_1, v_1] \times \cdots \times [u_n, v_n]$, and

$$sgn(\mathbf{a}) = \begin{cases} 1, & \text{if } a_i = u_i \text{ for an even number of } i's, \\ -1, & \text{if } a_i = u_i \text{ for an odd number of } i's. \end{cases}$$

Equivalently,
$$V_C([\mathbf{u}, \mathbf{v}]) = \Delta_{\mathbf{u}}^{\mathbf{v}} C(\mathbf{t}) = \Delta_{u_n}^{v_n} \cdots \Delta_{u_1}^{v_1} C(\mathbf{t}),$$

where $\Delta_{u_k}^{v_k} C(\mathbf{t}) = C(t_1, \cdots, t_{k-1}, v_k, t_{k+1}, \cdots, t_n) - C(t_1, \cdots, t_{k-1}, u_k, t_{k+1}, \cdots, t_n)$, $k = 1, \cdots, n$.

Note that above three conditions ensure that the range of C is I. By Sklar's theorem [16], any n random variables $X_1, \cdots, X_n$ can be connected by an n-copula via the equation
$$F(x_1, \cdots, x_n) = C(F_1(x_1), \cdots, F_n(x_n)),$$

where F is the joint distribution function of $X_1, \cdots, X_n$, F_i is the marginal distribution functions of X_i, $i = 1, \cdots, n$. In addition, if $X_1, \cdots, X_n$ are continuous, then the copula C is unique.

There are three important functions for n-copulas defined respectively by

$$M_n(\mathbf{u}) = \min\{u_1, \cdots, u_n\},$$

$$\Pi_n(\mathbf{u}) = \prod_{i=1}^{n} u_i,$$

and

$$W_n(\mathbf{u}) = \max\{u_1 + \cdots + u_n - n + 1, 0\},$$

for all $\mathbf{u} \in I^n$. Functions M_n and Π_n are n-copulas for all $n \geq 2$, but W_n is not an n-copula for any $n \geq 3$. M_n and W_n are called the *Fréchert-Hoeffding upper bound* and *lower bound* of n-copulas respectively since for any n-copula C, we have $W_n \leq C \leq M_n$.

Let $H : I^n \to \mathbb{R}$ be a function. The functions $H_{i_1 i_2 \cdots i_k} : I^k \to \mathbb{R}$ are called *k-dimensional marginals* of H defined by

$$H_{i_1 i_2 \cdots i_k}(u_{i_1}, \cdots, u_{i_k}) = H(v_1, \cdots, v_n),$$

where $v_j = u_{i_l}$ if $j = i_l$ for some $l = 1, 2, \cdots, k$, otherwise, $v_j = 1$.

Any n-copula C defines a function $\overline{C} : I^n \to I$ by

$$\overline{C}(\mathbf{u}) = 1 + \sum_{k=1}^{n}(-1)^k \sum_{1 \leq i_1 < \cdots < i_k \leq n} C_{i_1 i_2 \cdots i_k}(u_{i_1}, \cdots, u_{i_k}). \tag{1}$$

It is called the *survival function* of C. For more details about copulas theory, see Nelsen's book [11].

Now let's recall some existing results. In 2004, Rodríguez-Lallena and Úbeda-Flores [12] considered the following family of bivariate copulas,

$$C_\theta(u, v) = uv + \theta f(u)g(v), \tag{2}$$

where $f, g : [0, 1] \to \mathbb{R}$ are two functions, $\theta \in \mathbb{R}$ is a parameter. This family is a generalization of the well-known bivariate *Farlie-Gumble-Morgenstern* (or FGM, for short) family,

$$C_\theta(u, v) = uv + \theta uv(1 - u)(1 - v),$$

where $u, v \in [0, 1]$ and $\theta \in [-1, 1]$. In 2011, Kim et al. [7] extended Rodríguez-Lallena and Úbeda-Flores's work to the family,

$$C(u, v) = C^*(u, v) + \theta f(u)g(v), \tag{3}$$

where C^* is a known bivariate copula, $f, g : [0, 1] \to \mathbb{R}$ are two functions, θ is a parameter. In 2013 and 2015, Durante et al. [4] and Mesiar et al. [10] considered more general cases,

$$C(u, v) = C^*(u, v) + H(u, v), \tag{4}$$

where C^* is a known bivariate copula, $H : [0, 1] \times [0, 1] \to \mathbb{R}$ is a function.

3 Constructions of Multivariate Copulas

The constructions of all above results are adding some *perturbation functions* to a given bivariate copula. In fact, any n-copula C can be represented by a perturbation of the *independent* copula Π_n [19]. Based on this idea, we are going to extend these bivariate results to multivariate cases.

Firstly, for any given n-copula $C^* : I^n \to [0, 1]$, we consider the construction,

$$C(u_1, u_2, \cdots, u_n) = C^*(u_1, u_2, \cdots, u_n) + H(u_1, u_2, \cdots, u_n), \qquad (5)$$

where $H : I^n \to \mathbb{R}$ is a function, called a *perturbation function*. C is called a *perturbation* of C^* by H.

Theorem 1 *Let C^* be an n-copula, $H : I^n \to \mathbb{R}$ be a function. C is defined by (5). Then C is an n-copula if and only if H satisfies the following three conditions,*

(i) $H(0, u_2, \cdots, u_n) = \cdots = H(u_1, \cdots, u_{n-1}, 0) = 0$ *for all* $(u_1, \cdots, u_n) \in I^n$,
(ii) *There are* $1 \leq i < j \leq n$ *such that*

$$H(u_1, \cdots, u_{i-1}, 1, u_{i+1}, \cdots, u_n) = H(u_1, \cdots, u_{j-1}, 1, u_{j+1}, \cdots, u_n) = 0,$$

(iii) $V_{C^*}([\boldsymbol{u}, \boldsymbol{v}]) + V_H([\boldsymbol{u}, \boldsymbol{v}]) \geq 0$ *for all n-box* $[\boldsymbol{u}, \boldsymbol{v}]$ *in* I^n.

Proof The conditions (i) and (ii) ensure that C is grounded, and its one-dimensional marginals are uniform distributed, respectively. The n-increasing property of C is guaranteed by the condition (iii).   □

Next we provide a necessary and sufficient condition on H under which C defined by (5) is an absolutely continuous n-copula.

Theorem 2 *Let C^* be an absolutely continuous n-copula with the density c^*, $H : I^n \to \mathbb{R}$ be a non-zero absolutely continuous function with the Radon-Nikodym derivative h with respect to the Lebesgue measure on I^n. C is defined by (5) is an absolutely continuous n-copula if and only if H satisfies the following conditions.*

(i) $H(0, u_2, \cdots, u_n) = \cdots = H(u_1, \cdots, u_{n-1}, 0) = 0$ *for all* $(u_i, \cdots, u_n) \in I^n$,
(ii) *There are* $1 \leq i < j \leq n$ *such that*

$$H(u_1, \cdots, u_{i-1}, 1, u_{i+1}, \cdots, u_n) = H(u_1, \cdots, u_{j-1}, 1, u_{j+1}, \cdots, u_n) = 0,$$

(iii) $c^* + h \geq 0$ *almost surely.*

Proof Firstly, the boundary conditions of copulas are ensured by the condition (i) and (ii).

Next, we show that the condition (iii) is equivalent to the n-increasing property of C. On the one hand, suppose that C is n-increasing. If $c^* + h$ is not non-negative almost surely, then there exist $\mathbf{u} < \mathbf{v} \in I^n$ such that $c^* + h < 0$ on $[\mathbf{u}, \mathbf{v}]$. Note that $V_C([\mathbf{u}, \mathbf{v}]) = \int_{[\mathbf{u}, \mathbf{v}]} (c^* + h)(\mathbf{t}) d\mathbf{t}$. So $V_C([\mathbf{u}, \mathbf{v}]) < 0$. It contradicts the n-increasing property of C. On the other hand, if $c^* + h \geq 0$ almost surely, we must have $V_C([\mathbf{u}, \mathbf{v}]) \geq 0$ for all $\mathbf{u}, \mathbf{v} \in I^n$ with $\mathbf{u} \leq \mathbf{v}$. □

Now let's consider a special case of (5) as follows, which are multivariate extensions of the result in [7].

$$C(u_1, u_2, \cdots, u_n) = C^*(u_1, u_2, \cdots, u_n) + \prod_{i=1}^{n} f_i(u_i), \qquad (6)$$

where C^* is an n-copula, $f_i : [0, 1] \to \mathbb{R}$ is a function, $i = 1, 2, \cdots, n$.

The following theorem give us a sufficient condition under which C defined by (6) is an n-copula.

Theorem 3 *Let C^* be an n-copula, $f_i : [0, 1] \to \mathbb{R}$ be a function, $i = 1, 2, \cdots, n$. $C : [0, 1]^n \to \mathbb{R}$ is defined by (6) is an n-copula if $f_1, \cdots, f_n$ satisfy the following conditions,*

(i) $f_1(0) = \cdots = f_n(0) = 0$, *and there exist at least two functions f_i and f_j such that $f_i(1) = f_j(1) = 0$, $1 \leq i, j \leq n$,*
(ii) f_i *is absolutely continuous,*
(iii) $\min(B) \geq \sup \left\{ -\dfrac{V_{C^*}([\boldsymbol{u}, \boldsymbol{v}])}{\Delta(\boldsymbol{u}, \boldsymbol{v})} : \boldsymbol{u}, \boldsymbol{v} \in [0, 1]^n, \boldsymbol{u} < \boldsymbol{v} \right\}$,
where $B = \{\alpha_{i_1} \cdots \alpha_{i_k} \beta_{j_1} \cdots \beta_{j_{n-k}} : 1 \leq k \leq n, k \text{ is odd}, i_1, \cdots, i_k \text{ and } j_1, \cdots, j_{n-k}$ are pairwise distinct$\}$, $\alpha_i = \inf\{f_i'(u_i) : u_i \in A_i\} < 0$, $\beta_i = \sup\{f_i'(u_i) : u_i \in A_i\} > 0$, $A_i = \{u_i \in [0, 1] : f'(u_i) \text{ exists}\}$, $i = 1, \cdots, n$, and $\Delta(\boldsymbol{u}, \boldsymbol{v}) = (v_1 - u_1) \cdots (v_n - u_n)$.

Proof Firstly, if there is $f_i = 0$, then $C = C^*$ is an n-copula. So without loss of generality, we may assume that f_i is non-zero, $i = 1, \cdots, n$.

Since C^* is an n-copula, C is grounded and its marginals are uniformly distributed if and only if C satisfies the above condition (i). Next we are going to show that C is n-increasing if C satisfies the conditions (ii) and (iii).

Suppose that C satisfies conditions (ii) and (iii). By Lemma 2.2 in [12], it holds that for any $\mathbf{u}, \mathbf{v} \in I^n$ with $\mathbf{u} < \mathbf{v}$,

$$\frac{(f_1(v_1) - f_1(u_1)) \cdots (f_n(v_n) - f_n(u_n))}{(v_1 - u_1) \cdots (v_n - u_n)} \geq -\frac{V_{C^*}([\mathbf{u}, \mathbf{v}])}{\Delta(\mathbf{u}, \mathbf{v})},$$

i.e.,

$$V_C([\mathbf{u}, \mathbf{v}]) = V_{C^*}([\mathbf{u}, \mathbf{v}]) + (f_1(v_1) - f_1(u_1)) \cdots (f_n(v_n) - f_n(u_n)) \geq 0,$$

so C is n-increasing.   □

Based on the construction (6), we introduce the following parametric families of n-copulas, which is a multivariate extension of (3).

$$C(u_1, u_2, \cdots, u_n) = C^*(u_1, u_2, \cdots, u_n) + \theta \prod_{i=1}^{n} f_i(u_i), \qquad (7)$$

where C^* is an n-copula, $f_i : [0, 1] \to \mathbb{R}$ is a function, $i = 1, 2, \cdots, n, \theta \in \mathbb{R}$.

Corollary 1 *Let C^* be an n-copula, $f_i : [0, 1] \to \mathbb{R}$ be a function, $i = 1, 2, \cdots, n$. $C : [0, 1]^n \to \mathbb{R}$ is defined by (6) is an n-copula if $f_1, \cdots, f_n$ and θ satisfy the following conditions,*

(i) *$f_1(0) = \cdots = f_n(0) = 0$, and there exist at least two functions f_i and f_j such that $f_i(1) = f_j(1) = 0, 1 \leq i, j \leq n$,*

(ii) *f_i is absolutely continuous,*

(iii) $\sup\{-\dfrac{V_{C^*}([\boldsymbol{u}, \boldsymbol{v}])}{\Delta(\boldsymbol{u}, \boldsymbol{v})} : \boldsymbol{u}, \boldsymbol{v} \in [0, 1]^n, \boldsymbol{u} < \boldsymbol{v}\}\dfrac{1}{\max(B')} \leq \theta \leq \sup\{-\dfrac{V_{C^*}([\boldsymbol{u}, \boldsymbol{v}])}{\Delta(\boldsymbol{u}, \boldsymbol{v})} :$

$\boldsymbol{u}, \boldsymbol{v} \in [0, 1]^n, \boldsymbol{u} < \boldsymbol{v}\}\dfrac{1}{\min(B)},$

where B is the same as Theorem 3, $B' = \{\alpha_{i_1} \cdots \alpha_{i_k} \beta_{j_1} \cdots \beta_{j_{n-k}} : 1 \leq k \leq n, k$ is even, $i_1, \cdots, i_k$ and $j_1, \cdots, j_{n-k}$ are pairwise distinct\}, $\alpha_i = \inf\{f_i'(u_i) : u_i \in A_i\} < 0, \beta_i = \sup\{f_i'(u_i) : u_i \in A_i\} > 0, A_i = \{u_i \in [0, 1] : f'(u_i) \ exists\}, i = 1, \cdots, n,$ and $\Delta(\boldsymbol{u}, \boldsymbol{v}) = (v_1 - u_1) \cdots (v_n - u_n)$.

Remark 1 Conditions in Theorem 3 and Corollary 1 are sufficient but may not be necessary. Consider the Fréchert-Hoeffding upper bound of n-copulas, $M_n(u_1, \cdots, u_n) = \min\{u_1, \cdots, u_n\}$. For any $\mathbf{u}, \mathbf{v} \in [0, 1]^n$ such that $\mathbf{u} < \mathbf{v}$, it can be shown that

$$V_{M_n}([\mathbf{u}, \mathbf{v}]) = \max\{\min\{v_1, \cdots, v_n\} - \max\{u_1, \cdots, u_n\}, 0\}.$$

Thus,

$$\sup\left\{-\frac{V_{C^*}([\mathbf{u}, \mathbf{v}])}{\Delta(\mathbf{u}, \mathbf{v})} : \mathbf{u}, \mathbf{v} \in [0, 1]^n, \mathbf{u} < \mathbf{v}\right\} = 0.$$

So functions $f_1, \cdots, f_n$ that satisfy conditions in Theorem 3 or Corollary 1 for M_n must be zero, i.e., $f_1 = \cdots = f_n = 0$.

Next we provide a stronger sufficient condition on $f_1, \cdots, f_n$ to ensure that C defined by (6) is an n-copula. Example 2.1 in [12] shows that the condition is not necessary.

Theorem 4 *Let C be defined by (6). C is an n-copula if $f_1, \cdots, f_n$ satisfy the following conditions,*

(i) $f_1(0) = \cdots = f_n(0) = 0$, and there exist at least two functions f_i and f_j such that $f_i(1) = f_j(1) = 0, 1 \leq i, j \leq n$,

(ii) f_i satisfies the Lipschitz condition,

$$|f_i(v) - f_i(u)| \leq M_i|v - u|,$$

for all $u, v \in I$, such that $M_i > 0, i = 1, \cdots, n$, and

$$\prod_1^n M_i \leq \inf\left\{\frac{V_{C^*}([u, v])}{\Delta(u, v)} : u, v \in [0, 1]^n, u \leq v\right\}.$$

Proof By the condition (i), C is grounded and one-dimensional marginals of C are uniformly distributed. For any $\mathbf{u}, \mathbf{v} \in I^n$ with $\mathbf{u} < \mathbf{v}$, by the condition (ii), we have

$$-\frac{(f_1(v_1) - f_1(u_1)) \cdots (f_n(v_n) - f_n(u_n))}{(v_1 - u_1) \cdots (v_n - u_n)} \leq \frac{|f_1(v_1) - f_1(u_1)| \cdots |f_n(v_n) - f_n(u_n)|}{|v_1 - u_1| \cdots |v_n - u_n|}$$

$$\leq \prod_1^n M_i$$

$$\leq \inf\left\{\frac{V_{C^*}([\mathbf{u}, \mathbf{v}])}{\Delta(\mathbf{u}, \mathbf{v})} : \mathbf{u}, \mathbf{v} \in [0, 1]^n, \mathbf{u} \leq \mathbf{v}\right\}.$$

So

$$\frac{(f_1(v_1) - f_1(u_1)) \cdots (f_n(v_n) - f_n(u_n))}{(v_1 - u_1) \cdots (v_n - u_n)} \geq \sup\left\{-\frac{V_{C^*}([\mathbf{u}, \mathbf{v}])}{\Delta(\mathbf{u}, \mathbf{v})} : \mathbf{u}, \mathbf{v} \in [0, 1]^n, \mathbf{u} \leq \mathbf{v}\right\}.$$

Thus, as the proof of Theorem 3, C is n-increasing.   □

4 Properties of New Families

In this section, we are going to study some non-parametric copula-based measures of multivariate association, some dependence concepts for copulas defined in Sect. 3 and some properties of those families.

Firstly, recall that the multivariate generalizations of *Kendall's tau*, *Spearman's rho*, and *Blomqvist's beta* (see [13, 18] for details) are given by

$$\tau_n(C) = \frac{1}{2^{n-1} - 1}\left[2^n \int_{I^n} C(\mathbf{u})dC(\mathbf{u}) - 1\right], \tag{8}$$

$$\rho_n(C) = \frac{n+1}{2^n - n - 1} \left[2^{n-1} \left(\int_{I^n} C(\mathbf{u}) d\Pi_n(\mathbf{u}) + \int_{I^n} \Pi_n(\mathbf{u}) dC(\mathbf{u}) \right) - 1 \right], \quad (9)$$

$$\beta_n(C) = \frac{2^{n-1} \left[C\left(\frac{1}{2}\mathbf{1}_n\right) + \overline{C}\left(\frac{1}{2}\mathbf{1}_n\right) \right] - 1}{2^{n-1} - 1}, \quad (10)$$

where $\mathbf{1}_n$ is the vector $(1, \cdots, 1)' \in \mathbb{R}^n$.

Theorem 5 *Let C be an n-copula defined by (5), then the Kendall's tau, Spearman's rho, and Blomqvist's beta of C are given by*

$$\tau_n(C) = \tau_n(C^*) + \tau_n(H) + a_1, \quad (11)$$

$$\rho_n(C) = \rho_n(C^*) + \rho_n(H) + \frac{n+1}{2^n - n - 1}, \quad (12)$$

$$\beta_n(C) = \beta_n(C^*) + \beta_n(H) + \frac{1 - 2^{n-1}}{2^n - 1}, \quad (13)$$

where $a_1 = \frac{1}{2^{n-1} - 1} \left[2^n \int_{I^n} C^*(\mathbf{u}) dH(\mathbf{u}) + 2^n \int_{I^n} H(\mathbf{u}) dC^*(\mathbf{u}) + 1 \right]$.

Proof Firstly, by the definition of τ_n,

$$\tau_n(C) = \frac{1}{2^{n-1} - 1} \left[2^n \int_{I^n} C(\mathbf{u}) dC(\mathbf{u}) - 1 \right]$$

$$= \frac{1}{2^{n-1} - 1} \left[2^n \int_{I^n} C^*(\mathbf{u}) + H(\mathbf{u}) d(C^*(\mathbf{u}) + H(\mathbf{u})) - 1 \right]$$

$$= \frac{1}{2^{n-1} - 1} [2^n \int_{I^n} C^*(\mathbf{u}) dC^*(\mathbf{u}) + 2^n \int_{I^n} H(\mathbf{u}) dH(\mathbf{u})$$

$$+ 2^n \int_{I^n} C^*(\mathbf{u}) dH(\mathbf{u}) + 2^n \int_{I^n} H(\mathbf{u}) dC^*(\mathbf{u}) - 1]$$

$$= \tau_n(C^*) + \tau_n(H) + a_1.$$

Secondly, by the definition of ρ_n,

$$\rho_n(C) = \frac{n+1}{2^n - n - 1} \left\{ 2^{n-1} \left[\int_{I^n} C(\mathbf{u}) d\Pi_n(\mathbf{u}) + \int_{I^n} \Pi_n(\mathbf{u}) dC(\mathbf{u}) \right] - 1 \right\}$$

$$= \frac{n+1}{2^n - n - 1} \left\{ 2^{n-1} \left[\int_{I^n} C^*(\mathbf{u}) + H(\mathbf{u}) d\Pi_n(\mathbf{u}) + \int_{I^n} \Pi_n(\mathbf{u}) d(C^*(\mathbf{u}) + H(\mathbf{u})) \right] - 1 \right\}$$

$$= \frac{n+1}{2^n - n - 1} \{ 2^{n-1} [\int_{I^n} C^*(\mathbf{u}) d\Pi_n(\mathbf{u}) + \int_{I^n} H(\mathbf{u}) d\Pi_n(\mathbf{u}) + \int_{I^n} \Pi_n(\mathbf{u}) dC^*(\mathbf{u})$$

$$+ \int_{I^n} \Pi_n(\mathbf{u}) dH(\mathbf{u})] - 1 \}$$

$$= \rho_n(C^*) + \rho_n(H) + \frac{n+1}{2^n - n - 1}.$$

Lastly, by the definition of survival functions, for any $\mathbf{u} \in I^n$,

$$\overline{C}(\mathbf{u}) = 1 + \sum_{k=1}^{n}(-1)^k \sum_{1 \le i_1 < \cdots < i_k \le n} C_{i_1 i_2 \cdots i_k}(u_{i_1}, \cdots, u_{i_k})$$

$$= 1 + \sum_{k=1}^{n}(-1)^k \sum_{1 \le i_1 < \cdots < i_k \le n} \left[C^*_{i_1 i_2 \cdots i_k}(u_{i_1}, \cdots, u_{i_k}) + H_{i_1 i_2 \cdots i_k}(u_{i_1}, \cdots, u_{i_k}) \right]$$

$$= 1 + \sum_{k=1}^{n}(-1)^k \sum_{1 \le i_1 < \cdots < i_k \le n} C^*_{i_1 i_2 \cdots i_k}(u_{i_1}, \cdots, u_{i_k})$$

$$+ 1 + \sum_{k=1}^{n}(-1)^k \sum_{1 \le i_1 < \cdots < i_k \le n} H_{i_1 i_2 \cdots i_k}(u_{i_1}, \cdots, u_{i_k}) - 1$$

$$= \overline{C^*}(\mathbf{u}) + \overline{H}(\mathbf{u}) - 1.$$

Thus,

$$\beta_n(C) = \frac{2^{n-1}\left[C\left(\frac{1}{2}\mathbf{1}_n\right) + \overline{C}\left(\frac{1}{2}\mathbf{1}_n\right) \right] - 1}{2^{n-1} - 1}$$

$$= \frac{2^{n-1}\left[C^*\left(\frac{1}{2}\mathbf{1}_n\right) + H\left(\frac{1}{2}\mathbf{1}_n\right) + \overline{C^*}\left(\frac{1}{2}\mathbf{1}_n\right) + \overline{H}\left(\frac{1}{2}\mathbf{1}_n\right) - 1 \right] - 1}{2^{n-1} - 1}$$

$$= \frac{2^{n-1}\left[C^*\left(\frac{1}{2}\mathbf{1}_n\right) + \overline{C^*}\left(\frac{1}{2}\mathbf{1}_n\right) \right] - 1 + 2^{n-1}\left[H\left(\frac{1}{2}\mathbf{1}_n\right) + \overline{H}\left(\frac{1}{2}\mathbf{1}_n\right) \right] - 1 + 1 - 2^{n-1}}{2^{n-1} - 1}$$

$$= \beta_n(C^*) + \beta_n(H) + \frac{1 - 2^{n-1}}{2^n - 1}.$$

$\square$

Remark 2 In the above theorem, although the perturbation function H is not a copula, we still use $m(H)$ to denote the corresponding values of H, where $m = \tau_n, \rho_n,$ or β_n, and use $\overline{H}$ to denoted the corresponding function of H defined by (1). The similar notations are used in the following context.

Remark 3 As n increasing, we can see that

$$\tau_n(C) \approx \tau_n(C^*) + \tau_n(H) + 2\int_{I^n} C^*(\mathbf{u})dH(\mathbf{u}) + 2\int_{I^n} H(\mathbf{u})dC^*(\mathbf{u}),$$

$$\rho_n(C) \approx \rho_n(C^*) + \rho_n(H),$$

and

$$\beta_n(C) \approx \beta_n(C^*) + \beta_n(H).$$

Corollary 2 *Let C be an n-copula defined by (7), then the Kendall's tau, Spearman's rho, and Blomqvist's beta of C are given by*

$$\tau_n(C) = \tau_n(C^*) + \tau_n(\theta \prod_{i=1}^{n} f_i) + a_2, \tag{14}$$

$$\rho_n(C) = \rho_n(C^*) + \rho_n(\theta \prod_{i=1}^{n} f_i) + \frac{n+1}{2^n - n - 1}, \tag{15}$$

$$\beta_n(C) = \beta_n(C^*) + \beta_n(\theta \prod_{i=1}^{n} f_i) + \frac{1 - 2^{n-1}}{2^n - 1}, \tag{16}$$

where $a_2 = \dfrac{1}{2^{n-1} - 1} \left[2^n \int_{I^n} \theta C^*(\boldsymbol{u}) \prod_{i=1}^{n} f_i'(u_i) d\boldsymbol{u} + 2^n \int_{I^n} \theta \prod_{i=1}^{n} f_i(u_i) dC^*(\boldsymbol{u}) + 1 \right].$

In 2013, Tasena et al. [17] defined a measure of *multivariate complete dependence* as follows. Let C be an n-copula of random variables $X_1, \cdots, X_n$. Define

$$\delta_i(X_1, \cdots, X_n) = \delta_i(C) = \frac{\int (\partial_i C - \pi_i C)^2}{\int \pi_i C(1 - \pi_i C)},$$

where $\pi_i C : I^{n-1} \to I$ is defined by

$$\pi_i C(u_1, \cdots, u_{n-1}) = C(u_1, \cdots, u_{i-1}, 1, u_i, \cdots, u_{n-1}), \qquad i = 1, 2, \cdots, n.$$

By Theorem 3.6 in [17], δ_i satisfies following properties,

(i) $0 \leq \delta_i(C) \leq 1$,
(ii) $\delta_i(C) = 1$ if and only if $(X_1, \cdots, X_{i-1}, X_{i+1}, \cdots, X_n)$ is a function of X_i. For details, see [17].

Theorem 6 *Let C be an n-copula defined by (5). If*

$$H(u_1, \cdots, u_{i-1}, 1, u_{i+1}, \cdots, u_{n_1}) = 0,$$

then

$$\delta_i(C) = \delta(C^*) - \frac{\int 2\partial_i H(\partial_i C^* - \pi_i C^*) - (\partial_i H)^2}{\int \pi_i C^*(1 - \pi_i C^*)}.$$

Proof By the definition,

$$
\begin{aligned}
\delta_i(C) &= \frac{\int (\partial_i C - \pi_i C)^2}{\int \pi_i C(1 - \pi_i C)} \\
&= \frac{\int (\partial_i (C^* + H) - \pi_i (C^* + H))^2}{\int \pi_i (C^* + H)[1 - \pi_i (C^* + H)]} \\
&= \frac{\int (\partial_i C^* + \partial_i H - \pi_i C^* - \pi_i H)^2}{\int (\pi_i C^* + \pi_i H)(1 - \pi_i C^* - \pi_i H)}.
\end{aligned}
$$

If $\pi_i H(u_1, \cdots, u_{n_1}) = H(u_1, \cdots, u_{i-1}, 1, u_{i+1}, \cdots, u_{n_1}) = 0$, then

$$
\begin{aligned}
\delta_i(C) &= \frac{\int (\partial_i C^* + \partial_i H - \pi_i C^* - \pi_i H)^2}{\int (\pi_i C^* + \pi_i H)(1 - \pi_i C^* \pi_i H)} \\
&= \frac{\int (\partial_i C^* + \partial_i H - \pi_i C^*)^2}{\int \pi_i C^*(1 - \pi_i C^*)} \\
&= \frac{\int \left[(\partial_i C^* - \pi_i C^*)^2 - 2\partial_i H(\partial_i C^* - \pi_i C^*) + (\partial_i H)^2 \right]}{\int \pi_i C^*(1 - \pi_i C^*)} \\
&= \frac{\int (\partial_i C^* - \pi_i C^*)^2}{\int \pi_i C^*(1 - \pi_i C^*)} - \frac{\int 2\partial_i H(\partial_i C^* - \pi_i C^*) - (\partial_i H)^2}{\int \pi_i C^*(1 - \pi_i C^*)} \\
&= \delta(C^*) - \frac{\int 2\partial_i H(\partial_i C^* - \pi_i C^*) - (\partial_i H)^2}{\int \pi_i C^*(1 - \pi_i C^*)}.
\end{aligned}
$$

$\square$

Corollary 3 *Let C be an n-copula defined by (7). If $f_i(1) = 0$, then*

$$
\delta_i(C) = \delta_i(C^*) - \frac{\int 2\theta f_i' \prod_{j \neq i} f_j(\partial_i C^* - \pi_i C^*) - (\theta f_i' \prod_{j \neq i} f_j)^2}{\int \pi_i C^*(1 - \pi_i C^*)}.
$$

Now, let's recall some dependence concepts of copulas. For details, see [6, 11]. Let C_1 and C_2 be two n-copulas. If $C_1 \geq C_2$ ($\overline{C}_1 \geq \overline{C}_2$ resp.), i.e., $C_1(\mathbf{u}) \geq C_2(\mathbf{u})$ ($\overline{C}_1(\mathbf{u}) \geq \overline{C}_2(\mathbf{u})$ resp.) for all $\mathbf{u} \in I^n$, then we say that C_1 is *more positive lower* (*upper* resp.) *orthant dependent* (PLOD) (PUOD resp.) than C_2. C_1 is *more positive orthant dependent* (POD) than C_2 if $C_1 \geq C_2$ and $\overline{C}_1 \geq \overline{C}_2$ hold.

The following results give us some dependence relations between C and C^*. The proof is trivial.

Proposition 1 *Let C_1 and C_2 be two n-copulas defined by (5). If they share the same n-copula C^* and may have different perturbation functions H_i $i = 1, 2$, then*
   (i) *C_1 more PLOD than C_2 if and only if $H_1 \geq H_2$,*
   (ii) *C_1 more PUOD than C_2 if and only if $\overline{H}_1 \geq \overline{H}_2$,*
   (iii) *C_1 more POD than C_2 if and only if $H_1 \geq H_2$ and $\overline{H}_1 \geq \overline{H}_2$.*

Proposition 2 *Let C_1 and C_2 are two n-copulas defined by (7). If they share the same known n-copula C^* and may have different perturbation functions $f_{j1}, \cdots, f_{jn}$, and parameters θ_j, $j = 1, 2$ respectively, then*

(i) C_1 more PLOD than C_2 if and only if $\overline{\theta_1 \prod\limits_{i=1}^{n} f_{1i}} \geq \overline{\theta_2 \prod\limits_{i=1}^{n} f_{2i}}$,

(ii) C_1 more PUOD than C_2 if and only if $\underline{\theta_1 \prod\limits_{i=1}^{n} f_{1i}} \geq \underline{\theta_2 \prod\limits_{i=1}^{n} f_{2i}}$,

(iii) C_1 more POD than C_2 if and only if $\underline{\theta_1 \prod\limits_{i=1}^{n} f_{1i}} \geq \underline{\theta_2 \prod\limits_{i=1}^{n} f_{2i}}$ and $\overline{\theta_1 \prod\limits_{i=1}^{n} f_{1i}} \geq$

$\overline{\theta_2 \prod\limits_{i=1}^{n} f_{2i}}$.

The next theorem give us a property of the construction (6).

Theorem 7 *Let $(U_1^*, \cdots, U_n^*)$ and $(U_1, \cdots, U_n)$ be random vectors with uniform marginals on $[0, 1]$ and connected by copulas C^* and C respectively. C and C^* satisfy conditions of Theorem 3. Suppose that $f_i(1) = f_j(1) = 0$, $1 \leq i < j \leq n$.*

(i) If there is $1 \leq l \leq n$ such that $l \neq i, j$ and $f_l(1) = 0$, then $P\{U_i < U_j\} = P\{U_i^* < U_j^*\}$,

(ii) If $f_l(1) \neq 0$ for all $l \neq i, j$ and $f_i = f_j$, then $P\{U_i < U_j\} = P\{U_i^* < U_j^*\}$.

Proof (i) Let c and c^* be the densities of C and C^* respectively, then we have

$$c(\mathbf{u}) = \frac{\partial^n C(\mathbf{u})}{\partial u_1 \cdots \partial u_n} = c^*(\mathbf{u}) + \prod_{i=1}^{n} f_i'(u_i).$$

Then

$$P\{U_i < U_j\} = \int_0^1 \cdots \int_0^{u_j} \cdots \int_0^1 c(u_1, \cdots, u_i, \cdots, u_n) du_1 \cdots du_i \cdots du_n$$

$$= \int_0^1 \cdots \int_0^{u_j} \cdots \int_0^1 c^*(u_1, \cdots, u_i, \cdots, u_n) du_1 \cdots du_i \cdots du_n$$

$$+ \int_0^1 \cdots \int_0^{u_j} \cdots \int_0^1 f_1'(u_i) \cdots f_i'(u_i) \cdots f_n'(u_n) du_1 \cdots du_i \cdots du_n$$

$$= P\{U_i^* < U_j^*\} + \prod_{k \neq i, j}^{n} (f_k(1) - f_k(0)) \int_0^1 \int_0^{u_j} f_j'(u_j) f_i'(u_i) du_i du_j$$

$$= P\{U_i^* < U_j^*\},$$

since $f_l(0) = f_l(1) = 0$ and $l \neq i, j$.

(ii) Similarly, if $f_l(1) \neq 0$ for all $l \neq i, j$,

$$P\{U_i < U_j\} = P\{U_i^* < U_j^*\} + \prod_{k \neq i,j}^{n} (f_k(1) - f_k(0)) \int_0^1 \int_0^{u_j} f_j'(u_j) f_i'(u_i) du_i du_j$$

$$= P\{U_i^* < U_j^*\} + \prod_{k \neq i,j}^{n} f_k(1) \int_0^1 \int_0^{u_j} f_j'(u_j) f_i'(u_i) du_i du_j$$

$$= P\{U_i^* < U_j^*\} + \prod_{k \neq i,j}^{n} f_k(1) \int_0^1 f_i(u_j) f_j'(u_j) du_j.$$

Since $f_i = f_j$,

$$\int_0^1 f_i(u_j) f_j'(u_j) du_j = f_j(1) f_i(1) - f_j(0) f_i(0) - \int_0^1 f_j(u_j) f_i'(u_j) du_j$$

$$= -\int_0^1 f_i(u_j) f_i'(u_j) du_j = -\int_0^1 f_i(u_j) f_j'(u_j) du_j,$$

and hence $\int_0^1 f_i(u_j) f_j'(u_j) du_j = 0$. So $P\{U_i < U_j\} = P\{U_i^* < U_j^*\}$. $\qquad\square$

The following example shows that the converse of the above result (ii) in Theorem 7 may not hold in general. Moreover, it shows that Theorem 3 in [7] is incorrect.

Example 1 Let (U^*, V^*) and (U, V) be random vectors with uniform marginals on $[0, 1]$. Suppose that (U^*, V^*) is connected by the independent copula, i.e., $C^*(u, v) = uv$, and (U, V) is connected by $C(u, v) = C^*(u, v) + f(u)g(v)$, where $f(u) = u(1 - u)$, $g(v) = \frac{1}{2}v(1 - v)$. Then f and g satisfy the conditions in Theorem 3. In fact, C belongs to the bivariate FGM family.

As the proof of the above theorem, we have

$$P\{U < V\} = \int_0^1 \int_0^v c(u, v) du dv = \int_0^1 \int_0^v c^*(u, v) + f'(u)g'(v) du dv$$

$$= P\{U^* < V^*\} + \int_0^1 \int_0^v f'(u)g'(v) du dv = P\{U^* < V^*\} + \int_0^1 f(v)g'(v) dv.$$

where

$$\int_0^1 f(v)g'(v) dv = \int_0^1 \frac{1}{2}v(1 - v)(1 - 2v) dv = 0.$$

Thus $P\{U < V\} = P\{U^* < V^*\}$, but $f \neq g$.

5 Examples

In this section, we provide two examples. The given copula C^* in the first example is the simplest one, the independent copula. To emphasis multivariate and for simplicity, we will only consider 3-copulas, but results could be extended to n-copulas. In the second example, C^* is nontrivial. Also for simplicity, we will only consider 2-copulas.

Example 2 Let C^* be the independent 3-copula, i.e., $C^*(u, v, w) = uvw$. Let $f(x) = x(1 - x^k)$, where $u, v, w, x \in I, k \in \mathbb{N}$, the set of all positive integers. Consider the 3-copula family,

$$C(u, v, w) = C^*(u, v, w) + \theta f(u) f(v) f(w)$$
$$= uvw + \theta uvw(1 - u^k)(1 - v^k)(1 - w^k),$$

where $\theta \in \mathbb{R}$.

It is clear that $f(x)$ satisfies the conditions (i) and (ii) of Corollary 1. Next we will use the condition (iii) of Corollary 1 to find the range of the parameter θ for each k. Firstly, it is easy to see that $\dfrac{V_{C^*}([\mathbf{u}, \mathbf{v}])}{\Delta(\mathbf{u}, \mathbf{v})} = 1$ for any $\mathbf{u}, \mathbf{v} \in [0, 1]^3$ with $\mathbf{u} < \mathbf{v}$.
Secondly, $f'(x) = 1 - (k + 1)x^k$, so

$$\alpha = \inf\{f'(x) : x \in I\} = f'(1) = 1 - (k + 1) = -k,$$

and

$$\beta = \sup\{f'(x) : x \in I\} = f'(0) = 1.$$

Thus, as the notations in Theorem 3, $B = \{-k, -k^3\}$, $B' = \{k^2\}$. So by the condition (iii) of Corollary 1, the range of θ is

$$-\frac{1}{\max(B')} \le \theta \le -\frac{1}{\min(B)},$$

i.e.,

$$-\frac{1}{k^2} \le \theta \le \frac{1}{k^3}.$$

So, we can see that the range of θ is shrinking as k increasing. Specifically, if $k = 1, -1 \le \theta \le 1$. If $k = 2, -\dfrac{1}{4} \le \theta \le \dfrac{1}{8}$. If $k = 3, -\dfrac{1}{9} \le \theta \le \dfrac{1}{27}$.

Next, let's compute three measures discussed in Sect. 4 for these 3-copulas. By the definition of τ_n,

$$\tau_3(\theta f(u) f(v) f(w)) = \frac{1}{3} \left[8\theta(\frac{k + 1}{2k + 3} - \frac{k + 2}{k + 3} + \frac{1}{3}) - 1 \right].$$

$$a_2 = \frac{1}{3}\left[8\int_{I^3}\theta C^*(\mathbf{u})\prod_{i=1}^3 f_i'(u_i)d\mathbf{u} + 8\int_{I^3}\theta\prod_{i=1}^3 f_i(u_i)dC^*(\mathbf{u}) + 1\right]$$

$$= \frac{1}{3}\left[-\frac{\theta k^3}{(k+2)^3} + \frac{\theta k^3}{(k+2)^3} + 1\right] = \frac{1}{3}.$$

So by Corollary 2,

$$\tau_3(C) = \tau_3(C^*) + \tau_3(\theta f(u)f(v)f(w)) + a_2$$

$$= 0 + \frac{1}{3}\left[80\left(\frac{k+1}{2k+3} - \frac{k+2}{k+3} + \frac{1}{3}\right) - 1\right] + \frac{1}{3}$$

$$= \frac{80}{3}\left(\frac{k+1}{2k+3} - \frac{k+2}{k+3} + \frac{1}{3}\right).$$

So the range of $\tau_3(C)$ is

$$\frac{8}{3k^3}\left(\frac{k+1}{2k+3} - \frac{k+2}{k+3} + \frac{1}{3}\right) \leq \tau_3(C) \leq -\frac{8}{3k^2}\left(\frac{k+1}{2k+3} - \frac{k+2}{k+3} + \frac{1}{3}\right).$$

By the definition of ρ_n,

$$\rho_3(\theta f(u)f(v)f(w)) = 4\left[\frac{\theta k^3}{8(k+2)^3} - \frac{\theta k^3}{8(k+2)^3}\right] - 1 = -1.$$

So

$$\rho_3(C) = \rho_3(C^*) + \rho_3(\theta f(u)f(v)f(w)) + \frac{3+1}{2^3-3-1}$$

$$= 0 - 1 + 1 = 0.$$

By the definition of survival function (1),

$$\overline{\theta\prod_{i=1}^3 f_i}\left(\frac{1}{2},\frac{1}{2},\frac{1}{2}\right) = 1 - \frac{\theta}{8}\left(1-\frac{1}{2^k}\right)^3.$$

So

$$\beta_3(\theta f(u)f(v)f(w)) = \frac{2^2\left[\theta\prod_{i=1}^3 f_i\left(\frac{1}{2},\frac{1}{2},\frac{1}{2}\right) + \overline{\theta\prod_{i=1}^3 f_i}\left(\frac{1}{2},\frac{1}{2},\frac{1}{2}\right)\right] - 1}{2^2 - 1}$$

$$= \frac{4\left[\frac{\theta}{8}\left(1-\frac{1}{2^k}\right)^3 + 1 - \frac{\theta}{8}\left(1-\frac{1}{2^k}\right)^3\right] - 1}{3} = 1.$$

Thus,

$$\beta_3(C) = \beta_3(C^*) + \beta_3(\theta f(u) f(v) f(w)) + \frac{1 - 2^{3-1}}{2^3 - 1}$$

$$= 0 + 1 - \frac{3}{7} = \frac{4}{7}.$$

Lastly, since $f(u)f(v)f(w) = uvw(1 - u^k)(1 - v^k)(1 - w^k) \geq 0$ for all $(u, v, w) \in I^3$, we have that C is more PLOD than Π_3 if and only if $\theta \geq 0$ and Π_3 is more PLOD than C if and only if $\theta \leq 0$.

Remark 4 From the above example, we can see that this 3-copulas family, $C(u, v, w) = C^*(u, v, w) + \theta uvw(1 - u^k)(1 - v^k)(1 - w^k)$, is interesting. As long as this C is a 3-copula, $\rho_3(C)$ and $\beta_3(C)$ are free of θ. Specifically, we always have $\rho_3(C) = \rho_3(C^*)$ and $\beta_3(C) = \beta_3(C^*) + \frac{4}{7}$.

Example 3 Let C^* be a *Frank's* copula [2, 5] defined by

$$C^*(u, v) = \ln\left[1 + \frac{(e^u - 1)(e^v - 1)}{e - 1}\right].$$

Let

$$H = \theta(1 - u)(1 - e^u)(1 - v)(1 - e^v),$$

where $\theta \geq 0$. Define a bivariate function C by $C = C^* + H$. We will use Theorem 2 to find the range of θ such that C is a copula.

Firstly, it is easy to see that $H(0, v) = H(u, 0) = H(1, v) = H(u, 1) = 0$.

Secondly, we can find that

$$c^*(u, v) = \frac{(e - 1)(u + v)}{[e - 1 + (e^u - 1)(e^v - 1)]^2},$$

and

$$h(u, v) = \theta(ue^u - 1)(ve^v - 1).$$

It can be shown that minimum values of $c = c^* + h$ occur at $(0, 1)$ and $(1, 0)$. So $c \geq 0$ if and only if $c(0, 1) = c(1, 0) = c^*(0, 1) + h(0, 1) = \frac{1}{e - 1} - \theta(e - 1) \geq 0$.

Thus $C = C^* + H$ is a copula if $\theta \leq \frac{1}{(e - 1)^2}$.

Acknowledgements The authors would like to thank referees for their valuable comments.

References

1. Aguiló I, Suñer J, Torrens J (2013) A construction method of semicopulas from fuzzy negations. Fuzzy Set Syst 226:99–114
2. Balakrishnan N, Lai CD (2009) Continuous bivariate distributions, 2nd edn. Springer, New York
3. Durante F, Foscolo E, Rodríguez-Lallena JA, Úbeda-Flores M (2012) A method for constructing higher-dimensional copulas. Statistics 46(3):387–404
4. Durante F, Sánchez JF, Úbeda-Flores M (2013) Bivariate copulas generated by perturbations. Fuzzy Set Syst 228:137–144
5. Frank M (1979) On the simultaneous associativity of $F(x, y)$ and $x + y - F(x, y)$. Aequationes Math 19(1):194–226
6. Joe H (1997) Multivariate models and dependence concepts. CRC Press, Boca Raton
7. Kim JM, Sungur EA, Choi T, Heo TY (2011) Generalized bivariate copulas and their properties. Model Assist Stat Appl 6:127–136
8. Liebscher E (2008) Construction of asymmetric multivariate copulas. J Multivar Anal 99: 2234–2250
9. Mesiar R, Komorník J, Komorníková M (2013) On some construction methods for bivariate copulas. In: Aggregation functions in theory and in practise, pp 39–45
10. Mesiar R, Komorníková M, Komorník J (2015) Perturbation of bivariate copulas. Fuzzy Set Syst 268:127 140
11. Nelsen RB (2006) An introduction to copulas, 2nd edn. Springer, New York
12. Rodríguez-Lallena JA, Úbeda-Flores M (2004) A new class of bivariate copulas. Stat Probab Lett 66(3):315–325
13. Schmid F, Schmidt R, Blumentritt T, Gaißer S, Ruppert M (2010) Copula-based measures of multivariate association. In: Copula theory and its applications, pp 209–236
14. Shan Q, Wongyang T, Wang T, Tasena S (2015) A measure of mutual complete dependence in discrete variables through subcopula. Int J Apporx Reason 65:11–23
15. Siburg KF, Stoimenov PA (2009) A measure of mutual complete dependence. Metrika 71: 239–251
16. Sklar A (1959) Fonctions de répartition á n dimensions et leurs marges. Publ Inst Statist Univ Paris 8:229–231
17. Tasena S, Dhompongsa S (2013) A measure of multivariate mutual complete dependence. Int J Apporx Reason 54:748–761
18. Úbeda-Flores M (2005) Multivariate versions of Blomqvist's beta and Spearman's footrule. Ann Inst Statist Math 57(4):781–788
19. Victor H, Ibragimov R, Sharakhmetov S (2006) Characterizations of joint distributions, copulas, information, dependence and decoupling, with applications to time series. In: Optimality, pp 183–209
20. Wei Z, Wang T, Nguyen PA (2015) Multivariate dependence concepts through copulas. Int J Apporx Reason 65:24–33

Plausibility Regions on the Skewness Parameter of Skew Normal Distributions Based on Inferential Models

Xiaonan Zhu, Ziwei Ma, Tonghui Wang and Teerawut Teetranont

Abstract Inferential models (IMs) are new methods of statistical inference. They have several advantages: (1) They are free of prior distributions; (2) They rely on data. In this paper, $100(1 - \alpha)\%$ plausibility regions of the skewness parameter of skew-normal distributions are constructed by using IMs, which are the counterparts of classical confidence intervals in IMs.

1 Introduction

In practical applications, the skew data sets occur in many diverse fields of our life, such as economics, finance, biomedicine, environment, demography, and pharmacokinetics. Usually for mathematical convenience, they are assumed to be normally distributed. This restrictive assumption, however, may result in not only a lack of robustness against departures of the normal distribution and but also invalid statistical inferences, especially when data are skewed. To fix this issue, one solution is *skew-normal* distributions defined by Azzalini in 1985 [2]. Theoretically, the skew normal family is an extension from normal distribution family, which shares a number of formal properties of normal distribution, such as $Z^2 \sim \chi^2$, if Z is a centered skew normal random variable. In practical, skew normal is suitable for the analysis

X. Zhu · Z. Ma · T. Wang (✉)
Department of Mathematical Sciences, New Mexico State University, Las Cruces, USA
e-mail: twang@nmsu.edu

X. Zhu
e-mail: xzhu@nmsu.edu
Z. Ma
e-mail: ziweima@nmsu.edu

T. Teetranont
Faculty of Economics, Chiang Mai University, Chiang Mai, Thailand
e-mail: teerawut.t@cmu.ac.th

© Springer International Publishing AG 2017
V. Kreinovich et al. (eds.), *Robustness in Econometrics*,
Studies in Computational Intelligence 692, DOI 10.1007/978-3-319-50742-2_16

of data which is unimodal empirical distributed but with some skewness [1, 8]. In past three decades, the family of skew-normal distributions, including multivariate skew-normal distributions, has been studied by many authors, e.g. Azzalini [3, 5, 6], Wang et al. [16], Ye et al. [17].

However, estimations of the skewness parameter is challenging, since for a random sample $Z_1, \ldots, Z_n \sim SN(0, 1, \lambda)$, its maximum likelihood function of the skewness parameter λ may be unbounded, and the method of moments may not give us good estimators. As stated in Azzalini and Capitanio's work [4], "... *there are cases where the likelihood shape and the MLE are problematic. We are not referring here to difficulties with numerical maximization, but to the intrinsic properties of the likelihood function, not removable by change of parameterizations. In case of this sort, the behavior of the MLE appears quite unsatisfactory, and an alternative estimation method is called for ...*". Recently, some researchers tried to solve this issue, such as Azzalini and Capitanio [4], Sartori [15], Liseo and Loperfido [9], Debarshi [7] and Mameli et al. [14].

In this paper, we make inferences about the skewness parameter of skew-normal distributions in three cases by using *Inferential Models* (IMs for short). IMs are new methods of statistical inference introduced by Martin and Liu [10, 12]. Comparing with Fisher's fiducial inference, Dempster-Shafer theory of belief functions and Bayesian inference, IMs have several advantages: (1) IMs are free of prior distributions; (2) IMs depend on the observed data. For details of IMs, see Martin et al.'s work [10–13].

The paper is organized as follows. In Sect. 2, some necessary concepts and definitions of skew-normal distributions and IMs are reviewed briefly. In Sect. 3, the skewness parameter is estimated by using IMs in three different cases. For each case, the plausibility function and the $100(1 - \alpha)\%$ plausibility region are given, which is the counterpart of classical confidence interval in IMs. Simulation studies and one example are provided in Sect. 4.

2 Preliminaries

2.1 Skew-Normal Distributions

A random variable Z is said to be *skew-normal* with the *skewness parameter* λ, if its density function is

$$f(z; \lambda) = 2\phi(z)\Phi(\lambda z),$$

where $\lambda, z \in \mathbb{R}$, ϕ and Φ are the density function (p.d.f.) and distribution function (c.d.f.) of standard normal $N(0, 1)$. Z is called a *centered skew normal* random variable, denoted by $Z \sim SN(0, 1, \lambda)$ [5].

For any $\mu \in \mathbb{R}$ and $\sigma > 0$, let

$$X = \mu + \sigma Z,$$

then X is said to be *skew-normal* with the *location parameter* μ, *scale parameter* σ and *skewness parameter* λ, denoted by $X \sim SN(\mu, \sigma^2, \lambda)$. The density function of X is

$$f(x; \mu, \sigma^2, \lambda) = \frac{2}{\sigma} \phi\left(\frac{x - \mu}{\sigma}\right) \Phi\left(\frac{\lambda(x - \mu)}{\sigma}\right).$$

There is an alternative representation of $Z \sim SN(0, 1, \lambda)$ [5],

$$Z = \delta|Z_0| + \sqrt{1 - \delta^2} Z_1, \qquad \delta \in (-1, 1), \tag{1}$$

where $Z_0, \ Z_1 \sim N(0, 1)$, Z_0 and Z_1 are independent and the relations between λ and δ are $\delta(\lambda) = \dfrac{\lambda}{\sqrt{1 + \lambda^2}}$ and $\lambda(\delta) = \dfrac{\delta}{\sqrt{1 - \delta^2}}$. For more details of skew-normal distributions, see Azzalini's book [5].

Multivariate skew-normal distributions were defined by Azzalini and Dalla Valle [6] and Azzalini and Capitanio [4]. A random vector $\mathbf{X} \in \mathbb{R}^k$ is said to have a multivariate skew-normal distribution, denoted by $\mathbf{X} \sim SN_k(\Sigma, \boldsymbol{\alpha})$, if its density function is

$$f(\mathbf{x}) = 2\phi_k(\mathbf{x}; \Sigma)\Phi(\alpha^T \mathbf{x}),$$

where Σ is a positive definite $k \times k$ matrix, $\alpha \in \mathbb{R}^k$, α^T is the transpose of α, $\phi_k(\mathbf{x}; \Sigma)$ is the k-dimensional normal density with mean zero and covariance matrix Σ, and $\Phi(\cdot)$ is the c.d.f. of standard normal distribution $N(0, 1)$.

As univariate skew-normal distributions, there is another method to construct multivariate skew-normal distributions. Let $\mathbf{Y} = (Y_1 \ldots, Y_k)^T$ be a k-dimensional normal random vector distributed with $\mathbf{Y} \sim N_k(0, \Psi)$, independent of $Y_0 \sim N(0, 1)$. Then

$$\begin{pmatrix} Y_0 \\ \mathbf{Y} \end{pmatrix} \sim N_{k+1}\left\{0, \begin{pmatrix} 1 & 0 \\ 0 & \Psi \end{pmatrix}\right\},$$

where Ψ is a $k \times k$ covariance matrix. If $\delta_i \in (-1, 1), i = 1, \ldots, k$, and let

$$Z_i = \delta_i|Y_0| + \sqrt{1 - \delta_i^2} Y_i,$$

then $Z_i \sim SN(0, 1, \lambda(\delta_i)), i = 1, \ldots, k$, and $\mathbf{Z} = (Z_1 \ldots, Z_k)^T \sim SN_k(\Sigma, \alpha)$, where

$$\alpha^T = \frac{\lambda^T \Psi^{-1} \Delta^{-1}}{(1 + \lambda^T \Psi^{-1} \lambda)^{\frac{1}{2}}}, \quad \Delta = \mathrm{diag}((1 - \delta_1^2)^{\frac{1}{2}}, \ldots, (1 - \delta_k^2)^{\frac{1}{2}}),$$

$$\lambda = (\lambda(\delta_1), \ldots, \lambda(\delta_k))^T, \quad \Sigma = \Delta(\Psi + \lambda\lambda^T)\Delta.$$

In 2009, Wang et al. [16] defined another family of multivariate skew-normal distributions as following, which are extensions of Azzalini and Dalla Valle's definition.

Let $\mathbf{Z} \sim SN_k(\mathbf{I}_k, \alpha)$, where $\mathbf{I}_k$ is the $k \times k$ identity matrix. The distribution of $\mathbf{Y} = \mu + B^T \mathbf{Z}$ is called a *multivariate skew-normal* random vector with the *location parameter* $\mu \in \mathbb{R}^n$, *scale parameter* B, a $k \times n$ matrix, and *shape parameter* $\alpha \in \mathbb{R}^k$, and is denoted by $Y = \mathbf{SN}_n(\mu, B, \alpha)$. Note that

$$SN_k(\Sigma, \alpha) = \mathbf{SN}_k(0, \Sigma^{\frac{1}{2}}, \Sigma^{\frac{1}{2}}\alpha).$$

Next, we propose some results of multivariate skew-normal distributions, which we will use to obtain main results of the paper.

Proposition 1 *Suppose that* $(X_1, X_2, \ldots, X_{2n})^T \sim \mathbf{SN}_{2n}(\mu \mathbf{1}_{2n}, \sigma^2 \mathbf{I}_{2n}, \lambda \mathbf{1}_{2n})$, *where* $\mu, \lambda \in \mathbb{R}, \sigma > 0, \mathbf{1}_{2n}$ *is the column vector* $(1, \ldots, 1)^T \in \mathbb{R}^{2n}, \mathbf{I}_{2n}$ *is the* $2n \times 2n$ *identity matrix. Let* $T = \sum_{i=1}^{n}(X_{2i-1} - X_{2i})^2$, *then* $\dfrac{(1+\lambda^2)T}{2\sigma^2} \sim \chi_n^2$.

Proof By the above discussion of multivariate skew-normal distributions, we have

$$X_i = \mu + \sigma(\delta|Y_0| + \sqrt{1 - \delta^2}Y_i), \qquad i = 1, \ldots, 2n,$$

where $\delta = \dfrac{\lambda}{\sqrt{1 + \lambda^2}}$, and $Y_0, Y_1, \ldots, Y_{2n}$ are independent distributed with $N(0, 1)$.
So for each $i = 1, \ldots, n$,

$$X_{2i-1} - X_{2i} = \sigma\sqrt{1 - \delta^2}(Y_{2i-1} - Y_{2i}) = \sigma\sqrt{2(1 - \delta^2)}Z_i = \sigma\sqrt{\frac{2}{1 + \lambda^2}}Z_i,$$

where $Z_1, \ldots, Z_n$ are independent distributed with $N(0, 1)$. So

$$\frac{(1+\lambda^2)}{2\sigma^2}(X_{2i-1} - X_{2i})^2 = Z_i^2 \sim \chi_1^2.$$

Thus

$$\frac{(1+\lambda^2)T}{2\sigma^2} = \frac{(1+\lambda^2)}{2\sigma^2}\sum_{i=1}^{n}(X_{2i-1} - X_{2i})^2 = \sum_{i=1}^{n}Z_i^2 \sim \chi_n^2. \quad \Box$$

Now let's consider the representation of $SN(0, 1, \lambda)$ in (1). We can make some connection between skew normal distributions and Cauchy distributions.

Recall that a random variable Z is said to be Cauchy $(0, 1)$ (or standard Cauchy) if its density function is

$$f(z) = \frac{1}{\pi(1 + z^2)}, \qquad \text{for } z \in \mathbb{R}.$$

There is a well known representation of standard Cauchy random variable is

$$Z = \frac{Z_1}{Z_0} \sim \text{Cauchy } (0, 1),$$

where Z_1 is a standard normal distributed random variable, and Z_0 is a half normal distributed random variable, i.e., its p.d.f. is

$$f_{Z_0}(z_0) = 2 \exp \left\{ -\frac{z_0^2}{2} \right\},$$

and Z_0 and Z_1 are independent. Furthermore, let $W = \eta + \omega Z$, then $W \sim \text{Cauchy } (\eta, \omega)$, where η and ω are location and scale parameter respectively.

Based on the above definition and facts, we have the following results.

Proposition 2 *Suppose that* $X = \delta Z_0 + \sqrt{1 - \delta^2} Z_1$, *i.e.,* $X \sim SN(0, 1, \lambda)$, *where* $\lambda = \dfrac{\delta}{1 - \delta^2}$, *then* $\dfrac{X}{\sqrt{1 - \delta^2} Z_0} \sim \text{Cauchy } (\lambda, 1)$.

Proof The proof is straightforward based on the density functions and independent assumption of Z_0 and Z_1. $\square$

Consequently, we could extend the above result to a more general case.

Proposition 3 *Suppose that* $Y = \mu + \sigma X$, *i.e.,* $Y \sim SN(\mu, \sigma^2, \lambda)$, *then* $\dfrac{Y - \mu}{\sqrt{1 - \delta^2} Z_0} \sim \text{Cauchy } (\sigma \lambda, \sigma)$.

Proof Notice that $\dfrac{Y - \mu}{\sqrt{1 - \delta^2} Z_0} = \sigma \dfrac{X}{\sqrt{1 - \delta^2} Z_0}$ and $\dfrac{X}{\sqrt{1 - \delta^2} Z_0} \sim \text{Cauchy } (\lambda, 1)$, which verifies the statement. $\square$

2.2 Inference Models

In this section, we provide a brief review of IMs and the section refers to Martin and Liu's works [10, 12].

Let X be an observable random sample with a probability distribution $P_{X|\theta}$ on a sample space $\mathbb{X}$, where θ is an unknown parameter, $\theta \in \Theta$, a parameter space. Let U be an unobservable *auxiliary variable* on an auxiliary space $\mathbb{U}$, where although U is unobservable, we assume that U and $\mathbb{U}$ are well-known. An *association* is a map $a : \mathbb{U} \times \Theta \rightarrow \mathbb{X}$ such that

$$X = a(U, \theta).$$

An IM consists of three steps based on a fixed association.

Association Step (A-step) Suppose we have an association $X = a(U, \theta)$ and an observation $X = x$, where x could be a scalar or vector, then the unknown θ must satisfy

$$x = a(u^*, \theta),$$

for some unobserved u^* of U. So from the observation $X = x$, we can construct sets of solutions

$$\Theta_x(u) = \{\theta \in \Theta : x = a(u, \theta)\}, \quad x \in \mathbb{X}, \quad u \in \mathbb{U}.$$

Prediction Step (P-step) Since the true u^* is unobservable, to make a valid inference, the key point is to predicate u^* validly.

Let $u \to S(u)$ be a set-value map from $\mathbb{U}$ to $\mathbb{S}$, a collection of P_U-measurable subsets of $\mathbb{U}$. Then the random set $S : \mathbb{U} \to \mathbb{S}$ is called a *predictive random set* of U with distribution $P_S = P_U \circ S^{-1}$. We will use S to predict u^*.

Combination Step (C-step) Define

$$\Theta_x(S) = \bigcup_{u \in S} \Theta_x(u).$$

For any assertion A of θ, i.e., $A \subseteq \Theta$, the *belief function* and *plausibility function* of A with respect to a predictive random set S are defined by,

$$\text{bel}_x(A; S) = P_S\{\Theta_x(S) \subseteq A : \Theta_x(S) \neq \emptyset\};$$

$$\text{pl}_x(A; S) = P_S\{\Theta_x(S) \nsubseteq A^c : \Theta_x(S) \neq \emptyset\}.$$

Note that

$$\text{pl}_x(A; S) = 1 - \text{bel}_x(A^c; S), \quad \text{bel}_x(A; S) + \text{bel}_x(A^c; S) \leq 1, \quad \text{for all } A \subseteq \Theta.$$

To make a good inference for assertions, we need some concepts of *validity*.

Definition 1 Let X and Y be two random variables. We say that X is stochastically no smaller than Y, denoted by $X \geq_{st} Y$, if $P(X > a) \geq P(Y > a)$, for all $a \in \mathbb{R}$.

Definition 2 ([12]) A predictive random set S is *valid* for predicting the unobserved auxiliary variable U if $\gamma_S(U)$, as a function of $U \sim P_U$, is stochastically no smaller than Uniform$(0, 1)$, where γ_S is called the *contour function* of S defined by

$$\gamma_S(u) = P_S(S \ni u), \quad u \in U.$$

If $\gamma_S(U) \sim$ Uniform$(0, 1)$ then S is *efficient*.

A simple way to construct a valid predictive random set was provided by Martin and Liu [12] as follows.

Theorem 1 *Let $\mathbb{S}$ be a collection of subsets of $\mathbb{U}$. If $\mathbb{S}$ and P_U satisfy following conditions,*

(i) $\mathbb{S}$ *is nested, i.e., all elements of $\mathbb{S}$ can be ordered by inclusions,*
(ii) There is some $F \in \mathbb{S}$ such that $P_U(F) > 0$,
(iii) All closed subsets of $\mathbb{U}$ are P_U-measurable,
(iv) $\mathbb{S}$ *contains $\emptyset$, $\mathbb{U}$, and all of the other elements are closed subsets, and define a predictive random set S, with distribution P_S, supported on $\mathbb{S}$, such that*

$$P_S\{F \subseteq K\} = \sup_{F \in \mathbb{S}; F \subseteq K} P_U(F), \qquad K \subseteq \mathbb{U},$$

then S is valid.

Definition 3 ([12]) Suppose $X \sim P_{X|\theta}$ and let A be an assertion of interest. Then the IM with a belief function $\mathsf{bel}_x(\cdot\ ; S)$ is *valid* for A if

$$\sup_{\theta \notin A} P_{X|\theta}\{\mathsf{bel}_X(A; S) \geq 1 - \alpha\} \leq \alpha, \qquad \text{for all } \alpha \in (0, 1).$$

The IM is *valid* if it is valid for all A.

Theorem 2 ([12]) *Suppose the predictive random set S is valid and $\Theta_x(S) \neq \emptyset$ with P_S-probability 1 for all x. Then the IM is valid.*

Remark 1 (a) It is easy to see that the IM is valid if and only if

$$\sup_{\theta \in A} P_{X|\theta}\{\mathsf{pl}_X(A; S) \leq \alpha\} \leq \alpha, \qquad \text{for all } \alpha \in (0, 1), \ A \subseteq \Theta.$$

(b) By the definition of validity of IMs, for any false assertion A, i.e., the true $\theta_0 \notin A$,

$$\sup_{\theta \notin A} P_{X|\theta}\{\mathsf{bel}_X(A; S) \geq 1 - \alpha\} \leq \alpha,$$

and for any true assertion A, i.e., the true $\theta_0 \in A$,

$$\sup_{\theta \in A} P_{X|\theta}\{\mathsf{pl}_X(A; S) \leq \alpha\} \leq \alpha.$$

It means that if the IM is valid, the Type I error and Type II error of inferences can be controlled by the plausibility function and the belief function respectively.

Given an IM, a $100(1 - \alpha)\%$ *plausibility region* is defined by

$$\Pi_x(\alpha) = \{\theta \in \Theta : \mathsf{pl}_x(\theta; S) > \alpha\},$$

which is an IM-based counterpart of classical confidence intervals.

3 Plausibility Regions on the Skewness Parameter of Skew Normal Distributions

In this section, we are going to use IMs to find belief functions and plausibility functions of the skewness parameter of skew-normal distributions, and construct plausibility regions on the parameter in three different cases. For simplicity, we may assume $\mu = 0$ and/or $\sigma^2 = 1$, and $\lambda > 0$, but our results should still hold when μ and σ^2 are known.

3.1 Samples are Identical Distributed but not Independent

Suppose that $X_{11}, X_{12}, X_{21}, \ldots, X_{n1}, X_{n2}$ are identical distributed samples from $SN(0, 1, \lambda)$ but not independent such that $(X_1, X_2, \ldots, X_{2n})^T \sim \mathbf{SN}_{2n}(\mathbf{0}_{2n}, \mathbf{I}_{2n}, \lambda \mathbf{1}_{2n})$.

A-step If we define a statistic $T = \sum_{i=1}^{n}(X_{i1} - X_{i2})^2$, then by Proposition 1 we have an association

$$T = \frac{2}{1 + \lambda^2} F_{\chi_n^2}^{-1}(U),$$

where $U \sim \text{Uniform}(0, 1)$, and $F_{\chi_n^2}$ is the c.d.f. of χ_n^2. For any observation $t > 0$ and $u \in [0, 1]$,

$$\Theta_t(\lambda) = \left\{ \lambda : t = \frac{2}{1 + \lambda^2} F_{\chi_n^2}^{-1}(u) \right\} = \left\{ \left(\frac{2 F_{\chi_n^2}^{-1}(u)}{t} - 1 \right)^{\frac{1}{2}} \right\}.$$

P-step The parameter set

$$\Theta_t(\lambda) = \left\{ \left(\frac{2 F_{\chi_n^2}^{-1}(u)}{t} - 1 \right)^{\frac{1}{2}} \right\} \neq \emptyset$$

requires $\frac{2 F_{\chi_n^2}^{-1}(u)}{t} - 1 > 0$, so we have $u > F_{\chi_n^2}\left(\frac{t}{2}\right)$. Hence for the auxiliary variable U, we use an *elastic predictive random set* S defined as following to predicate it. See Chap. 5 in [12] for details of elastic predictive random sets.

$$S(u) = \begin{cases} [1 - F_{\chi_n^2}(\frac{t}{2}), F_{\chi_n^2}(\frac{t}{2})], & \text{if } F_{\chi_n^2}(\frac{t}{2}) > 0.5 \text{ and } 1 - F_{\chi_n^2}(\frac{t}{2}) < u < F_{\chi_n^2}(\frac{t}{2}), \\ [0.5 - |u - 0.5|, 0.5 + |u - 0.5|], & \text{otherwise}, \end{cases}$$

for all $u \in [0, 1]$. By Theorem 5.1 in [12], the predictive random set

$$
S(F_{\chi_n^2}^{-1}(u)) = \begin{cases} [F_{\chi_n^2}^{-1}(1 - F_{\chi_n^2}(\frac{t}{2})), \frac{t}{2}], & \text{if } F_{\chi_n^2}(\frac{t}{2}) > 0.5, \text{ and } 1 - F_{\chi_n^2}(\frac{t}{2}) < u < F_{\chi_n^2}(\frac{t}{2}), \\ [F_{\chi_n^2}^{-1}(0.5 - |u - 0.5|), F_{\chi_n^2}^{-1}(0.5 + |u - 0.5|)], & \text{otherwise,} \end{cases}
$$

is valid for inference about every assertion of λ.

C-step By the P-step, we have

$$
\Theta_t(S) = \begin{cases} \{0\}, & \text{if } F_{\chi_n^2}(\frac{t}{2}) > 0.5, \text{ and } 1 - F_{\chi_n^2}(\frac{t}{2}) < u < F_{\chi_n^2}(\frac{t}{2}), \\ \left[\left(\frac{2F_{\chi_n^2}^{-1}(0.5 - |u - 0.5|)}{t} - 1 \right)^{\frac{1}{2}}, \left(\frac{2F_{\chi_n^2}^{-1}(0.5 + |u - 0.5|)}{t} - 1 \right)^{\frac{1}{2}} \right], & \text{otherwise,} \end{cases}
$$

so we can use the above IM to get the following result.

Proposition 4 *For any singleton assertion* $A = \{\lambda\}$,

$$
bel_t(A; S) = 0,
$$

$$
pl_t(A; S) = 1 - \left| 2F_{\chi_n^2}(\frac{t(1 + \lambda^2)}{2}) - 1 \right|,
$$

and the $100(1 - \alpha)\%$ *plausibility region* $\Pi_t(\lambda)$ *is*

$$
\left(\max\left\{ \frac{2F_{\chi_n^2}^{-1}(\frac{\alpha}{2})}{t} - 1, 0 \right\} \right)^{\frac{1}{2}} < \lambda < \left(\max\left\{ \frac{2F_{\chi_n^2}^{-1}(1 - \frac{\alpha}{2})}{t} - 1, 0 \right\} \right)^{\frac{1}{2}}.
$$

Proof It is clear that $\{\Theta_x(S) \subseteq A\} = \emptyset$, so $bel_t(A; S) = 0$.

$$
pl_t(A; S) = 1 - bel_t(A^c; S) = 1 - P_S(\Theta_t(S) \subseteq A^c)
$$

$$
= 1 - P_U\left(F_{\chi_n^2}^{-1}(0.5 - |U - 0.5|) > \frac{t(1 + \lambda^2)}{2} \right) - P_U\left(F_{\chi_n^2}^{-1}(0.5 + |U - 0.5|) < \frac{t(1 + \lambda^2)}{2} \right)
$$

$$
= 1 - P_U\left(|U - 0.5| < |F_{\chi_n^2}(\frac{t(1 + \lambda^2)}{2}) - 0.5| \right)
$$

$$
= 1 - \left| 2F_{\chi_n^2}\left(\frac{t(1 + \lambda^2)}{2} \right) - 1 \right|.
$$

By the definition, $\Pi_t(\alpha) = \{\lambda : pl_t(\lambda; S) > \alpha\}$. Let

$$
pl_t(\lambda; S) = 1 - |2F_{\chi_n^2}(\frac{t(1 + \lambda^2)}{2}) - 1| > \alpha
$$

Table 1 Simulation results of average lower bounds (AL), average upper bounds (AU), average lengths (ALength) and coverage probabilities (CP) of 95% plausibility regions for identical distributed but not independent samples from $SN(0, 1, \lambda)$ with size $2n$

$\lambda = 1$					$\lambda = 2$				
$2n$	AL	AU	ALength	CP	$2n$	AL	AU	ALength	CP
10	0.0759	2.4467	2.3708	0.9501	10	0.4276	4.1293	3.7017	0.9460
30	0.1994	1.7478	1.5484	0.9519	30	1.1144	3.0232	1.9087	0.9488
50	0.3206	1.5598	1.2392	0.9505	50	1.3268	2.7630	1.4362	0.9504
100	0.5405	1.3924	0.8519	0.9500	100	1.5229	2.5193	0.9964	0.9524
$\lambda = 5$					$\lambda = 10$				
$2n$	AL	AU	ALength	CP	$2n$	AL	AU	ALength	CP
10	2.2145	9.6246	7.4100	0.9511	10	4.8120	19.3389	14.5269	0.9470
30	3.3062	7.1770	3.8708	0.9487	30	6.7628	14.2924	7.5296	0.9555
50	3.6766	6.6352	2.9585	0.9514	50	7.4313	13.1619	5.7305	0.9478
100	4.0404	6.1036	2.0631	0.9514	100	8.1409	12.1451	4.0042	0.9500

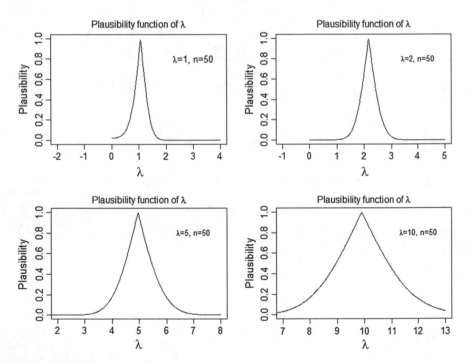

Fig. 1 Graphs of plausibility functions of identical distributed but not independent simulative data from $SN(0, 1, \lambda)$ with size $2n = 50$

and then solve it for λ. We have

$$\left(\max \left\{ \frac{2F_{\chi_n^2}^{-1}(\frac{\alpha}{2})}{t} - 1, 0 \right\} \right)^{\frac{1}{2}} < \lambda < \left(\max \left\{ \frac{2F_{\chi_n^2}^{-1}(1 - \frac{\alpha}{2})}{t} - 1, 0 \right\} \right)^{\frac{1}{2}} . \quad \Box$$

In Table 1 and Fig. 1 we provide a simulation study of $SN(0, 1, \lambda)$ when $\lambda = 1, 2, 5$ and 10.

3.2 Samples Are i.i.d. from $SN(0, \delta_\lambda^2, \lambda)$ Where δ_λ^2 Is a Nonnegative Monotone Function of λ

In this subsection, we suppose that the population is $SN(0, \delta_\lambda^2, \lambda)$ with unknown δ_λ^2 and λ, but δ_λ^2 is a nonnegative monotone function of λ, i.e., there is a nonnegative monotone function g such that $\delta_\lambda^2 = g(\lambda)$. Without loss of generality, we may assume that g is monotone increasing.

Now, suppose that we have a random sample $X_1, \ldots, X_n$ from $SN(0, \delta_\lambda^2, \lambda)$, i.e., $X_1, \ldots, X_n$ are i.i.d. with $SN(0, \delta_\lambda^2, \lambda)$.

A-step Since $X_i \sim SN(0, \delta_\lambda^2, \lambda)$, $\dfrac{X_i}{\delta_\lambda} \sim SN(0, 1, \lambda)$. So $\dfrac{X_i^2}{\delta_\lambda^2} \sim \chi_1^2, i = 1, \ldots, n$. As the above case, if we let $T = \sum_{i=1}^{n} X_i^2$, then we have

$$\frac{T}{\delta_\lambda^2} = \frac{T}{g(\lambda)} = F_{\chi_n^2}^{-1}(U),$$

where $U \sim \text{Uniform}(0, 1)$, and $F_{\chi_n^2}$ is the c.d.f. of χ_n^2. Thus we have an association of λ

$$T = g(\lambda) F_{\chi_n^2}^{-1}(U).$$

For any $t > 0$ and $u \in [0, 1]$,

$$\Theta_t(\lambda) = \left\{ g^{-1} \left(\frac{t}{F_{\chi_n^2}^{-1}(u)} \right) \right\}.$$

P-step For the auxiliary variable U, we also choose the default predictive random set $S(U) = [0.5 - |U - 0.5|, 0.5 + |U - 0.5|]$. Then

$$S(F_{\chi_n^2}^{-1}(U)) = [F_{\chi_n^2}^{-1}(0.5 - |U - 0.5|), F_{\chi_n^2}^{-1}(0.5 + |U - 0.5|)].$$

It is valid for every assertion of λ.

C-step By the P-step, we have

$$\Theta_t(S) = \left[g^{-1}\left(\frac{t}{F_{\chi_n^2}^{-1}(0.5 + |U - 0.5|)} \right), g^{-1}\left(\frac{t}{F_{\chi_n^2}^{-1}(0.5 - |U - 0.5|)} \right) \right].$$

Thus we have the following result.

Proposition 5 *For any singleton assertion* $A = \{\lambda\}$,

$$bel_t(A; S) = 0,$$

$$pl_t(A; S) = 1 - \left|2F_{\chi_n^2}\left(\frac{t}{g(\lambda)}\right) - 1\right|,$$

and the $100(1 - \alpha)\%$ *plausibility region* $\Pi_t(\lambda)$ *is*

$$g^{-1}\left(\frac{t}{F_{\chi_n^2}^{-1}(1 - \frac{\alpha}{2})} \right) < \lambda < g^{-1}\left(\frac{t}{F_{\chi_n^2}^{-1}(\frac{\alpha}{2})} \right).$$

Proof It is clear that $\{\Theta_x(S) \subseteq A\} = \emptyset$, so $bel_t(A; S) = 0$.

$$pl_t(A; S) = 1 - bel_t(A^c; S) = 1 - P_S(\Theta_x(S) \subseteq A^c)$$

$$= 1 - P_U\left(F_{\chi_n^2}^{-1}(0.5 + |U - 0.5|) < \frac{t}{g(\lambda)} \right) - P_U\left(F_{\chi_n^2}^{-1}(0.5 - |U - 0.5|) > \frac{t}{g(\lambda)} \right)$$

$$= 1 - P_U\left(|U - 0.5| < \left| F_{\chi_n^2}\left(\frac{t}{g(\lambda)}\right) - 0.5 \right| \right)$$

$$= 1 - \left|2F_{\chi_n^2}\left(\frac{t}{g(\lambda)}\right) - 1\right|.$$

By the definition, $\Pi_t(\alpha) = \{\lambda : pl_t(\lambda; S) > \alpha\}$. Let

$$pl_t(\lambda; S) = 1 - |2F_{\chi^2}\left(\frac{t}{g(\lambda)}\right) - 1| > \alpha$$

and then solve it for λ. We have

$$g^{-1}\left(\frac{t}{F_{\chi_n^2}^{-1}(1-\frac{\alpha}{2})}\right) < \lambda < g^{-1}\left(\frac{t}{F_{\chi_n^2}^{-1}(\frac{\alpha}{2})}\right).$$

□

Remark 2 (a) In this case, since we assume that the scale parameter is a one-to-one function of the skewness parameter, we can estimate the scale parameter instead of the skewness parameter as we did above. Actually, when μ and λ of a skew-normal population $SN(\mu, \sigma^2, \lambda)$ are known, we can use the same method to estimate σ^2.

(b) In these two cases, since we used the default predictive random sets $[0.5 - |U - 0.5|, 0.5 + |U - 0.5|]$, the $100(1 - \alpha)\%$ plausibility regions are identical with classical equal-tailed $100(1 - \alpha)\%$ confidence intervals.

Example 1 Suppose that the population $X \sim SN(0, 1 + \lambda^2, \lambda)$ with unknown $\lambda > 0$, we have

$$X = \lambda|Z_0| + Z_1,$$

where $Z_0, Z_1 \sim N(0, 1)$, and Z_0 and Z_1 are independent. So $g(x) = 1 + x^2$ and $g^{-1}(x) = \sqrt{x - 1}$.

In this case, by Proposition 5, for any singleton assertion $A = \{\lambda_0\}$,

$$\mathsf{bel}_t(A; S) = 0,$$

$$\mathsf{pl}_t(A; S) = 1 - \left|2F_{\chi_n^2}\left(\frac{t}{1 + \lambda_0^2}\right) - 1\right|,$$

Table 2 Simulation results of average lower bounds (AL), average upper bounds (AU), average lengths (ALength) and coverage probabilities (CP) of 95% plausibility regions for random sample from $SN(0, 1 + \lambda^2, \lambda)$ with size n

$\lambda = 1$					$\lambda = 2$				
n	AL	AU	ALength	CP	n	AL	AU	ALength	CP
10	0.2349	2.1872	1.9522	0.9518	10	1.1016	3.6923	2.5906	0.9500
30	0.4612	1.5802	1.1189	0.9478	30	1.4555	2.7897	1.3341	0.9499
50	0.5824	1.4277	0.8452	0.9470	50	1.5659	2.5768	1.0109	0.9498
100	0.7187	1.2922	0.5735	0.9491	100	1.6825	2.3861	0.7035	0.9511
$\lambda = 5$					$\lambda = 10$				
n	AL	AU	ALength	CP	n	AL	AU	ALength	CP
10	3.3181	8.6669	5.3487	0.9485	10	6.7712	17.1714	10.4002	0.9474
30	3.9118	6.6819	2.7700	0.9486	30	7.9088	13.2980	5.3892	0.9491
50	4.1279	6.2315	2.1035	0.9515	50	8.3034	12.3855	4.0821	0.9505
100	4.3509	5.8133	1.4624	0.9465	100	8.7422	11.5827	2.8404	0.9523

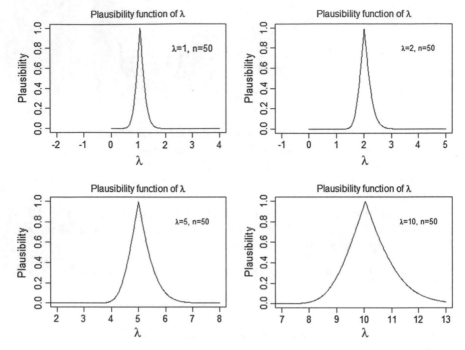

Fig. 2 Graphs of plausibility functions of i.i.d. simulative data from $SN(0, 1 + \lambda^2, \lambda)$ with size $n = 50$

and the $100(1 - \alpha)\%$ plausibility region $\Pi_t(\lambda)$ is

$$\left(\max \left\{ \frac{t}{F_{\chi_n^2}^{-1}(1 - \frac{\alpha}{2})} - 1, 0 \right\} \right)^{\frac{1}{2}} < \lambda < \left(\max \left\{ \frac{t}{F_{\chi_n^2}^{-1}(\frac{\alpha}{2})} - 1, 0 \right\} \right)^{\frac{1}{2}}.$$

In Table 2 and Fig. 2 we provide a simulation study of $SN(0, 1 + \lambda^2, \lambda)$ when $\lambda = 1, 2, 5$ and 10.

3.3 Samples with some extra information

From the stochastic representation (1) of skew-normal distributions, suppose that we can observe the ratio $\frac{X}{\sqrt{1 - \delta^2}|Z_0|}$, then by Proposition 2 we have

$$\frac{X}{\sqrt{1 - \delta^2}|Z_0|} = \lambda + \frac{Z_1}{|Z_0|} \sim \text{Cauchy } (\lambda, 1).$$

Based on this fact, we can build up an IM to make inferences on λ, which is the location parameter of a Cauchy random variable $\frac{X}{\sqrt{1-\delta^2}|Z_0|}$.

Now suppose that $X_1, X_2, \ldots, X_n$ are a random sample from a Cauchy population with the scale parameter 1, and an unknown location parameter λ, i.e., $X_i \sim$ Cauchy $(\lambda, 1)$ for $i = 1, 2, \ldots n$ and X_i's are independent.

Since Cauchy distributions are stable, sample mean $\bar{X} \sim$ Cauchy $(\lambda, 1)$, which means $\bar{X}$ can only provide information for location parameter λ as much as a single sample. But we know that order statistics are minimal sufficient statistics of location parameters, and the median of a random sample is a good estimator for λ. Thus, we will use sample median to construct IMs for λ as follows.

A-step Since $X_1, X_2, \ldots, X_n$ are independent distributed with Cauchy $(\lambda, 1)$, we have

$$X_i = \lambda + U_i,$$

where $U_1, \ldots, U_n$ are independent distributed with Cauchy $(0, 1)$.

Now let M be the sample median of $X_1, X_2, \ldots, X_n$ defined by

$$M = \begin{cases} X_{(k)}, & \text{if } n = 2k + 1, \\ \frac{1}{2} \left(X_{(k)} + X_{(k+1)} \right), & \text{if } n = 2k, \end{cases}$$

where $X_{(1)}, \ldots, X_{(n)}$ are order statistics of $X_1, \ldots, X_n$. The association is

$$M = \lambda + U_{\text{med}} = \lambda + G^{-1}(U),$$

where U_{med} is the median of $U_1, U_2, \ldots, U_n$, $U \sim$ Uniform $(0, 1)$ and G is the c.d.f. of the median U_{med}.

For any $m \in \mathbb{R}$, and $u \in [0, 1]$,

$$\Theta_m(\lambda) = \left\{ \lambda : m = \lambda + G^{-1}(u) \right\} = \left\{ m - G^{-1}(u) \right\}.$$

P-step Since we assume $\lambda > 0$, we must have $m - G^{-1}(u) > 0$, i.e., $u < G(m)$. So we use an elastic predictive random set

$$S(u) = \begin{cases} [G(m), 1 - G(m)], & \text{if } G(m) < u < 1 - G(m), \\ \left[\frac{1}{2} - \left| u - \frac{1}{2} \right|, \frac{1}{2} + \left| u - \frac{1}{2} \right| \right], & \text{otherwise.} \end{cases}$$

C-step Combine above two steps, we obtain the expanded set

$$\Theta_m(S) = \begin{cases} \{0\}, & \text{if } G(m) < u < 1 - G(m), \\ \left[m - G^{-1} \left(\frac{1}{2} + \left| u - \frac{1}{2} \right| \right), m - G^{-1} \left(\frac{1}{2} - \left| u - \frac{1}{2} \right| \right) \right], & \text{otherwise.} \end{cases}$$

So we can use the above IM to get the following result.

Proposition 6 *For any singleton assertion* $A = \{\lambda\}$,

$$bel_m (A; S) = 0,$$

$$pl_m (A; S) = 1 - |2G (m - \lambda) - 1|,$$

and the $100(1 - \alpha)\%$ *plausibility region* $\Pi_\lambda(\alpha)$ *is*

$$\Pi_\lambda (\alpha) = \left\{ \lambda : m - G^{-1} \left(1 - \frac{\alpha}{2} \right) < \lambda < m - G^{-1} \left(\frac{\gamma}{2} \right) \right\}.$$

Proof It is clear that $\{\Theta_x (S) \subseteq A\} = \emptyset$, so $bel_m (A; S) = 0$.

$$pl_m (\lambda) = 1 - bel_m \left(A^c; S \right)$$

$$= 1 - P_U \left(m - G^{-1} \left(\frac{1}{2} + \left| u - \frac{1}{2} \right| \right) > \lambda \right) - P_U \left(m - G^{-1} \left(\frac{1}{2} - \left| u - \frac{1}{2} \right| \right) < \lambda \right)$$

$$= 1 - P_U (1 - G (m - \lambda) < u < G (m - \lambda)) - P_U (G (m - \lambda) < u < 1 - G (m - \lambda))$$

$$= 1 - |2G (m - \lambda) - 1|.$$

By the definition, $\Pi_m(\alpha) = \{\lambda : pl_m(\lambda; S) > \alpha\}$. Let $pl_m(\lambda; S) = 1 - |2G (m - \lambda) - 1| > \alpha$ and solve it, then we have $\Pi_\alpha (\lambda) = \{\lambda : m - G^{-1} \left(1 - \frac{\alpha}{2} \right) < \lambda < m - G^{-1} \left(\frac{\alpha}{2} \right)\}$. $\square$

In Table 3 and Fig. 3 we provide a simulation study of $SN(0, 1, \lambda)$ when $\lambda = 1$, 2, 5 and 10.

Remark 3 The p.d.f. of median is as follows.

(a) If $n = 2k + 1$, $g (u) = \frac{n!}{(k)!^2} F^k (u) (1 - F (u))^k f (u)$, where $F (u) = \frac{1}{2} + \frac{1}{\pi} \arctan (u)$ and $f (u) = \frac{1}{\pi} \left(1 + x^2 \right)^{-1}$, which are c.d.f. and p.d.f. of Cauchy $(0, 1)$ respectively, thus the c.d.f. of median is

$$G (u) = (k + 1) \binom{n}{k + 1} \int_0^{F(u)} t^k (1 - t)^k \, dt.$$

(b) If $n = 2k$, the p.d.f. of median is

$$g (u) = \int_{-\infty}^{\infty} \frac{n!}{(k - 1)!^2} (F (2u - v) (1 - F (v)))^{k-1} f (2u - v) f (v) \, dv,$$

thus the c.d.f. of median is $G (u) = \int_{-\infty}^u g (s) \, ds$.

Table 3 Simulation results of average lower bounds (AL), average upper bounds (AU), average lengths (ALength) and coverage probabilities (CP) of 95% plausibility regions when some extra information is given

$\lambda = 1$					$\lambda = 2$				
n	AL	AU	ALength	CP	n	AL	AU	ALength	CP
11	-0.1025	2.1115	2.2140	0.9522	11	0.8897	3.1037	2.2140	0.9464
31	0.4090	1.5890	1.1800	0.9447	31	1.4139	2.5939	1.1800	0.9543
51	0.5485	1.4485	0.9000	0.9505	51	1.5528	2.4528	0.9000	0.9514
101	0.6879	1.3079	0.6200	0.9502	101	1.6881	2.3081	0.6200	0.9486
$\lambda = 5$					$\lambda = 10$				
n	AL	AU	ALength	CP	n	AL	AU	ALength	CP
11	3.8896	6.1036	2.2140	0.9471	11	8.8950	11.1090	2.2140	0.9486
31	4.4073	5.5873	1.1800	0.9502	31	9.4129	10.5929	1.1800	0.9523
51	4.5513	5.4513	0.9000	0.9546	51	9.5484	10.4484	0.9000	0.9501
101	4.6931	5.3131	0.6200	0.9474	101	9.6873	10.3073	0.6200	0.9474

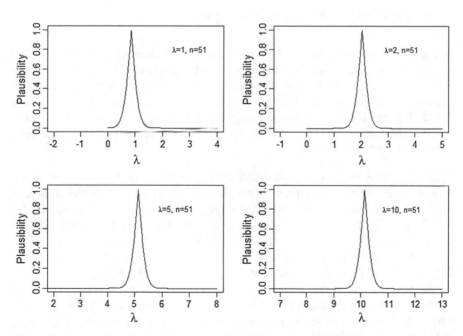

Fig. 3 Graphs of plausibility functions of i.i.d. simulative data from $SN(0, 1, \lambda)$ with size $n = 51$ when some extra information are given

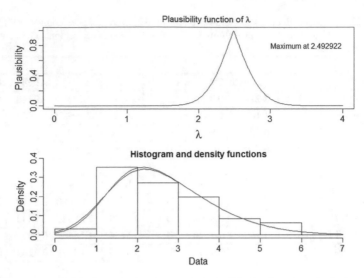

Fig. 4 Histogram, estimated density functions and the plausibility function of λ for Example 2

4 Simulation Study and an Example

In this section, we perform a simulation study to compare average lower bounds (AL), average upper bounds (AU), average lengths (ALength) and coverage probabilities (CP) of 95% plausibility regions of three cases discussed above. We choose samples sizes of 10, 30, 50 and 100. For each simple size, we simulated 10,000 times for $\lambda = 1, 2, 5$ and 10. We also provide graphs of plausibility functions for each case based on simulative data with sample size $n = 50$. Lastly, we provide an example for illustration of our results.

Example 2 The data set was obtained from a study of leaf area index (LAI) of robinia pseudoacacia in the Huaiping forest farm of Shannxi Province from June to October in 2010 (with permission of authors). The LAI is given in Table 4 of Appendix. By Ye et al.'s work [18], the LAI is approximately distributed as $SN(1.2585, 1.8332^2, 2.7966)$ via MME. We use our first result in Sect. 3.1 to explore the data again. In Fig. 4, the top graph is the plausibility function of λ based on the data set, where $\mathsf{pl}(2.492922) = 1$ and the bottom graph is the histogram of the data set and estimated density functions, where the red curve is the density of $SN(1.2585, 1.8332^2, 2.7966)$ via MME and the blue curve is the density of $SN(1.2585, 1.8332^2, 2.492922)$ by using IMs. Here since in this paper we suppose that the location and scale parameters are known, we used MME $\hat{\mu} = 1.2585$ and $\hat{\sigma} = 1.8332$ to standardize the data set, and then we used our result to obtain $\mathsf{pl}(2.492922) = 1$.

Acknowledgements The authors would like to thank referees for their valuable comments.

Appendix

Table 4 The observed values of LAI

LAI (Y)			
June (Y_1)	July (Y_2)	September (Y_3)	October (Y_4)
4.87	3.32	2.05	1.50
5.00	3.02	2.12	1.46
4.72	3.28	2.24	1.55
5.16	3.63	2.56	1.27
5.11	3.68	2.67	1.26
5.03	3.79	2.61	1.37
5.36	3.68	2.42	1.87
5.17	4.06	2.58	1.75
5.56	4.13	2.56	1.81
4.48	2.92	1.84	1.98
4.55	3.05	1.94	1.89
4.69	3.02	1.95	1.71
2.54	2.78	2.29	1.29
3.09	2.35	1.94	1.34
2.79	2.40	2.20	1.29
3.80	3.28	1.56	1.10
3.61	3.45	1.40	1.04
3.53	2.85	1.36	1.08
2.51	3.05	1.60	0.86
2.41	2.78	1.50	0.70
2.80	2.72	1.88	0.82
3.23	2.64	1.63	1.19
3.46	2.88	1.66	1.24
3.12	3.00	1.62	1.14

References

1. Aronld BC, Beaver RJ, Groenevld RA, Meeker WQ (1993) The nontruncated marginal of a truncated bivariate normal distribution. Psychometrica 58(3):471–488
2. Azzalini A (1985) A class of distributions which includes the normal ones. Scand J Stat 12(2):171–178
3. Azzalini A (1986) Further results on a class of distributions which includes the normal ones. Statistica 46(2):199–208
4. Azzalini A, Capitanio A (1999) Statistical applications of the multivariate skew normal distribution. J R Stat Soc 61(3):579–602

5. Azzalini A, Capitanio A (2013) The skew-normal and related families, vol 3. Cambridge University Press, Cambridge
6. Azzalini A, Dalla Valle A (1996) The multivariate skew-normal distribution. Biometrika 83(4):715–726
7. Dey D (2010) Estimation of the parameters of skew normal distribution by approximating the ratio of the normal density and distribution functions. PhD thesis, University of California Riverside
8. Hill M, Dixon WJ (1982) Robustness in real life: a study of clinical laboratory data. Biometrics 38:377–396
9. Liseo B, Loperfido N (2006) A note on reference priors for the scalar skew-normal distribution. J Stat Plan Inference 136(2):373–389
10. Martin R, Liu C (2013) Inferential models: a framework for prior-free posterior probabilistic inference. J Am Stat Assoc 108(501):301–313
11. Martin R (2014) Random sets and exact confidence regions. Sankhya A 76(2):288–304
12. Martin R, Liu C (2015) Inferential models: reasoning with uncertainty, vol 145. CRC Press, New York
13. Martin R, Lingham RT (2016) Prior-free probabilistic prediction of future observations. Technometrics 58(2):225–235
14. Mameli V, Musio M, Sauleau E, Biggeri A (2012) Large sample confidence intervals for the skewness parameter of the skew-normal distribution based on Fisher's transformation. J Appl Stat 39(8):1693–1702
15. Sartori N (2006) Bias prevention of maximum likelihood estimates for scalar skew-normal and skew-t distributions. J Stat Plan Inference 136(12):4259–4275
16. Wang T, Li B, Gupta AK (2009) Distribution of quadratic forms under skew normal settings. J Multivar Anal 100(3):533–545
17. Ye R, Wang T, Gupta AK (2014) Distribution of matrix quadratic forms under skew-normal settings. J Multivar Anal 131:229–239
18. Ye R, Wang T (2015) Inferences in linear mixed models with skew-normal random effects. Acta Mathematica Sinica, English Series 31(4):576–594

International Yield Curve Prediction with Common Functional Principal Component Analysis

Jiejie Zhang, Ying Chen, Stefan Klotz and Kian Guan Lim

Abstract We propose an international yield curve predictive model, where common factors are identified using the common functional principal component (CFPC) method that enables a comparison of the variation patterns across different economies with heterogeneous covariances. The dynamics of the international yield curves are further forecasted based on the data-driven common factors in an autoregression framework. For the 1-day ahead out-of-sample forecasts of the US, Sterling, Euro and Japanese yield curve from 07 April 2014 to 06 April 2015, the CFPC factor model is compared with an alternative factor model based on the functional principal component analysis.

Keywords Yield curve forecasting · Common factors

JEL classification C32 · C53 · E43 · E47

1 Introduction

Yield curve, also known as term structure of interest rates, illustrates the relationship between interest rate and time to maturity. Yield curve forecasting is important in economies. It is not only used by households, firms and financial institutions as primary input factors in making many economic and financial decisions, but also used by central banks to conduct monetary policy in order to achieve policy goals on

J. Zhang (✉) · Y. Chen
Department of Statistics and Applied Probability, National University of Singapore,
Singapore, Singapore
e-mail: jiejiezhang@u.nus.edu

Y. Chen
e-mail: stacheny@nus.edu.sg

S. Klotz
Center for Mathematics, Technische Universität München, Munich, Germany
e-mail: stefan-klotz@gmx.net

K.G. Lim
Lee Kong Chian School of Business, Singapore Management University, Singapore, Singapore
e-mail: kgl@smu.edu.sg

© Springer International Publishing AG 2017   287
V. Kreinovich et al. (eds.), *Robustness in Econometrics*,
Studies in Computational Intelligence 692, DOI 10.1007/978-3-319-50742-2_17

social investment, price stability and employment. Various forecast approaches have been proposed, which can be generally categorized into three strands [18]: the class of no-arbitrage and equilibrium models with theoretical underpinnings, factor models driven by economic theory, and reduced-form factor models driven by statistical data. The former class is widely used in studying risk premia and pricing derivatives. Yet, it is found to forecast poorly, compared with a simple random walk model [11, 12]. The second class can be fitted to multiple shapes of yield curves, e.g. humps, S shapes and monotonic curves such as in [23]. However it assumes uniform factor loadings and may misrepresent unique characteristics of some economies. The latter class has evolved from univariate factor models to multivariate factor models, and to functional factor models in recent advances. The modeling approach developed in this paper falls into the third class, and we show how it successfully forecasts the international yield curves compared to a bunch of alternative data-driven models.

Compared with individual interest rates, yield curve provides richer information about the whole spectrum of interest rates and is capable of reflecting markets' expectation on monetary policy and economic conditions. Numerous studies have detected predictive power in yield curves, see e.g. [14, 24]. It is however challenging to explore the cross-sectional dependence of international yield curves of multiple economies given the large dimensionality and simultaneously unique features. This motivates a proper factor extraction via dimension reduction to reduce the complexity of forecast models.

Factor models provide a balance between model complexity and forecast accuracy by representing the cross-sectional dependence with a few number of factors. The famous [23] model utilizes three factors—level, slope and curvature—within exponential factor loadings. Diebold and Li [9] took into account the temporal evolution of yield curves and developed the Dynamic Nelson-Siegel (DNS) model, which not only kept the parsimony and goodness-of-fit of the Nelson-Siegel interpolation, but also forecasted well compared with the traditional statistical models; see also [6].

The NS exponential factor loadings, though supported by economic theory, are less flexible to characterize features of international yield curves that are specific to the particular economies. This prompts identifying factors in a data-driven way as in [21]. Knez et al. [20] employed the principal component analysis (PCA) approach for analyzing fixed income data of five sectors and detected three to four common factors. Diebold et al. [10] extended the DNS model to multiple countries and identified two factors for the US, UK, German and Japanese zero-coupon government yields. Egorov et al. [13] discovered two factors for the US, LIBOR and EURIBOR term structures. All these studies assumed common covariance structure of the international yield curves of different economies, which ignores the heterogeneity among various groups. For multi-groups with individual covariance matrices, [16] developed Common Principal Component Analysis (CPC) in the multivariate setting that employed a common eigenstructure of these positive definite covariance matrices across several groups but yet allowed individual eigenvalues of each group to reflect heterogeneity.

In most studies, the international interest rates at various maturities are considered as multiple discrete variables. The PCA and CPC methods extract (common) factors based on covariance estimation. The covariance estimation again faces the challenge of high dimensionality that further influences the accuracy of factor identification. The recent development of functional data analysis paves the way for an accurate estimation. The multivariate interest rates can eventually be considered as a functional yield curve that is defined over a continuous interval of the time to maturity. Functional principal component analysis (FPC) reduces dimensionality of functional data based on the covariance operator estimation, which is an extension of the traditional multivariate PCA in functional domain as in [4]. Ferraty and Vieu [15] showed that FPC is a useful tool to quantify proximities between functional objects (e.g. yield curves) in reduced dimension, see also [8, 22, 26, 28, 30]. Coffey et al. [7] stated that the Common FPC (CFPC) is a useful tool for multi-groups of patients. Hyndman and Yasmeen [19] used a partial common principal component model to forecast age-specific mortality rates in Australia. When yield curves are considered, [5] used the FPC and identified three factors of the US yield curves. Benko et al. [3] derived multiple hypothesis tests based on the Common Functional Principal Component (CFPC) Analysis that is to verify the additional insight of using the common functional principal components. They found that the CFPC reduced the number of factors by half with a higher estimation precision than the FPC.

We propose an international yield curve predictive model, where common factors are identified using the CFPC method that enables a comparison of the variation patterns across different economies with heterogeneous covariances. The dynamics of the international yield curves are further forecasted based on the data-driven common factors in an autoregression framework. For the 1-day ahead out-of-sample forecasts of the US, Sterling, Euro and Japanese yield curve from 07 April 2014 to 06 April 2015, the CFPC model is compared with an alternative factor model based on the FPC method.

The reminder of the paper is structured as follows. In Sect. 2, we describe the data used. In Sect. 3, we present the CFPC method, and detail the estimation. In Sect. 4, we implement the proposed method to the international yield curves and report the forecast accuracy. Finally, Sect. 5 concludes.

2 Data

Since the credit crisis in 2007, most banks started using the Overnight Indexed Swaps (OIS) as risk-free interest rate when valuing collateralized derivatives. In a swap, a leg of fixed interest rate is exchanged for a leg of the floating rate that is the geometric mean of the overnight rate. There is a payment from the fixed-rate payer to the floating-rate payer if the geometric average of daily rates is less than the fixed rate for the period; otherwise the payment is vice versa. The fixed rate in the swap is referred to as the OIS rate, while the floating side replicates the aggregated interest that occurs when daily loans are sequentially rolled over at the overnight rate.

We consider the daily Bloomberg OIS data of four economies from July 09, 2012, until April 06, 2015: the US Effective Federal Funds Rate (USEFFR), the

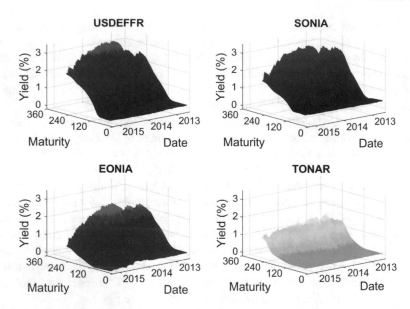

Fig. 1 3D plots of the OIS curves with maturities in months for USEFFR, SONIA, EONIA, and TONAR capturing the movement of the yield curves over time

Sterling OverNight Index Average (SONIA), the Euro OverNight Index Average (EONIA) and the Tokyo OverNight Average Rate (TONAR). On each day, there are 21 maturities of each economy. Two missing values in the data set were identified and replaced by using the average yield of the OIS with the same maturities of the two adjacent trading days. Moreover, four data irregularities have been discovered and corrected using the average yield with the same maturities of the two adjacent trading days. Figure 1 shows the time evolutions of the four OIS yield curves of USEFFR, SONIA, EONIA, and TONAR. It can be seen that while the levels and the slopes of the yield curves differ considerably, there is a common decline in the long-term-maturity yields.

The shape, level and serial dependence of the international yield curves vary over time. As an illustration, Fig. 2 displays the OIS yield curves of the four economies on different days. It shows that the yield curves are generally steeper in the past e.g. on May 31, 2013 and March 31, 2014 than in recent times, e.g. on January 30, 2015. Among others, the EONIA yield curves flatten from the mid-term-maturity over time, though the level continuously drops. The international yield curves exhibited different movements, which vary the dependence among the groups. For example, the level of EONIA was considerably higher than that of TONAR at the beginning of the sample in 2012. However, on January 30, 2015, the EONIA and TONAR curves dropped to a similar level.

The sample covariance surfaces of the international yield curves are displayed in Fig. 3. The covariance surfaces show a common shape for the four economies with greater values among larger maturities and smaller among short maturities. However, the variance-covariances of different economies are distinct in terms of

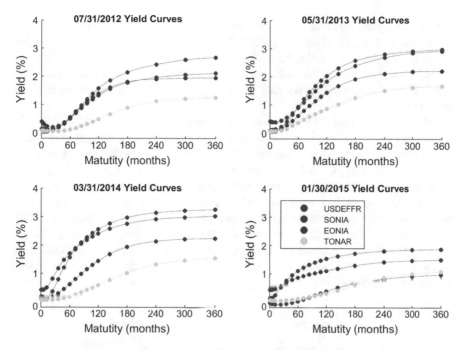

Fig. 2 OIS yield curves for USEFFR, SONIA, EONIA, and TONAR for the selected dates July 31, 2012, May 31, 2013, March 31, 2014, and January 30, 2015

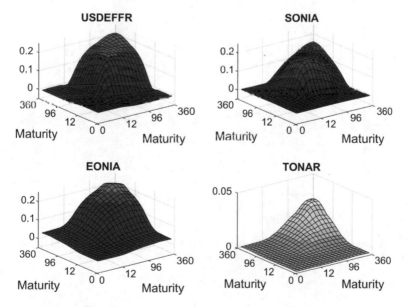

Fig. 3 Sample covariance surfaces of the USEFFR, SONIA, EONIA, and TONAR yield curves with maturities in months from July 09, 2012 to April 06, 2015

magnitude. For example the range of the covariances of USEFFR is -0.0134–0.2067, while that of TONAR is 0.0000–0.0285. For this reason, we adopt a common factor extraction approach rather an analysis of the combined data set. The CPC and CFPC are appropriate for the multi-groups with heterogeneity that characterize common characteristics while preserving unique features of yield curves across different economies.

Moreover, serial dependence is exhibited in the OIS curves. Figure 4 illustrates the sample autocorrelations of the OIS rates at three representative maturities of short-term (1-month), mid-term (12-month) and long-term (120-month). The autocorrelations of SONIA, and TONAR decline slower than those of USEFFR and EONIA. The long-term maturity rates are persistent with the least decline across all the four groups. All the autocorrelations are significant up to a lag order of 60 days. Among them, the minimums are 0.8105 for the USEFFR (12-month), 0.9106 for the SONIA (1-month), 0.7747 for the EONIA (12-month), and 0.9094 for the TONAR (1-month). As the dimension reduction methods are implemented under independence assumption, the serial dependence will be carried forward to the factors and this motivates the use of autoregressive model.

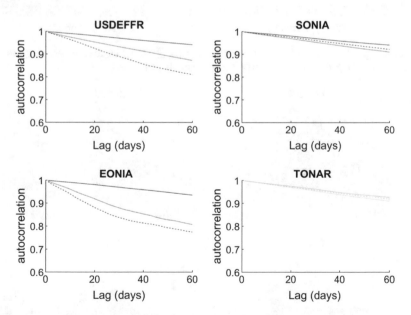

Fig. 4 Autocorrelation function of OIS time series of 1-month ($\cdots$), 12- (- - -) and 120-month (—) maturity for USEFFR, SONIA, EONIA, and TONAR

3 Method

In this section, we present the Common Functional Principal Component Analysis (CFPC) method that is used to characterize the multi-group yield curves in a data-driven way.

Let $X : (\Omega, \mathcal{A}, \mathbb{P}) \to (L^2(\mathcal{T}), \mathcal{B}_{L^2(\mathcal{T})})$ be a measurable function, where $(\Omega, \mathcal{A}, \mathbb{P})$ is a probability space and $\mathcal{B}_{L^2(\mathcal{T})}$ is a Borel field defined on the functional Hilbert space $L^2(\mathcal{T})$); see [2].

The Karhunen-Loève expansion of a random function X has the form:

$$X(\tau) = \mu(\tau) + \sum_{j=1}^{\infty} \xi_j \phi_j(\tau). \tag{1}$$

which converges in L^2 to X uniformly in $\tau \in [0, 1]$; see [1]. The Karhunen-Loève expansion forms the theoretical basis of analyzing the random function X by examining the corresponding orthonormal eigenfunctions ϕ_j and the principal component scores ζ_j; see [17].

Suppose there is only one group, then the Functional Principal Component Analysis (FPC) method is to find orthonormal functions $\phi_1, \phi_2, \ldots$ such that the variances of the principal scores are maximal. Thus, the problem of finding factors translates to an optimization problem on the covariance function $\nu(\pi, \tau) : L^2(\mathcal{T}) \to \mathbb{R}$, $\pi, \tau \in [0, 1]$, given by $\nu(\pi, \tau) = \text{Cov}(X(\pi), X(\tau)) = \mathbb{E}\{(X(\pi) - \mu(\pi))(X(\tau) - \mu(\tau))\}$, where $\mu(\tau)$ denotes the mean function:

$$\arg\max_{\phi_j} \text{Var}(\xi_j) = \arg\max_{\phi_j} \int_{\mathcal{T}} \int_{\mathcal{T}} \phi_j(\pi)\nu(\pi, \tau)\phi_j(\tau)d\pi d\tau, \tag{2}$$

subject to the constraint

$$\langle \phi_j, \phi_l \rangle = \delta_{jl}, \tag{3}$$

where, the j-th principal component score is defined as

$$\xi_j = \int_{\mathcal{T}} (X(\pi) - \mu(\pi)) \phi_j(\pi)d\pi. \tag{4}$$

Due to the orthogonality of ϕ_j and ϕ_l for $j \neq l$, it follows that $\mathbb{E}(\xi_j \xi_l) = 0$ for $j \neq l$. We have $\mathbb{E}(\xi_j^2) = \lambda_j$ and $\mathbb{E}(\xi_j) = 0$, see [3]. The covariance operator $\Upsilon : L^2(\mathcal{T}) \to L^2(\mathcal{T})$ associated with the covariance function ν is defined as

$$(\Upsilon z)(\tau) = \int_{\mathcal{T}} \nu(\pi, \tau)z(\pi)d\pi, \tag{5}$$

for a continuous function $z(\pi)$, $\pi \in [0, 1]$. The Cauchy-Schwarz inequality implies that the problem (2) transforms to the eigenequation [25]:

$$\int_T \nu(\pi, \tau)\phi_j(\pi)d\pi = \lambda_j\phi_j(\tau), \tag{6}$$

where ϕ_j represents the eigenfunction and λ_j the corresponding eigenvalue of the covariance operator Υ. Or equivalently,

$$(\Upsilon\phi_j)(\tau) = \lambda_j\phi_j(\tau). \tag{7}$$

3.1 Common Functional Principal Component Analysis

Now we extend the dimension reduction to multi-groups, where each economy g is an element of G. The CFPC assumes a common eigen-structure across different economies. In particular, the covariance operators Υ_g have common orthonormal eigenfunctions ϕ_j across the G groups, i.e. $\phi_{g_1,j} = \phi_{g_2,j}$ for $1 \leq g_1, g_2 \leq G$, with different eigenvalues $\lambda_{g,j}$ to represent the heterogeneity among the groups. The covariance function of group g is defined as:

$$\nu_g(\pi, \tau) = \sum_{j=1}^{\infty} \lambda_{g,j}\phi_j(\pi)\phi_j(\tau). \tag{8}$$

The eigenequation (2) in the CFPC translates to

$$\int_T \nu_g(\pi, \tau)\phi_j(\pi)d\pi = \lambda_{g,j}\phi_j(\tau) \Longleftrightarrow (\Upsilon_g\phi_j)(\tau) = \lambda_{g,j}\phi_j(\tau), \tag{9}$$

subject to the constraint

$$\langle\phi_j, \phi_l\rangle = \delta_{jl}. \tag{10}$$

3.2 Estimation

In our estimation, we shall approximate integrals over continuous general functions $f(s)$. The integrals $\int_T X_{g,t}(\pi)\phi_j(\pi)d\pi$ can be approximated with numeric quadrature techniques:

$$\int_T f(s)ds \approx \sum_{l=1}^{L} w_l f(s_l), \tag{11}$$

where L is the number of discrete arguments or quadrature points s_l, and w_l are quadrature weights. The application of quadrature techniques of type (11) to the left side of the eigenequation (9) yields an approximation of the form:

$$\int_{\mathcal{T}} \nu_g(\pi, \tau)\phi_j(\pi)d\pi \approx \sum_{l=1}^{L} w_l \nu_g(\pi_l, \tau)\phi_j(\pi_l)$$
$$= \bar{\nu}_g(\tau)^T w * \bar{\phi}_j, \quad (12)$$

where the vectors are set to $\bar{\nu}_g(\tau) = (\nu_g(\pi_1, \tau), \ldots, \nu_g(\pi_L, \tau))^T$, $w = (w_1, \ldots, w_L)^T$, $\bar{\phi}_j = (\phi_j(\pi_1), \ldots, \phi_j(\pi_L))^T$ and $*$ is a Hadamard product. The symmetry of the covariance function leads to:

$$\Upsilon\phi_j = V_g W\bar{\phi}_j, \quad (13)$$

with the $(L \times L)$ matrix $V_g = (\nu_g(\pi_l, \pi_k))_{l,k}$ containing the values of the covariance function at the quadrature points, $\bar{\phi}_i = (\phi_j(\pi_1), \ldots, \phi_j(\pi_L))^T$, and W being the diagonal $(L \times L)$ matrix having the weights w_l as elements. Thus, the eigenequation (9) is refined to

$$V_g W\bar{\phi}_j = \lambda_{g,j}\bar{\phi}_j, \quad (14)$$

subject to the orthonormality condition

$$\bar{\phi}_j W\bar{\phi}_l^T = \delta_{jl}. \quad (15)$$

By assuming positive weights, as most quadrature techniques do, the approximated eigen-equation can be transformed to a standard form,

$$W^{1/2} V_g W^{1/2} u_j = \lambda_{g,j} u_j, \quad (16)$$

under the constraint

$$\langle u_j, u_l \rangle = \delta_{jl}, \quad (17)$$

and we have $u_j = W^{1/2}\bar{\phi}_j$.

Suppose the interval $\mathcal{T}$ is divided into $L - 1$ equal-sized intervals and the function is evaluated at the boundaries s_l with $1 \leq l \leq L$ of these sub-intervals. The trapezoidal rule works

$$T(h) = \int_{\mathcal{T}} f(s)ds = h\left\{\frac{f(s_1)}{2} + \sum_{l=2}^{L-1} f(s_l) + \frac{s_L}{2}\right\}, \quad (18)$$

where $h = \frac{|T|}{L-1}$ is the width of the sub-intervals. Hence, the weight vector is $w = (w_1, \ldots, w_L)^T = (h/2, h, \ldots, h, h/2)^T$. The error of the approximation for the twice differentiable function f in the interval T is controlled [27]:

$$\left\| \int_T f(s)ds - T(h) \right\| \leq \frac{h^3}{12} \max_{t \in T} |f''(t)|. \tag{19}$$

When applying the trapezoidal rule Eq. (16) translates to

$$\text{diag}\,(h/2, h, \ldots, h, h/2)^{1/2}\, V_g \text{diag}\,(h/2, h, \ldots, h, h/2)^{1/2}\, u_j = \lambda_{g,j} u_j, \tag{20}$$

where $u_j = \text{diag}\,(h/2, h, \ldots, h, h/2)^{1/2}\, \bar{\phi}_j$ is subject to the constraint

$$\langle u_j, u_l \rangle = \delta_{jl}. \tag{21}$$

3.3   Factor AR Model

For each of the obtained single common factor, denoted as $\xi_{g,jt}$ for the j-th factor of the g-th economy, an autoregressive model of order 1 (AR(1)) is employed to perform the forecast.

$$\xi_{g,jt} = \alpha_{g,j0} + \alpha_{g,j1}\xi_{g,jt-1} + \epsilon_{g,jt}, \tag{22}$$

where $\alpha_{g,j0}$ and $\alpha_{g,j1}$ are the unknown parameters, and $\{\epsilon_{g,jt}\}$ is white noise satisfying the usual definition: $E(\epsilon_{g,jt}) = 0$ for all t, $E(\epsilon_{g,jt}^2) = \sigma_{\epsilon_{gj}}^2$, $\epsilon_{g,jt}$ and $\epsilon_{g,js}$ are independent for $t \neq s$. In the AR(1) approach, the parameters are estimated via maximum likelihood estimation under Gaussianity. Yield curve forecast is directly obtained based on the forecasts of these common factors.

4   Real Data Analysis

In this section, we apply the common functional factor model to perform out-of-sample forecasts of the yield curves and elaborate on its accuracy in a comparison with an alternative functional factor model based on the FPC method.

We use the yield data as described in Sect. 2: the four OIS data sets of the USEFFR, SONIA, EONIA, and TONAR from July 09, 2012 to April 06, 2015. There are 716 daily yield curves in each economy. We perform factor extraction and forecast 1-day ahead yield curves with the two specifications: FPC assuming homogeneous covariance operators among different economies and CFPC incorporating unique features of the international yield curves.

We perform out-of-sample forecast in real time with the forecast beginning on April 07, 2014, till April 06, 2015. The first factor model is estimated with the first 455 observations from July 09, 2012, until April 4, 2014, and is used to forecast the international interest rates for April 07, 2014. We move forward one day at a time to redo estimation and forecast till the end of the sample. In total, we obtain 261 out-of-sample one-step-ahead forecasts.

4.1 Forecasting Performance Measures

The forecasting power is evaluated using the root mean squared error (RMSE) which is common in forecast study, e.g. [29]. The RMSE is defined through

$$\mathrm{RMSE}(\tau) = \sqrt{\frac{1}{m_0} \sum_{t=1}^{m_0} \left(y_{t+h,\tau} - \hat{y}_{t+h|t,\tau}\right)^2},$$

where the forecast error $(y_{t+h,\tau} - \hat{y}_{t+h|t,\tau})$ at time $(t + h)$ for time to maturity τ is derived using observed score value y_{t+h} of the h-step-ahead forecast for the score and m_0 is the number of periods being forecasted.

4.2 Forecast Results

We selected three functional factors via the FPC method. The explanatory power of the three factors sums up to 99.8%, with the first explaining 90.4% variation, the second 9.1%, and the third 0.3% respectively. Figure 5a displays the estimated functional principal component scores. The first factor has an exponential increasing trend, representing the general shape of the OIS yield curve residuals. The second one shows an increasing section up to the short-term around 48-month maturity, followed by a decline. It captures the curvature behavior of the residual curves. The third factor is twisting from a hump around 24-month maturity to a valley around 108 months, and it is increasing thereafter.

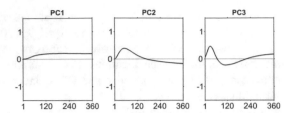

(a) Functional principal component scores against the maturities (months).

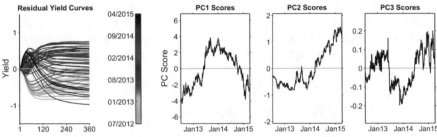

(b) Functional principal components of the USEFFR curves.

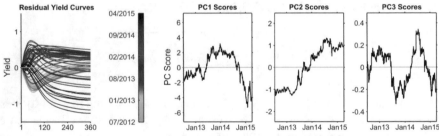

(c) Functional principal components of the SONIA curves.

Fig. 5 **a** Functional principal component scores against the maturities (months). **b** Functional principal components of the USEFFR curves. **c** Functional principal components of the SONIA curves. **d** Functional principal components of the EONIA curves. **e** Functional principal components of the TONAR curves

Figure 5b–e display the time evolution of the functional factors. The factors of the USEFFR and SONIA curves share several common characteristics. For example, the first factor shows the simultaneous increase in summer 2013 and decrease from mid-2014 onward. Meanwhile, these two economies exhibit obvious difference when the second factor is compared. The second factor of the USEFFR is more pronounced than that of the SONIA. Likewise, the second factor of the EONIA shows a more volatile evolution than that of the TONAR.

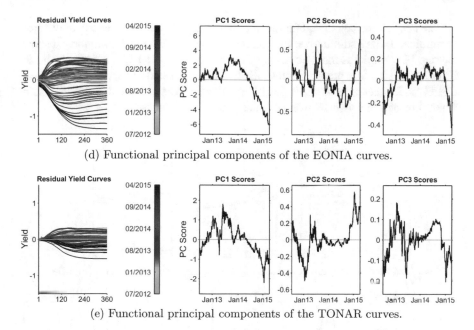

(d) Functional principal components of the EONIA curves.

(e) Functional principal components of the TONAR curves.

Fig. 5 (continued)

Common Functional Principal Component Model

The CFPC on the other hand incorporates the heterogeneous covariance matrices and extracts common functional factors. Figure 6a illustrates the common functional component scores. The CFPC scores are similar to those of FPC and thus share the similar interpretations. Again, the first common functional factor is slowly increasing up to maturities of around 144 months, and thereafter stays flat but with a slight decrease for 360 months. In contrast, the second common factor is increasing up to 48 months maturity and thereafter declines faster. The third one shows a strong increase up to 36-month maturity, followed by a minimum at 108 months. Figure 6b–e illustrate the time series of the common functional factors of the CFPC method.

Table 1 reports the explained variation of the common functional factors for each economy. The first factors capture between 81.3% of the variation for the SONIA curves and 98.5% for the EONIA curves, the second one 1.1% (EONIA) and 18.0% (SONIA), and the third one represents between 0.1% (USEFFR) and 1.1% (TONAR) respectively.

The RMSEs are calculated for the international yield curves forecasts of different economies. Table 2 summarizes the prediction power of the four economies. It shows that the FPC model is more accurate when considering the EONIA curves, and the CFPC model is more precise for the USEFFR, SONIA and TONAR. In general, the CFPC model seems to be more flexible for the OIS forecasts and is a good choice when forecasting a diverse group of international yield curves.

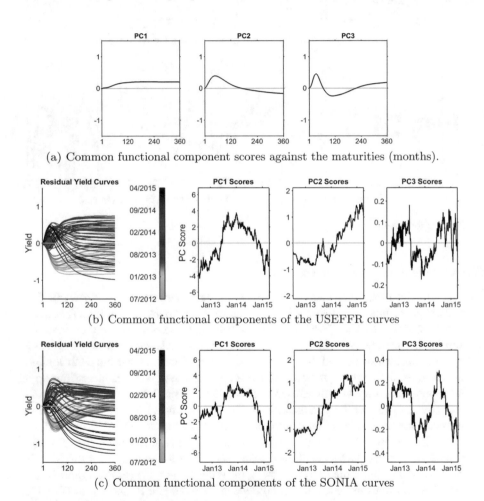

(a) Common functional component scores against the maturities (months).

(b) Common functional components of the USEFFR curves

(c) Common functional components of the SONIA curves

Fig. 6 **a** Common functional component scores against the maturities (months). **b** Common functional components of the USEFFR curves. **c** Common functional components of the SONIA curves. **d** Common functional components of the EONIA curves. **e** Common functional components of the TONAR curves

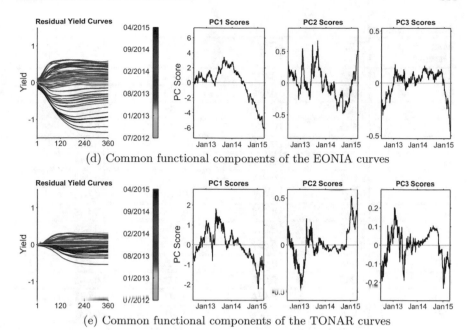

(d) Common functional components of the EONIA curves

(e) Common functional components of the TONAR curves

Fig. 6 (continued)

Table 1 Explained variation of the first three common functional factors

	USEFFR (%)	SONIA (%)	EONIA (%)	TONAR (%)
CFPC1	89.7	81.3	98.5	92.7
CFPC2	9.9	18.0	1.1	4.8
CFPC3	0.1	0.6	0.3	1.1
Total	99.7	99.9	99.9	98.6

Table 2 RMSE for one-step-ahead interest rates forecasts, evaluated at the given maturities; lower error measures are highlighted in boldface indicating better accuracy

Maturity (months)	FPC	CFPC	FPC	CFPC
	USEFFR		SONIA	
1	0.025	**0.020**	0.030	**0.012**
2	0.026	**0.024**	0.031	**0.015**
3	0.029	**0.028**	0.033	**0.020**
4	**0.033**	0.034	0.038	**0.029**
5	**0.038**	0.040	0.040	**0.034**
6	**0.043**	0.046	0.039	**0.036**

(continued)

Table 2 (continued)

Maturity (months)	FPC	CFPC	FPC	CFPC
12	**0.064**	0.077	**0.041**	0.045
24	**0.046**	0.056	0.033	**0.033**
36	0.036	**0.036**	0.038	**0.038**
48	**0.049**	0.050	0.044	**0.044**
60	**0.051**	0.052	0.046	**0.046**
72	0.049	**0.049**	0.045	**0.045**
84	0.047	**0.047**	0.044	**0.044**
96	0.045	**0.045**	0.044	**0.044**
108	0.045	**0.044**	0.045	**0.045**
120	0.046	**0.045**	0.046	**0.046**
144	0.047	**0.047**	0.045	**0.045**
180	**0.048**	0.049	0.043	**0.043**
240	**0.045**	0.046	0.042	**0.042**
300	0.045	**0.044**	0.042	**0.041**
360	0.045	**0.044**	0.041	**0.039**
	EONIA		TONAR	
1	**0.083**	0.098	0.009	**0.008**
2	**0.077**	0.089	0.009	**0.007**
3	**0.073**	0.082	0.009	**0.007**
4	**0.071**	0.076	0.008	**0.006**
5	**0.069**	0.072	0.009	**0.006**
6	0.067	**0.066**	0.009	**0.005**
12	0.061	**0.046**	0.013	**0.007**
24	0.039	**0.034**	0.020	**0.019**
36	**0.014**	0.019	**0.015**	0.018
48	0.025	**0.018**	**0.008**	0.009
60	0.031	**0.030**	0.015	**0.015**
72	**0.030**	0.032	**0.025**	0.027
84	**0.027**	0.030	**0.025**	0.026
96	**0.027**	0.030	**0.021**	0.023
108	**0.028**	0.029	**0.017**	0.018
120	0.030	**0.029**	0.018	**0.018**
144	0.035	**0.033**	0.026	**0.026**
180	0.041	**0.041**	**0.029**	0.030
240	**0.033**	0.034	0.025	**0.025**
300	0.034	**0.034**	0.024	**0.024**
360	**0.039**	0.040	0.032	**0.032**

5   Conclusion and Outlook

We proposed the CFPC model to characterize the common functional factors underlying multiple groups of yield curves. We found that the CFPC model captured between 98.6% (TONAR) and 99.7% (USEFFR) of the variation and was useful for explaining yield curves across different economic regions. In the AR framework, the factor model delivers reasonable forecasting performance. For the Japanese economy, i.e. the TONAR yield curves, the CFPC model is superior to the alternative.

Our study contributes to the existing literature on yield curve modeling and forecasting by using the CFPC method. It is data-driven and thus can be safely used in other applications of groups with heterogeneity in covariance matrices.

References

1. Ash RB, Gardner MF (1975) Topics in stochastic processes, Probability and mathematical statistics series. Academic Press, New York
2. Benko M (2006) Functional data analysis with applications in finance. Phd thesis, Humboldt-Universität, Berlin
3. Benko M, Härdle WK, Kneip A (2009) Common functional principal components. Ann Stat 37(1):1–34
4. Boente G, Rodriguez D, Sued M (2010) Inference under functional proportional and common principal component models. J Multivar Anal 101(2):464–475
5. Chen Y, Li B (2011) Forecasting yield curves in an adaptive framework. Cent Eur J Econ Model Econometrics 3(4):237–259
6. Chen Y, Niu L (2014) Adaptive dynamic nelson-siegel term structure model with applications. J Econometrics 180(1):98–115
7. Coffey N, Harrison AJ, Donoghue OA, Hayes K (2011) Common functional principal components analysis: a new approach to analyzing human movement data. Hum Mov Sci 30(6): 1144–1166
8. Di C-Z, Crainiceanu CM, Caffo BS, Punjabi NM (2009) Multilevel functional principal component analysis. Ann Appl Stat 3(1):458–488
9. Diebold FX, Li C (2006) Forecasting the term structure of government bond yields. J Econometrics 130(2):337–364
10. Diebold FX, Li C, Yue VZ (2008) Global yield curve dynamics and interactions: A dynamic nelson-siegel approach. J Econometrics 146(2):351–363
11. Duffee GR (2002) Term premia and interest rate forecasts in affine models. J Finance 57(1): 405–443
12. Duffee GR (2011) Forecasting with the term structure: the role of no-arbitrage restrictions. The Johns Hopkins University, Department of Economics, Economics Working Paper Archive (576)
13. Egorov AV, Li H, Ng D (2011) A tale of two yield curves: modeling the joint term structure of dollar and euro interest rates. J Econometrics 162(1):55–70
14. Estrella A, Hardouvelis GA (1991) The term structure as a predictor of real economic activity. J Finance 46(2):555
15. Ferraty F, Vieu P (2006) Nonparametric functional data analysis: theory and practice, Springer series in statistics. Springer, New York
16. Flury BN (1988) Common principal components and related multivariate models, Wiley series in probability and mathematical statistics. Wiley, New York

17. Hall P, Müller H-G, Wang J-L (2006) Properties of principal component methods for functional and longitudinal data analysis. Ann Stat 34(3):1493–1517
18. Hays S, Shen H, Huang JZ (2012) Functional dynamic factor models with application to yield curve forecasting. Ann Appl Stat 6(3):870–894
19. Hyndman RJ, Yasmeen F (2014) Common functional principal component models for mortality forecasting. Contributions in infinite-dimensional statistics and related topics, Societa Editrice Esculapio, Bologna, pp 19–24
20. Knez PJ, Litterman RB, Scheinkman J (1994) Explorations into factors explaining money market returns. J Finan 49(5):1861–1882
21. Litterman RB, Scheinkman J (1991) Common factors affecting bond returns. J Fixed Income 1(1):54–61
22. Müller H-G, Sen R, Stadtmüller U (2011) Functional data analysis for volatility. J Econometrics 165(2):233–245
23. Nelson CR, Siegel AF (1987) Parsimonious modeling of yield curves. J. Bus 60(4):473–489
24. Piazzesi M (2010) Affine term structure models. In: Handbook of financial econometrics. Handbooks in finance, vol 1. North-Holland, Boston, pp 691–766
25. Ramsay JO, Dalzell CJ (1991) Some tools for functional data analysis. J Roy Stat Soc Ser B (Stat Methodol) 53(3):539–572
26. Ramsay JO, Ramsey JB (2002) Functional data analysis of the dynamics of the monthly index of nondurable goods production. J Econometrics 107(1–2):327–344
27. Stoer J (2005) Numerische Mathematik: Eine Einführung, Springer-Lehrbuch, 9th edn. Springer, Berlin
28. Wang Z, Sun Y, Li P (2014) Functional principal components analysis of Shanghai stock exchange 50 index. Discrete Dyn Nat Soc 2014(3):1–7
29. Wooldridge JM (2013) Introductory econometrics: a modern approach, 5th edn. South-Western Cengage Learning, Mason
30. Yao F, Müller H-G, Wang J-L (2005) Functional data analysis for sparse longitudinal data. J Am Stat Assoc 100(470):577–590

An Alternative to p-Values in Hypothesis Testing with Applications in Model Selection of Stock Price Data

Hien D. Tran, Son P. Nguyen, Hoa T. Le and Uyen H. Pham

Abstract In support of the American Statistical Association's statement on p-value in 2016, see [8], we investigate, in this paper, a classical question in model selection, namely finding a "best-fit" probability distribution to a set of data. Throughout history, there have been a number of tests designed to determine whether a particular distribution fit a set of data, for instance, see [6]. The popular approach is to compute certain test statistics and base the decisions on the p values of these test statistics. As pointed out numerous times in the literature, see [5] for example, p values suffer serious drawbacks which make it untrustworthy in decision making. One typical situation is when the p value is larger than the significance level α which results in an inconclusive case. In many studies, a common mistake is to claim that the null hypothesis is true or most likely whereas a big p value merely implies that the null hypothesis is statistically consistent with the observed data; there is no indication that the null hypothesis is "better" than any other hypothesis in the confidence interval. We notice this situation happens a great deal in testing goodness of fit. Therefore, hereby, we propose an approach using the Akaike information criterion (AIC) or the Bayesian information criterion (BIC) to make a selection of the best fit distribution among a group of candidates. As for applications, a variety of stock price data are processed to find a fit distribution. Both the p value and the new approach are computed and compared carefully. The virtue of our approach is that there is always a justified decision made in the end; and, there will be no inconclusiveness whatsoever.

H.D. Tran (✉)
Tan Tao University, Duc Hoa, Long An, Vietnam
e-mail: hien.tran@ttu.edu.vn

S.P. Nguyen · H.T. Le · U.H. Pham
University of Economics and Law, Ho Chi Minh City, Vietnam
e-mail: sonnp@uel.edu.vn

H.T. Le
University of Science, VietNam National University of Ho Chi Minh City,
Ho Chi Minh City, Vietnam

© Springer International Publishing AG 2017
V. Kreinovich et al. (eds.), *Robustness in Econometrics*,
Studies in Computational Intelligence 692, DOI 10.1007/978-3-319-50742-2_18

Keywords p-Values · AIC · BIC · Hypothesis testing · Goodness of fit · Stock prices

1 Introduction

p values date back to the early work of the founders of modern statistics, namely the early twentieth century with Fisher, Neyman and Pearson. The methods marked the beginning of applications of the theory probability to statistics in order to create systematic procedures for utilizing observed data to verify assumptions. It has been widely adopted to a great number of applications in diverse scientific fields. Although p value has already been proven worthy in many cases, a lot of warnings from leading researchers have been issuing against misunderstandings and misuses of p values over several decades. These misinterpretations of what kind of information p values provide have led to inappropriate conclusions with severe consequences. Aware of the seriousness of the issue, the American Statistical Association officially announced a statement about the usages of p values in 2016. To the extreme is the ban of p values by the editors of the journal "Basic and Applied Social Psychology" in 2015.

There are a number of commonly seen mistakes in interpretations of p values. For a summary, see [5]. The main criticism is that p-values are not the probabilities that null hypotheses are true. Therefore, a p-value by itself should not be used to perform any statistical testing. Along this line is the fallacious conclusion that the null hypothesis holds when the p value is greater than the significance level α. This error and a way to avoid it are the focus of the present paper.

Note that, although a number of alternatives for p-value have been suggested, many researchers in statistics and data analysis believe there is no quick fix for all the problems of p values. The following excerpt by Andrew Gelman is extracted from [4]

> In summary, I agree with most of the ASAs statement on p-values but I feel that the problems are deeper, and that the solution is not to reform p-values or to replace them with some other statistical summary or threshold, but rather to move toward a greater acceptance of uncertainty and embracing of variation.

In this paper, on studying the question of choosing a good probability distribution to fit stock price data, we frequently encounter the inconclusive situation where p values of our predetermined candidate distributions are all greater than the significance level α. In our case, we must choose a model for subsequent questions, therefore, we are in need of a mechanism to make a justified choice of probability distribution. After having done a wide range of experiments with data, we realize that by using a combination of the Akaike Information criterion (AIC) and the Bayesian Information criterion (BIC), one can make an excellent selection from the set of prescribed distributions.

1.1 Summary of Our Contributions

In this paper, we propose an alternative method for statistical hypothesis testing utilizing the Akaike Information criterion and the Bayesian Information criterion. Each method has its own strength and those two strengths complement each other.

For application, we compute and compare the p values, the BIC and the AIC for a wide range of stock price data.

2 The Akaike Information Criterion (AIC)

The AIC was designed to find the probability model that gives the best prediction among a set of candidates. One important point is that no candidate is assumed to be the true unknown model. In practice, the set-up of the AIC is as follows:

Suppose we have models $M_1, \ldots, M_k$ where each model is a set of densities:

$$M_j = \left\{ p(y; \theta_j) : \theta_j \in \Theta_j \right\}$$

Suppose further that we have collected a sample $Y_1, \ldots, Y_n$ from an unknown density f.

First step is to use the maximum likelihood estimator to obtain $\hat{\theta}_i$ for each model M_i. The result is a set of estimates $\hat{p}_j(y) = \{p(y; \hat{\theta}_j)\}$, $\forall j$.

The quality of $\hat{p}_j(y)$ as an estimate of f can be measured by the Kullback–Leibler (KL) distance:

$$K(f, \hat{p}_j) = \int f(y) \log \left(\frac{f(y)}{\hat{p}_j(y)} \right) dy$$
$$= \int f(y) \log(f(y)) dy - \int f(y) \log \hat{p}_j(y) dy$$

To minimize the KL distance, we just have to maximize the quantity

$$K_j = \int f(y) \log \hat{p}_j(y) dy$$

Let $\hat{\ell}_j(\theta_j)$ be the log-likelihood function for model j. Akaike, see [1], showed that the sample mean

$$\bar{K}_j = \frac{1}{n} \sum_{i=1}^{n} \log \hat{p}_j(y_i) = \frac{\hat{\ell}_j(\hat{\theta}_j)}{n}$$

is a biased estimator of K_j. Therefore, we will compute the unbiased estimator

$$\hat{K}_j = \bar{K}_j - \frac{d_j}{n}$$

where d_j is the dimension of the parameter space Θ_j, or in other words, d_j is the number of free parameters in model j.

AIC(j) is a rescale of the KL distance, namely

$$\text{AIC}(j) = 2n\hat{K}_j = 2\ell_j(\hat{\theta}_j) - 2d_j$$

3 Bayesian Information Criterion (BIC) and Bayes Factor Approximation

The BIC assumes that one of the candidate distributions is the true model. It tries to find the best model using the Bayesian framework.

Let p_j be the prior probability of model j. Moreover, on each parameter space Θ_j, assume a prior distribution $\pi_j(\theta_j)$.

By Bayes' theorem,

$$P(M_j \mid Y_1, \ldots, Y_n) \propto P(Y_1, \ldots, Y_n \mid M_j)p_j$$

Furthermore,

$$P(Y_1, \ldots, Y_n \mid M_j) = \int P(Y_1, \ldots, Y_n \mid M_j, \theta_j)\pi_j(\theta_j)d\theta_j = \int L(\theta_j)\pi_j(\theta_j)d\theta_j$$

To maximize $P(M_j \mid Y_1, \ldots, Y_n)$, we can equivalently maximize the quantity

$$\log \int L(\theta_j)\pi_j(\theta_j)d\theta_j + \log p_j$$

Using some Taylor series approximation, we obtain

$$\log \int L(\theta_j)\pi_j(\theta_j)d\theta_j + \log p_j \approx \ell_j(\hat{\theta}_j) - \frac{d_j}{2}\log n$$

To be comparable with AIC, we can define

$$\text{BIC}(j) = 2\ell_j(\hat{\theta}_j) - d_j \log n$$

In view of model selection problem using Bayesian statistics, Bayes factor can be defined by

$$\text{Bayes factor}(M_k, M_j) = \frac{P(Y_n \mid M_k)}{P(Y_n \mid M_j)},$$

which measures the evidence for the model M_k versus model M_j based on the data information. The Bayes factor chooses the model with the largest value of marginal likelihood among a set of candidate models.

Noting that

$$\frac{P(M_k|Y_n)}{P(M_j|Y_n)} = \frac{P(Y_n|M_k)}{P(Y_n|M_j)} \times \frac{P(M_k)}{P(M_j)},$$

$$\left[\text{Posterior odds}(M_k, M_j) = \text{Bayes factor}(M_k, M_j) \times \text{Prior odds}(M_k, M_j)\right]$$

the Bayes factor is also given as the ratio of posterior odds and prior odds

$$\text{Bayes factor}(M_k, M_j) = \frac{\text{Posterior odds}(M_k, M_j)}{\text{Prior odds}(M_k, M_j)}.$$

Note that the BIC gives a rough approximation to the logarithm of the Bayes factor, see [2]

$$\log\left[\text{Bayes factor}(M_k, M_j)\right] = \log\left[P(Y_n|M_k)\right] - \log\left[P(Y_n|M_j)\right] \approx (BIC_j - BIC_k)/2.$$

4 Some Remarks AIC and BIC

Firstly, in both AIC and BIC, the minimum value is the desired value. However, since the penalty on the number of free parameters of models on BIC is much larger than that of the AIC due to the dependence on the sample size n, BIC and AIC usually behave differently on the same study. BIC prefers a simpler model with fewer parameters while AIC tends to choose complex model. So, if researchers would like to minimize errors for future predictions of data, AIC will be the suitable criterion, but, if a particular best fit model is more important than prediction then the BIC will be better.

Second, the next remark is about the differences between values of the criteria. Since there might be unexplained variations in variables, two models with similar criterion values should receive the same ranking in evaluating models.

In practice, we use the following tables from [3].

For AIC,

$AIC_j - AIC_{min}$	Level of empirical support for model j
0–2	Substantial
4–7	Considerably less
>10	Essentially none

For BIC,

$BIC_j - BIC_{min}$	Evidence against model j
0–2	Not worth more than a bare mention
2–6	Positive
4–7	Strong
>10	Very strong

Third, it should be emphasized that AIC or BIC only helps select the most suitable model among a list of candidates. Therefore, in some situations, if all the candidate models are badly supported by the observed data, the model chosen by AIC or BIC is simply the "least bad" model to fit data. It might still be of very poor quality. Therefore, one should take extra precaution when using the winning model to perform any statistical inferences.

Last, AIC and BIC provide ranking systems to evaluate candidate models. However, they do not reveal how well the observed data are consistent with any particular model, the kind of information routinely extracted from p values.

5   Applications

We use for best-fitting test of data. The variable h represents the decision in hypothesis testing problem, in

  $h = 1$: Reject the null hypothesis at the default 0.01 significance level,

  $h = 0$: Do not reject the null hypothesis at the default 0.01 significance level.

5.1   Closing Prices Data of the Company Apple

We consider the closing prices of Apple company from 06-Nov-2012 to 12-Aug-2013 from 200 observations.

Data transformation is the closing price today divided by the closing price yesterday.

5.1.1   Chi Square and Jacque Bera Test

Figure 1 shows the histogram of data and then using Chi square test, the result on distribution is shown in Table 1.

According to descriptive statistics such as the histogram and QQ plot, we can see that data seems to follow a normal distribution.

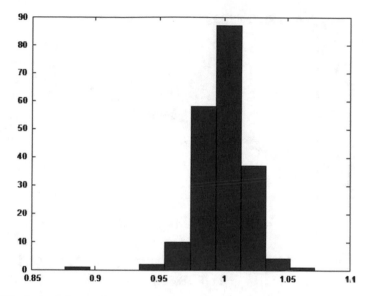

Fig. 1 The closing price graph

Table 1 Chi square test

Distribution	Normal	Gamma	Generalized beta	Student
h	0	0	1	0
p-value	0.0179218	0.013424	2.95E-22	0.0741824

Table 2 Jarque Bera test

h	1
p-value	0.001

However, from the Table 1 we cannot conclude that data follow normal, Gamma or Student distribution at 0.01 significance level, because p-value > 0.01. In the mean time, the generalized Beta distribution should not be a good candidate for fitting.

In addition, with analysis of Table 2 at 0.01 significance level, Jarque Bera test shows that the data does not follow a normal distribution.

5.1.2 AIC, BIC and Bayes Factor

From Sects. 3 and 4, we will use AIC, BIC and Bayes factor for model selection in terms of model distribution.

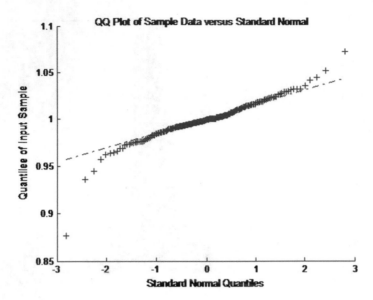

Fig. 2 QQ plot

Table 3 AIC, BIC, Bayes factor

Distribution	Normal	Gamma	Generalized beta	Student	Distribution min	Bayes factor
AIC	−992.782	−989.834	−795.547	375.138	Normal	
BIC	−986.186	−983.238	−788.950	385.033	Normal	4.366218

According to the Table 3, the data follows a normal distribution. Furthermore, Bayes factor coefficient 4.366 in the Table 3 (which is more than 3) can tell the data fits a normal distribution better than other distributions (Fig. 2).

In the similar method, we consider several other datum.

5.2 Closing Prices Data of the Company Google

We consider the closing prices of Google company from 07-Dec-11 to 11-Sep-12 from 200 observations.

5.2.1 Chi Square and Jarque Bera Test

Figure 3 shows the histogram of data is approximately symmetric. In addition, a normal distribution appears from the QQ plot in Fig. 4.

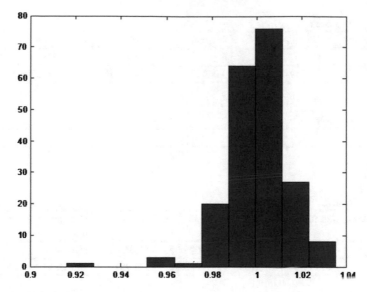

Fig. 3 The closing price graph

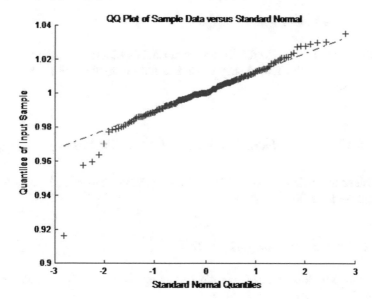

Fig. 4 QQ plot

Just like in the case of Apple company, from the Table 4 we can not conclude data follows normal, Gamma nor Student distribution at 0.01 significance level.

In addition, with analysis of Table 5 at 0.01 significance level, Jarque Bera test shows that the data does not follow a normal distribution with p-value <0.01.

Table 4 Chi square test

Distribution	Normal	Gamma	Generalized beta	Student
h	0	0	1	0
p-value	0.0498823	0.0434246	2.425E-16	0.8561765

Table 5 Jarque Bera test

h	1
p-value	0.001

Table 6 AIC, BIC, Bayes factor

Distribution	Normal	Gamma	Generalized beta	Student	Distribution min	Bayes factor
AIC	−1140.45	−1138.04	−1009.67	375.0561	Normal	
BIC	−1133.86	−1131.45	−1003.07	384.951	Normal	3.335081

5.2.2 AIC, BIC and Bayes Factor

According to the Table 6, the data follows a normal distribution. Furthermore, Bayes factor coefficient in the Table 6 shows that data fits a normal distribution better than other distributions.

5.3 Closing Prices Data USD/CHF Exchange Rate

We consider the closing prices of USD/CHF exchange rate from 12-Nov-2014 to 20-Aug-2015 from 200 observations.

5.3.1 Chi Square and Jarque Bera Test

Figure 5 shows the histogram of data and the Chi square test for different distributions results in Table 7.

Again, descriptive statistics make normal distribution a plausible choice. However, from the Table 7, we cannot conclude data follow normal, Gamma or Student distribution at 0.01 significance level.

In addition, with analysis of Table 8 at 0.01 significance level, Jarque Bera test shows that the data does not follow the normal distribution (Fig. 6).

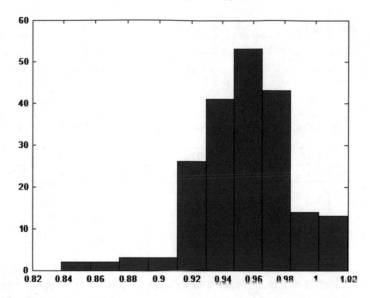

Fig. 5 The closing price graph

Table 7 Chi square test

Distribution	Normal	Gamma	Generalized beta	Student
p-value	0.1742495	0.1769862	1.63E-09	0.3069157

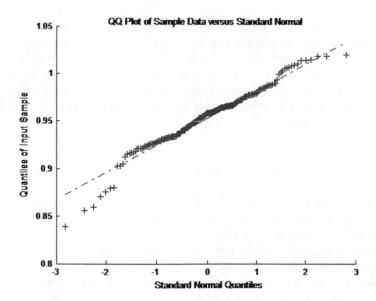

Fig. 6 QQ plot

Table 8 Jarque Bera test

h	1
p-value	0.001021

Table 9 AIC

Distribution	Normal	Gamma	Generalized beta	Student	Distribution min	Bayes factor
AIC	−829.771	−827.134	−757.99	374.9986	Normal	
BIC	−823.174	−820.537	−751.394	384.8936	Normal	3.738204

5.3.2   AIC, BIC and Bayes Factor

According to the Table 9, the data follows a normal distribution. Furthermore, Bayes factor coefficient in the Table 9 shows that the data fits a normal distribution better than other distributions.

5.4   Simulated Data from a Normal Distribution with Mean 2 and Standard Deviation 0.5

By simulating data from a normal distribution with mean 2 and standard deviation 0.5, we obtained six sets of simulation data with 250 observations for each.

All the data from previous sections have shown that Jarque Bera test always disagree with Chi square goodness-of-fit test. For illustration purposes, simulated data have been used for two reasons: first, to show that Jarque Bera test will work for ideal cases (e.g. simulated data), and then to confirm the strength of AIC and BIC in model selection without p values.

5.4.1   Chi Square and Jarque Bera Test

Based on chi-square test, Table 10 shows that the data do not follow the generalized Beta distribution, and we can not conclude that data follows normal, Gamma nor Student distribution at 0.01 significance level.

Following the analysis of Table 11, at 0.01 significance level, none of these data sets come out from a normal distribution.

Unlike in case of the experimental data, Jarque Bera test and Chi square test give the same conclusions, that is no basis to conclude that the data follows a normal distribution or not.

Table 10 Chi square test

Distribution	Normal	Gamma	Generalized beta	Student
Data 1: h	0	0	1	0
Data 1: p-value	0.464447	0.013139	9.03E-05	0.33762
Data 2: h	0	0	1	0
Data 2: p-value	0.378353	0.061114	1.07E-06	0.257894
Data 3: h	0	0	1	0
Data 3: p-value	0.81861	0.40886	6.74E-05	0.71038
Data 4: h	0	0	1	0
Data 4: p-value	0.2631097	0.0136797	8.58E-04	0.1796478
Data 5: h	0	0	1	0
Data 5: p-value	0.9791309	0.1178657	1.71E-08	0.9350344
Data 6: h	0	1	1	0
Data 6: p-value	0.401851	0.0041582	5.60E-06	0.2864364

Table 11 Jarque Bera test

Data	1	2	3	4	5	6
h	0	0	0	0	0	0
p-value	0.5	0.5	0.5	0.260869	0.465078	0.346738

Table 12 AIC

Distribution and AIC	1	2	3	4	5	6
Normal	372.4261	359.9674	382.5945	371.5895	370.1364	358.3627
Gamma	384.7519	375.7018	388.9089	381.0509	404.0931	381.1214
Generalized beta	430.6064	456.1119	448.8263	419.3184	496.9188	438.2722
Student	471.0849	472.6507	467.6781	472.8842	469.7819	467.9981
Distribution min	Normal	Normal	Normal	Normal	Normal	Normal

5.4.2 AIC, BIC and Bayes Factor

According to the Table 12, the data follows a normal distribution.

Bayes factor coefficient in the Table 13 (which is greater than 3) shows the data fit a normal distribution better than other distributions.

Table 13 BIC and Bayes factor

Distribution and BIC	1	2	3	4	5	6
Normal	379.469	367.0103	389.6374	378.6327	377.1793	365.4057
Gamma	391.7949	382.7447	395.9519	388.0938	411.136	388.1643
Generalized beta	437.6493	463.1548	455.8692	426.3614	503.9618	445.3151
Student	481.6492	483.2151	478.2424	483.4486	480.3463	478.5625
Distribution min	Normal	Normal	Normal	Normal	Normal	Normal
Bayes factor	474.8135	2610.171	23.50536	113. 3589	23638475	87494.11

6 Conclusion

In this paper, we present the BIC and AIC criteria together with the usage suggestions in practice. The main contribution is that AIC and BIC resolve the second error in the list of twelve common errors of p values as found in [5]. On one hand, we propose using BIC and AIC as alternatives for p value in model selection, in association with approximated Bayes factor as a confirmation. On the other hand, we also highlight the similarities and differences between the two criteria which serve as the guidelines on when to use which criteria.

As for applications, we collect various stock price data and show that the classical p value approach results in inconclusiveness in most of the time. In contrast, BIC or AIC will offer a justified decision in all these cases.

Acknowledgements We would like to express our deep gratitude to professor Hung T. Nguyen of New Mexico State University/Chiang Mai university for his generous help in our research, for his encouragements, and for numerous discussions.

References

1. Akaike H (1973) Information theory and an extension of the maximum likelihood principle. In: 2nd international symposium on information theory. Academiai Kiado
2. Ando T (2010) Bayesian model selection and statistical modeling. CRC Press
3. Burnham KP, Anderson DR (2002) Model selection and multimodel inference. Springer, New York
4. Gelman A (2012) p values and statistical practice. Epidemiology 24(1):69–72
5. Goodman S (2008) A dirty dozen: twelve p-value misconceptions. In: Seminar in hematology. Elsevier

6. Jarque CM, Anil KB (1980) Efficient tests for normality, homoscedasticity and serial independence of regression residuals. Econ Lett 6(3):255–259
7. Martin R, Chuanhai L (2014) A note on p-values interpreted as plausibilities. Statistica Sinica 24:1703–1716
8. Wasserstein RL, Nicole AL (2016) The ASA's statement on p-values: context, process, and purpose. Am Stat 70(2):225

Confidence Intervals for the Common Mean of Several Normal Populations

Warisa Thangjai, Sa-Aat Niwitpong and Suparat Niwitpong

Abstract This paper proposes a novel approach for confidence interval estimation for the common mean of several normal populations. This will be achieved by using the concept of an adjusted method of variance estimates recovery approach. The Monte Carlo simulation was used to evaluate the coverage probability and average length. Simulation results are presented to compare the performance from the proposed approach with that of existing approaches. The promising simulation results indicated that the proposed approach should be considered as an alternative to the interval estimation for the common mean.

Keywords Normal distribution · Generalized confidence interval · Adjusted method of variance estimates recovery

1 Introduction

The mean of a normal distribution has been used for statistical estimation in many fields of applied research covering areas of social and behavioral sciences and clinical trials. For example, Meier [7] provided the ability to estimate the mean percentage of albumin in the plasma protein with four different methods to obtain data. For other examples see Eberhardt et al. [3] and Skinner [9].

For multiple sample cases, it is a common practice to replicate an experiment or collect data at different settings. Therefore inference procedures regarding several normal means are of interest. The paper by Krishnamoorthy and Lu [5] presented procedures for hypothesis testing and interval estimation of the common mean of

W. Thangjai (✉) · S.-A. Niwitpong · S. Niwitpong
Faculty of Applied Science, Department of Applied Statistics,
King Mongkut's University of Technology North Bangkok, Bangkok 10800, Thailand
e-mail: wthangjai@yahoo.com

S.-A. Niwitpong
e-mail: sa-aat.n@sci.kmutnb.ac.th

S. Niwitpong
e-mail: suparat.n@sci.kmutnb.ac.th

© Springer International Publishing AG 2017
V. Kreinovich et al. (eds.), *Robustness in Econometrics*,
Studies in Computational Intelligence 692, DOI 10.1007/978-3-319-50742-2_19

several normal populations based on inverting weighted linear combinations of the generalized pivotal quantities which use concepts of generalized p-values and generalized confidence interval. A new generalized pivotal based on the best linear unbiased estimator of the common mean was proposed by Lin and Lee [6].

Inference procedures referring to common mean based on several independent normal samples are of practical and theoretical importance. The goal of this paper is to develop a novel approach for confidence interval estimation of the common normal mean derived from several independent samples. This paper investigates the concept method of variance estimates recovery (MOVER) confidence interval which is called adjusted MOVER confidence interval, and then compares the results with the exiting confidence intervals: the generalized confidence interval (GCI) was proposed by Lin and Lee [6] and the large sample confidence interval. The MOVER approach was inspired by the score interval approach proposed by Bartlett [1]. Many researchers have successfully used the MOVER approach to construct confidence interval for the common parameter; for example, see Zou and Donner [12], Zou et al. [13], Donner and Zou [2], Suwan and Niwitpong [10], and Niwitpong and Wongkhao [8]. However, no adjusted MOVER approach exists for the common mean on several independent normal samples. Therefore, this paper proposes the adjusted MOVER approach focusing on the common mean of several normal populations.

This paper is organized as follows. The proposed approach and computational procedures to construct confidence interval for the common mean of several normal populations are presented in Sect. 2. Two existing approaches including GCI approach and large sample approach are addressed in Sect. 2. Section 3, simulation results are presented to evaluate the empirical coverage probabilities and average lengths of the proposed approach are compared to the existing approaches. Section 4, illustrates the proposed approaches with real examples. And finally, Sect. 5 summarizes this paper.

2   Confidence Intervals for the Common Mean of Several Normal Populations

2.1   The Generalized Confidence Interval Approach (GCI)

Suppose that $\underset{\sim}{X} = (X_1, X_2, \ldots, X_n)$ is a random sample from a distribution which depends on a vector of parameters $\theta = \left(\theta, \underset{\sim}{v}\right)$ where θ is the parameter of interest and $\underset{\sim}{v}$ is a vector of nuisance parameters. Weerahandi [11] defines a generalized pivot $R\left(\underset{\sim}{X}, \underset{\sim}{x}, \theta, \underset{\sim}{v}\right)$ for interval estimation, where $\underset{\sim}{x}$ is an observed value of $\underset{\sim}{X}$, as a random variable having the following two properties:

(1) $R\left(\underset{\sim}{X}, \underset{\sim}{x}, \theta, \underset{\sim}{v}\right)$ has a distribution free of the vector of nuisance parameters $\underset{\sim}{v}$.

(2) The observed value of $R\left(\underset{\sim}{X}, \underset{\sim}{x}, \theta, \underset{\sim}{v}\right)$ is θ.

Let R_α be the 100α-th percentile of R. Then R_α becomes a $100\,(1-\alpha)\,\%$ lower bound for θ and $\left(R_{\alpha/2}, R_{1-\alpha/2}\right)$ becomes the $100\,(1-\alpha)\,\%$ two-side generalized confidence interval for θ.

Consider k independent normal populations with a common mean θ. Let X_{i1}, $X_{i2}, \ldots, X_{in_i}$ be a random sample from the i-th normal population as follows:

$$X_{ij} \sim N\left(\mu_i, \sigma_i^2\right); i = 1, 2, \ldots, k.$$

Let $\bar{X}_i$ and S_i^2 denote sample mean and sample variance for normal data for the i-th sample and let $\bar{x}_i$ and s_i^2 denote observed sample mean and observed sample variance respectively. From

$$\bar{X}_i \sim N\left(\mu_i, \frac{\sigma_i^2}{n_i}\right), U_i = \frac{(n_i - 1)\,S_i^2}{\sigma_i^2} = \frac{V_i}{\sigma_i^2} \sim \chi_{n_i-1}^2; i - 1, 2, \ldots, k,$$

where $\chi_{n_i-1}^2$ denotes the chi-square distribution with $n_i - 1$ degrees of freedom.

Let v_i be the observed values of V_i,

$$V_i = (n_i - 1)\,S_i^2, v_i = (n_i - 1)\,s_i^2; i = 1, 2, \ldots, k.$$

According to Lin and Lee [6], the generalized pivotal quantity to estimate the common mean θ based on the best linear unbiased estimator of common mean θ is

$$R_\theta = \frac{\sum_{i=1}^k \frac{n_i \bar{x}_i U_i}{v_i} - Z\sqrt{\sum_{i=1}^k \frac{n_i U_i}{v_i}}}{\sum_{i=1}^k \frac{n_i U_i}{v_i}} \tag{1}$$

where Z denotes the standard normal distribution.

Therefore, the $100\,(1-\alpha)\,\%$ confidence interval for the common mean θ based on the generalized confidence interval approach is

$$(R_\theta\,(\alpha/2),\, R_\theta\,(1-\alpha/2)) \tag{2}$$

The following algorithm is used to construct the generalized confidence interval:

Algorithm 1

Step 1: Generate $X_{i1}, X_{i2}, \ldots, X_{in_i}$ from $N\left(\mu_i, \sigma_i^2\right), i = 1, 2, \ldots, k$, and calculate the observed values of $\bar{x}_i$ and s_i^2.

Step 2: Generate $Z \sim N\,(0, 1)$.

Step 3: Generate $U_i \sim \chi^2_{n_i-1}$ and then calculate v_i and σ_i^2.

Step 4: Repeat step 3, calculate R_θ following (1) for $i = 1, 2, \ldots, k$.

Step 5: Repeat step 2–4 a total of m times and obtain an array of R_θ's. Rank this
         array of R_θ's from small to large.

Let $R_\theta (\alpha)$ be the 100α-th percentile of R_θ. Then $R_\theta (\alpha)$ is an estimate of
the lower bound of the $100 (1 - \alpha) \%$ one-sided confidence interval for θ and
$(R_\theta (\alpha/2) , R_\theta (1 - \alpha/2))$ is a $100 (1 - \alpha) \%$ two-sided generalized confidence interval for θ.

2.2 The Large Sample Approach

The large sample estimate of the mean of normal distribution is a pooled estimate of
mean defined as

$$
\hat{\theta} = \frac{\sum_{i=1}^{k} \frac{\hat{\theta}^{(i)}}{var\left(\hat{\theta}^{(i)}\right)}}{\sum_{i=1}^{k} \frac{1}{var\left(\hat{\theta}^{(i)}\right)}},
\tag{3}
$$

where $\hat{\theta}^{(i)} = \bar{x}_i$ and $var\left(\hat{\theta}^{(i)}\right) = s_i^2/n_i$.

Since the distribution of $\hat{\theta}$ is close to normal for a sufficiently large sample. The
quantiles of the normal distribution are used to gain a confidence interval. Therefore,
the $100 (1 - \alpha) \%$ confidence interval for the common mean θ based on the large
sample approach is

$$
\left(\hat{\theta} - z_{1-\alpha/2} \sqrt{\frac{1}{\sum_{i=1}^{k} \frac{1}{var\left(\hat{\theta}^{(i)}\right)}}}, \hat{\theta} + z_{1-\alpha/2} \sqrt{\frac{1}{\sum_{i=1}^{k} \frac{1}{var\left(\hat{\theta}^{(i)}\right)}}} \right),
\tag{4}
$$

where $z_{1-\alpha/2}$ is the $1 - \alpha/2$ quantile of the standard normal distribution.

2.3 The Adjusted Method of Variance Estimates Recovery
Approach (Adjusted MOVER)

Let $\theta^{(1)}, \theta^{(2)}, \cdots , \theta^{(k)}$ be k parameters of interest. The common mean defined as

$$\theta = \frac{\sum_{i=1}^{k} \frac{\theta^{(i)}}{var(\theta^{(i)})}}{\sum_{i=1}^{k} \frac{1}{var(\theta^{(i)})}}.$$

Suppose it is of interest to construct a $100\,(1-\alpha)\,\%$ two-sided confidence interval (L, U) for common mean θ. In order to motivate the MOVER approach, introduced by Zou et al. [13], with confidence intervals for $\theta^{(1)}, \theta^{(2)}, \cdots, \theta^{(k)}$ are given by $(l_1, u_1), (l_2, u_2), \cdots, (l_k, u_k)$, respectively.

The MOVER approach is based on central limit theorem. The $100\,(1-\alpha)\,\%$ two-sided confidence interval (L, U) for the parameter $\theta^{(1)} + \theta^{(2)}$ under the assumption of independence between the point estimates $\hat{\theta}^{(1)}$ and $\hat{\theta}^{(2)}$. The lower limit L and the upper limit U are given by

$$L = \hat{\theta}^{(1)} + \hat{\theta}^{(2)} - z_{\alpha/2}\sqrt{\widehat{var}\left(\hat{\theta}^{(1)}\right) + \widehat{var}\left(\hat{\theta}^{(2)}\right)},$$

and

$$U = \hat{\theta}^{(1)} + \hat{\theta}^{(2)} + z_{\alpha/2}\sqrt{\widehat{var}\left(\hat{\theta}^{(1)}\right) + \widehat{var}\left(\hat{\theta}^{(2)}\right)},$$

where $z_{\alpha/2}$ is the $\alpha/2$ quantile of the standard normal distribution.

Motivating the MOVER approach to construct a $100\,(1-\alpha)\,\%$ two-sided confidence interval (L, U) for $\theta^{(1)} + \theta^{(2)} + \cdots + \theta^{(k)}$, where the estimates $\hat{\theta}^{(1)}, \hat{\theta}^{(2)}, \ldots,$ and $\hat{\theta}^{(k)}$ are independent. Using the central limit theorem, a lower limit L is given by

$$L = \hat{\theta}^{(1)} + \ldots + \hat{\theta}^{(k)} - z_{\alpha/2}\sqrt{\widehat{var}\left(\hat{\theta}^{(1)}\right) + \ldots + \widehat{var}\left(\hat{\theta}^{(k)}\right)},$$

where $z_{\alpha/2}$ is the $\alpha/2$ quantile of the standard normal distribution. The lower limit L is not readily applicable because $var\left(\hat{\theta}^{(i)}\right)$ $(i = 1, 2, \ldots, k)$ is unknown.

Suppose that a $100\,(1-\alpha)\,\%$ two-sided confidence interval for $\theta^{(i)}$ is given by (l_i, u_i), where $i = 1, 2, \ldots, k$. The lower limit L is in the neighborhood of $l_1 + l_2 + \ldots + l_k$. Inspired by the score interval approach; see Bartlett [1], the procedure to estimate the variances for lower limit L at $\theta^{(1)} + \theta^{(2)} + \ldots + \theta^{(k)} = l_1 + l_2 + \ldots + l_k$, i.e., $\theta^{(i)} = l_i$. The central limit theorem

$$l_i = \hat{\theta}^{(i)} - z_{\alpha/2}\sqrt{\widehat{var}\left(\hat{\theta}^{(i)}\right)},$$

which gives a variance estimate for $\hat{\theta}^{(i)}$ at $\theta^{(i)} = l_i$ of

$$\widehat{var}\left(\hat{\theta}^{(i)}\right) = \frac{\left(\hat{\theta}^{(i)} - l_i\right)^2}{z_{\alpha/2}^2}. \tag{5}$$

Therefore, the lower limit L for $\theta^{(1)} + \theta^{(2)} + \ldots + \theta^{(k)}$ is given by

$$L = \hat{\theta}^{(1)} + \ldots + \hat{\theta}^{(k)} - z_{\alpha/2}\sqrt{\widehat{var}\left(\hat{\theta}^{(1)}\right) + \ldots + \widehat{var}\left(\hat{\theta}^{(k)}\right)} \tag{6}$$

$$= \hat{\theta}^{(1)} + \ldots + \hat{\theta}^{(k)} - z_{\alpha/2}\sqrt{\frac{\left(\hat{\theta}^{(1)} - l_1\right)^2}{z_{\alpha/2}^2} + \ldots + \frac{\left(\hat{\theta}^{(k)} - l_k\right)^2}{z_{\alpha/2}^2}}$$

$$= \hat{\theta}^{(1)} + \ldots + \hat{\theta}^{(k)} - \sqrt{\left(\hat{\theta}^{(1)} - l_1\right)^2 + \ldots + \left(\hat{\theta}^{(k)} - l_k\right)^2}.$$

By performing similar steps with the idea that $u_1 + u_2 + \ldots + u_k$ is close to upper limit U, and the variance estimate at $\theta^{(i)} = u_i$ is

$$\widehat{var}\left(\hat{\theta}^{(i)}\right) = \frac{\left(u_i - \hat{\theta}^{(i)}\right)^2}{z_{\alpha/2}^2}, \tag{7}$$

the upper limit U as

$$U = \hat{\theta}^{(1)} + \cdots + \hat{\theta}^{(k)} + z_{\alpha/2}\sqrt{\widehat{var}\left(\hat{\theta}^{(1)}\right) + \cdots + \widehat{var}\left(\hat{\theta}^{(k)}\right)} \tag{8}$$

$$= \hat{\theta}^{(1)} + \cdots + \hat{\theta}^{(k)} + z_{\alpha/2}\sqrt{\frac{\left(u_1 - \hat{\theta}^{(1)}\right)^2}{z_{\alpha/2}^2} + \cdots + \frac{\left(u_k - \hat{\theta}^{(k)}\right)^2}{z_{\alpha/2}^2}}$$

$$= \hat{\theta}^{(1)} + \cdots + \hat{\theta}^{(k)} + \sqrt{\left(u_1 - \hat{\theta}^{(1)}\right)^2 + \cdots + \left(u_k - \hat{\theta}^{(k)}\right)^2}.$$

Since

$$l_i = \bar{x}_i - t_{1-\alpha/2}\frac{s_i}{\sqrt{n_i}}; i = 1, 2, \ldots, k \tag{9}$$

and

$$u_i = \bar{x}_i + t_{1-\alpha/2}\frac{s_i}{\sqrt{n_i}}; i = 1, 2, \ldots, k. \tag{10}$$

From the i-th sample, the maximum likelihood estimator of common mean θ is

$$\hat{\theta}^{(i)} = \bar{x}_i. \tag{11}$$

The adjusted MOVER uses the concepts of large sample approach and MOVER approach are defined in (3)–(8); the common mean θ is weighted average of mean $\hat{\theta}^{(i)}$ based on k individual samples as; see Graybill and Deal [4]

$$\hat{\theta} = \frac{\sum_{i=1}^{k} \frac{\hat{\theta}^{(i)}}{var\left(\hat{\theta}^{(i)}\right)}}{\sum_{i=1}^{k} \frac{1}{var\left(\hat{\theta}^{(i)}\right)}}, \tag{12}$$

which gives a variance estimate for $\hat{\theta}^{(i)}$ at $\theta^{(i)} = l_i$ and $\theta^{(i)} = u_i$ of

$$var\left(\hat{\theta}^{(i)}\right) = \frac{1}{2} \left(\frac{\left(\hat{\theta}^{(i)} - l_i\right)^2}{z_{\alpha/2}^2} + \frac{\left(u_i - \hat{\theta}^{(i)}\right)^2}{z_{\alpha/2}^2} \right). \tag{13}$$

Therefore, the lower limit L for the common mean θ is given by

$$L = \hat{\theta} - z_{1-\alpha/2} \sqrt{\frac{1}{\sum_{i=1}^{k} \frac{z_{\alpha/2}^2}{\left(\hat{\theta}^{(i)} - l_i\right)^2}}}. \tag{14}$$

Similarly, the upper limit U for the common mean θ is given by

$$U = \hat{\theta} + z_{1-\alpha/2} \sqrt{\frac{1}{\sum_{i=1}^{k} \frac{z_{\alpha/2}^2}{\left(u_i - \hat{\theta}^{(i)}\right)^2}}}. \tag{15}$$

Therefore, the $100\,(1 - \alpha)\,\%$ confidence interval for the common mean θ based on adjusted MOVER approach is

$$\left(\hat{\theta} - z_{1-\alpha/2} \sqrt{\frac{1}{\sum_{i=1}^{k} \frac{z_{\alpha/2}^2}{\left(\hat{\theta}^{(i)} - l_i\right)^2}}}, \hat{\theta} + z_{1-\alpha/2} \sqrt{\frac{1}{\sum_{i=1}^{k} \frac{z_{\alpha/2}^2}{\left(u_i - \hat{\theta}^{(i)}\right)^2}}} \right). \tag{16}$$

The following algorithm is used to estimate the coverage probability and average length:

Algorithm 2

Step 1: Generate $X_{i1}, X_{i2}, \ldots, X_{in_i}$ from $N\left(\mu_i, \sigma_i^2\right), i = 1, 2, \ldots, k$, and calculate the observed values of $\bar{x}_i$ and s_i^2.

Step 2: For each approach, construct confidence intervals and record whether or not all the values of μ fall in their corresponding confidence intervals.

Step 3: Repeat steps 1–2 a total of M times. Then, for each approach, the fraction of times that all μ are in their corresponding confidence intervals provides an estimate of the coverage probability.

3 Simulation Studies

In this section, simulation studies are carried out to evaluate the performance of the proposed approach for the common mean of several normal populations, comparison studies are also conducted using the GCI approach and large sample approach. The performance of these three approaches was evaluated through the empirical coverage probabilities and the average lengths. In particular, the confidence interval is satisfactory when the coverage probability is greater than or close to the nominal confidence level $(1 - \alpha)$ and the shortest average length.

In the simulation, each confidence interval is computed at the nominal confidence level of 0.95. Following Lin and Lee [6], the number of samples used are $k = 2$ with the sample sizes $(n_1, n_2) = (10,10), (15,15), (30,10)$ and $(10,30)$, respectively, the population mean of normal data within each sample $\mu_1 = \mu_2 = 1$, and the population variance $(\sigma_1^2, \sigma_2^2) = (5,5), (5,10), (5,15), (5,20), (5,30), (5,40)$ and $(5,50)$. For each parameter setting, 10000 random samples are generated and thus 2500 R_θ's are obtained for each of the random samples.

Table 1 presents the simulated coverage probabilities and average lengths of 95% two-sided confidence intervals for mean, respectively. The coverage probabilities of the GCI approach and the proposed approach are much closer to the nominal confidence level of 0.95 than that of the large sample approach. The average lengths of the large sample confidence interval are shorter than those of the GCI and the proposed confidence interval. Overall, the GCI approach and the proposed approach provide much better confidence interval estimation than the large sample approach.

4 An Empirical Application

In this section, two examples were previously considered by Krishnamoorthy and Lu [5]. The first data, originally given by Meier [7], comes from four experiments which estimate the mean percentage of albumin in the plasma protein of normal human subjects: experiment A, experiment B, experiment C and experiment D. The summary statistics are $\bar{x}_1 = 62.3$, $\bar{x}_2 = 60.3$, $\bar{x}_3 = 59.5$, $\bar{x}_4 = 61.5$, $s_1^2 = 12.986$, $s_2^2 = 7.840$, $s_3^2 = 33.433$, $s_4^2 = 18.513$, $n_1 = 12$, $n_2 = 15$, $n_3 = 7$ and $n_4 = 16$. The 95% generalized confidence intervals for the population mean percentage of albumin in the plasma protein are $(60.0104, 64.5896), (58.7494, 61.8506), (54.1524,$

Table 1 Empirical coverage probabilities (CP) and average lengths (AL) of approximately 95% of two-side confidence bounds for mean of the normal distribution: 2 sample cases

n_1	n_2	σ_1^2	σ_2^2	GCI		Large sample		Adjusted MOVER	
				CP	AP	CP	AP	CP	AP
10	10	5	5	0.9420	2.0474	0.9038	1.8249	0.9420	2.1062
		5	10	0.9421	2.3902	0.9048	2.1205	0.9431	2.4475
		5	15	0.9415	2.5578	0.9055	2.2593	0.9429	2.6077
		5	20	0.9446	2.6607	0.9111	2.3385	0.9436	2.6991
		5	30	0.9420	2.7926	0.9086	2.4439	0.9445	2.8207
		5	40	0.9461	2.8643	0.9158	2.4937	0.9493	2.8782
		5	50	0.9458	2.9072	0.9125	2.5258	0.9474	2.9152
15	15	5	5	0.9438	1.6536	0.9199	1.5326	0.9435	1.6771
		5	10	0.9408	1.9169	0.9203	1.7731	0.9416	1.9403
		5	15	0.9466	2.0413	0.9266	1.8825	0.9485	2.0600
		5	20	0.9452	2.1235	0.9240	1.9538	0.9468	2.1381
		5	30	0.9458	2.2107	0.9233	2.0290	0.9463	2.2204
		5	40	0.9527	2.2632	0.9331	2.0724	0.9530	2.2679
		5	50	0.9500	2.2988	0.9316	2.1013	0.9536	2.2994
10	30	5	5	0.9437	1.4216	0.9264	1.3387	0.9467	1.4336
		5	10	0.9438	1.7984	0.9220	1.6776	0.9443	1.8217
		5	15	0.9417	2.0214	0.9162	1.8733	0.9425	2.0526
		5	20	0.9418	2.1731	0.9161	2.0021	0.9437	2.2080
		5	30	0.9479	2.3723	0.9231	2.1629	0.9487	2.4069
		5	40	0.9467	2.5011	0.9231	2.2633	0.9500	2.5332
		5	50	0.9421	2.5833	0.9151	2.3227	0.9440	2.6114
30	10	5	5	0.9441	1.4203	0.9258	1.3368	0.9425	1.4317
		5	10	0.9446	1.5237	0.9283	1.4405	0.9447	1.5269
		5	15	0.9429	1.5683	0.9292	1.4863	0.9442	1.5683
		5	20	0.9504	1.5900	0.9341	1.5076	0.9502	1.5870
		5	30	0.9499	1.6136	0.9386	1.5333	0.9514	1.6098
		5	40	0.9473	1.6226	0.9360	1.5443	0.9484	1.6190
		5	50	0.9514	1.6345	0.9406	1.5569	0.9515	1.6308

64.8476) and (59.2073, 63.7927) for experiment A, experiment B, experiment C and experiment D respectively. Using the generalized confidence interval approach, the 95% generalized confidence interval for the overall mean is (59.9106, 62.1268) with the length of interval 2.2163. The 95% confidence interval by the large sample approach is (60.0038, 61.9860) with the length of interval 1.9821. In comparison, the 95% confidence interval by the proposed approach is (59.9030, 62.0958) with the length of interval 2.1928.

The second data, originally given by Eberhardt et al. [3], comes from four different analytical methods which estimate the mean selenium content in nonfat milk power:

atomic absorption spectrometry (method 1), neutron activation instrumental (method 2), radiochemical (method 3) and isotope dilution mass spectrometry (method 4). The summary statistics are $\bar{x}_1 = 105.00$, $\bar{x}_2 = 109.75$, $\bar{x}_3 = 109.50$, $\bar{x}_4 = 113.25$, $s_1^2 = 85.711$, $s_2^2 = 20.748$, $s_3^2 = 2.729$, $s_4^2 = 33.640$, $n_1 = 8$, $n_2 = 12$, $n_3 = 14$ and $n_4 = 8$. The 95% generalized confidence intervals for the population mean selenium content in nonfat milk power are (97.2601, 112.7399), (106.8559, 112.6441), (108.5462, 110.4538) and (108.4011, 118.0989) for method 1, method 2, method 3 and method 4 respectively. Using the generalized confidence interval approach, the 95% generalized confidence interval for the overall mean is (108.7193, 110.5284) with the length of interval 1.8090. The 95% confidence interval by the large sample approach is (108.8045, 110.3996) with the length of interval 1.5950. In comparison, the 95% confidence interval by the proposed approach is (108.7047, 110.4740) with the length of interval 1.7693. As a result, the length of the large sample confidence interval is shorter than that of the GCI and the proposed confidence interval. These results confirm the simulation results in the previous section.

5 Discussion and Conclusions

Lin and Lee [6] proposed the GCI approach on the common mean of several normal populations, based on the best linear unbiased estimator of mean. This is better than the existing approaches in terms of having the shortest average lengths.

The simulation studies indicated that the GCI approach and the adjusted MOVER approach both provide much better confidence interval estimates than the large sample approach. However, confidence interval based on the adjusted MOVER approach is also easier to use than the confidence interval based on GCI which is a computational approach. As a result, the adjusted MOVER approach should be considered as an alternative to estimate the confidence interval for the common mean.

Acknowledgements The first author gratefully acknowledges the financial support from Science Achievement Scholarship of Thailand.

References

1. Bartlett MS (1953) Approximate confidence intervals. 2. more than one unknown parameter. Biometrika 40:306–317
2. Donner A, Zou GY (2010) Closed-form confidence intervals for function of the normal standard deviation. Stat Methods Med Res 86–89
3. Eberhardt KR, Reeve CP, Spiegelman CH (1989) A minimax approach to combining means, with practical examples. Chemom Intell Lab Syst 5:129–148
4. Graybill FA, Deal RB (1959) Combining unbiased estimators. Biometrics 15:543–550
5. Krishnamoorthy K, Lu Y (2003) Inferences on the common mean of several normal populations based on the generalized variable method. Biometrics 59:237–247

6. Lin SH, Lee JC (2005) Generalized inferences on the common mean of several normal populations. J Stat Plan Inference 134:568–582
7. Meier P (1953) Variance of a weighted mean. Biometrics 9:59–73
8. Niwitpong S, Wongkhao A (2016) Confidence intervals for the difference between inverse of normal means. Adv Appl Stat 48:337–347
9. Skinner JB (1991) On combining studies. Drug Info J 25:395–403
10. Suwan S, Niwitpong S (2013) Estimated variance ratio confidence interval of nonnormal distributions. Far East J Math Sci 4:339–350
11. Weerahandi S (1993) Generalized confidence intervals. J Am Stat Assoc 88:899–905
12. Zou GY, Donner A (2008) Construction of confidence limits about effect measures: a general approach. Stat Med 27:1693–1702
13. Zou GY, Taleban J, Huo CY (2009) Confidence interval estimation for lognormal data with application to health economics. Comput Stat Data Anal 53:3755–3764

A Generalized Information Theoretical Approach to Non-linear Time Series Model

Songsak Sriboochitta, Woraphon Yamaka, Paravee Maneejuk
and Pathairat Pastpipatkul

Abstract The limited data will bring about an underdetermined, or ill-posed problem for the observed data, or for regressions using small data set with limited data and the traditional estimation techniques are difficult to obtain the optimal solution. Thus the approach of Generalized Maximum Entropy (GME) is proposed in this study and applied it to estimate the kink regression model under the limited information situation. To the best of our knowledge, the estimation of kink regression model using GME has been not done yet. Hence, we extend the entropy linear regression to non-linear kink regression by modifying the objective and constraint functions under the context of GME. We use both Monte Carlo simulation and real data study to evaluate the performance of our estimation from Kink regression and found that GME estimator performs slightly better compared to the traditional Least squares and Maximum likelihood estimators.

Keywords Kink regression · Maximum entropy · GDP/Debt ratio

1 Introduction

In information theory, there is a classical maximum entropy (ME) principle which was introduced by Jaynes [11] and is based on Shannons entropy measure. Then, Golan, Judge and Miller, [6] developed a generalized entropy estimation (GME) for constructing distributions based on limited information. They adapt the entropy approach into the regression context by transforming the estimated parameters of the model to be described by a discrete probability distribution defined on a certain interval or support bound. They, then maximize these entropies to estimate unknown probabilities of parameters and error term subject to the constraints imposed by the data. Nowadays, this estimation approach has found wide-spread applications in various fields of science such as engineering, communication and information, physics, chemistry, biology, political science as well as economics. However, in this

S. Sriboochitta · W. Yamaka (✉) · P. Maneejuk · P. Pastpipatkul
Faculty of Economics, Chiang Mai University, Chiang Mai, Thailand
e-mail: woraphon.econ@gmail.com

© Springer International Publishing AG 2017
V. Kreinovich et al. (eds.), *Robustness in Econometrics*,
Studies in Computational Intelligence 692, DOI 10.1007/978-3-319-50742-2_20

study, we will focus our study in the economic problem. We found from literature review that the entropy estimation has been employed in linear regression framework such as the work of Golan, Perloff, and Shen [9], Al-Nasser [1, 2] and Wu [14]. For more detail from a brief review on the ME econometrics, refer to Golan [4].

Although GME has been widely employed to estimate linear regression models and was applied successfully to economic data, there is no study on non-linear regression models. Many previous researchers and practitioners often find that the economic data seems to have a non-linear structure and the usual linear regression fails to explain this non-linear relationship. We then introduce our contribution, generalized entropy-based approach to fitting economic data within the non-linear model framework.

Recently, nonlinear regression models have been widely applied in economics and one of the most interesting nonlinear regression models is kink regression. The kink regression model was first modified and popularized by Card, Lee, Pei and Weber [3] and then developed a theory of least square (LS) estimation and inference by Hansen [8]. In this paper, we do not emphasize in this models structure but the estimation method is what we want to develop. In the area of macroeconomic study, one of the common problems in analyzing macroeconomic model is the limitation of data, such as from the lack of high frequency data or long period data, since many countries have a poor data collection system, especially in the less developed countries. Thus, in the estimation of the unknown parameters of the model, the limited data will bring about an underdetermined, or ill-posed problem for the observed data, or for regressions that face with a small data sets, i.e. the number of unknown parameters are larger than the number of data points [7, 10]. In the last decade, the traditional Maximum likelihood (ML) and least squares (LS) estimation methods have been proposed and employed in various models. However, the validity of the normal assumption has been questioned and, in the case of the ML estimation, there is difficulty to construct an appropriate likelihood function. Without a likelihood function, it is not possible to do an estimation and testing. Although the Bayesian method is another traditional estimation that is proposed to deal with the small sample data by adding a prior distribution on model parameters, this prior distribution assumption has been questioned in many studies. To overcome these problems, this study proposes a GME estimation method to the kink regression framework. Mittelhammer, Judge, Miller [13] suggested that GME estimation is the most suitable alternative option available to the model estimation in order to avoid making any parametric assumptions.

In previous literature, we found some evidences that the entropy estimation is outperforms those two traditional estimations. Golan et al. [5] did an experiment study on the Tobit model and compare the GME with ML estimator when the error term is not normal. They found that GME estimator was more efficient than the ML estimator with small data set since it showed a lower Mean Square Error (MSE). The comparison between GME and ML estimator was also presented in Wu [14] who also did an experiment study to investigate the performance of GME estimator when the errors are generated from either normal or non-normal distributions. The result showed that the proposed GME estimator provides superior performance under various error distributions.

As we mentioned above, GME estimator can be employed as an alternative to LS and ML estimators. Yet, to the best of our knowledge, the estimation of kink regression using GME estimator has not been proposed in the literature, and hence the study we conducted is the first concerned with applying the GME estimator to kink regression model. Thus, our objective is to develop an estimation model with kink regression when the data is limited or has small samples.

The kink regression model structure and the GME approach are presented in Sect. 2. In Sect. 3, the experiment study and results are presented. The application of regression kink model, where GDP growth is the dependent variable and the debt/GDP ratio is the regressor, are presented in Sect. 4. Conclusion and discussion is provided in Sect. 5.

2 Kink Regression Model

2.1 Model Structure

The two-regime kink regression model can be written as

$$
\begin{aligned}
Y_t = {} & \beta_1^-(x'_{1,t} \le \gamma_1)_- + \beta_1^+(x'_{1,t} > \gamma_1)_+ +, \cdots, + \beta_K^-(x'_{K,t} \le \gamma_K)_- \\
& + \beta_K^+(x'_{K,t} > \gamma_K)_+ + \varepsilon_t
\end{aligned}
\tag{1}
$$

where Y_t is $[T \times 1]$ continuous dependent variable at time t, $x'_{k,t}$ is a matrix of $(T \times K)$ continuous independent variables at time t, and β is a matrix of $(T \times K \times 2)$ unknown parameters where $(\beta_1^-, \ldots, \beta_K^-)$ and $(\beta_1^+, \ldots, \beta_K^+)$ are the regressor coefficients with respect to variable $x'_{k,t}$ for value of $(x'_{k,t} \le \gamma_k)_-$ (lower regime) and with respect to variable $x'_{k,t}$ for value of $(x'_{k,t} > \gamma_k)_+$ (upper regime) respectively. Following, Hansen [8], the regressor variables are subject to regime-change at unknown kink point or threshold variable $(\gamma_1, \ldots, \gamma_K)$ and thereby separating these regressors into two regimes. These threshold variables are compact and strictly in the interior of the support of $(x_{1,t}, \ldots, x_{K,t})$. In addition, the error term of the model ε_t is $[T \times 1]$ dimensional vector which may represent one or more sources of noise in the observed system, such as errors in the data and modeling errors.

2.2 Generalized Maximum Entropy Approach

In this study, we proposed to use a maximum entropy estimator to estimate our unknown parameters in Eq. 1. Before, we discuss this estimator for kink regression and its statistical properties, let us describe briefly the concept about the entropy approach. The maximum entropy concept is about inferring the probability distribution that maximizes information entropy given a set of various constraints. Let p_k be a proper probability mass function on a finite set α where $\alpha = \{a_1, \ldots, a_K\}$. Shannon [12] developed his information criteria and proposed classical entropy, that is

$$H(p) = -\sum_{k=1}^{K} p_k \log p_k \qquad (2)$$

where $0 \log 0 = 0$ and $\sum_{k=1}^{K} p_k = 1$. The entropy measures the uncertainty of a distribution and reaches a maximum when p_k has uniform distribution [14]. To apply this concept to be an estimator in our model, we generalize the maximum entropy into the inverse problem to the kink regression framework. Rather than searching for the point estimates $(\beta_1^-, \ldots, \beta_k^-)$ and $(\beta_1^+, \ldots, \beta_K^+)$, we can view these unknown parameters as expectations of random variables with M support value for each estimated parameter value (k), $Z = [z_1, \ldots, z_K]$ where $z_k = [\underline{z}_{k1}, \ldots, \bar{z}_{km}]$ for all $k = 1, \ldots K$. Note that $\underline{z}_{k1}$ and $\bar{z}_{km}$ denote the lower and upper bound, respectively, of each support z_k. Thus we can express parameter β_k^- and β_k^+ as

$$\begin{aligned} \beta_k^- &= \sum_m p_{km}^- z_{km}^-, \quad x_{k,t} \le \gamma_k \\ \beta_k^+ &= \sum_m p_{km}^+ z_{km}^+, \quad x_{k,t} > \gamma_k \end{aligned} \qquad (3)$$

where p_{km}^- and p_{km}^+ are the M dimensional estimated probability distribution defined on the set z_{km}^- and z_{km}^+, respectively. For the threshold γ_k, we also view each (k) element of γ as a discrete random variable with M support value, $q_k = [\underline{q}_{k1}, \ldots, \bar{q}_{km}]$, where $\underline{q}_{k1}$ and $\bar{q}_{km}$ are the lower and upper bounds of γ_k

$$\gamma_k = \sum_m h_{km} q_{km} \qquad (4)$$

Next, similar to the above expression, ε_t is also constructed as the mean value of some random variable v. Each ε_t is assumed to be a random vector with finite and discrete random variable with M support value, $v_t = [v_{t1}, \ldots, v_{tM}]$. Let w_t be an M dimension proper probability weights defined on the set v_t such that

$$\varepsilon_t = \sum_m w_{tm} v_{tm} \qquad (5)$$

Using the reparameterized unknowns β_k^-, β_k^+, γ_k, and ε_t, we can rewrite Eq. 1 as

$$Y_t = \sum_m p_{1m}^- z_{1m}^- (x'_{1,t} \leq \sum_m h_{1m} q_{1m})_- + \sum_m p_{1m}^+ z_{1m}^+ (x'_{1,t} > \sum_m h_{1m} q_{1m})_+ +, \ldots,$$

$$+ \sum_m p_{Km}^- z_{Km}^- (x'_{K,t} \leq \sum_m h_{Km} q_{Km})_- + \sum_m p_{Km}^+ z_{Km}^+ (x'_{K,t} > \sum_m h_{Km} q_{Km})_+$$

$$+ \sum_m w_{tm} v_{tm} \tag{6}$$

where the vector support z_{km}^-, z_{km}^+, q_{Km}, and v_{tm}, are convex set that is symmetric around zero with $2 \leq M < \infty$. Then, we can construct our Generalized Maximum Entropy (GME) estimator as

$$H(p, h, w) = \arg\max_{p,h,w} \{H(p) + H(h) + H(w)\} \equiv -\sum_k \sum_m p_{km}^- \log p_{km}^-$$

$$- \sum_k \sum_m p_{km}^+ \log p_{km}^+ - \sum_k \sum_m h_{km} \log h_{km} - \sum_t \sum_m w_{tm} \log w_{tm} \tag{7}$$

subject to

$$Y_t - \sum_m p_{1m}^- z_{1m}^- (x'_{1,t} \leq \sum_m h_{1m} q_{1m})_- + \sum_m p_{1m}^+ z_{1m}^+ (x'_{1,t} > \sum_m h_{1m} q_{1m})_+ +, \ldots,$$

$$+ \sum_m p_{Km}^- z_{Km}^- (x'_{K,t} \leq \sum_m h_{Km} q_{Km})_- + \sum_m p_{Km}^+ z_{Km}^+ (x'_{K,t} > \sum_m h_{Km} q_{Km})_+$$

$$+ \sum_m w_{tm} v_{tm} \tag{8}$$

$$\sum_m p_{km}^+ = 1, \sum_m p_{km}^- = 1, \sum_m h_{km} = 1, \sum_m w_{tm} = 1 \tag{9}$$

where p, h, and w are on the interval $[0, 1]$. Consider repressor $(k = 1)$, this optimization problem can be solved using the Lagrangian method which takes the form as

$$L = H(p, h, w) + \lambda'_1 (Y_t - \sum_m p_{1m}^- z_{1m}^- (x'_{1,t} \leq \sum_m h_{1m} q_{1m})_-$$

$$+ \sum_m p_{1m}^+ z_{1m}^+ (x'_{1,t} > \sum_m h_{1m} q_{1m})_+ + \sum_m w_{tm} v_{tm}) + \lambda'_2 (1 - \sum_m p_{km}^-)$$

$$+ \lambda'_3 (1 - \sum_m p_{km}^+) + \lambda'_4 (1 - \sum_m h_{km}) + \lambda'_5 (1 - \sum_m w_{tm}) \tag{10}$$

where λ'_i, $i = 1, \ldots, 5$ are the vectors of Lagrangian multiplier. Thus, the resulting first-order conditions are

$$\frac{\partial L}{p_{1m}^-} = -\log(p_{1m}^-) - \sum_m \lambda_{1m} z_{1m}^- (x_{1,t}' \leq \sum_m h_{1m} q_{1m})_- - \lambda_{2i} = 0 \tag{11}$$

$$\frac{\partial L}{p_{1m}^+} = -\log(p_{1m}^+) - \sum_m \lambda_{1m} z_{1m}^+ (x_{1,t}' > \sum_m h_{1m} q_{1m})_+ - \lambda_{3i} = 0 \tag{12}$$

$$\frac{\partial L}{w_{tm}} = -\log(w_{tm}) - \sum_m \lambda_{1m} \, v_{tm}) - \lambda_{5i} = 0 \tag{13}$$

$$\frac{\partial L}{h_{1m}} = -\log(h_{1m}) - \sum_m \lambda_{1m} p_{1m}^- z_{1m}^- (x_{1,t}' \leq \sum_m q_{1m})_- \\ + \sum_m \lambda_{1m} p_{1m}^+ z_{1m}^+ (x_{1,t}' > \sum_m q_{1m})_+ - \lambda_{4i} = 0 \tag{14}$$

$$\frac{\partial L}{\lambda_1} = -(Y_t - \sum_m p_{1m}^- z_{1m}^- (x_{1,t}' \leq \sum_m h_{1m} q_{1m})_- \\ + \sum_m p_{1m}^+ z_{1m}^+ (x_{1,t}' > \sum_m h_{1m} q_{1m})_+ + \sum_m w_{tm} \, v_{tm}) = 0 \tag{15}$$

$$\frac{\partial L}{\lambda_2} = 1 - \sum_m p_{1m}^- = 0 \tag{16}$$

$$\frac{\partial L}{\lambda_3} = 1 - \sum_m p_{1m}^+ = 0 \tag{17}$$

$$\frac{\partial L}{\lambda_4} = 1 - \sum_m h_{1m} = 0 \tag{18}$$

$$\frac{\partial L}{\lambda_5} = 1 - \sum_m w_{tm} = 0 \tag{19}$$

Thus, we have

$$p_{1m}^- = \exp(-\lambda_{2i} - \sum_m \lambda_{1m} z_{1m}^- (x_{1,t}' \leq \sum_m h_{1m} q_{1m})_-) = 1, \tag{20}$$

$$p_{2m}^+ = \exp(-\lambda_{2i} - \sum_m \lambda_{1m} z_{1m}^+ (x_{1,t}' > \sum_m h_{1m} q_{1m})_+) = 1, \tag{21}$$

$$w_{tm} = \exp(-\lambda_{5i} - \sum_m \lambda_{1m} v_{tm}) = 1, \tag{22}$$

$$h_{1m} =$$

$$\exp\left(-\lambda_{4i} - \left(\sum_m \lambda_{1m} p_{1m}^- z_{1m}^- (x'_{1,t} \le \sum_m q_{1m})_- + \sum_m \lambda_{1m} p_{1m}^+ z_{1m}^+ (x'_{1,t} > \sum_m q_{1m})_+\right)\right) = 1 \tag{23}$$

Then, by setting $\lambda = 0$, this optimization yields

$$\widehat{p}_{1m}^- = \frac{\exp(-z_{1m}^- \sum_t \widehat{\lambda}_{1t} (x'_{1,t} \le \sum_m h_{1m} q_{1m})_-)}{\sum_m \exp(-z_{1m}^- \sum_t \widehat{\lambda}_{1t} (x'_{1,t} \le \sum_m h_{1m} q_{1m})_-)}, \tag{24}$$

$$\widehat{p}_{1m}^+ = \frac{\exp(-z_{1m}^+ \sum_t \widehat{\lambda}_{1t} (x'_{1,t} > \sum_m h_{1m} q_{1m})_+)}{\sum_m \exp(-z_{1m}^+ \sum_t \widehat{\lambda}_{1t} (x'_{1,t} > \sum_m h_{1m} q_{1m})_+)}, \tag{25}$$

$$\widehat{w}_{tm} = \frac{\exp(-\widehat{\lambda}_{1t} v_{1m})}{\sum_m \exp(-\widehat{\lambda}_{1t} v_{1m})}, \tag{26}$$

$$\widehat{h}_{1m} =$$

$$\frac{\exp\left(-\left(\sum_t \widehat{\lambda}_{1t} p_{1m}^- z_{1m}^- (x'_{1,t} \le \sum_m q_{1m})_- + \sum_t \widehat{\lambda}_{1t} p_{1m}^+ z_{1m}^+ (x'_{1,t} > \sum_m q_{1m})_+\right)\right)}{\sum_m \exp\left(-\left(\sum_t \widehat{\lambda}_{1t} p_{1m}^- z_{1m}^- (x'_{1,t} \le \sum_m q_{1m})_- + \sum_t \widehat{\lambda}_{1t} p_{1m}^+ z_{1m}^+ (x'_{1,t} > \sum_m q_{1m})_+\right)\right)} \tag{27}$$

Summing up the above equations, we maximize the joint-entropy objective, Eq. 7 subject to the Kink regression Eq. 8, with adding restrictions Eq. 9. The solution to this maximization problem is unique by forming the Lagrangean and solving for the first-order conditions to obtain the optimal solution $\widehat{p}_{1m}^-$, $\widehat{p}_{1m}^+$, $\widehat{h}_{1m}$, and $\widehat{w}_{tm}$. Then these estimated probabilities are used to derive the point estimates for the Kink regression coefficients and error term, see Eqs. 3, 4, and 5.

2.3 Testing for a Threshold Effect and Goodness of Fit Test

Since the non-linear structure of the model has been proposed in this study, we develop an entropy-ratio test, which is introduced in Golan and Dose [7], to check whether or not kink regression model is significant relative to the linear regression model. It corresponds to the likelihood ratio test which measures the entropy discrepancy between the constrained and unconstrained model. Consider the constrained simple linear repression model with one regressor.

$$Y_t = \beta_1(x'_{1,t}) + \varepsilon_t \tag{28}$$

The null hypothesis of entropy test is present in Eq. 28 with the restriction $\beta_1 = \beta_1^- = \beta_1^+$. Under this hypothesis the kink or threshold variable γ_1 does not exist. While the alternative unconstraint hypothesis can be presented by

$$Y_t = \beta_1^-(x'_{1,t} \le \gamma_1)_- + \beta_1^+(x'_{1,t} > \gamma_1)_+ + \varepsilon_t \tag{29}$$

Hence, the entropy ratio statistic can be defined as

$$ER = 2\left|H_U(\beta_1^- \neq \beta_1^+) - H_R(\beta_1 = \beta_1^- = \beta_1^+)\right| \tag{30}$$

where $R^* = 0$ implies no informational value of the data set, and $R^* = 1$ implies perfect certainty or perfect in-sample prediction. And $H_u(\beta_1^- \neq \beta_1^+)/H_R(\beta_1 = \beta_1^- = \beta_1^+)$ is directly related to the normalized entropy measure [6, 9].

3   Experiment Study

In this section, we conduct a simulation and experiment study to investigate the finite sample performance of the propose GME method. In the first part of these experiments, the purposed of these experiments is to compare the performance of the estimation when the number of support varies in its values as $M = 3, 5, 7$. For example, for three point support $z_{1m} = [-z, 0, z]$ or for five points support $z_{1m} = [-z, -z/2, 0, z/2, z]$. The support space of the β_1, γ_1, and ε_t are chosen to be uniformly symmetric around zero. In this study, several choices of the support for each of GME unknown parameters and errors are used. The support space of β_1 are specified to be in the interval $[-z, z]$ for $z = 3, 6$, and 9. The support space of γ_1 is specified to be in the interval $[-h, h]$ for $h = 5, 10$, and 15. Lastly, the interval $[-v, v]$, for $v = 1, 2$, and 3, are set for model errors ε_t.

In the second part, we also conduct a Monte Carlo simulation study to show the performance of our proposed entropy estimation on kink regression model with the least square (LS) and maximum likelihood (ML) estimation. To compare these methods, the study used Bias and Mean Squared Error (MSE) approaches.

Our sampling experiment model is based on

$$Y_t = \alpha + \beta_1^- 0 + \beta_1^+(x'_{1,t} > \gamma_1)_+ + \varepsilon_t \tag{31}$$

we simulate $x'_{1,t} \sim Unif[-3, 5]$ and threshold value is $\gamma_1 = 3$. The true value for parameter α, β_1^-, and β_1^+ are set to be 1, 2, and -1, respectively. To make a fair comparison, the error terms are generated from $\chi^2(2)$, $N(0, 1)$, $t(0, 1, 4)$, and $Unif(-2, 2)$. In this Monte Carlo simulation, we consider sample size $n = 20$ and $n = 40$. Then, we assess the performance of our proposed method through the Bias and Mean Squared Error (MSE) of each parameter in which the Bias and MSE of each parameter are given by

$$Bias = N^{-1} \sum_{r=1}^{N} (\tilde{\phi}_r - \phi_r),$$

and

$$MSE = M^{-1} \sum_{r=1}^{M} (\tilde{\phi}_r - \phi_r)^2.$$

where $N = 100$ is the number of bootstrapping; and $\tilde{\phi}_r$ and ϕ_r are the estimated value and true value, respectively.

3.1   Experiment Results

In the first experiment study, the choice of number of support points (M) and value of support space z, h and v are considered in this experiment. According to Heckelei, Mittelhammer, and Jansson [10], the choice of reference prior probabilities on support points and the number of support points are complicated, composite, and difficult to be defined. Thus, we examine the performance of the GME estimator by varying its support space and the number of support points. We use the same simulated data when the error is generated from $N(0, 1)$ to investigate the performance of the proposed estimator for all cases. The estimated results for this experiment are shown in Table 1. We observe that when we increase the number of support points, the MSE and Bias of the estimated parameters are decrease. However, with regard to the value of support space, the increase in the value of bound does not represent a lower Bias and MSE in some cases. Thus, we can conclude that regarding to the number of support points, this experiment showed that the GME estimation performs better when the number of support points increase. For the support value, our experiment results seem not be quite stable, we suspect that the value of the support might not large enough.

3.2   Comparison of the Estimation Results

Next we turn our attention to the second experiment to investigate the performance of GME estimator when the errors are generated from either normal or non-

Table 1 My caption

Support space N = 20

z	h	v	M	Bias of parameters				MSE of parameters			
				α	β_1^-	β_1^+	γ_1	α	β_1^-	β_1^+	γ_1
[−3,3]	[−5,5]	[−1,1]	3	−0.1539	−0.1282	0.0885	−0.1999	0.2419	0.1648	0.0788	0.3999
[−3,3]	[−5,5]	[−1,1]	5	−0.1295	−0.0776	0.0515	0.0794	0.2421	0.093	0.0471	0.2413
[−3,3]	[−5,5]	[−1,1]	7	−0.1473	−0.0564	0.0398	−0.012	0.3503	0.0509	0.0349	0.3025
[−6,6]	[−10,10]	[−2,2]	3	−0.2017	−0.0541	0.0729	−0.1794	0.4428	0.0332	0.0599	0.4121
[−6,6]	[−10,10]	[−2,2]	5	−0.1641	−0.0508	0.0586	0.1852	0.3628	0.0312	0.051	0.4381
[−6,6]	[−10,10]	[−2,2]	7	−0.1452	−0.03	0.0286	0.0915	0.3342	0.0161	0.0217	0.1816
[−9,9]	[−15,15]	[−3,3]	3	−0.0911	−0.0213	−0.0252	−0.0128	0.0141	0.0112	0.0112	0.3306
[−9,9]	[−15,15]	[−3,3]	5	−0.1312	0.0055	0.0271	0.0901	0.3661	0.0023	0.0259	0.2311
[−9,9]	[−15,15]	[−3,3]	7	−0.0681	0.0083	0.0072	−0.0557	0.1668	0.0028	0.0098	0.1573

Support space N = 40

z	h	v	M	Bias of parameters				MSE of parameters			
				α	β_1^-	β_1^+	γ_1	α	β_1^-	β_1^+	γ_1
[−3,3]	[−5,5]	[−1,1]	3	−0.0791	−0.0834	0.0543	0.134	0.0989	0.0993	0.0403	0.2885
[−3,3]	[−5,5]	[−1,1]	5	0.0557	0.0543	0.0236	0.1334	0.0557	0.0543	0.0233	0.1334
[−3,3]	[−5,5]	[−1,1]	7	−0.0179	−0.0092	0.0086	0.0002	0.0082	0.004	0.0011	0.0069
[−6,6]	[−10,10]	[−2,2]	3	−0.1114	−0.0357	0.0458	0.0624	0.1723	0.0194	0.0285	0.0856
[−6,6]	[−10,10]	[−2,2]	5	−0.0522	−0.0136	0.0229	0.0331	0.0846	0.0085	0.0136	0.0667
[−6,6]	[−10,10]	[−2,2]	7	−0.0059	0.0023	0.0035	−0.0088	0.0029	0.00001	0.0001	0.0018
[−9,9]	[−15,15]	[−3,3]	3	−0.1085	0.0091	0.0293	−0.1081	0.1651	0.0014	0.0155	0.2172
[−9,9]	[−15,15]	[−3,3]	5	−0.0302	0.0045	0.0087	−0.0339	0.055	0.0004	0.0026	0.0485
[−9,9]	[−15,15]	[−3,3]	7	−0.0105	0.0028	0.0045	−0.0138	0.0039	0.0001	0.0002	0.0031

Source: Calculation

normal distributions. In this experiment, we consider two more estimators, LS and ML, to compare with GME estimator on kink regression model. For GME estimator, we employ seven-point support $z = [-9, -4.5, -2.25, 0, 2.25, 4.5, 9]$, $h = [-15, -7.5, -3.25, 0, 3.25, 7.5, 15]$, and $v = [-3, -1.5, -0.75, 0, 0.75, 1.5, 3]$. Using the same sample simulation design explained above, we generate our data from $\chi^2(2)$, $N(0, 1)$, $t(0, 1, 4)$, and $Unif(-2, 2)$; and the results are shown in Tables 2, 3, 4, and 5, respectively.

According to the above Tables, the similar results are obtained from different cases. We found that GME estimator seems to outperform LS and ML estimators in terms of lower MSE and Bias of parameters. Whereas, the parameters in kink regression model, especially α, provide a strong accuracy when compared with the other two estimators. In addition, considering the sample size of this experiment,

Table 2 Kink regressions with $\chi^2(2)$ errors

N = 20	GME		LS		ML	
	Bias	MSE	Bias	MSE	Bias	MSE
α	0.1664	0.4672	1.0598	1.5575	1.1346	1.6454
β_1^-	-0.0359	0.0167	0.2909	0.9606	-0.0374	0.0145
β_1^+	0.0429	0.0223	0.0215	0.0221	0.0078	0.0032
γ_1	0.1788	0.5294	0.3584	1.9196	0.4675	0.3906
N = 40	GME		LS		ML	
	Bias	MSE	Bias	MSE	Bias	MSE
α	-0.0723	0.0526	1.2323	1.1651	1.2659	1.7246
β_1^-	-0.1669	0.2791	0.392	1.819	0.0537	0.0342
β_1^+	0.0895	0.0802	0.0169	0.0043	-0.0117	0.0066
γ_1	0.1999	0.3999	-0.0956	0.2846	0.1503	0.1183

Source: Calculation

Table 3 Kink regressions with $N(0, 1)$ errors

N = 20	GME		LS		ML	
	Bias	MSE	Bias	MSE	Bias	MSE
α	-0.0681	0.1668	0.2364	0.5312	-0.357	0.2671
β_1^-	0.0083	0.0028	-0.0323	0.0024	-0.0279	0.0017
β_1^+	0.0072	0.0098	-0.0684	0.0802	-0.0199	0.0143
γ_1	-0.0557	0.1573	0.0486	0.2466	-0.0371	0.0611
N = 40	GME		LS		ML	
	Bias	MSE	Bias	MSE	Bias	MSE
α	-0.0105	0.0039	-0.0579	0.0586	0.0297	0.0467
β_1^-	0.0028	0.0001	-0.0166	0.1045	0.0179	0.0205
β_1^+	0.0045	0.0002	0.0498	0.0267	0.0194	0.0013
γ_1	-0.0138	0.0031	-0.2366	0.2657	-0.0916	0.0376

Source: Calculation

Table 4 Kink regressions with $t(0, 1, 4)$ errors

N = 20	GME		LS		ML	
	Bias	MSE	Bias	MSE	Bias	MSE
α	−0.0551	0.0054	0.4439	6.4992	−0.1075	0.0646
β_1^-	0.001	0.0005	0.0082	0.0054	0.0147	0.0005
β_1^+	−0.0056	0.0005	−0.1838	6.1044	−0.0176	0.0061
γ_1	−0.0055	0.0012	0.3329	5.3791	−0.0773	0.0164
N = 40	GME		LS		ML	
	Bias	MSE	Bias	MSE	Bias	MSE
α	−0.0271	0.032	0.0745	0.0697	0.2726	0.152
β_1^-	−0.0235	0.027	0.0011	0.0013	0.0719	0.0128
β_1^+	0.0165	0.0131	−0.0276	0.0031	−0.0127	0.0026
γ_1	0.0388	0.08	0.0185	0.0336	−0.1401	0.0743

Source: Calculation

Table 5 Kink regressions with $Unif(-2, 2)$ errors

N = 20	GME		LS		ML	
	Bias	MSE	Bias	MSE	Bias	MSE
α	−0.0727	0.1667	0.687	2.8672	0.8237	0.9328
β_1^-	−0.0687	0.0098	0.0468	0.0087	0.0448	0.008
β_1^+	0.003	0.0105	−0.0442	0.0233	−0.0546	0.0081
γ_1	0.0031	0.2609	0.2429	0.8261	0.3161	0.1994
N = 40	GME		LS		ML	
	Bias	MSE	Bias	MSE	Bias	MSE
α	−0.0914	0.0994	−0.1981	0.1254	−0.7941	0.8471
β_1^-	−0.114	0.1364	0.0138	0.0028	0.043	0.6761
β_1^+	0.0646	0.0449	0.091	0.0099	0.1726	0.0317
γ_1	0.1579	0.2877	−0.2178	0.0924	−0.4565	0.376

Source: Calculation

we found that the performance is quite similar between n = 20 and n = 40 for GME estimator since the values of Bias and MSE are not much different. In contrast, the larger sample size can bring a lower Bias and MSE of the parameters in the case of LS and MLE. This indicates that the larger sample size cannot reduce uncertainty and its performance seems to be a little affected. Consider our key threshold or kink parameter γ_1, the interesting result is obtained. We found that the values of Bias and MSE for GME estimator increase when the error of the model is assumed to have non-normal distribution.

In summary, we can conclude that the overall performance of the GME estimator applied to kink regression is quite good. The GME estimators produce a lowest bias estimates when compared with LS and MLE when the errors are generated from normal and non-normal distribution.

4 Empirical Illustration

Finally, the method presented in this paper as well as LS and MLE are applied to the real data analysis. We consider the growth/debt regression problem which was introduced in Hansen [8]. The paper suggested that the growth of economy tends to slow when the government debt relative to GDP exceeds a threshold. Thus, we employed this model specification and applied it in Thai economy and our model can be written as

$$GDP_t = \beta_1^- (Debt/GDP_t \leq \gamma_1)_- + \beta_1^+ (Debt/GDP_t > \gamma_1)_+ + \varepsilon_t$$

Our dataset consists of yearly data GDP growth and Debt/GDP ratio for Thailand. The sample period spans from 1993 to 2015, covering 22 observations. The data are collected from Thomson Reuter DataStream, Faculty of Economic, Chiang Mai University. The data is plotted in Fig. 1. Prior to estimating the kink regression model, it is necessary to check whether there exists a kink or threshold effect or not. The study conducted an Entropy ratio test as discussed in Sect. 2.3 and the result is shown in Table 5. The Entropy ratio test statistic is significant at 1% level. This result led us to reject the null hypothesis of linear regression (H_{linear}) for the relationship between GDP growth and Debt/GDP ratio for Thai economy, which means that the kink regression model is better to describe this non-linear relationship.

4.1 Application Result

The kink regression model is then estimated by three estimators, GME, LS, and MLE, and the estimated results are shown in Table 6. Similar results are obtained in this study, as we can see a negative effect of Debt/GDP on the economic growth in both two regimes. We are surprised that our result show negative sign in both of the two regime which seem not correspond to the Keynesian economics approaches that

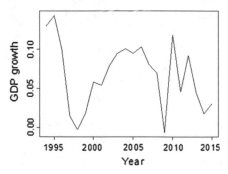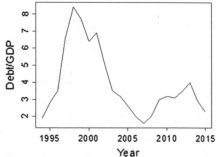

Fig. 1 GDP growth and Debt/GDP ratio plot

Fig. 2 Kink plot of GDP
growth and Debt/GDP

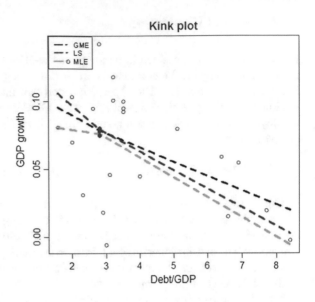

suggested Debt/GDP ratio should have a positive effect during economic recession
(regime 1). We reckon that the Thai economy may not change too much along our
sample period and the Thai economic structure is different from those of advanced
countries. However, we can observe that there exists a significant kink point and the
fitted kink regression line is plotted in Fig. 2. The result illustrates a steep negative
slope for low GDP-Debt ratio, with a kink or threshold point (γ_1) around 3.47% (For
GME), switching to a low negative slope for GDP-Debt ratio above that point.

Table 6 Coefficients (standard errors) from Kink regression

Parameter	GME	LS	MLE
α	0.0782***	0.0790***	0.0757*
	(−0.0127)	(−0.0172)	(−0.0395)
β_1^- (regime 1)	−0.0143	−0.0223	−0.0039
	(−0.0245)	(−0.0183)	(−0.0197)
β_1^+ (regime 2)	−0.0104***	−0.0136**	−0.0145*
	(−0.005)	(−0.0072)	(−0.0084)
γ_1	2.8001***	3.5008***	3.9686
	(−1.2011)	(−1.7383)	(−3.7315)
sigma			0.0354***
			(−0.0053)
RMSE	0.0347	0.0365	0.0357

Source: Calculation

Note: ***, **, * are significant at 1, 5, and 10% level, respectively

Table 7 Kink regression application result

	{H}_{linear}	{H}_{kink}	Entropy ratio
Regime 1 vs. Regime 2	26.35955	50.55449	47.999***

Source: Calculation
Note: *** is significant at 1% level

Consider the performance of GME estimator, Table 7 also provide the Root Mean Square Error (RMSE) of each estimator and confirms that RMSE value of the entropy kink regression model is lower than those of the other two estimators. These results gave us a confidence that entropy kink regression technique can at least better than the traditional methods, namely Maximum likelihood (ML) and least squares (LS) estimation.

5 Conclusion

This paper developed an estimation of kink regression model using the Generalized Maximum Entropy (GME) estimator and applied it in both experiment and application studies. An interesting contribution is the model can capture a non-linear relationship between response variable and its covariate when the data set is too small. The results of experiment presented in this study provide a strong evidence about the robust performance of the GME compared with the traditional estimation by Maximum likelihood (ML) and least square (LS). It is evident that with various sample sizes, number of support points and support space for unknown parameters and disturbance, the Bias and MSE of GME estimators are quite smaller when the number of support increase but the Bias and MSE values of the parameters are not quite stable when we consider various support space and number of observations. We, then compare the GME estimator with LS and ML estimators and found that the values of Bias and MSE are lower for GME. Therefore, the GME can provide a better alternative for the Kink regression model compared with the other parametric methods. In addition, we apply our method to study the effect of Debt/GDP ratio on the economic growth and found that Debt/GDP ratio provide a negative effect on economic growth for both regimes and the value of RMSE of GME estimator is slightly better than those of others.

References

1. Aigner DJ, Lovell CAK, Schmidt P (1977) Formulation and estimation of stochastic frontier production function models. J Econ 6:21–37
2. Battese GE, Coelli TJ (1992) Frontier production functions, technical efficiency and panel data: with application to paddy farmers in India. J Product Anal 3:153–169

3. Card D, Lee D, Pei Z, Weber A (2012) Nonlinear policy rules and the identification and estimation of causal effects in a generalized regression kind design. NBER Working Paper 18564
4. Furkov A, Surmanov K (2011) Stochastic frontier analysis of regional competitiveness. Metody Ilociowe w Badaniach Ekonomicznych 12(1):67–76
5. Garelli S (2003) Competitiveness of nations: the fundamentals. IMD World competitiveness yearbook, pp 702–713
6. Greene WH (2003) Simulated likelihood estimation of the normal-gamma stochastic frontier function. J Prod Anal 19:179–190
7. Hansen BE (2000) Sample splitting and threshold estimation. Econometrica 68(3):575–603
8. Hansen BE (2015) Regression kink with an unknown threshold. J Bus Econ Stat (Just-accepted)
9. Nelsen RB (2013) An introduction to copulas. Springer, New York
10. Noh H, Ghouch AE, Bouezmarni T (2013) Copula-based regression estimation and inference. J Am Stat Assoc 108(502):676–688
11. Schwab K, Sala-i-Martin X (2015) World Economic Forums Global Competitiveness Report, 2014–2015. In: World Economic Forum, Switzerland. Accessed from. http://reports.weforum.org/global-competitiveness-report-2014-2015/
12. Simm J, Besstremyannaya G, Simm MJ (2016) Package rDEA
13. Smith MD (2008) Stochastic frontier models with dependent error components. Econometrics J 11(1):172–192
14. Wang H, Song M (2011) Ckmeans. 1d. dp: optimal k-means clustering in one dimension by dynamic programming. R J 3(2):29–33

Predictive Recursion Maximum Likelihood of Threshold Autoregressive Model

Pathairat Pastpipatkul, Woraphon Yamaka and Songsak Sriboonchitta

Abstract In the threshold model, it is often the case that an error distribution is not easy to specify, especially when the error has a mixture distribution. In such a situation, standard estimation yields biased results. Thus, this paper proposes a flexible semiparametric estimation for Threshold autoregressive model (TAR) to avoid the specification of error distribution in TAR model. We apply a predictive recursion-based marginal likelihood function in TAR model and maximize this function using hybrid PREM algorithm. We conducted a simulation data and apply the model in the real data application to evaluate the performance of the TAR model. In the simulation data, we found that hybrid PREM algorithm is not outperform Conditional Least Square (CLS) and Bayesian when the error has a normal distribution. However, when Normal-Uniform mixture error is assumed, we found that the PR-EM algorithm produce the best estimation for TAR model.

Keywords Threshold autoregressive · PR-EM algorithm · Predictive recursion marginal likelihood · Stock market

1 Introduction

Most financial time series data have non-linear movements reflecting different economic behaviors over time. Thus it is not appropriate to assume linearity in econometric modeling attempts. The threshold autoregressive (TAR) model developed by Tong [11] is one of the most popular nonlinear time series models appear in the literature and TAR has become influential as well in the fields of economics and econometrics. This model allows regime shift in the time series data, which is determined by a threshold value. Moreover, it is typically applied to time series data as an extension of autoregressive models, in order to capture the regime switching in economics data. In the recent years, TAR model has been widely applied in various research fields such as ecology, agriculture, economics, and finance.

P. Pastpipatkul · W. Yamaka (✉) · S. Sriboonchitta
Faculty of Economics, Chiang Mai University, Chiang Mai, Thailand
e-mail: woraphon.econ@gmail.com

© Springer International Publishing AG 2017

349

V. Kreinovich et al. (eds.), *Robustness in Econometrics*,
Studies in Computational Intelligence 692, DOI 10.1007/978-3-319-50742-2_21

It is interesting to highlight that the estimation of TAR model has been estimated by many approaches such as conditional least squares (e.g., [5, 7]), Maximum Likelihood (e.g., [3, 8]) and Bayesian estimation by Geweke and Terui [4], Chen and Lee [2], and Koop and Potter [2]. However, these approaches assume the density function to have normal distribution with mean zero and variance, σ^2; thus, if the density is not normal, then these conventional models might not be appropriate for estimating the parameters in TAR model. Ang [1] suggested that the Normal distribution presents a symmetric and zero excess kurtosis which might not be correct to model the financial time series data. Recently, alternative non-normal density estimation procedures such as nonparametric maximum likelihood and Bayes have been proposed to deal with these problems. Nevertheless, Martin and Han [9] claimed that nonparametric Maximum Likelihood is generally rough and the optimization is unstable while nonparametric Bayes needs a large round of iterations if the prior is not proper. Thus, to solve these problems, they proposed a general scale mixture model for the error distribution using Predictive Recursion marginal likelihood $L_{PR}(\theta)$ and developed the hybrid Predictive Recursion-Expected Maximization (PREM) algorithm based on the scale mixture of normals model for the error terms for maximizing $L_{PR}(\theta)$ in the linear regression model.

This main purpose of this paper is to extend the estimation technique of Martin and Han [9] to estimate the parameters and purpose an alternative way for TAR estimation. To the best as our knowledge, we are the first ever to estimate TAR model using this method. Therefore the model becomes improved and has more flexible where the error distribution is taken to be a general scale mixture of normal with unknown mixing distribution. The remainder of this paper is organized as follows: The next section introduces the TAR model and estimation approach. Section 3 consists of simulations and demonstrates estimation of parameters compared to some existing methods. An empirical application is provided in Sect. 4 while conclusions is given in Sect. 5.

2 Methodology

2.1 Threshold Autoregressive Model

In this paper, we present a simplest class of TAR model, namely Self Exciting Threshold Autoregressive (SETAR), which consist of lag order (p) of autoregressive (AR) and k regimes. This model is a piecewise linear models or regime switching model The paper focused on the SETAR models because it is the most popular in forecasting and predicting and it is easily to estimated using PREM algorithm Tong [11] introduced SETAR with 2 regimes and specified as the following:

$$y_t = \alpha_0 + \alpha_i \sum_{i=1}^{p_1} y_{t-p_1} + \varepsilon_{1,t} \quad if \ y_{t-d} > \omega \tag{1}$$

$$y_t = \beta_0 + \beta_i \sum_{i=1}^{p_2} y_{t-p_2} + \varepsilon_{2,t} \quad if \ y_{t-d} \leq \omega$$

where α_i and β_i are the estimated coefficients, $\varepsilon \sim N(0, \sigma^2)$, is threshold value and is an $n \times 1$ vector of independent and identically distributed (iid) errors with normal distribution. The movement of the observations between the regimes is controlled by Y_{t-d}, the delay variable. If Y_{t-d} is greater or lower than ω, the separated observations can be estimated as AR process then the AR order can be vary across regime. Thus, it is important to select the appropriated lag delay (d) of delay variable. In the TAR model Eq. (1) is separated into two regimes which the non-linear model depends on below or above the threshold parameter. In addition the different lag order of autoregressive in each regime is allowed to be different. For identification of regimes, we interpret first regime as high growth regime while second regime is interpreted as low growth regime.

2.2 Predictive Recursion Marginal Likelihood

Consider the TAR model Eq. (1), errors are assumed to have normal distribution and iid. But in this case, the error distribution is taken to be a general scale mixture of normal with unknown mixing distribution. As we mentions before, the assumption of normal distribution might not be appropriate in some cases. Following, Martin and Han [9], they purposed that the density function of the model should have a heavy tail thus the estimated parameters can be sensitive to the extreme observations. The general form of the error density as a mixture can be written as

$$f(\varepsilon_i) = \int_0^\infty N(\varepsilon_i \,|\, 0, u^2) \Psi(du) \tag{2}$$

$$f(\varepsilon_i) = \int_0^\infty N(y_i - x_i \beta \,|\, 0, u^2) \Psi(du) \tag{3}$$

Then, applying the PR algorithm which is a recursive estimation of mixing distribution to estimate the mixing density [10]. Therefore, in TAR model, we can extend the general form of the error density and apply the PR algorithm to the residuals to estimate the mixing density, let $x_i = y_{t-1}, \ldots, y_{i,t-p}$ thus the PR marginal likelihood for β and α can be written as

$$L_{PR}(\theta) = \sum_{j=1}^{2} \left(\prod_{i=1}^{n} f_{j,i-1,\theta}(\varepsilon_{j,i}) \right) \tag{4}$$

where $\theta = (\beta, \alpha)$ based on $y_1, \ldots, y_k$ and $x_1, \ldots, x_k$, $k = 1, \ldots n$ repeated step in PR algorithm. Ψ is an unspecified mixing distribution with supported on U [Umin,Umax], where Umin is fix at 0.00001 and Umax can be specified as positive integer, $[0, \infty]$. However, there are many possibilities of density of the Ψ, in this study, the uniform density on U is preferred as an initial start of the PR-EM algorithm.

To estimate the β and α, normally, the numerical optimization technique is used to maximize the likelihood function to obtain the estimated parameters in the model. However, the study employed hybrid PREM algorithm as proposed in Martin and Han [9] as an optimization algorithm of this model since it work well in the estimation of the parameter in semiparametric mixture model.

2.3 Hybrid PREM Algorithm

Martin and Han [9] proposed a hybrid PREM algorithm taking an advantage of the latent scale parameter structure $U_1, \ldots, U_n$ in the mixture model. Note that $x_i = y_{t-1}, \ldots, y_{i,t-p}$ then, we take a logarithm in Eq. (4), we get

$$
\begin{aligned}
&\log L_{PR}(\theta) \\
&= \sum_{i=1}^{n} \log \int N(y_i - x_i\beta \mid 0, u^2)\Psi_{i-1,\beta} + \sum_{i=1}^{n} \log \int N(y_i - x_i\alpha \mid 0, u^2)\Psi_{i-1,\alpha} \\
&= \left[\sum_{i=1}^{n} \log N(y_i - x_i\beta \mid 0, U_i^2) - \sum_{i=1}^{n} \log \left\{ \frac{N(y_i - x_i\beta \mid 0, U_i^2)}{f_{i-1,\beta}(y_i - x_i\beta)} \right\} \right] \\
&\quad + \left[\sum_{i=1}^{n} \log N(y_i - x_i\alpha \mid 0, U_i^2) - \sum_{i=1}^{n} \log \left\{ \frac{N(y_i - x_i\alpha \mid 0, U_i^2)}{f_{i-1,\alpha}(y_i - x_i\alpha)} \right\} \right]
\end{aligned}
\tag{5}
$$

Then, integrate out U_i with respect to the density $\Psi_{\tilde{\beta}}^{\beta}(u)$ and $\Psi_{\tilde{\alpha}}^{\alpha}(u)$, thus

$$
\begin{aligned}
&\log L_{PR}(\theta) \\
&= \left[\sum_{i=1}^{n} \log \int N(y_i - x_i\beta \mid 0, U_i^2) - \sum_{i=1}^{n} \log \left\{ \frac{N(y_i - x_i\beta \mid 0, U_i^2)}{f_{i-1,\beta}(y_i - x_i\beta)} \right\} \Psi_{\tilde{\beta}}^{\beta}(u)d(u) \right] \\
&\quad + \left[\sum_{i=1}^{n} \log \int N(y_i - x_i\alpha \mid 0, U_i^2) - \sum_{i=1}^{n} \log \left\{ \frac{N(y_i - x_i\alpha \mid 0, U_i^2)}{f_{i-1,\alpha}(y_i - x_i\alpha)} \right\} \Psi_{\tilde{\alpha}}^{\alpha}(u) \right]
\end{aligned}
\tag{6}
$$

Where $\widehat{\alpha}$ and $\widehat{\beta}$ are some estimate. Let rewrite Eq. (6) as $Q_1(\beta, \alpha \,|\, \widehat{\beta}, \widehat{\alpha}) +$ $Q_2(\beta, \alpha \,|\, \widehat{\beta}, \widehat{\alpha})$ where

$$Q_1(\beta, \alpha \,|\, \widehat{\beta}, \widehat{\alpha}) = \left[0.5 \sum_{i=1}^{n} w_{1,i}(y_i - x_i \beta)^2 + \beta_0 \right] + \left[0.5 \sum_{i=1}^{n} w_{2,i}(y_i - x_i \alpha)^2 + \alpha_0 \right]$$

Where w_i is a weight which depend on mixing distribution and error term in each step and can be computed by

$$\widehat{w}_{1,i} = \int u^{-2} \Psi_{i,\alpha}^{\alpha}(u) du \qquad if \ y_{t-d} > \omega$$

$$\widehat{w}_{2,i} = \int u^{-2} \Psi_{i,\beta}^{\beta}(u) du \qquad if \ y_{t-d} \leq \omega \tag{7}$$

These weights is required to help the PR algorithm not too sensitive to the outlier. To maximize the PR marginal likelihood, we conducted PR-EM algorithm by maximizing Q_1 correspond to weighted least square.

2.4 Bayesian Inference on Threshold Autoregressive Model

The posterior estimation can be computed by combining these prior with the likelihood function using Bayes theorem in order to compute the posterior. Let $\Theta = \alpha, \beta, \sigma_1^2, \sigma_2^2$, the posterior estimation can be formed as following

Posterior probability $\propto$ likelihood x Prior probability

$$\Pr(\Theta, \omega \,|\, y_t, x_t) = \Pr(y_t, x_t \,|\, \Theta, \omega) \Pr(\Theta, \omega) \tag{8}$$

Where the likelihood function is compute form the summation of normal density distribution with mean equal $x_i \alpha$ for regime 1 and $x_i \beta$ for regime 2.

Following Chen and Lee [2], The Metropolis-Hasting (MH) sampler has been employed to sample the initial parameters In this study, we use normal distribution for parameter coefficient and threshold parameter. The distribution for variance is assumed to be inverse gamma. To sample the initial parameters, Θ^0 and ω^0. The MH algorithm is conducted to sampler from the posterior density, $\Pr(\Theta, \omega)$. The proportion distribution for each parameters are generate be Gaussian distribution and uniform distribution.

2.4.1 Conditional Least Square on Threshold Autoregressive Model

The conditional least square (CLS) in the estimation method that minimize the sum square errors of the model to obtain the estimated parameters [5]. Let $E|Y_t < \infty$, $t = 1, ...N$, Then the objection function for minimize sum square errors can be written as:

$$\text{Min } Er(\alpha, \beta, \omega) = \sum_{i=1}^{N} [Y_j - E_\theta (Y_j | B_{j-1})]^2$$

With respect to α, β, ω.

3 Simulation Study

In this section, we carry out the simulation study to evaluate the performance and accuracy of the PR-EM algorithm. We proceed the simulation to examine the estimation performance of this estimation applying in TAR model. The simulation is the realization of a simple two regime SETAR model. In this simulation, we generate random data from this model specification:

$$y_t = 0.2 + 0.6y_{t-1} + \varepsilon_{1,t} \quad if \ y_{t-1} > 0.4$$
$$y_t = 0.1 + 0.4y_{t-1} + \varepsilon_{1,t} \quad if \ y_{t-1} > 0.4$$

Where the error terms are assumed to be normal distribution. We generate 200, 500, and 1000 observations using the specified parameters. Figure 1 shows the stationary path of the simulated data from the specified parameter. In addition, we compare our estimation method with parametric Conditional least square (CLS) and Bayesian methods for robustness.

Tables 1, 2 and 3 present the estimation result for the specified parameter in TAR model, considering 3 sets of sample sizes. The result shows that TAR model can be estimated well using PR-EM algorithm. For, N = 200, the estimated values of the threshold is 0.3962, the lower regime coefficient are 0.910 and −0.3181 and the upper regime coefficients are 0.2488 and 0.6442. These estimated parameters are very close to the true values of 0.1 and −0.4 for lower regime and 0.2 and 0.6 for upper regime. In addition, another 2 sets of sample size also present a closely value of estimated parameters and true value. In this study, we also show the performance of this estimation by comparing with other two methods namely, Conditional least square (CLS) and Bayesian. The result show that the PR-EM algorithm is not the best method when compare with CLS and Bayesian methods. Although, some of standard

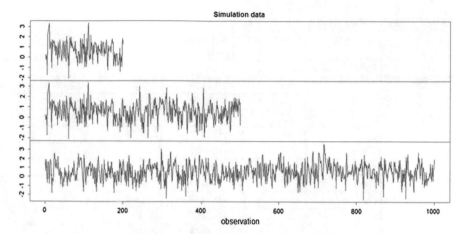

Fig. 1 Simulated path of the simulated data n = 200, 500, 1000

Table 1 The estimation result based on 200 sample sizes

	Parameter	True value	Estimates parameter	S.E.	RMSE
	PR-EM estimation				
N = 200	β_0	0.1	0.0910	0.1286	0.7911
	β_1	−0.4	−0.3181	0.2295	
	α_0	0.2	0.2488	0.5644	
	α_1	0.6	0.6442	0.1107	
	Threshold		0.3962		
	CLS estimation				
	β_0	0.1	0.0750	0.1113	0.7793
	β_1	−0.4	−0.5526	0.1877	
	α_0	0.2	0.1737	0.1445	
	α_1	0.6	0.6212	0.1107	
	Threshold		0.2849		
	Bayesian estimation				
	β_0	0.1	0.2096	0.1200	0.7841
	β_1	−0.4	−0.3786	0.2097	
	α_0	0.2	0.1426	0.1496	
	α_1	0.6	0.6424	0.1088	
	Threshold		0.4327		

Source: Calculation

Table 2 The estimation result based on 500 sample sizes

	Parameter	True value	Estimates parameter	S.E.	RMSE
	PR-EM estimation				
N = 500	β_0	0.1	0.1119	0.0742	0.7854
	β_1	−0.4	−0.2307	0.1038	
	α_0	0.2	0.1783	0.093	
	α_1	0.6	0.5969	0.0737	
	Threshold		0.3963		
	CLS estimation				
	β_0	0.1	0.1439	0.0568	0.7823
	β_1	−0.4	−0.2733	0.0949	
	α_0	0.2	0.2292	0.1143	
	α_1	0.6	0.5623	0.0862	
	Threshold		0.4501		
	Bayesian estimation				
	β_0	0.1	0.1353	0.0705	0.7825
	β_1	−0.4	−0.285	0.1184	
	α_0	0.2	0.1961	0.0931	
	α_1	0.6	0.5834	0.0712	
	Threshold		0.4476		

Source: Calculation

error (S.E.) of estimated parameters from PR-RM algorithm are lower than other methods. The Root Mean Square of Error (RMSE) of PR-EM algorithm is a bit higher. Interestingly, the PR-EM does not show evidence of completely competitive with other methods. However, it is provide a good estimation and could be an alternative estimation for TAR model. We expected that the assumption of normal distribution on the error term in this simulation study may not give us the expected competitive of PR-EM algorithm with other methods since these simulations are still assume a normal error. Thus, we simulate the sample data with non-standard normal scale mixture with respect to a uniform distribution which is illustrated in Fig. 2. We could see that the semiparametric PR-EM algorithm present the lowest of S.E. and RMSE. Therefore, we can conclude that if the error density function is non-normal, PR-EM algorithm is outperform other two method (Table 4).

Table 3 The estimation result based on 1000 sample sizes

	Parameter	True value	Estimates parameter	S.E.	RMSE
	PR-EM estimation				
N = 1000	β_0	0.1	0.0541	0.0542	1.1525
	β_1	−0.4	−0.3144	0.0391	
	α_0	0.2	0.1696	0.0700	
	α_1	0.6	0.5885	0.0527	
	Threshold		0.3963		
	CLS estimation				
	β_0	0.1	0.0801	0.0429	1.1482
	β_1	−0.4	−0.3572	0.0710	
	α_0	0.2	0.2164	0.0802	
	α_1	0.6	0.5571	0.0592	
	Threshold		0.4419		
	Bayesian estimation				
	β_0	0.1	0.0806	0.0499	1.1301
	β_1	−0.4	−0.3558	0.0833	
	α_0	0.2	0.1901	0.0722	
	α_1	0.6	0.5733	0.0532	
	Threshold		0.4190		

Source: Calculation

Fig. 2 Normal-Uniform mixture density

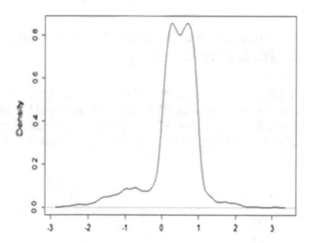

Table 4 The estimation result of TAR based mixture distribution

	Parameter	True value	Estimates parameter	S.E.	RMSE
	PR-EM estimation				
N = 200	β_0	0.1	0.6045	0.0749	0.4676
	β_1	−0.4	−0.1543	0.1527	
	α_0	0.2	0.5873	0.0847	
	α_1	0.6	0.5075	0.0660	
	Threshold		0.646		
	CLS estimation				
	β_0	0.1	0.4662	0.0915	0.4687
	β_1	−0.4	−0.1598	0.1670	
	α_0	0.2	0.5884	0.1305	
	α_1	0.6	0.4687	0.1012	
	Threshold		0.6463		
	Bayesian estimation				
	β_0	0.1	0.6148	0.1129	0.5132
	β_1	−0.4	−0.2035	0.1699	
	α_0	0.2	0.5291	0.1610	
	α_1	0.6	0.5070	0.1157	
	Threshold		0.4190		

Source: Calculation

4 Application on Bangkok Stock Exchange Index Forecasting

In this section apply the PR-EM to real data analysis which unknown any information about error distribution. We demonstrate the use of PR-EM algorithm using Bangkok Stock Exchange Index (SET). The data set is a daily time series data from January 2014 to October 2015 covering 443 observations (Fig. 3).

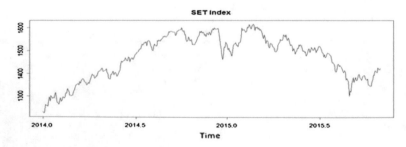

Fig. 3 Normal-Uniform mixture density

In the estimation of TAR model, it is important to find out an appropriate threshold value in the data series. If the value of threshold is known, the estimation of TAR model is surely available. Thus, with the unknown information the study proposed a pooled AIC of Gonzalo and Pitarakis [6] for testing the threshold value in the data series. Figure 4 illustrates the grid search method and shows that the appropriate autoregressive lag order for upper and lower regimes is lag 1 and the threshold value takes place between 0.000 and −0.001, as presented by black dot plot. Thus, this confirms that the data exhibit a non-linear trend.

According to the results of estimation presented in Table 5, we observe that PR-EM algorithm outperforms the CLS and Bayesian methods. The standard errors of estimated parameters from PR-EM algorithm show the lowest value when compared with the two parametric methods. Thus, we can conclude that PR-EM algorithm can improve the estimation in TAR model. Moreover, we plot PR-EM weights (w_i) which are used for adjustment of the outlier data in Fig. 5b. The observations with low weights are presented as a low influence on the fitting of the regression. Conversely, high weights are presented as a high influence. The PR-EM algorithm works well with these weights. Figure 5a also shows the histogram of residuals and we can observe that the residuals have normal distribution. Surprisingly, the results seem to be a bit different from simulation study with normal distribution case because the standard error and RMSE for PR-EM algorithm are completely less than those from CLS and Bayesian methods. We can say that though it is not the best performer in the simulation study, it can be the alternative estimation for TAR model. Then, we plot

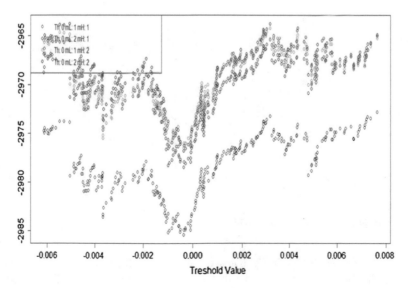

Fig. 4 Threshold value Pooled AIC

Table 5 The estimation TAR result based on SET index

Parameter	Estimates parameter	S.E.	RMSE
PR-EM estimation			
β_0	−0.0004	0.0001	0.0075
β_1	−0.0707	0.0015	
α_0	0.0003	0.0001	
α_1	0.0615	0.0013	
Threshold	−0.0003		
CLS estimation			
β_0	−0.0033	0.0568	0.0075
β_1	−0.2787	0.0949	
α_0	0.0008	0.1143	
α_1	0.0046	0.0862	
Threshold	−0.0036		
Bayesian estimation			
β_0	−0.0011	0.0014	0.0075
β_1	−0.1203	0.1363	
α_0	0.0005	0.0008	
α_1	0.0396	0.0938	
Threshold	−0.001		

Source: Calculation

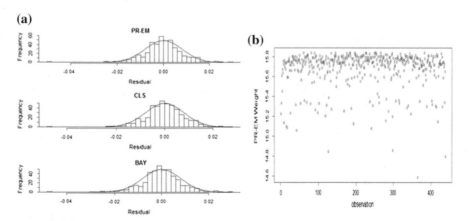

Fig. 5 Histogram fromresiduals from TAR

the forecasts for SET index in Fig. 6 in order to evaluate the forecasting performance of the estimation. The results show that the TAR model based PR-EM algorithm produces a good forecast and it is competitive with other methods.

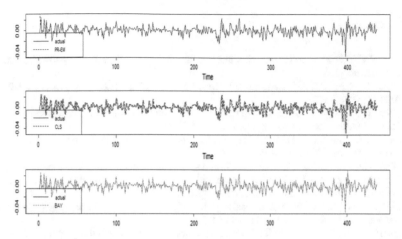

Fig. 6 Forecasting SET index verses actual

5 Concluding Remarks

This paper proposes a flexible semiparametric estimation for Threshold autoregressive model (TAR) where the error distribution is taken to be a general scale mixture of normals with unknown mixing distribution. To estimate the parameters in the TAR model, we maximize the PR-based likelihood function using hybrid PREM algorithm which purposed by Martin and Han [9]. We conducted a simulation study to evaluate the performance of the TAR model base PR-EM algorithm and found that it is not outperform in all cases, yet it is not dominated by other methods, namely Conditional Least Square (CLS) and Bayesian. However, when we conduct another simulated data based on Normal-Uniform mixture density, we found that the PR-EM algorithm produce the best estimation in term of standard error and root mean square error. In the last section, we apply the model to real data, Bangkok Stock Exchange Index (SET). The result form this estimation is perform well.

For further study, in this study does not examine the PR-EM algorithm with other mixing error distribution. We suggest to apply this estimation with other mixing distribution. Furthermore, this estimation method can be apply to other model such as a switching model, logit and probit regression model.

Acknowledgements The authors are grateful to Puay Ungphakorn Centre of Excellence in Econometrics, Faculty of Economics, Chiang Mai University for the financial support.

References

1. Ang V (2015) Financial interaction analysis using best-fitted probability distribution. IFC Bulletins chapters, 39
2. Chen CW, Lee JC (1995) Bayesian inference of threshold autoregressive models. J Time Ser Anal 16(5):483–492
3. Chan NH, Kutoyants YA (2012) On parameter estimation of threshold autoregressive models. Statistical inference for stochastic processes 15(1):81–104
4. Geweke J, Terui N (1993) Bayesian threshold autoregressive models for nonlinear time series. J Time Ser Anal 14(5):441–454
5. Gibson D, Nur D (2011) Threshold autoregressive models in finance: a comparative approach
6. Gonzalo J, Pitarakis J (2002) Estimation and model selection based inference in single and multiple threshold models. J Econometrics 110:319 352
7. Li D, Li WK, Ling S (2011) On the least squares estimation of threshold autoregressive moving-average models. Stat Interface 4(1):183–196
8. Qian L (1998) On maximum likelihood estimators for a threshold autoregression. J Stat Plann Inference 75(1):21–46
9. Martin R, Han Z (2016) A semiparametric scale-mixture regression model and predictive recursion maximum likelihood. Comput Stat Data Anal 94:75–85
10. Newton MA (2002) On a nonparametric recursive estimator of the mixing distribution. Sankhya Ser A 64(2):306322
11. Tong H (1983) Threshold Models in Non-linear Time Series Analysis. In: Krickegerg K (ed) Lecture Notes in Statistics, vol 21. Springer, New York

A Multivariate Generalized FGM Copulas and Its Application to Multiple Regression

Zheng Wei, Daeyoung Kim, Tonghui Wang and Teerawut Teetranont

Abstract We introduce a class of multivariate non-exchangeable copulas which generalizes many known bivariate FGM type copula families. The properties such as moments, affiliation, association, and positive lower orthant dependent of the proposed class of copula are studied. The simple-to-use multiple regression function and multiple dependence measure formula for this new class of copulas are derived. Several examples are given to illustrate the main results obtained in this paper.

1 Introduction

Copulas have been successfully applied to many areas and become much more widely used in recent years. The study of copula functions has become a major phenomenon in constructing joint distribution functions and modelling real multivariate data. In particular, there is a great need in developing the multivariate nonexchangeable copulas in modeling the asymmetric and nonlinear dependence data. Several researchers have put considerable efforts into developing multivariate asymmetric copulas [1, 2, 7, 9, 20, 23]. The purpose of this paper is to propose a class of non-exchangeable multivariate copulas which generalizes many known copulas such as bivariate Farlie-Gumbel-Morgenstern (FGM) family of copulas, the bivariate copulas proposed in [19], the bivariate copulas with quadratic sections in [15] and with cubic sections in [11], among others.

Z. Wei (✉)
Department of Mathematics and Statistics, University of Massachusetts Amherst,
Amherst, USA
e-mail: wei@math.umass.edu

D. Kim
Department of Mathematics and Statistics, Organization University of
Massachusetts Amherst, Amherst, USA
e-mail: daeyoung@math.umass.edu

T. Wang · T. Teetranont
Faculty of Economics, Chiang Mai University, Chiang Mai, Thailand
e-mail: twang@nmsu.edu

T. Teetranont
e-mail: teerawut.t@cmu.ac.th

© Springer International Publishing AG 2017
V. Kreinovich et al. (eds.), *Robustness in Econometrics*,
Studies in Computational Intelligence 692, DOI 10.1007/978-3-319-50742-2_22

Recently, the study of copula-based regression has drawn attention in different fields. Sungur [17, 18] studied the directional dependence through the copula based regression in bivariate case. In order to study the asymmetric dependence/interaction between the variables of interest, the asymmetric copulas should be utilized. [14] proposed the semiparametric estimation of the copula based regression. In non-experimental field, the copula based regression has been applied in the research context such as deficit hyperactivity disorder [4], gene expression network [3], and the aggression in adolescence development analysis [5]. In this paper, we derive the simple-to-use multiple regression formula based on the proposed class of non-exchangeable multivariate copulas.

Copulas, multivariate distributions with standard uniform marginals, contain the most information on the multivariate dependence structure independent of the marginals of the variables. References for a detailed overview of copula theory and applications, see [2, 13, 21, 22, 24].

Definition 1.1 [13] A $(k + 1)$-*dimensional* **copula** is a function $C : [0, 1]^{k+1} \mapsto [0, 1]$ satisfying following properties:

(a) C is grounded, i.e., if at least one $u_i = 0$, $C(u_0, u_1, \ldots, u_k) = 0$;

(b) For every $u_i \in [0, 1]$, $C(1, \ldots, 1, u_i, 1, \ldots, 1) = u_i$;

(c) C is $(k + 1)$-increasing in the sense that, for any $J = \prod_{i=0}^{k} [u_i, v_i] \subseteq [0, 1]^{k+1}$

with $u_i, v_i \in [0, 1]$,

$$volC(J) = \sum_{a} sgn(a)C(a) \geq 0,$$

where the summation is over all vertices a of J, and for $a = (a_0, a_1, \ldots, a_k)^T$, with $(a_0, a_1, \ldots, a_k)^T$ is the transpose of $(a_0, a_1, \ldots, a_k)^T$, and $a_i = u_i$ or v_i,

$$sgn(a) = \begin{cases} 1, & \text{if } a_i = v_i \quad \text{for an even number of } i's, \\ -1, & \text{if } a_i = v_i \quad \text{for an odd number of } i's. \end{cases}$$

The exchangeability is a type of symmetric assumption commonly used in the copula literatures [13].

Definition 1.2 [8]   A $(k + 1)$-copula C is **exchangeable** if it is the distribution function of a $(k + 1)$-dimensional exchangeable uniform random vector $U = (U_0, U_1, \ldots, U_k)^T$ satisfying $C(u_0, u_1, \ldots, u_k) = C(u_{\sigma(0)}, u_{\sigma(1)}, \ldots, u_{\sigma(k)})$ for all $\sigma \in \Gamma$, where Γ denotes the set of all permutations on the set $\{0, 1, \ldots, k\}$.

2   The Family of Multivariate Generalized FGM Copulas

[19] describes a wide class of bivariate copulas depending on two univariate functions of the following form, $C(u, v) = uv + f(u)g(v)$. It generalizes many known

bivariate families such as the Farlie-Gumbel-Morgenstern distributions. We propose a multivariate non-exchangeable copula by extending the work of [19].

Proposition 2.1 *Let f_i, $i = 0, 1, \ldots, k$, be $(k + 1)$ non-zero real functions defined on $[0, 1]$. Let C be the function on $[0, 1]^{k+1}$ given by*

$$C(u_0, u_1, \ldots, u_k) = \prod_{i=0}^{k} u_i + \prod_{i=0}^{k} f_i(u_i). \tag{1}$$

Then C is a copula if and only if
(1) $f_i(0) = f_i(1) = 0$, $i = 0, 1, \ldots, k$;
(2) f_i is absolutely continuous, $i = 0, 1, \ldots, k$; and
(3) $\min\left\{\prod_{i=0}^{k} f_i'(u_i)\right\} \geq -1$; i.e., $\displaystyle\min_{\ell \leq k : \ell \text{ is odd}} \left\{K_{i_1}, K_{i_1,i_2,i_3}, \ldots, K_{i_1,\ldots,i_\ell}\right\} \geq -1$, where $K_{i_1,\ldots,i_t} = \prod_{j \in \{i_1,\ldots,i_t\}} \alpha_j \times \prod_{j \notin \{i_1,\ldots,i_t\}} \beta_j$, $t \leq k$ is an odd number, $\alpha_i = \inf\{f_i'(u_i); u_i \in A_i\} < 0$, and $\beta_i = \sup\{f_i'(u_i); u_i \in A_i\} > 0$, with $A_i = \{u_i \in I; f_i'(u_i) \text{ exists}\}$ and $f_i'(u_i)$ denotes the derivative of f_i, for $i = 0, 1, \ldots, k$. Furthermore, in such a case, C in Eq. (1) is absolutely continuous.

Proof It is easy to say that the function given by Eq. (1) satisfies the boundary conditions in definition of copula if and only if $f_i(0) = f_i(1) = 0$, $i = 0, 1, \ldots, k$. We will show that the copula C is $(k + 1)$-increasing if and only if (2) and (3) holds. First, we assume the function C defined by Eq. (1) is an n-copula, and we want to show (2) and (3) holds. Let $F_i(x, y)$, $i = 0, 1, \ldots, k$, denote the functions defined on the set $T = \{(x, y) \in [0, 1]^2 | x < y\}$ by

$$F_i(x, y) = \frac{f_i(y) - f_i(x)}{y - x}, \quad \text{for} \quad i = 0, 1, \ldots, k.$$

Then, C is $(k + 1)$-increasing if and only if

$$-1 \leq \prod_{i=1}^{n} F_i(u_i, v_i). \tag{2}$$

Hence, C is $(k + 1)$-increasing if and only if the following inequalities holds:

$$-1 \leq L_{i_1}, -1 \leq L_{i_1 i_2 i_3}, \ldots, -1 \leq L_{i_1 \ldots i_\ell},$$

where $L_{i_1 \ldots i_t} = \prod_{j \in \{i_1,\ldots,i_t\}} \gamma_j \times \prod_{j \notin \{i_1,\ldots,i_t\}} \delta_j$, $t \leq n$ is an odd number, $\gamma_i = \inf\{F_i(u_i, v_i) : u_i < v_i, f_i(u_i) > f_i(v_i)\}$ and $\delta_i = \sup\{F_i(u_i, v_i) : u_i < v_i, f_i(u_i) < f_i(v_i)\}$, $i = 0, 1, \ldots, k$. Since $f_i(0) = f_i(1) = 0$ and f_i's are non-zero, the sets above are non-empty. Also, since Eq. (2) holds for all $(u_i, v_i) \in T$, we know that $F_i(u_i, v_i)$ is bounded and therefore $f_i(u_i)$ is absolute continuous for $i = 0, 1, \ldots, k$.

Furthermore, we have

$$
\begin{aligned}
\gamma_i &= \inf\{F_i(u_i, v_i) : u_i < v_i, f_i(u_i) > f_i(u_i)\} = \inf\{F_i(u_i, v_i) : u_i < v_i\} \\
&= \inf\{f_i'(u), u \in A\} = \alpha_i < 0, \\
\delta_i &= \sup\{F_i(u_i, v_i) : u_i < v_i, f_i(u_i) < f_i(u_i)\} = \sup\{F_i(u_i, v_i) : u_i < v_i\} \\
&= \sup\{f_i'(u), u \in A\} = \beta_i > 0.
\end{aligned}
$$

In summary, we have shown that if C is a copula, then (1), (2), and (3) are true. Conversely, the proof follows the same steps backwards, which completes the proof.

Remark 2.1 The copula given in Eq. (1) is exchangeable if and only if there exists constant K_i such that $f_i(u) \equiv K_i f_0(u)$ for all $u \in [0, 1]$, and $i \in \{1, \dots, k\}$.

In the following, the set of copulas characterized in Proposition 2.1 will be called the **multivariate non-exchangeable generalized FGM copula**, denoted by $\mathcal{C}$. The following corollary, whose proof is straightforward, provides the multivariate non-exchangeable FGM parametric copula.

Corollary 2.1 *Let f_i, $i = 0, 1, \dots, k$, be $(k + 1)$ non-zero real functions defined on* $[0, 1]$. *Let C be the function on $[0, 1]^{k+1}$ given by*

$$
C(u_0, u_1, \dots, u_k) = \prod_{i=0}^{k} u_i + \theta \prod_{i=0}^{k} f_i(u_i). \tag{3}
$$

Then C is a copula if and only if

$$
-1 / \max \left\{ \prod_{i=0}^{k} f_i'(u_i) \right\} \leq \theta \leq -1 / \min \left\{ \prod_{i=0}^{k} f_i'(u_i) \right\}.
$$

In the following, we provide a trivariate non-exchangeable copula which extends the Example 4.1. given in [19].

Example 1 Let $C_1(u_0, u_1, u_2) = u_0 u_1 u_2 + f_0(u_0) f_1(u_1) f_2(u_2)$, where $f_i(u_i) = u_i^{p_i}(1 - u_i)^{q_i}$ for $i = 0, 1, 2$. After some calculations, we get $\inf\{f_i'(u_i)\} = \alpha_i$, and $\sup\{f_i'(u_i)\} = \beta_i$, for $i = 0, 1, 2$, where

$$
\begin{aligned}
\alpha_i &= -\left(\frac{p_i}{p_i + q_i}\right)^{p_i - 1} \left(1 + \sqrt{\frac{q_i}{p_i(p_i + q_i - 1)}}\right)^{p_i - 1} \left(\frac{q_i}{p_i + q_i}\right)^{q_i - 1} \\
&\quad \times \left(1 - \sqrt{\frac{p_i}{q_i(p_i + q_i - 1)}}\right)^{q_i - 1} \sqrt{\frac{p_i q_i}{p_i + q_i - 1}},
\end{aligned}
$$

and

$$\beta_i = \left(\frac{p_i}{p_i + q_i}\right)^{p_i - 1} \left(1 - \sqrt{\frac{q_i}{p_i(p_i + q_i - 1)}}\right)^{p_i - 1} \left(\frac{q_i}{p_i + q_i}\right)^{q_i - 1}$$

$$\times \left(1 + \sqrt{\frac{p_i}{q_i(p_i + q_i - 1)}}\right)^{q_i - 1} \sqrt{\frac{p_i q_i}{p_i + q_i - 1}}.$$

From Proposition 2.1, C_1 is a copula if and only if $p_i \geq 1, q_i \geq 1$, for $i = 0, 1, 2$.

When the associated copula is given by $C_2(u_0, u_1, u_2) = u_0 u_1 u_2 + \theta f_0(u_0) f_1(u_1) f_2(u_2)$, where $f_i(u_i) = u_i^{p_i}(1 - u_i)^{q_i}$, $p_i \geq 1, q_i \geq 1$, by Corollary 2.1, C_2 is a copula if and only if $-1/\max\left\{\prod_{i=0}^{k} f_i'(u_i)\right\} \leq \theta \leq -1/\min\left\{\prod_{i=0}^{k} f_i'(u_i)\right\}$. For instance, let $f_0(u_0) = u_0(1 - u_0)$, $f_1(u_1) = u_1^2(1 - u_1)$, and $f_2(u_2) = u_2(1 - u_2)^2$. It is easy to check $\min\left\{\prod_{i=0}^{k} f_i'(u_i)\right\} = -1$ and $\max\left\{\prod_{i=0}^{k} f_i'(u_i)\right\} = 1$. Therefore, in this case we have $\theta \in [-1, 1]$.

3 Properties of the New Class of Multivariate Generalized FGM Copulas

In this section, we study several properties of the multivariate generalized FGM copula proposed in Eq. (1). The joint product moments and multivariate version of association measures formulas are obtained. Furthermore, the multivariate dependence properties are investigated, such as affiliation, left tail decreasing, right tail increasing, stochastically increasing and positive lower orthant dependent.

First, the joint product moments below follows immediately from the definition of the copulas in Eq. (1).

Theorem 3.1 *Let $\boldsymbol{U} = (U_0, U_1, \ldots, U_k)^T$ be a random vector of $(k + 1)$ uniform random variables on $[0, 1]$ with associated copula $C \in \mathcal{C}$ Eq. (1). Then, for any $n_i \geq 1$,*

$$E\left(\prod_{i=0}^{k} U_i^{n_i}\right) = \prod_{i=0}^{k} \frac{1}{n_i + 1} + \prod_{i=0}^{k} n_i \int_0^1 u_i^{n_i - 1} f_i(u_i) du_i.$$

Continuence of Example 1 Consider the copula $C(u_0, u_1, u_2) = u_0 u_1 u_2 + f_0(u_0) f_1(u_1) f_2(u_2)$, where $f_i(u_i) = u_i^{p_i}(1 - u_i)^{q_i}$ and $p_i \geq 1, q_i \geq 1$, for $i = 0, 1, 2$. From Theorem 3.1, for any $n_i \geq 1$,

$$E\left(U_0^{n_0} U_1^{n_1} U_2^{n_2}\right) = \prod_{i=0}^{2} \frac{1}{n_i + 1} + \prod_{i=0}^{2} n_i Beta(n_i + p_i, q_i + 1),$$

where $Beta(a, b) = \int_0^1 t^{a-1}(1 - t)^{b-1} dt$.

The most common non-parametric measures of association between the components of multivariate random vector are Kendall's tau and Spearman's rho. These measures depend only on their associated copula. The population version of the multivariate of Spearman's rho ([16]) is

$$\rho_C = h_\rho(k+1) \left(2^{k+1} \int_{I^n} C(u_0, u_1, \ldots, u_k) du_0 \ldots du_k - 1 \right),$$

where $h_\rho(n) = \frac{n+1}{2^n - (n+1)}$. Multivariate Kendalls tau has the form (see [12]),

$$\tau_C = h_\tau(k+1) \left(\int_{I^n} C(u_0, u_1, \ldots, u_k) dC(u_0, u_1, \ldots, u_k) - \frac{1}{2^{k+1}} \right),$$

where $h_\tau(n) = \frac{2^n}{2^{n-1}-1}$. The following theorem provides the expressions of multivariate concordance measures for the proposed family of copulas in Eq. (1).

Theorem 3.2 *Let $U = (U_0, U_1, \ldots, U_k)^T$ be a random vector of $(k+1)$ uniform random variables on $[0, 1]$ with associated copula $C \in \mathcal{C}$ in Eq. (1). Then the values of multivariate Spearman's rho and multivariate Kendall's tau are respectively given by,*

$$\rho_C = h_\rho(k+1) 2^{k+1} \prod_{i=0}^{k} \int_0^1 f_i(u_i) du_i,$$

and

$$\tau_C = h_\tau(k+1)((-1)^{k+1} + 1) \prod_{i=0}^{k} \int_0^1 f_i(u_i) du_i.$$

Proof By the definition of Spearman's rho,

$$\rho_C = h_\rho(k+1) \left(2^{k+1} \int_{I^n} C(u_0, u_1, \ldots, u_k) du_0 \ldots du_k - 1 \right)$$

$$= h_\rho(k+1) 2^{k+1} \left(\frac{1}{2^{k+1}} + \prod_{i=0}^{k} \int_0^1 f_i(u_i) du_i - 1 \right)$$

$$= h_\rho(k+1) 2^{k+1} \prod_{i=0}^{k} \int_0^1 f_i(u_i) du_i.$$

For multivariate Kendall's tau, we observe that $\int_0^1 f_i(u_i) f_i'(u_i) du_i = 0$, then

Table 1 Spearman's rho ρ for $p_0 = q_0 = 1$.

p_1, q_1	p_2, q_2								
	1, 1	1, 1.5	1, 2	1.5, 1	1.5, 1.5	1.5, 2	2, 1	2, 1.5	2, 2
1, 1	.037	.025	.019	.025	.016	.011	.019	.011	.007
1, 1.5	.025	.017	.013	.017	.011	.008	.013	.008	.005
1, 2	.019	.013	.009	.013	.008	.006	.009	.006	.004
1.5, 1	.025	.017	.013	.017	.011	.008	.013	.008	.005
1.5, 1.5	.016	.011	.008	.011	.007	.005	.008	.005	.003
1.5, 2	.011	.008	.006	.008	.005	.003	.006	.003	.002
2, 1	.019	.013	.009	.013	.008	.006	.009	.006	.004
2, 1.5	.011	.008	.006	.008	.005	.003	.006	.003	.002
2, 2	.007	.005	.004	.005	.003	.002	.004	.002	.001

$$
\tau_C = h_\tau(k+1)\left(\int_{I^n} C(u_0, u_1, \ldots, u_k)dC(u_0, u_1, \ldots, u_k) - \frac{1}{2^{k+1}}\right)
$$

$$
= h_\tau(k+1)\left(\int_{I^n} (\prod_{i=0}^{k} u_i + \prod_{i=0}^{k} f_i(u_i))(1 + \prod_{i=0}^{k} f_i'(u_i))du_0 \ldots du_k - \frac{1}{2^{k+1}}\right)
$$

$$
= h_\tau(k+1)((-1)^{k+1} + 1)\prod_{i=0}^{k}\int_0^1 f_i(u_i)du_i.
$$

Continuence of Example 2 1 Consider the copula given in Example 1,

$$
C(u_0, u_1, u_2) = u_0 u_1 u_2 + u_i^{p_0}(1 - u_0)^{q_0} u_1^{p_1}(1 - u_1)^{q_1} u_2^{p_2}(1 - u_2)^{q_2}.
$$

From Theorem 3.2, $\rho_C = 8 \prod_{i=0}^{2} Beta(p_i + 1, q_i + 1)$ and $\tau_C = 0$, where $Beta(a, b) = \int_0^1 t^{a-1}(1 - t)^{b-1}dt$. Table 1 shows the Spearman's rho for various combinations of p_i and q_i.

In the followings, we give the definitions of several well-known multivariate dependence properties and study them for the proposed family of copulas in Eq. (1).

Definition 3.1 [13] A $(k + 1)$-copula C is said to be **positive lower orthant dependent** (PLOD) if $C(u_0, u_1, \cdots, u_k) \geq \prod_{i=0}^{k} u_i$ holds for all $\boldsymbol{u} = (u_0, u_1, \ldots, u_k) \in [0, 1]^{k+1}$. For the case of $k = 1$, C is called **positively quadrant dependent** (PQD).

Theorem 3.3 *Let* $U = (U_0, U_1, \ldots, U_k)^T$ *be a random vector of* $(k+1)$ *uniform random variables on* $[0, 1]$ *with associated copula* $C \in \mathcal{C}$ *in Eq.* (1). *Then* U *is PLOD if and only if* $\prod_{i=0}^{k} f_i(u_i) \geq 0$.

Note that the copula $C_1(u_0, u_1, u_2) = u_0 u_1 u_2 + u_i^{p_0}(1 - u_0)^{q_0} u_1^{p_1}(1 - u_1)^{q_1} u_2^{p_2}$ $(1 - u_2)^{q_2}$ given in Example 1 is PLOD. Furthermore, the parametric copula $C_2(u_0, u_1, u_2) = u_0 u_1 u_2 + \theta u_i^{p_0}(1 - u_0)^{q_0} u_1^{p_1}(1 - u_1)^{q_1} u_2^{p_2}(1 - u_2)^{q_2}$ given in Example 1 is PLOD if and only if $\theta \geq 0$.

Definition 3.2 [2] Let $U = (U_0, U_1, \ldots, U_k)^T$ be a $(k+1)$-dimensional random vector and let the sets A and B be a partition of $\{0, 1, \ldots, k\}$, $U_A = (U_i : i \in A)$ and $U_B = (U_i : i \in B)$.

(1) U_B is **left tail decreasing** in U_A (denoted by $LTD(U_B|U_A)$) if $P(U_B \leq u_B|U_A \leq u_A)$ is non-increasing in each component of u_A for all u_B.
(2) U_B is **right tail increasing** in U_A (denoted by $RTI(U_B|U_A)$) if $P(U_B > u_B|U_A > u_A)$ is non-decreasing in each component of u_A for all u_B.
(3) U_B is **stochastically increasing** in U_A (denoted by $SI(U_B|U_A)$) if $P(U_B \leq u_B|U_A = u_A)$ is non-decreasing in each component of u_A for all u_B.

The next result shows the multivariate positive dependence properties for the proposed family of copulas in Eq. (1).

Theorem 3.4 *Let* $U = (U_0, U_1, \ldots, U_k)^T$ *be a random vector of* $(k+1)$ *uniform random variables on* $[0, 1]$ *with associated copula* $C \in \mathcal{C}$ *given in Eq.* (1), *where* $f_i(u_i)$, $i = 0, 1, \ldots, k$, *in* C *are assumed to be non-negative. Let the sets* A *and* B *be a partition of* $\{0, 1, \ldots, k\}$, $U_A = (U_i : i \in A)$ *and* $U_B = (U_i : i \in B)$. *Then,*

(i) $LTD(U_B|U_A)$ *if and only if* $f_i(u_i) \geq u_i f_i'(u_i)$ *for all* $u_i \in [0, 1]$ *and for every* $i \in A$.
(ii) $RTI(U_B|U_A)$ *if and only if* $f_i(u_i) \geq (u_i - 1) f_i'(u_i)$ *for all* $u_i \in [0, 1]$ *and for every* $i \in A$.
(iii) $SI(U_B|U_A)$ *if and only if* $(-1)^{k+1} f_i''(u_i) \prod_{j \in A \setminus \{i\}} f_j'(u_j) \geq 0$, *for all* $u_i, u_j \in [0, 1]$ *and for every* $i \in A$.

Proof (i) We observe that all the m-dimensional margins of the copula C_2 are independent copulas. We have

$$P(U_B \leq u_B|U_A \leq u_A) = \prod_{i \in B} u_i + \prod_{i \in B} f_i(u_i) \prod_{i \in A} [f_i(u_i)/u_i].$$

For any $j \in A$,

$$\frac{\partial P(U_B \leq u_B|U_A \leq u_A)}{\partial u_j} = \prod_{i \in B} f_i(u_i) \prod_{i \in A \setminus \{j\}} [f_i(u_i)/u_i] \left(\frac{f_j'(u_j)u_j - f_j(u_j)}{u_j^2} \right).$$

Thus, $f_j(u_j) \geq u_j f_j'(u_j)$ for all $u_j \in [0, 1]$ implies $P(U_B \leq u_B | U_A \leq u_A)$ is non-increasing in u_j, which completes the proof for (i).

(ii) The survival function $\bar{C}$ of U is $P(U_0 > u_0, U_1 > u_1, \ldots, U_k > u_k) = \prod_{i=0}^{k}(1 - u_i) + (-1)^{k+1} f_i(u_i)$. We have,

$$P(U_B > u_B | U_A > u_A) = \prod_{i \in B}(1 - u_i) + \left(\prod_{i \in B} f_i(u_i)\right)\left(\prod_{j \in A} \frac{f_j(u_j)}{(1 - u_j)}\right).$$

For any $j \in A$,

$$\frac{\partial P(U_B > u_B | U_A > u_A)}{\partial u_j} =$$

$$\left(\prod_{i \in B} f_i(u_i)\right)\left(\prod_{i \in A\setminus\{j\}} \frac{f_i(u_i)}{1 - u_i}\right)\left(\frac{f_j'(u_j)(1 - u_j) + f_j(u_j)}{u_j^2}\right).$$

Thus, $f_j(u_j) \geq (1 - u_j) f_j'(u_j)$ for all $u_j \in [0, 1]$ implies $P(U_B > u_B | U_A > u_A)$ is non-decreasing in u_j, which completes the proof for (ii).

(iii) Since the survival function $\bar{C}$ of U is $P(U_0 > u_0, U_1 > u_1, \ldots, U_k > u_k) = \prod_{i=0}^{k}(1 - u_i) + (-1)^{k+1} f_i(u_i)$, we have

$$P(U_B > u_B | U_A = u_A) =$$

$$(-1)^{|A|}\prod_{i \in B}(1 - u_i) + (-1)^{k+1}\left(\prod_{i \in B} f_i(u_i)\right)\left(\prod_{i \in A} f_i'(u_i)\right).$$

Thus, for any $j \in A$,

$$\frac{\partial P(U_B > u_B | U_A = u_A)}{\partial u_j} = (-1)^{k+1}\left(\prod_{i \in B} f_i(u_i)\right) f_j''(u_j)\left(\prod_{i \in A\setminus\{j\}} f_i'(u_i)\right)$$

which completes the proof for (iii).

Continuence of Example 3 1 Consider the copula $C_1(u_0, u_1, u_2) = u_0 u_1 u_2 + f_0(u_0) f_1(u_1) f_2(u_2)$, where $f_i(u_i) = u_i^{p_i}(1 - u_i)^{q_i}$ and $p_i \geq 1, q_i \geq 1$, for $i = 0, 1, 2$. Let $A = \{1, 2\}$ and $B = \{0\}$. From Theorem 3.4, it is easy to see that C is $LTD(U_B | U_A)$ if and only if $p_1 = p_2 = 1$, and C is $RTI(U_B | U_A)$ if and only if $q_1 = q_2 = 1$. Furthermore, $f_i'(u_i) = u_i^{p_i-1}(1 - u_i)^{q_i-1}(p_i - (p_i + q_i)u_i)$ could be negative or positive for any p_i and q_i, therefore, C is not $SI(U_B | U_A)$ for any $p_i \geq 1$ and $q_i \geq 1$. The same results apply for the parametric copula proposed in Corollary 2.1, $C_2(u_0, u_1, u_2) = u_0 u_1 u_2 + \theta u_i^{p_0}(1 - u_0)^{q_0} u_1^{p_1}(1 - u_1)^{q_1} u_2^{p_2}(1 - u_2)^{q_2}$ when $\theta \geq 0$.

Note that the copulas C_1 and C_2 are exchangeable if and only if $p_0 = p_1 = p_2$ and $q_0 = q_1 = q_2$.

In the following definition of affiliated copulas, for column vectors $x = (x_0, \ldots, x_k)^T$, $y = (y_0, \ldots, y_k)^T \in R^{k+1}$, let $x \vee y = (\max\{x_0, y_0\}, \ldots, \max\{x_n, y_n\})^T$ and $x \wedge y = (\min\{x_0, y_0\}, \ldots, \min\{x_k, y_k\})^T$. A function $f : R^n \to R$ is said to be multivariate totally positive of order two(MTP2) if it satisfies $f(x)f(y) \leq f(x \vee y)f(x \wedge y)$, for $x, y \in R^n$. MTP2 is an interesting property studied by [6]. The importance of the affiliation properties in application of auction theory can be found in [10]. It is true that the random vector X is affiliated if and only if its corresponding copula is affiliated. The affiliation is a strong positive dependence property among the elements of a random vector, a copula C is affiliated implies C is associated and PLOD [22].

Definition 3.3 A copula $C(u_0, \cdots, u_k)$ is said to be **affiliated** if for all $u = (u_0, \ldots, u_k)^T$ and $v = (v_0, \ldots, v_k)^T$ in $[0, 1]^{k+1}$,

$$c(u)c(v) \leq c(u \vee v)c(u \wedge v) \tag{4}$$

holds where $c(u_0, \cdots, u_k) = \frac{\partial^{k+1} C(u_0, \cdots, u_k)}{\partial u_0 \ldots \partial u_k}$ is the copula density function.

The following result gives a characterization of the affiliation property for the proposed family of copulas in Eq. (1).

Theorem 3.5 *Let $U = (U_0, U_1, \ldots, U_k)^T$ be a random vector of $(k + 1)$ uniform random variables on $[0, 1]$ with associated copula $C \in \mathcal{C}$ in Eq. (1), where $f_i(u_i)$, $i = 0, 1, \ldots, k$, in C are non-negative. Then, U is affiliated if and only if for any partition A and B of $\{0, 1, \ldots, k\}$, $U_A = (U_i : i \in A)$ and $U_B = (U_i : i \in B)$,*

$$f_i''(u_i) \prod_{j \in A \setminus \{i\}} f_j'(u_j) \quad and \quad f_i''(u_i) \prod_{j \in B \setminus \{i\}} f_j'(u_j)$$

are both positive or both negative for all $u_i, u_j \in [0, 1]$ and for every i in A and B.

Proof Let $u = (u_0, \ldots, u_k)^T$ and $v = (v_0, \ldots, v_k)^T$ in $[0, 1]^{k+1}$. Let $A = \{i \mid \max (u_i, v_i) = u_i\}$ and $B = \{i \mid \max(u_i, v_i) = v_i\}$ be the partition for $\{0, 1, \ldots, k\}$. We can rewrite the density functions as follows,

$$c(u \vee v) = 1 + \prod_{\substack{i \in A \\ j \in B}} f_i'(u_i)f_j'(v_j) \text{ and } c(u \wedge v) = 1 + \prod_{\substack{i \in A^C \\ j \in B^C}} f_i'(u_i)f_j'(v_j).$$

Thus,

$$c(\boldsymbol{u} \vee \boldsymbol{v})c(\boldsymbol{u} \wedge \boldsymbol{v}) - c(\boldsymbol{u})c(\boldsymbol{v})$$

$$= \left(1 + \prod_{i \in A, j \in B} f_i'(u_i)f_j'(v_j)\right)\left(1 + \prod_{i \in A^C, j \in B^C} f_i'(u_i)f_j'(v_j)\right)$$

$$- \left(1 + \prod_{i=1}^{n} f_i'(u_i)\right)\left(1 + \prod_{i=1}^{n} f_i'(v_i)\right)$$

$$= \left(1 + \prod_{i=1}^{n} f_i'(u_i)f_i'(v_i) + \prod_{i \in A^C, j \in B^C} f_i'(u_i)f_j'(u_j) + \prod_{i \in A, j \in B} f_i'(u_i)f_j'(u_j)\right)$$

$$- \left(1 + \prod_{i=1}^{n} f_i'(u_i)f_i'(v_i) + \prod_{i=1}^{n} f_i'(u_i) + \prod_{i=1}^{n} f_i'(v_i)\right)$$

$$= \left(\prod_{i \in A, j \in B} f_i'(u_i)f_j'(v_j) + \prod_{i \in A^C, j \in B^C} f_i'(u_i)f_j'(v_j) - \prod_{i=1}^{n} f_i'(u_i) - \prod_{i=1}^{n} f_i'(v_i)\right)$$

$$= \left(\prod_{i \in A} f_i'(u_i) - \prod_{i \in A} f_i'(v_i)\right)\left(\prod_{j \in B} f_j'(v_j) - \prod_{j \in B} f_j'(u_j)\right)$$

Thus, $f_i''(u_i) \prod_{j \in A \setminus \{i\}} f_j'(u_j)$ and $f_i''(u_i) \prod_{j \in B \setminus \{i\}} f_j'(u_j)$ are both positive or both negative for all $u_i, u_j \in [0, 1]$ and for every i in A and B if and only if $c(\boldsymbol{u} \vee \boldsymbol{v})c(\boldsymbol{u} \wedge \boldsymbol{v}) - c(\boldsymbol{u})c(\boldsymbol{v}) \geq 0$, which completes the proof.

Continuence of Example 4 1 Consider the copula $C(u_0, u_1, u_2) = u_0 u_1 u_2 + f_0(u_0) f_1(u_1) f_2(u_2)$, where $f_i(u_i) = u_i^{p_i}(1 - u_i)^{q_i}$ and $p_i \geq 1, q_i \geq 1$, for $i = 0, 1, 2$. Let $A = \{0\}$ and $B = \{1, 2\}$. Note that the second derivative of $f_i(u_i)$ is $f_i''(u_i) = u_i^{p_i - 2}(1 - u)^{q_i - 2}\{[(p_i - 1) - (p_i + q_i - 2)u_i][p_i - (p_i + q_i)u_i] - (p_i + q_i)u_i(1 - u_i)\}$. It is easy to see if $p_i = q_i = 1$ for $i = 1, 2$, $f_1''(u_1)f_2'(u_2) = -2(1 - 2u_2)$ could be either positive or negative. Therefore, C is not affiliated in this case. Also note that $[(p_i - 1) - (p_i + q_i - 2)u_i][p_i - (p_i + q_i)u_i] - (p_i + q_i)u_i(1 - u_i)$ is a quadratic function of u_i which could be either positive or negative for all cases when $p_i > 1, q_i > 1$, $p_i = 1, q_i > 1$, and $p_i > 1, q_i = 1$. Thus, by Theorem 3.5, the copula C is not affiliated for any $p_i \geq 1$ and $q_i \geq 1$.

The following Example is a trivariate extension of the Example 2.1 given in [19].

Example 2 Let m_0, m_1, m_2 be positive real numbers such that $m_0 m_1 m_2 > 1$, and let

$$f_0(u_0) = \begin{cases} m_0 u_0, & 0 \leq u_0 \leq \frac{1}{m_0 m_1 m_2}, \\ \frac{1 - u_0}{m_1 m_2}, & \frac{1}{m_0 m_1 m_2} \leq u_0 \leq 1, \end{cases} \qquad f_1(u_1) = \begin{cases} m_1 u_1, & 0 \leq u_1 \leq \frac{1}{m_0 m_1 m_2}, \\ \frac{1 - u_1}{m_0 m_2}, & \frac{1}{m_0 m_1 m_2} \leq u_1 \leq 1, \end{cases}$$

and

$$f_2(u_2) = \begin{cases} m_2 u_2, & 0 \leq u_2 \leq \frac{1}{m_0 m_1 m_2}, \\ \frac{1-u_2}{m_0 m_1}, & \frac{1}{m_0 m_1 m_2} \leq u_2 \leq 1. \end{cases}$$

In this case, $\alpha_0 = -\frac{1}{m_1 m_2}$, $\alpha_1 = -\frac{1}{m_0 m_2}$, $\alpha_2 = -\frac{1}{m_0 m_1}$, and $\beta_i = m_i$ for $i = 0, 1, 2$. By Proposition 2.1, we know that the function defined by $C(u_0, u_1, u_2) = u_0 u_1 u_2 + f_0(u_0) f_1(u_1) f_2(u_2)$ is a copula. Furthermore, by Theorem 3.5, we know that C is affiliated.

4 Multiple Regression Using Non-exchangeable Multivariate Copula

Sungur [17, 18] studied the directional dependence through the biavriate non-exchangeable copula-based regression. However, the analysis of directional dependence are often needed in the setting of multivariate data. This section shows how the generalized multivariate FGM copulas can be used in multivariate regression setting and proposes the simple-to-use multiple regression formulas based on the proposed generalized multivariate FGM copulas.

We first define the copula-based multiple regression function. For a $(k + 1)$-copula $C(u_0, u_1, \ldots, u_k)$ of the $(k + 1)$ uniform random variates U_0 and $\boldsymbol{U} = (U_1, \ldots, U_k)^T$, the **copula-based multiple regression function** of U_0 on $\boldsymbol{U} = \boldsymbol{u}$ is defined by

$$r^C_{U_0|\boldsymbol{U}}(\boldsymbol{u}) \equiv E(U_0|\boldsymbol{U} = \boldsymbol{u}) = \int_0^1 u_0 \frac{c(u_0, u_1, \ldots, u_k)}{c_{\boldsymbol{U}}(\boldsymbol{u})} du_0 \qquad (5)$$

$$= 1 - \frac{1}{c_{\boldsymbol{U}}(\boldsymbol{u})} \int_0^1 C_{U_0|\boldsymbol{U}}(u_0) du_0,$$

where $C_{\boldsymbol{U}}(\boldsymbol{u}) = C(1, u_1, \ldots, u_k)$ is the marginal distribution of $\boldsymbol{U}$, and

$$c(u_0, u_1, \ldots, u_k) = \frac{\partial^{k+1} C(u_0, u_1, \ldots, u_k)}{\partial u_0 \partial u_1 \cdots \partial u_k}, \quad c_{\boldsymbol{U}}(\boldsymbol{u}) = \frac{\partial^k C_{\boldsymbol{U}}(\boldsymbol{u})}{\partial u_1 \cdots \partial u_k},$$

$$C_{U_0|\boldsymbol{U}}(u_0) = \frac{\partial^k C(u_0, u_1, \ldots, u_k)}{\partial u_1 \cdots \partial u_k}$$

are the joint copula density of U_0 and $\boldsymbol{U}$, the marginal copula density of $\boldsymbol{U}$ and the conditional distribution of U_0 given $\boldsymbol{U}$, respectively.

The following proposition gives a basic property of the copula-based multiple regression function for uniform random variates.

Proposition 4.1 *[23] Let $r^C_{U_0|U}(\boldsymbol{u})$ be the copula-based multiple regression function of U_0 on $\boldsymbol{U}$. Then*

$$E[r^C_{U_0|U}(\boldsymbol{U})] = \frac{1}{2}.$$

In order to quantitatively measure the directional dependence given the copula-based multiple regression, say $r^C_{U_0|U}(\boldsymbol{\omega})$, [23] proposed the following multiple dependence measure,

$$\rho^2_{(U \to U_0)} = \frac{Var(r^C_{U_0|U}(\boldsymbol{U}))}{Var(U_0)} = \frac{E[(r^C_{U_0|U}(\boldsymbol{U}) - 1/2)^2]}{1/12} \tag{6}$$
$$= 12E[(r^C_{U_0|U}(\boldsymbol{U}))^2] - 3.$$

If $U_0, U_1, \ldots, U_k$ are independent, then the copula-based multiple regression function in Eq. (5) is equal to $1/2$ (e.g., $r^C_{U_0|U}(\boldsymbol{u}) = 1/2$) and so the corresponding dependence measure in Eq. (6) is zero.

Using the proposed multivariate non-exchangeable copula in Eq. (1), the theorem below gives the closed forms of copula-based regression functions.

Theorem 4.1 *For the proposed copula $C \in \mathcal{C}$ given in Eq. (1), the copula regression function is*

$$r^C_{U_0|U}(\boldsymbol{u}) = \frac{1}{2} - \prod_{i=1}^k f'_i(u_i) \int_0^1 f_0(u_0) du_0.$$

Proof For the generalized FGM copula $C \in \mathcal{C}$ family given in (1), we have

$$\frac{\partial^k C(u_0, \ldots, u_k)}{\partial u_1 \ldots \partial u_k} = \frac{\partial^k}{\partial u_1 \ldots \partial u_k}\left(\prod_{i=0}^k u_i + \prod_{i=0}^k f_i(u_i)\right) = u_0 + f_0(u_0)\prod_{i=1}^k f'_i(u_i).$$

Note that the marginal copula density $c_U(\boldsymbol{u}) = 1$. Therefore, the copula based regression function is

$$r^C_{U_0|U}(u_1, \ldots, u_k) = 1 - \frac{1}{c_U(\boldsymbol{u})}\int_0^1 \left(u_0 + f_0(u_0)\prod_{i=1}^k f'_i(u_i)\right) du_0$$
$$= \frac{1}{2} - \prod_{i=1}^k f'_i(u_i)\int_0^1 f_0(u_0) du_0.$$

Continuence of Example 5 1

Let $f_i(u_i) = u_i^{p_i}(1 - u_i)^{q_i}$, $i = 0, 1, 2$. Consider the copula C is given by

$$C(u_0, u_1, u_2) = u_0 u_1 u_2 + u_0^{p_0}(1 - u_0)^{q_0} u_1^{p_1}(1 - u_1)^{q_1} u_2^{p_2}(1 - u_2)^{q_2}.$$

From Theorem 4.1, the closed forms of the copula-based regression functions are

$$r_{U_0|U_1,U_2}^C (u_1, u_2) = \frac{1}{2} - Beta(p_0 + 1, q_0 + 1) \prod_{i=1}^{2} (1 - u_i)^{q_i-1}[p_i - (p_i + q_i)u_i],$$

$$r_{U_1|U_0,U_2}^C (u_0, u_2) = \frac{1}{2} - Beta(p_1 + 1, q_1 + 1) \prod_{i=0,2} (1 - u_i)^{q_i-1}[p_i - (p_i + q_i)u_i],$$

$$r_{U_2|U_0,U_1}^C (u_0, u_1) = \frac{1}{2} - Beta(p_2 + 1, q_2 + 1) \prod_{i=0}^{1} (1 - u_i)^{q_i-1}[p_i - (p_i + q_i)u_i].$$

We also obtain the closed-form formulas for the dependence measures in Eq. (6) for the copula-based regression functions derived above,

$$\rho_{(U_1,U_2 \to U_0)}^2 = 12 Beta(p_0 + 1, q_0 + 1)^2 g(p_1, q_1) g(p_2, q_2),$$
$$\rho_{(U_0,U_2 \to U_1)}^2 = 12 Beta(p_1 + 1, q_1 + 1)^2 g(p_0, q_0) g(p_2, q_2),$$
$$\rho_{(U_0,U_1 \to U_2)}^2 = 12 Beta(p_2 + 1, q_2 + 1)^2 g(p_0, q_0) g(p_1, q_1),$$

where $g(p, q) = p^2 Beta(2p - 1, 2q + 1) - 2pq Beta(2p, 2q) + q^2 Beta(2p + 1, 2q - 1)$. Tables 2 and 3 show the dependence measures of $\rho_{(U_1,U_2 \to U_0)}^2$ for various p_i and q_i.

Table 2 $\rho_{(U_1,U_2 \to U_0)}^2$ for $p_1 = q_1 = p_2 = q_2 = 1$.

q_0	p_0		
	1	1.5	2
1	0.037	0.017	0.009
1.5	0.017	0.007	0.003
2	0.009	0.003	0.002

Table 3 $\rho^2_{(U_1,U_2 \rightarrow U_0)}$ for $p_0 = q_0 = 1$.

p_1, q_1	p_2, q_2								
	1, 1	1, 1.5	1, 2	1.5, 1	1.5, 1.5	1.5, 2	2, 1	2, 1.5	2, 2
1, 1	.037	.021	.015	.021	.008	.005	.015	.005	.002
1, 1.5	.021	.012	.008	.012	.005	.003	.008	.003	.001
1, 2	.015	.008	.006	.008	.003	.002	.006	.002	.001
1.5, 1	.021	.012	.008	.012	.005	.003	.008	.003	.001
1.5, 1.5	.008	.005	.003	.005	.002	.003	.003	.001	.001
1.5, 2	.005	.003	.002	.003	.001	.001	.002	.001	0
2, 1	.015	.008	.006	.008	.003	.002	.006	.002	.001
2, 1.5	.005	.003	.002	.003	.001	.001	.002	.001	0
2, 2	.002	.001	.001	.001	.001	0	.001	0	0

References

1. Dudford T, Cooke RM (2002) Vines: a new graphical model for dependent random variables. Ann Stat 1031–1068
2. Joe H (1997) Multivariate models and dependence concepts. Chapman & Hall, London
3. Kim JM, Jung YS, Sungur EA, Han KH, Park C, Sohn I (2008) A copula method for modeling directional dependence of genes. BMC Bioinformatics 9(1):1–12
4. Kim D, Kim J (2014) Analysis of directional dependence using asymmetric copula-based regression models. J Stat Comput Simul 84:1990–2010
5. Kim S, Kim D (2015) Directional dependence analysis using skew-normal copula- based regression. In: Statistics and Causality: methods for Applied Empirical Research. John Wiley & Sons
6. Karlin S, Rinott Y (1980) Classes of orderings of measures and related correlation inequalities. I. Multivariate totally positive distributions. J Multivar Anal 10(4):467–498
7. Liebscher E (2008) Construction of asymmetric multivariate copulas. J Multivar Anal 99(10):2234–2250
8. Mai J, Scherer M (2014) Financial engineering with copulas explained. Springer, Palgrave Macmillan
9. McNeil AJ, Nešlehová J (2010) From archimedean to liouville copulas. J Multivar Anal 101(8):1772–1790
10. Milgrom PR, Weber RJ (1982) A theory of auctions and competitive bidding. Econometrica J Econometric Soc 1089–1122
11. Nelsen RB, Quesada Molina JJ, Rodriguez-Lallena JA (1997) Bivariate copulas with cubic sections. J Nonparametr Statist 7:205–220
12. Nelsen RB (2002) Concordance and copulas: a survey, in distributions with given marginals and statistical modelling. Kluwer Academic Publishers, Dordrecht
13. Nelsen RB (2007) An introduction to copulas. Springer, New York
14. Noh H, El Ghouch A, Bouezmarni T (2013) Copula-based regression estimation and inference. J Am Stat Assoc 108:676–688
15. Quesada Molina JJ, Rodriguez-Lallena JA (1995) Bivariate copulas with quadratic sections. J Nonparametr Stat 5:323–337
16. Schmid F, Schmidt R (2007) Multivariate extensions of Spearman's rho and related statistics. Stat Probab Lett 77(4):407–416
17. Sungur EA (2005a) A note on directional dependence in regression setting. Commun Stat Theor Methods 34:1957–1965

18. Sungur EA (2005b) Some observations on copula regression functions. Dempster Commun Stat Theor Methods 34:1967–1978
19. Úbeda-Flores M (2004) A new class of bivariate copulas. Stat Probab Lett 66(3):315–325
20. Wei Z, Kim S, Kim D (2016) Multivariate skew normal copula for non-exchangeable dependence. Procedia Comput Sci 91:141–150
21. Wei Z, Wang T, Panichkitkosolkul W (2014) Dependence and association concepts through copulas. In: Modeling dependence in econometrics. Springer International Publishing
22. Wei Z, Wang T, Nguyen PA (2015) Multivariate dependence concepts through copulas. Int J Approximate Reasoning 65:24–33
23. Wei Z, Wang T, Kim D (2016) Multivariate non-exchangeable copulas and their application to multiple regression-based directional dependence. In: Causal inference in econometrics. Springer
24. Wei Z, Wang T, Li B (2016) On consistency of estimators based on random set vector observations. In: Causal inference in econometrics. Springer

Part III
Applications

Key Economic Sectors and Their Transitions: Analysis of World Input-Output Network

T.K. Tran, H. Sato and A. Namatame

Abstract In the modern society, all major economic sectors have been connected tightly in an extremely complicated global network. In this type of network, a small shock occurred at certain point can be spread instantly through the whole network and may cause catastrophe. Production systems, traditionally analyzed as almost independent national systems, are increasingly connected on a global scale. The world input-output database, only recently becoming available, is one of the first efforts to construct the global and multi-regional input-output tables. The usual way of identifying key sectors in an economy in Input-output analysis is using Leontief inverse matrix to measure the backward linkages and the forward linkages of each sector. In other words, evaluating the role of sectors is performed by means of their centrality assessment. Network analysis of the input-output tables can give valuable insights into identifying the key industries in a world-wide economy. The world input-output tables are viewed as complex networks where the nodes are the individual industries in different economies and the edges are the monetary goods flows between industries. We characterize a certain aspect of centrality or status that is captured by the network measure. We use an α-centrality modified method to the weighted directed network. It is used to identify both how a sector could be affected by other sectors and how it could infect the others in the whole economy. The data used is the world input-output table, part of the world input-output database (WIOD) funded by European Commission from 1995 to 2011. We capture the transition of key industries over years through the network measures. We argue that the network structure captured from the input-output tables is a key in determining whether and how microeconomic expansion or shocks propagate throughout the whole economy and shape aggregate outcomes. Understanding the network structure of world input-output data can better inform on how the world economy grows as well as how to prepare for and recover from adverse shocks that disrupt the global production chains. Having analyzed these

T.K. Tran · H. Sato · A. Namatame (✉)
Department of Computer Science, National Defense Academy, Yokosuka, Japan
e-mail: em54051@nda.ac.jp

© Springer International Publishing AG 2017   381
V. Kreinovich et al. (eds.), *Robustness in Econometrics*,
Studies in Computational Intelligence 692, DOI 10.1007/978-3-319-50742-2_23

results, the trend of these sectors in that range of time will be used to reveal how the world economy changed in the last decade.

Keywords Production network · α Centrality · Amplification Index · Vulnerability Index · Key industrial sector · World input output network

1 Introduction

In the modern society, all major economic sectors have been connected tightly in an extremely complicated global network. In this type of network, a small shock occurred at certain point can be spread instantly through the whole network and may cause catastrophe. The usual way of identifying key sectors in an economy in Input-output analysis is using Leontief inverse matrix to measure the backward linkages and the forward linkages of each sector. The input-output table initially formalized by Leontief [12] has used extensively by economists, environmentalists, and policy makers. By keeping track of the inter-industrial relationships, the input-output table offers a reasonably accurate measurement of the response of any given economy in the face of external shocks or policy interventions.

The fundamental underlying relationship of input-output analysis proposed by Leontief is that the amount of a product (good or service) produced by a given sector in the economy is determined by the amount of that product that is purchased by all the users of the product. By its nature, input-output analysis encompasses all the formal market place activity that occurs in an economy, including the service sector which is frequently poorly represented. Consequently, input-output analysis frequently plays a fundamental role in the construction of the national accounts. In effect, an input-output model provides a snapshot of the complete economy and all of its industrial interconnections at one time. The power of the model is that it can show the distribution of overall impacts. A column of the total requirements table indicates which sectors in the region will be affected and by what magnitude. This can be used to make important policy decisions when translated into income and employment effects. Policy makers can use the information derived from the model to identify an industrial growth target and others.

Today input-output analysis has become important to all the highly-industrialized countries in economic planning and decision making because of this flow of goods and services that it traces through and between different industries. Input-output analysis is capable of simulating almost any conceivable economic impact. The nature of input-output analysis makes it possible to analyze the economy as an interconnected system of industries that directly and indirectly affect one another, tracing structural changes back through industrial interconnections. This is especially important as production processes become increasingly complex, requiring the interaction of many different businesses at the various stages of a product's processing. Input-output techniques trace these linkages from the raw material stage to the sale of the product as a final, finished good. This allows the decomposition analysis to account for the fact that

a decline in domestic demand. In analyzing an economy's reaction to changes in the economic environment, the ability to capture the indirect effects of a change is a unique strength of input-output analysis. One of the interests in the field of input-output economics lies with the fact that it is very concrete in its use of empirical data.

Alternatively, Acemoglu et al. [1] and Carvalho [10] argue that the structure of the production network is a key in determining whether and how microeconomic shocks propagate throughout the economy and shape aggregate outcomes. Therefore, understanding the structure of the production network can better inform on the origins of aggregate fluctuations and policymakers on how to prepare for and recover from adverse shocks that disrupt these production chains. The usual way of identifying key sectors in an economy in Input-output analysis is using Leontief inverse matrix to measure the backward linkages and the forward linkages of each sector. Alternatively, they evaluate the role of sectors by means of network measures such as degree centrality and α-centrality.

All changes in the endogenous sectors are results of changes in the exogenous sectors. The input-output analysis also allows a decomposition of structural change which identifies the sources of change as well as the direction and magnitude of change. Most importantly, an input-output based analysis of structural change allows the introduction of a variable which describes changes in producer's recipes—that is, the way in which industries are linked to one another, in input-output language, called the "technology" of the economy. It enables changes in output to be linked with underlying changes in factors such as exports, imports, domestic final demand as well as technology. This permits a consistent estimation of the relative importance of these factors in generating output and employment growth. In a general sense, the input-output technique allows insight into how macroeconomic phenomena such as shifts in trade or changes in domestic demand correspond to microeconomic changes as industries respond to changing economic conditions.

Production systems, traditionally analyzed as almost independent national systems, are increasingly connected on a global scale. As the global economy becomes increasingly integrated, an isolated view based on the national input-output table is no longer sufficient to assess an individual economy's strength and weakness, not to mention finding solutions to global challenges such as climate change and financial crises. Hence, a global and multi-regional input-output data is needed to draw a high-resolution representation of the global economy. Only recently becoming available, the World Input-Output Database (WIOD) is one of the first efforts to construct the global multi-regional input-output (GMRIO) tables. By viewing the world input-output system as an interdependent network where the nodes are the individual industries in different economies and the edges are the monetary goods flows between industries. Cerina et al. [11] analyzed the network properties of the so-called world input-output network (WION) and investigate its evolution over time. At global level, we find that the industries are highly but asymmetrically connected, which implies that micro shocks can lead to macro fluctuations. We also propose the network-based measures and these can give valuable insights into identifying the key industries.

In the modern economy, industry sectors have specific roles in an extremely complicated linked network despite of their size or range of effect. Since the linkage structure in the economy is considered to be dominated by a small group of sectors (key sectors) that connect to other different sectors in different supply chains, even a small shock originated from any firm could be conducted through the network and cause the significant impacts to the whole economy. Hence, to identify the sectors that belong to the such kind of hub group in the economy, it is not only based on the sector's output production or how much resource is used, but also its influence to all other nodes throughout the whole economy network, as well as its own external impact. These sectors play very important role in the whole economy since knowing them will help the policymakers actively preparing for and recovering from the impact of them to the economy. Traditionally, some network measurements are used to identified the key sectors such as the high forward and backward linkages with the rest of the economy, and most of these methodologies consider only the direct input or output coefficient (weight) of the sectors as the basis to determine sector's importance. There are two examples that use these methods to identify the key sectors of the economy; one is Alatriste-Contreras [2] used forward and backward to identify the major sectors of EU economy; and the other is Botri [8] who identified the key sectors of the Croatian economy.

In regard to key sectors, the first thought is that they are very important to the whole economy. However, some sectors or firms will mostly influence to the other sectors, and in the same way some of them might be the mostly affected from the other sectors. The economy is the very complicated linkages of different supply chains, which involve companies, people, activities, information producing, handling and/or distributing a specific product to the end users (customers). These supply chains are being connected together by means of some very specific industries. That is, if there is any economy shock originated from these key sectors, it will propagate throughout the economy and influences to the production of all other firms [10]. These key industries are also known as the hub sectors that shorten the distance between unrelated sectors in the economy. They provide the bridges for the separated parts which do not have direct trade inputs entire the economy. Therefore, the aggregate performance of the network also could be contributed by these kind of sectors as the shock from anywhere in the network may be conducted via them.

This paper aims to provide the different methods to identify the key economic sectors that most contribute to the economy based on the sector's influence scores to other nodes. These influence scores are calculated regarding the supply and consume from the input-output network. These scores do not depend much on the economic sector's direct transactions, but its relationship with the others throughout the whole network and its own external influence. In general, if there is any shock originating from one of these key economy sectors, it will be propagated through the entire the economy network via its links to the others whether its transaction is high or not. The introduced methods are developed based on the measurement of α-centrality. Two types of measurement are proposed: Amplification Index (AI) and Vulnerability Index (VI). The AI score is a measurement of influence to others, that is, how each economic sector influences the other economic sectors. The VI score measures influences from the other economic sectors, that is, it measures the impact that a sector receives from all other sectors. These scores also vary according to the value of a specific parameter that is the capital coefficient. A dataset of the world input-output network, which conducted by a project of European Commission, were used to demonstrate these methods. This dataset is the collection of intermediate matrixes that contains relationships between the industries in each economy and between the economies in the world in 17 years. For each year data, AI and VI are calculated, and then from those scores, a list of key economic sectors of the world economy is identified and their transitions over years are also traced. These results are compared with the results from other well-known measurement such as eigenvector centrality.

2   A Model of Input-Output Network

Consider an economy where production takes place at N distinct nodes, each specializing in a different good. These goods can be used as an intermediate input to be deployed in the production of other goods. A natural interpretation for these production nodes is to equate them with the different sectors of an economy. They assume that the production process at each of these sectors is well approximated by a Cobb-Douglas technology with constant returns to scale, combining a primary factor—which in this case is labor—and inter-mediate inputs. The output of sector i is then given by: Let's begin with the networks of input flows. In an economy, an industry's production Y is computed based on the investment in capital K and labor L. The Cobb-Douglas production is defined as

$$Y = F(L, K) = AK^{\alpha}L^{\beta} \tag{1}$$

Where:

  Y: total production (the real value of all goods produced in a year)
  L: labor input (the total number of person-hours worked in a year)
  K: capital input (the real value of all machinery, equipment, and buildings)
  A: total factor productivity

α and β are the output elasticity of capital (K) and labor (L), respectively. These values are constants determined by available technology.

  The basic input-output analysis assumes constant returns to scale, the change of output subsequent to a proportional change in all inputs. The input-output model assumes that the same relative mix of inputs will be used by an industry to create output regardless of quantity. Therefore in this case, $\alpha + \beta = 1$. The different values of α and A are selected depends on the specific economy and its current status. For example, in 2014, in the top positions of businesses listed in Tokyo Stock Exchange, this formula above was used in regard to about 1000 manufacturing industries, α is estimated as 0.121 and A = 0.081 [3].

  Acemoglu et al. [1] and Carvalho [10] develop a unified framework for the study of how network interactions can function as a mechanism for propagation and amplification of microeconomic shocks. The framework nests various classes of games over networks, models of macroeconomic risk originating from microeconomic shocks, and models of financial interactions. Under the assumption that shocks are small, they provide a fairly complete characterization of the structure of equilibrium, clarifying the role of network interactions in translating microeconomic shocks into macro-economic outcomes. Using Cobb-Douglas production function in Eq. 1, Acemoglu et al. [1] obtained the output of an economic sector i as:

$$x_i = (z_i l_i)^{1-\alpha} \left(\prod_{j=1}^{N} x_{ji}^{\omega_{ji}} \right)^{\alpha} \tag{2}$$

The first term in Eq. 2 shows the contribution from primary factors to production. The amount of labor hired by sector i is given by l_i, z_i is a sector specific productivity disturbance, and $1 - \alpha$ is the share of labor in production and α is the share of capital.

These interconnections between production nodes come into play with the second term of the production function, which reflects the contribution of intermediate inputs from other sectors. Thus, the term x_{ij} denotes the amount of good j used in the production of good i. The exponent ω_{ij} (≥ 0) in the production function gives the share of good j in the total intermediate input used by sector i. For a given sector i, the associated list of ω_{ij}'s thus encodes a sort of production recipe. Each nonzero element of this list singles out a good that needs to be sourced in order to produce good i. Whenever a ω_{ij} is zero, we are simply stating that sector i cannot usefully incorporate j as input in production, no matter what input prices sector i is currently facing. Note further that all production technologies are, deliberately, being kept largely symmetric: all goods are equally valued by final consumers and all production technologies are equally labor-intensive (specifically, they all share the same α). The only difference across sectors then lies in the bundle of intermediate inputs specified by their production recipe—that is, which goods are necessary as inputs in the production process of other goods.

When we stack together all production recipes in the economy, we obtain a collection of N lists, or rows, each row giving the particular list of ω_{ij}'s associated with the production technology in sector i. This list-of-lists is nothing other than an input-output matrix, W, summarizing the structure of intermediate input relations in this economy. The production network, W, which is the central object of this paper, is then defined by three elements: (i) a collection of N vertices or nodes, each vertex corresponding to one of the sectors in the economy; (ii) a collection of directed edges, where an edge between any two vertices denotes an input-supplying relationship between two sectors; and (iii) a collection of weights, each of which is associated with a particular directed edge and given by the exponent ω_{ij} in the production function.

In this paper; we focus on this matrix to find out the list of what it is called the hub-like unit or key economic sector.

3 Centrality Measures

One of the key concepts in network analysis is the notion of node centrality, which defines as the importance of a node due to its structural position in the network as a whole. Several centrality measures have been defined. Identifying the central input-supplying technologies and ranking their roles in an economy requires applying an appropriate measure of "node centrality" to the production network. While network analysis has developed a variety of centrality measures, here we will focus on so-called "influence measures" of centrality, where nodes are considered to be relatively more central in the network if their neighbors are themselves well-connected nodes.

The best known of these recursively defined centrality measures is called "eigenvector centrality." One of the best-known types of centrality is eigenvector centrality [4]. The eigenvector captures a certain aspect of centrality or status that is not captured by other measures. The idea here is that a node that is connected to nodes that are themselves well connected should be considered more central than a node that is connected to an equal number of less connected nodes. For instance, consider two firms, each with ten customers. Suppose industry A's directly connected industries have many direct connection industries of their own, and those industries have many direct connection industries and so on. Economic sector A's actions potentially affect a great number of other industries downstream. In contrast, if industry B's directly connected industries do not have many direct connection industries of their own, B's actions could have much less effect on the economic system as a whole. Thus, the eigenvector concept takes into account both direct and indirect influences. Variants of eigenvector have been deployed in the sociology literature, notably Eigenvector centrality [4] and Katz Centrality [9], in computer science with Google's PageRank algorithm [5]. Thus, as in the example above, an industry's centrality need not be dictated by its out-degree (or in-degree) alone, but will also be determined by its direct connections' out-degree.

Bonacich et al. [6] introduced α-centrality to address a problem of evaluating key nodes using eigenvalue centrality with an asymmetric network. Unlike eigenvector centrality, α centrality is also appropriated for certain classes of directed networks. In this measure, each node is considered having its own exogenous source that does not depend on other individual in the network. α-centrality expresses the centrality of a node as the number of paths linking it to other nodes, exponentially attenuated by their length. It is defined as Eq. (3) and matrix notation is given in Eq. (4).

$$x_i = \alpha A^T x_i + e \tag{3}$$

$$x = (I - \alpha A^T)^{-1} e \tag{4}$$

if node I does not have a tie to node j, node I still influence node j via other intermediate nodes between them. Therefore, we can also rewrite this centrality as an accumulation of its centrality along with time:

$$x = \left(\sum_{i=0}^{\infty} \alpha^i A^{Ti} \right) e = (1 + \alpha A^T + (\alpha A^T)^2 + \ldots + (\alpha A^T)^t + \ldots)e \qquad (5)$$

In these equations, x_i is node's α centrality or influence of node i. e is a vector of exogenous source of information, A^T is the transpose matrix. For example, $(\alpha A^T)^t$ considers the direct influence vertices expanded through t steps. The parameter α has 2 different roles in this centrality measurement. First, it is an attenuation parameter or a probability to influence others throughout the network. α-centrality measures the relative influence of not only a node within its network but also a node through intermediate paths of network. It also represents a trade-off between the exogenous source and endogenous or the possibility that each node's status may also depend on information that comes from outside the network or that may regard solely the member. Low value of α makes α-centrality probes only the local structure of the network and a range of nodes contributes to the centrality score of a given node is increased with the increase of α. The rank obtained using α-centrality can be considered as the steady state distribution of information spread process on a network, with probability α to transmit a message or influence along a link.

Based on the structure of the input-output network that we are considering (weighted and directed network), when applying α-centrality measurement, we can divide it into 2 different cases, Amplification Index (AI) and Vulnerability Index (VI). The idea of AI is to calculate the infection of a sector (or industry), or how the sector infects other nodes in the network. In the economy, the impact of an industry can be measured by a transaction between it and other industries. We measure the total influence (both directly and indirectly) that a sector gives to all other sectors.

$$x_i = \alpha \sum_j \omega_{ij} x_j + e_i \qquad (6)$$

w_{ij} represents the flow from the sector i to sector j, e_i is the exogenous factor of sector i. In the framework of the Cobb-Douglas production in Eq. (1) or (2), α is the output elasticity of the capital or the share of capital. The vector of these measurements of all sectors defined as the Amplification Index (AI), which is obtained

$$AI = (I - \alpha W)^{-1} e \qquad (7)$$

In some cases a sector may not have any direct connection to other sectors in the economy, it still indirectly impact them via other intermediate sectors. This could be done if we measure its influence in a period of time. Hence, the formula (7) could be rewritten as an accumulative one

$$AI = \left(\sum_{i=0}^{\infty} \alpha^i W^i \right) e = (I + \alpha W + (\alpha W)^2 + \cdots + (\alpha W)^t + \cdots) e \qquad (8)$$

Another measurement is the total influence (both directly and indirectly) that a sector receives from all other sectors, which is obtained as

$$x_i = \alpha \sum_j \omega_{ji} x_j + e_i \qquad (9)$$

The vector of these measurements of all sectors defined as the Vulnerability Index (VI), which is obtained as

$$VI = (I - \alpha W^T)^{-1} e \qquad (10)$$

Similarly to the accumulative AI, we represent the formula (10) as

$$VI = \left(\sum_{i=0}^{\infty} \alpha^i W^{Ti} \right) e = (I + \alpha W^T + (\alpha W^T)^2 + ... + (\alpha W^T)^t + ...) e \qquad (11)$$

In the next section, we will obtain AI and VI values using the input-output database of the world economy and identify some key industries in the world economy.

4 Applying to the World Economy Data

Ever since Leontief formalized its structure, the input-output table has been used extensively. By keeping track of the inter-industrial relationships, the input-output table offers a reasonably accurate measurement of the response of any given economy in the face of external shocks or policy interventions. However, as the global economy becomes increasingly integrated, an isolated view based on the national input-output table is no longer sufficient to assess an individual economy's strength and weakness, not to mention finding solutions to global challenges such as climate change and financial crises. Hence, a global multi-regional input-output (GMRIO) framework is needed to draw a high-resolution representation of the global economy.

Cerina et al. [11] constructed the WION based on the World Input-Output Database (WIOD). The empirical counterpart to a network of production technologies consisting of nodes that represent different sectors and directed flows these capture input transactions between sectors is given by input-output data. To investigate the network structure of sector-to-sector input flows, we use WIOD. At the time of writing, the WIOD input-output tables cover 35 industries for each of the 40 economies

(27 EU countries and 13 major economies in other regions) plus the rest of the world (RoW) and the years from 1995 to 2011. For each year, there is a harmonized global level input-output table recording the input-output relationships between any pair of industries in any pair of economies. The relationship can also be an industry to itself and within the same economy. The numbers in the WIOD are in current basic (producers') prices and are expressed in millions of US dollars.

We will take as nodes in the sector input-network. Each nonzero (i, j) entry is a directed edge of this network—that is, a flow of inputs from supplying sector j to customer i. For some of the empirical analysis below, we will be focusing only on properties of the extensive margin of input trade across sectors. To do this, we use only the binary information contained in this input-output data—that is, who sources inputs from whom—and disregard the weights associated with such input linkages.

Recognizing a network structure or the complexity of a network can help us understanding more the world economic behaviors. The data we are trying to work around is the world input-output table (WIOT). This dataset is a part of the world input-output database (WIOD), which was funded by the European Commission. Although WIOD main data tables contains 4 different tables, namely, world input-output Tables, National input-output tables, Socio Economic Accounts and Environmental Accounts, we only take an advantage on WION since this table contains 40 countries' economy transaction value to find out which country's industries are the most important to the world economy. WIOT is provided in current prices, denoted in millions of dollars, and covers 27 EU countries and 13 other major countries in the world, which contains 35 main industries each. While the data is available from 1995 to 2011, we will mainly focus on the latest year's dataset (2011) and make use of the others as an addition trend analysis. This table includes the flows between the industries of 40 countries and 1 group (Rest of the World). We considered only the transactions of those 40 major countries' industries; hence we have 40*35 = 1,400 sectors as nodes in the sectorial input-output network. Let's have a glance at some network's characterization at the regional level first. This world input-output network consists of 1,400 nodes with about 908,587 nonzero edges out of possible 1400^2 edges; therefore, this network is dense with the network density

$$\rho = \frac{m}{n * (n - 1)} = \frac{908587}{1400 * 1399} = 0.46 \qquad (12)$$

Regarding the transaction volume, there were 52 sectors that had the total transactions (both input and output transactions) larger than 500 billion dollars. In the top 10 highest total transaction sectors illustrated in Fig. 1, all of them were from China and the USA.

All 27 EU countries plus 13 other major countries are divided into 4 main groups, European Union, North America, Latin America and Asia and Pacific group, as shown in Table 1.

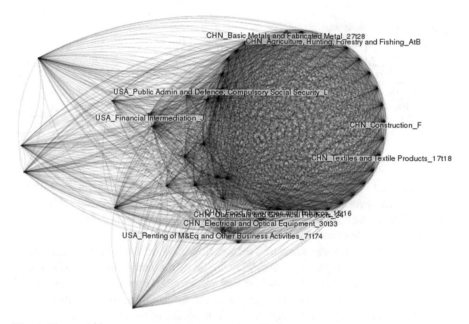

Fig. 1 The world input-output network (year 2011). 52 sectors had total transaction greater than 500 billion dollars

Table 1 Major countries and their groups in WIOT

European Union								
Austria	Germany	Netherlands	Belgium	Greece	Poland	Bulgaria	Hungary	
Portugal	Cyprus	Ireland	Romania	Czech Republic	Italy	Slovak Republic		Denmark
Lavia	Slovenia	Estonia	Lithuania	Spain	Finland	Luxembourg	Sweden	France
Malta	United Kingdom							
North America								
Canada	United States							
Latin America								
Brazil	Mexico							
Asia and Pacific								
China	India	Japan	South Korea	Australia	Taiwan	Turkey	Indonesia	Russia

We consider these groups as the sub-networks of the whole world economy network. We calculate the network density of each sub-network (group), and compare their economic connectivity. Each sub-networks (or groups), European Union (EU), North America (NA), Latin America (LA), and Asia and Pacific (AP), has a very high network density, 0.9, 0.95, 0.91, and 0.7 respectively. This fact indicates high linkage among the countries within the same sub-network (group). It is also unsurprising that the group North America, consists of the two strong economy countries, namely USA

Table 2 Network parameters of 4 groups

Group	Nodes	Non-zero edges	Density	Average inner transaction
European Union	945	808722	0.91	15685.28
North America	70	4635	0.96	169905.61
Latin America	70	4415	0.91	30522.54
Asia and Pacific	315	71056	0.72	84787.40

and Canada, has very high intra transactions, and leads this comparison list with the total inner-transaction is about 170 billion dollars. The following position belongs to Asia and Pacific group in the presence of China and Russia (about 87,787 million dollars). These values are summarized in Table 2.

We now identify the key economic sectors (industries) among 40 best economies of the world by obtaining their AI and VI from the world input-output table. Through the lenses of our model, sectors such as real estate, management of companies and enterprises, advertising, wholesale trade, telecommunications, iron and steel mills, truck transportation, and depository credit intermediation alongside a variety of energy-related sectors—petroleum refineries, oil and gas extraction, and electric power generation and distribution—are seemingly key to U.S. aggregate volatility as they sit at the center of the production network. When applying these equations in the real economic input-output network (ION), we see that the intermediate table of this network is a directed and weighted network. Each element of this intermediate matrix (W) represents the trade volume either between 2 commodities or a node itself, measured by a unit of million dollars. Using two measurements from Eqs. (8) and (11), assuming time is infinite, the measurements' results will be diverge if the values of each element in the matrix W is larger than 1. Hence, to overcome this problem, each element of W is divided by the maximum value of the matrix element. We denote the normalized input-output matrix as $M = W/\max(W)$, and we define VI as

$$VI = (I + (\alpha M^T) + (\alpha M^T)^2 + ... + (\alpha M^T)^t + ...)e \qquad (13)$$

Similarly, AI matrix is defined as

$$AI = (I + (\alpha M) + (\alpha M)^2 + ... + (\alpha M)^t + ...)e \qquad (14)$$

Table 3 shows the top five economic sectors with the highest AI from the input-output data in 2011 with the different α values. In most cases, the top economic sectors were from China, which leading by the "Basic metals and fabricated metals" sector, and "Electrical and optical equipment" with the high value of α. These sectors from China were the greatest impact to the world economy in that period of time. However, if lower the range of sectors affected by a given sector, or reduce the value of α, the U.S' sector "Renting of M&Eq and Other Business Activities" replaced the

Table 3 The top 5 economic sectors with high amplification index (AI) in 2011

α	Rank	Sector	AI
1	1	(CHN) Basic metals and fabricated metal	0.233
1	2	(CHN) Electrical and optical equipment	0.090
1	3	(CHN) Mining and quarrying	0.068
1	4	(CHN) Electricity, gas and water supply	0.053
1	5	(CHN) Chemicals and chemical products	0.043
0.85	1	(CHN) Basic metals and fabricated metal	0.233
0.85	2	(CHN) Electrical and optical equipment	0.090
0.85	3	(CHN) Mining and quarrying	0.068
0.85	4	(CHN) Electricity, gas and water supply	0.053
0.85	5	(CHN) Chemicals and chemical products	0.043
0.5	1	(CHN) Basic metals and fabricated metal	0.004
0.5	2	(USA) Renting of M&Eq and other business activities	0.004
0.5	3	(USA) Financial intermediation	0.003
0.5	4	(CHN) Electrical and optical equipment	0.0026
0.5	5	(CHN) Chemicals and chemical products	0.0019
0.25	1	(USA) Renting of M&Eq and other business activities	0.0016
0.25	2	(CHN) Basic metals and fabricated metal	0.0015
0.25	3	(USA) Financial intermediation	0.0014
0.25	4	(CHN) Electrical and optical equipment	0.0013
0.25	5	(CHN) Chemicals and chemical products	0.0011

sector "Basic metals and Fabricated metals" of China to become the most influenced industry. These results also indicate the evidence that USA and China enjoyed the largest economy in the world in 2011.

Similarly, in the Table 4 below, the top five economic sectors with the highest vulnerability index in the different cases of the value of α in 2011 are pointed out. The top most be influenced economic sectors were still belong to China, which leading by the "Electrical and Optical Equipment" and "Basic metals and fabricated metals" sector, despite of the change of the value of α. Even a small change of any other industries may also lead to a fluctuation of this sector's transaction.

Comparing to the result of World Input-Output network analysis by Federica Cerina et al. [11] in 2011, the authors used 4 different parameters to evaluate the industries. The first calculation was produced by the Laumas method of backward linkages (w), next was the eigenvector method of backward linkages e, the third and the

Table 4 The top 5 economic sectors with high vulnerability index (VI) in 2011

α	Rank	Sector	AI
1	1	(CHN) Electrical and optical equipment	0.200
1	2	(CHN) Basic metals and fabricated metal	0.076
1	3	(CHN) Construction	0.075
1	4	(USA) Financial intermediation	0.055
1	5	(CHN) Machinery, nec	0.053
0.85	1	(CHN) Electrical and optical equipment	0.200
0.85	2	(CHN) Basic metals and fabricated metal	0.076
0.85	3	(CHN) Construction	0.075
0.85	4	(USA) Financial intermediation	0.055
0.85	5	(CHN) Machinery, nec	0.054
0.5	1	(CHN) Electrical and optical equipment	0.004
0.5	2	(CHN) Basic metals and fabricated metal	0.003
0.5	3	(CHN) Construction	0.003
0.5	4	(USA) Financial intermediation	0.002
0.5	5	(USA) Public admin and defence; compulsory social security	0.002
0.25	1	(CHN) Electrical and optical equipment	0.0014
0.25	2	(CHN) Construction	0.0013
0.25	3	(CHN) Basic metals and fabricated metal	0.0013
0.25	4	(USA) Financial intermediation	0.0011
0.25	5	(USA) Public admin and defence; compulsory social security	0.0011

fourth were PageRank centrality PR and the community coreness measure—dQ—respectively. In the Table 5, we compare the results (top 5 sectors) implemented by Cerina et al. (2011) and our measurements with the different values of α. According to this table, the results got from backward linkages method (w) and Vulnerability Index (with the different α values) are almost identical. That is, there is an existence of the same sectors from China and the USA in both methods such as China's Construction (CHN_Cst), "Public Admin and Defence; Compulsory Social Security" from the USA (USA_Pub), etc. However, the results generated by Amplification Index measurement are different to the other methods since an approach of this implement is based on the outward link while the others use the inward link as main factor to measure the centrality.

Table 5 The comparison of results implemented by AI, VI and other methods conducted by Cerina et al. [11]

Rank	w	e	PR	—dQ—	AI(α:1)	AI(α:0.5)	AI(α:0.25)	VI(α:1)	VI(α:0.5)	VI(α:0.25)
1	CHN-Cst	CHN-Tpt	GRB-Hth	CHN-Cst	CHN-Met	CHN-Met	USA-Obs	CHN-Elc	CHN-Elc	CHN-Elc
2	USA-Pub	CHN-Tex	DEU-Tpt	USA-Obs	CHN-Elc	USA-Obs	CHN-Met	CHN-Met	CHN-Met	CHN-Cst
3	USA-Hth	CHN-Elc	USA-Pub	CHN-Met	CHN-Min	USA-Fin	USA-Fin	CHN-Cst	CHN-Cst	CHN-Met
4	USA-Est	CHN-Rub	CHN-Elc	USA-Pub	CHN-Ele	CHN-Elc	CHN-Elc	USA-Fin	USA-Fin	USA-Fin
5	CHN-Elc	CHN-Lth	USA-Hth	USA-Est	CHN-Chm	CHN-Chm	CHN-Chm	CHN-Mch	USA-Pub	USA-Pub

Abbreviation	
CHN: China	USA: the USA
GRB: Great Britain	DEU: Germany
Cst: Construction	Tpt: Transport Equipment
Hth: Health and Social Work	Met: Basic Metals and Fabricated Metal
Obs: Renting of M&Eq and Other Business Activities	Elc: Electrical and Optical Equipment
Mch: Machinery, NEC	Tex: Textiles and Textile Products
Min: Mining and Quarrying	Fin: Financial Intermediation
Est: Real Estate Activities	Rub: Rubber and Plastics
Chm: Chemicals and Chemical Products	Pub: Public Admin and Defence; Compulsory Social Security
Ele: Electricity, Gas and Water Supply	Lth: Leather, Leather and Footwear

5 Transitions of Important Industries

By mixing the top 20 sectors of the highest AI and the top 20 sectors of the highest VI, we get a list of 27 sectors sorted by AI value and VI. We will try to examine whether or not the relationship between an AI and VI value of a sector and its input and output strength in 2011, then take a deeper look at these differences throughout the period of 17 years.

With $\alpha = 1$, we pick out some sectors to analysis that have both high AI and VI value such as "Basic metals and Fabricated Metal" (CHN_Met), "Electrical and Optical Equipment" (CHN_Elc), "Mining and Quarrying" (CHN_Min) from China, the two sectors "Financial Intermediation" (USA_Fin) and "Renting of M&Eq and Other Business Activities" (USA_Obs) from the US.

Firstly, we will examine the change of AI value of these sectors in the period from 1995 to 2011 (Fig. 2). It can be easily seen that the only two sectors from USA were leading the remain with the fluctuation of their AI values in the first 16 years before dropping down and being replaced by sectors from China in the year 2011. In the first 16-year period, the USA's sectors had very high values of AI compared to the ones of China. The very important milestone, the financial crisis of 2007–2008 or global financial crisis, caused by the collapse of Lehman Brothers, also affected to the AI value of these two USA's sectors. After 2008, their reactions to this event were quite different. In this year, while the AI of the sector "Financial Intermediation" felt below

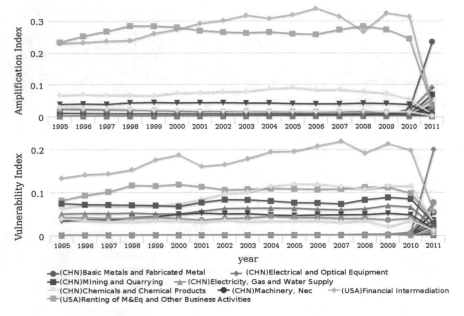

Fig. 2 The transitions of the important sectors in terms of AI and VI ($\alpha = 1$) in the periods 1995–2011

the value of the other the U.S.' sector before bouncing back to the higher value in 2009, the AI value of the sector "Financial Intermediation" of the United States had increased gradually to reach a peak of 0.28 first time since 2000. As we are considering the influence of sector through the entire network of world economy (α is 1), it seems that there was a prediction of this crisis from the reducing value of the sector "Financial Intermediation" since 2006. A year later, both of these sectors had the sharp declines and bottom out around the AI value 0.03 in 2011, and were replaced by the sectors from China. One thing to note is that, while in the previous 16 years, the AI value of these top China's sector were very small compared to other sectors, in the last year of this period (2011), their AI value dramatically rose up to nearly 0.24 and 0.1 corresponding to the sector CHN_Met and CHN_Elec respectively. Similar to the change of AI, the change of VI of these VI had the same trend. While almost the high VI value are of the sectors from the U.S. in the first 16 years, in the year 2011, the sector CHN_Elec from China had a sudden leap to the VI value of 0.2 after having a slight change from 0.002 in 2008 to 0.007 in 2010. However, these changes can be seen considering the total degree of these sectors in the year 2011.

In the other hand, considering the sectors with the highest VI value in 2011, the leading is the sector Electrical and Optical Equipment(CHN_Elc) of China, followed by the sector CHN_Met and the sector Construction (CHN_Cst). Based on the WION, we see that the sector CHN_Elc, itself consumed its products valued about 660 billion dollars, had imported approximately 198 billion dollars mostly from the same industry type of the foreign countries (mostly from Taiwan, Japan and Korea). In China's

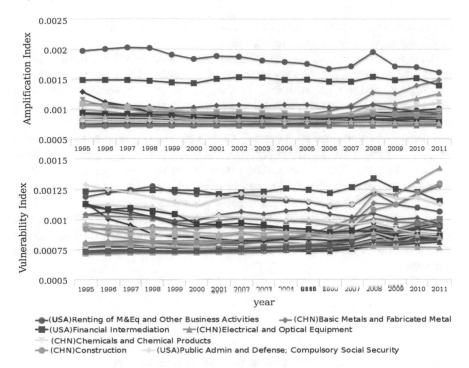

Fig. 3 The transitions of the important sectors in terms of AI and VI ($\alpha = 0.25$) in the periods 1995–2011

local market, products from the sector CHN_Met were mostly used (about 278 billion dollars), the sectors "Rubber and Plastics", "Wholesale Trade and Commission Trade, Except of Motor Vehicles and Motorcycles" and "Chemicals and Chemical Products" were the three following sectors that provided much products to the sector CHN_Elc with 87, 70 and 61 billion dollars respectively. Moreover, these directly supported sectors had a very high ranking of AI value in the top sectors with the highest AI value. This may be one possible explanation of this sector's high VI value.

If we reduce the range of effect to the other sectors by reducing α value, it is clearly seen that any sector had the more direct investment, the higher AI value it got. Similarly, high volume transaction of direct supported sectors had more influence to the VI value of the target sector. For example, with α is 0.25 (Fig. 3), the sector "Renting of M&Eq and Other Business Activities" (or USA Obs) from the United State of America became the top most AI value sector followed by the other sectors from China and the U.S, namely "Basic metals and Fabricated Metal" and "Financial Intermediation" respectively. In 2011, this sector had the highest total-strength and the highest Out-strength (with nearly 2,429 billion dollars), According to the National Accounts Main Aggregates Database of United Nations Statistics Division, the United States was the largest consumer market of the world. Hence, despite of the fact that this sector's output mostly to the USA's local market, it still had the very high AI value comparing to the other industries.

In terms of VI value, the sector CHN_Elc from China was still the most be influenced sectors since it has the very high imported products from other industries of both regional and foreign countries. However, the sector "Construction" of China (CHN_Cst) consumed more products that the sector CHN_Elc from the other China's industries. From the sector "Other Non-Metallic Mineral" of China, about 375 billion dollars was consumed by the sector CHN_Cst. The others were from the sector CHN_Met with 367 billion dollars and CHN_Elc with only 96 billion dollars. Although the sector "Other Non-Metallic Mineral" did not have high AI, in this case of small range of affect ($\alpha = 0.25$), it still had enough influence to make the sector CHN_Cst become more vulnerability than the other sectors.

To conclude this complicated relationship, it is very hard to decide which sectors have high influence or most being affected if based only on their transaction. The use of AI and VI with the varied value of α might make the keys sector evaluation more precisely.

6 Conclusion

In the modern society, all major economic sectors have been connected tightly in an extremely complicated global network. In this type of network, a small shock occurred at certain point can be spread instantly through the whole network and may cause catastrophe. Production systems, traditionally analyzed as almost independent national systems, are increasingly connected on a global scale. Only recently becoming available, the world input-output database is one of the first efforts to construct the global and multi-regional input-output tables. The network measures can give valuable insights into identifying the key industries. By viewing the world input-output tables as complex networks where the nodes are the individual industries in different economies and the edges are the monetary goods flows between industries, we characterize a certain aspect of centrality or status that is captured by the α-centrality measure of the world input-output network. We also capture their evolution of over years. We also argue that the network structure captured from the input-output data is key in determining whether and how microeconomic impacts or shocks propagate throughout the economy and shape aggregate outcomes. Understanding the network structure of world input-output data can better inform on how the world economy grows as well as how to prepare for and recover from adverse shocks that disrupt the global production chains.

The discussion in this paper has attempted to introduce another way to look for the key sectors in the world economy. Applying the method based on the AI and VI, we identified the sectors that could be considered as key, or the major, industries in the world economy in the period from 1995 to 2011. In short, these measurements are defined as:

- Amplification Index, AI, is used to measure the total influence that a sector could affect to other sectors in a long time.
- Vulnerability Index, VI, is, on the other hand, a cumulative impact that a sector could receive from other sectors in a period of time.

Using of the two methods heavily depends on the value of the trade-off parameter α. The value of α determines how far influence could be spread through the network. The higher value of α, the further nodes that impact could reach to. If α is chosen correctly according to the considering economy and the research scale of the economists, AI and VI might be the useful measurements for the economist to evaluate the influence of the key sectors in that economy.

Since there are some traditional ways to analyze key sectors in the economy such as finding Forward links and Backward links, these introduced methods may be contributed to the policy makers' toolkit to help them in analyzing the economy easily, and also preparing and recovering from adverse shocks that disrupt the production chains.

References

1. Acemoglu D, Vasco MC, Asuman O, Alireza T (2012) The network origins of aggregate fluctuations. Econometrica 80(5):1977–2016
2. Alatriste-Contreras M (2015) The relationship between the key sectors in the European Union economy and the intra-European Union trade. J Econ Struct. doi:10.1186/s40008-015-0024-5
3. Aoyama H et al. (2007) Pareto firms (in Japanese). Nihon Kezai Hyoronsha Chapter 3, pp 91–147. ISBN 978-4-8188-1950-4
4. Bonacich P, (1972) Factoring and weighting approaches to status scores and clique identification. J Math Sociol 2(1):113–120
5. Page L, Brin S, Motwani R, Winograd T (1999) The pagerank citation ranking: bringing order to the web. Technical Report 1999-66, Stanford InfoLab
6. Bonacich P, Lloyd P (2001) Eigenvector-like measures of centrality for asymmetric relations. Soc Netw 23:191–201
7. Borgatti SP, Li X (2009) On the social network analysis in a supply chain context. J Supply Chain Manage 45:5–22
8. Botri V (2013) Identifying key sectors in croatian economy based on input-output tables. Radni materijali EIZ (The Institute of Economics, Zagreb) - a EIZ Working Papers EIZ-WP-1302
9. Katz L (1953) A new status index derived from sociometric analysis. Psychometrika 18(1):39–43
10. Carvalho VM (2014) From micro to macro via production networks. J Econ Perspect 28:23–48
11. Cerina F, Zhu Z, Chessa A, Riccaboni M (2015) World input-output network. PLoS One. doi:10.1371/journal.pone.0134025
12. Leontief W (1986) Input-output economics. Oxford University Press, Oxford

Natural Resources, Financial Development and Sectoral Value Added in a Resource Based Economy

Ramez Abubakr Badeeb and Hooi Hooi Lean

Abstract This chapter vestigates the effects of natural resource dependence and financial development on the sectoral value added in a resource based economy, Yemen. We allow the effect of these two factors to be different for the growth of agricultural, manufacturing and service sectors respectively. We remark on one hand that natural resource curse hypothesis is strongly supported. The agricultural and manufacturing sectors are affected by this phenomenon which implies the existence of Dutch disease symptoms in Yemen. On the other hand, financial sector development does not play an important role in fostering real sectors activities. The service sector is the only sector that benefit from the financial sector development in Yemen. This finding opens up a new insight for Yemeni economy to sustain sectoral growth by controlling the level of natural resource dependence and proactiveness sectoral strategy for financial sector development.

Keywords Natural resource curse · Financial development · Sectoral value added · Republic of Yemen

1 Introduction

Natural resource dependence interacts with and alters various social, political and economic factors; and thus slower the economic growth and development [17, 18, 21, 28, 36–38, 43]. This is called natural resource curse, a phenomenon of slow economic growth that caused by a series of negative effects from the excessive dependence on natural resources in a country [40]. On the other hand, endogenous growth economists claim that enhancement in productivity can lead to a faster pace of innovation and

R.A. Badeeb · H.H. Lean (✉)
Economics Program, School of Social Sciences, Universiti Sains Malaysia,
11800 Penang, Malaysia
e-mail: hooilean@usm.my

© Springer International Publishing AG 2017
V. Kreinovich et al. (eds.), *Robustness in Econometrics*,
Studies in Computational Intelligence 692, DOI 10.1007/978-3-319-50742-2_24

extra investment in human capital. Given that, financial intermediaries that promotes and allocates resources to the development of new innovation may have a positive role in fostering economic growth [22, 24]. What is unclear, however, is whether these potential effects of natural resource dependence and financial development are being reaped in the Yemeni economy, and how these factors can impact the sectoral growth specifically.

Previous studies that related to the resource-growth nexus and finance-growth nexus focus on the ultimate effect on aggregate growth and ignore the sectoral growth in the economy. Therefore, one must dig much deeper to understand the processes and its effects on different economic sectors. To this end, this study contributes to the literature by investigating the impacts of financial development and natural resource dependence on three main economic sectors i.e. agriculture, manufacturing and service in Yemen.

We situate the study in Yemen because the country has a wide spectrum of economic potential for a number of industries[1] and shows a bright and promising future for its agriculture and fishery sectors, besides the political chaotic and military conflicts. Yemen has one of the best natural harbors in the world that has a unique strategic geographical location linking the East and the West [14]. Despite these potentials, it has been widely observed that the economic performance of Yemen is accompanied by a surge in dependence on oil, where the economy is dominated by the production and export of oil, which generates around 70% (as an average in the 10 years) of government's revenues, contributes about 80–90% of its exports, and is responsible for building up most of the country's foreign exchange reserves. Moreover, Yemen is a suitable example of a country that has experienced early financial sector activities relative to other developing countries.[2]

The rest of the paper is organized as follows: the literature review is presented in Sect. 2. In Sect. 3, we overview the Yemeni economy. Section 4 focuses on the data and methodology, and the empirical results and discussion are presented in Sect. 5. Finally, Sect. 6 concludes with policy implications.

2 Literature Review

In the last decades, economists observed that resource-rich countries, especially many African, Latin American and Arab nations tend to grow at slower rates than countries with fewer natural resources. These countries suffer from what economists call the "resource curse" [7]. This situation usually led to two sorts of reversals, economic and political reversals. The highly dependence on natural resources can hurt the economic growth indirectly by releasing forces that hamper the development of

---

[1] Other regional countries such as Saudi Arabia, Oman and Kuwait do not have other important real sectors in their economy besides oil

[2] The first bank was established in the north of Yemen in 1962

national economy through the Dutch disease,[3] the price volatility of natural resources and rent seeking, in addition to other economic and political reasons. The curse can be exemplified by empirical findings that have been presented by many economists like [6, 17, 21, 28, 30, 36–38, 43].[4]

Sachs and Warner [38] followed a large panel of natural resource economies between 1970 and 1989 and found that natural resource dependence was negatively correlated with economic growth. Following their influential studies, a large volume of subsequent research has been inspired to examine the direct and indirect relationships between natural resource dependence and economic growth. Gylfason [17, 18], Mehlum et al. [28] have argued that since 1970, countries that have based their economies on natural resources have tended to be examples of development failure. For oil, in particular, [30] find that oil exporting countries have witnessed a fall of average per capita income of 29% over the period 1975–2000. This compares to the rest of the world whose average per capita income increased by 34% over the same twenty-five year period. Finally Apergis and Payne [5] found a negative relationship between oil rents and agriculture value added in the long run in oil-producing Middle East and North African (MENA) countries. The author attributed this result to a resource movement effect from other economic sectors to the booming oil sector in these countries.

Along the same line, the role of financial development in growth has also attracted a widespread attention in the past decades and in particular since the emergence of endogenous growth theory in 1980s [12, 26, 35].

The theoretical links between financial development and economic growth can be traced back to Schumpeter [39] who was the earliest economist and highlighted the importance of finance in the process of economic development. A lot of complementary propositions have been put forward to the positive role of financial development on economic growth. The central idea of these studies is that financial development can encourage economic growth by channeling resources to high productive investments [27, 32, 41]. Those studies asserted that the financial intermediation (banks) has an important role in economy by raising saving and capital accumulation. This idea has been supported with modern analysis. A large body of empirical evidences supporting this "Schumpeterian" logic [13, 19, 22, 25, 31]. Robinson [34] in contrast pointed out that the economic growth promotes financial development by creating the demand for financial services, and the financial sector responds to this demand. However, Abu-Bader and Abu-Qarn [1], Demetriades and Hussein [16], Singh [42] argued that financial development and economic growth cause each other. Financial development promotes growth by allowing a higher rate of return to be earned on capital and economic growth, which in turn provides the means to implement well-developed financial structures. Contrary to all previous perspectives, Lucas [26]

---

[3]The Dutch disease phenomena works when natural resources booms increase domestic income and, consequently the demand for goods, which generate inflation and appreciation of the real exchange rate making much of the manufacturing industry uncompetitive in the world market

[4]See [6, 9] for recent literature survey of the curse of natural resources

claimed that the relationship between financial development and economic growth does not exist and considered the role of the financial sector in economic growth to be "over-stressed".

These theoretical discussions reveal that there is not a consensus on the role of natural resource dependence and financial development, on economic growth. However, the debate whether these factors lead sectoral growth has important policy implications for both developed and developing countries. Thus, in contrast to previous studies, this paper provides evidence for Yemen where the effect of these factors on sectoral growth has not been studied.

3 Yemen Economy Review

Yemen is one of the poorest countries in the Middle East and Arab region, where nearly 40% of its population live below the poverty line [46]. The Republic of Yemen was established in May 1990, after unification between the Yemen Arab Republic (YAR) and the Marxist People's Democratic Republic of Yemen (PDRY). In 1990, the new country confronted a difficult task by unifying two countries with different economic systems. During this period, however, the GDP growth averaged 6.2% because of constantly increasing oil revenue. In 1995, Yemeni authorities initiated an economic reform program to achieve two sets of goals: (1) stabilization to restore macroeconomic balance and reduce the inflation rate, and (2) structural reform to foster economic growth and thereby reduce the high poverty rates. Since initiating this reform program Yemen has become one of the most open and trade-liberalized economies in the MENA region United Nation Development Programme [44]. However, economic liberalization has not been transformed into tangible benefits for the majority of the population. The economy remained vulnerable to price and demand fluctuation of oil exports, which became the main GDP contributor and exceeded the contribution of the agriculture sector.

The structure of the economy underwent fundamental changes, with the share of key sectors changing substantially. Figure 1 shows that the GDP shares (at current prices) have increased for industry (including oil and gas) and services. For industry, the fluctuations are strong, in part due to changing in oil world prices. Manufacturing and agriculture have declined strongly. In addition to its increased relative importance, the nature of the service sector has changed from being mostly involved in supporting agriculture and manufacturing in 1990 toward responding to demands fuelled by oil revenues [2]. In general, since 1990, all sectors have increased their real output, with the strongest increase for services; however, for industry, output has declined strongly in recent years even though it remains higher than in 1990.[5]

As in the case of the most developed countries, the financial system in Yemen is dominated by the banking sector, with no existence of a stock market, and

---

[5]Manufacturing is a subsector of industry sector; however, we focus on its separate contribution due to its importance in our study

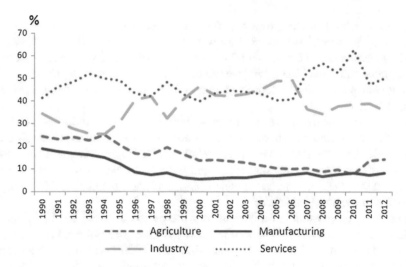

Fig. 1 GDP share by sector at current prices

marginal roles for non-bank financial institutions such as insurance companies, money-changers and pension funds.[6] After achieving the unity in 1990 and the integration of the banking sector in both countries, several problems have appeared in the financial sector, such as bad loans, loan and client concentration, lack of investment opportunities, short-term contracts, and weak regulatory and institutional frameworks [44]. Therefore, in line with the early mentioned economic reform program, the Yemeni government elaborated a reform program focusing on the financial sector. Following this reform program, the total assets of commercial banks increased from YR179 billion in 1996 (24% of GDP), to YR310 billion in 2000 and YR1323 billion in 2012 (30% of GDP). The deposits also witnessed a significant rise from YR120 billion in 1996, to YR250 billion in 2000 (16% of GDP) and YR1799 billion in 2012 (23% of GDP). However, the credit to the private sector represents only 29% of these deposits and around only 5% of GDP.

4 Data, Model and Methodology

4.1 Measurement and Data Sources

Natural resource dependence refers to the degree that an economy relies on resource revenues. Therefore, to gauge the reliance of the economy on natural resources,

---

[6]According to the Central Statistical Organization of Yemen, there were eighteen Yemeni and international commercial and Islamic banks, thirty exchange companies, and nineteen insurance corporations and pension funds operating in Yemen in 2012

the ratio of natural resource export to GDP has been used widely in relevant literature since Sachs and Warner [38]. However, in Yemen, in addition to the export of the government's share of natural resources, natural resources revenue includes other grants and taxes.[7] Therefore, we use natural resource revenue (which includes the components mentioned above) relative to GDP as a proxy for natural resource dependence.

For the financial development measures, this study uses two indicators related to financial intermediaries.[8] These two proxies are, domestic credit to private sector as share of GDP, and the size of the deposits relative to GDP. According to Wurgler [47], the most natural way to measure sector and industry growth is to use the sectoral value added. Therefore, the value added output of each sector to GDP is used to proxy for the sectoral value added.

The study employs data for Yemen over the period of 1980–2012.[9] Following Vetlov and Warmedinger [45] for the Germany case, we use Northern Yemen data for the period prior to 1990 and united Yemen data after 1990, combined with a dummy variable to account for the unification (see Badeeb and Lean [8]).[10,11] Data are sourced from International Financial Statistics (IFS), World Development Indicator (WDI) for Republic of Yemen (1990–2012). The data of natural resources revenues is sourced from the Yemeni Ministry of Finance and IMF country reports.

4.2 Models

Three models have been constructed to capture the effects of natural resource dependence, and financial development on sectoral growth. The equations are inspired from Apergis and Payne [5] after adding the control variables such as trade openness and government expenditure:

---

[7]Oil revenues in Yemen include the concession commissions that the government receives from oil production companies, tax charges on foreign oil companies that operate in Yemen, and grants that the government receives from oil companies after signing contracts (Yemeni Ministry of Finance)

[8]We focus on financial intermediaries due to the absence of stock market in Yemen

[9]Yemeni unification took place on May 22, 1990, when the People's Democratic Republic of Yemen (also known as South Yemen) was united with the Yemen Arab Republic (also known as North Yemen), forming the Republic of Yemen (known as simply Yemen)

[10]Angelini and Marcellino [4] argued that this simple treatment of the unification problem has been used widely in empirical macroeconomic analyses in Europe. It is based on the economic reasoning that East Germany's economy represented very small portion of the unified Germany economy in real GDP terms in 1991

[11]Evidence of the validity of this treatment comes from the fact that the economy of former Southern Yemen accounted for only 17.3% of real GDP of united Yemen. Additionally, the economy of united Yemen is largely based on the market system which was followed by the Northern part before unification

Model 1:

$$AGR_t = \delta_0 + \delta_1 NR_t + \delta_2 FD_t + \delta_3 TO_t + \delta_4 GOV_t + \delta_5 Dum_t + \Psi_t \qquad (1)$$

Model 2:

$$MNF_t = \eta_0 + \eta_1 NR_t + \eta_2 FD_t + \eta_3 TO_t + \eta_4 GOV_t + \eta_6 Dum_t + \kappa_t \qquad (2)$$

Model 3:

$$SRV_t = \omega_0 + \omega_1 NR_t + \omega_2 FD_t + \omega_3 TO_t + \omega_4 GOV_t + \omega_6 Dum_t + \zeta_t \qquad (3)$$

where AGR is agriculture value added of GDP measured in percentage, MNF is manufacturing value added of GDP measured in percentage, SRV is service value added of GDP measured in percentage, NR is natural resource dependence, FD is financial development indicators proxy by credit to private sector and deposit, TO is trade openness, the sum of exports and imports of GDP measured in percentage, GOV is government expenditure of GDP measured in percentage; Ψ, κ and ζ are error terms. All variables have been transformed into natural logarithm before the estimation. Dum is the dummy variable for unification period; it will take 1 if the observation is in the period of 1990–2012 and 0 if the observation is in the period of 1980–1989.

Including trade openness to the model because it is widely accepted that trade openness is an important growth determinant (see Barro and Sala-i Martin [11], Chang et al. [15]). It is also important because Yemen's economy is well integrated with the globe. Therefore, any change of trade openness is expected to affect the economic sector's performance. We also include government expenditure to the model because of its important role in the natural resource-based economies [36].

4.3 Methodology

To test the long-term relation of the variables, we adopt the auto-regressive distributed lag (ARDL) bound testing approach to cointegration by Pesaran et al. [33]. Most recent studies indicated that an ARDL model is more preferable in estimating the cointegration relation because it is reliable and applicable irrespective of whether the underlying regressors are I(0) or I(1). In addition, this approach is better and performs well for a small sample size.

To estimate the ARDL approach to cointegration, short-run dynamics are added into the long run equation. The ARDL models in error correction form are written as Eqs. (4)–(6) respectively.

$$\Delta AGR_t = \delta_0 + \delta_1 AGR_{t-1} + \delta_2 NR_{t-1} + \delta_3 FD_{t-1} + \delta_4 TO_{t-1} + \delta_5 GOV_{t-1}$$
$$+ \sum_{i=1}^{0} \delta_6 \Delta AGR_{t-i} + \sum_{i=0}^{p} \delta_7 \Delta NR_{t-i} + \sum_{i=0}^{q} \delta_8 \Delta FD_{t-i}$$
$$+ \sum_{i=0}^{r} \delta_9 \Delta TO_{t-i} + \sum_{i=0}^{s} \delta_{10} \Delta GOV_{t-i} + \delta_{11} Dum_t + \epsilon_t \qquad (4)$$

$$\Delta MNF_t = \eta_0 + \eta_1 MNF_{t-1} + \eta_2 NR_{t-1} + \eta_3 FD_{t-1} + \eta_4 TO_{t-1} + \eta_5 GOV_{t-1}$$
$$+ \sum_{i=1}^{0} \eta_6 \Delta MNF_{t-i} + \sum_{i=0}^{p} \eta_7 \Delta NR_{t-i} + \sum_{i=0}^{q} \eta_8 \Delta FD_{t-i}$$
$$+ \sum_{i=0}^{r} \eta_9 \Delta TO_{t-i} + \sum_{i=0}^{s} \eta_{10} \Delta GOV_{t-i} + \eta_{11} Dum_t + \epsilon_t \qquad (5)$$

$$\Delta SRV_t = \omega_0 + \omega_1 SRV_{t-1} + \omega_2 NR_{t-1} + \omega_3 FD_{t-1} + \omega_4 TO_{t-1} + \omega_5 GOV_{t-1}$$
$$+ \sum_{i=1}^{0} \omega_6 \Delta SRV_{t-i} + \sum_{i=0}^{p} \omega_7 \Delta NR_{t-i} + \sum_{i=0}^{q} \omega_8 \Delta FD_{t-i}$$
$$+ \sum_{i=0}^{r} \omega_9 \Delta TO_{t-i} + \sum_{i=0}^{s} \omega_{10} \Delta GOV_{t-i} + \omega_{11} Dum_t + \epsilon_t \qquad (6)$$

The coefficients of the first portion of the model measure the long-term relation, whereas the coefficients of the second portion that attach with $\sum$ represent the short-term dynamics. The F-statistic is used to test the existence of a long-term relation among the variables. We test the null hypothesis $H_0 : \delta_1 = \delta_2 = \delta_3 = \delta_4 = \delta_5$ in $(4)^{12}$ that there is no cointegration among the variables. The F-statistics is then compared with the critical value provided by Narayan [29], which is more suitable for a small sample. If the computed F-statistic is greater than the upper bound critical value, we reject the null hypothesis of no cointegration and conclude that steady state equilibrium exists among the variables. If the computed F-statistic is less than the lower bound critical value, the null hypothesis of no cointegration cannot be rejected. However, if the computed F-statistic lies between the lower and upper bounds' critical values, the result is inconclusive.

Finally, this study adopts two stages of robustness check. First, we use different measure of financial development on the same estimation technique (i.e., ARDL). In the second stage, we use new natural resource dependence and financial development measures on the alternative estimation techniques (i.e., fully modified OLS and dynamic OLS).

5 Empirical Findings and Discussion

To test the integration order of the variables, ADF and PP tests are employed. We do not report the results in order to conserve space. Overall, the unit root tests suggest that all variables are stationary in its first differences besides government expenditure.

The ARDL bounds testing approach starts with F-test to confirm the existence of the cointegration between the variables in the model. Lags up to two years have been imposed on the first difference of each variable. We use Schwarz Bayesian Criterion (SBC) to suggest the optimum lag for our ARDL models. Given the sample size of 33 observations, the critical values of Narayan [29] for the bounds F-test are employed. The result of ARDL bound test of cointegration is tabulated in Table 1.

---

[12] The same approach is applied to Eqs. (5) and (6) respectively

Table 1 Result from ARDL cointegration test

Sector	Model	Max. lag	SBC optimum lag	F-Statistic	ECT_{t-1} (t-ratio)
Agriculture	Credit	2	(1,1,0,0,0)	2.1475	−0.596 (−5.669)***
	Deposit	2	(1,0,0,0,0)	4.1167*	−0.435 (−4.709)***
Manufacturing	Credit	2	(1,0,0,0,0)	5.1457**	−0.2121 (3.098)***
	Deposit	2	(1,0,0,0,0)	4.3123*	−0.235 (−2.292)**
Service	Credit	2	(1,0,0,1,0)	3.6777	−0.641 (−3.967)***
	Deposit	2	(1,0,0,0,0)	3.7375	−0.727 (−4.943)***

Note ***, ** and * denotes the significance at 1, 5 and 10% levels respectively. Critical values bounds are from [29] with unrestricted intercept and no trend (Case III)

The result of cointegration test in Table 1 shows that the F-statistics is greater than its upper bound critical value of 1% level in manufacturing sector models, indicating the existence of long run relationship. Moreover, the coefficient of lagged error correction term (ECT_{t-1}) is significant and negative which confirms the existence of long run relationship. However, for the case of agriculture and service sectors, the results are inconclusive. Therefore, we need to show the cointegration among the variables by using the alternative way by testing the coefficient of ECT_{t-1} which is considered by Kremers et al. [23] as a more efficient way of establishing cointegration. Kremers et al. [23] argued that the significant and negative coefficient for ECT_{t-1} will indicate the adjustment of the variables towards equilibrium hence the cointegration. So, the cointegration of these two sectors was supported by significant and negative coefficient for ECT_{t-1}.

As there is cointegration among the variables, we can derive the long-term coefficient as the estimated coefficient of the one lagged level independent variable divided by the estimated coefficient of the one lagged level dependent variable and multiply it with a negative sign. Conversely, the short-term coefficients are calculated as the sum of the lagged coefficient of the first differenced variables.

Table 2 Panel A provides the long run estimation results for the three sectors which will be discussed as follows:

Agriculture Sector

Our findings for agriculture sector reveal that there is no significant impact of financial development on agricultural sector for the case deposit model. However, surprisingly, the result reveals significant negative relationship between financial development and agriculture value added for the case of credit to private sector. This negative result is against expectation and could be linked with the fact that the agricultural shocks i.e. drought, flooding and other hazards, negatively affect the ability of agricultural borrowers to repay their loans and advances which left farmers in dept trap. Additionally, the findings reveal the negative role of natural resource dependence on agricultural value added for the both cases of financial development indicators. This

Table 2 Long run and short run analysis

	Agriculture		Manufacturing		Service	
	FD		FD		FD	
	Credit	Deposit	Credit	Deposit	Credit	Deposit
Panel A. Long run results						
C	6.0414***	5.7428	2.9188	3.9284	1.9371**	0.6927
	(10.1683)	(0.9109)	(0.9241)	(0.7607)	(2.3091)	(0.6597)
FD	−0.4316***	−0.2436	0.8012	1.9471	0.3550**	0.4737**
	(−0.1306)	(−0.9581)	(0.6567)	(1.6164)	(2.0537)	(2.6395)
NR	−0.1306***	−0.1375**	−0.4873***	−0.3994*	−0.1257	−0.1145
	(−3.2748)	(−2.1923)	(−3.1570)	(−1.7775)	(1.6173)	(−1.1921)
TO	−0.5094***	−0.8115***	−0.3330***	−1.3837***	−0.0745	0.4771**
	(−3.6078)	(−3.2571)	(−3.0924)	(−3.1832)	(−0.4462)	(2.6606)
GOV	0.0812	0.3899	−0.2480	−1.2876	0.3912	−0.0097
	(0.4406)	(1.3630)	(−0.2682)	(−1.2034)	(1.5559)	(−0.0522)
Dum	−0.4968***	−0.1806	1.3171	1.1613	0.5775**	0.3851**
	(−3.2312)	(−0.9109)	(0.8049)	(1.5547)	(2.4374)	(2.7324)
Panel B. Short run results						
Δ FD	−0.1052	−0.1059	0.1699	0.4583***	0.2276**	0.3445***
	(−0.9422)	(−0.8766)	(1.1260)	(3.1392)	(2.6089)	(3.1093)
Δ NR	−0.0778***	−0.0597*	−0.1033**	0.0940**	−0.0806	–
	(−2.7660)	(−1.9686)	(−2.4200)	(−2.5526)	(−1.5107)	0.08323**
						(−2.7489)
Δ TO	−0.3035***	−0.3526**	−0.0706	−0.3257*	0.1538	0.3470**
	(−3.8014)	(−2.5016)	(−0.5160)	(−1.8889)	(1.2850)	(2.9869)
Δ GOV	0.04842	0.1694	−0.0525	−0.3031*	0.2508*	−0.0070
	(0.4485)	(1.3096)	(−0.2760)	(−(1.7162)	(1.7621)	(−0.0522)
Dum	−0.2960**	−0.0785	0.2793	0.2734**	0.3703***	0.2801***
	(−2.435)	(−0.8497)	(1.5303)	(2.4306)	(2.8566)	(3.1292)
ECT_{t-1}	−0.5959***	−0.4345***	−0.2121***	−0.2354**	−0.6412***	–
	(−5.6696)	(−4.7091)	(−3.0985)	(−2.2924)	(−3.9674)	0.7272***
						(−4.9427)
Panel C. Diagnostic test						
Serial correlation	0.1043	0.1349	1.7017	2.3135	0.3553	0.73063
	(0.747)	(0.713)	(0.192)	(0.128)	(0.551)	(0.393)
Functional form	2.267	1.5570	0.2983	0.6274	1.4708	1.1595
	(0.132)	(0.137)	(0.585)	(0.428)	(0.225)	(0.282)
Normality	2.5871	3.0172	2.9999	0.9722	3.5715	9.0371
	(0.274)	(0.221)	(0.223)	(0.615)	(0.168)	(0.011)
Heteroscedasticity	0.729	1.5264	0.0117	0.7415	0.2740	0.4158
	(0.393)	(0.217)	(0.914)	(0.389)	(0.601)	(0.519)

Note ***,** and *denotes the significance at 1, 5 and 10% levels respectively. t statistics in parenthesis

infers that natural resource dependence leads to a reallocation of resources from the agricultural sector to other sectors i.e. natural resource sector. In fact, this result in line with Apergis and Payne [5] and provides evidence of natural resource curse hypothesis through Dutch disease mechanism. A clear manifestation of this was the rural-urban migration in which resources were mopped-up from the rural areas in Yemen and deposited in urban areas. Young graduates migrated to the cities in search of jobs with associated abandonment of investment in agriculture.

Additionally, Dutch disease phenomenon works when natural resources booms increase domestic income and, consequently the demand for goods, which generate inflation and appreciation of the real exchange rate. Hence, the relative price of all non-natural resource commodities increases, so the exports become expensive relative to world market prices and thereby decreases the competitiveness of these commodities as well as investment in these sectors. Furthermore, the table reports a negative and significant long-run relationship between trade openness and agriculture value added. This is so as the lower prices of imported agricultural products could suppress the price of domestically produced products and discourage domestic production. This mean trade openness in Yemen excreted competitive pressure on the producers of agricultural goods. Especially, weakening the protection for agricultural products such as the fruits and vegetable, has reduced its export, thus affected negatively the welfare of those peasants who produce these goods.

Finally, the dummy variable of unification period inters the model significantly negative. This sheds more light on the declining role of agriculture sector in recent years in Yemen due to lack of resources over time, climate change, social conflicts, and lack of security are also having a significant impact on the agriculture sector.

Manufacturing Sector

For the case of manufacturing sector the result reveal that there is no role of financial development on fostering manufacturing value added in Yemen. These results are not surprising because many Yemeni small entrepreneurs that are unable to provide collateral have to create small plants that require minimal capital and rely on family members and relatives. The main private clients of the banking system are the large family-owned enterprises that dominate the formal business sector. Often they are also the main bank shareholders. Banks are reluctant to lend outside of this limited group of companies because of the difficulties they face in ensuring repayment or exercising rights over collateral [44]. Likewise the role of natural resource dependence on agriculture sector, the negative and significant role of natural resource dependence on manufacturing sector has also appeared in the Table. A 10% increase in natural resource dependence is expected to decrease manufacturing value added by between 4 and 4.9%. Therefore, it is fair to say that Yemen experienced the same kind of outcome of other natural resource dominated economies suffering from the "Dutch disease". The results of manufacturing sector model also revealed the negative role of trade openness on this which confirms the competitive pressure on the producers of manufacturing goods. Intuitively, this result is not surprising as manufacturing sector in Yemen exhibits low-quality export basket. When countries have specialized in low-quality products they are more likely to experience negative trade impact Huchet-Bourdon et al. [20].

Service Sector

Unlike the case of agriculture and manufacturing sectors, the service sector has a different story. The table reports a positive and significant relationship between financial development and service value added. A 10% increase in credit to private sector is expected to increase service value added by 3.6% and a 10% increase in the deposits is expected to increase service value added by 4.7%. This implies that service sector growth in Yemen relies on the development of financial sector. These findings confirm that financial sector in Yemen more involved in less risky modes of investment in those sectors that less volatile than other such as service sector. Furthermore, one of the main characteristics of this sector and have been revealed also by the findings is the absence of significant negative effect of natural resource dependence on this sector in the long-run which make the investment in this sector less risky than agriculture and manufacturing sectors.

Finally, the significant positive sign of the dummy variable reflects the growing importance of the service sector in Yemeni economy in recent years associated with deteriorating of the importance of the agriculture and manufacturing sectors after unification.

Short run estimation results in error-correction representation are provided in Table 2 Panel B. The table reveals the absence of any role of financial development on agriculture and manufacturing sectors, whereas the positive role on service sector is still exist. The negative effect of natural resource dependence on economic sectors has appeared in the case of the agriculture and manufacturing sectors also which confirm again the natural resource curse phenomenon in Yemeni economy through Dutch disease mechanism. The short run analysis also reveals the negative role of trade openness on agriculture sector. Furthermore, the significance coefficients of the dummy variable come in line with our long run analysis.

Finally, the coefficient of the estimated error correction model is negative and significant, this confirm the existence of long run relationship among our variables. In addition, the coefficient suggests that a deviation from the long-run equilibrium following a short-run shock is corrected by about 59 and 43% per year for agriculture sector, 21 and 24% per year for manufacturing sector and 65 and 72% per year for service sector.

Table 2 Panel C tabulates the result of some major diagnostic statistics such as the LM statistics which is test for serial correlation; the misspecification is checked by Ramsey Reset test, Heteroscedasticity and normality tests. The stability of coefficients by testing the CUSUM and CUSUMQ test are also examined.[13] Based on the results the null hypothesis of normality of residuals, null hypothesis of no misspecification of functional form null hypothesis of no first order serial correlation and null hypothesis of no heteroscedasticity are accepted. Furthermore, stability of the model was supported because the plots of both of both CUSUM and CUSUMQ fell

---

[13] Figures of CUSUM and CUSUMQ are available upon request

inside the critical bounds of 5% significance level. Finally, the size of the adjusted R^2 indicated a good fit.

The natural resource dependence plays a significant negative effect on the growth of agriculture and manufacturing sectors. These findings are clear manifestation that natural resource curse in Yemen works through Dutch disease mechanism.[14] Agriculture has declined in importance, from over 24% of GDP in 1990 to about 8% in 2012. Manufacturing has also shriveled, from about 19% to only about seven per cent. A low-paid, low-skilled urban service sector, increasingly tied to natural resource sector instead of agriculture and manufacturing, continues to account for about 40% of GDP.

The result revealed that the service sector in Yemen is the only sector that benefit from financial sector development, whereas agricultural and manufacturing sector have different story. The study found that financial development in Yemen has a negative or no effect on agriculture sector. The reason behind this result can be attributed to fact that farmers are very unlikely to borrow from the formal sector and thereby around 80% of farmers have no outstanding loans [44]. Financial sector imposes high interest rates on loans to farmers by virtue of sector having a longer production period as compared with other sectors. Furthermore, the uncertain nature of agricultural output whose risks include, uncertain prices, high input costs, climatic conditions, affect the production of this sector and thereby the ability of farmers to repay their loans. On the other hand the findings revealed that the manufacturing sector is still far from being influenced by financial sector development due to difficulties in obtaining bank credit in high interest rate environment and high level of collaterals.

Robustness Check

As a robustness check, we re-estimate the models using alternative proxy for financial development and also a new natural resource dependence proxy. In addition, we utilize two other econometric approaches i.e. FMOLS and DOLS for estimation. In general, our previous findings with the ARDL approach are robust.

The alternative financial development proxy is FD_{PCA}. This is a new proxy constructed with the previous two financial development proxies and the M2 to GDP[15] based on the Principal Components Analysis approach. This proxy is able to capture most of the information from the original dataset of these proxies (see Ang and McKibbin [3]).

---

[14]A simple univariate regression applied by IMF (2013) of the real effective exchange rate on real oil prices for the period 1995–2012 suggested that a 1% increase in oil prices leads to a real appreciation of Yemeni Rial about 0.3%

[15]This measure is considered to be the broadest measure of financial intermediation and includes three types of financial institutions: the central bank, deposit money banks and other financial institutions. Although this measure does not represent the effectiveness of the financial system, but by assuming that the size of the financial intermediary system is positively correlated with the financial system activities, this can be used for constructing PCA as a measure of financial development for robustness check

Table 3 Robustness check

	FMOLS			DOLS		
	Agriculture	Manufacturing	Service	Agriculture	Manufacturing	Service
C	5.7458***	4.4533***	2.7878***	3.6730***	4.1617***	2.0357
	(7.2668)	(5.9379)	(7.6423)	(3.7369)	(4.5688)	(1.5421)
FD_{PCA}	−0.1578**	0.0016	0.1135***	−0.3104***	−0.0087	0.0517***
	(−2.5683)	(0.0268)	(4.0041)	(−7.4820)	(−0.2251)	(3.9280)
NR_{dum}	−0.3924***	−0.0906***	−0.0972	−1.0634***	−0.4130***	−0.2663
	(−2.8252)	(−3.2748)	(−1.5164)	(−4.4108)	(−2.8484)	(−0.8225)
TO	−0.3030	−0.5506**	0.2105*	−0.2324	−0.8365*	−0.1347
	(−1.3404)	(−2.5675)	(2.0181)	(−0.5694)	(−2.2118)	(−0.2457)
GOV	−0.2361	−0.0291	−0.0011	0.3382	0.3468	0.5973
	(−0.9199)	(−0.1195)	(−0.0092)	(0.9207)	(1.0189)	(1.2110)
Dum	−0.8872***	0.0221	0.3671***	−0.7004***	0.4239**	0.5522**
	(−4.07530)	(0.1072)	(3.6552)	(−4.6420)	(3.0311)	(2.7250)

Note ***,** and *denotes the significance at 1%, 5% and 10% levels respectively. t statistics in parenthesis

Additionally, we follow Badeeb et al. [10] by using an alternative proxy for natural resource dependence. This proxy is a dummy variable for the number of years where the oil rent is greater than 10% of GDP. The results of our robustness check, which are tabulated in Table 3, are in line and in agreement with our main models that confirm our stated argument and conclusion.

6   Conclusion

This paper empirically examines the impact of natural resource dependence, financial development and trade openness on Yemeni sectoral growth. The paper found that the natural resource dependence plays a significant negative effect on the growth of agriculture and manufacturing sectors. These findings are clear manifestation that natural resource curse in Yemen works through Dutch disease mechanism. These results provide evidence on how economic structure in resource based countries is shaped by natural resource dependence. Also, the results revealed how financial sector in these countries tend to involve in less risky activities in those sectors that less volatile than other such as service sector.

This findings offer several policy implication in Yemen. On one hand, since natural resource dependence is a key obstacle to the development of sectoral growth, the country needs to rebalance its economy away from natural resource sector in order to reduce the level of natural resource dependence. The government should try to promote the service sectors where it enjoys a potential comparative advantage for example, in tourism and sea transport. The infrastructure for both services remains weak. On the other hand, strengthening the role of the financial sector in financial intermediary through accelerating the establishment of a stock market and boosting the confidence in the banking system and reforms is important step toward more economic diversification. Government should play a more proactive role in encouraging credit to enable financial sector to play a more efficient intermediary role in mobilizing domestic savings and channeling them to private productive investment across economic sectors.

References

1. Abu-Bader S, Abu-Qarn AS (2008) Financial development and economic growth: the Egyptian experience. J Policy Model 30(5):887–898
2. Albatuly A, Al-Hawri M, Cicowiez M, Lofgren H, Pournik M (2011) Assessing development strategies to achieve the MDGs in the Republic of Yemen. Country study, July 2011
3. Ang JB, McKibbin WJ (2007) Financial liberalization, financial sector development and growth: evidence from Malaysia. J Dev Econ 84(1):215–233
4. Angelini E, Marcellino M (2011) Econometric analyses with backdated data: unified Germany and the Euro area. Econ Model 28(3):1405–1414
5. Apergls N, Payne JE (2014) The oil curse, institutional quality, and growth in MENA countries: evidence from time-varying cointegration. Energy Econ 46:1–9
6. Arezki R, van der Ploeg F (2011) Do natural resources depress income per capita? Rev Dev Econ 15(3):504–521
7. Auty R (2002) Sustaining development in mineral economies: the resource curse thesis. Taylor & Francis, New York. ISBN: 9781134867905
8. Badeeb RA, Lean HH (2016) Financial development, oil dependence and economic growth: evidence from Republic of Yemen. Studies in economics and finance (to appear). Forthcoming
9. Badeeb RA, Lean HH, Clark J (2017) The evolution of the natural resource curse thesis: a critical literature survey. Resour Policy 51:123–134
10. Badeeb RA, Lean HH, Smyth R (2016) Oil curse and financegrowth nexus in Malaysia. Energy Econ 57:154–165
11. Barro RJ, Sala-i Martin X (2004) Economic growth. Advanced series in economics. McGraw-Hill, New York. ISBN: 9780262025539
12. Barro RJ (1989) A cross-country study of growth, saving, and government. Working paper 2855, National Bureau of Economic Research, February 1989
13. Bittencourt M (2012) Financial development and economic growth in Latin America: is Schumpeter right? J Policy Model 34(3):341–355
14. Burrowes RD (2010) Historical dictionary of Yemen., Asian/Oceanian historical dictionaries. Scarecrow Press, Lanham
15. Chang R, Kaltani L, Loayza N (2005) Openness can be good for growth: the role of policy complementarities. Working paper 11787, National Bureau of Economic Research, November 2005
16. Demetriades PO, Hussein KA (1996) Does financial development cause economic growth? Time-series evidence from 16 countries. J Dev Econ 51(2):387–411
17. Gylfason T (2001) Natural resources, education, and economic development. Eur Econ Rev 45(46):847–859 ISSN: 0014-2921. 15th annual congress of the European Economic Association
18. Gylfason T (2006) Natural resources and economic growth: from dependence to diversification. Springer, Berlin, pp 201–231
19. Hassan MK, Sanchez B, Yu J-S (2011) Financial development and economic growth: new evidence from panel data. Q Rev Econ Finan 51(1):88–104 ISSN: 1062-9769

20. Huchet-Bourdon M, Le Mouël CLM, VIJIL M (2011) The relationship between trade openness and economic growth: some new insights on the openness measurement issue. In: XIIIème Congrès de l'Association Européenne des Economistes Agricoles (EAAE), Zurich, August 2011
21. Kim D-H, Lin S-C (2015) Natural resources and economic development: new panel evidence. Environ Resour Econ 1–29. ISSN: 1573-1502
22. King RG, Levine R (1993) Finance and growth: Schumpeter might be right. Q J Econ 108:717–737
23. Kremers JJM, Ericsson NR, Dolado JJ (1992) The power of cointegration tests. Oxf Bull Econ Stat 54(3):325–348 ISSN: 1468-0084
24. Levine R (1997) Financial development and economic growth: views and agenda. J Econ Lit. 35(2):688–726
25. Levine R, Loayza N, Beck T (2000) Financial intermediation and growth: causality and causes. J Monet Econ 46(1):31–77 ISSN: 0304-3932
26. Lucas RE (1988) On the mechanics of economic development. J Monet Econ 22(1):3–42 ISSN: 0304-3932
27. McKinnon RI (1973) Money and capital in economic development. Brookings Institution, Washington, DC
28. Mehlum H, Moene K, Torvik R (2006) Institutions and the resource curse*. Econ J 116(508):1–20 ISSN: 1468-0297
29. Narayan PK (2005) The saving and investment nexus for China: evidence from cointegration tests. Appl Econ 37(17):1979–1990
30. Nili M, Rastad M (2007) Addressing the growth failure of the oil economies: the role of financial development. Q Rev Econ Finan 46(5):726–740 ISSN: 1062-9769
31. Pagano M (1993) Financial markets and growth. Eur Econ Rev 37(2):613–622 ISSN: 0014-2921
32. Patrick HT (1966) Financial development and economic growth in underdeveloped countries. Econ Dev Cult Change 14(2):174–189 ISSN: 00130079, 15392988
33. Pesaran MH, Shin Y, Smith RJ (2001) Bounds testing approaches to the analysis of level relationships. J Appl Econ 16(3):289–326 ISSN: 1099-1255
34. Robinson J (1952) The rate of interest and other essays. Am Econ Rev 43(4):636–641 ISSN: 00028282
35. Romer P (1990) Endogenous technological change. J Polit Econ 98:71102
36. Sachs JD, Warner A (1999) The big push, natural resource booms and growth. J Dev Econ 59(1):43–76
37. Sachs JD, Warner A (2001) The curse of natural resources. Eur Econ Rev 45(4–6):827–838
38. Sachs JD, Warner AM (1995) Natural resource abundance and economic growth. Working paper 5398, National Bureau of Economic Research, December 1995
39. Schumpeter JA (1934) The theory of economic development: an inquiry into profits, capital, credit, interest, and the business cycle. Economics third world studies. Transaction Books, New Brunswick
40. Shao S, Yang L (2014) Natural resource dependence, human capital accumulation, and economic growth: a combined explanation for the resource curse and the resource blessing. Energy Policy 74:632–642 ISSN: 0301-4215
41. Shaw ES (1974) Financial deepening in economic development. J Finan 29(4):1345–1348 ISSN: 00221082, 15406261
42. Singh T (2008) Financial development and economic growth nexus: a time-series evidence from India. Appl Econ 40(12):1615–1627
43. Torvik R (2009) Why do some resource-abundant countries succeed while others do not? Oxf Rev Econ Policy 25(2):241–256
44. United Nation Development Programme UNDP (2006) Macroeconomic policies for growth, employment and poverty reduction. Technical report, Sub-regional Resource Facility for Arab States (SURF-AS), Beirut

45. Vetlov I, Warmedinger T (2006) The German block of the ESCB multi-country model. Working paper 654, European Central Bank, Frankfurt am Main, Germany
46. World Bank (2013) Project appraisal document on a proposed grant to the Republic of Yemen. Working paper 78624YE, Finance and Private Sector Development Department Middle East and North Africa Region
47. Wurgler J (2000) Financial markets and the allocation of capital. J Financ Econ 58(1):187–214

Can Bagging Improve the Forecasting Performance of Tourism Demand Models?

Haiyan Song, Stephen F. Witt and Richard T. Qiu

Abstract This study examines the forecasting performance of the general-to-specific (GETS) models developed for Hong Kong through the bootstrap aggregating method (known as bagging). Although the literature in other research areas shows that bagging can improve the forecasting performance of GETS models, the empirical analysis in this study does not confirm this conclusion. This study is the first attempt to apply bagging to tourism forecasting, but additional effort is needed to examine the effectiveness of bagging in tourism forecasting by extending the models to cover more destination-source markets related to destinations other than Hong Kong.

Keywords Bagging · General-to-specific modeling · Tourism demand · Hong Kong

1 Introduction

Tourism demand modeling and forecasting plays a crucial role in the process of decision making among tourism stakeholders in both the public and private sector. As policy makers and private practitioners base their decisions largely on tourism demand forecasts, efforts to improve the accuracy of tourism demand forecasts are ongoing.

H. Song (✉) · R.T. Qiu
School of Hotel and Tourism Management, The Hong Kong Polytechnic University,
17 Science Museum Road, TST East, Kowloon, Hong Kong SAR
e-mail: haiyan.song@polyu.edu.hk

R.T. Qiu
e-mail: richard.tr.qiu@connect.polyu.hk

S.F. Witt
School of Hospitality and Tourism Management, University of Surrey,
Guildford, Surrey GU2 7XH, UK
e-mail: stephen_f_witt@hotmail.com

© Springer International Publishing AG 2017   419
V. Kreinovich et al. (eds.), *Robustness in Econometrics*,
Studies in Computational Intelligence 692, DOI 10.1007/978-3-319-50742-2_25

In the field of tourism demand forecasting, the general-to-specific (GETS) modeling procedure has proved an effective tool due to its ease of specification and robustness in model estimation [23]. In contrast to its counterpart, the specific-to-general approach, the GETS procedure starts with a general model that contains all possible influencing factors, and reduces the model to its final form by eliminating insignificant factors recursively using t-statistics [4, 11, 21, 22]. However, as it suffers from an unstable decision rule (the rule for eliminating insignificant factors using t-statistics) in the model reduction process, the final forecasts may not be "optimal".

One possible way to overcome this "unstable decision rule" problem is the bootstrap aggregating (bagging) method proposed by Breiman [2] and Bühlmann and Yu [3]. Inoue and Kilian [9] and Rapach and Strauss [15, 16] demonstrated its effectiveness using estimations of U.S. inflation and U.S. national and regional employment growth, respectively. In this paper, we try to apply the bagging procedure to GETS forecasting based on Hong Kong tourism data to investigate whether GETS-bagging can overcome the "unstable decision rule" problem in demand forecasting for the Hong Kong tourism industry.

The rest of the paper is structured as follows. Section 2 briefly introduces the literature related to Hong Kong tourism demand forecasting, GETS, and the GETS-bagging procedure. Section 2 discusses the GETS-bagging method and the dataset used in the study, Sect. 4 shows the results of the forecasting exercise, and Sect. 5 concludes.

2 Literature Review

Hong Kong is one of the world's most popular tourist destinations. According to the Hong Kong Tourism Board [8], Hong Kong received 59.3 million visitors in 2015, putting it among the most popular tourist destinations in the world. Tourism has been and remains the second largest source of foreign currency in Hong Kong, and the income generated has contributed around 6% of Hong Kong's GDP over the last decade. Many businesses such as retailing, catering, accommodation, and entertainment are directly and indirectly influenced by the growth of tourism in Hong Kong.

Over the past two decades, growing research attention has been drawn to the modeling and forecasting of Hong Kong's tourism demand. A few studies have focused on modeling the trends and business cycles of tourism demand in Hong Kong based on time-series forecasting techniques [31]. Among others, Hiemstra and Wong [6] and Song and Wong [25] identified the key factors affecting Hong Kong's tourism demand and analyzed the demand elasticity based on econometric models. Song et al. [20] and [19] investigated the forecasting performance of alternative time-series and econometric forecasting techniques, while Song et al. [26] predicted the future growth of Hong Kong's tourism demand from key source markets. Song et al. [18, 24] developed a system to generate reliable forecasts of Hong Kong's tourism demand.

In the last category, the authors developed a web-based system called the Hong Kong Tourism Demand Forecasting System (HKTDFS) which uses the GETS procedure. The factors considered in the HKTDFS include the income level of tourists from the source markets, the prices of tourism products/services in Hong Kong (measured by Hong Kong's CPI relative to that of the source markets adjusted by the exchange rate between the Hong Kong dollar and the source market currencies), the prices of substitute destinations (also adjusted by the relevant exchange rates), the marketing expenditure of such destinations, and so on [23]. The system starts with an autoregressive distributed lag model (ADLM) incorporating all possible factors that may affect the demand for Hong Kong tourism together with their lagged values (four-period lagged values for each variable as the system uses quarterly data) and dummy variables (including seasonal dummies and one-off event dummies). This model is then estimated using the OLS method. The insignificant variable with the largest p-value is eliminated, and the OLS estimation process repeated until all variables left in the model are both statistically and economically significant (that is, the coefficients of the variables have the correct signs according to economic theory). Song et al. [17] showed that the ADLM used in the HKTDFS produces relatively accurate forecasts. However, they also mentioned that the model can generate relatively large forecasting errors for volatile markets such as mainland China and Taiwan (with a mean absolute percentage error greater than 10%). Furthermore, using t-statistics as a decision rule to eliminate variables can be problematic as the explanatory variables are correlated [2]. This "unstable decision rule" also prevents the forecasting system from being fully automated. The investigation in this paper can therefore be considered an extension of the HKTDFS, as it explores an alternative way to reduce the forecasting error and automate the forecasting process.

The term "bagging" was introduced by Breiman [2] to stand for "**b**ootstrap **agg**regating". It is an ensemble method combining multiple predictors. To improve the accuracy of the model, it trains multiple models on different samples (data splits) and averages their predictions. A large number (e.g. B) of bootstrap samples are first drawn, B predictions are then generated by applying the model to these samples, and the bagging predictor is finally calculated by averaging these predictions. This is based on the concept that the *"averaging of misclassification errors on different data splits gives a better estimate of the predictive ability of a learning method"* [32]. Experiments show that the bagging predictor works well for unstable learning algorithms, and a reduction of 21 to 46% can be made in mean squared errors (MSE) when bagging is applied to the regression tree [2].

3 The Model and Data

The GETS procedure was used in the web-based HKTDFS by Song et al. [24]. As a potential extension of that study, the same procedure is adopted in this investigation. The model starts with a general ADLM in the form of

$$q_{i,t+h}^{h} = \alpha_i + \sum_{j=1}^{k} \beta_{i,j} X_{i,j,t} + \varepsilon_{i,t+h}^{h}, \tag{1}$$

where $q_{i,t+h}^{h}$ is the demand for Hong Kong tourism (measured by total tourist arrivals) among residents in country i at time $t + h$, t is the time index with a maximum of T, h is the forecast horizon, $X_{i,j,t}$ are the vectors of k explanatory variables, including the lagged values of the independent and dummy variables, $\varepsilon_{i,t+h}^{h}$ is the error term, and α and βs are the parameters to be estimated.

Due to the data requirements for the bagging procedure, monthly data is used in this investigation instead of the quarterly data used in the HKTDFS. For the same reason, only relative price, substitute price, and GDP per capita (together with their lagged values) are considered as independent variables in this study. These variables are proven to be the most important factors determining tourism demand [10, 29]. Thus, the GETS procedure in this investigation is more of a "lag selector" than a "variable selector" in the HKTDFS. The model then becomes

$$q_{i,t+h}^{h} = \alpha_i + \sum_{j=0}^{12} \beta_{i,j} RP_{i,t-j} + \sum_{m=0}^{12} \gamma_{i,m} SP_{i,t-m} + \sum_{n=0}^{12} \varphi_{i,n} GDP_{i,t-n}$$

$$+ \sum_{k=2}^{12} ds_{i,k} DS_k + \sum_{p=1}^{x} de_{i,p} DE_p + \varepsilon_{i,t+h}^{h}, \tag{2}$$

where $RP_{i,t}$ are the relative prices, $SP_{i,t}$ are the substitute prices, $GDP_{i,t}$ are the GDP per capita in particular source markets, DS_ks are the seasonal dummies (with the first dummy for January being omitted to avoid collinearity), DE_ps are x one-off event dummies, and α, β, γ, φ, ds, and de are the parameters to be estimated.

To carry out the GETS procedure, this general model is then estimated using OLS. The estimates for all of the coefficients are then sorted by their t-statistics in ascending order. The variable associated with the first coefficient (the coefficient with the smallest t-statistics or largest p-value) is eliminated from the model if it is statistically insignificant. Here, the elimination of insignificant variables is done in a recursive manner instead of as a one-off act, as suggested by Song et al. [23]. The above procedure is repeated until all variables left in the model are significant (or all variables are dropped). The treatment of seasonal dummies is worth mentioning: as seasonality always has a considerable influence on tourism demand, all seasonal dummies are excluded from the variable elimination procedure. The GETS forecasts are calculated as

$$\hat{q}_{i,t+h}^{h,GETS} = \hat{\alpha}_i + \sum_{j=0}^{12} \hat{\beta}_{i,j} I_{i,j} RP_{i,t-j} + \sum_{m=0}^{12} \hat{\gamma}_{i,m} I_{i,m} SP_{i,t-m}$$

$$+ \sum_{n=0}^{12} \hat{\varphi}_{i,n} I_{i,n} GDP_{i,t-n} + \sum_{k=2}^{12} \hat{ds}_{i,k} DS_k + \sum_{p=1}^{x} \hat{de}_{i,p} I_{i,p} DE_p, \tag{3}$$

where $I_{i,j}$, $I_{i,m}$, $I_{i,n}$, and $I_{i,p}$ are relevant dummies that take the value of one if the associated coefficient is significant, and zero otherwise, and $\hat{\alpha}_i$, $\hat{\beta}_{i,j}$, $\hat{\gamma}_{i,m}$, $\hat{\varphi}_{i,n}$, $\hat{ds}_{i,k}$, and $\hat{de}_{i,p}$ are the OLS estimators of the model.

In the bagging procedure, a large number ($B = 100$ in this investigation) of bootstrap samples are generated from the original dataset. As the dataset contains time-series data, the moving-block bootstrap is used to maintain the structure of the data. For each draw, a block of 12 observations (as monthly data is used) is picked from the dataset (with replacement). After $\lceil T12 \rceil$ draws, a sample of ($\lceil T12 \rceil \times 12$) observations is generated, and the first T observations from this sample are used as one bootstrap sample. For each bootstrap sample (indexed by b), a series of GETS forecasts can be calculated using Eq. (3). The GETS-bagging forecasts can then be calculated as the average of these GETS forecasts,

$$\hat{q}_{i,t+h}^{h,GB} = \frac{1}{B} \sum_{b=1}^{B} \hat{q}_{i,t+h,b}^{h,GB}, \tag{4}$$

where $\hat{q}_{i,t+h,b}^{h,GB}$ is the GETS forecast for bootstrap sample b and $\hat{q}_{i,t+h}^{h,GB}$ is the GETS-bagging forecast for the total number of arrivals of tourists from country i at time t+h.

The data used for this investigation include the total arrivals in Hong Kong ($q_{i,t}$) from three source markets, namely mainland China, the U.S., and the U.K., where mainland China represents a short-haul market and the U.S. and U.K. represent long-haul markets. Australia was considered a long-haul sample from Oceania but was later excluded due to data availability. These data are obtained from statistical reports of the Hong Kong Tourism Board [7].

The relative price is the price of Hong Kong tourism relative to that of the source markets ($RP_{i,t}$), it is defined as

$$RP_{i,t} = \frac{CPI_{HK,t}/EX_{HK,t}}{CPI_{i,t}/EX_{i,t}}, \tag{5}$$

where $CPI_{i,t}$ is the consumer price index for Hong Kong (or the origin country i), and $EX_{i,t}$ is the exchange rate between the Hong Kong dollar (or currency of origin country i) and the U.S. dollar.

The substitute price is the price of tourism in substitute destinations relative to Hong Kong ($SP_{i,t}$), and is defined as

$$SP_{i,t} = \sum_{j=1}^{J} \frac{CPI_{j,t}}{EX_{j,t}} w_{i,j,t}, \qquad (6)$$

where $J = 6$, representing the 6 substitute destinations including mainland China, Taiwan, Singapore, Thailand, Korea, and Japan [25], and $w_{i,j,t}$ is the share of international tourist arrivals at region j, calculated as

$$w_{i,j,t} = \frac{q_{i,j,t}}{\sum_{j=1}^{J} q_{i,j,t}}. \qquad (7)$$

where $q_{i,j,t}$ is the total number of arrivals of tourists from country i to country j at time t. Notice that when mainland China is the origin country under examination, it is excluded from the calculation of the substitute price ($J = 5$ in this case).

The CPIs of mainland China, Korea, Japan, the U.K., and the U.S. are obtained from the OECD database [14], the CPI of Hong Kong is obtained from the Census and Statistics Department of the Hong Kong SAR Government [5], the CPI of Taiwan is obtained from National Statistics, Republic of China (Taiwan) [12], the CPI of Singapore is obtained from Statistics Singapore [27], and the CPI of Thailand is obtained from the Bank of Thailand [1]. The exchange rate data for all countries/regions above are obtained from OANDA fxTrade™ [13]. The data on tourists arrivals are retrieved from UNWTO [28].

The model includes three one-off event dummies. The first represents the effect of the 9/11 attack, which takes the value of one from September, 2001 to December, 2001 and zero otherwise. The second represents the effect of the Beijing Olympics in 2008, which takes the value of one from July, 2008 to December, 2008 and zero otherwise. The third represents the effect of the subprime mortgage crisis starting from 2008, which takes the value of one from January, 2008 to December, 2010 and zero otherwise.

4   The Forecasting Results

The GETS-bagging forecasts are generated using Eqs. (3) and (4) and the data described in Sect. 3. To compare the forecast accuracy of the GETS-bagging procedure with that of the pure GETS procedure, a series of GETS forecasts are also generated using Eq. (3) and the original dataset.

Figure 1 shows the results for the 1-period-ahead ($h = 1$) total arrival forecasts of all three countries. The results for $h = 2$ to 12 are available up request.

From Fig. 1, we can see that the forecasts of both procedures work similarly, with the GETS-bagging procedure responds more to variations in the explanatory variables. This "overreaction" problem downgrades the performance of the GETS-bagging procedure. Both procedures work poorly in the early stage of the forecasts,

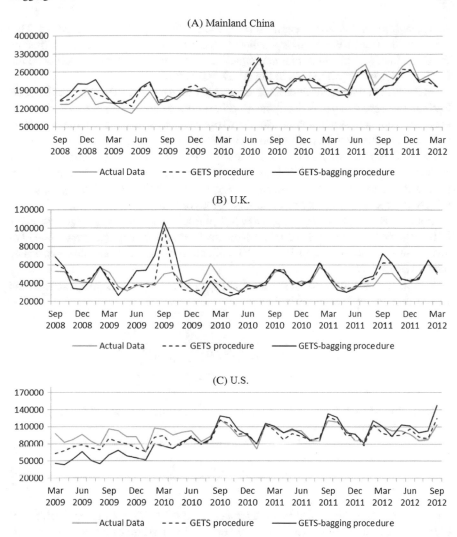

Fig. 1 1-period-ahead forecasts

caused by significant changes in the explanatory variables after 2008 due to the sub-prime mortgage crisis, although it is already controlled by the dummy. This phenomenon is more significant in the case of the U.S., which suffered the most during the crisis.

In the case of the U.K., a spike is forecast by both procedures around September, 2009. This is caused by a significant increase in the exchange rate of the pound sterling (GBP) against the U.S. Dollar (USD), which increased from 0.589 USD/GBP in October, 2008 to a peak of 0.704 USD/GBP in March, 2009, and recovered to 0.604 USD/GBP in August, 2009. The GETS-bagging procedure responds more to this shock and generates less accurate forecasts than does the GETS procedure. This spike in the forecasts becomes smaller as the forecast horizon increases (results are available from the authors upon request), but the improvement is smaller in the GETS-bagging procedure than in the GETS procedure.

In the case of mainland China, the forecasts become less accurate when the forecast horizon increases, whereas the opposite occurs in the case of the U.K. The forecast accuracy improves as h increases from 1 to 6 but worsens afterwards for the U.S. Thus it seems that the performance of both procedures has little to do with the forecast horizon. This phenomenon is also identifiable in the forecasting accuracy in later sections. In general, the GETS-bagging procedure does not outperform the GETS procedure as expected; therefore improving the HKTDFS by switching from the GETS to the GETS-bagging procedure is not an option.

Four measures of forecasting accuracy, namely the root mean squared error (RMSE), mean absolute error (MAE), mean absolute percentage error (MAPE), and mean absolute scaled error (MASE), are calculated to compare the forecasts of the two procedures.

RMSE is a quadratic scoring rule which measures the average magnitude of the forecasting errors. It is calculated by

$$RMSE_i^h = \sqrt{\frac{\sum_{t=1}^{T}(\hat{q}_{i,t+h}^{h,GB} - q_{i,t+h}^h)^2}{T}}. \tag{8}$$

MAE is the average magnitude of the forecasting errors without considering their direction. It is calculated by

$$MAE_i^h = \frac{1}{T}\sum_{t=1}^{T}\left|\hat{q}_{i,t+h}^{h,GB} - q_{i,t+h}^h\right|. \tag{9}$$

As errors are squared before being averaged in the RMSE, this gives a relatively high weight to large forecasting errors whereas the MAE gives equal weight to all forecasting errors. Together, the difference between RMSE and MAE can be used to diagnose the variation in the forecasting errors of both the GETS-bagging and GETS procedures.

The MAPE and MASE are measures of forecasting accuracy at the percentage level. They are calculated by

$$MAPE_i^h = \frac{1}{T}\sum_{t=1}^{T}\left|\frac{\hat{q}_{i,t+h}^{h,GB} - q_{i,t+h}^h}{q_{i,t+h}^h}\right|, \tag{10}$$

and

$$MASE_i^h = \frac{\displaystyle\sum_{t=1}^{T} \left| \hat{q}_{i,t+h}^{h,GB} - q_{i,t+h}^h \right|}{\dfrac{T}{T-1} \displaystyle\sum_{t=2}^{T} \left| q_{i,t+h}^h - q_{i,t+h-1}^h \right|}, \tag{11}$$

respectively. As these two measures are scale-free error metrics, they can be used not only to compare the GETS-bagging and GETS procedures in this investigation, but also to compare these two procedures with other procedures in future investigations.

Table 1 shows the RMSE of the two procedures for all three countries. The numbers in parentheses for the RMSE of the GETS-bagging procedure are the percentages of the RMSE of the GETS-bagging procedure relative to that of the GETS procedure. Thus, an improvement in the GETS-bagging procedure is shown if the number is below 100%. However, all these numbers are above 100%. This means that, according to the RMSE in this investigation, the GETS-bagging procedure is outperformed by the GETS procedure. The same conclusion can be drawn from the MAE, MAPE, and MASE comparisons (see Appendix).

As mentioned above, the difference between the RMSE and MAE can be used to diagnose the variation in the forecasting errors. The percentage of this difference is calculated by

$$\Delta\% = \frac{RMSE - MAE}{MAE}. \tag{12}$$

Table 2 compares this difference for both procedures for all countries.

Among the 36 groups of comparisons, 12 show the GETS-bagging differences to be smaller than the GETS differences. That is, in these 12 groups, the GETS-bagging procedure generates less variation in forecasting errors. Interestingly, of these 12 groups, 7 are forecasts for mainland China, the most volatile source market among the three. In these cases, the GETS-bagging procedure does serve to reduce the variance in forecasting error. However, the increase in bias spoils the forecasts so that the GETS-bagging procedure is outperformed by the GETS procedure. The variance reduction becomes more obvious when the source market is volatile. It is possible that, with a highly volatile source market, the reduction in variance exceeds the effects of increased bias, and the GETS-bagging procedure may then outperform the GETS procedure.

5   Concluding Remarks

The GETS-bagging procedure did not yield the expected results in our investigation. Although it reduced the variance in forecasting error to some extent in the case of mainland China, the forecasting error itself was increased compared with the GETS procedure. Furthermore, the interpretable structure is also lost in the process of bagging. However, the failure of the GETS-bagging procedure in this investigation does not imply a general failure of the procedure in tourism demand forecasting. As mentioned in Song et al. [17], the HKTDFS generates considerably accurate forecasts, and Breiman [2] indicated that "bagging can improve only if the unbagged is not optimal". This may be one of the reasons the GETS-bagging procedure was outperformed by the GETS procedure in the HKTDFS. Also, the linear regression using all variables is a fairly stable procedure, but the stability may decrease as the number of variables used in the predictor decreases. While the GETS-bagging procedure did not improve the HKTDFS, improvements in other aspects remain possible. As the model reduction process is sensitive to the sequence of removing insignificant variables, the process can vary from researcher to researcher. The final model may thus not be the "optimal" one due to this subjective influence. Judgmental adjustments with input from experts proposed by Song et al. [18] were able to reduce the forecasting error. However, these adjustments prevent the model from being automated.

Table 1 The RMSE of both procedures for all three countries

h	Mainland China		U.K.		U.S.	
	GETS	GETS-bagging	GETS	GETS-bagging	GETS	GETS-bagging
1	316992.29	347383.78 (109.6%)	9401.16	13944.74 (148.3%)	11836.58	21395.19 (180.8%)
2	342914.17	365484.59 (106.6%)	6026.12	15518.18 (257.5%)	11997.23	21530.27 (179.5%)
3	370967.52	382118.47 (103.0%)	6063.99	15194 (250.6%)	12187.47	20762.62 (170.4%)
4	363281.29	409763.51 (112.8%)	6166.34	14072.01 (228.2%)	13308.6	20842.35 (156.6%)
5	360874.27	429463.34 (119.0%)	5941.2	11482.92 (193.3%)	11069.28	20177.43 (182.3%)
6	408599.73	481190.18 (117.8%)	6196.81	10049.01 (162.2%)	11974.54	18972.67 (158.4%)
7	452876.15	579940.73 (128.1%)	5718.32	7830.49 (136.9%)	11974.12	21652.69 (180.8%)
8	424726.69	561524.24 (132.2%)	5706.74	7899.73 (138.4%)	12046.05	24285.48 (201.6%)
9	431886.14	530321.82 (122.8%)	7140.06	8432.27 (118.1%)	11011.98	25755.24 (233.9%)
10	398612.85	519010.39 (130.2%)	7593.48	8484.53 (111.7%)	14024.57	27406.42 (195.4%)
11	425273.43	482090.82 (113.4%)	6620.05	8537.44 (129.0%)	10757.54	33441.7 (310.9%)
12	458713.61	472027.7 (102.9%)	6951.79	8804.68 (126.7%)	11120.63	37875.14 (340.6%)

Table 2 The RMSE exceeds the MAE in percentages

h	Mainland China		U.K.		U.S.	
	GETS (%)	GETS-bagging (%)	GETS (%)	GETS-bagging (%)	GETS (%)	GETS-bagging (%)
1	127.6	125.6	178.9	149.2	124.3	129.6
		(98.5)		(83.4)		(104.3)
2	121.6	119.3	128.9	144.8	135.6	133.0
		(98.1)		(112.4)		(98.1)
3	122.8	124.1	126.5	144.7	127.6	133.7
		(101.1)		(114.4)		(104.8)
4	119.7	124.0	130.7	149.6	124.4	126.9
		(103.6)		(114.5)		(102.0)
5	120.6	124.3	117.1	143.8	122.9	127.9
		(103.0)		(122.9)		(104.1)
6	122.2	118.4	120.2	137.6	123.2	127.1
		(96.9)		(114.4)		(103.2)
7	121.9	113.4	122.3	126.1	122.6	132.4
		(93.0)		(103.2)		(108.0)
8	126.9	118.0	125.0	128.1	126.6	128.8
		(93.0)		(102.4)		(101.7)
9	116.1	121.6	122.0	127.6	124.3	129.8
		(104.8)		(104.6)		(104.4)
10	113.9	117.0	119.5	124.7	118.7	129.7
		(102.8)		(104.4)		(109.2)
11	125.5	122.2	122.3	126.1	122.0	119.6
		(97.4)		(103.1)		(98.1)
12	125.3	120.9	126.0	125.0	124.4	113.9
		(96.5)		(99.1)		(91.5)

Another alternative is to reconsider the dropped variables each time an insignificant variable is dropped. In the GETS process, whenever an insignificant variable is dropped, the already dropped variables can be reintroduced into the model to seek forecasting error reduction. This extra step can, to some extent, correct the "bad" drops. More importantly, it can be automated by computer. A Bayesian estimation is also an alternative. Wong et al. [30] showed that imposing this prior to the VAR model can improve model performance and reduce the forecasting error. Given that the t-statistics used in the GETS procedure can be problematic due to serial correlation among the explanatory variables, using Bayesian factors instead may improve the forecasting results.

Appendix: Tables of the MAE, MAPE, and MASE of Both Procedures for all Three Countries

See Appendix Tables 3, 4, 5.

Table 3 MAE of both procedures for all three countries

h	Mainland China		U.K.		U.S.	
	GETS	GETS-bagging	GETS	GETS-bagging	GETS	GETS-bagging
1	248514.61	276615.46 (111.3%)	5253.77	9345.38 (177.9%)	9520.478	16505.54 (173.4%)
2	282037.25	306362.47 (108.6%)	4676.06	10714.25 (229.1%)	8845.252	16186.01 (183.0%)
3	302180.56	307888.23 (101.9%)	4795.33	10498.54 (218.9%)	9551.193	15530.67 (162.6%)
4	303385.15	330366.69 (108.9%)	4717.01	9403.36 (199.3%)	10694.32	16419.15 (153.5%)
5	299147.38	345496.76 (115.5%)	5075.21	7982.88 (157.3%)	9004.759	15774.76 (175.2%)
6	334481.12	406464.95 (121.5%)	5154.13	7302.96 (141.7%)	9720.252	14929.27 (153.6%)
7	371528.4	511454.96 (137.7%)	4676.93	6208.69 (132.8%)	9766.333	16353.55 (167.4%)
8	334608.97	475861.52 (142.2%)	4563.59	6168.62 (135.2%)	9515.025	18855.38 (198.2%)
9	372134.15	436171.47 (117.2%)	5850.82	6608.72 (113.0%)	8858.444	19840.88 (224.0%)
10	350112.39	443452.18 (126.7%)	6356.59	6805.44 (107.1%)	11813.87	21133.23 (178.9%)
11	338785.28	394423.24 (116.4%)	5412.29	6770.56 (125.1%)	8819.396	27954.85 (317.0%)
12	366183.62	390310.34 (106.6%)	5515.32	7046.35 (127.8%)	8939.031	33260.36 (372.1%)

Table 4 MAPE of both procedures for all three countries

h	Mainland China		U.K.		U.S.	
	GETS (%)	GETS-bagging (%)	GETS (%)	GETS-bagging (%)	GETS (%)	GETS-bagging (%)
1	13	15	12	21	10	17
		(115.2)		(181.4)		(176.1)
2	15	17	11	24	9	17
		(110.8)		(228.3)		(185.2)
3	16	17	11	24	10	16
		(102.8)		(218.2)		(165.1)
4	16	17	11	21	11	17
		(106.3)		(197.0)		(155.9)
5	14	17	12	18	9	17
		(115.9)		(154.3)		(175.8)
6	17	19	12	17	10	17
		(114.2)		(138.7)		(156.0)
7	18	25	11	15	10	17
		(137.3)		(130.3)		(167.1)
8	16	22	11	14	10	19
		(136.6)		(131.3)		(195.0)
9	18	19	13	15	9	20
		(108.3)		(113.7)		(217.2)
10	16	19	14	15	12	22
		(115.9)		(107.0)		(175.1)
11	15	16	12	15	9	29
		(109.0)		(124.1)		(303.6)
12	16	16	12	16	9	34
		(96.9)		(127.5)		(363.1)

Table 5 MASE of both procedures for all three countries

h	Mainland China		U.K.		U.S.	
	GETS (%)	GETS-bagging (%)	GETS (%)	GETS-bagging (%)	GETS (%)	GETS-bagging (%)
1	84	93 (111.3)	71	126 (177.9)	79	137 (172.2)
2	95	103 (108.6)	63	144 (229.1)	74	134 (180.9)
3	102	104 (101.9)	65	141 (218.9)	80	129 (162.2)
4	102	111 (108.9)	63	127 (199.3)	89	133 (148.9)
5	101	116 (115.5)	68	107 (157.3)	75	121 (160.5)
6	113	137 (121.5)	69	98 (141.7)	81	114 (140.7)
7	125	172 (137.7)	63	84 (132.8)	82	129 (157.6)
8	113	160 (142.2)	61	83 (135.2)	79	149 (187.0)
9	125	147 (117.2)	79	89 (113.0)	74	159 (214.7)
10	118	149 (126.7)	86	92 (107.1)	99	165 (167.4)
11	114	133 (116.4)	73	91 (125.1)	74	228 (309.4)
12	123	131 (106.6)	74	95 (127.8)	75	278 (371.9)

References

1. Bank of Thailand (2016) Statistical reports, various issues. http://www.bot.or.th/English
2. Breiman L (1996) Bagging predictors. Mach Learn 24(2):123–140
3. Büchlmann P, Yu B (2002) Analyzing bagging. Ann Stat 30(4):927–961
4. Campos J, Ericsson NR, Hendry DF (2005) General-to-specific modeling: an overview and selected bibliography. FRB International Finance Discussion Paper No. 838
5. Census Statistics Department (2016) Statistical reports, various issues. http://www.censtatd.gov.hk/
6. Hiemstra S, Wong KKF (2002) Factors affecting demand for tourism in Hong Kong. J Travel Tourism Market 13(1–2):41–60
7. Hong Kong Tourism Board (2016) Annual reports. http://www.discoverhongkong.com/eng/about-hktb/annual-report/index.jsp

8. Hong Kong Tourism Board (2016) Tourism statistics 12 2015. http://partnernet.hktb.com/filemanager/intranet/ViS_Stat/ViS_Stat_E/ViS_E_2015/Tourism_Statistics_12_2015_0.pdf
9. Inoue A, Kilian L (2008) How useful is bagging in forecasting economic time series? A case study of U.S. consumer price inflation. J Am Stat Assoc 103(482):511–522
10. Li G, Song H, Witt SF (2005) Recent developments in econometric modeling and forecasting. J Travel Res 44(1):82–99
11. Narayan PK (2004) Fijis tourism demand: the ARDL approach to cointegration. Tourism Econ 10(2):193–206
12. National Statistics (2016) Statistical reports, various issues. http://eng.stat.gov.tw
13. OANDA (2016) fxTrade^{TM}. Historical exchange rates | OANDA. https://www.oanda.com/lang/cns/currency/historical-rates/
14. OECD Statistics (2016). https://stats.oecd.org/
15. Rapach DE, Strauss JK (2010) Bagging or combining (or both)? An analysis based on forecasting U.S. employment growth. Econometric Rev 29(5–6):511–533
16. Rapach DE, Strauss JK, Forecasting US (2012) State-level employment growth: an amalgamation approach. Int J Forecast 28(2):315–327
17. Song H, Dwyer L, Li G, Cao Z (2012) Tourism economics research: a review and assessment. Ann Tourism Res 39(3):1653–1682
18. Song H, Gao Z, Zhang X, Lin S (2012) A web-based Hong Kong tourism demand forecasting system. Int J Networking Virtual Organ 10(3–4):275–291
19. Song H, Li G, Witt SF, Athanasopoulos G (2011) Forecasting tourist arrivals using time-varying parameter structural time series models. Int J Forecast 27(3):855–869
20. Song H, Li G, Witt SF, Fei B (2010) Tourism demand modelling and forecasting: how should demand be measured? Tourism Econ 16(1):63–81
21. Song H, Lin S, Zhang X, Gao Z (2010) Global financial/economic crisis and tourist arrival forecasts for Hong Kong. Asia Pac J Tourism Res 15(2):223–242
22. Song H, Witt SF (2003) Tourism forecasting: the general-to-specific approach. J Travel Res 42(1):65–74
23. Song H, Witt SF, Li G (2008) The advanced econometrics of tourism demand. Routledge, New York
24. Song H, Witt SF, Zhang X (2008) Developing a web-based tourism demand forecasting system. Tourism Econ 14(3):445–468
25. Song H, Wong KKF (2003) Tourism demand modeling: a time-varying parameter approach. J Travel Res 42(1):57–64
26. Song H, Wong KKF, Chon KKS (2003) Modelling and forecasting the demand for Hong Kong tourism. Intl J Hosp Manage 22(4):435–451
27. Statistics Singapore (2016) Statistical reports, various issues. http://www.singstat.gov.sg
28. UNWTO (2016) Statistical reports, various issues. http://unwto.org
29. Witt SF, Witt CA (1995) Forecasting tourism demand: a review of empirical research. Int J Forecast 11(3):447–475
30. Wong KK, Song H, Chon KS (2006) Bayesian models for tourism demand forecasting. Tour Manag 27(5):773–780
31. Wong KKF (1997) The relevance of business cycles in forecasting international tourist arrivals. Tour Manag 18(8):581–586
32. Zhao Y, Cen Y (2013) Data mining applications with R. Academic Press, Amsterdam

The Role of Asian Credit Default Swap Index in Portfolio Risk Management

Jianxu Liu, Chatchai Khiewngamdee and Songsak Sriboonchitta

Abstract This paper aims at evaluating the performance of Asian Credit Default Swap (CDS) index in risk measurement and portfolio optimization by using several multivariate copulas-GARCH models with Expected Shortfall and Sharpe ratio. Multivariate copula-GARCH models consider the volatility and dependence structures of financial assets so that they are conductive to accurately predict risk and optimal portfolio. We find that vine copulas have better performance than other multivariate copulas in model estimation, while the multivariate T copulas have better performance than other kinds of copulas in risk measurement and portfolio optimization. Therefore, the model estimation, risk measurement, and portfolio optimization in empirical study should use different copula models. More importantly, the empirical results give evidences that Asian CDS index can reduce risk.

Keywords Gold · Crude oil · Bond · Expected shortfall · Copula · Sharpe ratio

1 Introduction

Since the global financial crisis in 2008, most countries around the world still have suffered from the economic slowdown. Therefore, policymakers have tried to stimulate their own economy by using economic stimulus packages, the so called Quantitative Easing. These packages, however, have caused the global financial market to get more volatile as the policies were changed more often and hence created more uncertainty. It also created higher risks in stock, commodities and bonds markets. Thus, it is more difficult for investors and corporates to manage their risks from market volatility.

Investors are now seeking ways to reduce the risks of the portfolios as they normally position their portfolio into asset class—equities, fixed-income, cash and commodities—because asset class is more liquid than other assets, for example real estate. This causes investors to face greater risks of their portfolios. Fortunately,

J. Liu (✉) · C. Khiewngamdee · S. Sriboonchitta
Faculty of Economics, Chiang Mai University, Chiang Mai, Thailand
e-mail: getliecon@gmail.com

© Springer International Publishing AG 2017
V. Kreinovich et al. (eds.), *Robustness in Econometrics*,
Studies in Computational Intelligence 692, DOI 10.1007/978-3-319-50742-2_26

credit default swap (CDS) provides very effective hedging strategies and has been widely used for risk management in terms of reducing risk exposure. It also has proved itself to be a helpful tool to manage the risk of the portfolio. Compared to American and European markets, CDS index spreads in Asia yield a higher average return after the crisis. Moreover, CDS index in Asia has grown rapidly over the last decade as investors have moved their investment to emerging markets in Asia.

The risk measurement of the CDS has been widely studied in several papers. For instance, Amato [1] estimated CDS risk premia and default risk aversion. Oh and Patton [2] modeled the systemic risk of CDS measurement. Additionally, there also are studies that analyze the risk of the portfolio consisting CDS. For example, Schönbucher [3], Kiesel and Scherer [4] and Kim et al. [5] measured the valuation losses on CDS portfolio. Pedersen [6] also evaluated the valuation and risk of CDS option portfolio. Furthermore, Bo and Capponi [7] considered the optimal investment in CDS portfolio under contagion risk. However, only a few studies, for example Raunig and Scheicher [8], have compared the risk between investing in CDS and other assets. In this paper, we bridge this gap by analyzing the risk of the CDS portfolio compared with another portfolio excluding CDS. In order to measure the risk of the portfolio, we introduce copula-based volatility models which take into account the joint dependence between asset returns.

Copula concept recently has been used to improve the accuracy of the risk measurement in finance such as Value at Risk (VaR) and Expected Shortfall (ES). Many studies applied copula based GARCH model and found that these copulas based models are more reliable than thus superior to ordinary GARCH model. For example, Hürlimann [9], Wei and Zhang [10], He and Gong [11] and Wang et al. [12] employed multivariate copula based GARCH models to evaluate the risk of the portfolio using VaR and ES. Their results show that copula based GARCH model is practicable and more effective in capturing the dependence structure between asset returns and hence accurately measure the risks.

Furthermore, other studies relaxed standard copula assumptions by applying vine copulas which allow different dependency structures between variables and more flexible in high dimensional data. For instance, Emmanouil and Nikos [13], Weiß and Supper [14], Sriboonchitta et al. [15] and Zhang et al. [16] estimated VaR and ES by using vine copulas based GARCH models and found that vine copula based models accurately forecast risk and return of the portfolios. Hence, vine copulas become a useful tool to estimate risks (Guegan and Maugis [17]; Low et al. [18]).

In this paper, we are interested in examining the effect of Asian CDS index on portfolio risk and return. In other words, we investigate whether including Asian CDS index to portfolio can reduce the risk or increase the return of the portfolio. In order to capture the main asset classes, we conduct our portfolio consisting of stock, commodities, and bond as they are traded widely in global market. We employ copula based GARCH models and Expected Shortfall to evaluate the risk and return of the portfolio. Firstly, we filter the data by using GARCH model and employ our copulas approach to estimate the dependence parameters between asset returns. We apply both standard and high dimensional copula based volatility models, namely multivariate copula, C-vine and D-vine copulas. Then, we estimate the risk of the equally weighted portfolio by applying the Expected Shortfall concept. We also construct another identical portfolio but includes Asian CDS index and re-estimate the risk of this portfolio. Finally, by the best fitting copula model, we compare the risk and return of these two portfolios using Sharpe ratio and confirm our result by employing robustness test.

The remainder of this paper is organized as follows. In Sect. 2, we describe the dataset. Section 3 presents the methodologies used in this study. Section 4 provides the empirical results and final section gives conclusions.

2 Data

In this paper, we use iTraxx Asia ex-Japan CDS index, Hang Seng future Index, 10-year US government bond, gold future price, and Brent crude oil future price to measure risk and return in financial market. Our sample covers the period from April 1st, 2011, to March 31st, 2016, with 1186 daily observations totally. All data are obtained from Thomson Reuter Database. We partition the data into two parts:

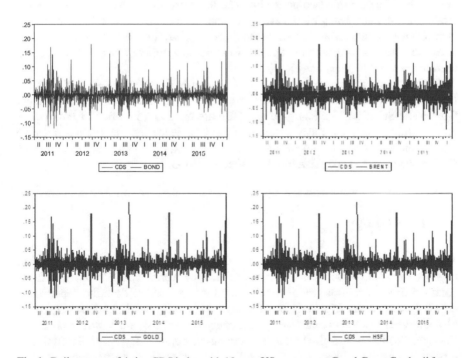

Fig. 1 Daily returns of Asian CDS index with 10-year US government Bond, Brent Crude oil future price, Gold future price and Hang Seng future index. We can find that CDS has stronger volatility than others, and all assets show phenomenon of volatility clustering, thereby implying that using GARCH model is appropriate in our study

Table 1 The estimate results of Kendall's tau

Kendall's tau	CDS	HSF	Bond	Crude oil	Gold
CDS	1	−0.3553***	0.1915***	−0.1215***	−0.03502***
HSF	−0.3553***	1	−0.0919***	0.0903***	0.0129
Bond	0.1915***	−0.0919***	1	−0.1751***	0.0761***
Crude oil	−0.1215***	0.0903***	−0.1751***	1	0.1089***
Gold	−0.0350*	0.0129	0.0761***	0.1089***	1

*, **, and *** denote rejection of the null hypothesis at the 10%, 5%, and 1% significance levels, respectively

in sample and out of sample. The in-sample data from April 1st, 2011 to March 31st, 2015, with 710 observations is used to estimate the parameters of the copula-based GARCH models. Thereafter, the 250 observations in the out of sample are used to forecasts the optimal portfolios and ES by using the principle of the daily rolling window forecasting, and to perform robust check whether the ES with copula-based GARCH models being used in our paper are adequate. To answer the question whether Asian CDS index can reduce risk or make more profit for portfolio or not. We divide the assets into two portfolios. The first portfolio includes 4 assets: bond, crude oil, gold, and Hang Seng index, we name it non-CDS portfolio. The second portfolio we include Asian CDS index to those 4 assets, this portfolio is called CDS portfolio. The risk measurement and portfolio optimization of two portfolios can be used to answer this question (Fig. 1).

Table 1 reports the estimate results of Kendall's tau. Firstly, we find that the Kendall's tau for the most of pairs are statistically significant at 5% level except Gold and HSF. It implies that gold and HSF are unrelated in terms of rank correlation, while other pairs have certain relationship. Secondly, there exists negative correlation between CDS and HSF, CDS and crude oil, CDS and gold, which means CDS probably can reduce risk for diversified portfolio.

3 Methodology

This study combines copula functions and GARCH model to capture volatility and dependence structure between asset returns. In other words, GARCH model reflects volatility of asset returns, while copula functions are used to describe dependence between asset returns. In general, considering the volatility and dependence structures of financial assets are conductive to accurately predict risk and optimal portfolio. The specification of marginal models, multivariate copulas, Expected Shortfall and robustness check are introduced in the following sub-sections, respectively.

3.1 Specification for Marginal Models

According to Bollerslev [19], the basic GARCH (1,1) model can be expressed as

$$r_t = \sigma_t z_t, \tag{1}$$

$$\sigma_t^2 = \omega + \alpha \varepsilon_{t-1}^2 + \beta \sigma_{t-1}^2, \tag{2}$$

where r_t is the return of time series r at time t, z_t is the standardized residual which can be assume as any distribution. σ_t^2 is the conditional variance of r_t, which depends on the previous value of squared error term and the previous value of the conditional variance. In order to capture the characteristics of heavy tail and asymmetry for the marginal Von Rohr and Hoeschele [20], we assume that the distribution for the standardized residual, z_t, follows a skewed Student-t distribution:

$$f_{skt}(z_t | \upsilon, \gamma) = \frac{2}{\gamma + \gamma^{-1}} \left(f_\upsilon(\frac{z_t}{\gamma}) I_{[0,\infty)}(z_t) + f_\upsilon(\gamma z_t) I_{(-\infty,o)}(z_t) \right), \tag{3}$$

where $f_\upsilon(\cdot)$ is the density of the Student-t distribution with υ degrees of freedom, is the indicator function to capture the asymmetry effect of the bull and bear markets, and γ is the skewness parameter ranging from 0 to ∞. If $\gamma > 1$, the distribution is skewed to the right; if $\gamma = 1$, it is symmetric; and if $0 < \gamma < 1$, the distribution is skewed to the left.

3.2 Specification for Multivariate Copulas

The essential characteristic of copula model is that any multivariate distribution function can be decomposed into the marginal distributions, which describe the return pattern of each asset individually, such as peak kurtosis and heavy tail, etc., and the copula, which fully captures the dependence between the asset returns Wu et al. [21]. In multivariate copulas, multivariate Gaussian and T copulas are generally used to analyze financial risk because of their easy implementation. Three-dimensional multivariate Gaussian and T copulas can be expressed as

$$c_{Gau}(u_1, u_2, u_3 | R) = \Phi_R(\Phi^{-1}(u_1), \Phi^{-1}(u_2), \Phi^{-1}(u_3)), \tag{4}$$

$$c_T(u_1, u_2, u_3 | R, v) = T_R(T_v^{-1}(u_1), T_v^{-1}(u_2), T_v^{-1}(u_3)), \tag{5}$$

respectively, where Φ^{-1} represents an inverse standard normal distribution function. Obviously, u_i is the values of the marginal probability distribution functions, $i = 1, 2, 3$. T_v^{-1} is an inverse student-t distribution with degree of freedom v. R is a correlation matrix, and there are several forms, exchangeable (ex), Toeplitz (toep), and unstructured (un), which are commonly used to describe dependence structures

as follows: $\begin{pmatrix} 1 & \rho & \rho \\ \rho & 1 & \rho \\ \rho & \rho & 1 \end{pmatrix}$, $\begin{pmatrix} 1 & \rho_1 & \rho_2 \\ \rho_1 & 1 & \rho_1 \\ \rho_2 & \rho_1 & 1 \end{pmatrix}$ and $\begin{pmatrix} 1 & \rho_1 & \rho_2 \\ \rho_1 & 1 & \rho_3 \\ \rho_2 & \rho_3 & 1 \end{pmatrix}$.

Vine copulas are a kind of multivariate copulas which use bivariate copulas to construct multivariate distributions, and specify the dependence and conditional dependence of selected pairs of random variables and all marginal distribution functions. A $d-$dimensional vine copula is decomposed into $d(d-1)/2$ pair-copulas, and the densities of multivariate vine copulas can be factorized in terms of pair-copulas and margins. Bedford and Cooke [22, 23] proposed two subclasses of pair-copula constructions (PCC), which are called as C-vine and D-vine copulas. Following Aas et al. [24], the densities of C-vines and D-vines are, respectively,

$$f(x_1, x_2, ..., x_d) = \prod_{k=1}^{d} f(x_k) \times \prod_{j=1}^{d-1} \prod_{i=1}^{d-j} c_{j,i+j|1,...,j-1}(F(x_j|x_{1:j-1}), F(x_{i+j}|x_{i+j|1:j-1})), \quad (6)$$

$$f(x_1, x_2, ..., x_d) = \prod_{k=1}^{d} f(x_k)$$
$$\times \prod_{j=1}^{d-1} \prod_{i=1}^{d-j} c_{j,i+j|i+1,...,i+j-1}(F(x_i|x_{i+1:i+j-1}), F(x_{i+j}|x_{i+1|i+1:i+j-1})), \quad (7)$$

where $F(\cdot|\cdot)$ is the conditional distribution which can be got from the first deviation of copulas, for example,

$$F(x_1|x_2) = \frac{\partial C_{x_1,x_2}(F(x_1), F(x_2))}{\partial F(x_2)}, \quad (8)$$

is the conditional distribution, given one variable. The marginal conditional distributions of given multivariate variables can be expressed by the form $F(x|v)$,

$$F(x|v) = \frac{\partial C_{x,v_j|v_{-j}}(F(x|v_{-j}), F(v_j|v_{-j}))}{\partial F(v_j|v_{-j})}, \quad (9)$$

where v stands for all the conditional variables. The copula family used in our work includes Gaussian copula, T copula, Clayton copula, Gumbel copula, Frank copula, BB1, BB7, BB8, and rotate copulas. If we select different copulas for all pairs in C-vine or D-vine copula, then this kind of C-vine or D-vine copula is called mixed C-vine or D-vine copula. Different copulas have different characteristics. In order to accurately capture the dependency, we apply copulas as much as possible. In addition, there are few studies that used vine copula with only Clayton or T copula to optimize portfolio strategies and measure risk. For example, Low et al. [18] employed the C-vine copula with only Clayton copula to examine several portfolio strategies, while only Brechmann et al. [25] used the vine copula with only T copula to measure the performance of several VaRs. Therefore, we also applies vine copulas with Clayton and T copulas models in this study thereby comparing them with mixed C-vine and D-vine copulas in terms of risk measurement and portfolio return.

Moreover, we use two-stage estimation method to estimate multivariate copula-based GARCH model. This method is well known as inference functions for margins (IFM). At first stage, we use maximum likelihood method to estimate GARCH with skewed student-t distribution model, and then we substitute marginals into multivariate copula functions, and estimate dependence parameters by maximum likelihood estimation method. Estimators by IFM are close to and asymptotically efficient to the maximum likelihood estimator under some regularity conditions (Joe [26]).

3.3 Expected Shortfall and Robustness Check

ES satisfies the property of subadditivity and provides a more conservative measure of losses relative to VaR which is a quantile. Following Rockafellar and Uryasev [27], the equation of ES can be expressed as

$$ES_\beta(r|w, \beta) = \alpha + \frac{1}{q(1 - \beta)} \sum_{k=1}^{q} [-w^T r_k - \alpha]^+, \tag{10}$$

where q represents the number of samples generated by Monte Carlo simulation, α represents VaR. β represents the threshold value usually set at 97.5% as suggested in the revised version of Basel III in 2013, and r_k is the kth vector of simulated returns. In this study, we simulate 5,000 possible return values at $t + 1$ period for each variable. The process of Monte Carlo simulation with multivariate copula-based GARCH model is explained by several studies, Liu et al. [28] and Aas et al. [24], etc.

In order to evaluate the performance of all multivariate copula-based models, the Percentage of Failure Likelihood Ratio (PoFLR) is used to test the model is accurate or not. The PoFLR is a widely known test based on failure rates and VaR has been suggested by Kupiec [29]. Under the null hypothesis of the model being correct, large PoFLR value is indicative that the proposed model systematically understates or overstates the underlying level of risk. Since PoFLR is based on VaR model, while VaR model does not consider the size of the expected loss. So, a test with ES should be used to test the robustness of the copula-based models. Therefore, mean predictive squared error (MPSE) is employed to robustness check for all multivariate copula-based GARCH models. The MPSE is given as

$$M = \frac{1}{N} \sum_{t=1}^{N} 1 + (ES_{t+i} - R_{t+i})^2, \tag{11}$$

where M represents MPSE that implies the average loss. R denotes the equally weighted return. N is the number of out-of-sample. The more robust model is the less value of M.

4 Empirical Results

In this section, we first filter all assets series by using GARCH with skewed-student-t distribution model. We find that skewed student-t distribution fits very well. Asian CDS index is skewed to right, leptokurtic, and fat-tailed. Hang Seng future index and bond seemingly are symmetric. Crude oil and gold future price are skewed to left (as shown in Fig. 2).

Table 2 reports the values of AIC of multivariate Gaussian and T copulas, and C-vine and D-vine copulas. The results present that a mixed D-vine copula has a better performance than others for CDS portfolio, while, for non-CDS portfolio, a mixed C-vine copula is selected in terms of AIC. Moreover, C-vine and D-vine with only T copula also report a good performance as well as the unstructured multivariate Gaussian and T copulas due to the very small values of AIC compared to other copulas. Nevertheless, the Exchangeable and Toeplitz forms of Gaussian and T copulas underperform others for both CDS and non-CDS portfolios. Since most of multivariate copulas have a good performance except Gaussian and T with exchangeable

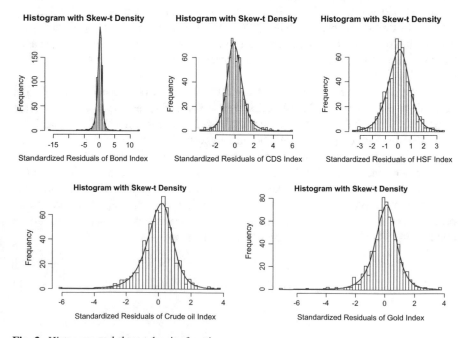

Fig. 2 Histogram and skew-t density function

Table 2 The values of AIC of multivariate copulas

Copulas	CDS Portfolio		Non-CDS Portfolio	
	Number of parameter	AIC	Number of parameter	AIC
Gaussian(ex)	1	−12.9923	1	−0.385
Gaussian(toep)	4	−118.544	3	−40.696
Gaussian(un)	10	−389.484	6	−141.563
T(ex)	2	−110.646	2	−56.619
T(toep)	5	−188.339	3	−84.350
T(un)	11	−428.753	7	−168.495
Mixed C-vine	14	−443.797	**9**	**−179.046**
T C-vine	20	−431.515	12	−172.395
Mixed D-vine	**15**	**−449.574**	10	−178.320
T D-vine	20	−434.200	12	−170.008

Table 3 The results of PoFLR and MPSE

Copulas	Number of violation	Expected number of violation	PoFLR	MPSE
CDS Portfolio				
Gaussian(un)	3	6	2.1393	1.000007
T(un)	**4**	**6**	**0.9504**	**1.000004**
Mixed C-vine	1	6	6.9471***	1.000011
T C-vine	1	6	6.9471***	1.000011
Mixed D-vine	**4**	**6**	**0.9504**	**1.000004**
T D-vine	3	6	2.1393	1.000005
Non-CDS Portfolio				
Gaussian(un)	6	6	0.0103	1.000122
T(un)	6	6	0.0103	1.000124
Mixed C-vine	6	6	0.0103	1.000138
T C-vine	6	6	0.0103	1.000137
Mixed D-vine	**6**	**6**	**0.0103**	**1.000119**
T D-vine	**6**	**6**	**0.0103**	**1.000115**

*, **, and *** denote rejection of the null hypothesis at the 10%, 5%, and 1% significance levels, respectively

Note T C-vine or T D-vine copula represents C-vine or D-vine with only T copulas

and Toeplitz forms, so we use them to calculate VaR and ES for an equally weighted portfolio, and then use PoFLR and MPSE to evaluate the risk for all models.

Table 3 presents the numbers of violation, and the results of PoFLR and MPSE. We find that, for CDS portfolio, the numbers of violation in mixed C-vine and T

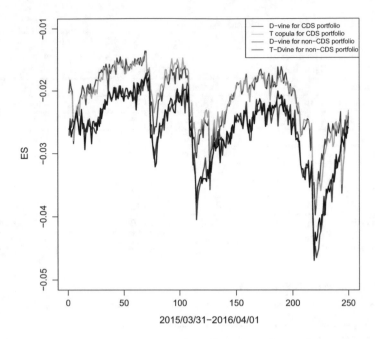

Fig. 3 ES of equally weighted portfolio for both CDS and non-CDS portfolios. It shows that the shapes of ES for these two portfolios are similar, which implies both portfolios face the same risk. Furthermore, the curve of ES for CDS portfolio is above the curve of ES for non-CDS portfolio. It implies that Asian CDS index reduces the risk of the portfolio, which is the reason that why CDS is the most widely used type of credit derivative

C-vine copula models equal to 1, and the PoFLR tests of both mixed C-vine and T C-vine models reject null hypothesis, which means the mixed C-vine and T C-vine copula models underestimate risk. The outperforming models are the unstructured multivariate T and mixed D-vine copulas which report the least violation from the expected violation number. Interestingly, for non-CDS portfolio, the expected violation is equal to the actual numbers of violation in all models, and the PoFLR test also accepts null hypothesis, which implies all multivariate copula models have good performance for forecasting VaR in terms of PoFLR test.

In addition, the results of MPSE show that unstructured multivariate T and mixed D-vine copulas are more robust than others due to the minimum MPSE value for CDS portfolio, while T D-vine and mixed D-vine copulas show more robust results than others for non-CDS portfolio. Therefore, we can conclude that the multivariate T copula and the mix D-vine copula models for CDS portfolio, and T D-vine copula and mixed D-vine copula model for non-CDS portfolio can predict ES accurately (Fig. 3).

Now we turn to portfolio optimization of CDS and non-CDS portfolio. We use maximum Sharpe ratio method to get optimal portfolio allocation, and calculate cumulative returns of these two portfolios. The Sharpe ratio measures the tradeoff

Table 4 The results of return based on maximum Sharpe ratio portfolio

Copulas	CDS Portfolio			Non-CDS Portfolio		
	Mean	S.D.	Cumulative return	Mean	S.D.	Cumulative return
Gaussian(un)	−0.0004	0.0104	0.8979	0.0003	0.0181	1.0395
T(un)	**0.0006**	**0.0125**	**1.1554**	−0.0004	0.0174	0.8542
Mixed C-vine	−0.0002	0.0121	0.9323	0.0003	0.0136	1.0519
T C-vine	0.0001	0.0141	1.0079	**0.0012**	**0.0155**	**1.3016**
Mixed D-vine	0.0001	0.0119	1.0062	−0.0002	0.0134	0.9227
T D-vine	−0.0002	0.0108	0.9442	0.0007	0.0157	1.1665

between risk and return for each portfolio. Since the earlier results show that Asian CDS index can reduce the risk for the portfolio, can Asian CDS index increase the portfolio returns? By employing maximum Sharpe ratio method, we can find out a portfolio on efficient frontier and hence investigate the effect of Asian CDS index in portfolio returns by comparing the return of CDS and non-CDS portfolios.

Table 4 presents the results of cumulative returns based on multivariate copula models for CDS and non-CDS portfolios. The results show that CDS portfolio has the maximum cumulative return of 15% which is obtained by unstructured T copula. Surprisingly, the mixed D-vine copula, which is the best performance copula in terms of AIC and MPSE, does not achieve good profit. For non-CDS portfolio, we obtain the maximum return of 30%, by using T C-vine copula model. This is much higher than the use of the mixed C-vine copula model which is the best one in terms of AIC, and the T D-vine copula model which is the best performance according to PoFLR and MPSE.

There are two critical conclusions that we can gain. First and foremost, Asian CDS index achieves risk reduction of the portfolio, however, it also reduces the portfolio returns. Second, if multivariate copulas have similar performance in model estimation in terms of AIC, we should use all of them to calculate VaR, ES, and portfolio optimization due to the uncertain performances among them.

5 Conclusion

Volatility and dependence structures of financial assets play a critical role in risk measurement and portfolio allocation. Negative correlation between financial assets also is deemed to serve as useful tools for strategic asset allocation and risk management. For these reasons, multivariate copula-GARCH models have attracted much attention among academics and institutional investors. In this study, the main purpose of using multivariate copula-GARCH models is to point out the role of Asian Credit

Default Swap (CDS) index in risk measurement and portfolio optimization of financial assets. To achieve the purpose, we calculate ES of equally weighted CDS and non-CDS portfolios corresponding to the best copula models, and then we calculate the cumulative returns of optimal portfolios using maximum Sharpe ratio method.

Our empirical results show that the performance of the multivariate copula-GARCH models are not consistent in model estimation, VaR, ES, and portfolio optimization. In model estimation, the mixed C-vine copula for non-CDS portfolio and mixed D-vine copula for CDS portfolio are selected in terms of AIC. In risk measurement, the unstructured multivariate T and the mixed D-vine copulas are more robust than others due to the minimum MPSE value for CDS portfolio, while T D-vine and mixed D-vine copulas are more robust than others for non-CDS portfolio. In portfolio optimization, the unstructured T copula and T C-vine copula show the highest cumulative returns for the portfolios of CDS and non-CDS portfolio, respectively. Therefore, in order to more accurately predict risk and portfolio allocations, most of multivariate copulas should be applied in empirical study. The most important finding is that Asian CDS index can reduce risk in investors' portfolio. However, from the maximum Sharpe ratio result, it also decreases the cumulative return of the portfolio. Therefore, risk-averse investors are willing to buy Asian CDS index for risk reduction while gain lower return.

Acknowledgements The financial support from the Puay Ungphakorn Centre of Excellence in Econometrics is greatly acknowledged. We would also like to express our gratitude to the many colleagues with whom, through the years, we have had the pleasure of discussing ideas on copulas and their applications.

References

1. Amato JD (2005) Risk aversion and risk premia in the CDS market. BIS Q Rev
2. Oh DH, Patton AJ (2016) Time-varying systemic risk: evidence from a dynamic copula model of CDS spreads. J Bus Econ Stat (just-accepted):1–47
3. Schönbucher P (2005) Portfolio losses and the term structure of loss transition rates: a new methodology for the pricing of portfolio credit derivatives. Working paper
4. Kiesel R, Scherer M (2007) Dynamic credit portfolio modelling in structural models with jumps. Preprint, Universität Ulm
5. Kim DH, Loretan M, Remolona EM (2010) Contagion and risk premia in the amplification of crisis: evidence from Asian names in the global CDS market. J Asian Econ 21(3):314–326
6. Pedersen CM (2003) Valuation of portfolio credit default swaptions. Lehman Brothers Quantitative Credit Research
7. Bo L, Capponi A (2014) Optimal investment in credit derivatives portfolio under contagion risk. Math Finan
8. Raunig B, Scheicher M (2008) A value at risk analysis of credit default swaps
9. Hürlimann W (2004) Multivariate Fréchet copulas and conditional value-at-risk. Int J Math Math Sci 2004(7):345–364
10. Wei YH, Zhang SY (2007) Multivariate Copula-GARCH model and its applications in financial risk analysis. Appl Stat Manage 3:008
11. He X, Gong P (2009) Measuring the coupled risks: a copula-based CVaR model. J Comput Appl Math 223(2):1066–1080

12. Wang ZR, Chen XH, Jin YB, Zhou YJ (2010) Estimating risk of foreign exchange portfolio: using VaR and CVaR based on GARCHEVT-Copula model. Physica A Stat Mech Appl 389(21):4918–4928
13. Emmanouil KN, Nikos N (2012) Extreme value theory and mixed canonical vine Copulas on modelling energy price risks. Working paper
14. Weiß GN, Supper H (2013) Forecasting liquidity-adjusted intraday Value-at-Risk with vine copulas. J Bank Finan 37(9):3334–3350
15. Sriboonchitta S, Liu J, Kreinovich V, Nguyen HT (2014) A vine copula approach for analyzing financial risk and co-movement of the Indonesian, Philippine and Thailand stock markets. In: Modeling dependence in econometrics. Springer International Publishing, pp 245–257
16. Zhang B, Wei Y, Yu J, Lai X, Peng Z (2014) Forecasting VaR and ES of stock index portfolio: a vine copula method. Physica A Stat Mech Appl 416:112–124
17. Guegan D, Maugis, PA (2010) An econometric study of vine copulas. SSRN 1590296
18. Low RKY, Alcock J, Faff R, Brailsford T (2013) Canonical vine copulas in the context of modern portfolio management: are they worth it? J Bank Finan 37(8):3085–3099
19. Bollerslev T (1986) Generalized autoregressive conditional heteroskedasticity. J Econometrics 31(3):307–327
20. Von Rohr P, Hoeschele I (2002) Bayesian QTL mapping using skewed Student-t distributions. Genetics Selection Evolution 34(1):1
21. Wu CC, Chung H, Chang YH (2012) The economic value of co-movement between oil price and exchange rate using copula-based GARCH models. Energy Econ 34(1):270–282
22. Bedford T, Cooke RM (2001) Monte Carlo simulation of vine dependent random variables for applications in uncertainty analysis. In: 2001 Proceedings of ESREL 2001, Turin, Italy
23. Bedford T, Cooke RM (2002) Vines-a new graphical model for dependent random variables. Ann Stat 30(4):10311068
24. Aas K, Czado C, Frigessi A, Bakken H (2009) Pair-copula construction of multiple dependence. Insur Math Econ 44:182198
25. Brechmann EC, Czado C, Paterlini S (2014) Flexible dependence modeling of operational risk losses and its impact on total capital requirements. J Bank Finan 40:271285
26. Joe H (2005) Asymptotic efficiency of the two-stage estimation method for copula based models. J Multivar Anal 94:401419
27. Rockafellar RT, Uryasev S (2002) Conditional value-at-risk for general loss distributions. J Bank Finan 26(7):1443–1471
28. Liu J, Sriboonchitta S, Phochanachan P, Tang J (2015) Volatility and dependence for systemic risk measurement of the international financial system. Lecture notes in artificial intelligence (Subseries of Lecture notes in computer science), vol 9376. Springer, Heidelberg, pp 403–414
29. Kupiec P (1995) Techniques for verifying the accuracy of risk measurement models. J Deriv 3:7384

Chinese Outbound Tourism Demand to Singapore, Malaysia and Thailand Destinations: A Study of Political Events and Holiday Impacts

Jianxu Liu, Duangthip Sirikanchanarak, Jiachun Xie
and Songsak Sriboonchitta

Abstract This chapter investigates the effects of Thailand's political turmoil and the Chinese Spring Festival on the dynamic dependence between the Chinese outbound tourism demand for Singapore, Malaysia and Thailand (SMT) using the bivariate and multivariate dynamic copula-based ARMAX-APARCH model with skewed Student's t-distribution and normal inverse Gaussian marginals. We selected political events and the Chinese Spring Festival as the forcing variables to explain the time-varying dependences, and also proposed a dynamic multivariate Gaussian copula to capture the dependence between the Chinese outbound tourism demand for Singapore, Malaysia and Thailand. The main empirical results show that Thailand's political turmoil and the Chinese Spring Festival, respectively, have negative and positive effects on Chinese tourist arrivals to SMT. Also, there does exist a high degree of persistence pertaining to the dependence structure among SMT. In addition, both the lagged one period of Thailand's political turmoil and the Chinese Spring Festival are found to have a positive influence on time-varying dependences. Lastly, we found that substitute effects exist between Thailand and Malaysia, while complementary effects prevail between Thailand and Singapore, and Singapore and Malaysia. The findings of this study have important implications for destination managers and travel agents as they help them to understand the impact of political events and holidays on China outbound tourism demand and provide them with a complementary academic approach on evaluating the role of dependencies in the international tourism demand model.

Keywords Copula · GARCH · Tourism · China

J. Liu · D. Sirikanchanarak · S. Sriboonchitta (✉)
Faculty of Economics, Chiang Mai University, Chiang Mai, Thailand
e-mail: songsakecon@gmail.com

J. Xie
School of Statistics and Mathematics, Yun Nan University of Finance and Economics,
Kun Ming 650221, China

© Springer International Publishing AG 2017
V. Kreinovich et al. (eds.), *Robustness in Econometrics*,
Studies in Computational Intelligence 692, DOI 10.1007/978-3-319-50742-2_27

1   Introduction

Since 2012, China's position as the largest outbound tourism market and the highest outbound tourism spender has been further consolidated. In 2013, 98.19 million Chinese traveled abroad, with the outbound expenditure reaching 128.7 billion USD, according to the China Tourism Academy. This figure is the greatest in the world in this regard. In this year 2013, Thailand, Singapore and Malaysia are the top three outbound destinations for Chinese tourists in the ASEAN region. In Asia, they are ranked first, third and fourth. The combination of the three countries—Singapore, Malaysia and Thailand (SMT)—is one of the most popular tour packages for the Chinese tourist. China has become the major tourism source market for SMT, as well. China has become the second largest tourists source country for Singapore, only behind Indonesia, since 2011; China is Malaysia's third largest source of tourists, following Indonesia and Singapore in the same year; for Thailand, China has been the largest passenger source market since 2012. The Chinese tourist arrivals to Singapore, Malaysia and Thailand totaled 4.71 million, 2.27 million and 1.79 million, respectively, in 2013. Thus, it can be seen that Thailand occupies the leading position among SMT. In addition, Chang et al. [2], Liu and Sriboonchitta [20] and Liu et al. [11] found that there exists a significant interdependence between SMT.

However, there is ongoing political turmoil in Thailand. Political shocks in Thailand are bound to have an impact on its economic development. Therefore, it might be that negative events can also affect the tourism economy by keeping Chinese tourists away from Thailand, thereby having a potential impact on the Chinese tourist arrivals in Singapore and Malaysia. Thus, one of the purposes of this paper is to investigate whether political turmoil in Thailand has an influence on the tourism demand for Thailand from China, besides examining whether there exist any potential spillover effects on tourism in Singapore and Malaysia. The second purpose is to estimate the extent of impact of political shocks on the dependence structure of SMT. Also, this paper attempts to investigate whether the Chinese Spring Festival can determine the tourism demand and the dependence structure of SMT. In addition, we attempt to answer the following questions: (1) Is the relationship between SMT as regards tourism demand constant or time-varying? (2) What kind of spillover effects do the negative shocks of Thailand have on Singapore and Malaysia: substitution effects or complementary effects?

The contributions of this article are threefold. Firstly, we propose the copula-based ARMAX-APARCH models to elastically describe the political turmoil effect, and holiday effect, volatility, leverage effect of SMT countries, as well as the dependence structure of tourism demand for Singapore and Malaysia (SM), Singapore and Thailand (ST), and Malaysia and Thailand (MT). We found that political events and holidays have negative and positive effects on tourist arrivals from China to SMT. The correlations between SM, ST and MT are not invariant, and political events and the Chinese Spring Festival do have an impact on the time-varying dependences of each pair. Secondly, we invented a time-varying multivariate Gaussian copula to capture the dynamic dependence of tourism demand for SMT and investigated the effects of political turmoil and the Chinese Spring Festival on their time-varying dependence. We observed that both the lagged one period of political events and the Chinese Spring Festival have significantly positive impact on the time-varying dependences between SMT. Lastly, some policy planning methods are recommended for the consideration of governments and travel agencies of SMT.

This study is organised as follows. Section 2 presents literature review. Section 3 reviews the data and methodology, including the marginal specification, the static and time-varying copulas, and the goodness of fit. The empirical findings are presented in Sect. 4. Policy planning is discussed in Sect. 5. Some concluding remarks are given in Sect. 6.

2 Literature Review

Several studies have been conducted to investigate the effects of political shocks on tourism demand [6, 13, 15, 19]. They all showed that political instability can hinder tourism development and damage economic growth. However, just a few papers used econometric models to investigate the effects of political instability on tourism demand, such as Neumayer [13] and Saha and Yap [15], using a panel data model. We conclude that most of the scholars just studied this problem in terms of qualitative analysis or macroeconomics.

In addition, there have been a few studies that focused on tourism demand by applying various econometric models to analyse the interdependencies of international tourism demand between destinations, such as Chan et al. [1], Shareef and McAleer [17], Seo et al. [16] and Chang et al. [2]. Chan et al. [1] investigated the interdependencies between major tourism source countries to Australia, namely Japan, New Zealand, the UK and the USA, using the symmetric Constant Conditional Correlation-Multivariate Generalised Autoregression Conditional Heteroscedasticity (CCC-MGARCH) model. Shareef and McAleer [17] studied the uncertainty of monthly international tourist arrivals from the eight major tourist source countries to the Maldives by using the symmetric CCC-MGARCH model as well. Seo et al. [16] employed the MGARCH model with the dynamic conditional correlation (DCC) specification to estimate the conditional correlation between the Korean tourism demand for Jeju Island, Thailand, Singapore and the Philippines, and the results showed that the conditional correlations are not constant but time-varying. Chang et al. [2] applied several models, such as VARMA-GARCH, VARMA-AGARCH and CCC models, to investigate the interdependence of the four leading destinations in the ASEAN region.

Nevertheless, the CCC-MGARCH, the DCC-MGARCH and the Vector Auto Regression Moving Average (VARMA)-GARCH were assumed to have a linear relationship with a multivariate Student's t or normal distribution [20, 23]. In many cases, however, these assumptions do not conform to the data because the distributions of the data are usually skewed (asymmetric), heavily tailed and leptokurtic, with different marginal distributions (Sriboonchitta et al. [20]). To deal with these drawbacks, this study applies the copula-based APARCH model, one of the most frequently used methods to estimate static or time-varying dependence, to analyse the time-varying relationships of tourism demand among SMT, and to estimate the impact of political turmoil and the Chinese Spring Festival on the dependencies between SMT. The copula-based GARCH models allow for better flexibility in joint distributions than bivariate normal or Students t-distributions. Patton [14] and Jondeau and Rockinger [10] independently proposed the copula-based GARCH model to analyse the dependencies of the exchange rate market, and an international stock market, respectively. Thereafter, the copula-based GARCH model has been widely used and developed. For example, Zhang and Guégan [24] proposed the GARCH processes with the time-varying copula model for option pricing; Huang et al. [5] presented an application of the copula-GARCH model in the estimation of a portfolios value at risk; Wu et al. [22, 23] proposed the dynamic copula-based GARCH model to describe the dependence structures of stock and bond returns, the oil price and the US dollar exchange rate; and Tang et al. [21] modeled the dependence between tourism demand and exchange rate using the copula-based GARCH model.

Based on previous research, this paper improves the copula-based GARCH model, for example, by substituting the ARMAX-APARCH model for the GARCH model, adding political events and the Chinese Spring Festival as the forcing variables in the time-varying equation, putting time-varying properties into multivariate Gaussian copulas etc. This study is applicable for destination competitive strategies and policy development.

3 Data and Methodology

3.1 Data

Our data set consists of Thailand political events, the Chinese Spring Festival and the number of tourist arrivals from China to SMT, where the Chinese Spring Festival and political events are dummy variables. The sample period covers January 1999 to September 2013. A total of 177 monthly observations for each country are obtained from EcoWin database. Our data set includes 15 observations for the Chinese Spring Festival, and 8 political events of Thailand. The 8 political events of Thailand are those that happened on December 2005, September 2006, October 2006, September 2008, December 2008, May 2009, April 2010 and May 2010. Figure 1 shows political events in Thailand, the Chinese Spring Festival and the number of tourist arrivals from China to SMT. The figure demonstrates that tourist arrivals from China to SMT increase over time, in general, and that there obviously exists non-linear correlation between SMT. We also can find that the number of tourist arrivals from China to SMT reaches the peak value at the Chinese Spring Festival, while the tourist volume becomes a trough at the time of Thailand political events. Obviously, Thailand political events and the Chinese Spring Festival have an influence on the tourist flow. Since the peak

Fig. 1 Political event of Thailand, the Chinese Spring festival, and the number of tourist arrivals from China to SMT

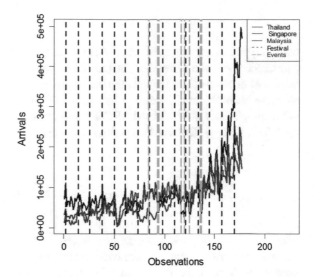

and the trough of tourist flow for SMT are concordant, Thailand's political events and the Chinese Spring Festival might have an impact on the dependence between SMT. We used a logarithmic transformation of the data to stabilize the increase in their volatility over time. Thereafter, this paper uses $r_{i,t} = \ln(Y_{i,t}/Y_{i,t-1})$ to measure the growth rates of the monthly tourist arrivals from China, where $i = 1, 2, 3$ represents the countries of SMT.

3.2 Methodology

3.2.1 Marginal Specifications

Ding et al. [3] proposed the Asymmetric Power ARCH (APARCH) model that allows for leverage and Taylor effects, and this model also nests several ARCH-type models, such as GARCH, GJR-GARCH and nonlinear ARCH etc. The ARMAX-APARCH model is given by

$$r_t = c + \sum_{i=1}^{p} AR_i r_{t-i} + \sum_{i=1}^{q} MA_i \varepsilon_{t-i} \sum_{i=1}^{k} \varphi_i X_{it} + \varepsilon_t, \tag{1}$$

$$\sigma_t^\delta = \omega + \sum_{j=1}^{Q} \alpha_j (|\varepsilon_{t-j}| - \lambda_j \varepsilon_{t-j})^\delta + \sum_{j=1}^{P} \beta_j \sigma_{t-j}^\delta, \tag{2}$$

$$\varepsilon_t = \sigma_t \eta_t, \tag{3}$$

$$\omega > 0, \delta > 0, \alpha_i \geq 0, -1 < \lambda_i < 1, \beta_i \geq 0.$$

where α_i and β_i are the standard ARCH and GARCH parameters, λ_i is the leverage parameter and δ the parameter for the power term. A positive λ_i implies negative information has stronger impact than the positive information on the volatility. In this study, we select Normal Inverse Gaussian [7, 12] and skewed Student's t-distribution [4] as marginals in ARMAX-APARCH model, namely, $\eta \sim NIG(0, 1, a, b)$ or $\eta \sim SSTD(0, 1, \upsilon, \gamma)$. The standard probability density function of NIG can be expressed as

$$f_{NIG}(\eta; 0, 1, a, b) = \frac{a}{\pi} \exp\left(\sqrt{a^2 - b^2} + b\eta\right) \frac{K_1\left(a\sqrt{1+\eta^2}\right)}{\sqrt{1+\eta^2}}, \qquad (4)$$

where $a > 0$ and $|b|/a < 1$; we define $\varsigma = \sqrt{a^2 - b^2}$ and $\xi = \frac{b}{a}$, the ς and ξ are called the parameters of shape and skewness, respectively. Noted that K_1 is a modified Bessel function of the third kind with index 1. The standard probability density function of SST is given as

$$f_{SSTD}(\eta \,|0, 1, \upsilon, \gamma) = \frac{2}{\gamma + \gamma^{-1}} \left\{ f_\upsilon\left(\frac{\eta}{\gamma}\right) I_{[0,\infty)}(\eta) + f_\upsilon(\gamma\eta) I_{(-\infty,0)}(\eta) \right\}, \quad (5)$$

where υ determines the shape, and γ determines the skewness. f_υ is standard Student's t density function. It is remarkable that normal-inverse Gaussian and skewed student's t-distribution are able to portray stochastic phenomena that have heavy tails and strongly skewed.

3.2.2 Static and Time-Varying Copulas

Copula methods have long been recognized and developed in various fields like econometrics, economics, financials, etc. Sklar [18] was the first person to give a definition of copula. If $x = (x_1, x_2, \ldots, x_n)$ is a random vector with joint distribution function H and marginal distribution $F_1, F_2, \ldots, F_n$, then there exists a function C called copula, which can be defined as

$$F(x_1, x_2, \ldots, x_n) = C(F_1(x_1), F_2(x_2), \ldots, F_n(x_n)). \qquad (6)$$

In the light of formula (6), the copula function can be expressed as

$$C(u_1, u_2, \ldots, u_n) = F(F_1^{-1}(u_1), F_2^{-1}(u_2), \ldots, F_n^{-1}(u_n)). \qquad (7)$$

If F_i is an absolutely continuous distribution which is strictly increasing, we have the density function as

$$f(x_1, \cdots\cdots x_n) = \frac{\partial F(x_1, \cdots\cdots, x_n)}{\partial x_1 \cdots \partial x_n}$$
$$= \frac{\partial C(u_1 \cdots, u_n)}{\partial u_1 \cdots \partial u_n} \times \prod \frac{\partial F(x_i)}{\partial x_i} \qquad (8)$$
$$= c(u_1 \cdots, u_n) \times \prod f_i(x_i),$$

where small letter c represents the density function of copula. An important feature of this result is that the marginal distributions do not need to be in any way similar to each other, nor is the choice of copula constrained by the choice of marginal distributions.

In this study, the bivariate copulas, such as Gaussian, T, Clayton, Frank, Gumbel, Joe, BB1, BB6, BB7, BB8 and survival copula, are considered for the analysis of the dependence structure. Also, the multivariate Gaussian, T, Clayton, Gumbel and Frank copulas are employed to study multivariate correlation. In addition, the Akaike Information Criteria (AIC) is used to select the preferable copula family. The bivariate Gaussian copula has the following form:

$$C_{Ga}(u_1, u_2|\rho) = \int_{-\infty}^{\Phi^{-1}(u_1)} \int_{-\infty}^{\Phi^{-1}(u_2)} \frac{1}{2\pi\sqrt{1-\rho^2}} \exp\left(-\frac{x_1^2 - 2\rho x_1 x_2 + x_2^2}{2(1-\rho^2)}\right) dx_1 dx_2$$
$$= \Phi_\rho\left(\Phi^{-1}(u_1), \Phi^{-1}(u_2)\right), \qquad (9)$$

where the ρ is the Pearson correlation belonging to $(-1, 1)$, both u_1 and u_2, which are the cumulative distribution functions (CDFs) of the standardized residuals from the marginal specifications, are uniformly distributed. Φ^{-1} is the inverse CDF of a standard normal and Φ_ρ is the CDF of the bivariate normal distribution with zero mean; the correlation equals to ρ. For a three-dimensional Gaussian copula, we have the form

$$C_{Gau}(u_1, u_2, u_3|R) = \Phi_R(\Phi^{-1}(u_1), \Phi^{-1}(u_2), \Phi^{-1}(u_3)), \qquad (10)$$

where R is the correlation matrix, and equals to $\begin{pmatrix} 1 & \rho & \rho \\ \rho & 1 & \rho \\ \rho & \rho & 1 \end{pmatrix}$. Most dependence struc-

tures for time series are, basically, not time-invariant. Time-varying copulas might be considered as the dynamic generalizations of a Pearson correlation or a Kendall's tau. However, it is still difficult to find causal variables to explain such dynamic characteristics. On the one hand, we consider the time-varying dependence structures between the growth rates of tourism demand of Singapore, Malaysia and Thailand. Simultaneously, we also guess that political instability in Thailand and the Chinese Spring Festival holidays might have an influence on the time-varying dependencies. Therefore, the time-varying bivariate Gaussian copula is constructed as follows;

$$\rho_t^* = w + \phi \cdot \rho_{t-1}^* + \delta \cdot |u_{i,t-1} - u_{j,t-1}| + \beta_1 x_{1,t-1} + \beta_2 x_{2,t-1}, \qquad (11)$$

where $i, j = 1, 2, 3, i \neq j$, represent different countries; $\rho_t^* = -\ln[(1 - \rho_t)/(1 + \rho_t)]$, which is used to ensure the correlation fall within $(-1, 1)$. The x_1 and x_2 in the equation are dummy variables which represent the Chinese Spring Festivals and Thailand's political events, respectively. The time-varying three-dimensional Gaussian copula is expressed as

$$\rho_t^* = w + \phi \cdot \rho_{t-1}^* + \frac{\delta}{2}(|u_{thai,t-1} - u_{\sin g,t-1}| + |u_{thai,t-1} - u_{malay,t-1}|) \quad (12)$$
$$+\beta_1 x_{1,t-1} + \beta_2 x_{2,t-1}.$$

Since Thailand occupies the leading position among SMT, we consider the average differences between Thailand and the other countries to be causal variable. The method of inference function for margins (IFM) is used to estimate the parameters of the copula-based APARCH models. Joe [8, 9] and Wu et al. [22, 23] showed that this estimator is close to and asymptotically efficient as the maximum likelihood estimator under some regularity conditions. The procedure of IFM can be described as having two steps: We firstly estimate the corresponding ARMAX-APARCH models using the maximum likelihood method, and then estimate the copula functions, given the parameters of the marginals.

3.2.3 Goodness of Fit

After we calculate the parameters for each family of copulas using IFM, the primary task becomes how to choose the most appropriate copula family. In this study, we use AIC as the criterion. The AIC is given by

$$AIC := -2 \sum_{t=1}^{T} \ln[c(\hat{u}_t; \Theta)] + 2k, \quad (13)$$

where $k = 1$ for one-parameter copulas; $k = 2$ for two-parameter copulas, such as t, BBX copulas, etc.; and $k = 5$ for time-varying copulas. $\hat{u}_t$ represents the matrix of marginal distributions, and Θ represents the vector of static or time-varying copulas. Excepting AIC, we also perform the likelihood ratio (LR) test to compare static copulas with the time-varying copulas. The likelihood ratio can be expressed as

$$LR = -2 \cdot [\log c(\hat{u}_t; \rho) - \log c(\hat{u}_t; \omega, \alpha, \gamma, \beta_1, \beta_2)], \quad (14)$$

where LR is a chi-Square distribution with four degrees of freedom. If we reject the null hypothesis, it implies that the time-varying copula gives better performance than static copula.

4   Empirical Results

4.1   Preliminary Study

Table 1 provides the descriptive statistics for this data. It shows that all data exhibit strongly positive skewness and excess kurtosis, which implies all series are heavy-tailed and right-skewed. Moreover, the Jarque-Bera test results show that the distribution of all the series strongly rejects the assumption of normality. Therefore, none of the series is normal distributed, implying the skewed Student's t-distribution and normal-inverse Gaussian distributions should be more appropriate than normal distribution in our study.

4.2   Estimation of Marginal Model

Table 2 presents the results of the ARMAX-APARCH model with different assumptions of marginal distribution for SMT. To investigate the political effects, holiday effects, volatility and leverage effects of the China outbound tourist demand to SMT, we employed ARMAX(12,4)–APARCH (1,1) with NIG distribution for Singapore, ARMAX (6,9)–APARCH (1,1) with SST distribution for Malaysia, and ARMAX (12,4)–APARCH (1,1) with NIG distribution for Thailand. The parameters of skewness and shape are significant at a 95% confidence level for all marginals, and none of the LM test were able to reject the hypothesis, which demonstrates that our models are appropriate. Firstly, the results show that all the estimated parameters of political turmoil and holiday effects were statistically significant with negative and positive effects, respectively, on tourism arrivals. Political turmoil in Thailand not only had an influence on itself but could also impact the tourism demand of Singapore and Malaysia. The holiday effect impact on Singapore was the greatest, while the holiday effect impact on Malaysia was the smallest. Secondly, the leverage parameter λ was positive and statistically significant, which implies that negative shocks

Table 1 Data description and statistics		Singapore	Malaysia	Thailand
	Mean	0.0591	0.0719	0.0640
	Maximum	1.5183	1.6126	1.4524
	Minimum	−0.8263	−0.7963	−0.7224
	Std. Dev.	0.3297	0.366	0.3374
	Skewness	1.1470	1.1277	0.9034
	Kurtosis	6.5119	5.2741	5.5368
	Jarque-Bera	129.0432	75.2348	71.1362
	Probability	0	0	0

Table 2 Results of ARMAX-APARCH Model

	Singapore		Malaysia		Thailand
Constant	0.0018***	Constant	0.0472***	Constant	0.0245***
	(0.0000)		(0.0071)		(0.0018)
Ar1	−0.0670***	Ar1	−0.9828***	Ar1	1.8619 ***
	(0.0000)		(0.0001)		(0.0002)
Ar2	−0.2784***	Ar2	0.1303***	Ar2	−1.9388***
	(0.0000)		(0.0002)		(0.0002)
Ar3	0.1274***	Ar3	0.5631***	Ar3	2.0428***
	(0.0001)		(0.0001)		(0.0001)
Ar4	−0.1064***	Ar4	0.0733***	Ar4	−1.3725***
	(0.0000)		(0.0001)		(0.0001)
Ar5	−0.0749***	Ar5	−1.0405***	Ar5	0.5834***
	(0.0000)		(0.0001)		(0.0001)
Ar6	−0.1872***	Ar6	−0.6768***	Ar6	−0.3931 ***
	(0.0003)		(0.0000)		(0.0001)
Ar7	−0.0760***	Ma1	0.8967***	Ar7	0.0935***
	(0.0001)		(0.0001)		(0.0008)
Ar8	0.0339***	Ma2	−0.2786***	Ar8	0.1154 ***
	(0.0001)		(0.0001)		(0.0013)
Ar9	0.0677***	Ma3	−0.7923***	Ar9	−0.1669
	(0.0001)		(0.0001)		(0.0007)
Ar10	−0.0924***	Ma4	−0.2416***	Ar10	−0.0018 ***
	(0.0000)		(0.0000)		(0.0002)
Ar11	−0.0613***	Ma5	1.2602***	Ar11	0.1337 ***
	(0.0000)		(0.0001)		(0.0047)
Ar12	0.5845***	Ma6	0.7698***	Ar12	−0.0865 ***
	(0.0002)		(0.0000)		(0.0032)
Ma1	−0.2683***	Ma7	−0.3117***	Ma1	−2.0585 ***
	(0.0000)		(0.0000)		(0.0003)
Ma2	0.0777***	Ma8	−0.3161***	Ma2	2.0906 ***
	(0.0000)		(0.0000)		(0.0003)
Ma3	−0.2841***	Ma9	−0.1031***	Ma3	−1.9767 ***
	(0.0001)		(0.0001)		(0.0002)

(continued)

Table 2 (continued)

	Singapore		Malaysia		Thailand
Ma4	0.2429***	Holiday effect	0.1663***	Ma4	1.0563 ***
	(0.0001)		(0.0002)		(0.0002)
Holiday effect	0.5319***	Political effect	−0.1511***	Holiday effect	0.4306***
	(0.0001)		(0.0409)		(0.0025)
Political effect	−0.0237***	ω	0.0003*	Political effect	−0.1914***
	(0.0003)		(0.0001)		(0.0670)
ω	0.2588***	α	0.1641***	ω	0.0069 ***
	(0.0675)		(0.0459)		(0.0054)
α	0.7458***	β	0.5683***	α	0.1800 ***
	(0.1007)		(0.1128)		(0.0596)
β	0	ψ	0.3506**	β	0.6779 ***
	(0.0064)		(0.1219)		(0.1112)
ψ	−0.3286**	υ	3.8194***	ψ	0.4696***
	(0.1100)		(0.0855)		(0.2778)
λ	0.2185*	υ	9.4759**	λ	2.0679***
	(0.0903)		(2.9894)		(0.6176)
ς	−0.2717*	γ	2.0486***	ς	−0.2870 ***
	(0.1059)		(0.2347)		(0.1202)
ξ	0.7197*	—	—	ξ	0.5366 ***
	(0.3034)				(0.1914)
LM-test	0.9855	LM-test	0.7221	LM-test	0.9453
P value		P value		P value	
LogL	63.8159	LogL	7.3100	LogL	23.9031
AIC	−0.4297	AIC	0.2010	AIC	0.0238
BIC	0.0386	BIC	0.6513	BIC	0.4921

Note Significant code: 0 '***' 0.001 '**' 0.01 '*' 0.05 '.' 0.1

have stronger impact than positive information on volatility for SMT. Since α is the ARCH parameter associated with ε_{t-1}^2 and β is the GARCH parameter associated with the volatility spillover effect from h_{t-1}, significantly positive estimates of both α and β imply that there exist both short-run and long-run persistence of shocks for Malaysia and Thailand. At the same time, the estimated parameter of β for Singapore tourism demand equals zero. This implies no long-run persistence of shocks for Singapore. In addition, the sum of the estimates of α and β equal 0.75, 0.72 and 0.86

Table 3 KS Test for Uniformity and Box-Ljung Test for Autocorrelation

KS Test of Three Margins for Uniformity

	Statistic	P value	Hypothesis
$u_{1,t}$	0.0529	0.7085	0 (acceptance)
$u_{2,t}$	0.0460	0.8510	0 (acceptance)
$u_{3,t}$	0.0597	0.5571	0 (acceptance)

Box-Ljung Test of Margins for Autocorrelation

		Chi-squared	P value
$u_{1,t}$	First moment	2.9177	0.9833
	Second moment	8.1753	0.6117
	Third moment	5.2056	0.8770
	Fourth moment	10.8725	0.3675
$u_{2,t}$	First moment	9.6388	0.4727
	Second moment	7.6270	0.6652
	Third moment	5.3204	0.8688
	Fourth moment	9.8930	0.4499
$u_{3,t}$	First moment	10.9961	0.3578
	Second moment	12.2648	0.2677
	Third moment	10.7407	0.3781
	Fourth moment	14.9657	0.1333

Note $u_{1,t} = F_{nig}(x_{thai,t})$, $u_{2,t} = F_{nig}(x_{sing,t})$, and $u_{3,t} = F_{skt}(x_{malay,t})$

for Singapore, Malaysia and Thailand, respectively. This implies that the impact of unexpected shock of Thailand as regards volatility has longer duration than in the case of the other two countries. Thirdly, the skewness parameters in the NIG and SST distributions are greater than zero, which implies that all the series are skewed to the right.

Before we conducted copula analysis, we had to test whether the marginal distribution for each series satisfied the serial independence assumption and the assumption that it is distributed as uniformly (0, 1). If any of these two assumptions is rejected, then the misspecification of the marginal distribution may cause incorrect-fit to the copulas. Thus, testing these two assumptions for the marginal distribution model specification is a critical step in constructing multivariate distribution models using copulas [14]. We employed the Box-Ljung statistic to test whether the marginal distribution for each series satisfied the serial independence assumption and adopted the Kolmogorov-Smirnov (KS) statistic to test whether it was distributed as uniform (0, 1). We report the results in Table 3. For the Box-Ljung test, we regressed $(u_{it} - \bar{u}_i)^k$ on 10 lags of each variable, for each series i, and for k = 1, 2, 3, 4. The LM test was of no autocorrelation under the null hypothesis. Table 3 shows that the probability values for both Box-Ljung and KS tests were greater than 10%, implying that the margins satisfy the assumptions of iid and uniform distribution, which, in turn, implies that our use of the ARMAX-APARCH with specified distributions to fit the margins is appropriate, as well.

4.3 Estimation of Copula Model

The 17 static bivariate copulas were firstly estimated by maximum likelihood estimation method, and AIC was used to select the preferable copula. Thereafter, the corresponding time-varying copula was estimated by ML as well. Figures 2, 3 and 4 show the AICs of static copulas between the growth rates of SM, ST and MT, respectively. Figures 2, 3 and 4 prove that the Gaussian copula had the best performance for each pair of tourism demand in terms of AIC values. This implies that there does not exist tail dependence, such as left tail dependence, right tail dependence or both sides, even if all the marginals were skewed to right. For SMT, we used the multivariate Gaussian, T, Frank, Gumbel and Clayton copulas to fit their joint distribution. Figure 5 demonstrates the AIC values of all the candidate copulas. According to AIC, the multivariate Gaussian copula exhibits the best performance. Therefore, the tourism demand of SMT also did not have tail dependence. These findings imply that extreme movements of tourism demand of Singapore, Malaysia and Thailand are not consistent with each other. These results also imply that introducing tail dependences between SMT does not add much to the explanatory ability of the models.

Table 4 shows the estimated results of static and dynamic Gaussian copulas with their comparison by LR test. Panel A of Table 4 shows the parameter estimates for the static Gaussian copula function. We can see that all the parameter estimates are statistically significant at the 0.1% confidence level. So, the tourism demands for SMT from China are related to each other, and these three countries also have some relationship on the whole. In addition, the Gaussian dependence structure between Singapore and Thailand demonstrates the biggest correlation. Panel B and Panel C of Table 4 present the parameter estimates for the dynamic Gaussian copula function

Fig. 2 The AIC values of static copulas between the growth rates of Thailand and Singapore. *Note* Note that, '1', '6', '7', and '8' represent BB1, BB6, BB7 and BB8 copulas, respectively; 'R1', 'R6', 'R7', and 'R8' represent the corresponding rotated BBX copulas by 180 degrees; 'Gau', 'T', 'Frk', 'Cl', 'G', and 'Joe' are Gaussian, student-t, Frank, Clayton, Gumbel and Joe copulas, respectively. While 'RCl', 'RG', and 'RJ' represent their rotated copulas by 180 degrees as well. Figures 3 and 4 have the same notation.

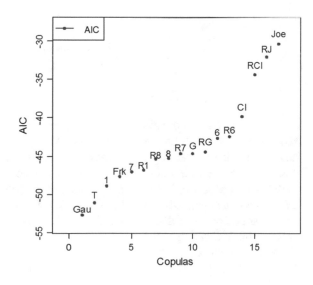

Fig. 3 The AIC values of static copulas between the growth rates of Thailand and Malaysia

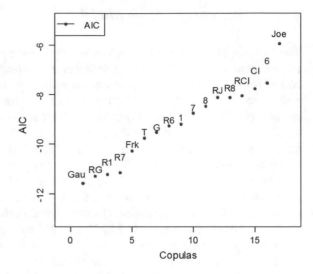

Fig. 4 The AIC values of static copulas between the growth rates of Singapore and Malaysia

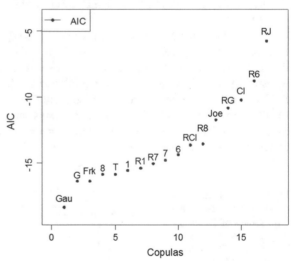

and the LR test for comparing the static copulas with the dynamic copulas. The AIC values in Panel B are smaller than the AIC values in Panel A, implying that the dynamic dependencies perform better than the invariant dependencies. Also, in Panel C, it can be observed that the LR tests do reject the null hypothesis, thereby implying that time-varying Gaussian copulas have better explanatory ability than static Gaussian copulas. Overall, it can be concluded that the dependencies between the tourism demands of SM, ST, MT and SMT are not constant but time-varying in terms of the AIC and the LR test.

We proceed to explain the estimate results of the time-varying Gaussian copula in Panel B of Table 4. First, we can observe that the autoregressive parameter ϕ is close

Fig. 5 The AIC values of static multivariate copulas between the growth rates of SMT

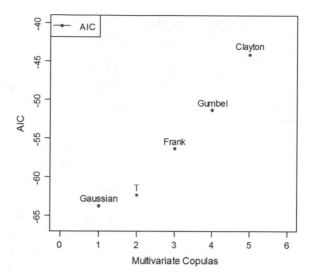

to 1, implying a high degree of persistence pertaining to the dependence structure between all the pairs, namely SM, ST and MT, as well as between SMT. The parameter δ also is significant for each time-varying Gaussian copula, which displays that the last period information is a meaningful measure. Second, the parameter β_1 for holiday effect is statistically significant and positive, and so the parameter β_2 for political events is the same. These results reflect that both the prior period to the Chinese Spring Festival and Thailand's political turmoil enlarge the interdependence of each pair. The estimated value of the parameter β_1 is greater than that of β_2, implying that holiday effects have a greater contribution to the dependence of each pair. Thus, Thailand's political events have a stronger effect on the dependences of ST and MT than the others; meanwhile, the estimates are also more significant. To sum up, political turmoil in Thailand and the Chinese Spring Festival not only impact the tourism demand for SMT from China, but also have a positive effect on the dependence structure between the pairs SM, ST and MT, as well as between SMT.

Figures 6, 7, 8 and 9 show the time-varying correlation between the SMT destinations for Chinese tourists. Firstly, we can observe that the correlations between the tourist arrivals to the different destination countries all have large fluctuation, and that the correlations are obviously different, with the no time variation in the Gaussian copula. The large fluctuation in the correlations reveals the instable relationship of tourism demand in SMT. This finding may serve as an indicator towards the need to strengthen cooperation and commutation between SMT, thereby improving relationship to achieve a win-win situation for all. Secondly, we observe that most of the correlations drop to minimum values during the Chinese Spring Festival. This may be because Chinese tourists prefer free-walking tours and not package tours of SMT during the Chinese Spring Festival. Maybe, we can ascribe this phenomenon to the exorbitant costs of package tours to SMT during the time of the

Table 4 The results of static and dynamic Gaussian copulas

	CTS	CTM	CSM	CSMT
Panel A: Static				
$\hat{\rho}$	0.4824***	0.2427***	0.2818***	0.3261***
	(0.0491)	(0.0617)	(0.0574)	(0.0394)
Kendalls tau	0.3204	0.1561	0.1819	—
LogL	27.3387	6.7922	6.5837	32.8600
AIC	−52.6775	−11.5845	−11.1675	−63.7200
Panel B: Dynamic				
W	−0.2722 ***	−0.4179***	−0.1768***	−0.2003*
	(0.0365)	(0.1178)	(0.0303)	(0.0849)
ϕ	0.9899 ***	0.9406***	0.9899***	0.9580***
	(0.0007)	(0.0423)	(0.1737)	(0.0447)
δ	0.7832 ***	1.0891**	0.4257*	0.4256.
	(0.2197)	(0.3435)	(0.1771)	(0.2218)
β_1	0.9938 ***	1.1731**	0.6239*	1.0716***
	(0.2980)	(0.4348)	(0.2838)	(0.2990)
β_2	0.3138 ***	0.5165**	0.1912*	0.2859*
	(0.0948)	(0.1782)	(0.0887)	(0.1288)
LogL	34.7947	15.1255	15.1234	42.3327
AIC	−59.5893	−20.2511	−20.2468	−74.6654
Panel C: LR test				
Statistics	14.9120	16.6670	17.0790	18.9450
Degree of freedom	4	4	4	4
Probability	0.0049**	0.0022**	0.0019**	0.0008***

Note Significant. codes: 0 '***' 0.001 '**' 0.01 '*' 0.05 '.' 0.1.
The numbers in the parentheses are the standard deviations

Chinese Spring Festival. The typical cost of a package tour of SMT in the time of the Chinese Spring Festival is about 11,000 RMB, whereas it costs only 6,000 RMB at other times of the year. As a result, the correlations become much stronger after the Chinese Spring Festival, which is caused by the low prices of package tours to SMT. Thirdly, the correlations are also found to be around minimum values during political events. On the one hand, Chinese tourists give up the classical package tour of SMT, while choosing to travel in Singapore and/or Malaysia, thereby giving rise to substitute effects. At the same time, these very Chinese tourists may cancel trips to Thailand and also give up the trip to Singapore and Malaysia, and choose to visit Japan, Korea, Vietnam or other destinations, or choose to not go on a trip. For example, in Fig. 7, the negative correlation between Thailand and Malaysia could be recognised as the 'substitute effect' during the political turmoil and during some Chinese holiday; whatever the reason—whether the Chinese Spring Festival or Thailand political turmoil—the correlations as regards tourism demand between

Fig. 6 The correlations between the growth rates of Thailand and Singapore

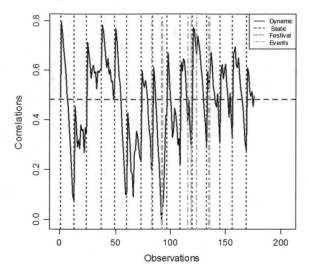

Fig. 7 The correlations between the growth rates of Thailand and Malaysia

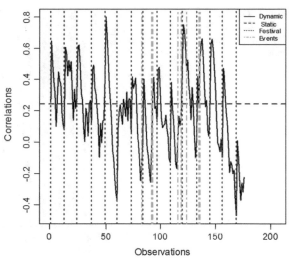

Thailand and Singapore are always positive, as demonstrated in Fig. 6. Therefore, we conclude that there exist complementary effects in Singapore and Thailand. In addition, Fig. 9 illustrates the correlations regarding the tourism demand of SMT. Political events and the Chinese Spring Festival prompt the correlations between SMT to become weak, which also demonstrates that the Chinese tourist prefers free-walking tours of one or two countries and not package tours of SMT due to the exorbitantly high costs of package tours. Lastly, the time-varying correlations between Singapore and Malaysia are illustrated in Fig. 8. It can be observed that Thailand political events also have an impact on the correlations between Singapore and Malaysia, and cause Chinese tourists to choose just one destination for travel.

Fig. 8 The correlations
between the growth rates of
Singapore and Malaysia

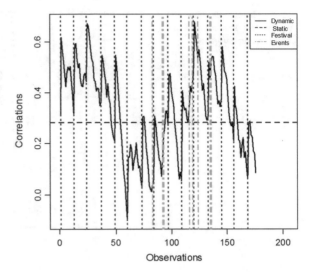

Fig. 9 The correlations
among the growth rates of
Thailand, Singapore and
Malaysia

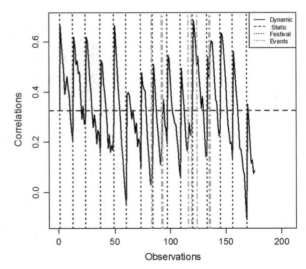

5　Policy Planning

The empirical findings of this study reveal that negative shocks to the tourism sector in
Thailand have statistically significant negative impact on tourist arrivals in Singapore
and Malaysia. At the same time, the Chinese Spring Festival reveals a positive effect
on tourist arrivals in SMT. The results also indicate that the last period of the Chinese
Spring Festival and Thailand's political events are able to explain the correlations of
Chinese outbound tourism demand for SMT. On the one hand, some Chinese tourists
may choose Malaysia as their destination instead of Thailand due to the high cost

of the trip for holidaying or due to Thailand's political instability, which results in substitution effects. On the other hand, some Chinese tourists may cancel trips to Thailand, and also give up the trips to Singapore or SMT package tours because of Thailand's political instability. According to the empirical findings, policy planning is recommended for the pairs of SMT as well as for the three countries (SMT) investigated in this study, which is given as follows:

1. **Singapore**. Since the Chinese Spring Festival has a positive effect on tourist arrivals in SMT, Singapore should consider moderate adjustment in the airline fare by introducing competition mechanisms during the Chinese Spring Festival. Cheaper airline fare would provide more opportunities for Chinese tourists to make trips, thereby realising the scale effects and the probit maximum for the tourism industry. Also, we suggest that Singapore combine with Malaysia and Thailand, marketing special tour packages for holidaying for Chinese tourists. For reducing the negative effects of Thailand's political events, the tour administration department of Singapore should strengthen communication with tour agencies or firms. Policy makers should provide some favourable policies, such as free visa or waiver of fees and more economy tour packages, to maintain the normal level of tourist arrivals if and when political unrest breaks out in Thailand.

2. **Malaysia**. Malaysia and Thailand share a competitive relationship at the time of the Chinese Spring Festival. First, Malaysia should reduce trip costs by introducing low-cost carriers, as well as provide visa fee relief, thereby improving the countrys competitive ability. Malaysia should also consider cooperating with Thailand and Singapore in terms of introducing new tour package plans, affordable tour packages, quicker visa processing, etc., to change the competitive relationship into a win-win relationship.

3. **Thailand**. Thailand offers a variety of landscapes, low-cost trips, traditional Buddhist culture and 'service with a smile' to the Chinese tourists. Thailand has the core status in SMT. In view of the negative effects of political turmoil, the government of Thailand should positively report physical truth regarding political events so as to reduce the negative effects of political instability. In terms of Thailand's political events, most of the political events break out only in Bangkok, so they have a minimal impact on popular tourist destinations such as Phuket, Chiang Mai, Chiang Rai etc. We suggest that the government disseminate this information to Chinese tourists, thereby encouraging and attracting them to visit Thailand. Travel agencies should introduce some special group packages (not including Bangkok) and low-cost trips to reduce the probability of the Chinese tourist not visiting Thailand. During the Chinese Spring Festival, travel agencies and airline companies should offer various travel package plans, such as honeymoon packages, winter vacation packages, visiting university packages etc.

In addition, we suggest that the SMT establish a department that focuses on dealing with political turmoil, formulating new tour packages, advertising the culture of SMT etc. Thus, tour operators and national tourism promotion authorities of SMT should collaborate closely in marketing and promoting joint tourism ventures and products.

6 Conclusion

This paper brings the bivariate and multivariate copula-based ARMAX-APARCH model into studying the SMT tourism demand. The copula-based models are used to investigate the effects of political events and holidays, describe the volatility and leverage effects, capture static and time-varying dependences, judge whether political events and holidays impact the dependencies or not, and scrutinise the relationship between two countries of SMT in order to ascertain whether there exist substitution effects or complementary effects. We found that the copula-based ARMAX-APARCH model is a complementary academic approach to analysing the spillover effects of political events in the international tourism demand model. Based on the method, we firstly found that Thailand political turmoil has a negative effect on tourist arrivals to SMT from China, while the Chinese Spring Festival has a positive effect. Secondly, there does not exist tail dependence for all the pairs of SMT, as well as for SMT, implying that the extreme movements of the Chinese tourist demand to SMT do not happen together. Thirdly, Thailand's political events and the Chinese Spring Festival are suitable for explaining the dynamic dependencies between SMT. This finding suggests that forcing variables of dynamic dependences should be specified according to the actual research problem. Also, there is a pair of countries, Malaysia and Thailand, which appears to have substitution effects. On the contrary, Malaysia and Singapore, and Singapore and Thailand have complementary effects. Lastly, we recommend some policies for each country of SMT. We always maintain that a competitive advantage can be developed for these three destinations for attracting Chinese outbound tourists if these three countries form a strategic partnership in developing tourism products and furthering cooperative destination promotion.

Acknowledgements The authors are very grateful to Professor Wing-Keung Wong and Professor Vicente Ramos for their comments. The authors wish to thank the Puey Ungphakorn Centre of Excellence in Econometrics and Bank of Thailand Scholarship for their financial supports.

References

1. Chan F, Lim C, Michael M (2005) Modelling multivariate international tourism demand and volatility. Tour Manag 26:459–471
2. Chang CC, Khamkaew T, Tansuchat R, Michael M (2011) Interdependence of international tourism demand and volatility in leading ASEAN destinations. Tour Econ 17(3):481–507(27)
3. Ding Z, Granger CWJ, Engle RF (1993) A Long memory property of stock market returns and a new model. J Empir Finan 1:83–106
4. Fernandez C, Steel MFJ (1998) On bayesian modeling of fat tails and skewness. J Am Statist As soc 93:359–371
5. Huang JJ, Lee KJ, Liang H, Lin W (2009) Estimating value at risk of portfolio by conditional copula-GARCH method Insurance. Math Econ 45:315–324
6. Issa IA, Altinay L (2006) Impacts of political instability on tourism planning and development: the case of Lebanon. Tour Econ 12(3):361–381
7. Jensen MB, Lunde A (2001) The NIG-S and ARCH model: a fat-tailed, stochastic, and autoregressive conditional heteroskedastic volatility model. Econometrics J 4:319–342

8. Joe H (1997) Multivariate models and dependence concepts. Chapman and Hall, London
9. Joe H (2005) Asymptotic efficiency of the two-stage estimation method for copula-based models. J Multivar Anal 94:401–419
10. Jondeau E, Rockinger M (2006) The Copula-GARCH model of conditional dependencies: an international stock market application. J Int Money Finan 25:827–853
11. Liu J, Sriboonchitta S, Nguyen HT, Kreinovich V (2014) Studying volatility and dependency of chinese outbound tourism demand in Singapore, Malaysia, and Thailand: a vine copula approach., Modeling dependence in econometrics. Springer, Heidelberg
12. Necula C (2009) Modeling heavy-tailed stock index returns using the generalized hyperbolic distribution. Romanian J Econ Forecasting pp 118–131
13. Neumayer E (2004) The impact of political violence on tourism: dynamic cross-national estimation. J Conflict Resol 48(2):259–281
14. Patton AJ (2006) Modelling asymmetric exchange rate dependence. Int Econ Rev 47(2):527–556
15. Saha S, Yap G (2014) The moderation effects of political instability and terrorism on tourism development a cross-country panel analysis. J Travel Res 53(4):509–521
16. Seo JH, Park SY, Yu L (2009) The analysis of the relationships of Korean outbound tourism demand: Jeju Island and three international destinations. Tour Manag 30:530–543
17. Shareef R, McAleer M (2007) Modeling the uncertainty in monthly international tourist arrivals to the Maldives. Tour Manag 28(1):23–45
18. Sklar M (1959) Fonctions de f epartition àn dimensions et leurs marges. Publ Inst Stat 8:229–231
19. Sönmez SF (1998) Tourism, terrorism and political instability. Ann Tourism Res 25(2):416–456
20. Sriboonchitta S, Nguyen HT, Wiboonpongse A, Liu J (2013) Modeling volatility and dependency of agricultural price and production indices of Thailand: static versus time-varying copulas. Int J approximate reasoning 54:793–808
21. Tang JC, Sriboonchitta S, Ramos V, Wong WK (2014) Modelling dependence between tourism demand and exchange rate using copula-based GARCH model. Current issues in method and practice (in press)
22. Wu CC, Liang SS (2011) The economic value of range-based covariance between stock and bond returns with dynamic copulas. J Empir Finan 18:711–727
23. Wu CC, Chung H, Chang YH (2012) Economic value of co-movement between oil price and exchange rate using copula-based GARCH models. Energy Econ 34(1):270–282
24. Zhang J, Guégan D (2008) Pricing bivariate option under GARCH processes with time-varying copula. Insur Math Econ 42:1095–1103

Forecasting Asian Credit Default Swap Spreads: A Comparison of Multi-regime Models

Chatchai Khiewngamdee, Woraphon Yamaka and Songsak Sriboonchitta

Abstract This paper aims to explore the best forecasting model for predicting the Credit Default Swap (CDS) index spreads in emerging markets Asia by comparing the forecasting performance between the multi-regime models. We apply threshold, Markov switching, Markov switching GARCH and simple least squares for structural and autoregressive modeling. Both in- and out-of-sample forecasts are conducted to compare the forecasting performance between models. The results suggest that Markov switching GARCH(1,1) structural model presents the best performance in predicting Asian Credit Default Swap (CDS) index spreads. We also check the preciseness of our selected model by employing the robustness test.

Keywords Credit default swaps · Threshold · Markov switching · Robustness

1 Introduction

Credit default swaps (CDS) contract, the most popular and powerful credit derivative in the world financial market, was first invented in 1994 by JPMorgan for the purpose of protecting its credit lending to Exxon. This contract is a tool for transferring the credit risk from credit owner to the CDS issuer. In other words, it is like an insurance since the buyer of CDS contract, who is a credit owner, will gain protection against the default of credit. The seller, on the other hand, will assume the credit risk by delivering principle and interest payments to the owner once the reference credit defaults in exchange for a protection fee or the so called spread. If there is no default event occurs, the seller of CDS contract can enjoy the profit from receiving the periodical payment fee from the buyer.

In order to simplify trading on CDS contracts, CDS contracts were initially developed into a CDS index contract in 2001, namely Markit CDX in North America. Then,

C. Khiewngamdee (✉) · W. Yamaka · S. Sriboonchitta
Faculty of Economics, Chiang Mai University, Chiang Mai, Thailand
e-mail: getliecon@gmail.com

W. Yamaka
e-mail: woraphon.econ@gmail.com

© Springer International Publishing AG 2017
V. Kreinovich et al. (eds.), *Robustness in Econometrics*,
Studies in Computational Intelligence 692, DOI 10.1007/978-3-319-50742-2_28

three years later, CDS index contracts trading spread into Europe and Asia, so-called Markit iTraxx. This development of CDS also brought greater liquidity, transparency and acceptance to the CDS market. As a result, CDS market has become more popular and grown rapidly since then. By the year 2004, the notional amount of the credit derivative market was totally $4.5 trillion. These CDS indexes are used by many licensed market makers including the world class investment banks such as Goldman Sachs, Citigroup, Deutsche Bank and UBS since it provides a very efficient way to take a view on the reference entity credit [1]. In fact, speculation has become a common function for CDS contracts as the value of CDS market is greater than the bonds and loans that the contracts reference.

Recently, as the world economy has not fully recovered from the crisis and most of the markets are still bearish, investors have been seeking for high return investment as well as hedging their risk exposures. Although government bonds seem to be the best investment asset during the economic turmoil, CDS contracts surprisingly provide a higher return and more efficiency than bonds. According to Bloomberg, trading bonds in huge volumes has become more difficult and more expensive but trading CDS in large scale has potentially increased the return by deploying leverage. Additionally, CDS market has higher liquidity than the associated bond market and hence speculators usually take advantage by using shorter trading horizons [2, 3]. Hence, CDS markets have returned to attract investors' attention once again. Particularly, CDS market in Asia with the exception of Japan has grown radically after the crisis which is much higher than CDS market in US, Europe and Japan (as shown in Fig. 1).

Predicting the trend of CDS spread changes if of interest to investors, financial institutions including policymakers since it provides the information on the future

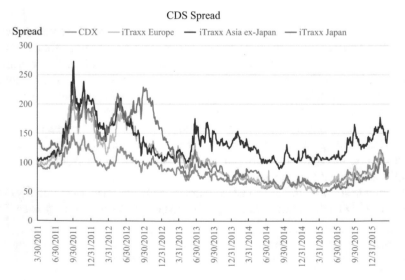

Fig. 1 Credit Default Swap (CDS) index spreads from major markets

credit performance in the market. A precise forecasting model will be extremely helpful to investors and financial institutions to make decisions on their hedging or trading strategy as well as the policymakers to foresee the future CDS spreads in order to apply an appropriate policy to stabilize the financial market and the economy. Moreover, since CDS market in Asia is still in its infancy, with the size of the markets quite small and trading in the markets not so liquid, there are still so much rooms to grow for Asian CDS market. Besides, policymakers in Asian countries practically focus on development of the local bond market and derivatives market. [4, 5]. Therefore, it leads to a rapid growth in CDS market in Asia.

While most of the CDS studies focused on forecasting CDS spreads only in US and EU markets, this paper aims to widen those studies by investigating the forecasting performance on the CDS index in emerging markets Asia by using advanced econometric models. In particular, we apply both linear and non-linear versions of autoregressive model and structural model. These models include simple model, threshold model, Markov switching model and Markov switching GARCH. We analyze both in- and out-of-sample forecasts for each model and compare the forecasting performance using root mean square forecast error (RMSE) and mean absolute error (MAE). Additionally, we perform a robustness test to check the validity of the selected model.

The remainder of the paper is organized as follows; Sect. 2 reviews the literature, Sect. 3 describes and analyzes the sample dataset, Sect. 4 presents the forecasting models used in this study, Sect. 5 provides empirical results, Sect. 6 discusses and compares the forecasting performance between the models and robustness test and Sect. 7 contains the conclusions of the paper.

2 Literature Review

For the past two decades, as an innovative financial instrument, CDS has drawn attention from many researchers who study the characteristics of CDS. Most of the studies focused on CDS forecasting and determinants of CDS. For example, Avino and Nneji [6] found that there are some evidences that iTraxx Europe CDS spreads can be predictable. They conducted the forecast by employing both linear and non-linear models. The results show that, in the out-of-sample forecast, the linear models are superior to non-linear Markov switching models.

Byström [7] and Alexander and Kaeck [8] studied first-order serial correlation (AR(1)) in iTraxx CDS spreads. They found that, during 2004–2006, the change of CDS spread shows a significantly positive sign in the first lagged correlation. Moreover, Byström [7] applied using a simple trading rule for his in-sample forecast and the result shows that it generates positive profits before transaction costs. Alexander and Kaeck [8] investigated the determinants of CDS spread in different regimes of the economy by using a Markov switching regression model. They found that, in a volatile regime, implied volatility for option is a main determinant of CDS spreads change. Nevertheless, in stable regime, stock market returns play more important role.

Collin-Dufresne et al. [9], Campbell and Taksler [10] and Cremers et al. [11] examined the determinants of credit spreads by using financial variables and found that there are limited explanatory power among those variables. However, Galil et al. [12] found that those financial market variables, especially stock returns, change in the volatility of stock return and change in CDS spread, can determine the CDS spread after controlling for firm-specific variables. They also suggested that adding financial market variables can improve the performance of the models to explain spread changes. Furthermore, Zhang et al. [13] and Ericsson et al. [14] found that, by employing a structural model, the structural variables have explanatory power for the determination of CDS spreads.

Several papers have used linear models, namely simple autoregressive or structural model, in order to forecast CDS spread changes. Thus, it raised a question whether advanced non-linear models can perform better in order to forecast CDS spread changes. In fact, there are several studies applied advanced non-linear models, namely threshold autoregressive (TAR) model, Markov switching model and Markov switching GARCH model, to forecast various financial assets. For instance, Domian et al. [15] employed threshold autoregressive (TAR) model to find the relation between stock returns and real economic activity. They found that positive stock returns slightly increase real economic activity, while negative stock returns radically decrease the growth of industrial production. Posedel and Tica [16] investigated the effect of exchange rate pass-through to inflation in Croatia by using TAR model. They found that, above the threshold, change in nominal exchange rate strongly affects inflation. On the contrary, below the threshold, nominal exchange rate has no effect on inflation. In addition, Nampoothiri and Balakrishna [17] suggested that TAR is a preferred model to estimate coconut oil prices at Cochin market, compared with a simple AR model.

Ailliot and Monbet [18] employed non-homogeneous Markov Switching Autoregressive (MS-AR) models to describe the changes of the wind speed in different weather types. They found that these models can provide good interpretations for the important properties of the changes of wind speed, for instance, the marginal distribution and the length of stormy and calm periods. Likewise, Piplack [19] suggested that a Markov switching model offers better interpretability than a linear model in forecasting the volatility of the S&P500 stock market index during 1962–2007. It also has less forecasting errors than linear model. However, Engel [20] argued that a Markov switching model performs worse than a random walk model in out-of-sample forecasting of exchange rates. He also suggested Markov switching model may perform better if it is allowed more than 2 regimes.

Erlwein and Muller [21] developed a Markov switching model by using a filtered-based EM-algorithm and hidden information from the Markov chain to optimize the parameter estimation of hedge fund returns. Their results show that the model becomes more flexible, since the parameters are updated everytime the market situation changed, and hence a forecasting performance of this model is more reliable.

A more advanced non-linear model, Markov switching GARCH, was proposed by combining Markov switching concept with GARCH model. This model has been used widely in the past decade. For instance, Brunetti et al. [22] employed a Markov switching GARCH model, consist of both the conditional mean and the conditional variance, to analyze Southeast Asia exchange rate turbulences. The results suggest that stock market returns, banking stock returns, money supply relative to reserve, real effective exchange rates and volatility are significant factors to describe exchange rate turbulences. Chitsazan and Keimasi [23] applied Markov switching GARCH model to investigate the volatility states of gold futures in Iran. They found that there is much volatility in gold future market and their volatility regimes change very often.

Several researches claimed Markov switching GARCH model is better than ordinary GARCH models. For example, Klaassen [24] showed that using GARCH model leads to the overestimation of US dollar exchange rate volatility in volatile period. Thus, in order to fix this problem, he proposed Markov switching GARCH model which is more flexible than ordinary GARCH model. The results show that, by using Markov switching GARCH, the problem has been solved and it also presents better out-of-sample forecasts.

Marcucci [25] compared out-of-sample prediction performance between a set of standard GARCH models, namely GARCH(1,1), EGARCH(1,1) and GJR(1,1), and Markov Switching GARCH models in order to forecast US stock market volatility. The results show that, under shorter horizons, Markov Switching GARCH models outperform all standard GARCH models while standard asymmetric GARCH models perform best under longer one. Similarly, Tang and Yin [26] compared the ability to forecast aluminium and copper prices between using a set of standard GARCH models and Markov switching GARCH model cover the period from 1993 to 2009. They found that a Markov switching GARCH model fits the data better than other standard GARCH models. It also outperforms standard GARCH models in out-of-sample forecasting.

3   Data and Data Analysis

The dataset used for this study were collected from Thomson Reuters database. The sample used is daily data which goes from April 1st, 2011 to March 31st, 2016 for a total of 1186 observations. The data variables consist of iTraxx Asia ex-Japan CDS Spreads, Hang Seng Future Index, 10-year US government bond yield and CBOE Market Volatility Index.

Table 1 reports the descriptive statistics for the variables, namely the changes in CDS spread, returns on Hang Seng Future index, changes in bond yield and changes in VIX index. According to the Jarque-Bera test, all variables are not normally distributed. It also shows, confirmed by the Augmented Dickey Fuller (ADF) statistic, that all variables are stationary. Besides, only the changes in CDS spread variable shows a positive mean value and it is the most volatile variable according to the standard deviation.

Table 1 Descriptive statistics

	ΔCDS_t	$\%\Delta HS_t$	$\Delta Bond_t$	ΔVIX_t
Mean	0.031197	(−0.002039)	(−0.00142)	(−0.003196)
Median	(−0.25)	0.02593	(−0.005)	(−0.08)
Maximum	31.25	5.85594	0.238	16
Minimum	(−30.83)	(−7.38861)	(−0.247)	(−12.94)
Std. dev.	4.688441	1.286005	0.053608	1.677486
Skewness	1.06774	(−0.155716)	0.0629	1.381229
Kurtosis	11.75315	5.891173	4.60257	19.22469
Jarque-Bera	4011.538[c]	417.9612[c]	127.6954[c]	13385.58[c]
ADF-stat	−33.5525[c]	−34.0693[c]	−36.7771[c]	−23.6903[c]

[a,b,c]indicate rejection of the null hypothesis at 10, 5 and 1%, respectively

4   The Forecasting Models

This study aims to investigate the performance of the forecasting linear and non-linear models for Asian CDS spreads. Since we are interested in whether CDS spreads can predict its own future. Hence, we use lagged CDS spreads to forecast future CDS spreads. Following Avino and Nneji [6], the simple AR(1) model is employed and suggested to investigate the CDS forecasting. Furthermore, we also use structural model in order to investigate whether macroeconomic and financial variables can predict the changes of Asian CDS spreads. We apply both linear and non-linear versions of these models namely simple model, threshold model, Markov switching model and Markov switching GARCH.

4.1   Autoregressive Model and Structural Regression Model

The simple AR(1) is expressed as follows:

$$y_t = \alpha + \beta_1 y_{t-1} + \varepsilon_t, \tag{1}$$

where y_t and y_{t-1} are the observed data and its lag, α and β are the estimated parameters of the model, and ε_t is assumed to be i.i.d. and normally distributed. For the structural model, the other explanatory variables are formed into the model in order to evaluate the forecasting ability of these variables in predicting future credit spreads. Thus the regression structural model can be expressed as

$$y_t = \alpha + \beta_1 y_{t-1} + \phi_i X'_t + \varepsilon_t, \tag{2}$$

where X'_t is $k \times n$ matrix of explanatory variables and ϕ_i is $1 \times k$ unknown parameter vector.

4.2 Threshold Autoregressive Model and Threshold Structural Regression Model

Since the introduction of Threshold model of Tong [27], the model has become popular in both statistics and econometrics and been employed in many application studies. The simple TAR(1) model with two regimes is considered for the study and it could be defined as follows

$$
\begin{aligned}
y_t &= \alpha_1 + \beta_1 y_{t-1} + \varepsilon_{1,t}, & y_{t-1} \leq \gamma, \\
y_t &= \alpha_2 + \beta_2 y_{t-1} + \varepsilon_{2,t}, & y_{t-1} > \gamma,
\end{aligned}
\tag{3}
$$

where γ is a threshold value and the threshold variable is set to be y_{t-1} in this case. The threshold process divides one dimensional Euclidean space into 2 regimes, with a linear AR(1) in each regime. For Threshold structural regression model, according to Yu [28], the general setup of threshold regression models (T-reg) is

$$
\begin{aligned}
y_t &= \alpha_1 + \beta_1 y_{t-1} + \phi_i X'_t + \varepsilon_{1,t}, & w \leq \gamma, \\
y_t &= \alpha_2 + \beta_2 y_{t-1} + \theta_i X'_t + \varepsilon_{2,t}, & w > \gamma,
\end{aligned}
\tag{4}
$$

where w is the threshold variable used to split the sample, γ is the threshold point.

4.3 Markov Switching Autoregressive Model and Markov Switching Structural Regression Model

Consider the following Gaussian Markov Switching Autoregressive model of Hamilton [29], the simple MS-AR(1) can be defined as

$$
y_t = \alpha_{s(t)} + \beta_{1,s(t)} y_{t-1} + \varepsilon_{t,s(t)},
\tag{5}
$$

where $\varepsilon_{s(t)} \sim i.i.d.N(0, \sigma^2_{s(t)})$, y_t is dependent variable. $s(t) = i$, $i = 1, ..., h$. $\alpha_{s(t)}$ and $\beta_{1,s(t)}$ are state dependent estimated parameters. However, in this study we focus on two regimes for being the most popular application in many works. For Markov Switching regression (MS-reg), we can extend Eq. (5) in the simplest form as

$$
y_t = \alpha_{s(t)} + \beta_{1,s(t)} y_{t-1} + \phi_i X_t + \varepsilon_{t,s(t)},
\tag{6}
$$

where y_t is a dependent variable and X_t is a matrix of explanatory variables. The model is postulated that the transition matrix (Q) is governed by a first order Markov chain, thus

$$p(s_t = j | s_{t-1} = i) = p_{ij}, \quad \sum_{j=1}^{h} p_{ij} = 1, \quad i = 1, ..., h. \tag{7}$$

In this study, we consider only 2-regime Markov switching model, thus, the first order Markov process could be written as:

$$\begin{aligned} p(s_t = 1 | s_{t-1} = 1) &= p_{11} \\ p(s_t = 1 | s_{t-1} = 2) &= p_{12} \\ p(s_t = 2 | s_{t-1} = 1) &= p_{21} \\ p(s_t = 2 | s_{t-1} = 2) &= p_{22} \end{aligned} \tag{8}$$

where p_{ij} are the transition probabilities from state j to state h.

4.4 Markov Switching Structural GARCH and Markov Switching Autoregressive GARCH

Bollerslev [30] noted that time series data generally exhibit variable volatility over time, thus tending to show GARCH (Generalized Autoregressive Conditionally Heteroscedastic) effects in the mean equation. As also snoted by Arango et al. [31] about the high-frequency financial series data, the assumption that the error sequence generated by the non-linear model for the conditional mean has a constant conditional variance is not realistic to explain the volatility in the data. Thus, we expect the GARCH effect in the model and extend Eq. (5) to be MS-AR(1)-GARCH(1,1) as

$$y_t = \alpha_{s(t)} + \beta_{1,s(t)} y_{t-1} + \varepsilon_{t,s(t)}, \tag{9}$$

$$h_{s(t)}^2 = \mu_{s(t)} + \delta_{s(t)} \varepsilon_{t-1,s(t)}^2 + \kappa_{s(t)} h_{t-1,s(t)}^2, \tag{10}$$

where Eqs. (9) and (10) are the mean and variance equations, respectively, and they are allowed to switch across regime. $h_{s(t)}^2$ is the state dependent conditional variance and $\mu_{s(t)}, \delta_{s(t)},$ and $\kappa_{s(t)}$ are state dependent estimated parameters which are restricted to be larger than zero. In this variance GARCH(1,1) specification, the state dependent unconditional variance can be computed by $\mu_{s(t)}/(1 - \delta_{s(t)} - \kappa_{s(t)})$. Consider MS-reg-GARCH(1,1), we can extend Eqs. (9) and (10) to be

$$\begin{aligned} y_t &= \alpha_{s(t)} + \beta_{1,s(t)} y_{t-1} + \phi_i X_t + \varepsilon_{t,s(t)}, \\ \varepsilon_{t,s(t)} &= h_{s(t)}^2 v_{s(t)}, \end{aligned} \tag{11}$$

$$h^2_{s(t)} = \mu_{s(t)} + \delta_{s(t)}\varepsilon^2_{t-1,s(t)} + \kappa_{s(t)}h^2_{t-1,s(t)}, \tag{12}$$

where $v_{s(t)}$ is a standardized residual. In this study, we proposed six different distributions of $v_{s(t)}$ consisting of normal, student-t, generalized error distribution (GED), skewed GED, skewed normal, and skewed student-t distributions. For selecting the best distribution, the lowest Akaiki Information criterion (AIC) is preferred. The probabilistic structure of the switching regime is defined as a first-order Markov process with constant transition probabilities $p_i j$ governing the latent state variable $s(t)$:

To estimate the parameter set in both MS-AR(1) and MS-AR(1)-GARCH(1,1), a maximum likelihood procedure is used to estimate the Markov switching model. The general form of the MS-AR(1) likelihood can be defined as

$$f(y_t|y_{t-1}, \alpha_{s(t)}, \beta_{1,s(t)}, \sigma^2_{s(t)}, p_{ij}) = f(y_t|y_{t-1}, \alpha_{s(t=1)}, \beta_{1,s(t=1)}, \sigma^2_{s(t=1)}, p_{ij})$$
$$\times f(y_t|y_{t-1}, \alpha_{s(t=2)}, \beta_{1,s(t=2)}, \sigma^2_{s(t=2)}, p_{ij}), \tag{13}$$

while the MS-AR(1)-GARCH(1,1) likelihood can be defined as

$$f(y_t|y_{t-1}, \alpha_{s(t)}, \beta_{1,s(t)}, \sigma^2_{s(t)}, p_{ij}, \Phi) = f(y_t|y_{t-1}, \alpha_{s(t=1)}, \beta_{1,s(t=1)}, \sigma^2_{s(t=1)}, p_{ij}, \Phi)$$
$$\times f(y_t|y_{t-1}, \alpha_{s(t=2)}, \beta_{1,s(t=2)}, \sigma^2_{s(t=2)}, p_{ij}, \Phi), \tag{14}$$

where Φ is a skew or degree of freedom parameter when the distribution of the standardized residuals is not normal. For MS-reg and MS-reg-GARCH(1,1), an explanatory variable (X_t) is taken into account in the likelihood function Eqs. (13) and (14).

5 Estimation Results

In this section, we analyze the estimation results of the CDS spread changes from both structural and AR(1) models. We employ simple linear model and three non-linear models namely, threshold model, Markov switching model and Markov switching GARCH. For non-linear models, we divide the estimated parameters into 2 regimes namely high market volatility regime and low market volatility regime. In other words, for threshold model, we divide by threshold into 2 regimes which are bull market regime and bear market regime. A bull market implies low market volatility and a bear market implies high market volatility [32, 33]. For regime switching models, we estimate the impact of selected explanatory variables which depend on whether the CDS market is in a high volatility or low volatility scenario. We also report log-likelihood and Akaike information criterion (AIC) values.

Under the error distributions assumption in GARCH model, namely, Normal Distribution, Student-t Distribution, Generalized Error Distribution (GED) and their skewed version, we consider Akaike information criterion(AIC) values in order to determine the best distribution that suits Markov switching-GARCH models. Table 2

Table 2 Markov switching GARCH(1,1) model selection

Distribution	Structural model	AR(1) model
	AIC	AIC
Normal	5851.309	6312.021
Student-T	5843.770	6333.789
Skewed normal	5841.275	**6295.586**
Skewed student-T	**5833.899**	6299.210
GED	5860.279	6310.076
Skewed GED	5850.157	6299.591

compares the Akaike information criterion(AIC) values of the Markov switching-GARCH models estimated from different error distributions. The result shows that, according to the lowest AIC value, Skewed Student-t Distribution fits the data best in Markov switching regression-GARCH(1,1) model and Skewed Normal is the best suitable distribution for Markov switching AR(1)-GARCH(1,1) model.

Table 3 shows the parameter estimation for structural models. In each model, most of the estimated parameters are significant and have the signs as we expected. Hang Seng Future index return and the changes in bond yield, for instance, show a negative impact on CDS spread changes. Since CDS is a substitution investment asset, when stock return or bond yield declines, investors tend to move their investment into the CDS market and hence increase in the changes of the CDS spread, and vice versa. Market Volatility Index (VIX) positively affects the changes of the CDS spread as when the market is more volatile, investors will hedge their risks by buying CDS contracts. For the lagged CDS spreads, it shows a negative effect on CDS spreads which means the changes of the CDS spread are not persistent.

For threshold regression model, the lagged CDS spreads, Hang Seng Futures return, bond yield and market volatility index significantly affect the change of CDS spread in low volatility regime. In high volatility regime, however, there is only market volatility index that has a significant effect on CDS spread change and the magnitude of the effect is higher than in low volatility regime. The performance of Markov switching regression model is more superior since all of the explanatory variables significantly affect the change of CDS spread in both low and high volatility regimes. The results also show that the impacts of all explanatory variables are much higher in high volatility period. Moreover, this Markov switching model has the lowest value of AIC compare to other models. Likewise, in Markov switching GARCH model with skewed Student t-distribution, explanatory variables all have a significant impact on CDS spread change but the size of impacts are smaller than in a Markov switching model. Nevertheless, bond yield shows no significant effect on the change of CDS spread in low volatility regime.

Table 4 presents the parameter estimation for AR(1) model. Interestingly, the results are mixed among models. In threshold model, the lag variable significantly affects the CDS spread change only in high volatility regime. On the contrary, the

Table 3 Parameter estimates for structural models

	Least square	Threshold		Markov switching		Markov switching-GARCH	
		Low volatile regime	High volatile regime	Low volatile regime	High volatile regime	Low volatile regime	High volatile regime
Constant	0.0188	−3.6100^c	0.3756^c	−0.1168^a	−0.0003	0.0965	−0.0366
	(−0.109)	(−0.4468)	(−0.1344)	(−0.071)	(−0.2481)	(−0.0709)	(−0.089)
CDS_{t-1}	−0.0767^b	−0.0740^b	−0.0381	−0.0730^b	−0.0882^b	−0.1007^c	−0.0873^c
	(−0.024)	(−0.0444)	(−0.0359)	(−0.029)	(−0.0354)	(−0.0605)	(−0.0268)
HS_t	−1.5871^c	−2.1999^c	−2.4212	−0.9310^c	−2.3244^c	−0.5250^c	−1.1546^c
	(−0.0901)	(−0.1384)	(−47.0725)	(−0.0683)	(−0.2041)	(−0.0402)	(−0.0812)
$Bond_t$	−10.3160^c	−16.8852^c	0.5495	−7.1515^c	−10.6548^c	−2.4377	−8.7246^c
	(−2.2841)	(−2.2868)	(−0.133)	(−1.7043)	(−5.0119)	(−1.8867)	(−1.8403)
VIX_t	0.7359^c	1.0781^c	1.4086^c	0.4415^c	1.3124^c	0.1653^c	0.6055^c
	(−0.0736)	(−0.123)	(−0.0812)	(−0.0471)	(−0.1844)	(−0.0427)	(−0.0588)
Threshold		0.1749^c					
		(−0.0444)					
σ	3.7531^c	3.9509^c	4.0060^c	1.8075^c	4.5004^c		
	(−0.0771)	(−0.0826)	(−0.4468)	(−0.0775)	(−0.2981)		
μ						0.1658	1.0033^c
						(−0.2735)	(−0.335)
δ						0.7318^c	0.0226^c
						(−0.2385)	(−0.0062)
κ						0.7202^c	0.9583^c
						(−0.0207)	(−0.0094)
df						3.7728^c	8.9105^c
						(−1.0063)	(−2.5907)
Skew						4.8651^c	0.8970^c
						(−1.9333)	(−0.0573)
P11				0.9761^c		0.7477^c	
				(−0.0072)		(−0.0684)	
P22				0.9570^c		0.2053	
				(−0.0169)		(−0.2066)	
Log-likelihood	−3248.734	−6635.382		−2896.437		−2894.95	
AIC	6509.467	13296.76		5820.875		5833.899	

^{a,b,c}indicate rejection of the null hypothesis at 10, 5 and 1%, respectively

Table 4 Parameter estimates for AR(1) models

	AR(1)	Threshold		Markov switching		Markov switching-GARCH	
		Low volatile regime	High volatile regime	Low volatile regime	High volatile regime	Low volatile regime	High volatile regime
Constant	0.034	0.1395	−0.9224[a]	−0.1568[a]	−0.0427	−0.0335	1.6942[b]
	(−0.1395)	(−0.1629)	(−0.5564)	(−0.0844)	(−0.3231)	(−0.0847)	(−0.7363)
CDS_{t-1}	0.0244	0.0451	0.1066[a]	0.0650[a]	0.0324	0.0447	−0.2354
	(−0.029)	(−0.0484)	(−0.0642)	(−0.0367)	(−0.0458)	(−0.0302)	(−0.1725)
Threshold		2.995					
σ		4.681	4.7039	2.0740[c]	6.1034[c]		
				(−0.109)	(−0.4476)		
μ						1.4450[c]	1.9761
						(−0.0302)	(−0.2381)
δ						0.1021[c]	0.2381[b]
						(−0.0271)	(−0.095)
κ						0.8148[c]	0.9354[c]
						(−0.0367)	(−0.0239)
Skew						1.2167[c]	2.2412[c]
						(−0.0746)	(−0.3774)
p_{11}				0.9571[c]		0.1077	
				(−0.0152)		(−0.1544)	
p_{22}				0.9724[c]		0.8844[c]	
				(−0.0082)		(−0.0296)	
Log-likelihood	−3511.82	−2860.387		−3179.702		−3133.793	
AIC	7029.65	3667		6375.403		6295.586	

[a,b,c] indicate rejection of the null hypothesis at 10, 5 and 1%, respectively

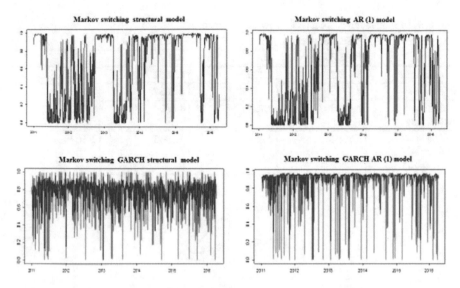

Fig. 2 Time series of regime probabilities for Asian CDS index. This figure shows the filtered probability of being in the high volatile regime estimated from Markov switching and Markov switching GARCH(1,1) models both in structural and AR(1) model

impact of the lag variable on the CDS spread change is significant only in low volatility regime for Markov switching model. Additionally, for simple AR(1) and Markov switching GARCH model with skewed Normal distribution, the lagged CDS spread change shows no significant effect on the change of CDS spread. Unlike the structural models, the AR(1) models suggest that the changes of the CDS spread are persistent since the lagged CDS spread change shows a positive effect on CDS spread change (Fig. 2).

6   In- and Out-of-Sample Forecasting Performance of the Models and Robustness Test

Economic forecasting typically differs from the actual outcome reflecting forecast uncertainty. In order to evaluate the forecasting accuracy, many studies employed a root mean square error (RMSE) and mean square error (MAE). They both combine the bias and the variance of the forecast inherently in its structure and simplify into the forecast error standard deviation and mean in the case of unbiased forecast (Stovicek [34]). In this section, the four types of model obtained from the previous section are investigated for the forecasting performance for both structural and AR(1) model. We compare the accuracy of these models by applying RMSE and MAE for both in- and out-of-sample forecast. The RMSE and MAE formulas are as follows:

Table 5 In-sample forecasting performance of the models

Model	Structural model		AR(1) model	
	RMSE	MAE	RMSE	MAE
Least square	3.7532	2.4768	4.6862	3.0245
Threshold	4.7765	3.1765	**4.683**	**3.0175**
Markov switching	**3.4434**	**2.2334**	4.6893	3.0202
Markov switching GARCH(1,1)	6.7791	4.2117	4.7946	3.1078

$$RMSE = \sqrt{\frac{1}{N}\sum_{j=1}^{N} e_j^2} \tag{15}$$

$$MAE = \frac{1}{N}\sum_{j=1}^{N} |e_j| \tag{16}$$

The in-sample performance of the forecasting models (as shown in Table 5) indicates that, based on the MRSE and MAE values, the Markov switching model clearly outperforms other models in structural type. Nevertheless, the results suggest that the Markov switching GARCH(1,1) model with skewed Student t-distribution is the worst performing model since its RMSE and MAE values are much higher than other models. For AR(1) forecasting models, the RMSE and MAE values are quit not much different among the models. The threshold model shows the best performance over the other models in forecasting the first order autoregressive process. Interestingly, Markov switching GARCH(1,1) model (with skewed Normal distribution) remains an underperforming model in this forecasting method.

For out-of-sampleforecast, we calculated a multi-step-ahead forecast, which is from March 1st, 2016 to March 31st, 2016, and compared our forecasting values with the real data. Then, we analyzed the out-of-sample forecasting performance of the proposed 4 different models both in structural and AR(1) model. According to Table 5, the results from the in-sample forecast indicated that the Markov switching model performs better than other models in structural model, while the threshold model shows the best forecasting performance in AR(1) model. However, consider the out-of-sample forecasting performance as shown in Table 6, the results show that Markov switching GARCH(1,1) model becomes an outperforming model over other models for structural models based on the RMSE and MAE criteria. Besides, for AR(1) models, Markov switching model becomes a superior model to forecast the change of CDS spread.

In addition, the robustness test is performed to evaluate the predictive performance and compare the out-of-sample predictive ability of our forecasting models. We

Table 6 Out-of-sample forecasting performance of the models

Model	Structural model		AR(1) model	
	RMSE	MAE	RMSE	MAE
Least square	5.8744	3.8182	5.8689	3.9702
Threshold	5.9275	3.8277	5.8814	4.0264
Markov switching	5.9176	3.8261	**5.8567**	**3.8743**
Markov switching GARCH(1,1)	**5.7475**	**3.6602**	6.4304	4.8271

conducted the conditional predictive ability test called GW-test which is proposed by Giacomini and White [35]. The approach is based on inference about conditional expectations of forecasts and forecast errors and it is valid under heterogeneity rather than stationarity data. The null hypothesis is $H_0 : Loss(M_a) - Loss(M_b) = 0$ which mean that two forecast models are equally accurate on average. The proposed test statistic for the GW-test can be calculated as:

$$GW = T\left(T^{-1}\sum_{t=1}^{T-\tau}\pi_t d_{t+\tau}^i\right)' \Omega_{t+\tau}^{-1}\left(T^{-1}\sum_{t=1}^{T-\tau}\pi_t d_{t+\tau}^i\right), \qquad (17)$$

where π_t are all information set available at time t, $\Omega_{t+\tau}^{-1}$ is the inverse covariance matrix which is a consistent HAC estimator for the asymptotic variance of $\pi_t d_{t+\tau}^i$. The significant testing of the test statistic follows a chi-squared distribution. The interpretation of the result can be explained by the sign of the test-statistic which indicates the superior forecasting model. In other words, a positive test-statistic indicates that M_a forecast produces larger average loss than the M_b forecast (M_b performs better than M_a), while a negative sign indicates the opposite interpretation.

Table 7 reports the results of the robustness test analysis by applying GW test. The estimated results are robust and consistent with the out-of-sample statistical performance of the forecasting models in previous section. In other words, for structural models, we can interpret that Markov switching GARCH(1,1) model slightly outperforms other models in out-of-sample forecast since the GW statistics of least square model, threshold model and Markov switching model against Markov switching GARCH(1,1) model are positive although they are not significant. The results also suggest that least square model obviously dominates threshold model and Markov switching model in out-of-sample forecast. The robustness test for AR(1) models presents much more apparent results. It clearly indicates that Markov switching model dominates all other models, followed by simple AR(1) model, threshold model and Markov switching GARCH(1,1), respectively, as the sign of GW statistics and they are all significant. Finally, in order to determine the best predictive model, we chose a dominant model from both structural and AR(1) models and compared these

Table 7 GW robustness test of forecasting performance for out-of-sample forecasts

GW tests for structural models	Test-statistic	GW tests for AR(1) models	Test-statistic
Least square against threshold	11.01^c (-)	**AR(1)** against threshold	22.00^c (-)
Least square against Markov switching	12.06^c (-)	AR(1) against **Markov switching**	18.59^c (+)
Least square against **Markov switching GARCH(1,1)**	2.67 (+)	**AR(1)** against Markov switching GARCH(1,1)	14.43^c (-)
Threshold against **Markov switching**	0.91 (+)	Threshold against **Markov switching**	5.99^c (+)
Threshold against **Markov switching GARCH(1,1)**	0.53 (+)	**Threshold** against Markov switching GARCH(1,1)	13.10^c (-)
Markov switching against **Markov switching GARCH(1,1)**	0.58 (+)	**Markov switching** against Markov switching GARCH(1,1)	15.85^c (-)
GW test between the best of Structural model and AR(1) model			**Test-statistic**
Markov switching AR(1) against **Markov switching GARCH(1,1) structural**			0.28 (+)

^{a,b,c}indicate rejection of the null hypothesis at 10, 5 and 1%, respectively

two models. We find that Markov switching GARCH(1,1) structural model slightly outperforms Markov switching AR(1) model which is consistent with the RMSE and MAE results.

7 Conclusion

This paper aims to investigate the forecasting performance on the CDS index in emerging markets Asia by using advanced econometric models. In particular, we apply ordinary least square model, threshold model, Markov switching model and Markov switching-GARCH model and the first lagged autoregressive (AR(1)) version of these models. We collect all the dataset from Thomson Reuters database. The variables consist of iTraxx Asia ex-Japan CDS Spreads, Hang Seng Futures Index, 10-year US government bond yield and CBOE Market Volatility Index. We consider daily data which goes from April 1st, 2011 to March 31st, 2016 for a total of 1186 observations.

We initially estimate the parameters of the models. The empirical results show that, for structural models, Hang Seng Futures index return and the changes in bond yield have a negative impact on the change of CDS spread since the CDS contracts are viewed as the substitution investment asset. Market Volatility Index (VIX), on the other hand, shows a positive effect on the CDS spread change as the CDS contracts are tools for hedging the risks. Furthermore, the lagged CDS spreads presents a negative effect on CDS spreads which implies that the changes of the CDS spread are not persistent. In addition, for 2-regime structural models, namely threshold model, Markov switching and Markov switching GARCH model, the impact size of those explanatory variables is higher in high volatility period. Interestingly, unlike structural models, the AR(1) models present that the changes of the CDS spread are persistent since the lagged CDS spread change shows a positive effect on CDS spread change and the estimated results are mixed among models.

The in- and out-of- sample forecasts are conducted to compare the forecasting performance of the proposed models both in structural model and AR(1) model using RMSE and MAE criteria. For in-sample forecasting performance, we find that Markov switching model outperforms other models in structural type, while threshold model performs better than other models in AR(1) variant. Moreover, Markov switching GARCH(1,1) model presents the worst performance in both structural and AR(1) models. However, the out-of-sample forecast demonstrates different results, that is, Markov switching GARCH(1,1) and Markov switching model become a superior model to forecast the change of CDS spread for structural model and AR(1) model, respectively. Additionally, we employ a robustness test called GW test to check the preciseness of the selected model. The results remain suggesting that Markov switching GARCH(1,1) and Markov switching model dominate other models in structural model and AR(1) model, respectively. Finally, we compare these two dominant models in order to determine the best predictive model. The result shows that Markov switching GARCH(1,1) structural model slightly outperforms Markov switching AR(1) model.

References

1. Alloway T (2015) Why would anyone want to restart the credit default swaps market? Bloomberg market
2. Lin H, Liu S, Wu C (2009) Determinants of corporate bond and CDS spreads. Working paper
3. Oehmke M, Zawadowski A (2014) The anatomy of the CDS market. SSRN 2023108
4. Gyntelberg J, Ma G, Remolona EM (2005) Corporate bond markets in Asia. BIS Q Rev 6(4):83–95
5. BIS (2009) The international financial crisis: timeline, impact and policy responses in Asia and the Pacific, prepared for the wrap-up conference of the Asian Research Programme in Shanghai, August 2009
6. Avino D, Nneji O (2014) Are CDS spreads predictable? An analysis of linear and non-linear forecasting models. Int Rev Finan Anal 34:262–274
7. Byström H (2006) CreditGrades and the iTraxx CDS index market. Finan Anal J 62(6):65–76
8. Alexander C, Kaeck A (2008) Regime dependent determinants of credit default swap spreads. J Bank Finan 32(6):1008–1021

9. CollinDufresne P, Goldstein RS, Martin JS (2001) The determinants of credit spread changes. J Finan 56(6):2177–2207
10. Campbell JY, Taksler GB (2003) Equity volatility and corporate bond yields. J Finan 58(6):2321–2350
11. Cremers M, Driessen J, Maenhout P, Weinbaum D (2008) Individual stock-option prices and credit spreads. J Bank Finan 32(12):2706–2715
12. Galil K, Shapir OM, Amiram D, Ben-Zion U (2014) The determinants of CDS spreads. J Bank Finan 41:271–282
13. Zhang BY, Zhou H, Zhu H (2009) Explaining credit default swap spreads with the equity volatility and jump risks of individual firms. Rev Finan Stud 22(12):5099–5131
14. Ericsson J, Jacobs K, Oviedo R (2009) The determinants of credit default swap premia. J Finan Quant Anal 44(1):109–132
15. Domian DL, Louton DA (1997) A threshold autoregressive analysis of stock returns and real economic activity. Int Rev Econ Finan 6(2):167–179
16. Posedel P, Tica J (2007) Threshold autoregressive model of exchange rate pass through effect: the case of Croatia. EFZG Working paper series/EFZG Serijalanaka u nastajanju, no 15, pp 1–12
17. Nampoothiri CK, Balakrishna N (2000) Threshold autoregressive model for a time series data. J Indian Soc Agric Stat 53:151–160
18. Ailliot P, Monbet V (2012) Markov-switching autoregressive models for wind time series. Environ Model Softw 30:92–101
19. Piplack J (2009) Estimating and forecasting asset volatility and its volatility: a Markov-switching range model. Discussion paper series/Tjalling C. Koopmans Research Institute 9(08)
20. Engel C (1993) Real exchange rates and relative prices: an empirical investigation. J Monetary Econ 32(1):35–50
21. Erlwein C, Mller M (2011) A regime-switching regression model for hedge funds. Fraunhofer-Institutfr Techno-und Wirtschaftsmathematik, Fraunhofer (ITWM)
22. Brunetti C, Scotti C, Mariano RS, Tan AH (2008) Markov switching GARCH models of currency turmoil in Southeast Asia. Emerg Markets Rev 9(2):104–128
23. Chitsazan H, Keimasi M (2014) Modeling volatility of gold futures market in Iran by switching GARCH models. Int J Econ Manage Soc Sci 3(11):703–707
24. Klaassen F (2002) Improving GARCH volatility forecasts with regime-switching GARCH. In: Advances in Markov-switching models. Physica-Verlag HD, Heidelberg, pp 223–254
25. Marcucci J (2005) Forecasting stock market volatility with regime-switching GARCH models. Stud Nonlinear Dyn Econometrics 9(4)
26. Tang, Yin S (2010) Forecasting metals prices with regime switching GARCH Models. The Chinese University of Hong Kong
27. Tong H (1983) Threshold models in non-linear time series analysis. Springer New York Inc., New York
28. Yu P (2012) Likelihood estimation and inference in threshold regression. J Econometrics 167(1):274–294
29. Hamilton JD (1989) A new approach to the economic analysis of nonstationary time series and the business cycle. Econometrica 57:357–384
30. Bollerslev T (1986) Generalized autoregressive conditional heteroscedasticity. J Econometrics 31(3):307–327 Amsterdam
31. Arango LE, Gonzlez A, Posada CE (2002) Returns and the interest rate: a non-linear relationship in the Bogota stock market. Appl Finan Econ 12(11):835–842
32. Li Q, Yang J, Hsiao C, Chang YJ (2005) The relationship between stock returns and volatility in international stock markets. J Empirical Finan 12(5):650–665

33. Dimitriou D, Simos T (2011) The relationship between stock returns and volatility in the seventeen largest international stock markets: a semi-parametric approach. Mod Econ 2(01):1
34. Stovicek K (2007) Forecasting with ARMA models: the case of Slovenian inflation. Bank of Slovenia
35. Giacomini R, White H (2006) Tests of conditional predictive ability. Econometrica 74:1545–1578

Effect of Helmet Use on Severity of Head Injuries Using Doubly Robust Estimators

Jirakom Sirisrisakulchai and Songsak Sriboonchitta

Abstract Causal inference based on observational data can be formulated as a missing outcome imputation and an adjustment for covariate imbalance models. Doubly robust estimators–a combination of imputation-based and inverse probability weighting estimators–offer some protection against some particular misspecified assumptions. When at least one of the two models is correctly specified, doubly robust estimators are asymptotically unbiased and consistent. We reviewed and applied the doubly robust estimators for estimating causal effect of helmet use on the severity of head injury from observational data. We found that helmet usage has a small effect on the severity of head injury.

Keywords Causal effect · Missing outcome imputation · Propensity score matching · Doubly robust estimators

1 Introduction

Motorcyclists and passengers are characterized as unprotected road users. Helmets are widely used to protect motorcyclists and passengers from fatal accidents and serious head injuries. To promote proper helmet usage policies, practitioners need to be able to measure the causal effect of helmet use on severity of head injury. The first objective of this paper is to reinvestigate the causal effect of helmet use on severity of head injuries by using doubly robust estimators. We consider the potential outcomes framework as an appropriate tool in addressing the causal effect estimation. The second objective the key contribution of this paper is to review and apply the doubly robust estimators, which are new to the economic literature.

J. Sirisrisakulchai (✉) · S. Sriboonchitta
Faculty of Economics, Chiang Mai University, Chiang Mai, Thailand
e-mail: sirisrisakulchai@hotmail.com

© Springer International Publishing AG 2017 491
V. Kreinovich et al. (eds.), *Robustness in Econometrics*,
Studies in Computational Intelligence 692, DOI 10.1007/978-3-319-50742-2_29

Table 1 Causal effect in the potential outcome framework

Unit	Potential outcomes		Covariate	Treatment assignment	Unit-level causal effect
	Treatment	Control			
	Y_1	Y_0	X_i	T_i	
1	$Y_1^{mis} = Y_{11}$	$Y_1^{obs} = Y_{01}$	X_1	0	$Y_{11} - Y_{01}$
2	$Y_2^{mis} = Y_{12}$	$Y_2^{obs} = Y_{02}$	X_2	0	$Y_{12} - Y_{02}$
3	$Y_3^{obs} = Y_{13}$	$Y_3^{mis} = Y_{03}$	X_3	1	$Y_{13} - Y_{03}$
$\vdots$	$\vdots$	$\vdots$	$\vdots$	$\vdots$	$\vdots$
i	$Y_i^{obs} = Y_{1i}$	$Y_i^{mis} = Y_{0i}$	X_i	1	$Y_{1i} - Y_{0i}$
$\vdots$	$\vdots$	$\vdots$	$\vdots$	$\vdots$	$\vdots$
n	$Y_n^{mis} = Y_{1n}$	$Y_n^{obs} = Y_{0n}$	X_n	0	$Y_{1n} - Y_{0n}$

The fundamental problem of causal effect estimation is the lack of observation/data of one of the potential outcomes. This can be viewed as a missing data problem. Suppose that researchers observe a random sample of unit $i = 1, 2, \ldots, n$ from a population. The outcome of interest is Y_i for each unit i. Let T_i be the assignment indicator for each unit i, so that $T_i = 1$ if unit i is assigned to a treatment group and $T_i = 0$ if unit i is assigned to a control group. There is a pair of potential outcomes associated with unit i: the potential outcome Y_{1i} that is realized if $T_i = 1$ and another potential outcome Y_{0i} that is realized if $T_i = 0$. For each unit i, we also observe the m-dimensional vector of covariates X_i for unit i. Let Y_i^{obs} be the observed potential outcome for unit i and Y_i^{mis} be the missing or the unobserved potential outcome for unit i. Thus, we have

$$Y_i^{obs} = T_i \cdot Y_{1i} + (1 - T_i) \cdot Y_{0i} \tag{1}$$

and

$$Y_i^{mis} = (1 - T_i) \cdot Y_{1i} + T_i \cdot Y_{0i}. \tag{2}$$

According to the potential outcomes framework [12], we can define a causal effect of the treatment on unit i as $Y_{1i} - Y_{0i}$. However, this effect is unobservable because we cannot observe the same unit at the same time in both the treatment and the control regimes. Table 1 summarizes the causal effect in the potential outcome framework. Notice that potential outcomes notation in randomized experiments was introduced by [6] and by [10] in non-randomized settings.

To handle the unobserved (missing) potential outcomes in randomized experiments, researchers usually model the relationship between all observed covariates and observed outcomes and use these models to predict the missing outcomes. As a result of this modelling, complete data table can be obtained and also the unit-level causal effects can be estimated.

In randomized experiments, a well-defined treatment is randomly assigned to a group of units, but it is not the case in non-randomized experiments or observational studies. Observational data is common and serves as the basis for economic and social science research. The causal effect obtained from observational data could be biased because of self-selection problems or some systematic judgement of treatment assignment mechanism by researchers. The difficulty arises because the treatment assignment mechanism and the outcome of interest may be related through some attributes. These correlations lead to an imbalance of treatment and control groups in those attributes.

Propensity score-matching methods are commonly used to adjust for covariate imbalance in observational data. Rosenbaum and Rubin [9] discussed the essential role of propensity score in correcting bias due to all observed covariates. In randomized experiments, we can directly compare the treatment and control groups because their units are likely to have the same characteristics. However, in observational data, the direct comparisons may be biased because the units exposed to treatment generally have different characteristics from the control units. Under the strong ignorability assumption, $\Pr(T|Y_1, Y_0, X) = \Pr(T|X, Y^{obs})$, and overlapping assumption, $0 < \Pr(T = 1|X, Y^{obs}) < 1$, [9] showed that this bias can be corrected by propensity score. The strong ignorability assumption implies that there is no confounding variable and all observed covariates contain all selection-bias information. The last assumption ensures the presence of bias-correction across the entire domain of X.

However, propensity score methods sometimes fail to correct this imbalance if the researchers misspecify the propensity score models. To cope with this problem, the researchers usually increase the complexity of the models until they find a sufficiently balanced solution.

To estimate causal effect of treatment from observational data, researchers have to simultaneously consider a missing outcome imputation and an adjustment for covariate imbalance. If researchers cannot specify two correct models, these misspecifications will lead to unbiased estimators of the causal effect of treatment. Over the last decade, doubly robust estimators have been developed in incomplete data analysis. These estimators can be viewed as a combination of imputation-based and inverse probability weighting estimators. When at least one of the two models is correctly specified, doubly robust estimators are asymptotically unbiased and consistent [13].

In this paper, we use the doubly robust procedures, which apply both the propensity score and the outcomes models simultaneously to produce a consistent estimate of causal effect. The additional protection provided by doubly robust estimators gives us much more confidence on our causal effect estimations. The paper proceeds as follows. In Sect. 2, we discuss the propensity score estimation. Then the doubly robust estimators are discussed in Sect. 3. Finally, we describe the data and causal effect estimations in Sects. 4 and 5.

2 Propensity Score Estimation

In this paper, we will assume the strong ignorability assumption. This assumption implies that the treatment assignment mechanism is unconfounded in the sense that Y_i and T_i are conditionally independent given the vector of covariate X_i [9]. Thus, the joint distribution of the complete data is

$$Pr(Y, T, X) = \prod_i \{Pr(Y_i|X_i) \times Pr(T_i|X_i) \times Pr(X_i)\} \tag{3}$$

The propensity score is the conditional probability of assignment to a treatment given a vector of the observed covariates [9]. The propensity score can be written as

$$e(X_i) = Pr(T_i = 1|X_i) = \pi_i. \tag{4}$$

By conditioning on observable covariates, this assignment mechanism can be taken as if it were random. Thus, by comparing two units with the same characteristics but different only in that one of whom was treated and another was not, is like comparing those two units in a randomized experiment [11].

To estimate the propensity score, discrete choice models, such as logit or probit models, can be used. The logit model is commonly used in application for the ease of estimation. The logit formula is

$$Pr(T_i = 1|X_i) = \frac{e^{\beta X_i}}{1 + e^{\beta X_i}}, \tag{5}$$

where X_i is a vector of covariates and β is a vector of parameters.

3 Doubly Robust Estimators

In this paper, we consider average treatment effect as a causal estimand or target parameter to be estimated. The average treatment effects can be expressed in two forms: the population average treatment effect (PATE) and the population average treatment effect on the treated (PATT). The PATE can be written as

$$\tau = E[Y_1|X] - E[Y_0|X], \tag{6}$$

and the PATT can be written as

$$\gamma = E[Y_1|X, T = 1] - E[Y_0|X, T = 1]. \tag{7}$$

Let us define the regression models $E[Y_1|X] = g_1(X)$ and $E[Y_0|X] = g_0(X)$. Robins et al. [7] described the doubly robust estimation framework for regression with some missing regressors. The applications of this framework for missing potential outcomes and subsequent literature are referred to the following articles and references therein: [1, 5, 8]. As we mentioned earlier, the doubly robust estimators apply both the propensity score and outcomes regression models to create the consistent estimate of parameters; thus, the estimate procedure consists of three components.

The first component is called inverse-probability weighting (IPW). Suppose that researchers have an estimate of propensity score $\hat{\pi}_i$. The IPW estimators can be written as follows:

$$\hat{\tau}^{IPW} = \sum_{T_i=1} \frac{\hat{\pi}_i^{-1}}{\sum_{T_i=1} \hat{\pi}_i^{-1}} Y_i - \sum_{T_i=0} \frac{(1 - \hat{\pi}_i^{-1})}{\sum_{T_i=0}(1 - \hat{\pi}_i)^{-1}} Y_i \qquad (8)$$

and

$$\hat{\gamma}^{IPW} = \sum_{T_i=1} \frac{1}{\sum_{T_i=1} T_i} Y_i - \sum_{T_i=0} \frac{\hat{\pi}_i(1 - \hat{\pi}_i^{-1})}{\sum_{T_i=0} \hat{\pi}_i(1 - \hat{\pi}_i)^{-1}} Y_i. \qquad (9)$$

The second component is just an outcomes regression model. Let $\hat{g}_0(X_i)$ and $\hat{g}_1(X_i)$ be the estimates of the mean potential outcomes. Thus the ordinary least square (OLS) estimators for PATE and PATT are as follows:

$$\hat{\tau}^{OLS} = \frac{1}{n} \sum_{i=1}^{n} (\hat{g}_1(X_i) - \hat{g}_0(X_i)) \qquad (10)$$

and

$$\hat{\gamma}^{OLS} = \frac{1}{\sum_{T_i=1} T_i} \sum_{i=1}^{n} (Y_i - \hat{g}_0(X_i)). \qquad (11)$$

Finally, a doubly robust estimator can be obtained by combining the two estimators as follows [14]:

$$\hat{\tau}^{DR} = \hat{\tau}^{OLS} + \sum_{T_i=1} \frac{\hat{\pi}_i^{-1}}{\sum_{T_i=1} \hat{\pi}_i^{-1}} (Y_i - \hat{g}_1(X_i)) - \sum_{T_i=0} \frac{(1 - \hat{\pi}_i^{-1})}{\sum_{T_i=0}(1 - \hat{\pi}_i)^{-1}} (Y_i - \hat{g}_0(X_i)) \qquad (12)$$

and

$$\hat{\gamma}^{DR} = \hat{\gamma}^{OLS} - \sum_{T_i=0} \frac{\hat{\pi}_i(1 - \hat{\pi}_i^{-1})}{\sum_{T_i=0} \hat{\pi}_i(1 - \hat{\pi}_i)^{-1}} (Y_i - \hat{g}_0(X_i)). \qquad (13)$$

The residual bias correction term is just the replacement of Y_i in the IPW estimators with the regression residual. The above DR estimators were first proposed by [2] and generalized by [7]. The DR estimator has the so-called double robustness property. If either the propensity score model or outcome regression model is correctly specified, then PATE and PATT are consistent [5].

Focusing on the estimation of PATT, [4] constructed doubly robust weighted least square estimators of PATT using entropy. This procedure is called the entropy balancing approach. Entropy balancing is performed by optimizing a set of weights over the control units and fixed equal weights for the treated units. [4] minimized the following problem to find the optimal weights (w_i):

$$\min_{w} \sum_{T_i=0} w_i \log w_i, \tag{14}$$

subject to

$$\sum_{T_i=0} w_i = 1, \quad w_i > 0 \quad \forall i, \tag{15a}$$

$$\sum_{T_i=0} w_i c_j(X_i) = \bar{c}_{1j}, \quad j = 1, 2, ..., m, \tag{15b}$$

where $c_j(X_i)$ denotes the moment function of the m-dimensional covariates. In practice, the balancing targets $\bar{c}_{1j}$ are set as

$$\bar{c}_{1j} = \frac{1}{\sum_{T_i=1} T_i} \sum_{T_i=1} c_j(X_i). \tag{16}$$

These weights w_i^{EB} can be used to estimate $E[Y_0|T=1] = \sum_{T_i=0} w_i^{EB} Y_i$. Thus, the entropy balancing estimator of PATT is

$$\hat{\gamma}^{EB} = \sum_{T_i=1} \frac{Y_i}{\sum_{T_i=1} T_i} - \sum_{T_i=0} w_i^{EB} Y_i. \tag{17}$$

Zhao and Percival [14] showed that $\hat{\gamma}^{EB}$ is a consistent estimator of PATT if the control outcome is linear in the same covariate moment constrained by the entropy balancing optimization.

Finally, we can combine entropy balancing with the outcome regression to get the doubly robust estimators. By replacing the IPW by w_i^{EB}, a doubly robust estimator of PATT can be obtained as follows:

$$\hat{\gamma}^{EB-DR} = \sum_{T_i=1} \frac{Y_i - \hat{g}_0(X_i)}{\sum_{T_i=1} T_i} - \sum_{T_i=0} w_i^{EB} (Y_i - \hat{g}_0(X_i)). \tag{18}$$

Zhao and Percival [14] also showed that $\hat{\gamma}^{EB-DR}$ is the same as $\hat{\gamma}^{EB}$ if $\hat{g}_0(X_i)$ is a linear model on the moments $c_j(X_i)$.

4 Data

The data used in this paper are the injury surveillance (IS) data of the accident victims who were sent to Vachira Phuket Hospital, Thailand. The data set was screened to select only the motorcycle accident victims with head injuries and those who were killed as a result of head injuries. The data consisted of a sample of 1,751 motorcycle accident victims involved in traffic crashes during the period from 2008 to 2012. The severity of head injuries was based on the abbreviated injury scale (AIS) criterion. The AIS criterion is a widely used clinical classification in which the severity of injury for each accident victim is classified by body region. There are 6 scores ranging from AIS-1 to AIS-6, where the highest level indicates the most serious injury (Hobbs and Hobbs, 1979).

About 18% of the victims were classified as helmet-wearing riders and 82% are classified as non-helmet riders. The average AIS score was about 2.25 for non-helmet users, and 2.18 for helmet-wearing riders. The description of the variables used in this paper and the main statistics are shown in Table 2. Figure 1 plots the histogram of severity of head injury.

Table 2 Description of variables and statistics

Variable	Description	Mean	SD
Helmet use	1 if victim had been wearing helmet; 0 otherwise	0.181	0.385
Severity of head injury	1 to 6 where 6 is the most serious injury	2.215	0.636
Age	Age of the victim	30.145	14.339
Male	1 if the victim is male; 0 otherwise	0.660	0.474
Passenger	1 if the victim was a passenger; 0 otherwise	0.173	0.378
M-crash	1 if the crash type is between motorcycle and another motorcycle; 0 otherwise	0.224	0.417
Night	1 if the accident occurred during 8.01 pm–6.00 am; 0 otherwise	0.482	0.500

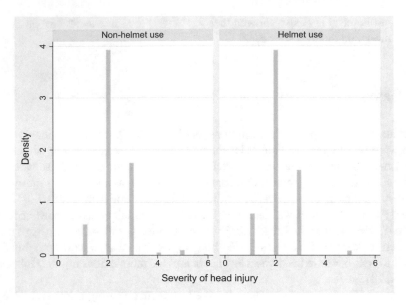

Fig. 1 Histogram of severity of head injury

5 Results and Discussion

In this section, we explore an application to estimating effect of helmet use to severity
of head injury. The propensity scores were estimated using logit model as described
in Sect. 2. The covariates included in the propensity score model consisted of Age,
Male, Passenger, and Night. These characteristics can be used to explain the decision
to wear helmets for each rider. Helmet use rate of males was lower than that of
females. There is a linear relationship between age and the propensity toward helmet
wearing. The older motorcyclists were more likely to wear a helmet when compared
with the younger motorcyclists. Helmet use for passengers was significantly lower
than for the rider position. During the night time, helmet use rate was lower than the
rate during the day. Figure 2 visualizes the histogram of estimated propensity scores
for helmet and non-helmet uses. The propensity score distributions indicate a bit of
imbalance between the two groups.

For outcome regression models, we added one more variable (i.e., M-crash), which
is the indicator for crash type. From the propensity score and outcome regression
models, we estimated $\hat{\gamma}^{IPW}$, $\hat{\gamma}^{OLS}$, $\hat{\gamma}^{DR}$, and $\hat{\gamma}^{EB}$. These estimated results are summarized in Table 3. All estimators are very similar. The standard error of $\hat{\gamma}^{DR}$ is the
smallest (in five-digit precision) in our case study. Notice that if a propensity score
is modelled correctly then $\hat{\gamma}^{DR}$ will have smaller variance than $\hat{\gamma}^{IPW}$. However, if
the outcome regression is modelled correctly, $\hat{\gamma}^{DR}$ may have larger standard error
than $\hat{\gamma}^{OLS}$ [3].

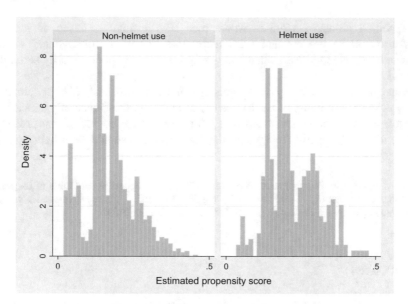

Fig. 2 Histogram of propensity score

Table 3 Population average treatment effect on the treated

PATT	Estimate	Sd. error
IPW	−0.07324	0.04223
OLS	−0.07904	0.04238
DR	−0.07313	0.04222
EB	−0.07026	0.04320

The use of motorcycle helmets had a small effect on the severity of head injury. The possible reasons why there was a small effect of helmet use on the severity of head injury is that we used the injury data from the Phuket city area in which the commuters use low speed of travel. This (surprising) result may also be due to the fact that many motorcyclists in Thailand wear cheap, low-quality helmets that do not protect the wearers' faces. Moreover, the Thai motorcyclists may not have used the helmets properly; for example, they might not have secured the chin strap, which could cause the helmet to fly off on impact.

The main concern and possible limitation in this study is the inability to control for all observed confounding variables. The obvious factors are the intrinsic motorcycle factors such as the size and type of motorcycle. The injury surveillance (IS) data did not include such variables. We believe that these factors would be related to the severity of head injury but not related to the helmet use. Thus the doubly robust estimators could give some protection against the misspecified model in our case. However, as pointed out by [5], the doubly robust estimators based on two misspecified models performed worse than IPW estimators based on incorrect propensity score models and OLS estimators based on misspecified models. Researchers have to be aware that two wrong models are not necessarily better than one [1].

Finally, one may ask why researchers have to estimate causal effect of helmet use on severity of head injury because physics and common sense generally indicate that helmets must provide some protection. Thus, even if the degree of that protection can be disputed, we still recommend that motorcyclists should wear helmets every time they ride.

Acknowledgements We graciously acknowledge the partial support of the Center of Excellence in Econometrics, Faculty of Economics, Chiang Mai University, Thailand. This work was also supported in part by the Thailand Research Fund grant MRG-5980209.

References

1. Bang H, Robins JM (2005) Doubly robust estimation in missing data and causal inference models. Biometrics 61(4):962–973
2. Cassel CM, Sarndal CE, Wretman JH (1976) Some results on generalized difference estimation and generalized regression estimation for finite populations. Biometrika 63(3):615–620
3. Funk MJ, Westreich D, Wiesen C, Davidian M (2011) Doubly robust estimation of causal effects. Am J Epidemiol 173:761–767
4. Hainmueller J (2011) Entropy balancing for causal effects: a multivariate reweighting method to produce balanced samples in observational studies. Polit Anal 20(1):25–46
5. Kang JD, Schafer JL (2007) Demystifying double robustness: a comparison of alternative strategies for estimating a population mean from incomplete data. Stat Sci 22(4):523–539
6. Neyman J (1923) On the application of probability theory to agricultural experiments: essays on principles, Section 9. Statist Sci 5(1990):465–480 Translated from the Polish and edited by Dabrowska DM, and Speed TP
7. Robins JM, Rotnitzky A, Zhao L (1994) Estimation of regression coefficients when some regressors are not always observed. J Am Stat Assoc 89:846–866
8. Robins JM, Wang N (2000) Inference for imputation estimators. Biometrika 87(1):113–124
9. Rosenbaum PR, Rubin DB (1983) The central role of the propensity score in observational studies for causal effects. Biometrika 70:41–55
10. Rubin DB (1974) Estimating causal effects of treatments in randomized and nonrandomized studies. J Educ Psychol 66:688–701
11. Rubin D (1977) Assignment to a treatment group on the basis of a covariate. J Educ Stat 2(1):1–26
12. Rubin DB (2005) Causal inference using potential outcomes: design, modeling, decisions. J Am Stat Assoc 100:322–331
13. Vermeulen K, Vansteelandt S (2015) Bias-reduced doubly robust estimation. J Am Stat Assoc 110(511):1024–1036
14. Zhao Q, Percival D (2015) Primal-dual covariate balance and minimal double robustness via entropy balancing. arXiv e-prints

Forecasting Cash Holding with Cash Deposit Using Time Series Approaches

Kobpongkit Navapan, Jianxu Liu and Songsak Sriboonchitta

Abstract The levels of cash holding and cash deposit for Thai banks have significantly increased over the past 10 years. This paper aims to forecast cash holding by using cash deposit. For banks, cash holding partially is from the cash deposited. In addition, accurate prediction on the cash holding would provide valuable information and indicators supervising bankers to control the levels of both cash holding and cash deposit effectively. In addition, the empirical relevance of cash holding and cash deposit is examined with three different models; linear model, ARIMA model and state space model. Experimental results with real data sets illustrate that state space model tends be the most accurate model compared to the other two models for prediction.

Keywords Cash holding · Cash deposit · Linear model · ARIMA model · State space model

1 Introduction

The amounts of cash holding and cash deposit always play a crucial role in cash management for banks. This is because the cost of raising external funds which is significantly higher than the internal funds [9]. Another crucial point lays on the issue that banks receive cash from their depositors. Thus, cash holding partially is from the cash deposited. Obviously the level of cash holding and the level of cash deposited are correlated. However, in the real practice cash holding can come from many financial resources like money invested by investors, financial profits from previous years, tax refunds, etc.

K. Navapan · J. Liu (✉) · S. Sriboonchitta
Faculty of Economics, Chiang Mai University, Chiang Mai, Thailand
e-mail: liujianxu1984@163.com

© Springer International Publishing AG 2017
V. Kreinovich et al. (eds.), *Robustness in Econometrics*,
Studies in Computational Intelligence 692, DOI 10.1007/978-3-319-50742-2_30

Since the time series data sets used in this study are shown in a sequence of points over time, it is appropriate to focus on traditional regression models, namely Linear model, Autoregressive integrated moving average model (ARIMA), and State Space for predictions. When using these three models in time series data, it is necessary to assume that all errors are not identical and independent distributed which can be referred to serial correlation or auto correlation. As a result, these first two models cannot fit the data leading to biased results and inaccurate predictions [15]. Moreover, it is found that State Space model provides more less error for prediction compared to ARIMA model. Some examples supporting this finding are the study of forecasting sale performance of retail sales of women footwear by Ramos et al. [13], and the study of forecasting the monthly commercial banks interest rate in Nigeria by Ologunde et al. [11].

This study aims to forecast cash holding with cash deposits for Thai banks. It provides a clearer picture to our understanding of these two variables on how they are relevant to each other. Then, accurate prediction on the cash holding would provide valuable information and indicators supervising bankers to control the levels of both cash holding and cash deposit effectively.

There has been so many great works of literature on cash holding but mainly focusing on theoretical issues. It would be appropriate to discuss theoretical issues first then follow with discussions of some works with respect to the mentioned models.

Bank as a financial intermediary has an important role in creating money in economy and establishing a connection between our lives and financial activities. The more money banks hold is the more money banks can lend in economy. In general, A central bank uses the monetary policy as a tool in order to control the amount of cash in the economy through interest rate or purchasing assets or quantitative easing (QE) [8]. Klein [7] also supported this point because the limitation on competitive behavior controlled by interest rate regulation can cause an effect on the level of cash holding.

Moreover, the degree of tightening of monetary policy can lever the level of cash holding. The tighter the monetary policy, the higher cash holding will be because of more constraints of external financing [6]. Ferreira and Vilela [4] introduced another different point. They concluded that the higher opportunity of investment is relatively connected with the higher amount of cash holding because of trade off model. It shows that the changing level of cash holding is considered by weighting the marginal cost and marginal benefits of holding cash. In addition, the main reason that banks have to be regulated and monitored closely by the central bank is that banking stability is subject to public confidence. Loss of confidence caused by one bank can extent to other banks causing financial instability to the entire economy [14]. Their study explained that banks and other depository institutions offer deposit and saving accounts to individuals and firms which they can be withdrawn at any time. Thus, they share liquidity risk together. At the level of withdrawals exceeding the amount of new deposits, they will go into liquidity trouble. Cash holding might not be a crucial issue to some firms as there are a number of ways to raise cash holding. For instance, by selling assets, raising funds in the capital market, limiting dividends and investment, and refinancing debt contracts [12]. On the contrary, a firm tends to hold cash because the cost of raising external funds is much higher than internal funds [9].

In another way Barclay and Smith [1] introduced scale economies showing that large firms seem to have a better scale economic. In the market large firms can raise money at the lower cost compared to small firms. Consequently, small firms tend to have more cash in order to avoid the cost. Meanwhile Bates et al. [2] focused on dividend payments. Firms with dividend payments would have better access to the capital markets, causing their levels of cash holding relatively low. Agency cost is also another factor. However, it can be reduced through bank loan [3].

With a different point of view, Gatev et al. [5] showed that transaction deposits can provide a great benefit to the bank in order to hedge liquidity risk from unused loan commitments. Furthermore, Naceur and Goaied [10] supported that a high level of deposit account relating to assets can be used as a determinant to measuring bank performance.

Up to this time, the next content focuses on some experimental predictions with three different models, linear model, ARIMA model and State Space model. Zhang [15] suggested that the first step for prediction is to follow the rule of thumb which focuses on the size of the sample. The minimum of the sample size at least should be equal to or larger than 104. They also found that it is impossible to avoid historical features of trend component, cyclical component and irregular component in financial data when traditional regression models are applied. As a result, ARIMA model is more fitted compared to linear model.

In addition, Ramos et al. [13] found that ARIMA model provide less accurate prediction compared to State Space model in their experiment. They illustrated that in one-step forecasts the RMSE and MAPE values of State Space model are smaller compared to those values of ARIMA. This shows that State Space predicts more correctly. These previous experiments are taken into account when designing methodologies used in the area of the study. Since commonly traditional regression models are cointegrating regression and ARIMA regression, this study places emphasis on State Space model in the methodology section.

The remainder of this paper is organized into five sections: Sect. 2 outlines state space methodology. Section 3 discusses data. Empirical results are presented in Sects. 4 and 5 concludes the paper.

2   Methodology

2.1   State Space Model

In linear analysis, the coefficient parameter is approximately constant over time, meanwhile most of time series data is varying over time. Dynamic linear model with state space approach is more flexible because it allows the coefficients are moving over time. It is more suitable to analysis time series data, and it assumes that their system properties vary over time. Additionally, it uses unobservable state variables to model the processes that drives the variation of observed data.

In general, the dynamic model consists of two parts, the first part consisting of State Space which represents of the data, and the second part representing the state of system. The sequence of the process has conditional dependence with only the previous step. The Kalman filter formulas help to estimate the state which is given by the data. General dynamic linear model can be written with a help of observation equation and state equation. That is,

$$Y_t = F_t X_t + V_t, V_t \sim N(0, V_t) \tag{1}$$

$$X_t = G_t X_{t-1} + W_t, W_t \sim N(0, W_t), \tag{2}$$

where Y_t are the observations at time t, with $t = 1, 2, \ldots, n$. Vector X_t of length m contains the unobserved states of the system that are assumed to evolve in time according to a linear system operator G_t (a m × m matrix). In time series settings X_t will have elements corresponding to various components of the time series process, like trend, seasonality, etc. We observe a linear combination of the states with noise and matrix F_t (mp) is the observation operator that transforms the model states into observations. Both observations end system equations can have additive Gaussian errors with covariance σ_v^2 and matrices σ_w^2.

2.2 Checking for Robustness of Predicted Ability

In order to check the robustness of predicted ability, we employ two benchmarks which are Mean absolute percentage error (MAPE) and Root mean square error (RMSE). MAPE measures accuracy as a percentage which can be expressed as

$$MAPE = \frac{1}{n} \sum_{t=1}^{n} \left| \frac{Y_t - Y_t'}{Y_t} \right|, \tag{3}$$

where, F_t is the predicted value, and A_t is the actual value. This measure tends to be biased when comparing the accuracy of prediction methods. The model with the lowest value of MAPE will be selected.

RMSE benchmark is used to measure the sample standard deviation, the differences between actual values and predicted values. RMSE is given as follows:

$$RMSE = \sqrt{\frac{1}{n} \sum_{i=1}^{n} (Y_t - Y_t')^2}. \tag{4}$$

RMSE is a good benchmark when only compare predicting error of the different models for specific variables. Compared with MAPE, RMSE has some disciplines due to large error.

3 Data

Cash holding and cash deposit data sets, 120 observations in each, are from the Bank of Thailand Statistic Database. The total amount of cash holding consists of cash in local currency and in foreign currencies held by Thai banks, whereas the total amount of cash deposits is cash from three different types of accounts, current, saving and fixed accounts held by them. The data is originally collected from all registered commercial banks (excluding their branches in other countries), all branches of foreign commercial bank and banks international banking facilities (BIBF) in Thailand. The data is also in the term of monthly basis, covering from January 2006 to December 2015.

After plotting the original data of both cash holding and cash deposit, it is found that there is an upward trend for both cash holding and cash deposit during the period of 2006–2016 as illustrated in Fig. 1. The graphical result of cash deposit also shows a rapid increase during 2011–2013. Moreover, after year 2013 it shows a slight drop before continuing to increase in later year.

This characteristic of data, monthly time series, can be adjusted by using the X12 procedure to remove seasonality. The procedure makes additive or multiplicative adjustments and creates an output data set containing the adjusted time series and intermediate calculations.

However, regarding to checking for robustness of predicted ability with MAPE and RMSE the data is organized as the followings. The data is divided into two parts, in sample and out of sample. The first 96 observations contained in sample are for the period January, 2006–December, 2013. The remained observations out of sample are from the period January, 2014 to December, 2015. The in sample observations are carefully used in the prediction calculation through MAPE and RMSE methods.

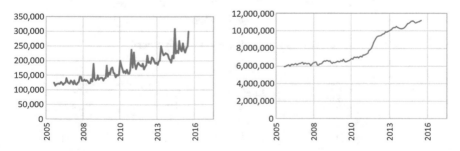

Fig. 1 The graph of series of cash holding (on the *left*) and cash deposit (on the *right*)

Then they are compared with the out of sample observations to find out the most accurate model for prediction.

4   Empirical Results

In this study, we can express the observation equation as follows:

$$Log(Cashholding_t) = C_t + \beta_{1t} \, Log(Cashdeposit_t) + \beta_{2t} \, Log(Cashdeposit_{t-1}) + V_{it},$$
$$V_{it} \sim N(0, \sigma_t^2),$$
(5)

Then followed by state equation as follows:

$$C_t = C_{t-1} + W_{1t}, \, W_{1t} \sim N\left(0, \sigma_{W_1}^2\right)$$
(6)

$$\beta_{1t} = \beta_{1t-1} + W_{2t}, \, W_{2t} \sim N\left(0, \sigma_{W_2}^2\right)$$
(7)

$$\beta_{2t} = \beta_{2t-1} + W_{3t}, \, W_{3t} \sim N\left(0, \sigma_{W_3}^2\right)$$
(8)

Figure 2 illustrates an upward trend in the relationship of log cash holding and log cash deposit. Then we implement unit root test in order to check if stationary of cash holding and cash deposit exists. The results of unit root test are demonstrated in Table 1 showing that both cash holding and cash deposit are nonstationary according to the figures with test different ways of testing, for a unit root, for a unit root with drift, and for a unit root with drift and deterministic time trend.

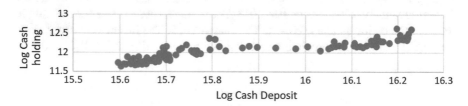

Fig. 2   The relationship between log cash holding and log cash deposits

Table 1   Unit root test

	Log(cash holding)	ΔLog(cash holding)	Log(cash deposit)	ΔLog(cash deposit)
ADF	5.9919	−17.67101***	3.861144	−3.569143***
PP	2.3587	−14.19103***	3.30591	−9.831593***

*** significant at 1% level ($p < 0.01$)

Table 2 Estimated parameters of the models

Parameters	Linear	ARIMA	State space
Constant (β_0)	−0.2137	–	3.4634
	−0.1669	–	−4.0026
$Cashdeposit_t(\beta_1)$	0.02689	0.0614	0.5808
	−0.0141	−0.1653	−0.2453
$Cashdeposit_{t-1}(\beta_2)$	0.9936	−0.6419	−0.0488
	−0.0169	−0.1266	−0.1021

All the previous findings can conclude that the time series data sets are not stationary in mean value. The results from unit root test also confirm this behavior. We then correct through appropriate differencing of the data. In this case, we applied ARIMA (1, 1, 0) model. Model parameter of linear model and ARIMA model are shown in Table 2.

After we run linear regression, we also consider whether there is a long run relationship between them or not. Therefore, the next step is test error of regression for a unit root. The result shows that the error contains a unit root indicating that they are cointegration which can be referred to the long-run relationship.

In order to compare Linear regression with ARIMA, we can see that $Cashdeposits_t(\beta_1)$ are positively related to cash holdings in both model. For $Cashdeposits_{t-1}(\beta_2)$ or cash deposits at time t − 1, it has a negative relationship with cash holding. On the other hand, $Cashdeposits_{t-1}(\beta_2)$ is positively related to cash holdings. For the State Space model, the coefficients from Table 2 can be used as final state which is 12th month 2012. The results show that $Cashdeposits_t(\beta_1)$ positively impacts on cash holdings. Meanwhile, $Cashdeposits_{t-1}(\beta_2)$ is negatively correlated to cash holdings.

After using Kalman smoothing of their coefficient parameters, the result shows that only their intercepts are varying over time. The rests are quite time-invariant coefficients, in other words they are not affected by time. Therefore, we will show only the varying-intercept that demonstrates a dynamic trend as illustrated in Fig. 3. Therefore, we can say that the relationship between cash deposits and cash in hand is quite stable. So the majority change in cash holding comes from intercept.

The next step is to use three models, linear, ARIMA, and state space models to forecast cash holding with cash deposit with respect to data divided into in sample and out of sample. The results in Table 3 show that state space model has the lowest errors in both MAPE and RMSE followed by ARIMA and linear models respectively.

The result also shows that the MAPE and RMSE values of all three models are significant different. The MAPE and RMSE values of state space model are 0.007–0.106 smallest respectively. The highest MAPE and RMSE values consisting of 0.360–4.592 belong to linear model compared to the values of ARIMA, 0.028–0.394 in that order.

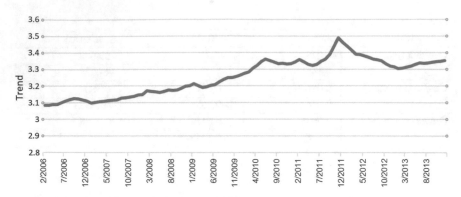

Fig. 3 Smoothing estimates of intercept (on the *right*)

Table 3 Forecasting criteria for the three models

Three models	Linear	ARIMA (1, 1, 0)	State space
MAPE	0.360	0.028	0.007***
RMSE	4.592	0.394	0.106***

Note *** lowest values compared to the other models

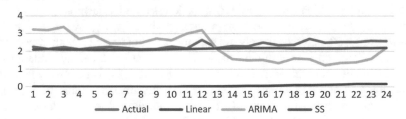

Fig. 4 Plotting actual variables with predicted values of all the models (on the *right*)

Overall state space model produces more accuracy for forecasting than the other models. As expected, the result has similarity to the finding of Ramos et al. [13]. Smallest differences from the calculation can refer to the model as the best model with smallest errors in prediction.

As shown in Fig. 4, it is obvious that the line generated by the predicted values of state space minus by the actual values moves almost the same pattern with the actual values of out of sample. This result is also supported by the results in Table 3. In addition, all the graphs also show some useful information. By using linear model, the line is lowest compared to the other lines, illustrating that the results generated by the model is underestimated.

However, state space model also produces the same, but its line is much closer to the actual line. The line generated by ARIMA model shows the different result. The line moves in the upper part of out of sample and then it dramatically drops and moves in the lower part for the rest.

5 Conclusions

In order to forecast cash holding with cash deposit for banks in Thailand with three different models linear model, ARIMA model and state space model, the empirical results show that the state space model tends to be the most accurate model for prediction, followed by the ARIMA model and the linear model respectively. The predicted values of the linear model at the beginning of time are very far from the actual values in the out of sample values. However, it might be an interesting issue if we forecast with longer period of time. Overall, this would be advantageous to bankers who need to monitor the level of cash holding and cash deposit because selecting the most appropriate model would help them manage cash holding effectively and efficiently.

For further study, as shown in Fig. 3 that the upward trend of the intercept from the state space analysis shows a peak in 2011. During that time, there was worst flooding in Thailand affecting all industrial sector and society. It has been claimed that this crisis covers about 90 billion square kilometers of land, accounted more two-thirds of the country. Focusing on more variables during the time with more advance econometric methods would provide more useful explanations with respect to this crisis effecting cash holding to bankers.

References

1. Barclay MJ, Smith CW (1995) The maturity structure of corporate debt. J Finan 50(2):609–631
2. Bates TW, Kahle KM, Stulz RM (2009) Why do US firms hold so much more cash than they used to? J Finan 64(5):1985–2021
3. Diamond DW (1984) Financial intermediation and delegated monitoring. Rev Econ Stud 51(3):393–414
4. Ferreira MA, Vilela AS (2004) Why do firms hold cash? Evidence from EMU countries. Eur Finan Manag 10(2):295–319
5. Gatev E, Schuermann T, Strahan PE (2009) Managing bank liquidity risk: how deposit-loan synergies vary with market conditions. Rev Finan Stud 22(3):995–1020
6. Jigao Z, Zhengfei L (2009) Monetary policies, enterprise'growth, and the change in the level of cash-holding. Manag World 3:019
7. Klein MA (1971) A theory of the banking firm. J Money Credit Bank 3(2):205–218
8. McLeay M, Radia A, Thomas R (2014) Money creation in the modern economy. Bank England Q Bull Q1
9. Myers SC, Majluf NS (1984) Corporate financing and investment decisions when firms have information that investors do not have. J Finan Econ 13(2):187–221
10. Naceur SB, Goaied M (2001) The determinants of the Tunisian deposit banks' performance. Appl Finan Econ 11(3):317–319

11. Ologunde AO, Elumilade DO, Asaolu TO (2006) Stock market capitalization and interest rate in Nigeria: a time series analysis
12. Opler T, Pinkowitz L, Stulz R, Williamson R (1999) The determinants and implications of corporate cash holdings. J Finan Econ 52(1):3–46
13. Ramos P, Santos N, Rebelo R (2015) Performance of state space and ARIMA models for consumer retail sales forecasting. Robot Comput Integr Manuf 34:151–163
14. Samad A, Hassan MK (1999) The performance of Malaysian Islamic bank during 1984–1997: an exploratory study. Int J Islam Finan Serv 1(3):1–14
15. Zhang R (2007) OLS regression? Auto-regression? Dynamic regression? A practical modeling example in financial industry SAS global forum

Forecasting GDP Growth in Thailand with Different Leading Indicators Using MIDAS Regression Models

Natthaphat Kingnetr, Tanaporn Tungtrakul and Songsak Sriboonchitta

Abstract In this study, we compare the performance between three leading indicators, namely, export, unemployment rate, and SET index in forecasting QGDP growth in Thailand using the mixed-frequency data sampling (MIDAS) approach. The MIDAS approach allows us to use monthly information of leading indicators to forecast QGDP growth without transforming them into quarterly frequency. The basic MIDAS model and the U-MIDAS model are considered. Our findings show that unemployment rate is the best leading indicator for forecasting QGDP growth for both MIDAS settings. In addition, we investigate the forecast performance between the basic MIDAS model and the U-MIDAS model. The results suggest that the U-MIDAS model can outperform the basic MIDAS model regardless of leading indicators considered in this study.

1 Introduction

Governments, financial institutions, and private sectors have put great attentions on economic time series for decades to anticipate future states of economy. An accurate prediction would help policy makers, economists, and investors determine appropriate policies and financial strategies. It is no doubt that the gross domestic product (GDP) is one of the most important economic variables that contain information about the state of economy. Therefore, numerous investigations have been done to forecast GDP.

In macroeconomic theory, the important factors of GDP growth are consumption, investment, government expenditure, export and import. According to the World Bank [15], Thailand's export is the main factor driving the country's economy and the proportion of Thailand's export to GDP has been more than 60% since 2,000. Many

N. Kingnetr (✉) · T. Tungtrakul · S. Sriboonchitta
Faculty of Economics, Chiang Mai University, Chiang Mai, Thailand
e-mail: natthaphat.kingnetr@outlook.com

© Springer International Publishing AG 2017
V. Kreinovich et al. (eds.), *Robustness in Econometrics*,
Studies in Computational Intelligence 692, DOI 10.1007/978-3-319-50742-2_31

studies suggested that export has effect on GDP growth [14, 16]. Moreover, Hsiao and Hsiao [12] found that there exists bidirectional causality between exports and GDP for the selected eight rapidly developing Asian countries including Thailand.

In addition, there is a huge body of literature including Estrella et al. [4] for the United States and Bellégo and Ferrara [1] for Euro area that uses the financial variables as leading indicators of GDP growth. Ferrara and Marsilli [5] concluded that the stock index could improve forecasting accuracy on GDP growth. Moreover, unemployment is also important variable which reflect the condition of ongoing economy and corresponds to output change [8]. Therefore, we consider these three variables such as export, stock index, and unemployment rate as leading indicators.

However, GDP is available quarterly, while other macroeconomic variables, such as export and unemployment rate, are recorded monthly. Moreover, other economic variables, stock price index for instance, are available at higher frequency as daily or even real time. The traditional GDP forecasting approaches assumed that all variables in a model of interest are sampled at the same frequency. Thus, they restricts specification of forecasting model to quarterly data. In most empirical works, the higher frequency is converted to the lower frequency by averaging [2]. This solution is not appealing because information of the monthly indicators have not been fully exploit for the prediction of GDP.

To enable the use of high frequency indicators to forecast low frequency variable, Ghysels et al. [9] proposed Mixed Data Sampling (MIDAS) model. The MIDAS model has been applied in various fields such as financial economics [11] and macroeconomics [2, 3, 13] to forecast GDP. Additionally, Clements and Galvão [3] concluded that the predictive ability of the indicators in comparison with an autoregression is stronger. It also allows the dependent variable and explanatory variables to be sampled at different frequencies and be in a parsimonious way of allowing lags of explanatory variables. Foroni et al. [7] employed MIDAS regression with unrestricted linear lag polynomials (U-MIDAS) model in macroeconomic applications and also showed that U-MIDAS is better than MIDAS model when the frequency mismatch is small. Therefore, it is interesting to see whether these results hold in the case of Thailand.

The objective of this paper to use the important leading indicators that are export, stock index, and unemployment rate to forecast the GDP growth. In addition to the forecasting performance of the selected indicators, we compare the forecasting performance between two MIDAS models. The result of study will be useful for government for imposing policies and strategies for stabilising the country's economy.

The organization of this paper is as follows. Section 2 describes the scope of the data used in this study. Section 3 provides the methodology of this study. Section 4 discusses the empirical results. Finally, conclusion and of this study is drawn in Sect. 5.

2 Data Analysis

In order to forecast the GDP growth rate using MIDAS models using export growth, unemployment rate, and stock index, the data during the period of 2001Q1–2016Q1 are obtained from different sources. Starting with low frequency series, the nominal GDP growth rate is measured quarterly and obtained from the Bank of Thailand. For the high frequency series, the monthly unemployment rate and Thailand's export are obtained from the Bank of Thailand. The Stock Exchange of Thailand (SET) index is used as the stock index series and obtained from the Stock Exchange of Thailand. Then, export and stock index are transformed into export growth rate and SET return respectively.

Figure 1 illustrates the plot of series used in this study. It can be noticed the negative shock in export growth rate and SET return, and positive shock in unemployment rate in response to economic downturn after the end of 2008Q1. In addition, there has been relatively high fluctuation in both the export growth rate and SET return, whereas unemployment rate has gradually declined as the GDP growth rate maintains positive most of the time.

We provide descriptive statistics for each variable in the study as shown in Table 1. According to skewness and kurtosis, it is possible to conclude that the monthly export growth data are normally distributed. In addition, the positive skewness can be seen with quarterly GDP and unemployment exhibiting right tail distribution. Meanwhile, SET return has a negative skewness meaning that it has left tail distribution

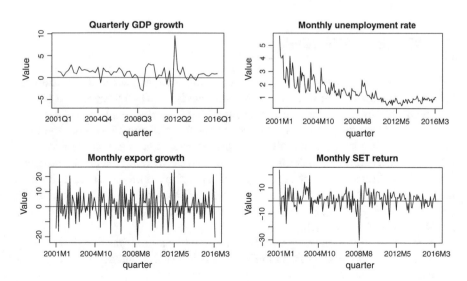

Fig. 1 Plots of series in this study during 2001Q1–2016Q1

Table 1 Descriptive statistics

	QGDP growth	Export growth	Unemployment rate	SET return
Mean	1.013	0.936	1.485	1.121
Median	1.128	0.610	1.220	1.381
Maximum	9.543	24.378	5.730	23.619
Minimum	−6.314	−22.040	0.390	−30.176
Std.Dev.	1.881	9.121	0.891	6.471
Skewness	0.312	0.130	1.666	−0.472
Kurtosis	11.865	2.893	6.368	6.093
Period	2001Q1–2016Q1	2001M1–2016M3		

3 Methodology

In this section, we explore the forecasting approaches that can take mixed-frequency data into account. We first discuss the basic version of Mixed Data Sampling (MIDAS) model introduced by Ghysels et al. [9] and the unrestricted version proposed by Foroni and Marcellino [6]. However, we consider the case of single high-frequency indicator for each regression in this study to reduce the curse of dimensionality and to avoid the estimation issues in the MIDAS framework [9].

3.1 The Basic MIDAS Model

Ghysels et al. [9] proposed a Mixed Data Sampling (MIDAS) approach to deal with various frequencies in univariate model. Particularly, a MIDAS regression tries to deal with a low-frequency variable by using higher frequency explanatory variables as a parsimonious distributed lag. Moreover, the MIDAS model also does not use any aggregation procedure and allow the use of long lags of explanatory variable with only small number of parameters that have to estimated. This can be achieved because the coefficient for each lag of explanatory variable is modelled as a distributed lag function instead [3]. Furthermore, Foroni and Marcellino [6] asserts that the parametrisation of the lagged coefficient in the parsimonious way is one of the key features of MIDAS approach.

Suppose that a low-frequency variable is measured quarterly and a high-frequency explanatory variable is measured monthly, the basic MIDAS model for h-step forecasting is then given by

$$y_t = \beta_0 + \beta_1 \left(\sum_{j=1}^{K} B\left(j; \theta\right) x_{t-h-(j-1)/m}^{(m)} \right) + \varepsilon_t \tag{1}$$

where y_t is a quarterly variable, $x^{(m)}_{t-h-(j-1)/m}$ is a monthly indicator measured at $j-1$ months prior to the last month of quarter $t-h$, h is the forcasting step, m is a frequency ratio, which is 3 in this case since a quarter consists of three months, K is a number of monthly data used to predict y_t, and $B(j;\theta)$ is weighting function.

In this study, we choose the normalized exponential Almon lag polynomial weighting function proposed by Ghysels et al. [10] that can be specified as

$$B(j;\theta) = \frac{\exp(\theta_1 j + \theta_2 j^2)}{\sum\limits_{i=1}^{K} \exp(\theta_1 i + \theta_2 i^2)} \tag{2}$$

where the parameters θ_1, θ_2 are also part of the estimation problem and can be influenced by the last K values of $x^{(m)}_t$. This weight specification gives more weight to the most recent observed period and is attractive due to its weight restriction that weights are non-negative and sum up to one. Additionally, it has been widely used in MIDAS literature.

To have a better view of how the basic MIDAS model with exponential Almon weight scheme are specified, we consider the case that $K = 3$ and $h = 1$. The MIDAS model can then be written as

$$y_t = \beta_0 + \beta_1 \left(\frac{\exp(\theta_1 + \theta_2)}{\sum\limits_{i=1}^{K} \exp(\theta_1 i + \theta_2 i^2)} \left(x^{(3)}_{t-1} \right) + \frac{\exp(2\theta_1 + 4\theta_2)}{\sum\limits_{i=1}^{K} \exp(\theta_1 i + \theta_2 i^2)} \left(x^{(3)}_{t-1-1/3} \right) + \frac{\exp(3\theta_1 + 9\theta_2)}{\sum\limits_{i=1}^{K} \exp(\theta_1 i + \theta_2 i^2)} \left(x^{(3)}_{t-1-2/3} \right) \right) + \varepsilon_t \tag{3}$$

If y_t is the GDP growth for the first quarter of 2015, then $x^{(3)}_{t-1}$ is a value of an indicator from December 2014, $x^{(3)}_{t-1-1/3}$ is from November 2014, and $x^{(3)}_{t-1-2/3}$ is from October 2014. $\beta_0, \beta_1, \theta_1,$ and θ_2 are parameters to be estimated through maximum likelihood (ML) or non-linear least squares (NLS). In addition, the number of estimated parameters are not influenced by the amount of lags in the regression model. Thus, the weighting function allows us to incorporate long historical data of the indicator while maintains parsimonious parameter estimation, hence, one of the key features of the MIDAS approach [9].

3.2 The Unrestricted MIDAS Model

Even though the basic MIDAS model can significantly reduce the coefficients via a specific weighting scheme, the strong assumption on the dynamics of the data or its process is needed. To overcome this issue, Foroni and Marcellino [6] proposed the unrestricted MIDAS (U-MIDAS) model. Suppose that low frequency data is measured quarterly, while the high frequency is measured monthly, the U-MIDAS model for h-step forecasting can be specified as

$$y_t = \beta_0 + \sum_{j=1}^{K} \beta_j x_{t-h-(j-1)/m}^{(m)} + \varepsilon_t \tag{4}$$

where y_t is a quarterly variable, $x_{t-h-(j-1)/m}^{(m)}$ is a monthly indicator measured at $j-1$ months prior to the last month of the quarter $t - h$, h is the forecasting step, m is a frequency ratio, which is 3 in our case, K is a number of monthly data used to predict y_t.

For demonstration purpose, consider the case that $K = 9$ and $h = 2$, then the U-MIDAS model can be specified as

$$
\begin{aligned}
y_t = {} & \beta_0 + \beta_1 x_{t-2}^{(3)} + \beta_2 x_{t-2-1/3}^{(3)} + \beta_3 x_{t-2-2/3}^{(3)} + \beta_4 x_{t-3}^{(3)} + \beta_5 x_{t-3-1/3}^{(3)} \\
& + \beta_6 x_{t-3-2/3}^{(3)} + \beta_7 x_{t-4}^{(3)} + \beta_8 x_{t-4-1/3}^{(3)} + \beta_9 x_{t-4-2/3}^{(3)} + \varepsilon_t
\end{aligned}
\tag{5}
$$

If y_t is the GDP growth for the first quarter of 2015, then $x_{t-2}^{(3)}$ is a value of an indicator from September 2014, $x_{t-2-1/3}^{(3)}$ is from August 2014, $x_{t-2-2/3}^{(3)}$ is from July 2014, $x_{t-3}^{(3)}$ is from June 2014 and so on.

It can be seen that the U-MIDAS approach is simply adding individual components of the higher frequency data to the linear regression, which allows one to find a separate coefficient for each high frequency component. This simplicity brings advantages of using the U-MIDAS model over the basic MIDAS model. The U-MIDAS model does not require functional distributed lag polynomials and can include the autoregressive term without common factor restriction. Additionally, one can use ordinary least squares (OLS) to estimates the individual coefficients unconstrained. Moreover, the basic MIDAS model with normalised exponential Almon weight may not lead to desirable outcome when the differences in sampling frequencies between variables in the study are small, quarterly-and-monthly data for instance. Lastly, the exponential Almon weight specification may not be general enough [7].

4 Empirical Results

In this section, we provide the results of forecasting performance between the basic MIDAS model and the U-MIDAS model. The data sample during the period of 2002Q1–2015Q1 is used for model specification and estimation, while the data during the period of 2015Q2–2016Q1 is used for forecast evaluation. For the MIDAS model specification, we allow the maximum possible data points of monthly indicators up to $K = 24$ (i.e., values of high frequency series from current period and from its lag up to 23 periods). This means the model can incorporate up to the last two years of monthly data for predicting the QGDP growth. The optimal specifications for MIDAS and U-MIDAS model are based on the Akaike's information criterion (AIC).

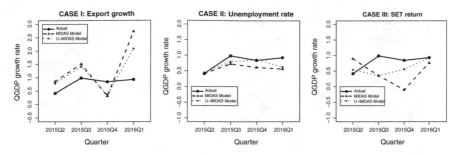

Fig. 2 Plots of actual and forecast values of QGDP 2015Q2–2016Q1

Table 2 Out-of-sample QGDP growth forecast of selected original MIDAS models

Leading indicator	Forcasting period				Selected lag for MIDAS model
	2015Q2	2015Q3	2015Q4	2016Q1	
Export growth	0.853	1.509	0.311	2.723	3
Unemployment rate	0.409	0.728	0.615	0.566	21
SET return	0.895	0.351	−0.102	0.759	5
Actual value	0.414	0.988	0.844	0.930	

Figure 2 gives us a simple comparison in forecasting performance for each leading indicator. We can see that forecast results using unemployment rate are closer to the actual value than the ones from export growth and SET return. In addition, it seems that the U-MIDAS model could perform better than the basic MIDAS model for this study since the plots of forecasts are nearer to the actual line compared to the basic MIDAS model for all three leading indicators.

We also provide the results of model selection and forecast in more detail. Table 2 shows that the optimal lag selection for export growth, unemployment rate, and SET return in the basic MIDAS model are 3, 21, and 5 respectively. As we have already discussed, the inclusion of high lag length without losing parsimonious parametrisation can not be done in the U-MIDAS framework. Thus, using the basic MIDAS model specification allows us to reach the lag length of 21 in the case of unemployment rate. On the other hand, Table 3 shows that the optimal lag selections in U-MIDAS model are 6, 4, and 3 for export growth, unemployment rate and SET return respectively. It can be seen that the selected lag length in the U-MIDAS approach is not as large as the one in the basic MIDAS approach.

Next, we calculate the out-of-sample root mean square error (RMSE) for each leading indicator. The results are shown in Table 4. Starting with export growth, it can be seen that the indicator exhibits the highest error among all indicators when considering 4 forecasting periods. According to macroeconomic theory, there are many factors that contribute to QGDP growth such as consumption, investment,

Table 3 Out-of-sample QGDP growth forecast of selected U-MIDAS models

Leading indicator	Forcasting period				Selected lag for U-MIDAS model
	2015Q2	2015Q3	2015Q4	2016Q1	
Export growth	0.779	1.415	0.366	2.083	6
Unemployment rate	0.451	0.795	0.875	0.629	4
SET return	0.539	0.353	0.558	0.925	3
Actual value	0.414	0.988	0.844	0.930	

Table 4 Out-of-sample forecast evaluation

Forecasting period	RMSE		RMSE ratio (r)
	MIDAS	U-MIDAS	
Panel I: Export growth			
1	0.438	**0.365**	0.832
2	0.482	**0.397**	0.824
3	0.499	**0.426**	0.852
4	0.996	**0.684**	0.687
Panel II: Unemployment rate			
1	**0.005**	0.037	7.343
2	0.184	**0.139**	0.756
3	0.200	**0.115**	0.575
4	0.251	**0.180**	0.718
Panel III: SET return			
1	0.480	**0.124**	0.259
2	0.564	**0.458**	0.811
3	0.714	**0.408**	0.572
4	0.625	**0.354**	0.566

Note:
1. $r < 1$ means the U-MIDAS model performing better than the basic MIDAS model.
2. The lower RMSE values are in **bold**.

government expenditure, and import. Hence, the information from export growth alone is not adequate to explain the QGDP growth.

In the case of unemployment rate, it is the best leading indicator based on RMSE. The reason is that unemployment rate is the main factor of GDP growth, thus it could reflect the QGDP growth directly. Moreover, it has a link with many factors particularly other macroeconomic factors. For instance, increase in export growth means that people are employed to produce goods for export and consequently, unemployment rate decreases. Therefore, we can conclude that unemployment rate is very important factor for forecasting QGDP growth.

The return of SET index as leading indicator does not explain the QGDP growth as well because Thailand's investment is very sensitive to the world economy. So, it does not reflect only the country's economy. Hence, unemployment rate contributes the most for forecasting QGDP growth regardless of the approaches in this study. Nevertheless, the results from both approaches confirm that the unemployment rate is the best leading indicator for predicting QGDP growth.

Thus, using export growth and SET price as leading indicators exhibits wider forecasting error. This indicates that using the proper leading indicators is very important for forecasting. In our case, the unemployment rate is the best leading indicator for forecasting GDP Growth because it reflects to current economic condition and corresponds to output change. Additionally, the RMSE is monotonically increasing with respect to the number of forecasting periods.

Lastly, we calculate the ratios of RMSEs of export growth rate between the basic MIDAS model and the U-MIDAS model for each leading indicator to see how much they differ in forecasting performance. The RMSE ratio (r) can be specified as

$$r = \frac{RMSE_{U-MIDAS}}{RMSE_{MIDAS}}$$

when the ratio r is lower than one, it indicates that the selected U-MIDAS model could outperform the selected basic MIDAS model.

The results in Table 4 show that the U-MIDAS regression models provide better forecast accuracy than the basic MIDAS model as most of RMSE ratios are less than 1. In the case of unemployment rate, it can be seen that the basic MIDAS model can only outperform the U-MIDAS for the first period, while the U-MIDAS model dominates the rest. In the case of Export growth and SET return, the U-MIDAS model improves the forecasting performance for every period. Although the improvement seems to be moderate, the U-MIDAS model still can improve overall forecasting accuracy. These results confirm the superiority in forecast performance of U-MIDAS when the difference in frequency sampling between variables is small. Another reason could be that the normalised exponential Almon weight specification in the basic MIDAS model is not appropriate choice to forecast QGDP growth given the indicators, resulting in imprecise forecasting performance. Therefore, it is possible to conclude that the U-MIDAS model could provide us a greater forecasting precision than basic MIDAS model.

5 Conclusion

In this paper, we compared the forecasting performance between different leading indicators namely, Export growth, unemployment rate, and, SET return for forecasting quarterly GDP growth in Thailand. The MIDAS model allow us to use monthly information of leading indicators directly to forecast without transforming it into quarterly frequency. Two types of MIDAS model are employed in the study, the basic MIDAS model and the U-MIDAS model. The period of data in this study is 2001Q1–2016Q1 where the data during 2015Q2–2016Q1 were left out for forecast evaluation. Our results showed that unemployment rate is the best leading indicator for forecasting GDP Growth because it reflects to current economic condition and corresponds to output change. Additionally, we investigated the forecasting performance between MIDAS and U-MIDAS model. In this particular setting, we found that the U-MIDAS model could outperform the basic MIDAS model. However, this study focused on four-period forecasting with only three leading indicators and two types of MIDAS model, the basic MIDAS model and the U-MIDAS model. Therefore, the recommendation for the future research would be the consideration of additional leading indicators and types of MIDAS model. In addition, one may be interested in investigating the performance of MIDAS approach in longer forecasting horizon which is remained to be seen in the context of macroeconomic variables in Thailand.

Acknowledgements The authors would like to thank the anonymous reviewer for useful suggestions which have greatly improved the quality of this paper. This research is supported by Puay Ungphakorn Centre of Excellence in Econometrics, Faculty of Economics, Chiang Mai University.

References

1. Bellégo C, Ferrara L (2009) Forecasting Euro-area recessions using time-varying binary response models for financial. Working papers 259, Banque de France
2. Clements MP, Galvão AB (2008) Macroeconomic forecasting with mixed-frequency data. J Bus Econ Stat 26(4):546–554
3. Clements MP, Galvão AB (2009) Forecasting US output growth using leading indicators: an appraisal using MIDAS models. J Appl Econ 24(7):1187–1206
4. Estrella A, Rodrigues AR, Schich S (2003) How stable is the predictive power of the yield curve? Evidence from germany and the united states. Rev Econ Stat 85(3):629–644
5. Ferrara L, Marsilli C (2013) Financial variables as leading indicators of GDP growth: Evidence from a MIDAS approach during the Great Recession. Appl Econ Lett 20(3):233–237
6. Foroni C, Marcellino M (2013) A survey of econometric methods for mixed-frequency data. Working Paper 2013/06, Norges Bank
7. Foroni C, Marcellino M, Schumacher C (2015) Unrestricted mixed data sampling (MIDAS): MIDAS regressions with unrestricted lag polynomials. J Roy Stat Soc: Ser A (Statistics in Society) 178(1):57–82
8. Gabrisch H, Buscher H (2006) The relationship between unemployment and output in post-communist countries. Post-Communist Econ. 18(3):261–276
9. Ghysels E, Santa-Clara P, Valkanov R (2004) The MIDAS touch: mixed data sampling regression models. CIRANO Working Papers 2004s-20, CIRANO
10. Ghysels E, Sinko A, Valkanov R (2007) Midas regressions: Further results and new directions. Econ Rev 26(1):53–90
11. Ghysels E, Valkanov RI, Serrano AR (2009) Multi-period forecasts of volatility: Direct, iterated, and mixed-data approaches. In: EFA 2009 Bergen Meetings Paper
12. Hsiao FS, Hsiao MCW (2006) FDI, exports, and GDP in East and Southeast Asia-Panel data versus time-series causality analyses. J Asian Econ 17(6):1082–1106

13. Kuzin V, Marcellino M, Schumacher C (2011) MIDAS vs. mixed-frequency VAR: Nowcasting GDP in the euro area. Int J Forecast 27(2):529–542
14. Liu X, Burridge P, Sinclair PJN (2002) Relationships between economic growth, foreign direct investment and trade: evidence from china. Appl Econ 34(11):1433–1440
15. World Bank: World development indicators (2015)
16. Xu Z (1996) On the causality between export growth and gdp growth: An empirical reinvestigation. Rev Int Econ 4(2):172–184

Testing the Validity of Economic Growth Theories Using Copula-Based Seemingly Unrelated Quantile Kink Regression

Pathairat Pastpipatkul, Paravee Maneejuk and Songsak Sriboonchitta

Abstract The distinct points of view about factors driving economic growth are introduced all the time in which some effectively useful suggestions then become the growth theories, which in turn lead to various researches on economic growth. This paper aims to examine the joint validity of the growth theories using our introduced model named copula based seemingly unrelated quantile kink regression as a key tool in this work. We concentrate exclusively on the experience of Thailand and found that the growth models can prove their validities for the Thai economy through this experiment.

Keywords Economic growth · Threshold effect · Nonlinear regression · Copulas

1 Introduction

Economic growth has been a critical issue in macroeconomics over several decades, whereas the sources of economic growth also have vexed economists for a long time. It is not easy to define the best measures stimulating the economy; therefore the vast amounts of researches have been conducted for the solutions and some accepted hypothesis, then, become the theories of economic growth. There are three main economic growth theories in the history the classical, neoclassical, and the new growth theories trying to explain the same thing, economic growth, using different exogenous variables. All the points of view about the growth are valuable in their own ways; but in some sense, we believe that these three theories may be related nontrivially. Therefore this research is conducted to examine a joint validity among

P. Pastpipatkul (✉) · P. Maneejuk (✉) · S. Sriboonchitta
Faculty of Economics, Chiang Mai University, Chiang Mai, Thailand
e-mail: ppthairat@hotmail.com

P. Maneejuk
e-mail: mparavee@gmail.com

© Springer International Publishing AG 2017
V. Kreinovich et al. (eds.), *Robustness in Econometrics*,
Studies in Computational Intelligence 692, DOI 10.1007/978-3-319-50742-2_32

these three theories as suggested by Sumer [16]. But, before we investigate the validity among these growth theories, we had better give a brief history and idea of each growth theory.

The first era of growth theory called 'the classical theory' was proposed by classical economists, such as Adam Smith, David Ricardo, and Thomas Malthus, in the eighteenth and early nineteenth centuries. The classical theory provides many of the basic ingredients for economic growth and also points out the existence of a steady state. The classical economists seem to concentrate much more on the steady state and the deviation from that state, than a variety of factors influencing the growth. However, this theory is still important since it is a good starting point for various extensions and spills over empirical researches.

The second growth theory is called 'the neoclassical theory'. It began in the mid nineteenth century. Within the framework of neoclassic theory, one economist named Robert Solow makes the important contribution to the growth theory which is known as the Solow model [3]. The key property of the Solow model is the neoclassical production function which focuses on two factors, capital and labor, and assumes constant returns to scale and diminishing marginal returns for each factor. This model tries to explain the growth via the factors of production, capital and labor, and it turns out to be useful for the nation to find the level of exact inputs (labor and capital) that can maintain the steady state. However, prior to the Solow model, there was the most common growth theory built on the model called the Harrod-Domar model. It was introduced independently by Roy Harrod in 1939 and Evsey Domar in 1946. This model concentrates especially on the role of capital; more capital accumulation can raise economic growth. However, the Harrod-Domar model does not concern the role of labor which in turn makes this model unrealistic. This point turns out to be a weakness of the pre-neoclassical growth model like Harrod-Domar model, and leads to a great opportunity for Solow to develop this model [1].

The third growth theory is called 'the new growth theory'. This theory points out the results of the driving force behind economic growth called endogenous factors such as research and development, human capital, innovation, and education that can generate the long-run economic growth. Paul Romer is the one of growth theorists who first omits the old growth theory; instead, he constructs a model that allows the endogenous factors to spill over into the economy [10]. Romer's most important work published in 1986 [14] suggested considering the impact of investment in human capital on economic growth since he experimentally found that the growth could be increasing over time due to the greater accumulation of knowledge and new researches. Many studies are generated after the discovery of Romer; however the results tend to be not very different from what Romer has suggested previously [10]. The new growth theory comes alive following another discovery of Robert Barro in 1990. His empirical study found that the public investment or government spending is the importantly supporting force in which the productive public investment, such as infrastructure and property right, is positively related to long-run economic growth [3].

2 Econometric Framework

To investigate the validity of the economic growth theories, we follow the study of Sumer [16] using four different economic growth models to stand for those three growth theories. The four growth models are shown as in the equations that follow in which the first equation is the Harrod-Domar model representing the pre-neoclassical economic growth theory. This model defines the role of capital as the driver of economic growth. The authors have shown that the growth of GDP is proportional to the change in investment. The second equation representing the neoclassical growth theory is the Solow model. As Solow argues with the Harrod-Domar model that the growth cannot be achieved through capital alone, he suggests considering both labor and capital as the forefronts of economic growth. Next, we consider the original work of endogenous growth theory, the Romer model, to illustrate the new growth theory. Romer indicates that investment in human capital, such as research and development (R&D) expenditures, is the key element behind the growth that we should take into account, and hence the Romer model is represented by the third equation. In addition, as the suggestion of Sumer [16], we can see that the Barro model also provides another effective way to explain the endogenous growth theory using the government expenditures which, in turn, is illustrated by the fourth equation. These four economic growth models can be formed as a system of equations below.

$$\ln(GDP_t) = \alpha_1 + \alpha_{11} \ln(Investment_t) + u_1 \tag{1}$$

$$\ln(GDP_t) = \alpha_2 + \alpha_{21} \ln(K_t) + \alpha_{22} \ln(L_t) + u_2 \tag{2}$$

$$\ln(GDP_t) = \alpha_3 + \alpha_{31} \ln(R \& D_t) + u_3 \tag{3}$$

$$\ln(GDP_t) = \alpha_4 + \alpha_{41} \ln(Gov_t) + u_4 \tag{4}$$

In addition, technically, Sumer [16] uses a well-known model for the system of equations called Seemingly Unrelated Regression (SUR) to examine the validity of growth theories. Why do we view the growth theories as an equation system? It is because we have many theorists that try to explain the same endogenous variable namely economic growth, but they have their own view about the sources of growth. Therefore, the different factors being economic growth in the model are set up for each theory, and then bring on the system of equations.

What is the SUR model? In brief, it is a system of equations that comprises several linear regressions. Each equation in the SUR model contains only exogenous regressors and this property makes the SUR model different from a simultaneous equation system. The key point of the SUR model is the disturbances which are assumed to correlate across equations, and hence we are able to estimate all equations jointly. (More details about the SUR model will be discussed later in the next section).

What is new in this paper? The empirical analysis in this paper still gains efficiency from the SUR model in which those four equations of the growth theories are estimated simultaneously. However, the SUR model here is different from what Sumer [16] used to find the validity of growth theories in 2012. Our previous work Pastpipatkul et al. [12] just found that the SUR model has a strong assumption of normally distributed residuals which in turn makes the model unrealistic; therefore we suggest using the Copulas which provides a joint cumulative distribution function for different marginal distributions to relax this normality assumption. This means each equation in the SUR model is allowed to have different marginal distributions of residuals which are also not necessary to be normally distributed. Then, Copulas can play a role as a joint distribution linking these marginal distributions together.

In addition, but importantly, we realize that we are dealing with the real economy which has complex processes. It is often that economists use the statistical inferences based on the conditional mean, but not always that the mean is a good representation for the whole economic processes. Therefore, following the studies of Chen et al. [4] and Jun and Pinkse [8], we decide to apply a quantile approach into the Copula based SUR model to capture the unequal impacts of the exogenous variables on economic growth across different quantiles. Furthermore, it is not only the quantile; empirical works in economic growth also suggest that the growth is a nonlinear process [5]. For example, in the growth path of any economy, if it has stagnation as an initial phase, then it will be followed by a take-off in which the growth rates are increasing, and eventually the economy will recover. This event implies that economy has different growth regimes; therefore the Kink regression approach with unknown threshold as introduced in Hansen [6] is applied to our base model to capture this consideration. Hence, we have the Seemingly Unrelated Quantile Kink Regression (SUQKR) as an econometric model for this work.

2.1 Modeling the Seemingly Unrelated Quantile Kink Regression

The base model here is the seemingly unrelated regression (SUR) which is proposed by Zellner [19]. The SUR model is a system of equations consisting of several linear regressions. The important assumption of the SUR model which let it gain the efficiency of estimation is that the error terms are assumed to correlate across equations. Thus, all equations are estimated jointly. The dependency of error terms will be thoroughly described in the next section. Suppose we have m regression equations where the term $y_{i,t}$ denotes dependent variable and $x_{ij,t}$ denotes k vector of independent variable at time t, where $i = 1, \ldots, m$ and $j = 1, \ldots, k$. As we apply the idea of quantile into the SUR model, the term $\varepsilon_{i,t}^{\tau}$ is m-dimensional unobserved error terms whose distributions depend on a quantile level, $\tau \in (0, 1)$, and these error terms are assumed to correlate across equations due to the property of SUR model. Hence, the structure of the system is given by

$$y_{1,t} = \beta_{11}^{\tau-}(x_{11,t} < \gamma_{11}) + \beta_{11}^{\tau+}(x_{11,t} > \gamma_{11}^{\tau}) + \cdots$$
$$+ \beta_{k1}^{\tau-}(x_{11,t} < \gamma_{k1}^{\tau}) + \beta_{k1}^{\tau+}(x_{k1,t} > \gamma_{k1}^{\tau}) + \varepsilon_{1,t}^{\tau}$$

$$\vdots \qquad (5)$$

$$y_{m,t} = \beta_{1m}^{\tau-}(x_{1m,t} < \gamma_{1m}^{\tau}) + \beta_{1m}^{\tau+}(x_{1m,t} > \gamma_{1m}^{\tau}) + \cdots$$
$$+ \beta_{km}^{\tau-}(x_{1m,t} < \gamma_{km}^{\tau}) + \beta_{km}^{\tau+}(x_{km,t} > \gamma_{km}^{\tau}) + \varepsilon_{m,t}^{\tau}.$$

As illustrated above, the system is consisting of m equations in which each equation contains the $m \times k$ matrix of regression parameters denoted by $\beta_{ij}^{\tau-}$ and $\beta_{ij}^{\tau+}$. As we can see that the matrices of regression parameters are split into two different parts depending on a certain level of the parameter γ_{ij}^{τ} which is called a kink point (a threshold). The coefficient matrix $\beta_{ij}^{\tau-}$ is for any values of $x_{ij,t}$ less than γ_{ij}^{τ} and the matrix $\beta_{ij}^{\tau+}$ is for any values of $x_{ij,t}$ greater than γ_{ij}^{τ}. We consider the idea of the kink point as introduced by Hansen [6] due to its capability to separate each independent variable into different regimes which may capture a non-linear process of economy. In addition, the work of Koenker and Bassett [9], who first mention about the concept of regression quantiles, also points out that estimated parameters $\beta_{ij}^{\tau-}$, $\beta_{ij}^{\tau+}$ and γ_{ij}^{τ} depending on the quantile level τ are useful for analysing the extremes value distribution, i.e. the tail behaviour of the distribution. Together, this system then can be viewed as the structure of the seemingly unrelated quantile kink regression or SUQKR model which is our proposed model for this work.

2.2 Dependence Measure

As we mentioned previously that the important assumption of the SUQKR model is that the error terms are assumed to correlate across equations; therefore, at this stage, we employ a well-known joint distribution named Copula to construct a dependence structure for the SQUKR model.

What is the Copula? It is a multivariate dependence function which is used to join two (or more) marginal distributions of random variables. In practice, we refer to the Sklar's theorem and assume that the terms $x_1, \ldots x_n$ are continuous random variables with marginal F_i, where $i = 1, 2, \ldots n$. Therefore, the n-dimension joint distribution $F(x_1, \ldots, x_n)$ or copula C exists such that for all $x_1, \ldots x_n \in R^2$ is given by

$$F(x_1, \ldots, x_n) = C(F_1(x_1), \ldots, F_n(x_n)), \qquad (6)$$

$$C(u_1, \ldots, u_n) = C(F_1^{-1}(u_1), \ldots, F_n^{-1}(u_n)), \qquad (7)$$

where $u_1, \ldots, u_n$ are n-dimensional cumulative distribution function of standardize residuals which have a uniform $[0, 1]$. Moreover, the paper considers two classes of copula namely Elliptical and Archimedean copulas to model the dependency among the error terms $\varepsilon_{i,t}^{\tau}$ in which the elliptical copula consists of Gaussian and Student-t copulas. The Archimedean copula consists of the four, well-known, families namely Gumbel, Joe, Frank, and Clayton copulas.

2.3 Estimation Technique

As we consider four economic growth models to illustrate three growth theories, the multivariate Copula is used to find the joint distribution among the continuous marginal distributions of residuals which will be described in this estimation part. Prior to the model estimation, the Augmented Dickey-Fuller test is conducted to check the stationaries of the data. Then, the Maximum likelihood estimation (MLE) is employed to estimate the unknown parameters. To illustrate the likelihood function of our model, we let the term $\Theta^{\tau} = \{\psi^{\tau}, \theta^{\tau}\}$, where ψ^{τ} denotes the set of parameters in the SUQKR model and θ^{τ} denotes the copula dependence parameter. Then, we can derive the likelihood function for the multivariate Copula based SUQKR model as

$$
L(\Theta^{\tau} \,|y_{i,t}, \; x_{ij,t}) = \prod_{t=1}^{T}\prod_{i=1}^{4} f(\psi^{\tau} \,|y_{i,t}, \; x_{ij,t}) \cdot f(\theta^{\tau}, F^{\tau}(u_{1,t}^{\tau}, u_{2,t}^{\tau}, u_{3,t}^{\tau}, u_{4,t}^{\tau})). \quad (8)
$$

As illustrated above, the likelihood function of the model contains two density functions shown in the right hand side of the likelihood in which the first density function, $\prod_{i=1}^{4} f(\psi^{\tau} \,|y_{i,t}, \; x_{ij,t})$, is the Asymmetric Laplace density of the four economic growth equations which takes a form as

$$
\prod_{i=1}^{4} f(\psi^{\tau} \,|y_{i,t}, \; x_{ij,t})
$$

$$
= \prod_{i=1}^{4} \frac{\alpha^{n}(1-\alpha)^{n}}{\sigma^{2}} \times \exp\left(\sum_{t=1}^{T} \frac{(1-\alpha)(y_{i,t} - \phi_{\alpha,i})}{\sigma^{2}}\right) \quad if \; y_{i,t} < \phi_{\alpha,i} \quad (9)
$$

$$
\prod_{i=1}^{4} \frac{\alpha^{n}(1-\alpha)^{n}}{\sigma^{2}} \times \exp\left(\sum_{t=1}^{T} \frac{(-\alpha)(y_{i,t} - \phi_{\alpha,i})}{\sigma^{2}}\right) \quad if \; y_{i,t} \geq \phi_{\alpha,i},
$$

where $\phi_{\alpha,i}$ is the mean of each equation which is given by

$$
\phi_{\alpha,i} = \beta_{1i}^{\tau-}(x_{1i,t} < \gamma_{1i}^{\tau}) + \beta_{1i}^{\tau+}(x_{1i,t} > \gamma_{1i}^{\tau}) + \cdots + \beta_{ki}^{\tau-}(x_{1i,t} < \gamma_{ki}^{\tau}) + \beta_{ki}^{\tau+}(x_{ki,t} > \gamma_{ki}^{\tau}).
$$

The second density function, $f(\theta^{\tau}, F^{\tau}(u_{1}^{\tau}, u_{2}^{\tau}, u_{3}^{\tau}, u_{4}^{\tau}))$, is the copula density that can be any families of the copula, i.e. Gaussian, Student's t, Joe, Clayton, Gumbel and Frank. The terms $u_{1}^{\tau}, u_{2}^{\tau}, u_{3}^{\tau}$ and u_{4}^{τ} represent uniform marginal distributions transformed from the probability of Asymmetric Laplace distribution function (ALD). The copula density functions are derived as follows: Hofert [7] and Wand [17]

(1) Following Patton's formula [13], the multivariate Gaussian copula density is given by

$$c(\theta_G^\tau, u_1^\tau, u_2^\tau, u_3^\tau, u_4^\tau) = (\sqrt{\det \theta^\tau})^{-1}.$$

$$\exp\left(\frac{1}{2}\left(\Phi_1^{-1}(u_1^\tau), \Phi_2^{-1}(u_2^\tau), \Phi_3^{-1}(u_3^\tau), \Phi_4^{-1}(u_4^\tau)\right) \cdot (\theta_G^{\tau-1} - I) \begin{pmatrix} \Phi_1^{-1}(u_1^\tau) \\ \Phi_2^{-1}(u_2^\tau) \\ \Phi_3^{-1}(u_3^\tau) \\ \Phi_4^{-1}(u_4^\tau) \end{pmatrix}\right)$$

(10)

where Φ_i is standard normal cumulative distribution at quantile, θ_G^τ is a dependence of Gaussian copula at quantile τ with $[-1, 1]$ interval.

(2) The multivariate Student-t copula density is given by

$$c(\theta_T^\tau, u_1^\tau, u_2^\tau, u_3^\tau, u_4^\tau)$$

$$= \frac{|\theta_T^\tau|^{-1/2} \Gamma\left(\frac{v+2}{2}\right) \Gamma\left(\frac{v}{2}\right)}{\left(\Gamma\left(\frac{v+2}{2}\right)\right)^2}$$

$$\left(1 + \begin{pmatrix} (t_v^{-1}(u_1^\tau)) \\ (t_v^{-1}(u_2^\tau)) \\ (t_v^{-1}(u_3^\tau)) \\ (t_v^{-1}(u_4^\tau)) \end{pmatrix} \theta_T^\tau \left((t_v^{-1}(u_1^\tau), t_v^{-1}(u_2^\tau), (t_v^{-1}(u_3^\tau), t_v^{-1}(u_4^\tau))\right) / v\right)^{\frac{-v+2}{2}}$$

(11)

where Γ is a gamma function, t_v^{-1} is the quantile function of a standard univariate Student-t distribution with degree of freedom v and θ_T^τ is a dependence of Student-T copula at quantile τ with $[-1, 1]$ interval.

(3) The multivariate Frank copula density is given by

$$c(\theta_F^\tau, u_1^\tau, u_2^\tau, u_3^\tau, u_4^\tau) = \left(\frac{\theta_F^\tau}{1 - \exp(-\theta_F^\tau)}\right)^{4-1} Li_{-(4)}(h_{F,\theta}(u_1^\tau, u_2^\tau, u_3^\tau, u_4^\tau))$$

$$\frac{\exp(-\theta_F^\tau \Sigma_{j=1}^4 u_i^\tau)}{h_{F,\theta}(u_1^\tau, u_2^\tau, u_3^\tau, u_4^\tau)}$$

(12)

where $h_{F,\theta}(u_1^\tau, u_2^\tau, u_3^\tau, u_4^\tau) = (1 - e^{-\theta_F^\tau})^{1-4} \prod_{j=1}^4 \left\{1 - \exp(-\theta_F^\tau u_j^\tau)\right\}$ and Li denotes the polylogarithm.

(4) The multivariate Clayton copula density is shown by

$$c(\theta_C^\tau, u_1^\tau, u_2^\tau, u_3^\tau, u_4^\tau)$$
$$= \prod_{k=0}^{4-1}(\theta_C^\tau k + 1)(\prod_{j=1}^{4} u_j^\tau)^{-(1+\theta_C^\tau)}(1 + t_{\theta_C^\tau}(u_1^\tau, u_2^\tau, u_3^\tau, u_4^\tau))^{-4(4+1/\theta_C^\tau)} \quad (13)$$

(5) The multivariate Gumbel copula density is shown by

$$c(\theta_{Gu}^\tau, u_1^\tau, u_2^\tau, u_3^\tau, u_4^\tau) = (\theta_{Gu}^\tau)^4 C_{Gu,\theta}(u_1^\tau, u_2^\tau, u_3^\tau, u_4^\tau)$$
$$\frac{\prod_{j=1}^{4}(-\log u_j^\tau)^{\theta_{Gu}^\tau - 1}}{t_{Gu,\theta}((u_1^\tau, u_2^\tau, u_3^\tau, u_4^\tau) \prod_{j=1}^{4} u_j^\tau} P_{G,n}(t_{Gu,\theta}((u_1^\tau, u_2^\tau, u_3^\tau, u_4^\tau)^{1/\theta_{Gu}^\tau}) \quad (14)$$

where $P_{G,n}(x) = \Sigma_{k=1}^{4} \kappa_{2k}^{G}(\alpha)x^k$.

(6) The multivariate Joe copula density is shown by

$$c(\theta_J^\tau, u_1^\tau, u_2^\tau, u_3^\tau, u_4^\tau) = (\theta_J^\tau) \frac{\prod_{j=1}^{4}(1 - u_j^\tau)^{\theta_J - 1}}{h_{J,\theta}(u_1^\tau, u_2^\tau, u_3^\tau, u_4^\tau)}$$
$$(1 - h_{J,\theta}(u_1^\tau, u_2^\tau, u_3^\tau, u_4^\tau))^\upsilon P_{n,\upsilon}^J(\frac{h_{J,\theta}(u_1^\tau, u_2^\tau, u_3^\tau, u_4^\tau)}{1 - h_{J,\theta}(u_1^\tau, u_2^\tau, u_3^\tau, u_4^\tau)}) \quad (15)$$

where $h_{J,\theta}(u_1^\tau, u_2^\tau, u_3^\tau, u_4^\tau) = \prod_{j=1}^{4}(1 - (1 - u_j^\tau)^{\theta_J})$ and $P_{n,\kappa}^J(x) = \kappa_{2k}^J(\alpha)x^k, \alpha = 1/\theta_J^\tau$. Note that the dependence parameters are restricted by $[0, +\infty)$ for Clayton, $[1, +\infty)$ for Gumbel and Joe, and $[0, +\infty)$ for Frank. Then, we employ the maximum likelihood estimator (MLE) to maximize the multivariate Copula based SUQKR likelihood function Eq. (8) in order to obtain the final estimation results.

3 Simulation Study

A simulation study was conducted to evaluate performance and accuracy of the Copula based SUQKR model. We simulated the data from the bivariate Copula-based SUQKR model and employed the Monte Carlo method to simulate cumulative distribution function of standardized residuals u_1 and u_2 from the dependence parameter of Copula families, i.e. Gaussian, Student-t, Clayton, Frank, Joe and Gumbel. We set the true values of parameters and copula which are shown in Tables 1, 2 and

3. To model the error term $\varepsilon_{i,t}^\tau$ of two equations, the quantile function of asymmetric Laplace distribution (ALD) was used to convert the simulated uniform u_1 and u_2 into the term $\varepsilon_{i,t}^\tau$, where $\varepsilon_{i,t}^\tau \sim ALD(0, 1)$ for three different quantile levels $\tau = (0.1, 0.5, 0.9)$. Finally, we constructed the Copula based SUQKR model using the specified parameter from Tables 1, 2 and 3 and obtained the simulated dependent and independent variables for our model at the specified quantile levels. Hence, the simulation model takes the following form:

$$
\begin{aligned}
y_{1,t} &= \alpha_1^\tau + \beta_{11}^{\tau-}(x_{11,t} < \gamma_{11}) + \beta_{11}^{\tau+}(x_{11,t} > \gamma_{11}^\tau) + \varepsilon_{1,t}^\tau \\
y_{2,t} &= \alpha_2^\tau + \beta_{12}^{\tau-}(x_{12,t} < \gamma_{12}^\tau) + \beta_{12}^{\tau+}(x_{12,t} > \gamma_{12}^\tau) + \varepsilon_{2,t}^\tau.
\end{aligned}
\tag{16}
$$

In this experiment, we simulated the independent variables $x_{1j,t}$ where $j = 1, 2$ from $N(\mu, 10)$ in which the mean μ is equal to the values of kink points. We set the value of the kink point γ_{11}^τ equal to 8 and γ_{12}^τ equal to 12. The values for intercept terms α_1^τ and α_2^τ are equal to 1 and for the coefficients $\beta_{11}^{\tau-}$, $\beta_{11}^{\tau+}$, $\beta_{12}^{\tau-}$, and $\beta_{12}^{\tau+}$ are equal to $-4, 0.5, -2$ and 1, respectively.

Tables 1, 2 and 3 show the results of the Monte Carlo simulation investigating the maximum likelihood estimation of the Copula based SUQKR model. We found that our proposed model can perform well through this simulation study. The overall mean parameters at different quantile levels are somewhat close to the true values with acceptable standard errors. For instance, Table 1 shows the results of the estimated parameters of the multivariate Copula based SUQKR model at quantile level 0.1. It is found that the mean value of the coefficient α_1^τ is 1.03 with a standard error equal to 0.0537 while the true value is 1. The estimate of the same parameter α_1^τ at quantile level 0.5 is equal to 0.82 with standard error equal to 0.0676 and equal to 1.05 with standard error equal to 0.1774 at quantile level 0.9. Overall, the Monte Carlo simulation suggests that our introduced model Copula based SUQKR is reasonably accurate.

4 Robustness Checks by Kullback-Leibler Divergence

The previous part shows the accuracy of our proposed model, the seemingly unrelated quantile kink regression (SUQKR), through the Monte Carlo simulation and we found that the SUQKR model is reasonably precise. However, how precise the result is must depend on the copula we choose as well, because we need the appropriate copula function to join the error terms in SUQKR model. If the true copula is known, the proposed model or even the estimator will be accurate. But in the case of copula misspecification as pointed out by Noh et al. [11]; the selection of wrong copula function will bring about a bias in the estimation of the model. Therefore, we employed the Kullback-Leibler divergence (KLD) which is a measure of the distance between two probability distributions, i.e. the true distribution and the alternative distribution, to test the robustness of the model.

Table 1 Results of multivariate Copula based SUQKR with $\tau = 0.1$

Copula	Parameter	True	Estimate	S.E.	Copula	Parameter	True	Estimate	S.E.
Gaussian	α_1^τ	1	1.03	0.05	Student-t	α_1^τ	1	1.21	0.01
	$\beta_{11}^{\tau-}$	−4	−4.01	0.01		$\beta_{11}^{\tau-}$	−4	−4.11	0.08
	$\beta_{11}^{\tau+}$	0.5	0.37	0.26		$\beta_{11}^{\tau+}$	0.5	0.62	0.01
	$\sigma_1.$	1	1.08	0.11		$\sigma_1.$	1	1.04	0.11
	α_2^τ	1	0.47	0.06		α_2^τ	1	1.00	0.18
	$\beta_{12}^{\tau-}$	−2	−2.06	0.00		$\beta_{12}^{\tau-}$	−2	−1.85	0.01
	$\beta_{12}^{\tau+}$	1	0.97	0.01		$\beta_{12}^{\tau+}$	1	1.17	0.38
	σ_2	1	1.15	0.11		σ_2	1	0.92	0.09
	γ_{11}^τ	8	7.65	0.44		γ_{11}^τ	8	8.17	0.03
	γ_{12}^τ	12	11.84	0.05		γ_{12}^τ	12	12.77	0.96
	θ^τ	0.5	0.51	0.07		θ^τ	0.5	0.32	0.09
Joe	α_1^τ	1	0.71	0.09	Clayton	α_1^τ	1	1.01	0.07
	$\beta_{11}^{\tau-}$	−4	−3.89	0.03		$\beta_{11}^{\tau-}$	−4	−3.96	0.02
	$\beta_{11}^{\tau+}$	0.5	0.50	0.01		$\beta_{11}^{\tau+}$	0.5	0.62	0.01
	$\sigma_1.$	1	0.98	0.09		$\sigma_1.$	1	1.01	0.08
	α_2^τ	1	0.36	0.33		α_2^τ	1	0.38	0.11
	$\beta_{12}^{\tau-}$	−2	−1.99	0.02		$\beta_{12}^{\tau-}$	−2	−2.00	0.02
	$\beta_{12}^{\tau+}$	1	1.22	0.03		$\beta_{12}^{\tau+}$	1	1.05	0.02
	σ_2	1	1.04	0.10		σ_2	1	1.14	0.10
	γ_{11}^τ	8	8.54	0.03		γ_{11}^τ	8	8.35	0.02
	γ_{12}^τ	12	12.23	0.06		γ_{12}^τ	12	12.11	0.08
	θ^τ	2	2.17	0.26		θ^τ	3	3.59	0.43
Gumbel	α_1^τ	1	1.05	0.17	Frank	α_1^τ	1	0.89	0.08
	$\beta_{11}^{\tau-}$	−4	−4.08	0.12		$\beta_{11}^{\tau-}$	−4	−3.96	0.05
	$\beta_{11}^{\tau+}$	0.5	0.44	0.06		$\beta_{11}^{\tau+}$	0.5	0.14	0.71
	$\sigma_1.$	1	1.02	0.10		$\sigma_1.$	1	1.13	0.13
	α_2^τ	1	1.19	0.05		α_2^τ	1	0.53	0.04
	$\beta_{12}^{\tau-}$	−2	−2.17	0.01		$\beta_{12}^{\tau-}$	−2	−2.16	0.01
	$\beta_{12}^{\tau+}$	1	0.93	0.09		$\beta_{12}^{\tau+}$	1	0.89	0.01
	σ_2	1	1.02	0.09		σ_2	1	0.80	0.06
	γ_{11}^τ	8	7.51	0.32		γ_{11}^τ	8	7.37	2.16
	γ_{12}^τ	12	11.37	0.04		γ_{12}^τ	12	11.90	0.03
	θ^τ	3	2.99	0.35		θ^τ	2	2.38	0.66

Source Calculation

To illustrate the Kullback-Leibler divergence, we denote F as the true distribution and $\widehat{F}$ as the alternative distribution. The KLD then is employed to measure the distance between these two distributions using the following formula

$$D(F, \widehat{F}) = \int_{-\infty}^{\infty} f(x) \log \frac{f(x)}{\widehat{f}(x)} dx \qquad (17)$$

Table 2 Results of multivariate Copula based SUQKR with $\tau = 0.5$

Copula	Parameter	True	Estimate	S.E.	Copula	Parameter	True	Estimate	S.E.
Gaussian	α_1^τ	1	0.82	0.06	Student-t	α_1^τ	1	0.77	0.05
	$\beta_{11}^{\tau-}$	-4	-3.96	0.01		$\beta_{11}^{\tau-}$	-4	-3.96	0.01
	$\beta_{11}^{\tau+}$	0.5	0.51	0.01		$\beta_{11}^{\tau+}$	0.5	0.45	0.00
	$\sigma_{1.}$	1	1.01	0.10		$\sigma_{1.}$	1	0.99	0.09
	α_2^τ	1	1.14	0.03		α_2^τ	1	0.73	0.11
	$\beta_{12}^{\tau-}$	-2	-2.12	0.00		$\beta_{12}^{\tau-}$	-2	-1.98	0.02
	$\beta_{12}^{\tau+}$	1	0.98	0.01		$\beta_{12}^{\tau+}$	1	1.07	0.01
	σ_2	1	1.15	0.11		σ_2	1	0.87	0.08
	γ_{11}^τ	8	7.96	0.04		γ_{11}^τ	8	8.07	0.02
	γ_{12}^τ	12	11.59	0.15		γ_{12}^τ	12	12.20	0.09
	θ^τ	0.5	0.48	0.07		θ^τ	0.5	0.41	0.08
Joe	α_1^τ	1	0.90	0.02	Clayton	α_1^τ	1	0.42	0.11
	$\beta_{11}^{\tau-}$	-4	-3.98	0.01		$\beta_{11}^{\tau-}$	-4	-3.98	0.01
	$\beta_{11}^{\tau+}$	0.5	0.61	0.03		$\beta_{11}^{\tau+}$	0.5	0.50	0.00
	$\sigma_{1.}$	1	0.85	0.07		$\sigma_{1.}$	1	1.04	0.09
	α_2^τ	1	0.94	0.08		α_2^τ	1	1.11	0.24
	$\beta_{12}^{\tau-}$	-2	-1.97	0.01		$\beta_{12}^{\tau-}$	-2	-1.97	0.02
	$\beta_{12}^{\tau+}$	1	1.01	0.01		$\beta_{12}^{\tau+}$	1	1.05	0.03
	σ_2	1	1.00	0.09		σ_2	1	1.13	0.09
	γ_{11}^τ	8	8.31	0.02		γ_{11}^τ	8	7.99	0.06
	γ_{12}^τ	12	12.15	0.04		γ_{12}^τ	12	12.11	0.16
	θ^τ	2	2.25	0.26		θ^τ	3	3.31	0.48
Gumbel	α_1^τ	1	1.08	0.26	Frank	α_1^τ	1	0.48	0.07
	$\beta_{11}^{\tau-}$	-4	-4.01	0.03		$\beta_{11}^{\tau-}$	-4	-3.96	0.02
	$\beta_{11}^{\tau+}$	0.5	0.46	0.02		$\beta_{11}^{\tau+}$	0.5	0.51	0.02
	$\sigma_{1.}$	1	0.99	0.09		$\sigma_{1.}$	1	1.10	0.11
	α_2^τ	1	1.55	0.05		α_2^τ	1	0.41	0.04
	$\beta_{12}^{\tau-}$	-2	-2.00	0.01		$\beta_{12}^{\tau-}$	-2	-2.06	0.01
	$\beta_{12}^{\tau+}$	1	0.97	0.01		$\beta_{12}^{\tau+}$	1	0.99	0.00
	σ_2	1	0.93	0.09		σ_2	1	0.99	0.09
	γ_{11}^τ	8	7.94	0.09		γ_{11}^τ	8	8.08	0.06
	γ_{12}^τ	12	11.86	0.02		γ_{12}^τ	12	11.85	0.04
	θ^τ	3	3.08	0.33		θ^τ	2	2.07	0.63

Source Calculation

where $\widehat{f}$ is a likelihood density function of $\widehat{F}$ and f is a likelihood density function of the true distribution F in which all parameters are known. By using this formula, we are able to calculate the distance between the true function and the approximation. This study conducts an experimental study and selects the true function from Joe, Student-t and Clayton copulas for quantile level 0.1, 0.5, and 0.9, respectively.

The result shown in Fig. 1 illustrates three panels of each quantile level. We can see that the approximations of Joe, Student-t and Clayton copulas based on the SUQKR model achieve their minimum and close to their true function lines (dashed lines). In addition, we also compared the performance of our purposed model with

Table 3 Results of multivariate Copula based SUQKR with $\tau = 0.9$

Copula	Parameter	True	Estimate	S.E.	Copula	Parameter	True	Estimate	S.E.
Gaussian	α_1^τ	1	1.05	0.17	Student-t	α_1^τ	1	0.98	0.40
	$\beta_{11}^{\tau-}$	−4	−3.90	0.19		$\beta_{11}^{\tau-}$	−4	−4.00	0.01
	$\beta_{11}^{\tau+}$	0.5	0.52	0.12		$\beta_{11}^{\tau+}$	0.5	0.46	0.05
	$\sigma_{1\cdot}$	1	0.85	0.08		$\sigma_{1\cdot}$	1	0.86	0.08
	α_2^τ	1	1.81	0.23		α_2^τ	1	1.08	0.07
	$\beta_{12}^{\tau-}$	−2	−2.06	0.02		$\beta_{12}^{\tau-}$	−2	−2.10	0.00
	$\beta_{12}^{\tau+}$	1	0.94	0.01		$\beta_{12}^{\tau+}$	1	1.33	0.12
	σ_2	1	0.95	0.10		σ_2	1	0.83	0.08
	γ_{11}^τ	8	7.97	0.65		γ_{11}^τ	8	7.98	0.09
	γ_{12}^τ	12	11.90	0.24		γ_{12}^τ	12	12.25	0.06
	θ^τ	0.5	0.48	0.08		θ^τ	0.5	0.40	0.08
Joe	α_1^τ	1	0.86	0.21	Clayton	α_1^τ	1	0.40	0.08
	$\beta_{11}^{\tau-}$	−4	−3.92	0.06		$\beta_{11}^{\tau-}$	−4	−3.90	0.01
	$\beta_{11}^{\tau+}$	0.5	0.64	0.03		$\beta_{11}^{\tau+}$	0.5	0.55	0.10
	$\sigma_{1\cdot}$	1	0.83	0.07		$\sigma_{1\cdot}$	1	0.99	0.09
	α_2^τ	1	0.94	0.08		α_2^τ	1	0.96	0.18
	$\beta_{12}^{\tau-}$	−2	−1.99	0.03		$\beta_{12}^{\tau-}$	−2	−1.84	0.09
	$\beta_{12}^{\tau+}$	1	0.96	0.04		$\beta_{12}^{\tau+}$	1	0.87	0.00
	σ_2	1	0.97	0.09		σ_2	1	0.95	0.09
	γ_{11}^τ	8	8.45	0.11		γ_{11}^τ	8	8.36	0.01
	γ_{12}^τ	12	12.11	0.05		γ_{12}^τ	12	12.12	0.13
	θ^τ	2	2.47	0.31		θ^τ	3	2.68	0.43
Gumbel	α_1^τ	1	0.97	0.06	Frank	α_1^τ	1	0.73	0.03
	$\beta_{11}^{\tau-}$	−4	−3.99	0.03		$\beta_{11}^{\tau-}$	−4	−3.94	0.02
	$\beta_{11}^{\tau+}$	0.5	0.47	0.02		$\beta_{11}^{\tau+}$	0.5	0.53	0.00
	$\sigma_{1\cdot}$	1	0.81	0.07		$\sigma_{1\cdot}$	1	1.01	0.09
	α_2^τ	1	1.06	0.17		α_2^τ	1	1.65	0.05
	$\beta_{12}^{\tau-}$	−2	−2.05	0.06		$\beta_{12}^{\tau-}$	−2	−2.05	0.01
	$\beta_{12}^{\tau+}$	1	0.99	0.01		$\beta_{12}^{\tau+}$	1	0.82	0.01
	σ_2	1	0.78	0.07		σ_2	1	1.17	0.11
	γ_{11}^τ	8	7.91	0.05		γ_{11}^τ	8	8.12	0.02
	γ_{12}^τ	12	11.88	0.16		γ_{12}^τ	12	11.31	0.00
	θ^τ	3	2.98	0.26		θ^τ	2	2.19	0.64

Source Calculation

the independence copula based SUQKR function denoted by M0 (independent error term), and then we found that our purposed model performs better than the independence copula SUQKR model since the distance of our proposed model is less than the model with independent error terms for all cases, i.e. $D(f_{\text{True}}, f_{Joe}) < D(f_{\text{True}}, f_{M0})$, $D(f_{\text{True}}, f_{Student-t}) < D(f_{\text{True}}, f_{M0})$, and $D(f_{\text{True}}, f_{Clayton}) < D(f_{\text{True}}, f_{M0})$. In the case of misspecified copula function, we can see from the result that the misspecification brings a larger deviation of the approximated SUQKR function from the true one. Therefore, these all results allow us to indicate that our proposed model is more

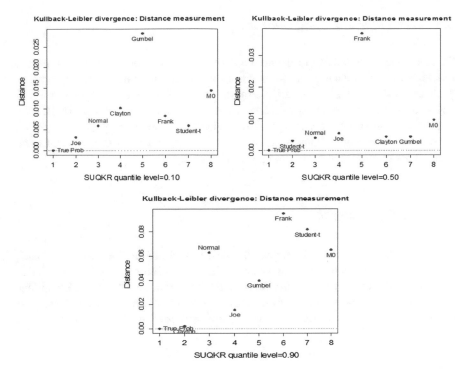

Fig. 1 The performance of Copula based SUQKR at different quantile levels

robust than the alternative model through this simulated data and the misspecification copula function will lead to the low accuracy of the model.

5   Application to Thailand's Economy

In this part, we aim to test a joint validity of economic growth theories by analysing the case of Thailand. Our reasoning for choosing Thailand as a case study is that Thailand provides a sufficiently long series of macroeconomic data and, most importantly, there is meager research done on the economic growth of Thailand. To the best of our knowledge, no research has been done to test the correctness of growth theories for the Thai economy while a large number of Thai economists usually follow the ideas of these four growth theories, i.e. Harrod-Domar, Solow, Romer, and Barro models, and use them to explain Thailand's economy. We think that this will be a good opportunity to conduct a research testing the validity of the growth theories for the Thai economy.

5.1 Variables and Data Sources

To investigate the validity of economic growth theories for Thailand, we used a quarterly data set related to those four growth models, specifically gross domestic product or GDP as a response variable for all growth theories and the covariate variables consisting of level of capital, labor force, investment, government expenditure, and research and development expenditure, spanning from 1996:Q1 to 2015:Q3. We derived the data from Thomson-Reuter DataStream, Financial Investment Center (FIC), Faculty of Economics, Chiang Mai University.

5.2 Model Selection

This section is constructed to choose the best model among the candidates using classical criteria for selecting model namely the Akaike Information Criterion (AIC) and also the Bayesian Information Criterion (BIC) to strengthen the result. We applied the well-known copula families from both Elliptical and Archimedean classes, i.e. Gaussian, Student t, Clayton, Gumbel, Frank, and Joe, as described in Sect. 2.2,

Table 4 Model selection

Two regimes (Kink effect)						
AIC/BIC	Gaussian	Student-t	Clayton	Gumbel	Joe	Frank
$\tau = 0.10$	**−2511.7**	−2144.6	−1969.3	1906.5	−1553.1	−1394.2
	−2470.6	−2045.9	−1878.2	1997.6	−1462.1	−1303.1
$\tau = 0.50$	**−2652.3**	−1710.2	−2093.2	2259.1	−2040.1	−1775.3
	−2561.2	−1611.5	−2002.1	2350.2	−1948.9	−1684.1
$\tau = 0.70$	**−2538.4**	−2059.8	−1964.1	2411.6	2270.7	−1764.6
	−2447.3	−1961.1	−1873.1	2502.7	2361.8	−1673.5
$\tau = 0.90$	**−2322.6**	−2076.5	−1718.2	2222.4	−2062.9	−1574.9
	−2231.5	−1977.8	−1627.1	2313.5	−1971.8	−1483.8
One regime (No kink effect)						
AIC/BIC	Gaussian	Student-t	Clayton	Gumbel	Joe	Frank
$\tau = 0.10$	−2460.2	−1302.2	−1525.1	−879.1	−649.2	−1302.2
	−2407.1	−1249.1	−1471.9	−825.9	−596.1	−1249.1
$\tau = 0.50$	−2571.6	−1510.9	−1320.8	−1244.3	181.6	−1650.9
	−2518.5	−1507.1	−1267.7	−1191.2	234.7	−1597.9
$\tau = 0.70$	−2220.4	−2116.9	−1909.8	−53.2	−1307.2	−114.2
	−2167.3	−2113.1	−1856.7	−0.1	−1254.1	−61.06
$\tau = 0.90$	−2303.6	−1718.5	375.5	−973.6	−1183.2	297.5
	−2250.5	−1714.7	428.7	−920.5	−1129.9	350.6

Source Calculation

to model a dependence structure of our SUQKR model at different quantile levels. Here we assume 4 different levels for quantile denoted by $\tau = 0.10, 0.50, 0.70, 0.90$, in which each quantile level is suspected to have the kink (threshold) effect on a relationship between response variable and its covariate. In this part, we are not only choosing the best-fit copula for the data, but also verifying whether or not the kink effect exists with respect to our model.

Table 4 shows the values of AIC and BIC of all candidate models in which the bold numbers display the lowest values of AIC and BIC at different quantile levels. Among the trial runs of several alternative copula functions, we found that Gaussian presents the lowest AIC and BIC for all considered quantile levels 0.10, 0.50, 0.70, and 0.90. Moreover, we also compared the two-regime SUQKR models with the single-regime counterpart (no kink effect) where the results are also shown in Table 4. It is found that the values of AIC and BIC of two-regime quantile model are lower than those of the single-regime quantile model, meaning that the SUQKR model with two regimes is favourable based on the AIC and BIC criteria.

5.3 Estimates of the SUQKR Model

To test the validity of the growth models, the Copula based SUQKR as chosen from the previous section was estimated by the maximum likelihood estimator (MLE) and the results are then reported in the Table 5. We found that the parameters are statistically significant at the 5% significance level with low and acceptable standard errors. The results show that the four growth models, namely Harrod-Domar, Solow, Romer, and Barro, can prove their validities for the Thai economy but the magnitudes of effects are different across the quantiles and regimes. This paper assumes four different quantile levels denoted by τ to be the 0.1, 0.5, 0.7, and 0.9 quantiles. The estimated results of the SUQKR model corresponding to those four growth models will be discussed respectively, but again those four growth models were estimated simultaneously.

The first equation is the Harrod-Domar model which takes the form as $\ln(GDP_t) = \alpha_1^\tau + \beta_{11}^{\tau-} \ln(Investment_t)^- + \beta_{11}^{\tau+} \ln(Investment_t)^+ + \sigma_1 u_1$. We found that the relationship between Thailand's GDP growth and investment is split into two regimes based on the significant kink points γ^τ which are different across quantile levels. In the lower regime denoted by $(.)^-$, it is found that the changes in investment lead to contrary impacts on the GDP growth for all quantiles, except the 0.1 quantile which is found that the higher investment the more economic growth. Conversely, in the upper regime denoted by $(.)^+$, there exists a positive relationship between GDP growth and investment in which the size of the effect is different across quantile levels, except the 0.1 quantile which is found to have a negative relationship among those variables.

Similarly, the second equation is the Solow growth model which takes the form as $\ln(GDP_t) = \alpha_2^\tau + \beta_{21}^{\tau-} \ln(K_t)^- + \beta_{21}^{\tau+} \ln(K_t)^+ + \beta_{22}^{\tau-} \ln(L_t)^- + \beta_{22}^{\tau+} \ln(L_t)^+ + \sigma_2 u_2$. The results shown in Table 5 prove that the variables: capital and labor as

Table 5 Estimated results of the SUQKR model

Parameter\Quantile	$\tau = 0.10$	$\tau = 0.50$	$\tau = 0.70$	$\tau = 0.90$
The Harrod-Domar model				
α_1^τ	0.0060	0.0152[a]	0.0115[a]	0.0136[a]
	(0.0011)	(0.0015)	(0.0005)	(0.0012)
$\beta_{11}^{\tau-}(I^-)$	0.0003[a]	−0.0001[a]	−0.0007[a]	−0.0001
	(0.0001)	(0.0000)	(0.0000)	(0.0001)
$\beta_{11}^{\tau+}(I^+)$	−0.0002[a]	0.0001[a]	0.0006[a]	0.0001
	(0.0001)	(0.0000)	(0.0000)	(0.0001)
γ_{11}^τ	12.7538[a]	12.7339[a]	12.7197[a]	12.7253[a]
	(0.041)	(2.3242)	(4.5655)	(0.0007)
σ_1	0.0039[a]	0.0076[a]	0.0079[a]	0.0101[a]
	(0.0003)	(0.0002)	(0.0002)	(0.0075)
The Solow model				
α_2^τ	0.0038[a]	0.0154[a]	0.0130[a]	0.0176[a]
	(0.0013)	(0.0016)	(0.0006)	(0.0016)
$\beta_{21}^{\tau-}(K^-)$	−0.0024[a]	−0.0030[a]	0.0024[a]	0.0063[a]
	(0.0006)	(0.0009)	(0.0009)	(0.0020)
$\beta_{21}^{\tau+}(K^+)$	0.0005[a]	0.0003	−0.0005[a]	0.0001[a]
	(0.0003)	(0.0003)	(0.0002)	(0.0004)
$\beta_{22}^{\tau-}(L^-)$	0.001	0.0001	−0.0003	0.0008
	(0.0004)	(0.0003)	(0.0004)	(0.0004)
$\beta_{22}^{\tau+}(L^+)$	−0.0001	−0.0009[a]	−0.0004	−0.0013[a]
	(0.0003)	(0.0004)	(0.0005)	(0.0007)
γ_{21}^τ	0.9643[a]	0.9690[a]	0.9683	0.9578[a]
	(0.4671)	(0.1582)	(0.8222)	(0.4107)
γ_{22}^τ	0.2684[a]	0.2809[a]	0.2794	0.2722
	(0.0813)	(0.1719)	(0.2788)	(0.2374)
σ_3	0.0047[a]	0.0060[a]	0.0079[a]	0.0097[a]
	(0.0005)	(0.0002)	(0.0002)	(0.0006)
The Romer model				
α_3^τ	0.0054[a]	0.0161[a]	0.0123[a]	0.0155[a]
	(0.0008)	(0.0019)	(0.0006)	(0.0015)
$\beta_{31}^{\tau-}(RD^-)$	0.0036	0.0214[a]	0.0382[a]	0.0399[a]
	(0.0632)	(0.0047)	(0.0021)	(0.0029)
$\beta_{31}^{\tau+}(RD^+)$	0.0043	0.0017	0.0182[a]	0.0131[a]
	(0.041)	(0.0004)	(0.0034)	(0.0051)
γ_{31}^τ	0.0081	0.0062	0.0074	0.0062
	(0.0339)	(0.0192)	(0.0067)	(0.0082)
σ_3	0.0041[a]	0.0074[a]	0.0078[a]	0.0100[a]
	(0.0003)	(0.0001)	(0.0002)	(0.0007)

(continued)

Table 5 (continued)

Parameter\Quantile	$\tau = 0.10$	$\tau = 0.50$	$\tau = 0.70$	$\tau = 0.90$
The Barro model				
α_4^{τ}	0.0035[a]	0.0146[a]	0.0107[a]	0.0127[a]
	(0.0001)	(0.0015)	(0.0009)	(0.0015)
$\beta_{41}^{\tau-}(G^-)$	−0.0191[a]	−0.0123[a]	−0.0043[a]	0.0028
	(0.0063)	(0.0063)	(0.0021)	(0.0024)
$\beta_{41}^{\tau+}(G^+)$	−0.0761	0.0049[a]	0.0142	0.0280[a]
	(0.0034)	(0.0022)	(0.008)	(0.0034)
γ_{41}^{τ}	0.2991[a]	0.0333	0.0354[a]	0.0364
	(0.0192)	(0.0714)	(0.0091)	(0.0271)
σ_4	0.0044[a]	0.0075[a]	0.0079[a]	0.0104[a]
	(0.0004)	(0.0002)	(0.0002)	(0.0008)
θ_G^{τ}	0.9900[a]	0.910[a]	0.9511[a]	0.9940[a]
	(0.0021)	(0.0009)	(0.0016)	(0.0011)

Source Calculation

Note [a]denotes the 5% significance level. The value in parenthesis is standard deviation

introduced in the Solow model are statistically significant but their effects in terms of coefficients are different across quantiles and regimes based on the kink points. For example, in the upper regime, Thailands GDP growth depends negatively on the labor variable but the size of the effect is different across quantiles. For example, the economic growth in the 0.9 quantile is found to be most negatively sensitive to the change in labor.

The third equation is the Romer model which is used to represent the new growth theory. This model can be formed as $\ln(GDP_t) = \alpha_3^{\tau} + \beta_{31}^{\tau-} \ln(RD_t)^- + \beta_{31}^{\tau+} \ln(RD_t)^+ + \sigma_3 u_3$. We can see that the kink points split the effect of R&D expenditure on Thailands GDP growth into two regimes. In the lower regime, GDP growth is much more positively sensitive to the change in R&D expenditure than in the upper regime, except the result in the 0.1 quantile which is not very different across regimes. Furthermore, we also observed that the impact of R&D expenditure on GDP growth grew quite dramatically as we move up through the conditional distribution of R&D expenditure.

The last equation is the Barro model which takes the form as $\ln(GDP_t) = \alpha_4^{\tau} + \beta_{41}^{\tau-} \ln(Gov_t)^- + \beta_{41}^{\tau+} \ln(Gov_t)^+ + \sigma_4 u_4$. The result demonstrates the unequal impacts of government expenditure on Thailands GDP growth which are split into two regimes based on kink points. In the lower regime, an increase in government expenditure causes the GDP growth to decline but the level of negative impact is somehow reducing as the level of quantile is moved up. On the contrary, in the upper regime it is found that if the government spending increases, then the GDP growth rises and the impacts are found to be different significantly across quantiles.

6 Conclusions

This paper attempts to test the joint validity of economic growth theories with special focus on the experience of Thailand since many works on economic growth in Thailand have been found to follow the idea of conventional growth theories without testing their correctness for the Thai economy. Motivated by this reasoning, we consider three main economic growth theories namely the classical, neoclassical, and the new growth theories and employ some important growth models that are the Harrod-Domar, Solow, Romer, and Barro models to represent the three eras of growth theory.

To investigate the validity of the growth theories, we introduce Copula based seemingly unrelated quantile kink regression (SUQKR) as a key tool in this work. Evidences from this study show that the four growth models can prove their validities for the Thai economy through the data set. We found that the investment variable of the Harrod-Domar model, the capital and labor variables of the Solow model, the R&D expenditure of the Romer model, and the government expenditure of the Barro model are statistically significant.

Furthermore, as a specific capability of our method, we are allowed to preserve the unequal and nonlinear impacts of those variables of interest on Thailand's GDP growth. That is the impacts are different across quantiles; some estimated coefficients dramatically increase as the level of quantile moves up, or have both positive and negative effects on the growth. Moreover, we also found that the impacts of those variables are nonlinear; they are split into two regimes i.e. the lower and upper regimes due to the kink effect. That is, for example, an increase in government spending tends to create a negative impact on the growth, but in the upper regime government spending is necessary to propel the Thai economy.

Acknowledgements The authors are grateful to Puey Ungphakorn Centre of Excellence in Econometrics, Faculty of Economics, Chiang Mai University for the financial support.

References

1. Acemoglu D (2008) Introduction to modern economic growth. Princeton University Press, Princeton
2. Barro R (1990) Government spending in a simple model of endogenous growth. J Polit Econ 98:S103–S125
3. Barro R, Sala-i-Martin X (2004) Economic growth, 2nd edn. The MIT Press, Cambridge
4. Chen CW, So MK, Chiang TC (2016) Evidence of stock returns and abnormal trading volume: a threshold quantile regression approach. Jpn Econ Rev 67(1):96–124
5. Fiaschi D, Lavezzi AM (2007) Nonlinear economic growth: some theory and cross-country evidence. J Dev Econ 84(1):271–290
6. Hansen BE (2017) Regression kink with an unknown threshold. J Bus EconStat (to appear)
7. Hofert M, Machler M, McNeil AJ (2012) Likelihood inference for Archimedean Copula in high dimensions under known margins. J Multivar Anal 110:133–150

8. Jun SJ, Pinkse J (2009) Efficient semiparametric seemingly unrelated quantile regression estimation. Econom Theory 25(05):1392–1414
9. Koenker R, Bassett G Jr (1978) Regression quantiles. Econom J Econom Soc 46:33–50
10. Minea A (2008) The role of public spending in the growth theory evolution. Rom J Econ Forecast 2:99–120
11. Noh H, Ghouch AE, Bouezmarni T (2013) Copula-based regression estimation and inference. J Am Stat Assoc 108(502):676–688
12. Pastpipatkul P, Maneejuk P, Sriboonchitta S (2015) Welfare measurement on Thai rice market: a Markov switching Bayesian seemingly unrelated regression. In: Integrated uncertainty in knowledge modelling and decision making. Springer International Publishing, pp 464–477
13. Patton AJ (2006) Modeling asymmetric exchange rate dependence. Int Econ Rev 47(2):527–556
14. Romer PM (1986) Increasing returns and long-run growth. J Polit Econ 94:1002–1037
15. Sklar M (1959) Fonctions de rpartition n dimensions et leurs marges. Universit Paris 8:229–231
16. Sumer K (2012) Testing the validity of economic growth theories with seemingly unrelated regression models: application to Turkey in 1980–2010. Appl Econom Int Dev 12(1):63–72
17. Wang Y (2012) Numerical approximations and goodness-of-fit of Copulas
18. Wichitaksorn N, Choy STB, Gerlach R (2006) Estimation of bivariate Copula-based seemingly unrelated Tobit models. Discipline of business analytics, University of Sydney Business School, NSW
19. Zellner A (1962) An efficient method of estimating seemingly unrelated regressions and tests for aggregation bias. J Am Stat Assoc 57(298):348–368

Analysis of Global Competitiveness Using Copula-Based Stochastic Frontier Kink Model

Paravee Maneejuk, Woraphon Yamaka and Songsak Sriboonchitta

Abstract The competitiveness is a considerable issue for nations who rely on the international trade and hence leads to the competitiveness evaluation. This paper suggests considering a country's productive efficiency to reflect the competitive ability. We introduce the copula-based nonlinear stochastic frontier model as a contribution to the competitiveness evaluation due to a special concern about the difference among countries in terms of size and structure of the economies. As a specific capability of this proposed model, we are able to find the different impact of inputs on output from the group of small countries to the group of large countries. Finally, this paper provides the efficiency scores according to our analysis and the overall ranking of global competitiveness.

Keywords Technical efficiency · Competitiveness · Nonlinear stochastic frontier · Kink regression · Copula

1 Introduction

The competitive ability is crucial for any economy or nation that relies on the international trade and is relevant to the modern economy since it is considered to be a key criterion for assessing the success of country. The competitiveness evaluation deserves this special attention because countries need to know their competitive powers, as well as the ability of other countries in the international market, to formulate the proper structural reforms to move their economy forward.

P. Maneejuk (✉) · W. Yamaka · S. Sriboonchitta
Faculty of Economics, Chiang Mai University, Chiang Mai, Thailand
e-mail: mparavee@gmail.com

W. Yamaka
e-mail: woraphon.econ@gmail.com

© Springer International Publishing AG 2017
V. Kreinovich et al. (eds.), *Robustness in Econometrics*,
Studies in Computational Intelligence 692, DOI 10.1007/978-3-319-50742-2_33

Because of the importance of competitiveness in contributing to the world economy, a variety of institutions take interests in the global competitiveness, trying to measure and make the rankings of countries based on their performances in the economic sphere. On the one hand, World Economic Forum (WEF) defines the term competitiveness as the ability to produce goods and services or the productivity of a country which is determined by twelve pillars namely institutions, infrastructure, macroeconomic environment, health and primary education, higher education and training, goods market efficiency, labor market efficiency, financial market development, technological readiness, market size, business sophistication, and innovation [13]. On the other hand, one section of International Institute for Management Development (IMD) named World Competitiveness Center also focuses on the competitive ability and explains that the competitiveness determines how country manages its competency to achieve a long-term growth or how much the country success in the international market [6]. The rankings of global competitiveness suggested by IMD are determined by the set of economic performance, government efficiency, business efficiency, and infrastructure of a country.

These two institutions, WEF and IMD, have been suggesting the use of the comprehensive reports which help appropriately policymakers and business leaders make long term decision. However, as there is no certain technique for measuring the competitive ability and no unique consensus on its definition has been reached yet, this paper intends to introduce an alternative way of the evaluation of the global competitiveness which is relevant to a country's productive efficiency. We aim to compare the competitive ability of countries through the levels of technical efficiency. That is, we will identify the performance (efficiency) of a country and then make the rankings based on their levels of efficiency. We deal with some specific indicators as suggested by Furkov and Surmanov [5] to represent the basic performance of the national economy with respect to the competitiveness definition, additionally, with special concern about income distribution of nations. Note that all economic indicators will be discussed later in the part of model specification.

The next section will explain about the methodology we use to evaluate the global competitiveness including the basic idea regarding technical efficiency (TE). In Sect. 3 we will do some Monte Carlo experiments and report the Robustness Checks by Kullback-Leibler Divergence. Section 4 explains the data and the model specification. Section 5 reports the empirical estimate of global competitiveness through the efficiency frontier and the competitiveness rankings. Section 6 contains the conclusions.

2 Methodology: An Introduction to the Nonlinear Stochastic Frontier Model

To evaluate a country's efficiency, we consider the idea of Stochastic Frontier Model (SFM) which is used generally to assess technical efficiency of production units and measure the impact of input on output. The SFM was first introduced by Aigner et al. [1] applied in the context of cross-sectional data. The use of SFM which is helpful for efficiency evaluation is for the separation of the inefficiency effect from the statistical noise (or normal error term). Technically, the original SFM can be viewed as a linear regression where its error term is composed of two uncorrelated terms, saying that the first term represents the statistical noise and the second term represents the inefficiency relative to the frontier.

Based on the idea of the conventional SFM, the output depends linearly on inputs with two independent error components. However, as in the literature of global competitiveness, we realize that the global production function may have nonlinear behaviors and processes [10]. Importantly, we use the term production function because, in this paper, we define the concept of global competitiveness in the context of growth theory in which the output is presented by the growth of GDP. Then, the level of countrys competitiveness is analyzed by some specific indicators that evaluate the performance of the countrys growth strategies. And, as countries have different size (in terms of GDP) and structure of the economies, we doubt that the relationship between output and inputs may differ across countries. For this reason, the originally linear SFM seems to no longer be appropriate for evaluation of the global competitiveness. Therefore, this paper proposes the nonlinear SFM as an innovational tool. To the best of our knowledge, the nonlinear SFM has not yet been explored and if our proposal worked, it would be useful for many applications of stochastic frontier analysis with structural change. To construct the nonlinear SFM, we apply the idea of the kink regression as introduced by Card et al. [3] and Hansen [8] to the conventional SFM. And hence, our paper proposes the stochastic frontier kink model. This model will be explained thoroughly in the following part, however, in brief, it is split into two (or more) parts based on a kink point. This specific ability allows the disparate impact of input on the output across countries which is so-called the nonlinear relationship.

2.1 Modelling the Stochastic Frontier Kink Model

The structure of the stochastic frontier kink model can simply take the form as the following equation. Suppose we have T different countries, the independent variable of each country is denoted by Y_t where $t = 1, \ldots, T$. The term x'_t is a matrix $(T \times k)$ of k regressors or input variables of country t where the coefficients are presented by a matrix β with dimension $(T \times k \times 2)$ for the case of two different regimes.

$$Y_t = \beta_1^- (x'_{1,t} - \gamma_1)_- + \beta_1^+ (x'_{1,t} - \gamma_1)_+ + \cdots \qquad (1)$$
$$\ldots + \beta_k^- (x'_{k,t} - \gamma_k)_- + \beta_k^+ (x'_{k,t} - \gamma_k)_+ + \varepsilon_t$$

Following the original work of kink regression, Hansen [9], we use $(x'_{k,t})_- = \min[x'_{k,t}, 0]$ and $(x'_{k,t})_+ = \max[x'_{k,t}, 0]$ to separate $x'_{k,t}$ into two regimes. The term regime often refers to the state of the economy, but in this study, it refers to the different size and structure of economy i.e. regime 1 refers to a small country and regime 2 refers to a large country. As we can see, Eq. (1) shows that the slope with respect to variable $x'_{k,t}$ or the estimated parameters β are split into two groups, for two regimes, in which $(\beta_1^-, \ldots, \beta_k^-)$ present the parameters in the lower regime and $(\beta_1^+, \ldots, \beta_k^+)$ present the parameters in the upper regime. The slope is equal to β_k^- for any value of $x'_{k,t} < \gamma_k$ and β_k^+ is for the case of $x'_{k,t} > \gamma_k$ where the term γ_k is called a kink point. According to this specific characteristic, Eq. (1) is said to be the stochastic frontier kink model since the relationship between the independent variable and its covariates has the kink effect at $x'_{k,t} = \gamma_k$ and the error term of the model ε_t consists of the two independent error terms expressed as $\varepsilon_t = V_t - W_t$. Again, the error term V_t which represents the statistical noise, is assumed to follow normal distribution while the term W_t representing the inefficiency is assumed to have non-negative distribution. Note that we assume a truncated normal distribution for W_t.

2.2 A Copula-Based Stochastic Frontier Kink Model

Apart from the advantage of the SFM, the assumption of independence between two error components of the conventional SFM is considered to be weak. It is argued that the inefficiency term at the present time may depend on the noise at the previous time [4]. Therefore, many researchers have worked on this problem in which one of the most influential works is the study of Smith [15]. He suggests allowing the two error components to be related by using a copula to fit the joint distribution. This suggestion has spread various extensions in which some can prove that the copula-based SFM perform better than the conventional one [17]. Thus, we decide to take the advantage of copula joining the two error components to relax this weak assumption, and propose the copula-based stochastic frontier kink model to evaluate a countrys efficiency.

According to the Sklar's theorem (See Nelsen [11]), let H be a joint distribution of W_t and V_t, a two dimensional distribution with marginals $F_1(W_t)$ and $F_2(V_t)$. Then, there exists a bivariate copula C such that

$$H(W_t, V_t) = C(F_1(W_t), F_2(V_t)). \tag{2}$$

The term C is bivariate copula distribution function of the two error components. Furthermore, if the marginals are continuous, then the copula C is unique. Otherwise, C is uniquely determined on $R(F_1) \times R(F_2)$ where $R(F_1)$ and $R(F_2)$ denote the range of the marginal F_1 and F_2, respectively. However, if F_1 and F_2 are univariate distribution, then the function $H : \bar{R}^n \to [0, 1]$ as defined in Eq. (2) is a joint distribution function of marginal distributions F_1 and F_2. If we have a continuous marginal distribution, saying that $F_1(W_t) = w_t$ and $F_2(V_t) = v_t$, the copula is determined by

$$C(w_t, v_t) = C(F_1^{-1}(W_t), F_2^{-1}(V_t)), \tag{3}$$

where F_i^{-1} is the quantile functions of marginal $i = 1, 2$; and w_t, v_t are uniform [0, 1]. The term C is copula distribution function of n dimensional random variable with uniform margin on [0, 1]. Using the chain-rule from calculus, the copula density c is obtained by differentiating Eq. (1); thus, we get probability distribution function (pdf) of W_t and V_t as shown below.

$$\frac{\partial^2}{\partial W_t \partial v_t} H(W_t, V_t) = \frac{\partial^2}{\partial W_t \partial V_t} C(F_1(W_t), F_2(V_t))$$
$$= f_1(W_t) f_2(V_t) c(F_1(W_t), F_2(V_t)) \tag{4}$$

The function f_i is the density function of each marginal distribution and c is the copula density. To extend copula to the SFM, the component error cannot be obtained directly from Eq. (1), thus we transform (W_t, V_t) to be (W_t, ε_t) where $\varepsilon_t = V_t - W_t$. Therefore, we can rewrite Eq. (4) as

$$f(W_t, \varepsilon_t) = f_1(W_t) f_2(W_t + \varepsilon_t) c(F_1(W_t), F_2(W_t + \varepsilon_t)). \tag{5}$$

According to the work of Smith [15], who first introduced the copula-based SFM, the pdf of ε_t is given by

$$f(\varepsilon_t) = \int_0^\infty f(w_t, \varepsilon_t) dw_t$$
$$= E_{W_t}(f_2(W_t + \varepsilon_t) c(F_1(W_t), F_2(W_t + \varepsilon_t) | \theta_C)) \tag{6}$$

where E_{W_t} is the expectation with respect to W_t and θ_C is the dependence parameter. We can see that Eq. (6) comprises two densities, i.e. $f(W_t + \varepsilon_t)$ and $c(F_1(W_t), F_2(W_t + \varepsilon_t))$. Therefore, we can construct the likelihood function of the copula-based stochastic frontier kink model by deriving those two densities. The first density is given by

$$Y_t = \beta_1^-(x'_{1,t} - \gamma_1)_- + \beta_1^+(x'_{1,t} - \gamma_1)_+ + \ldots + \beta_k^-(x'_{k,t} - \gamma_k)_- + \beta_k^+(x'_{k,t} - \gamma_k)_+ + \varepsilon_t$$
$$Y_t = \Omega + \varepsilon_t$$
$$Y_t = \Omega + V_t - W_t; \ \varepsilon_t = V_t - W_t \tag{7}$$
$$Y_t - \Omega = V_t - W_t$$

Since V_t is assumed to have normal distribution thus the density function of V_t is given by

$$f(V_t) = f(\varepsilon_t + W_t) = \frac{1}{\sqrt{2\pi \sigma_V^2}} \exp\left\{ \frac{(\varepsilon_t + W_t)^2}{2\sigma_V^2} \right\} \tag{8}$$

where W is simulated from the positive truncated normal distribution with mean $\mu = 0$ and variance σ_W^2. Consider the second density, the bivariate copula density for $W_t + \varepsilon_t$ and W_t is constructed by either Elliptical copulas or Archimedean copulas. In this study, we consider six well-known copula families namely Normal copula, Student-t copula, Frank copula, Clayton copula, Gumbel copula, and Joe copula (See Nelson [11]). Note that the joint distribution through the copula function needs the uniform marginal thus the simulated $W_t + \varepsilon_t$ and W_t are transformed by cumulative normal distribution and cumulative truncated normal distribution.

As discussed in Smith [15] and Wibiinpongse et al. [17], it is not easy to estimate the model directly from the maximum likelihood estimation since it has a closed-form expression. To overcome this problem, the Monte Carlo simulation is conducted to simulate the error ε_t and W_t. The expected log-likelihood function is obtained from Eq. (6) then yields

$$L(\beta_k^-, \beta_k^+, \sigma_V, \sigma_W, \theta_C) = \sum_{t=1}^{T} \log\left(\frac{1}{M} \sum_{i=1}^{M} f(W_{it} + \varepsilon_{it}) c(w_{it}, (w_{it} + \varepsilon_{it}) \,|\theta_C)\right).$$

(9)

The log likelihood function in Eq. (9) can then be maximized using the maximum simulated likelihood method (See Greene [7] and Wiboonpongse et al. [17]).

The main contribution of the stochastic frontier model is the technical efficiency (TE) which is the effectiveness with a given set of inputs used to produce an output. The technical efficiency is generally defined by the ratio of the observed output to the corresponding frontier output conditional on the levels of inputs used by a country. Then, a country is said to have technical efficiency if it can produce maximum output by using minimum amounts of inputs. Following Battese and Coelli [2], TE can be defined through Eq. (10) in which W is the inefficiency term. We can specify TE equation by using Monte Carlo simulation as

$$TE = E(\exp(-W) \,|\xi = \varepsilon_t)$$

$$= \frac{\displaystyle\sum_{i=1}^{M} \exp(-W_{it}) f(W_{it} + \varepsilon_t) c(F_1(W_{it}), F_2(W_{it}+\varepsilon_t) \,|\theta_C)}{\displaystyle\sum_{i=1}^{M} f(W_{it} + \varepsilon_t) c(F_1(W_{it}), F_2(W_{it}+\varepsilon_t) \,|\theta_C)}.$$

(10)

3   Simulation Study

The Monte Carlo simulation study is conducted to explore the performance and accuracy of our proposed model copula-based stochastic frontier kink. This simulation study consists of two sub-sections. First, we evaluate the performance of our model using the Bias and Mean Squared Error (MSE). Second, the Kullback–Leibler Divergence (KLD) is employed to measure the distance between the true model and alternative model to examine the effect of misspecified copula function.

3.1 General Specifications

The simulation model takes the form as the following equation where we set the true value for parameters α, β_1^-, and β_1^+ equal to 1, 0.5, and -2, respectively. Then, we simulate the variable $x_{1,t}'$ on the standard normal distribution with zero mean and variance equal to 1. The kink parameter γ_1 is equal to 6. The error term $\varepsilon_t = V_t - W_t$ is assumed to follow a normal distribution with $V_t \sim N(0, \sigma_V)$ and half-normal distribution with $W_t \sim N^+(0, \sigma_W)$, where $\sigma_V = 1$ and $\sigma_W = 1$, respectively.

$$Y_t = \alpha + \beta_1^-(x_{1,t}' - \gamma_1)_- + \beta_1^+(x_{1,t}' - \gamma_1)_+ + \varepsilon_t$$

For the joint distribution of the two error components, we apply the six copula families consisting of Gaussian, Student-t, Gumbel, Clayton, Joe and Frank copulas to model the nonlinear dependence structure between V_t and W_t. We set the true value for the dependence parameter θ_c equal to 0.5 for Gaussian, Student-t, and Clayton copulas, additionally; the degree of freedom for Student-t copula is set to be 4. For the case of Gumbel and Frank copulas, the dependence parameter θ_c is equal to 2 and 3 for Joe copula. We also vary the sample size to strengthen the accuracy of our model by generating N = 100, 200, and 300. For each data set, we obtain 100 bootstrap samples through a parametric resampling. Additionally, the Mean Squared Error (MSE) which is the difference between the estimator and the estimated value is computed for each parameter using the following formula:

$$MSE = M^{-1} \sum_{r=1}^{M} (\tilde{\phi}_r - \phi_r)^2$$

where M is the number of bootstrapping. The terms $\tilde{\phi}_r$ and ϕ_r represent the estimated parameter and the unknown parameter, respectively.

Table 1 shows the result of the Monte Carlo simulation investigating the maximum likelihood estimation of the copula-based stochastic frontier kink model. We explore six simulation experiments based on the different copula functions. As shown in this table, most of the mean parameters are very close to their true values and the MSE values of all parameters are low and gradually decrease as the sample size increases. Our model can perform well through the simulation study. Therefore, overall, the Monte Carlo simulation suggests that our proposed Copula-based stochastic frontier kink model is reasonably precise.

3.2 Robustness Checks by Kullback–Leibler Divergence

The previous part shows the accuracy of our proposed model, copula-based stochastic frontier kink model, through the Monte Carlo simulation. However, how precise the result is must depend on the copula we choose as well, because we need the

Table 1 Simulation study

Copula	Gaussian						Student-t					
	Estimated parameter			MSE			Estimated parameter			MSE		
N	100	200	300	100	200	300	100	200	300	100	200	300
α	1.266	0.621	0.683	0.361	0.196	0.245	1.078	0.995	0.886	0.295	0.514	1.065
β_1^-	0.484	0.479	0.49	0.012	0.002	0.036	0.488	0.484	0.498	0.047	0.003	0.151
β_1^+	−2.035	−1.949	−2.005	0.041	0.038	0.559	−1.965	−2.018	−2.011	0.635	0.022	0.24
γ_1	6.157	6.028	6.065	0.159	0.672	0.616	5.904	6.109	6.065	0.684	0.948	0.922
σ_u	1.045	1.077	1.004	0.362	0.139	0.233	0.774	1.191	1.34	0.29	0.176	0.105
σ_v	0.755	0.758	0.587	0.716	0.38	0.558	1.063	1.035	1.056	0.619	0.481	0.105
θ_c	0.295	0.608	0.718	0.335	0.418	0.139	0.472	0.602	0.803	0.365	0.21	0.171
df							4.1	5.289	4.539	0.255	0.296	0.111
Copula	Clayton						Gumbel					
	Estimated parameter			MSE			Estimated parameter			MSE		
N	100	200	300	100	200	300	100	200	300	100	200	300
α	1.269	1.122	0.873	0.158	0.272	0.248	1.435	1.198	0.85	0.669	0.794	0.251
β_1^-	0.346	0.57	0.495	0.003	0.002	0.001	0.486	0.075	0.468	0.004	0.003	0.001
β_1^+	−1.809	−2.022	−2.015	0.096	0.025	0.009	−2.008	−1.358	−2.144	0.064	0.027	0.009
γ_1	5.672	5.771	6.098	0.262	0.842	0.302	6.152	4.641	6.768	0.23	0.905	0.118
σ_u	0.845	0.921	1.036	0.16	0.147	0.356	1.104	1.0001	0.5	0.706	0.426	0.175
σ_v	0.933	1.029	0.8	0.351	0.454	0.412	1.163	1.763	1.208	1.401	1.641	0.481
θ_c	0.501	0.524	0.63	1.007	2.763	0.264	2.645	1.413	2.288	0.556	0.585	0.324
Copula	Frank						Joe					
	Estimated parameter			MSE			Estimated parameter			MSE		
N	100	200	300	100	200	300	100	200	300	100	200	300
α	0.819	1.165	1.138	0.605	0.532	0.206	0.91	1.296	1.19	0.273	1.032	0.296
β_1^-	0.491	0.727	0.511	0.003	0.022	0.001	0.481	0.4789	0.499	0.923	0.003	0.027
β_1^+	−2.02	−1.846	−1.998	0.083	0.027	0.018	−2.012	−1.989	−1.979	1.159	0.023	0.412
γ_1	6.093	4.689	6.004	0.195	1.143	0.121	6.094	6.09	5.913	0.196	1.016	0.523
σ_u	1.187	1	1.423	0.314	0.512	0.155	1.642	1.183	0.424	5.53	0.285	0.221
σ_v	0.853	1.527	1.117	1.183	0.798	0.335	1.606	1.307	1.212	1.102	0.185	0.609
θ_c	2.759	2.587	2.406	1.051	0.265	0.817	3.218	2.794	1.694	0.255	0.306	1.379

Source Calculation

appropriate copula function to join the two error components V_t and W_t. Selecting the proper copula is a crucial step since the inappropriate copula function will bring about a bias in the estimation [12]. Therefore, in this part, we employ the Kullback–Leibler divergence (KLD)—which is a measure of the distance between two probability distributions—to strengthen the model.

We do two experiment tests in which the true copula functions are set to be Student-t and Clayton copulas. The simulation procedure is based on the sample size of $N = 100$. Then, simulated data are estimated by the copula-based stochastic frontier kink model with six different copula functions namely Gaussian, Student-t, Gumbel, Clayton, Joe and Frank copulas. The distance between the true function and the alternative one is computed using the following formula of the KLD. Moreover, we not only examine the selection of appropriate copula; but also consider specially the selection of the right structure. That is, we are going to compare the copula-based

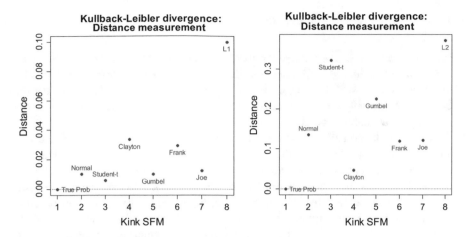

Fig. 1 The Kullback–Leibler Divergence: distance measurements

nonlinear SFM -or the copula-based stochastic frontier kink model that we proposed-
, with the copula-based linear SFM. Given the same true model, we decide to use
the same simulated data, which has a structure change, to observe the performance
of our model when the data exists with nonlinear structure.

For the true distributions denoted by P and the alternative distribution denoted
by Q of a continuous random variable, the Kullback–Leibler divergence is defined
to be the integral:

$$D(P, Q) = \int_{-\infty}^{\infty} p(x) \log \frac{p(x)}{q(x)} dx \qquad (11)$$

The term $p(x)$ is a true likelihood density function of P where all parameters
are known and $q(x)$ is an alternative likelihood density function of Q. According to
this KLD formula, we are able to calculate the distance between the true functions
i.e. Student-t and Clayton copulas based stochastic frontier kink model, and the
approximation of alternative models. The results are shown in Fig. 1.

Figure 1 shows the two panels of the KLD results. We can see from the Fig. 1
that the approximations of the Student-t and Clayton based stochastic frontier kink
models provide the minimum distances which are closest to their true function lines
(dashed lines). Considering the case of misspecified copulas, we see that the wrong
specification brings the larger deviation from the true one. In addition, as we aim to
compare the performance of our model to the copula-based linear SFM, we construct
the Student-t based linear SFM denoted by L1 and the Clayton based linear SFM
denoted by L2. The results show that our model performs much better than the
copula-based linear SFM; our model is much closer to the true probability density in
both two experiments. Overall, the KLD suggests that our proposed Copula-based
nonlinear SFM is more accurate than the alternative models through this simulated
data, and the misspecification of copula will lead to low accuracy.

4   Data and Model Specification

The cross-sectional data are derived from World Bank database in year 2014, covering 134 countries including 34 Asian countries, 6 North American countries, 10 South American countries, 52 European countries, and 32 African countries. Note that the data are not available for some countries in that year; therefore, we decide to use the nearest year available instead.

We define the specific factors representing the basic performance of national economy with respect to the definition of competitiveness by following the study of Furkov and Surmanov [5]. Together with a special concern about income distribution of nations, we speculate that income distribution is somehow necessary for the competitiveness evaluation. That is because; the competitive ability sometimes requires tapping from the inequality in which unequal distribution of income may provide an incentive for individual to work harder or for entrepreneur to expand a business, which in turn, increase a countrys competitive power. Therefore, the inequality is measured using the best known index called Gini and the copula-based stochastic frontier kink model in the context of global competitiveness is given by

$$
\begin{aligned}
\ln(GDP_t) = {} & \alpha + \beta_1 \ln(Capital_t) + \beta_2 \ln(Labour_t) + \beta_3 \ln(Consump_t) \\
& + \beta_4 \ln(Gini_t) + V_t - W_t
\end{aligned}
\tag{12}
$$

The independent variables consist of capital, labour, consumption, and the Gini coefficient. The variable $Capital_t$ refers to the ability of the country t to transform capital for further development measured by the gross fixed capital formation (USD millions). The variable $Labour_t$ refers to the number of people employed in various sectors (in thousands) and $Consump_t$ represents the household final consumption of the country t (in USD millions). This variable implies the purchasing power of the country t which directly relates to the competitiveness. The dependent variable is real GDP measured in million USD.

5   Empirical Results

Prior to evaluating the global competitiveness, we begin with an experimental test to verify a kink effect or the nonlinear structure. We employ the Likelihood ratio test (LR-test) to verify which model between the linear model and the nonlinear model is best-fit for the data [8]. Technically, we let the linear model be a null hypothesis and the kink (nonlinear) model is an alternative one. In this LR-test, we select a Gaussian copula function for the production estimate and present in this analysis. More specifically, the LR-test is defined by

$$
LR - test = 2[\log \max Lik(H_a) - \log \max Lik(H_0)].
$$

Table 2 Results of likelihood ratio test

	lnL (linear SFM)	lnL (kink SFM)	LR
Capital	−36.03	−29.47	13.12**
Labour	−251.58	−247.4	8.25*
Consumption	−59.85	−151.22	−182.74
Gini	−292.3	−305.18	−25.79

Source Calculation
Note *, and **, denote rejections of the null hypothesis at the 10% and 5% significance levels, respectively

This test expresses how many times the data are more likely to be under one model than another. Therefore, the model is considered to have the kink effect if the null hypothesis based on LR statistic is rejected. The probability distribution for this test statistic is chi-squared distribution with degrees of freedom equal to the difference of the number of parameters of the two models. The kink effect on a relationship between response variable and its covariate is examined as a pair test. This algorithm is kept using for each pair of the covariate and response variable and the result is shown in Table 2.

Table 2 shows the result of testing for the kink effect with respect to our model. We found that the null hypothesis of linearity based on the LR-test is rejected with a significance level at 5 and 10% for the pairs of labour and capital against GDP, respectively, whereas the null hypothesis is held for the cases of consumption and the Gini coefficient. This means only the relationship between capital and labour with GDP are in favor of the kink model. And hence, we can define the model with specified kink effect as in the following:

$$
\begin{aligned}
\ln(GDP_t) = {} & \alpha + \beta_1^- (\ln(Capital_t) - \gamma_1)_- + \beta_1^+ (\ln(Capital_{1t}) - \gamma_1) \\
& + \beta_2^- (\ln(Labour_t) - \gamma_2)_- + \beta_2^+ (\ln(Labour_t) - \gamma_2)_+ \qquad (13) \\
& + \beta_3 \ln(Consump_t) + \beta_4 \ln(Gini_t) + V_t - W_t.
\end{aligned}
$$

The model is set similarly to Eq. (12), but the coefficient terms β_i^- and β_i^+, $i = 1, 2$, represent the coefficients of regime 1 and regime 2, respectively. Here we treat the parameters γ_1 and γ_2 as the kink points which need to be estimated.

5.1 Selection of Copula

This part is about selecting a Copula that is best-fit for the data. Given a set of Copulas, Table 3 shows the values of AIC and BIC for each copula-based stochastic frontier kink model. According to both criteria, the best model is the one based on the Gumbel copula. It has the minimum values of AIC and BIC which are −**48.84**

Table 3 AIC and BIC criteria for each copula-based stochastic frontier kink model

Copula	Gaussian	Student-t	Clayton	Gumbel	Frank	Joe	SFM0
AIC	346.04	353.91	−15.56	**−48.84**	−24.12	−22.32	−19.4
BIC	349.52	357.68	29.99	**−3.29**	21.44	23.23	26.15

Source Calculation

Note SFM0 is the is the conventional Stochastic frontier model [2]

and −**3.29** (bold numbers), respectively, whereas the conventional stochastic frontier model (SFM0) is not chosen for this data set.

5.2 Estimates of Gumbel Copula-Based Stochastic Frontier Kink Model

The estimated parameters of the Gumbel copula-based stochastic frontier kink model are displayed in Table 4. We found that the model can perform well across the data sets in which most of the parameters are rightly signed and statistically significant. This model can capture a nonlinear effect of some variables, meaning that it can prove our hypothesis: the relationship between output and inputs may differ across countries depending on the size (in terms of GDP) and structure of the economy.

Table 4 displays the estimated parameters corresponding to Eq. (13), including the dependence parameters of copula and variances. As shown in Table 4, all the parameters are significant except the coefficient of labour in regime 1 (β_2^-). The parameters of capital and labour are split into two regimes based on the kink points and mildly different from one regime to the other. The first regime (β_i^-, $i = 1, 2$) refers to a group of small countries while the second regime (β_i^+) refers to a group of large countries.

The impact of capital on the output or GDP is more intense in the group of large countries such as US, UK, and China where success depends much on the physical capital. Their coefficients are approximately equal to 1.01, meaning that an additional 1% of capital leads to 1.01% increase in GDP. But, beyond the significant

Table 4 Estimated results

Parameter	α	β_1^-	β_1^+	β_2^-	β_2^+	β_3
Estimate	8.30***	0.37***	1.01***	−0.41	0.93***	0.54***
S.E.	0.42	0.08	0.04	0.28	0.32	0.04
Parameter	β_4	γ_1	γ_2	σ_v	σ_w	θ_c
Estimate	−0.17***	11.08***	7.89***	0.33***	0.46***	2.75***
S.E.	0.07	0.76	0.73	0.05	0.01	0.46

Source Calculation

Note ***, denotes rejections of the null hypothesis at the 1% significance levels

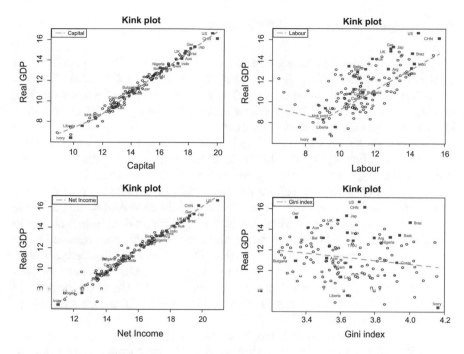

Fig. 2 Plot of the data fitting to lines estimated by Gumbel copula-based SF kink model

kink point (γ_1) around 11.08, there is a place of the small countries such as Liberia and Ivory in which the additional 1% of capital leads to just 0.37% increase in GDP. On the other hand, we failed to find the impact of labour on GDP of the small countries, only the parameters in regime 2 are statistically significant. It is found that an increase by 1% of labour tends to get 0.93% increase in GDP for the group of large countries. Additionally, the estimated parameter β_3 corresponding to the household final consumption variable has positive sign, meaning that this variable creates the positive impact on countrys competitiveness. Nevertheless, the estimated Gini coefficient surprisingly has negative sign. Evidence from this data set suggests that an increase in the Gini coefficient would reduce the countrys competitive ability rather than propel the economy. Additionally, technically, we found that the estimated parameter of the Gumbel copula is significant and equal to 2.75. This result confirms the dependence between two error components and justifies the use of the copula-based stochastic frontier kink model.

Figure 2 is generated corresponding to the estimated parameters shown in Table 4, to illustrate the position of each country, especially when we separate all countries into two groups based on the kink point. We plot the data of each input, i.e. capital (top left), labour (top right), consumption (bottom left), and Gini coefficient (bottom right), against the level of GDP. Note that the data are log-transformed. The lines are estimated from the copula-based stochastic frontier kink model.

5.3   Estimate of Technical Efficiency

This section provides the efficiency estimates and clustering of 134 countries based on the efficiency score. We calculate the technical efficiency (TE) using Eq. (10) where the TE value would be between 0 and 1. TE = 1 means perfectly efficient country whereas TE = 0 means perfectly inefficient country. Therefore, the more efficiency implies the more competitiveness.

Figure 3 displays the TE values of 134 countries. The efficiency score obtained from our model (denoted by Cop-SFK) varies between 0.932 and 0.995, while the average score is approximately 0.976. About half of the observations (66 countries) can produce at efficiency level above the mean while the remainders are below the mean. Additionally, we employ the conventional evaluation method namely Data Envelopment Analysis (DEA) which is a method that uses a linear programming to calculate the technical efficiency by measuring a country's efficiency compared with a reference performance. To calculate TE from DEA, we use the R-package 'rDEA' provided by Simm et al. [14]. We found that the range of efficiency score obtained from DEA is almost the same as our model that is 0.937–1.00, while the average score is 0.977. These identical results can help prove that the TE value obtained from our model is not overestimated.

Finally, the cluster analysis is employed to classify all counties based on their competitive abilities or TE scores. More technically, we conduct a dynamic programming algorithm for optimal one-dimensional clustering provided by Wang and Song [16]. We define the k-means problem, that is, the values of TE obtained from the previous experiment are separated into k groups so that the sum of squared within-cluster distances to each group mean is minimized. In this study, we assign k = 2,

Fig. 3 Technical efficiency plot

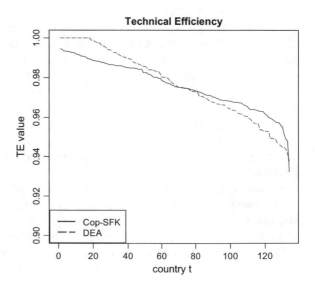

Fig. 4 Clustering of the efficiency scores

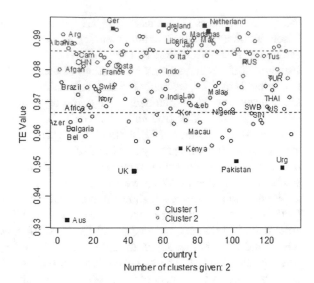

which are the first cluster for low TE value (low competitive ability) and the second cluster for high TE value (high competitive ability).

Figure 4 displays two clusters which are indicated using red and black circles with the two horizontal dashed lines representing the mean of each cluster. There are 69 countries form 134 countries are classified in the cluster 1 in which the five highest competitive countries according to our analysis are Ireland (0.9945), Madagascar (0.9942), Germany (0.9934), the Netherlands (0.9932), and Mexico (0.9925), respectively. The mean score of this cluster is 0.9861. On the other hand, the five lowest competitive countries are Australia (0.9325), UK (0.9479), Uruguay (0.9490), Pakistan (0.9551), and Kenya (0.9574), respectively, while the mean score of this cluster is 0.9665. We should notice that this clustering is based on the TE value measured by some specific factors i.e. capital, labour, consumption, and the Gini coefficient which are the basic factors reflecting a countrys economic performance. The efficiency score in this paper actually refers to the efficient use of inputs, which are those four factors, to produce the maximum possible level of output (GDP). Therefore, the lowest competitiveness here does not mean that the country is really weak; the efficiency is actually caused by various factors which should be addressed further.

6 Conclusion

Since competitive ability is important for every country relying on international trade, the competitiveness evaluation becomes a crucial problem for nations. This paper suggests considering a countrys productive efficiency in terms of TE to reflect the competitiveness. To measure the TE value, this paper introduces the copula-based nonlinear stochastic frontier model as a contribution to the competitiveness evaluation since we realize that the difference among countries such as size and structure of the economy may lead to the unequal impact of inputs on output across countries. Therefore, we apply the kink regression which is a nonlinear approach to the conventional SFM, and hence, we come up with an innovative tool called the copula-based stochastic frontier kink model.

Prior to the estimation of TE, we explore the performance of our model through the Monte Carlo simulation study and the Kullback–Leibler Divergence and we found that our model is reasonably accurate. Then, this model is applied to estimate the production function in which we found a kink effect on the relationship between GDP and capital. The result shows that capital creates a larger positive impact on GDP in the group of large countries such as US, UK, and China than the small countries. Additionally, we also found the kink effect on the relationship between labour and GDP but we failed to find the significant impact of labour on the GDP of the small countries. The variable consumption is found to have a positive impact on GDP whereas the impact of Gini coefficient is surprisingly negative. Finally, the estimate of TE value shows that the efficiency score varies between 0.932 and 0.995 where the most competitive country is Ireland with score 0.9945. It is necessary to note that these rankings and TE scores are true based on some specific factors with respect to the competitiveness definition. The result would not be the same if the data sets changed. However, this result should at least be useful for country leader, policymakers, and domestic businesses to know their competitive ability and position on the globe, to formulate the proper strategies and structural reforms to move their economy forward.

References

1. Aigner DJ, Lovell CAK, Schmidt P (1977) Formulation and estimation of stochastic frontier production function models. J Econ 6:21–37
2. Battese GE, Coelli TJ (1992) Frontier production functions, technical efficiency and panel data: with application to paddy farmers in India. J Product Anal 3:153–169
3. Card D, Lee D, Pei Z, Weber A (2012) Nonlinear policy rules and the identification and estimation of causal effects in a generalized regression kind design. NBER Working Paper 18564
4. Das A (2015) Copula-based stochastic frontier model with autocorrelated inefficiency. Cent Eur J Econ Model Econometrics 7(2):111–126
5. Furkov A, Surmanov K (2011) Stochastic frontier analysis of regional competitiveness. Metody Ilociowe w Badaniach Ekonomicznych 12(1):67–76
6. Garelli S (2003) Competitiveness of nations: the fundamentals. IMD World competitiveness yearbook, pp 702–713
7. Greene WH (2003) Simulated likelihood estimation of the normal-gamma stochastic frontier function. J Prod Anal 19:179–190
8. Hansen BE (2000) Sample splitting and threshold estimation. Econometrica 68(3):575–603
9. Hansen BE (2015) Regression kink with an unknown threshold. J Bus Econ Stat (just-accepted)

10. Lukas M, Nevima J (2011) Application of econometric panel data model for regional competitiveness evaluation of selected EU 15 countries. J Competitiveness 3(4):23–38
11. Nelsen RB (2013) An introduction to copulas, vol 139. Springer Science & Business Media, New York
12. Noh H, Ghouch AE, Bouezmarni T (2013) Copula-based regression estimation and inference. J Am Stat Assoc 108(502):676–688
13. Schwab K Sala-i-Martin X (2015) World economic forums global competitiveness report, 2014–2015. World Economic Forum, Switzerland. http://reports.weforum.org/global-competitiveness-report-2014-2015/
14. Simm J, Besstremyannaya G, Simm, MJ (2016) Package rDEA
15. Smith MD (2008) Stochastic frontier models with dependent error components. Econometrics J 11(1):172–192
16. Wang H, Song M (2011) Ckmeans.1d.dp: optimal k-means clustering in one dimension by dynamic programming. R Journal 3(2):29–33
17. Wiboonpongse A, Liu J, Sriboonchitta S, Denoeux T (2015) Modeling dependence between error components of the stochastic frontier model using copula: application to intercrop coffee production in Northern Thailand. Int J Approximate Reasoning 65:34–44

Gravity Model of Trade with Linear Quantile Mixed Models Approach

Pathairat Pastpipatkul, Petchaluck Boonyakunakorn
and Songsak Sriboonchitta

Abstract The Thai economy has mostly depended on exports, which has significantly declined in recent years. Hence, this paper is to investigate the determinants
affecting Thailands exports with its top ten trading partners by using gravity model
approach along with panel data. In panel data, there are different characteristics
between entities that account for unobserved individual effects. Previous studies
have only focused on estimating mean effect. Mixed models are relatively selected
as additional approach for panel data that accounts for individual heterogeneity in
terms of variance components. Another advantage is that they are suitable for dependent data which are likely to be similar as collected repeatedly on the same country.
We also take an interest in studying different magnitudes and directions of the effects
of determinants on different parts of the distribution of export values. Meanwhile,
Quantile regression (QR) allows study of different quantiles of the conditional distribution. In this study we combine the benefits of both mixed effects and quantile
estimator to study Thai exports and employ linear quantile mixed models (LQMMs)
with gravity model.

Keywords Mixed-models · Quantile regression · Heterogeneity · Gravity model

1 Introduction

Thai economy has been growing through exports which account for more than 60% of
Thailands GDP. However, Thailands export has declined in recent years. In January
2016, Thai exports values was 15.71 billion USD, decreasing by 8.91% compared
to the same month of the previous year. It was the lowest since November 2011
according to the statistics of the Ministry of Commerce of Thailand. The global
economy also plays a crucial role in Thai export contraction. Key trading markets
to Thailand, such as, the U.S., China, the EU, Japan as well as ASEAN have been
gaining more importance in recent years. Exports to China, as one of Thailand's

P. Pastpipatkul (✉) · P. Boonyakunakorn · S. Sriboonchitta
Faculty of Economics, Chiang Mai University, Chiang Mai, Thailand
e-mail: ppthairat@hotmail.com

© Springer International Publishing AG 2017
V. Kreinovich et al. (eds.), *Robustness in Econometrics*,
Studies in Computational Intelligence 692, DOI 10.1007/978-3-319-50742-2_34

largest trading partners, accounted for 11% of total export in 2014, followed by United States, the European Union, and Japan respectively.

In addition, there are many factors playing a vital role in Thai exports. For examples, according to Tulasombat et al. [19] exchange rate is the first factor followed by wage. If there is a depreciation in the value of a currency, it will make exports cheaper, meanwhile an increase in wages leads to higher cost in production, especially for small and medium sized industries. Thailand has consequently lost its competitive advantage of export. Moreover, import tariff can also have negative impacts on Thai exports. Thus, this paper is to study the determinants of Thailand's exports to its top ten trading partners by using gravity approach.

The majority of the empirical literatures on the gravity model are based on panel methods with fixed effects (FE) approach. FE treats heterogeneity only overall mean effects which can lead to an over-fitting of the results. Furthermore, there is dependence in panel data that violates the dependence assumption of FE model (Gibbons [10]). Mixed models are another approach for panel data which treat individual heterogeneity in terms of variance components. Moreover, Geraci and Bottai [8] proposed that random structure allows us to consider between individual heterogeneity to be correlated to explanatory variables. Therefore, mixed effects models are more flexible than FE approach.

In this study, we are also interested in studying different magnitudes and directions of the effects of determinants on different parts of the distribution. The standard approach considers only the conditional mean of the dependent variable, meanwhile Quantile regression (QR) allows study of different quantiles of the conditional distribution. It is selected and employed to investigate on how the impact of the determinants on the Thailand's exports varies with the conditional distribution of Thai exports. We combine the benefits of both mixed effects and quantile estimator in order to study Thai exports. Therefore, we relatively apply linear quantile mixed models (LQMMs) with gravity model.

To our best knowledge, this is the first gravity model applied with LQMMs. Our panel data set contains data of Thai exports and top ten largest trading partners, which account for over 70% of total trade value, based on the recorded value of the Ministry of Commerce of Thailand in 2014. These partners are China, Japan, Hong Kong, Malaysia, Australia, Vietnam, Singapore, Indonesia, the Philippines and the United States. This paper is divided into 5 sections as follows. Section 2 outlines relevant methods and methodology. Section 3 presents estimating the model. Section 4 shows the empirical results of gravity model. Conclusion will be in the final section.

2 Model and Methodology

2.1 Gravity Model

Gravity model has become one of the most popular approaches in economics in order to study the relationship of international trade. Tinbergen [18] firstly introduced the gravity model for international trade. Gravity model comes from a physics model based on the law of gravitation. This gravity law focuses on the relationship between two objects in proportional to their masses and inversely correlated to their distance, and consists of trade flows (T_{ij}) from country i to country j. It is proportional to the national income denoted by Y_i and Y_j, and inversely relative to their geographic distance (D_{ij}). Gravity model for international trade can be expressed as

$$T_{ij} = \alpha_0 Y_i^{\alpha_1} Y_j^{\alpha_2} D_{ij}^{\alpha_3}, \tag{1}$$

where $\alpha_0, \alpha_1, \alpha_2$, and α_3 are unknown parameters.

In order to simplify the application, Eq. (1) is transformed by taking the logarithmic. In the first place, estimation of gravity model with ordinary least squares (OLS) can be written as

$$\ln T_{ij} = \ln \alpha_0 + \alpha_1 \ln Y_i + \alpha_2 \ln Y_j + \alpha_3 \ln D_{ij} + \varepsilon_{ij}. \tag{2}$$

2.2 Gravity Model with Panel Data Approach

Recently, panel data has been increased in its importance in the economic field as it provides a greater number of observations that lead to more information. therefore, it improves the quality of the estimates. Gravity model is generally applied with panel data. This is a result of having different characteristics in panel data, which can generate different effects on their imports. However, we cannot observe country heterogeneity.

Gravity model has been commonly estimated with fixed effects (FE) approach. FE model treats heterogeneity only overall mean effects (Bell and Jones [2]). This can lead to weak results since FE is not appropriate for specifying heterogeneity. Cheng and Wall [4] also proposed that if the estimator is not taken into account heterogeneity, this causes misleading interpretation because of the parameter estimates being bias. Moreover, FE approach has some drawbacks, such as, losing a lot of degree of freedom (Chan et al. [3]). Furthermore, data in panel data are dependent. For instance, observations collected many times from the same country tend to be more similar in many characteristics than other observations from other different countries (Geraci and Bottai [8]). Therefore, it violates the independence assumption of FE model (Gibbons [10]).

Another way for panel data is to apply with mixed models, as the individual effects can be conceptualized as a linear combination of FE and random effects (RE) (Pinheiro [16], Richardson [17]). It treats individual heterogeneity in terms of variance components. According to (Hedeker and Mermelstein [6]), they stated that the variance components describe the structure of the correlation in panel data. It allows correlation of errors which has more flexibility in modeling the error covariance structure. Moreover, mixed models also consider data that has more complex dependence structures, as it considers the within-subject dependence of samples as a random effect.

2.3 Quantiles

Since we consider the heterogeneity and dependent data of panel data, we apply gravity model with mixed effects. Additionally, we take more interest in the left and right tails of the distribution indicating low and high values of exports. While standard regression models are linear mean regression model $Y = \theta X + \varepsilon$ where, $E(Y|X) = \theta X$. It studies the effect of the covariate X on the mean of Y. While quantiles can investigate on how X affects the small or large value of Y. Therefore, we use QR with mixed effect panel data.

2.3.1 Quantile Regression (QR) for Independent Data

Suppose that $X_i = (X_{i1}, \ldots, X_{ip})$ is a p-dimensional vector of covariates, countries at $i = 1, \ldots, n$, where $y_1, \ldots, y_n$ is a dependent variable and identically distributed random variable with conditional cumulative distribution function (CDF). $F_{y_i|x_i}$ which is assumed to be unknown. QR function can be defined as its inverse: $Q_{y_i|x_i} \equiv F_{y_i|x_i}^{-1}$, where $\tau, 0 < \tau < 1$. The linear quantile regression can be expressed as

$$Q_\tau(y_i|\beta, x_i) = x_i'\beta_\tau, \tag{3}$$

where $\beta_\tau \in R^p$ is denoted as a vector of unknown fixed parameter with length p, whereas τ is denoted for the level of quantile.

2.3.2 Quantile Regression (QR) for Panel Data

As data is collected several times on a sample of individuals across time, this invalids the independent assumption. We consider a conditional model to deal with dependent data. In order to account for the dependence between panel data the approach should be based on the inclusion in the predictor of sources of unobserved heterogeneity. For a given $\tau \in (0, 1)$, a conditional QR can be defined by

$$Q_\tau(y_{it}|b_i, \beta, x_{it}) = b_i + x_i'\beta_\tau, \tag{4}$$

which can be expressed as

$$y_{it} = b_i + x_{it}'\beta_\tau + \varepsilon_{it}, \tag{5}$$

where ε_{it} refers to an error term, $Q_\tau (\varepsilon_{it} | b_i, \beta, x_{it}) = 0$, β_τ concludes the correlation between the determinants X and the $\tau - th$ response quantile for an individual with its level that is equivalent to b_i. As the regression shares the same b_i, there is no longer independent.

According to Geraci and Bottai [9], for conditional QR, there are two approaches involved which are defined as distribution free and likelihood based methods. For the first approach, fixed individual intercepts are considered as pure location shift parameters common to all conditional quantiles. Initially, Koenker [13] proposed fixed effect quantile regression for panel data, followed by Harding and Lamarche [5] who expanded fixed effect to be more flexible for endogenous independent variables. Meanwhile, for the likelihood based methods, individual parameters are assumed to be independent and identical distributed random variables. Hence, differences can be explained in term of the response quantiles across entities (Geraci and Bottai [8]).

2.4 Mixed Models (MM)

Consider panel data in the form $(y_{ij}, z'_{ij}, x'_{ij})$, for $j = 1, .., n$, and $i = 1, \ldots, M$, where y_{ij} is the jth observation of the response variable in the ith subject x'_{ij} is a row of a known $n_i \times p$ matrix X_i, z'_{ij} is a row of a known $n_i \times q$ matrix Z_i.

A linear mixed effects function of response y_{ij} can be expressed as

$$y_{ij} = x'_{ij}\beta_\tau + z'_{ij}u_i + \varepsilon_{ij}, \tag{6}$$

where β and u_i are fixed and random effects respectively. y_i is assumed multivariate normal distribution; The random effects u_i help reduce the dependence among the observations within ith subject. It is shared by all observations within the same subject. The random effects and the within-subject errors are assumed to be independent for different subjects (Pinheiro and Bates [16]).

2.4.1 Linear Quantile Mixed Models (LQMMs) with Panel Data

At this stage, it is to assume that the $y_i = (y_{11}, \ldots, y_{1n_i})'$ for $i = 1, \ldots, M$, conditionally on a vector random effects (u_i), are independent distribution with respect to a joint asymmetric Laplace (AL) with location and scale parameters given by σ^τ and $\mu_i^{(\tau)} = x_i \theta_x^{(\tau)} + Z_i u_i$, where $\theta_x^{(\tau)} \in R^p$ is a vector of unknown fixed effects.

Suppose that $u = (u'_1, \ldots, u'_M)'$, $y = (y'_1, \ldots, y'_M)'$, $X = [X'_1], \ldots, [X'_M]'$, and $Z = \oplus_{i=1}^M Z_i$, $u^{(\tau)} = X\theta_x^{(\tau)} + Zu$, the joint density of (y, u) depends on M subjects for the LQMM is written as

$$p\left(y, u \left| \overset{(\tau)}{\underset{x}{\theta}}, \overset{(\tau)}{\sigma}, \overset{(\tau)}{\Psi} \right.\right) = p\left(y \left| \overset{(\tau)}{\underset{x}{\theta}}, \overset{(\tau)}{\sigma}, \overset{(\tau)}{\Psi} \right.\right) p\left(u \left| \overset{(\tau)}{\Psi} \right.\right) \tag{7}$$

$$= \prod_{i=1}^{M} p\left(y \left| \overset{(\tau)}{\underset{x}{\theta}}, \overset{(\tau)}{\sigma}, \overset{(\tau)}{\Psi} \right.\right) p\left(u \left| \overset{(\tau)}{\Psi} \right.\right),$$

where $\Psi^{(\tau)} \in S_{++}^{q}$, S_{++}^{q} is the set of real symmetric positive-definite $q \times q$ matrices. Later on LQMM is developed by Geraci and Bottai [8] in order to obtain ith contribution to the marginal likelihood by mixing the random effects, which is

$$L_i(\theta_x, \sigma, \Psi | y_i) = \int_{R^q} p(y_i, u_i, |\theta_x, \sigma, \Psi) du_i, \tag{8}$$

where R^q represents the q-dimension Euclidean space. The marginal log-likelihood is written as $l_i(\theta_x, \sigma, \Psi | y) = \log L_i(\theta_x, \sigma, \Psi | y), i = 1, .., M$.

2.5 Estimation

In order to estimate the interested parameter, to deal with unobserved random effects is processed by using Gaussian quadrature (Geraci and Bottai [7]). The interested integral for estimating for the marginal distribution of y_i in model (8) is expressed as

$$p_y(y_i | \theta_x, \sigma, \Psi) = \sigma_{n_i}(\tau) \int_{R^q} \exp\left\{-\frac{1}{\sigma}\rho_\tau(y_i - u_i)\right\} p(u_i | \Psi) du_i, \tag{9}$$

To choose a proper distribution for the random effects can be difficult. As robustness refers to the error model in general, at this stage it is used to refer to the random effects. Geraci and Bottai [8] suggested that apply of the symmetric Laplace is the robust alternative to the Gaussian choice. Hence, we emphasis on two types of the distribution of random effects, which are Gaussian and Laplacian. This is equivalent to apply a Gauss-Hermite and a Gauss–Laguerre quadrature to the integral in (9). The Gauss-Hermite quadrature is considered as normal random effects meanwhile, the Gauss–Laguerre quadrature is considered as robust random effects. The Gauss–Laguerre quadrature can be applied because in this each one-dimensional integral in (9). Therefore, we select normal an robust for random effects as the argument types of quadrature.

For the argument covariance, we consider three types of argument covariance. The first one is multiple of an identity (pdIdent), which has equal covariance. The second one is compound symmetry structure (pdCompSymm), that all the variances are the same and all the covariances are the same. These allow random effects to be correlated. The last one is diagonal structure (pdDiag), that variances are equal and covariances of zero assuming no correlation between the random effects (Geraci [7]). Kincaid [11] proposed that there still remains a question of how to choose the good covariance structure. This paper will be applied with the lqmm (Linear

Table 1 Summary table of the covariance structures and type of quadrature

Covariance matrix	Argument covariance	$\text{var}(u_i)$	$\text{cov}(u_l, u_l)$	m	Argument type
Multiple of an identify	pdIdent	ψ_u^2	0	1	Normal or robust
Compound symmetry	pdComp-Symm	ψ_u^2	ϕ	$1 = (q = 1)$ or $2(q>1)$	Normal
Diagonal	pdDiag	ψ_l^2	0	q	Normal or robust

Quantile Mixed Model) package in R, which is implemented by Marco Geraci [7] (Table 1).

3 Estimating the Model

3.1 Model Specification

Based on the convention of gravity model, the bilateral trade flow is determined by size of economy (measured by GDP), population, and the distance between two countries. More macroeconomic variables are added in order to provide more explanation of the exports flows between Thailand and its trading partners. These variables are exchange rate, wage, and import tariff rate. Gravity model of trade is written as

$$ln EXPORT_{ij,t} = \alpha_{ij} + \beta_1 ln GDP_{i,t} + \beta_2 ln GDP_{j,t} + \beta_3 ln POPULATION_{i,t} \quad (10)$$
$$+ \beta_4 ln POPULATION_{j,t} + \beta_5 ln EXCHANGE_{ij,t} + \beta_6 ln WAGE_{i,t}$$
$$+ \beta_7 ln DISTANCE_{ij,t} + \beta_8 ln TARIFF_{j,t} + u_{ij,t},$$

for $i = 1, \ldots, N$ where i refers to Thailand, $j = 1, \ldots, N$ where j refers to partner countries, $ij = 1, \ldots, N$ where ij refers to between two countries, $t = 1, \ldots, T$, where $EXPORT_{ij,t}$ corresponds to the value of Thai exports goods and services to top 10 trading partners at year t. The explanatory variables are defined as follows: $GDP_{i,t}$ and $GDP_{j,t}$ denote the nominal income of Thailand and partner countries respectively. It represents the economic size of country, $POPULATION_{i,t}$ and $POPULATION_{j,t}$ correspond to number of population in Thailand and partner countries at time t respectively, $EXCHANGE_{ij,t}$ denotes the exchange rates between two countries at time t, $WAGE_{i,t}$ denotes the Thai average wage at time t, $DISTANCE_{ij,t}$ is measured the distance between Thailand and partner countries, and $TARIFF_{j,t}$ corresponds to an average tax rate imposed on the import of goods applied by country j at time t.

3.2 The Covariance Structures and Type of Quadrature

We consider three types of argument covariance; pdIdent, pdCompSymm, and pdDiag with two argument types of quadrature; normal and robust (Geraci [7]). These three covariance structures are employed with two argument types together resulting in the total number of five models. The Model 1; pdIdent covariance with normal type, Model2; pdIdent covariance with robust type, Model 3; pdCompSymm covariance with normal type, Model 4; pdDiag covariance with normal type, and Model 5; pdDiag covariance with robust type. Then, the most appropriate model based on AIC criterion is selected.

3.3 Data

Panel data is applied in order to provide more meaningful information of Thailands export with data of top 10 trading partners. The period of the data is at the range of 2002–2014. These top ten trading partners selected with respect to their amounts recorded in 2014 contain China, Japan, Hong Kong, Malaysia, Australia, Vietnam, Singapore, Indonesia, the Philippines and the United States. The panel data produces 130 observations with no missing data. In addition, the data come from several sources. Thai exports data are from Foreign Trade Statistics of Thailand, whereas GDP and population data are from the World Bank database. The data of distances between Thailand to the other trading partners are obtained from the data base for the CEPII Geodist dyadic dataset (Mayer and Zignago [15]). Exchange rate and wage data are from Bank of Thailand (BOT) and tariff rates are from the World Banks WITS.

4 Empirical Results

4.1 Model Selection

In order to choose the most appropriate model for interpretation from five models, we use Akaikes information criterion (AIC). According to the types of argument, robust and normal, the results in Table 2 can be categorized into 2 groups. The first group consists of AIC model 2 and 5 with the type of robust. The second group contains the rest of the models with the type of normal. The first group illustrates lower AICs compared to the second group at all selected quantiles except 95th quantile. Moreover, the AIC difference between two groups is less than 3. Meanwhile, the second group of model 1, 3, and 4 shows the lower standard errors (SE) of Thai population and tariff rate variables at 95th quantile compared to SEs of the models in the first group. However, the SEs of the other variables in the first group are lower

Table 2 Reports the impact of each covariate on Thailands exports at different quantiles for Model 1 to Model 3

	5th	25th	50th	75th	95th
Model 1 covariance="pdIdent" type="normal"					
(Intercept)	−83.72	−83.745	−83.745	−83.745	−82.663
	(−21.710)	(−21.723)	(−21.724)	(−21.765)	(−21.85)
lnGDPi	0.297	0.385	0.691	0.385	0.27
	(−0.223)	(−0.176)	(−0.122)	(−0.19)	(−0.217)
lnGDPj	0.738	0.691	0.691	0.691	0.655
	(−0.153)	(−0.122)	(−1.918)	(−0.116)	(−0.140)
lnPOPULATIONi	9.079	9.103	9.103	9.103	9.091
	(−1.936)	(−1.912)	(−1.918)	(−1.914)	(−1.939)
lnPOPULATIONj	−0.19	−0.177	0.177	−0.177	−0.195
	(−0.152)	(−0.119)	(−0.120)	(−0.119)	(−0.156)
lnEXCHANGEij	−0.029	−0.003	−0.003	−0.003	0.012
	(−0.092)	(−0.066)	(−0.065)	(−0.065)	(−0.088)
lnWAGEi	−0.427	−0.541	−0.541	−0.541	−0.474
	(−0.247)	(−0.13)	(−0.128)	(−0.146)	(−0.241)
lnDISTANCEij	−0.762	−0.712	−0.712	0.712	0.652
	(−0.344)	(−0.349)	(−0.342)	(−0.346)	(−0.374)
lnTARIFFj	−3.439	−3.546	−3.546	−3.546	−3.17
	(−3.058)	(−3.036)	(−3.041)	(−2.943)	(−2.874)
Log-likelihood	322.3	263.3	300.3	191	271.3
AIC	11.52	19.69	−55.11	19.69	−62.78
Model2 covariance="pdIdent" type="robust"					
(Intercept)	−83.676	−83.75	−83.745	−83.745	−83.611
	(−22.508)	(−22.504)	(−22.491)	(−22.509)	(−22.481)
lnGDPi	0.444	0.385	0.385	0.385	0.264
	(−0.184)	(−0.150)	(−0.148)	(−0.143)	(−0.157)
lnGDPj	0.638	0.691	0.691	0.691	0.651
	(−0.114)	(−0.083)	(−0.084)	(−0.088)	(−0.100)
lnPOPULATIONi	8.963	9.103	9.103	9.103	9.204
	(−2.036)	(−1.989)	(−1.979)	(−1.980)	(−1.980)
lnPOPULATIONj	−0.141	−0.177	−0.177	−0.177	−0.191
	(−0.095)	(−0.070)	(−0.067)	(−0.068)	(−0.090)
lnEXCHANGEij	−0.014	−0.003	−0.003	−0.003	0.013
	(−0.054)	(−0.039)	(−0.032)	(−0.030)	(−0.041)
lnWAGEi	−0.509	−0.541	−0.541	−0.541	−0.471
	(−0.134)	(−0.118)	(−0.115)	(−0.118)	(−0.180)
lnDISTANCEij	−0.621	−0.712	−0.712	−0.712	−0.649
	(−0.196)	(−0.139)	(−0.138)	(−0.140)	(−0.161)

(continued)

Table 2 (continued)

	5th	25th	50th	75th	95th
lnTARIFFj	−3.625	−3.546	−3.546	−3.546	−3.445
	(−3.447)	(−3.274)	(−3.276)	(−3.276)	(−3.302)
Log-likelihood	390.7	346.2	348	280.6	257.6
AIC	−50.96	−57.2	−96.99	−63.72	−59.86
Model 3 covariance="pdCompSymm" type="normal"					
(Intercept)	−83.72	−83.75	−83.745	−83.745	−82.663
	(−18.695)	(−18.62)	(−18.633)	(−18.557)	(−18.596)
lnGDPi	0.297	0.385	0.385	0.385	0.27
	(−0.215)	(−0.124)	(−0.127)	(−0.128)	(−0.207)
lnGDPj	0.738	0.691	0.691	0.691	0.655
	(−0.122)	(−0.122)	(−0.117)	(−0.117)	(−0.160)
lnPOPULATIONi	9.079	9.103	9.103	9.103	9.091
	(−1.686)	(−1.678)	(−1.678)	(−1.669)	(−1.649)
lnPOPULATIONj	−0.19	−0.177	−0.177	−0.177	−0.195
	(−0.104)	(−0.097)	(−0.091)	(−0.087)	(−0.110)
lnEXCHANGEij	−0.029	−0.003	−0.003	−0.003	0.012
	(−0.099)	(−0.072)	(−0.069)	(−0.066)	(−0.084)
lnWAGEi	−0.427	−0.541	−0.541	−0.541	−0.652
	(−0.234)	(−0.138)	(−0.128)	(−0.137)	(−0.195)
lnDIATANCEij	−0.762	−0.712	−0.712	−0.712	−0.652
	(−0.230)	(−0.152)	(−0.149)	(−0.150)	(−0.195)
lnTARIFFj	−3.439	−3.546	−3.546	−3.546	−3.17
	(−2.684)	(−2.684)	(−2.687)	(−2.574)	(−3.023)
Log-likelihood	322.3	263.3	300.3	191	271.3
AIC	11.52	19.69	−55.11	19.69	−62.78

Note The numbers in the parentheses are the standard errors(SE)

than the variables in the second group. As a result, models 2 and 5 (robust type) are selected to present the relationship between Thai exports and covariates (Table 3).

4.2 Interpretation

Table 2 demonstrates coefficient estimates of all variables at theve different quantiles, 5th, 25th, 50th, 75th, and 95th quantiles, representing the different sizes of export flows. The extremely low of export flow is at 5th and followed by low, average, high export, and extremely high of export flows with 25th, 50th, 75th, and 95th respectively.

Firstly, we will consider the positive effect of the relationship between all covariates with Thai exports. The effects of Thailands GDP, partners GDPs and Thai population are positive at all different quantiles. The study of Anderson [1] also

Table 3 Reports the impact of each covariate on Thailand's exports at different quantiles for Model 4 and Model 5

	5th	25th	50th	75th	95th
Model 4 covariance="pdDiag" type="normal"					
(Intercept)	−83.720	−83.745	−83.745	−83.745	−82.663
	(−20.423)	(−18.620)	(−20.483)	(−20.579)	(−20.614)
lnGDPi	0.297	0.385	0.385	0.385	0.270
	(−0.274)	(−0.189)	(−0.211)	(−0.222)	(−0.227)
lnGDPj	0.738	0.691	0.691	0.691	0.655
	(−0.157)	(−0.120)	(−0.133)	(−0.130)	(−0.125)
lnPOPULATIONi	9.079	9.103	9.103	0.691	9.091
	(−1.812)	(−1.789)	(−1.782)	(−0.130)	(−1.811)
lnPOPULATIONj	−0.190	−0.177	−0.177	−0.177	−0.195
	(−0.151)	(−0.094)	(−0.093)	(−0.095)	(−0.108)
lnEXCHANGEij	−0.029	−0.003	−0.003	−0.003	0.012
	(−0.115)	(−0.076)	(−0.083)	(−0.071)	(−0.185)
lnWAGEi	−0.427	−0.541	−0.003	−0.541	−0.474
	(−0.254)	(−0.140)	(−0.083)	(−0.142)	(−0.220)
lnDISTANCEij	−0.762	−0.712	−0.712	−0.712	−0.652
	(−0.258)	(−0.194)	(−0.185)	(−0.178)	(−0.197)
lnTARIFFj	−3.439	−3.546	−3.546	−3.546	−3.17
	(−3.410)	(−3.423)	(−3.456)	(−3.413)	(−3.392)
Log-likelihood	322.3	263.3	300.3	191	271.3
AIC	11.52	19.69	−55.11	19.69	−62.78
Model 5 covariance="pdDiag" type="robust"					
(Intercept)	−83.676	−83.745	−83.745	−83.745	−83.611
	(−18.966)	(−18.955)	(−18.951)	(−18.951)	(−18.953)
lnGDPi	0.444	0.385	0.385	0.385	0.264
	(−0.162)	(−0.145)	(−0.127)	(−0.126)	(−0.139)
lnGDPj	0.638	0.691	0.691	0.691	0.651
	(−0.118)	(−0.095)	(−0.100)	(−0.112)	(−0.122)
lnPOPULATIONi	8.963	9.103	9.103	9.103	9.204
	(−1.682)	(−1.706)	(−1.698)	(−1.695)	(−1.692)
lnPOPULATIONj	−0.141	−0.177	−0.177	−0.177	−0.191
	(−0.094)	(−0.076)	(−7.690)	(−1.695)	(−0.083)
lnEXCHANGEij	−0.014	−0.003	−0.003	−0.003	0.013
	(−0.067)	(−0.053)	(−0.051)	(−0.053)	(−0.057)
lnWAGEi	−0.509	−0.541	−0.541	−0.541	−0.471
	(−0.145)	(−0.110)	(−0.113)	(−0.127)	(−0.184)
lnDISTANCEij	−0.621	−0.712	−0.712	−0.712	−0.649
	(−0.236)	(−0.190)	(−0.186)	(−0.194)	(−0.201)
lnTariffj	−3.625	−3.546	−3.546	−3.546	−3.445
	(−3.470)	(−3.529)	(−3.479)	(−3.484)	(−3.485)
Log-likelihood	390.7	346.2	348	280.6	257.6
AIC	−50.96	−57.20	−96.99	−63.72	−59.86

Note The numbers in the parentheses are the standard error

provided a strong support that the bilateral trade rises when the GDPs of both countries increase.

The effect of Thai GDP is stronger at the 5th quantile than the other quantiles, implying that it is a major determinant in explaining at extremely low export. In addition, the effect of low Thai GDP at the 5th quantile is significantly different, compared to the one at the 95th quantile. It indicates that the effect of Thai GDP, when the export flow is extremely low, is much higher than the one, when the export flow is extremely high. As partner GDP effect is positively stronger at the 25th, 50th, and 75th quantiles compared to the other quantiles, it implies that it is a key factor in explaining at low, average, and high exports. On the other hand, for Thai population effect, it is stronger at the 95th than at the others. This indicates that in order to explain at extremely high export Thai population is a major factor. The study of Koh [14] provided support to this result because more population tends to increase ability to create a wider variety of products which leads to have more opportunities to exports.

Secondly, we take another interest on the negative effect of the relation. The effects of partners population, Thai wage, distance, and partners tariff rates are negative at the whole selected quantiles. Meanwhile, exchange rate is the only one that is positive at extremely high quantile (95th) but negative at the other quantiles. This implies that exchange rate has a positive related to Thai export only extremely high exports.

The effect of partners population is stronger at the 95th quantile than the other quantiles, implying that it tends to be a major factor in explaining at extremely high export. At low, average, high Thai exports (25th, 50th, and 75th quantiles respectively) it shows that Thai wage has the higher effects than at extremely low and high exports. It implies that Thai wage is a key element in explaining at low and average, high exports. Partners tariff effects are stronger at the 5th than at other selected quantiles. This indicates that the impacts of tariff rates peak at very low exports and then begin to steadily decline when the exports increase, therefore partners tariff is a key factor to explain at extremely low export.

Overall, Thai population also has the positive highest effect at extremely high trade flow among other determinants. It implies that Thai exports are mostly influenced by Thai population at extremely high exports. Moreover, the effect of Thai population, when the export flow is extremely high, is much higher than the one, when the export flow is extremely low. For extremely exports, the partners tariff tends to the major determinants compared to the others as it has the highest negative impact on Thai exports. A possible explanation is that the partners tariff produces a negative effect to exports due to losing competitive advantages in price as demonstrated by Koczan and Plekhanov [12].

5 Conclusion

The objective of this paper is to study the determinants of Thailands exports to its trading partners by using gravity model. Linear quantile mixed models (LQMMs) is relatively employed because of the heterogeneous individuals, and complex dependence structures. To be more explainable for Thailands exports, more determinants, exchange rate, wage, and import tariff rate are added into the gravity model.

The models with the robust argument type are selected to present the results, as they provide the lower AICs compared to normal one at all selected quantiles except at 95th quantile. Thailands GDP, partners GDPs, and Thai population have positive impacts on Thai exports. In contrast, partners population, Thai wage, distance, and partners tariff rates have negative impacts on Thai exports. In the meantime, exchange rate effects are negative related to exports except at 95th quantile.

Considering at the extremely low Thai exports, the positive effect of Thai GDP on the exports appears the highest and begins to decline while the exports increase. At the same time, the negative effects of exchange rate and tariff rate peak at extremely low Thai exports and their effects are weaker as Thailands exports increase.

Moreover, at extremely high Thai exports, the effect of Thai population and partners population are stronger than at lower value of exports. It implies that the impact of them decreases as Thailands exports decrease. At low, average, and high Thai exports, partners GDP, Thai wage, and distance have the higher effects than at extremely low and high exports. Overall, Thai population has the highest impact on Thai exports compared to partners tariff which has the highest negative impact on exports.

Policy suggestions based on the empirical results are given as follows; the major issue places emphasis on tariff rate, which can be contributed by more cooperation between Thailand and major trading partners in order to relax tariff rates. In addition, exchange rate requires an appropriate maintenance at competitive levels, especially when Thai exports are very low. Meanwhile, Thai GDP is also important at the very low level of export because it can increase significant supply to boost export. Wage is another determinant that requires special attention when Thai exports are relatively low.

Acknowledgements The authors are very grateful to Professor Vladik Kreinovich and Professor Hung T.Nguyen for their comments. The authors wish to thank the Puey Ungphakorn Centre of Excellence in Econometrics.

References

1. Anderson JE (2014) Trade, Size, and Frictions: the Gravity Model. Mimeo Boston College
2. Bell A, Jones K (2015) Explaining fixed effects: Random effects modelling of time-series cross-sectional and panel data. Polit Sci Res Methods 3(01):133–153
3. Chan F, Harris MN, Greene, WH, Knya L (2014) Gravity models of trade: unobserved heterogeneity and endogeneity
4. Cheng IH, Wall HJ (2001) Controlling for heterogeneity in gravity models of trade. Federal Reserve Bank of St. Louis, Research Department
5. Harding M, Lamarche C (2009) A quantile regression approach for estimating panel data models using instrumental variables. Econ Lett 104(3):133–135
6. Hedeker D, Mermelstein RJ, (2012) Mixedeffects regression models with heterogeneous variance. Taylor and Francis

7. Geraci M (2014) Linear quantile mixed models: the LQMM package for Laplace quantile regression. J Stat Softw 57(13):1–29
8. Geraci M, Bottai M (2007) Quantile regression for panel data using the asymmetric Laplace distribution. Biostatistics 8(1):140–154
9. Geraci M, Bottai M (2014) Linear quantile mixed models. Stat Comput 24(3):461–479
10. Gibbons RD, Hedeker D, DuToit S (2010) Advances in analysis of longitudinal data. Ann Rev Clin Psychol 6:79
11. Kincaid C (2005) Guidelines for selecting the covariance structure in mixed model analysis, paper 198-30. In: Proceedings of the Thirtieth Annual SAS Users Group Conference, Inc., Portage, Michigan
12. Koczan Z, Plekhanov A (2013) How Important are Non-tariff Barriers? Complementarity of Infrastructure and Institutions of Trading Partners, European Bank for Reconstruction and Development
13. Koenker R (2004) Quantile regression for longitudinal data. J Multivar Anal 91(1):74–89
14. Koh W (2013) Brunei darussalams trade potential and ASEAN economic integration: a gravity model approach. SE Asian J Econ 1:67–89
15. Mayer T, Zignago S (2011) Notes on CEPIIs distances measures: The GeoDist database
16. Pinheiro JC, (2005) Linear mixed effects models for longitudinal data. Encycl Biostat
17. Richardson TE, (2013) Treatment heterogeneity and potential outcomes in linear mixed effects models. Doctoral dissertation, Kansas State University
18. Tinbergen J, (1962) Shaping the world economy; suggestions for an international economic policy. Books (Jan Tinbergen)
19. Tulasombat S, Bunchapattanasakda C, Ratanakomut S (2015) The effect of exchange rates on agricultural goods for export: a case of Thailand. Inf Manage Bus Rev 7(1):1

Stochastic Frontier Model in Financial Econometrics: A Copula-Based Approach

P. Tibprasorn, K. Autchariyapanitkul and S. Sriboonchitta

Abstract This study applies the principle of stochastic frontier model (SFM) to calculate the optimal frontier of the stock prices in a stock market. We use copula to measure dependence between the error terms in SFM by examining several stocks in Down Jones industrial. The results show that our modified stochastic frontier model is more applicable for financial econometrics. Finally, we use AIC for model selection.

Keywords Copulas · Gaussian quadrature rule · Optimization method · Price efficiency · Stochastic frontier model

1 Introduction

The production frontier is originally used in production analysis to estimate the efficient frontier. The idea behind this method is that the observed production of a single unit cannot exceed the production frontier. Such that, we can define the most efficient production process as set. According to Koopmans [1, 2] and Färe and Shawna [3], the production set is explained by a set of inputs used to generate an output which is described as the set of (x, y) given the following

$$\Psi = \{(x, y)|x \rightarrow y\} \tag{1}$$

P. Tibprasorn · S. Sriboonchitta
Faculty of Economics, Chiang Mai University, Chiang Mai 52000, Thailand
e-mail: phachongchit_t@cmu.ac.th

S. Sriboonchitta
e-mail: songsakecon@gmail.com

K. Autchariyapanitkul (✉)
Faculty of Economics, Maejo University, Chiang Mai 50290, Thailand
e-mail: Kittawit_a@mju.ac.th

© Springer International Publishing AG 2017
V. Kreinovich et al. (eds.), *Robustness in Econometrics*,
Studies in Computational Intelligence 692, DOI 10.1007/978-3-319-50742-2_35

where Ψ is the production frontier, y is an output from using inputs x. The efficient output function can be written as $\partial Y(x) = \max Y(x)$ where $Y(x)$ is a function of output Y in term of inputs x. Different other assumptions can be imposed on Ψ (see, Aragon et al. [4]).

There are two different methods have been developed in this area: the first one is the classical frontier model; given this model, all the data points in X_n are under Ψ. The second is the stochastic frontier model where random error allows some data points to be out of Ψ.

There are many studies on stochastic frontier model (SFM) and the reader can refer to the studies of Azadeh et al. [5], Greene [6, 7], Filippini et al. [8], Stevenson [9] and Wang et al. [10]. In their studies, they specified a model with a restrictive distributional structure for the error term, i.e. the non-negative error term. Generally, the distance between the frontier curve and the observation indicates an inefficiency. Then, the distance is given by the error components (see, Tibprasorn et al. [13], Kuosmanen et al. [11] and Sanzidur et al. [12]). The technique assumes that the non-negative error term and symmetric error term are i.i.d. Moreover, the non-negative error term is forced to be a half-normal distribution whilst a normal distribution is imposed for the symmetric error term.

The main concept of this paper is to estimate the efficient frontier of stock price which exhibits the highest possible price a stock can reach. It is noted that the present prices under the frontier curve show the inefficiency of the price in stock market. We can use the result as a strategy to manage the stock in the portfolio. Moreover, we can show the significance of a historical price and volume of the stock related to predicted stock prices. We indicate that the modified stochastic frontier model is more reasonable for financial analysis.

In applying the technical efficiency (TE) in financial context, Tibprasorn et al. [13] estimated the TE of several stocks in the Stock Exchange of Thailand (SET). In their study, they assumed normal and half-normal distributions in the model. Hasan et al. [14] used a stochastic frontier model to estimate the TE of Dhaka Stock Exchange and again, they assumed independence for the error components. Thus, we relaxed this assumption using copula to estimate the dependency between the error components (see, Tibprasorn et al. [13], Carta and Steel [15], Lai and Cliff [16] and Wiboonpongse et al. [17]). For the setting of the error components, we assumed an exponential distribution for the non-negative error term while a normal distribution is assumed for the symmetric error term.

We used the likelihood function to obtain the unknown parameters (see, Burns [18] and Green [19]). Additionally, in this study, we approximated an integral by using the *Gaussian quadrature rule*, and then, the optimization method was used to get these parameters.

The paper unfolds as follows. Section 2 lays out the concept of copula and stochastic frontier model with correlated errors. Section 3 presents the implementation to stock market. Section 4 shows the empirical results, and the final section draws the conclusion.

2 Copula and Stochastic Frontier Model

2.1 Copula

Suppose X and Y are two real valued random variables. The joint distribution of (X, Y) is denoted as, for $x, y \in \mathbb{R}$,

$$H(x, y) = p(X \leqslant x, Y \leqslant y) \tag{2}$$

The marginal distributions of the components X, Y are denoted, respectively, as

$$F(x) = P(X \leqslant x) = \lim_{y \to \infty} H(x, y) = H(x, \infty) \tag{3a}$$

$$G(x) = P(Y \leqslant y) = \lim_{y \to \infty} H(x, y) = H(\infty, y) \tag{3b}$$

Thus, the marginals can be obtained from the joint. Since F and G are non-decreasing functions and continuous, the random variables $F(X) = U_1, G(Y) = U_2$ are uniformly distributed on the unit interval $[0, 1]$. Thus,

$$H(x, y) = P(U_1 \leqslant F(x), U_2 \leqslant G(y)) \tag{4}$$

Then, the joint distribution of (U_1, U_2) is computed at the values $F(x), G(y) \in [0, 1]$. Thus, if we denote the joint distribution of (U_1, U_2) as $C(U_1, U_2)$, then

$$H(x, y) = C(F(x), G(y)) \tag{5}$$

In the above equation, the C is a copula function of two marginals $F(x)$, and $G(y)$. Let H be a bi-variate distribution with continuous marginals F and G. Then, the unique copula C is determined by

$$C(u_1, u_2) = H(F^{-1}(u_1), G^{-1}(u_2)) \tag{6}$$

where $F^{-1}(u_1), G^{-1}(u_2)$ are quantile functions defined as $F^{-1} : [0, 1] \to \mathbb{R}$ where $F^{-1}(\alpha) = inf\{x \in \mathbb{R} : F(x) \geqslant \alpha\}$ (in fact, the infimum here is a minimum).

2.2 Stochastic Frontier Model with Correlated Errors

The classical SFM is given as

$$Y = f(X; \beta) + (W - U), \tag{7}$$

where Y is an output given set of inputs X; β is the set of parameters; W represents the symmetric error term and U represents the non-negative error term. W and U are assumed to be uncorrelated. Following the above equation, the technical efficiency (TE) can be written as

$$TE = \exp(-U). \tag{8}$$

In this study, we modified the classical SFM to allow the error terms to be correlated, and we assumed an exponential distribution for the error term U while a normal distribution is assumed for the error term W. Thus, by Sklar's Theorem, the joint cumulative distribution function (cdf) of (U, W) is

$$H(u, w) = Pr(U \leqslant u, W \leqslant w) \tag{9a}$$
$$= C_\theta(F_U(u), F_W(w)), \tag{9b}$$

where $C_\theta(\cdot, \cdot)$ is denoted as the bi-variate copula with unknown parameter θ. Following Smith [20], transforming (U, W) to (U, ξ) as the probability density function (pdf) of (U, ξ) then we get

$$h(u, \varepsilon) = f_U(u)f_W(u + \varepsilon)c_\theta(F_U(u), F_W(u + \varepsilon)), \tag{10}$$

where $f_U(u)$ and $f_W(w)$ are the marginal density of $H(u, w)$, the composite error $\xi = \varepsilon(-\infty < \varepsilon < \infty), w = u + \varepsilon$ and $c_\theta(\cdot, \cdot)$ is the copula density of $C_\theta(\cdot, \cdot)$. Thus, the pdf of ε is obtained by

$$h_\theta(\varepsilon) = \int_0^\infty h(u, \varepsilon)du \tag{11a}$$

$$= \int_0^\infty f_U(u)f_W(u + \varepsilon)c_\theta(F_U(u), F_W(u + \varepsilon))du. \tag{11b}$$

Smith [20] argued that it is very difficult to find the closed-form expression for finding the pdf of ε because there are very few densities of ε for estimating the maximum likelihood. Thus, we employ the Gaussian quadrature rule mentioned in Tibprasorn et al. [13] to approximate integral of $h_\theta(\varepsilon)$. According to the Gaussian quadrature rule, this technique is designed to yield an exact result by a suitable choice of the points r_j and weight s_j for $j = 1, ..., J$ where J is the number of nodes for approximating. Let $u = 0.5 + r/(1 - r)$, where $r \in (-1, 1)$, we get

$$\int_{-1}^1 f_U(a)f_W(a + \varepsilon)c_\theta(F_U(a), F_W((a + \varepsilon)))dr = \int_{-1}^1 g(r, \varepsilon)dr, \tag{12}$$

where $a = 0.5 + r/(1 - r)$. Thus, the pdf of ε approximated by the Gaussian quadrature rule can be obtained by

$$h_\theta(\varepsilon) = \int_{-1}^{1} g(r, \varepsilon)dr \approx \sum_{j=1}^{J} s_j g(r_j, \varepsilon) \tag{13}$$

The likelihood function for copula-based stochastic frontier model is represented by

$$L(\beta, \sigma_w, \lambda, \theta) = \prod_{i=1}^{N} h_\theta(y_i - x_i'\beta) = \prod_{i=1}^{N} h_\theta(\varepsilon_i), \tag{14}$$

where σ_w and λ are the scale parameters of marginal distribution of W and U, θ is the parameter of copula and $i = 1, ..., N$ is the number of observations.

Taking a natural logarithm, the log-likelihood function becomes

$$\ln L(\beta, \sigma_w, \lambda, \theta) = \sum_{i=1}^{N} \ln h_\theta(\varepsilon_i) \approx \sum_{i=1}^{N} \ln \sum_{j=1}^{J} s_j g(r_j, \varepsilon_i). \tag{15}$$

Following Battese and Coelli [21], the technical efficiency of each copula (TE_θ) can be computed by

$$TE_\theta = E[\exp(-U)|\varepsilon] \tag{16a}$$

$$= \frac{1}{h_\theta(\varepsilon)} \int_{0}^{\infty} \exp(-u)h(u, \varepsilon)du \tag{16b}$$

$$= \frac{\int_{0}^{\infty} \exp(-u) f_U(u) f_W(u + \varepsilon) c_\theta(F_U(u), F_W(u + \varepsilon))du}{\int_{0}^{\infty} f_U(u) f_W(u + \varepsilon) c_\theta(F_U(u), F_W(u + \varepsilon))du}. \tag{16c}$$

and using Monte Carlo integration. Thus, we obtain

$$TE_\theta = \frac{\sum_{i=1}^{N} \exp(-u_i) f_U(u_i) f_W(u_i + \varepsilon_i) c_\theta(F_U(u_i), F_W(u_i + \varepsilon_i))du}{\sum_{i=1}^{N} f_U(u_i) f_W(u_i + \varepsilon_i) c_\theta(F_U(u_i), F_W(u_i + \varepsilon_i))du}. \tag{17}$$

Note that, in this paper, u_i follows the cumulative distribution of U assumed to be exponentially distributed. Since $U \geqslant 0$, the density function is given by

$$f_U(u; \lambda) = \lambda e^{-\lambda U}, U \geq 0, \lambda > 0. \tag{18}$$

And w_i follows the distribution of W assumed to be distributed as normal. Then, the density function follows

$$f_W(w; \sigma_w) = \frac{1}{\sqrt{2\pi \sigma_w^2}} \exp \left\{ -\frac{w^2}{2\sigma_w^2} \right\}. \tag{19}$$

3 Model Implementation

This paper proposes the technical inefficiency to captures the deviation of actual price from the efficient frontier a stock can reach by applying the stochastic frontier approach with the form of the compound interest. Given Eq. (7), the modification of SFM can be shown as:

$$P_{i,t} = f(P_{i,t-i}; V_{i,t-i}, \beta_0, \beta_1, \beta_2) + W_{i,t} - U_{i,t} \tag{20a}$$

$$P_{i,t} = P_{i,t-1} \exp \left\{ \beta_0 + \beta_1 ln\left(\frac{P_{i,t-1}}{P_{i,t-2}}\right) + \beta_2 \ln\left(\frac{V_{i,t-1}}{V_{i,t-2}}\right) \right\} + W_{i,t} - U_{i,t} \tag{20b}$$

$$\ln\left(\frac{P_{i,t}}{P_{i,t-1}}\right) = \beta_0 + \beta_1 \ln\left(\frac{P_{i,t-1}}{P_{i,t-2}}\right) + \beta_2 \ln\left(\frac{V_{i,t-1}}{V_{i,t-2}}\right) + W_{i,t} - U_{i,t}, \tag{20c}$$

where $i = 1, 2, ..., N$; $P_{i,t}$ represents the adjusted closing price of each stock at week t; $V_{i,t}$ exhibits the actual trade volume of each stock at week t; $W_{i,t}$ is the stochastic noise capturing measurement error of each stock at week t and $U_{i,t}$ is non-negative error capturing the inefficiency of each stock at week t. The value of inefficiency is defined in terms of price inefficiency which explains the failure of actual stock price to achieve the efficient price due to some uncontrollable events, and it shows the level of efficiency that stock price can reflect all information in market.

Copulas used in this study include Gumbel, Gaussian, Clayton, t and Frank copulas. We obtained all parameters using a likelihood function, and the Gaussian Quadrature algorithm is applied for the maximization process. Therefore, the price efficiency of each stock at week t for each copula $((PE_\theta)_{i,t})$ can be written as

$$(PE_\theta)_{i,t} = E[\exp(-U_{i,t})|\varepsilon]. \tag{21}$$

4 Empirical Results

We examined several stocks in Dow Jones industrial, namely, Microsoft Corporation (MSFT), Wal-Mart Stores Inc. (WMT), GNC Holdings Inc. (GNC), Helix Energy Solutions Group Inc. (HLX), Fluidigm Corporation (FLDM) and SciQuest Inc. (SQI). All the weekly observations are obtained from 2011 until 2015. Tables 1, 2, 3, 4, 5, 6 and 7 shows descriptive statistics of the variables.

Based on the AIC, the best choices of copula in this study are the one based on the Frank 2 copula with $\theta < 0$ for the case of SQI, and Clayton copula for the cases

Table 1 Descriptive statistics

	MSFT		WMT		FLDM	
	Price	Volume	Price	Volume	Price	Volume
No. Obs.	252	252	252	252	246	246
mean	0.0031	−0.0027	0.0008	−0.0008	−0.0011	−0.0006
SD	0.001	0.1033	0.0005	0.1067	0.0057	0.3164
max	0.1399	1.0901	0.0518	1.6152	0.306	1.6522
min	−0.1551	−0.9763	−0.1244	−1.2165	−0.4439	−2.5587
skewness	−0.2898	0.3022	−1.0931	0.6056	−1.0299	0.1617
kurtosis	7.044	3.9957	6.4971	6.4219	9.9282	4.698
	SQI		HLX		GNC	
	Price	Volume	Price	Volume	Price	Volume
No. Obs.	252	252	252	252	240	240
mean	−0.0003	0.0069	−0.0024	−0.0017	0.003	−0.0066
SD	0.0034	0.4482	0.0045	0.1243	0.0024	0.3061
max	0.1806	3.7777	0.1887	1.1437	0.1396	1.5519
min	−0.2395	−1.9605	−0.2643	−0.8392	−0.2059	−2.8595
skewness	−0.358	0.754	−0.6419	0.4744	−0.4107	−0.2001
kurtosis	4.9116	6.9486	5.4207	3.428	4.7745	5.5409

*All values are the growth rate of price and volume

Table 2 MFST's parameters estimation

Parameter	IID	Frank 1	Frank 2	Gaussian	T cop	Gumbel	Clayton
β_0	0.0201	0.0201	0.0361	0.0359	0.0395	0.0461	0.0371
	(0.0036)	(274.7127)	(0.0061)	(0.0068)	(0.0150)	(0.0071)	(0.0090)
β_1	−0.1095	0.1095	0.0950	0.1047	−0.0934	−0.1090	−0.0916
	(0.0607)	(32.2757)	(0.0613)	(0.0603)	(1.2760)	(0.0630)	(0.0484)
β_2	−0.0049	−0.0049	−0.0026	−0.0046	−0.0020	−0.0040	0.0012
	(0.0058)	(230.8108)	(0.0053)	(0.0140)	(0.0061)	(0.0057)	(0.0057)
σ_V	0.0267	0.0267	0.0408	0.0351	0.0408	0.0425	0.0428
	(0.0018)	(0.2824)	(0.0091)	(0.0032)	(0.0307)	(0.0071)	(0.0019)
λ	59.9791	59.9885	29.4483	30.8432	26.7258	22.8419	29.6819
	(12.2448)	(68.7530)	(5.7384)	(3.0281)	(11.0441)	(3.5626)	(5.4195)
θ		−0.0001	7.6739	0.6291	0.7623	2.2565	5.0156
		(0.0980)	(3.7548)	(0.0811)	(0.7757)	(0.5791)	(0.6929)
df(T cop)					2.0000		
					(3.1666)		
Lo gL	512.61	512.61	523.28	514.68	525.01	516.13	526.92
AIC	−1,015.22	−1,013.22	−1,034.57	−1,017.36	−1,036.02	−1,020.27	−1,041.84

Table 3 WMT's parameters estimation

Parameter	IID	Frank 1	Frank 2	Gaussian	T cop	Gumbel	Clayton
β_0	0.0175	0.0175	0.0187	0.0339	0.0338	0.0315	0.0190
	(0.0043)	(0.0033)	(0.0032)	(0.0448)	(1.4783)	(0.4218)	(0.0100)
β_1	−0.0055	−0.0055	−0.0079	−0.0054	−0.0054	−0.0140	−0.0045
	(1.5631)	(0.1596)	(0.6139)	(0.0553)	(0.1209)	(0.0625)	(0.5402)
β_2	−0.0032	−0.0032	−0.0036	−0.0034	−0.0034	−0.0041	−0.0034
	(0.0234)	(0.0315)	(0.1821)	(0.0038)	(0.1705)	(0.0320)	(0.0316)
σ_V	0.0152	0.0152	0.0160	0.0286	0.0286	0.0240	0.0158
	(0.0058)	(0.0011)	(0.0589)	(0.0559)	(0.0948)	(0.0026)	(0.0105)
λ	59.7189	59.7225	55.5525	30.1561	30.1736	32.4815	54.7416
	(5.6130)	(7.8743)	(27.1714)	(40.9690)	(7.0163)	(3.4267)	(9.5915)
θ		−0.0006	0.7841	0.8090	0.8086	1.9767	0.2335
		(0.0086)	(3.3267)	(0.9055)	(0.2646)	(0.2692)	(0.2298)
df(T cop)					4929.46		
					(6.9264)		
Lo gL	605.71	605.71	605.81	606.33	606.33	606.07	605.92
AIC	−1,201.41	−1,199.41	−1,199.61	−1,200.65	−1,198.65	−1,200.15	−1,199.84

Table 4 GNC's parameters estimation

Parameter	IID	Frank 1	Frank 2	Gaussian	T cop	Gumbel	Clayton
β_0	0.0308	0.0308	0.0493	0.0556	0.0403	0.0621	0.0542
	(0.0076)	(4.4000)	(0.0088)	(0.0194)	(0.0055)	(0.0814)	(0.0206)
β_1	−0.1040	−0.1040	−0.1158	−0.1118	−0.1242	−0.1097	−0.1084
	(0.3865)	(7.8969)	(0.3259)	(0.0684)	(0.1094)	(0.5189)	(0.0317)
β_2	0.0048	0.0048	0.0075	0.0051	0.0073	0.0064	0.0057
	(0.0528)	(0.9897)	(0.0222)	(0.0051)	(0.0066)	(0.0189)	(0.0436)
σ_V	0.0396	0.0396	0.0551	0.0539	0.0480	0.0577	0.0578
	(0.0032)	(0.0631)	(0.0219)	(0.0281)	(0.0068)	(0.0171)	(0.0220)
λ	36.1204	36.1195	21.3638	19.0663	26.3939	16.7120	19.6570
	(20.6323)	(189.7732)	(4.1584)	(5.6967)	(6.4038)	(25.9514)	(1.2023)
θ		−0.0001	5.3075	0.6483	0.4169	1.9019	2.8334
		(0.3279)	(1.9538)	(0.2121)	(0.1542)	(2.1392)	(0.3374)
df(T cop)					2.0001		
					(0.7608)		
Lo gL	387.11	387.11	391.28	388.48	391.92	388.44	391.76
AIC	−764.22	−762.22	−770.57	−764.97	−769.85	−764.88	−771.51

Table 5 HLX's parameters estimation

Parameter	IID	Frank 1	Frank 2	Gaussian	T cop	Gumbel	Clayton
β_0	0.0384	0.0384	0.0715	0.0733	0.0665	0.0908	0.0677
	(0.0046)	(0.0494)	(0.0145)	(0.0092)	(0.1731)	(0.3361)	(0.0073)
β_1	−0.0592	−0.0592	−0.0383	−0.0549	−0.0328	−0.0522	−0.0305
	(0.0784)	(0.2637)	(0.0582)	(0.1641)	(2.2720)	(0.0839)	(0.0620)
β_2	−0.0195	−0.0195	−0.0252	−0.0195	−0.0248	−0.0215	−0.0217
	(0.0121)	(0.1895)	(0.0108)	(0.0113)	(0.1416)	(0.0328)	(0.0102)
σ_V	0.0500	0.0500	0.0769	0.0690	0.0635	0.0821	0.0644
	(0.0032)	(0.0363)	(0.0154)	(0.0091)	(0.0075)	(0.0210)	(0.0055)
λ	24.3222	24.3224	13.2318	13.1631	14.2751	10.5283	14.1912
	(1.9554)	(8.6335)	(2.8112)	(1.6205)	(50.1502)	(1.8228)	(1.4151)
θ		−0.0001	6.7352	0.6603	0.5803	2.2871	1.9660
		(0.0048)	(2.5802)	(0.0851)	(1.6144)	(0.7162)	(0.2380)
df(T cop)					2.0001		
					(0.4279)		
Lo gL	334.53	334.53	342.40	336.52	343.68	337.71	342.76
AIC	−659.07	−657.07	−672.81	−661.03	−673.35	−663.41	−673.51

Table 6 FLDM's parameters estimation

Parameter	IID	Frank 1	Frank 2	Gaussian	T cop	Gumbel	Clayton
β_0	0.0434	0.0434	0.0831	0.0863	0.0856	0.1063	0.0828
	(0.0850)	(0.0055)	(0.0125)	(0.0131)	(9.5038)	(0.0123)	(0.0040)
β_1	0.0131	0.0131	0.0415	0.0146	0.0501	0.0209	0.0400
	(0.0451)	(0.0360)	(0.0496)	(0.0462)	(1.1175)	(0.0506)	(0.0378)
β_2	−0.0127	−0.0127	−0.0085	−0.0122	−0.0086	−0.0101	−0.0107
	(0.1002)	(0.0129)	(0.0068)	(0.0061)	(24.0378)	(0.0072)	(0.0076)
σ_V	0.0555	0.0555	0.0923	0.0819	0.0845	0.0957	0.0862
	(0.0070)	(0.0063)	(0.0056)	(0.0104)	(8.4114)	(0.0089)	(0.0045)
λ	22.4195	22.4199	11.5208	11.4629	11.3324	9.1053	11.9070
	(2.4534)	(0.1199)	(1.8978)	(1.5219)	(944.8902)	(1.1263)	(0.6049)
θ		−0.0001	8.3491	0.7133	0.7505	2.5149	3.7889
		(0.0330)	(0.6944)	(0.0665)	(10.6989)	(0.3301)	(0.0401)
df(T cop)					2.0000		
					(140.16)		
Lo gL	302.40	302.40	315.49	305.90	317.41	308.11	318.19
AIC	−594.81	−592.81	−618.98	−599.81	−620.82	−604.23	−624.39

Table 7 SQI's parameters estimation

Parameter	IID	Frank 1	Frank 2	Gaussian	T cop	Gumbel	Clayton
β_0	0.0333	0.0333	0.0601	0.0617	0.0533	0.0752	0.0596
	(0.0069)	(2.5076)	(2.6309)	(1.5965)	(0.0580)	(0.0112)	(0.3010)
β_1	−0.0982	−0.0981	−0.0505	−0.1015	−0.0559	−0.0872	−0.0711
	(0.0288)	(0.9401)	(5.1181)	(0.2816)	(0.1712)	(0.6299)	(0.2176)
β_2	−0.0008	−0.0008	0.0000	−0.0005	0.0006	−0.0004	0.0009
	(0.0247)	(0.5601)	(1.2231)	(0.4138)	(0.0152)	(0.0124)	(0.0120)
σ_V	0.0466	0.0466	0.0727	0.0619	0.0589	0.0728	0.0619
	(0.0047)	(0.0117)	(2.5735)	(0.0123)	(0.0373)	(0.0327)	(0.1476)
λ	29.4236	29.4242	16.2544	16.0476	18.2917	13.0092	16.6509
	(2.2857)	(5.5058)	(33.5619)	(5.9284)	(19.8508)	(2.3613)	(2.5362)
θ		−0.0001	6.9277	0.6239	0.5238	2.0899	2.1066
		(0.0027)	(49.5111)	(0.1906)	(1.5295)	(0.3856)	(0.5823)
df(T cop)					2.0000		
					(4.5061)		
Lo gL	362.60	362.60	367.85	363.73	368.32	364.50	367.69
AIC	−715.20	−713.20	−723.70	−715.45	−722.64	−716.99	−723.37

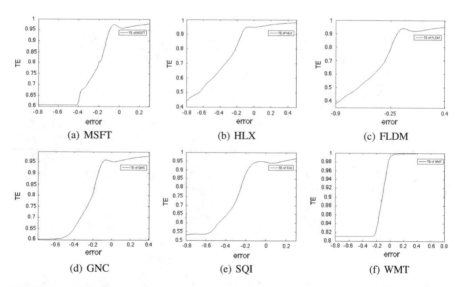

Fig. 1 Price efficiencies for Clayton copula **a, b, c, d,** Flank copula with $\theta < 0$ **e** and the independent copula **f** based models

of MFST, HLX, FLDM and GNC. The independence copula (IID) is more suitable for WMT whereas Frank1 copula with $\theta > 0$ is the worst result for MFST, HLX, FLDM, GNC and SQI. T-copula performs the worst result for WMT. Figure 1 presents the price efficiencies using the best models based on Frank 2 copula with $\theta < 0$, Clayton and the independence copula based model. The price efficiency range will vary depending on individual stock. According to the results, the price efficiencies are 0.60–0.99 (average 0.80), 0.45–0.99 (average 0.70), 0.49–0.95 (average 0.70), 0.6–0.99 (average 0.80), 0.54–0.96 (average 0.75) and 0.54–0.96 (average 0.75) for MFST, HLX, FLDM, GNC, SQI and WMT, respectively. This finding can be used strategically to choose stock into the portfolio. We can conclude that increasing in positive dependence between the error components can increase more the price efficiency. On the contrary, the large negative value between the error components can reduce the price efficiency.

5 Conclusions

This paper modified the stochastic frontier model that is usually used for production efficiency analysis in agricultural science problems. Its additions to the conventional SFM include (1) the substitution of PE equation in a production function with the compound interest equation for financial analysis. The strong assumption of classical SFM is relaxed by using copula approach to present the dependence structure of error components, w and u. (2) the use of Gaussian quadrature rule to approximate the probability distribution function (pdf) of composite error. (3) the model selection by using AIC and that it is also used to show the dependence between the random variable and inefficiency of stock price. The copulas such as t, Gaussian, Gumbel, Frank, and Clayton are applied to combine with a stochastic frontier model. In this study, the marginal distributions of the error components were assigned to be normal and exponential, but different distributions could be examined as well, and checked jointly with the copulas.

Finally, we investigated the prices of several stocks in Dow Jones industrial market by using SFM and copula. The choice of copulas was chosen using the AIC, and the results indicated Frank 2 copula with $\theta < 0$ for the case of SQI, Clayton copula for the cases of MFST, HLX, FLDM and GNC. The independence copula (iid) is more suitable for WMT whereas Frank1 copula with $\theta > 0$ is the worst result for MFST, HLX, FLDM, GNC and SQI. T-copula performs the worst result for WMT. We considered the prices of the selected stocks to be not underestimated or overestimated in terms of pricing efficiency.

References

1. Koopmans D, Tjalling C (1951) Activity analysis of production and allocation, 13th edn. Wiley, New York
2. Koopmans D, (1951) Stochastic non-parametric frontier analysis in measuring technical efficiency: a case study of the north American dairy industry
3. Färe R, Shawna G (2004) Modeling undesirable factors in efficiency evaluation: comment. Eur J Oper Res 157(1):242–245

4. Aragon Y, Abdelaati D, Christine TA (2005) Nonparametric frontier estimation: a conditional quantile-based approach. Econom Theory 21(2):358–389
5. Azadeh A, Asadzadeh SM, Saberi M, Nadimi V, Tajvidi A, Sheikalishahi M (2011) A neuro-fuzzy-stochastic frontier analysis approach for long-term natural gas consumption forecasting and behavior analysis: the cases of Bahrain, Saudi Arabia, Syria, and UAE. Appl Energy 88(11):3850–3859
6. Greene W (2005) Reconsidering heterogeneity in panel data estimators of the stochastic frontier model. J Econom 126(2):269–303
7. Greene W (2005) Fixed and random effects in stochastic frontier models. J Prod Anal 23(1):7–32
8. Filippini M, Hunt LC (2011) Energy demand and energy efficiency in the OECD countries: a stochastic demand frontier approach. Energy J 32(2):59–80
9. Stevenson RE (1980) Likelihood functions for generalised stochastic frontier estimation. J Econom 13:57–66
10. Wang H-J, Ho C-W (2010) Estimating fixed-effect panel stochastic frontier models by model transformation. J Econom 157(2):286–296
11. Kuosmanen T, Kortelainen M (2012) Stochastic non-smooth envelopment of data: semi-parametric frontier estimation subject to shape constraints. J Prod Anal 38(1):11–28
12. Sanzidur R, Wiboonpongse A, Sriboonchitta S, Chaovanapoonphol Y (2009) Production efficiency of Jasmine rice producers in Northern and North-Eastern Thailand. J Agric Econ 60(2):419–435
13. Tibprasorn P, Autchariyapanitkul K, Chaniam S, Sriboonchitta S (2015) A copula-based stochastic frontier model for financial pricing. In: Integrated uncertainty in knowledge modelling and decision making. Springer International Publishing, Cham, pp 151–162
14. Hansan MZ, Kamil AA, Mustafa A, Baten MA (2012) Stochastic frontier model approach for measuring stock market efficiency with different distributions. PLos ONE 7(5):e37047. doi:10.1371/journal.pone.0037047
15. Carta A, Steel MF (2012) Modelling multi-output stochastic frontiers using copulas. Comput Stat Data Anal 56(11):3757–3773
16. Lai H, Huang CJ (2013) Maximum likelihood estimation of seemingly unrelated stochastic frontier regressions. J Prod Anal 40(1):1–14
17. Wiboonpongse A, Liu J, Sriboonchitta S, Denoeux T (2015) Modeling dependence between error components of the stochastic frontier model using copula: application to intercrop coffee production in Northern Thailand. Int J Approx Reason 65:34–44
18. Burns R (2004) The simulated maximum likelihood estimation of stochastic frontier models with correlated error components. The University of Sydney, Sydney
19. Greene WA (2010) Stochastic frontier model with correction for sample selection. J Prod Anal 34(1):15–24
20. Smith MD (2008) Stochastic frontier models with dependent error components. Econom J 11(1):172–192
21. Battese GE, Coelli TJ (1988) Prediction of firm-level technical efficiencies with a generalized frontier production function and panel data. J Econom 38(3):387–399

Quantile Forecasting of PM10 Data in Korea Based on Time Series Models

Yingshi Xu and Sangyeol Lee

Abstract In this chapter, we analyze the particulate matter PM10 data in Korea using time series models. For this task, we use the log-transformed data of the daily averages of the PM10 values collected from Korea Meteorological Administration and obtain an optimal ARMA model. We then conduct the entropy-based goodness of fit test for the obtained residuals to check the departure from the normal and skew-t distributions. Based on the selected skew-t ARMA model, we obtain conditional quantile forecasts using the parametric and quantile regression methods. The obtained result has a potential usage as a guideline for the patients with some respiratory disease to pay more attention to health care when the conditional quantile forecast is beyond the limit values of severe health hazards.

Keywords ARMA model · Goodness of fit test · PM10 · Quantile regression · Quantile forecasting · Value-at-risk

1 Introduction

The air pollution, especially, the particulate matter (PM) has been a critical social issue in recent years. PM stands for a complex mixture of extremely small solid particles and liquid droplets found in the air, which can be only detected using an electronic microscope. These particles form as a result of complex reactions of chemicals such as sulfur dioxide, emitted from power plants, industries and automobiles, and once inhaled, can get into the heart, lungs and bloodstream and cause serious health problems (https://www.epa.gov).

It was first demonstrated in the early 1970s that many of deaths and health diseases are associated with the PM pollution. Thereafter, it has been reported that particulate

Y. Xu · S. Lee (✉)
Department of Statistics, Seoul National University, Seoul 08826, Korea
e-mail: sylee@stats.snu.ac.kr

© Springer International Publishing AG 2017   587
V. Kreinovich et al. (eds.), *Robustness in Econometrics*,
Studies in Computational Intelligence 692, DOI 10.1007/978-3-319-50742-2_36

matter smaller than about $10\,\mu m$ (PM10) has serious effects on lung cancer [14], cardiovascular disease [4], etc. The IRAQ and WHO designated PM a Group 1 carcinogen. In 2013, the ESCAPE study showed that for every increase of $10\,\mu g/m^3$ in PM10, the lung cancer rate rises by 22% [14]. People with heart or lung diseases, children, and seniors are the most likely to be affected by particle pollution exposure (https://www.epa.gov), and thereby, most governments have created regulations for the particulate concentration. Korean government also set up the limits for particulate in the air: the annual average of PM10 should not exceed $50\,\mu g/m^3$ and the daily average should not exceed $100\,\mu g/m^3$ (http://www.airkorea.or.kr).

Owing to its importance, the task of predicting the amount of particulate in the air has been a core issue among researchers and practitioners. Many statistical models have been developed in the literature in order to cope with this issue. For example, multivariate linear regression (MLR) is widely used for PM10 forecasting [16], and artificial neural networks (ANNs) are also used for daily forecasting, often showing better results than the MLR [2, 10]. Besides these, ARIMA (autoregressive integrated moving average) models [1], CART (classification and regression trees) [17], GAM (generalized additive model) [12], and SVM (support vector machine) methods [15] are used for the purpose of PM10 forecasting.

All those methods are mainly designed to forecast the daily or monthly averages of PM10 levels. However, the average forecasting may lack information useful for certain group of people. For example, the patients with lung disease, who are much more sensitive to the variations of PM10 concentration, might need information on its upper quantiles rather than averages, because if the upper quantile forecasting value appears to exceed the limit values of PM10 concentration, it can trigger a signal to take more careful actions. Motivated by this, we study the conditional quantile forecasting method for PM10 concentration based on time series models. Among the estimating methods, we adopt the parametric method based on time series models, wherein the error distribution is assumed to belong to a specific distribution family, and the semiparametric method based on the quantile regression: see [5, 13] who consider the quantile regression method for financial time series. The quantile regression method is broadly appreciated as a functional device to estimate the value-at-risk (VaR), which is the most popular risk measurement for asset processes, and is particularly well known to be robust against outliers and a model bias: see [11].

This paper is organized as follows. Section 2 introduces the PM10 data and selects the optimal ARMA model for the log-transformed PM10 data. An entropy-based goodness of fit test is also conducted to identify the underlying distribution family of error terms. Section 3 presents the conditional quantile forecasting methods taking the aforementioned parametric and semiparametric approaches. Their performance is compared through some back-testing methods. Section 4 provides concluding remarks.

2 Data Description and Model Building

In this section, we describe the PM10 data collected in Seoul, Korea, from January 1, 2012 to December 31, 2015, consisting of 1461 observations. The data is obtained from Korea Meteorological Administration (http://data.kma.go.kr), which provides hourly mean concentration data of PM10 obtained from 28 monitoring stations in Korea including Seoul. In our analysis, we use the mean of the hourly data PM10 values to obtain the daily mean concentration PM10 data. In this procedure, missing values of hourly data are imputed with the daily averages of the data. Moreover, the two missing daily averages are imputed with the averages of the previous and following days' data values.

2.1 ARMA Model Fitting for the PM10 Data

Figure 1 plots the daily average PM10 concentration, denoted by X_t at time t, from January 1, 2012 to December 31, 2015. Since the data shows some spikes and fluctuations, we apply the log-transformation to the data. Figure 2 shows the plot of log-transformed PM10 concentration time series data, denoted by Y_t. The plot shows a pattern seemingly stationary: the Dicky-Fuller test suggests that the log-transformed time series have no unit roots. Then, comparing the AIC values, we obtain ARMA(1,3) model as the most appropriate, which is confirmed by the

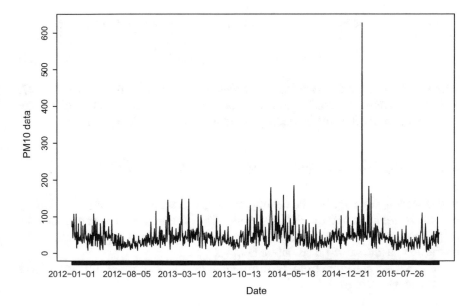

Fig. 1 The time series plot of the PM10 data

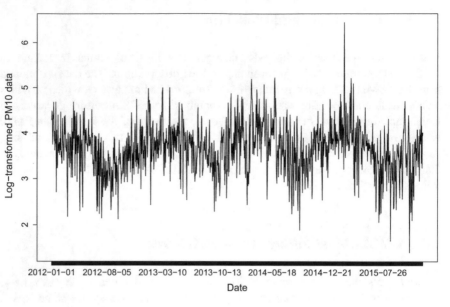

Fig. 2 The time series plot of the log-transformed PM10 data

Portmanteu test. Our analysis shows that the estimated ARMA(1,3) model for the log-transformed PM10 data, Y_t, is given as follows:

$$Y_t = 0.0647 + 0.9824 \, Y_{t-1} + \epsilon_t - 0.3359 \, \epsilon_{t-1} - 0.4254 \, \epsilon_{t-2} - 0.1041 \, \epsilon_{t-3}. \quad (1)$$

Based on this model, one can also obtain the one-step-ahead forecasts, that is,

$$\hat{Y}_{t+1} = \hat{\delta} + \hat{\phi}_1 Y_t + \hat{\theta}_1 \hat{\epsilon}_{t-1} + \hat{\theta}_2 \hat{\epsilon}_{t-2} + \hat{\theta}_3 \hat{\epsilon}_{t-3}, \quad (2)$$

where $\hat{\phi}_1, \hat{\theta}_i, \hat{\epsilon}_{t-i}$ are obtained using 1000 day moving window from September 27, 2014 to December 31, 2015. The results can be used to check whether or not the PM10 concentration would exceed the preassigned limits in the next day.

2.2 Entropy-Based Goodness of Fit Test for Residuals

In this section, we carry out the entropy-based goodness of fit test of [7] to identify the error distribution of the underlying model. The information on error distributions helps improve the accuracy of the quantile forecasting. Here, we particularly focus on the normal and skew-t distribution families.

To check the departure from the distribution family $\{F_\theta\}$, we set up the null and alternative hypotheses:

$$\mathcal{H}_0 : F \in \{F_\theta : \theta \in \Theta^d\} \quad \text{versus} \quad \mathcal{H}_1 : not \ \mathcal{H}_0. \quad (3)$$

Suppose that $X_1, \ldots, X_n$ are observed. To implement the test, we check whether the transformed random variable $\hat{U}_t = F_{\hat{\theta}_n}(X_t)$ follows a uniform distribution on $[0, 1]$, say, $U[0, 1]$, where $\hat{\theta}_n$ is an estimate of θ. We set $\hat{F}_n(r) = \frac{1}{n}\sum_{i=1}^n I(F_{\hat{\theta}_n}(X_i) \leq r)$, $0 \leq r \leq 1$, where $\hat{\theta}_n$ is any consistent estimator of true parameter θ_0 under the null, for example, the maximum likelihood estimator (MLE).

As in [9], we generate independent and identically distributed (i.i.d.) random variables w_{ij}, $j = 1, \cdots, J$, from $U[0, 1]$, where J is a large integer, say, 1,000, such that $\tilde{w}_{ij} = \frac{w_{ij}}{w_{1j} + \cdots + w_{mj}}$ and $s_i = i/m$, $i = 1, \cdots, m$, and use the test:

$$\hat{T}_n = \sqrt{n} \max_{1 \leq j \leq J} \left| \sum_{i=1}^m \tilde{w}_{ij}\left(\hat{F}_n\left(\frac{i}{m}\right) - \hat{F}_n\left(\frac{i-1}{m}\right)\right) \times \log m\left(\hat{F}_n\left(\frac{i}{m}\right) - \hat{F}_n\left(\frac{i-1}{m}\right)\right)\right|, \quad (4)$$

wherein $m = n^{1/3}$ is used because this choice consistently produces reasonably good results as seen in [9]. Since the entropy test has a limiting distribution depending upon the choice of $\hat{\theta}_n$, we obtain the critical values through Monte Carlo simulations (i.e. the bootstrap method) as follows:

(i) From the data $X_1, \ldots, X_n$, obtain the MLE $\hat{\theta}_n$.
(ii) Generate $X_1^*, \ldots, X_n^*$ from $F_{\hat{\theta}_n}(\cdot)$ to obtain $\hat{T}_n$, denoted by $\hat{T}_n^*$, with the pre-assigned m in (4) based on these random variables. Here, for the empirical distribution, we use $F_n^*(r) = \frac{1}{n}\sum_{i=1}^n I(F_{\hat{\theta}_n}(X_i^*) \leq r)$, where $\hat{\theta}_n^*$ is the estimator obtained from the bootstrap sample.
(iii) Repeat the above procedure B times, and for a preassigned $0 < p < 1$, calculate the $100(1 - p)\%$ percentile from the obtained B number of $\hat{T}_n^*$ values.
(iv) Reject $\mathcal{H}_0$ if the value of $\hat{T}_n$ obtained from the original observations is larger than the $100(1 - p)\%$ percentile obtained in (iii).

According to [9], the bootstrap entropy test is 'weakly consistent', which justifies its usage in practice. Applying the test to the residuals $\hat{\epsilon}_t$, we implement a goodness of fit test for the normal and skew-t distributions: the latter has the density function of the form [6]:

$$p(y|\mu, \sigma, \nu, \gamma) = \frac{2}{\gamma + \frac{1}{\gamma}}\frac{\Gamma(\frac{\nu+1}{2})}{\Gamma(\frac{\nu}{2})(\pi\nu)^{1/2}}\sigma^{-1} \times \left[1 + \frac{(y-\mu)^2}{\nu\sigma^2}\left\{\frac{1}{\gamma^2}I_{[0,\infty)}(y-\mu) + \gamma^2 I_{(-\infty,0)}(y-\mu)\right\}\right]^{-(\nu+1)/2}$$

with location parameter $\mu = -0.003$, scale parameter $\sigma = 0.406$, shape parameter $\nu = 7.1$, and skewness parameter $\gamma = 0.882$.

The entropy test rejects the null hypothesis of the normal distribution and to accept the skew-t distribution at the nominal level of 0.05. Figure 3 shows the histogram of the residuals and corresponding theoretical normal and skew-t densities. Apparently, the skew-t is a better fit to the residuals. Figure 4 shows the Q-Q plot of the residual quantiles and fitted skew-t quantiles. These two coincide except for some extreme cases, which are speculated to affect the quantile forecasting as seen below.

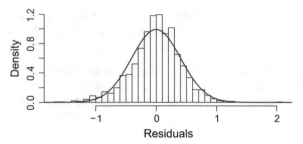

(a) The histogram of theresiduals and theoretical normal density

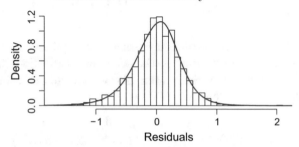

(b) The histogram of theresiduals and theoretical skew-t density

Fig. 3 The histograms of the residuals and theoretical densities

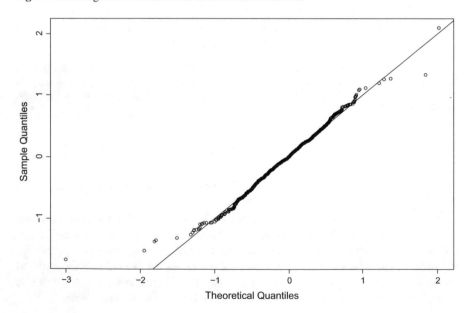

Fig. 4 The Q-Q plot of the residual and theoretical skew-t quantiles

3 Conditional Quantile Forecasting

3.1 Parametric Method

Since the log-transformed PM10 data follows ARMA(1,3) model and the residuals follow a skew-t distribution, we use the parametric approach to obtain conditional quantile forecasts [5]. Since Y_t follows an ARMA(1,3) model:

$$Y_t = \delta + \phi_1 Y_{t-1} + \epsilon_t - \theta_1 \epsilon_{t-1} - \theta_2 \epsilon_{t-2} - \theta_3 \epsilon_{t-3}, \tag{5}$$

where ϵ_t are i.i.d. skew-t random variables, the one-step-ahead conditional quantile forecast based on the information up to time t is given by:

$$\widehat{VaR}_t = \hat{\delta} + \hat{\phi}_1 Y_{t-1} + Q_\tau(\epsilon) - \hat{\theta}_1 \hat{\epsilon}_{t-1} - \hat{\theta}_2 \hat{\epsilon}_{t-2} - \hat{\theta}_3 \hat{\epsilon}_{t-3}, \tag{6}$$

where $Q_\tau(\epsilon)$ denotes the τ-th quantile of the skew-t distribution. Here, notation VaR is used because this conventionally indicates the value-at-risk in the financial time series context.

Based on this, we obtain the one-step-ahead conditional quantile forecasts using a moving window of size 1000. For example, we estimate the parameters using the 1000 days' log-transformed PM10 data from January 1, 2012 to September 26, 2014 in order to get the one-step-ahead conditional quantile for September 27, 2014. We repeat this procedure until December 31, 2015. Figure 5 shows the one-step-ahead conditional quantile forecasts at the level of 0.9.

3.2 Quantile Regression Method

In this subsection, we use the quantile regression method to get the conditional quantile forecasts [13]. Based on the ARMA(1,3) model, we express the τ-th conditional quantile function of Y_t given past information $\mathcal{F}_{t-1}$ as follows:

$$q_t(\theta) = \phi_1 Y_{t-1} + \xi(\tau) - \theta_1 \epsilon_{t-1} - \theta_2 \epsilon_{t-2} - \theta_3 \epsilon_{t-3}, \ 1 \leq t \leq n, \tag{7}$$

where the $\xi(\tau) = \delta + F_\epsilon^{-1}(\tau)$ (F_ϵ denotes the distribution of ϵ_t) emerges as a new parameter. In this case, the parameter vector is denoted by $\beta = (\xi, \phi_1, \theta_1, \theta_2, \theta_3)^T$, and then, the τ-th quantile regression estimator of true parameter is defined by:

$$\hat{\beta}_n(\tau) = \arg \min_{\beta \in \mathcal{B}} \frac{1}{n} \sum_{t=1}^{n} \rho_\tau(Y_t - q_t(\tau)), \tag{8}$$

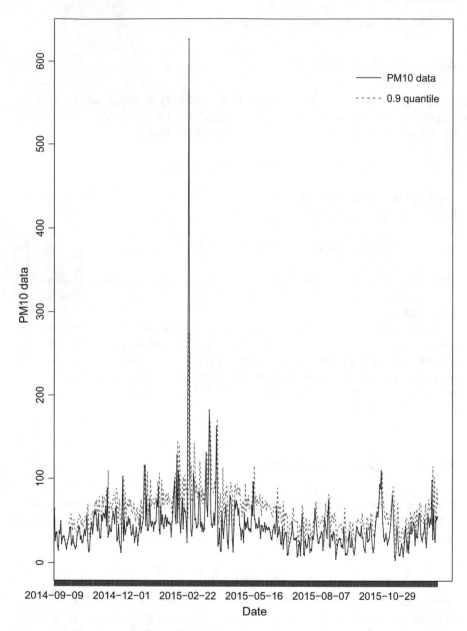

Fig. 5 The PM10 conditional quantile forecasts based on the parametric method

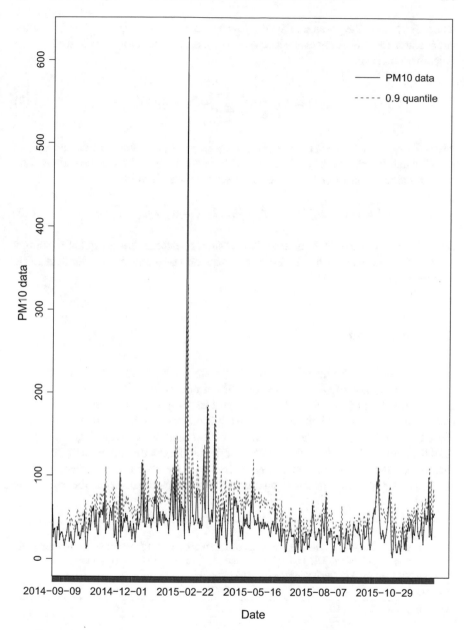

Fig. 6 The PM10 conditional quantile forecasts based on the quantile regression method

where $\mathcal{B} \subset \mathbb{R}^5$ is the parameter space, $\rho_\tau(u) = u(\tau - I(u < 0))$, and $I(\cdot)$ denotes an indicator function. However, since $\epsilon_{t-1}, \epsilon_{t-2}, \epsilon_{t-3}$ are unknown, we introduce the objective function:

$$\hat{\beta}_n(\tau) = \arg\min_{\beta \in \mathcal{B}} \frac{1}{n} \sum_{t=1}^{n} \rho_\tau(Y_t - \hat{q}_t(\tau)), \qquad (9)$$

where $\hat{q}_t(\tau) = \phi_1 Y_{t-1} + \xi(\tau) - \theta_1 \hat{\epsilon}_{t-1} - \theta_2 \hat{\epsilon}_{t-2} - \theta_3 \hat{\epsilon}_{t-3}$ with initial values $Y_t = \epsilon_t = 0$ when $t \leq 0$, and the residuals are obtained from the whole observations. The one-step-ahead conditional quantile forecasts are then given by:

$$\widehat{VaR}_t(\tau) = \hat{\phi}_{1n} Y_{t-1} + \hat{\xi}_n - \hat{\theta}_{1n} \hat{\epsilon}_{t-1} - \hat{\theta}_{2n} \hat{\epsilon}_{t-2} - \hat{\theta}_{3n} \hat{\epsilon}_{t-3}. \qquad (10)$$

As in the previous case, we forecast the one-step-ahead conditional quantile using a moving window of size 1000. Figure 6 shows the one-step-ahead conditional quantile forecasts at the level of 0.9.

3.3 Evaluation

In order to evaluate the quantile forecasting methods, we conduct the unconditional coverage (UC) test of [8] and the conditional coverage (CC) test of [3]. Table 1 lists the p values of the UC and CC tests and the proportions of the Y_t's below the estimated one-step-ahead quantile. The result indicates that the two methods all pass the UC and CC tests, and the quantile regression method produces larger p values than the parametric method. We reason that the outliers in the tail part (Fig. 3) affect the performance of the parametric method in a negative way. The quantile regression method is well known to be robust against outliers or a model bias than the parametric method.

Table 1 The evaluation of quantile estimates of log-transformed PM10 concentrate: the figures denote the p values of the UC and CC tests and the proportion of the log-transformed PM10 concentrates below the estimated one-step-ahead quantile

Method		0.75	0.90
Parametric model	UC	0.570	0.334
	CC	0.296	0.399
	Proportion	0.761	0.913
Quantile regression	UC	0.726	0.889
	CC	0.647	0.885
	Proportion	0.757	0.898

4 Concluding Remarks

In this study, we analyzed the PM10 dataset based on an ARMA model and discovered that the best fit is the ARMA(1,3) model with skew-t errors. The conditional quantile forecasting is conducted at the levels of $\tau = 0.75$ and 0.9, which revealed that the quantile regression method slightly outperforms the parametric method based on the MLE. Our quantile forecasting procedure can be potentially used to alarm a group of people, sensitive to the air pollution, to pay more attention to their health treatment. Although not handled in this study, one may consider clinical experiments to determine a proper quantile τ that can be used as a guideline for a group of people or patients who should mandatorily avoid the exposure to high degree PM10 concentration. The forecasting value can be used as a warning signal particularly when it exceeds the preassigned limit values. For the ordinary people, $\tau = 0.5$ may be good enough for such a purpose, whereas higher τ's would be appropriate for some patients at a high risk. In general, the air pollution problem is strongly linked with health economics that aims to understand the behavior of the individuals, health care providers, hospitals, health organizations, and governments in decision-making, and to evaluate market failure and improper allocation of resources. Economic cost of the health impact of air pollution is one of the important issues in health economics. We leave all the relevant issues as our future project.

Acknowledgement This research is supported by Basic Science Research Program through the National Research Foundation of Korea (NRF) funded by the Ministry of Science, ICT and future Planning (No. 2015R1A2A2A010003894).

References

1. Andy GP, Chan T, Jaiswal N (2006) Statistical models for the prediction of respirable suspended particulate matter in urban cities. Atmos Environ 40(11):2068–2077
2. Chaloulakou A, Grivas G, Spyrellis N (2003) Neural network and multiple regression models for pm10 prediction in athens: a comparative assessment. J Air Waste Manage Assoc 53(10):1183–1190
3. Christoffersen P, Hahn J, Inoue A (2001) Testing and comparing value-at-risk measures. J Empirical Finan 8(3):325–342
4. Cohen AJ, Ross Anderson H, Ostro B, Pandey KD, Krzyzanowski M, Künzli N, Gutschmidt K, Pope A, Romieu I, Samet JM et al (2005) The global burden of disease due to outdoor air pollution. J Toxicol Environ Health Part A 68(13–14):1301–1307
5. Engle RF, Manganelli S (2004) Caviar: conditional autoregressive value at risk by regression quantiles. J Bus Econ Stat 22(4):367–381
6. Fernández C, Steel MFJ (1998) On bayesian modeling of fat tails and skewness. J Am Stat Assoc 93(441):359–371
7. Ilia V, Lee S, Karagrigoriou A (2011) A maximum entropy type test of fit. Comput Stat Data Anal 55(9):2635–2643
8. Kupiec PH (1995) Techniques for verifying the accuracy of risk measurement models. J Deriv 3(2):73–84
9. Lee S, Kim M (2016) On entropy-based goodness-of-fit test for asymmetric student-t and exponential power distributions. J Stat Comput Simul 87(1):1–11

10. McKendry IG (2002) Evaluation of artificial neural networks for fine particulate pollution (pm10 and pm2. 5) forecasting. J Air Waste Manage Assoc 52(9):1096–1101
11. McNeil AJ, Frey R, Embrechts P (2015) Quantitative risk management: concepts, techniques and tools. Princeton University Press, Princeton
12. Munir S, Habeebullah TM, Seroji AR, Morsy EA, Mohammed AMF, Saud WA, Abdou AEA, Awad AH (2013) Modeling particulate matter concentrations in makkah, applying a statistical modeling approach. Aerosol Air Qual Res 13(3):901–910
13. Noh J, Lee S (2015) Quantile regression for location-scale time series models with conditional heteroscedasticity. Scand J Stat 43(3):700–720
14. Raaschou-Nielsen O, Andersen ZJ, Beelen R, Samoli E, Stafoggia M, Weinmayr G, Hoffmann B, Fischer P, Nieuwenhuijsen MJ, Brunekreef B et al (2013) Air pollution and lung cancer incidence in 17 european cohorts: prospective analyses from the european study of cohorts for air pollution effects (escape). Lancet Oncol 14(9):813–822
15. Sánchez AS, Nieto PJG, Fernández PR, del Coz Díaz JJ, Iglesias-Rodríguez FJ (2011) Application of an svm-based regression model to the air quality study at local scale in the avilés urban area (spain). Math Comput Model 54(5):1453–1466
16. Shahraiyni HT, Sodoudi S (2016) Statistical modeling approaches for pm10 prediction in urban areas; a review of 21st-century studies. Atmosphere 7(2):15
17. Slini T, Kaprara A, Karatzas K, Moussiopoulos N (2006) Pm10 forecasting for thessaloniki, Greece. Environ Model Softw 21(4):559–565

Do We Have Robust GARCH Models Under Different Mean Equations: Evidence from Exchange Rates of Thailand?

Tanaporn Tungtrakul, Natthaphat Kingnetr and Songsak Sriboonchitta

Abstract This study investigates the exchange rate volatility of Thai baht using GARCH, TGARCH, EGARCH and PGARCH models and examines the robustness of these models under different mean equation specifications. The data consisted of monthly exchange rate of Thai baht with five currencies of leading trade partners during January 2002–March 2016. The results show that the GARCH model is well-fitted for Chinese yuan and US dollar exchange rate, while TGARCH model is suitable to be selected for Japanese yen, Malaysian ringgit and Singapore dollar. For the model sensitivity, the findings indicate that the GARCH model is robust for the cases of Chinese yuan and US dollar, while TGARCH model is robust only for Malaysian ringgit. Therefore, We conclude that the selection of GARCH models is sensitive to mean equation specification. This confirms that researchers should pay attention to mean equation specifications when it comes to volatility modelling.

Keywords Robust · GARCH · EGARCH · TGARCH · PGARCH · Exchange rate · Thailand

1 Introduction

Since the introduction of the Autoregressive Conditional Heteroskedasticity (ARCH) model [13] and the Generalized Autoregressive Conditional Heteroskedasticity (GARCH) model [6], a large number of volatility models have been developed to estimate the conditional volatility of stock return, one of the important factors in financial investment [27]. Application of volatility models has been extended to international economics for modelling and forecasting volatility of macroeconomic factors involved in international trade [21, 24]. Trade is one of important factors

T. Tungtrakul (✉) · N. Kingnetr · S. Sriboonchitta
Faculty of Economics, Chiang Mai University, Chiang Mai, Thailand
e-mail: tanapornecon@gmail.com

© Springer International Publishing AG 2017
V. Kreinovich et al. (eds.), *Robustness in Econometrics*,
Studies in Computational Intelligence 692, DOI 10.1007/978-3-319-50742-2_37

contributing to economic development [28]. Studying the characteristics of its determinants would not only help to understand how economic fluctuation occurs through trade but also contribute to several implications such as trade forecasting for investors, long and short-run policy making, and international policy coordination [22]. As far as determinant of trade is concerned, Auboin and Ruta [5] pointed out that exchange rate and its volatility play an important role in fluctuations in international trade across countries around the world.

According to the Observatory of Economic Complexity (OEC), Thailand is among the top 30 exporters and importers in the world [25]. The crucial partners of Thailand are China, Japan, United States, Malaysia, and Singapore, which were the top five trade partners in 2012–2015 [9]. After the Asian financial crisis in July 1997, Bank of Thailand decided to change the exchange rate system from fixed exchange scheme to managed floating exchange rate scheme, causing the exchange rate of Thai baht to fluctuate since then.

Many studies focused on the impact of exchange rate volatility on trade flows. They used different econometric models to measure exchange rate volatility. Rahmatsyah et al. [23] studied the effect of exchange rate volatility on the bilateral trade of Thailand with the US by using the Moving Average Standard Deviation (MASD) and the GARCH models to estimate Thailands exchange rate volatility. Their results indicate that there exists a negative effect from the exchange rate volatility on both export and import under the MASD framework, whereas only negative effect on import was found for the case of the GARCH model. Hooy and Baharumshah [16] also studied exchange rate volatility in six selected East Asian countries, including Thailand, trading with the US by employing the EGARCH model. In the case of Thailand, they found that the exchange rate volatility has only positive effect on Thailands imports; however, evidence for impact on export could not be found. From these studies, it can be seen that employing different volatility models leads to completely different results.

The conflicting predictions from the theoretical models and the failure of empirical studies on the effects of exchange rate volatility on trade have led to the development of various kinds of volatility model to facilitate as many situations as possible [26]. Engle [13] was the first to develop such a model, the so-called ARCH model. Later on, Bollerslev [6] generalized the ARCH model into the GARCH model. Even though these models were developed decades ago, they still have a great influence on the volatility literature nowadays. However, both the models require the assumption that effects of positive and negative shocks are symmetric. To relax this assumption, Nelson [18] developed the Exponential Generalized Autoregressive Conditional Heteroskedasticity (EGARCH) model. Glosten et al. [14] and Zakoian [30] proposed the Threshold Generalized Autoregressive Conditional Heteroskedasticity (TGARCH) as an alternative approach. These two models are superior to the original GARCH model because the asymmetric effects of shock can now be captured with these two models.

Lee [17] concluded that the performance of volatility models depends on the information criteria for model selection. Pilbeam and Langeland [20] suggested determining the efficiency of the exchange rate volatility model by comparing the forecasts from different volatility models. Bollerslev [6] found that the GARCH model outperforms the ARCH model. Hansen and Lunde [15] showed that there are no models in the GARCH family capturing asymmetric shock effects, that can outperform the original GARCH model. Donaldson and Kamstra [12] found that the simple GARCH is the best for predicting stock return volatility. However, Brownlees and Gallo [7] found that the EGARCH model has better forecast precision. Ali [2] concluded that the TGARCH model outperforms other GARCH models under wider tail distribution. However, asymmetric ARCH models have been successfully fitted for the exchange rate data [8, 29].

Nevertheless, many researchers in volatility literature have overlooked the importance of mean equation specification. Asteriou and Hall [4] pointed out these volatility models may be sensitive to mean equation specification such that it could lead to undetermined or misleading results. Using an inappropriate GARCH model could cause wrong estimates of exchange rate volatility which world then be used to investigate the determinants of international trade, resulting in misleading conclusion and policy implication.

Therefore, there are two objectives in this study, Firstly, we examine the exchange rate volatility in Thailand in relation to leading trade partners, using different univariate symmetric and asymmetric GARCH models. The Akaike information criterion (AIC) is employed for model selection in this study. Secondly, we investigate the robustness of the model in terms of how responsive the GARCH model selection is to mean equation specification. The analysis of volatility in exchange rate is useful not only for policy makers to understand Thailands exchange rate volatility behaviour and movement, but also to international traders and investors that require the fitted volatility model in order to make appropriate decisions.

The organization of this paper is as follows. The data analysis and model specification used in this study are presented in Sect. 2. In Sect. 3, we provide the methodology employed in this study. Section 4 shows the empirical results, followed by the robustness checks of the GARCH models in Sect. 5. Finally, the conclusion of this study is drawn in Sect. 6.

2　Data Analysis

To investigate the volatility models of the exchange rate of Thai baht, the monthly of the exchange rates of Thai baht over the period from January 2002 to March 2016 collected from Bank of Thailand are employed. Five different currency exchanges are considered in the study, based on the currencies of the top five trade partners of Thailand, and these currencies are, Chinese yuan (CHY), Japanese yen (JPY), US dollar (USD), Malaysian ringgit (MYR), and Singapore dollar (SGD). All exchange rates are measured as Thai baht per 1 unit of the respective currency and transformed into growth rate (as a percentage change) using the following formula:

$$g_{i,t} = \frac{e_{i,t} - e_{i,t-1}}{e_{i,t-1}} \times 100, \tag{1}$$

where $g_{i,t}$ is the growth rate of the exchange rate of Thai baht with currency i at time t; $e_{i,t}$ and $e_{i,t-1}$ are the exchange rate of Thai baht with currency i in the current month and the previous month, respectively.

The periods prior to 2002 are not considered in this study since Thailand was using a fixed exchange rate scheme, causing the rate to be almost fixed during that time. Therefore, only the period from 2002 to 2016 is of concern. Figure 1 shows the growth rate of the exchange rates during the period of study. It can be seen that the series has been fluctuating over time.

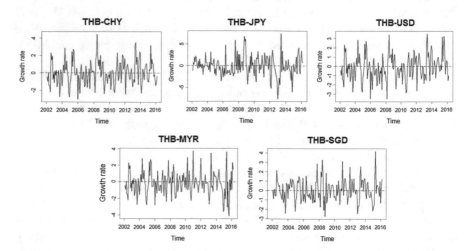

Fig. 1 Plots of monthly exchange rate during the period from January 2002 to March 2016

Table 1 Descriptive statistics of growth rate of exchange rates

Currency	THB–CHY	THB–JPY	THB–USD	THB–MYR	THB–SGD
Mean	0.0198	−0.0101	−0.1208	−0.1594	0.0473
Median	−0.0628	−0.1349	−0.1993	−0.1668	−0.0291
Maximum	4.4461	7.3836	3.5171	3.7487	4.1996
Minimum	−3.0407	−7.5639	−3.5487	−4.1218	−2.7289
Std. dev.	1.3514	2.2894	1.3811	1.4018	1.1389
Skewness	0.3431	0.2252	0.1597	0.1167	0.4410
Kurtosis	3.2255	4.2146	3.0766	3.6453	3.7658
Jarque-Bera test	3.6952	11.888	0.7641	3.3354	9.6656

The descriptive statistics for each exchange rate is provided in Table 1. We can see that the growth rates of THB–JPY and THB–SGD exhibit positive skewness, which may imply right-tail distribution. The other three cases suggest normal distribution since the levels of skewness are moderate and the kurtosis values are close to three.

3 Methodology

In this section, we describe the brief information regarding the unit roots tests and the volatility models for the exchange rate. Four univariate conditional volatility models are considered in this study, namely GARCH(1,1), EGARCH(1,1), TGARCH(1,1), and PGARCH(1,1).

3.1 Stationarity Testing

When it comes to modelling time series data, one of the necessary steps is to check whether the series is stationary. This study will employ the Augmented Dickey-Fuller (ADF) test [10], with the following model specification:

$$\Delta y_t = \alpha_0 + \alpha_1 y_{t-1} + \sum_{i=1}^{p} \alpha_{2i} \Delta y_{t-i} + \varepsilon_t \tag{2}$$

where y_t is the time series being tested and ε_t stands for residuals. The hypothesis testing can be specified as $H_0 : \alpha = 0$ for non-stationary against $H_1 : |\alpha| < 1$ for stationary. Nimanussornkul et al. [19] pointed out that the variance will become infinitely large if the series is non-stationary. However, the conclusion made from the conventional ADF test is based on the use of p-value and test statistic. However, there have been arguments recently that it may not be appropriate to use p-value for making conclusion. Therefore, this study employed the ADF test through the use of the Akaike information criterion (AIC) which was introduced by Anderson et al. [3]. The idea is to convert the null hypothesis testing problem of the ADF test into a model selection problem. Thereafter, we obtain the Akaike weight based on AIC and calculate the probability for each hypothesis as

$$Prob(H_i|data) = \frac{\exp(-\frac{1}{2}\Delta_i)}{\exp(-\frac{1}{2}\Delta_0) + \exp(-\frac{1}{2}\Delta_a)} \tag{3}$$

where $i = 0$ for H_0, $i = a$ for H_a and $\Delta_i = AIC_i - \min AIC$. The hypothesis with higher probability will be chosen.

3.2 GARCH(1,1) Model

To specify the appropriate equations. If the mean equation follows Autoregressive moving average with p autoregressive terms and q moving average terms ARMA(p,q) model can be written as

$$y_t = c + \sum_{i=1}^{p} \varphi_i y_{t-i} + \sum_{i=1}^{q} \theta_i \mu_{t-i} \qquad (4)$$

For variance equation, the Autoregressive Conditional Heteroscedasticity (ARCH) model introduced by Engle [13] assumes that positive and negative shocks affect the volatility equally. Bollerslev [6] extended ARCH into the GARCH model which can be specified as

$$\sigma_t^2 = \gamma + \alpha \sigma_{t-1}^2 + \beta u_{t-1}^2 \qquad (5)$$

where γ represents the predicted variance by the weighted average of a long term average; α represents the forecast variance from the previous period (GARCH effect); and β represents the information regarding the observed volatility in the previous period (ARCH effect). For ensuring that the conditional variance (σ_t^2) is greater than zero, $\gamma > 0, \alpha \geq 0$, and $\beta \geq 0$ are required. The GARCH(1,1) model shows that volatility is affected by previous shocks and its own past. The assumption of this model is that positive shocks ($u_t > 0$) and negative shocks ($u_t < 0$) exhibit the same impact on the conditional variance (σ_t^2). The general specification of the GARCH(m,n) model is

$$\sigma_t^2 = \gamma + \sum_{i=1}^{m} \alpha_i \sigma_{t-i}^2 + \sum_{j=1}^{n} \beta_j u_{t-j}^2 \qquad (6)$$

where m is the number of lagged σ_t^2 terms and n is the number of lagged u_t^2 terms.

3.3 EGARCH(1,1) Model

The exponential GARCH(EGARCH) model was introduced by Nelson [18]. This model allow the impact of positive shocks and negative shocks to be asymmetric. Additionally, this approach ensures that the conditional variance remains positive. According to Asteriou and Hall [4], the EGARCH(1,1) model can be written as

$$\log(\sigma_t^2) = \gamma + \alpha \log(\sigma_{t-1}^2) + \beta \left| \frac{u_{t-1}}{\sigma_{t-1}} \right| + \delta \frac{u_{t-1}}{\sigma_{t-1}}, \qquad (7)$$

where the leverage parameter (δ) reflects the symmetry effect of the shocks. If $\delta = 0$, then the effect is symmetric. When $\delta < 0$, it implies that the negative shocks have a greater contribution to volatility than the positive shocks. In macroeconomic analysis, negative shocks usually imply bad news that leads to more uncertain and unpredictable future [1]. The logarithm of conditional volatility in EGARCH model allow the estimates of the conditional variance to remain positive. For the general case, the EGARCH(m,n) is specified as follows:

$$\log(\sigma_t^2) = \gamma + \sum_{i=1}^{m} \alpha_i \log(\sigma_{t-i}^2) + \sum_{j=1}^{n} \beta_j \left| \frac{u_{t-j}}{\sigma_{t-j}} \right| + \sum_{j}^{n} \delta \frac{u_{t-j}}{\sigma_{t-j}} \qquad (8)$$

3.4 TGARCH(1,1) Model

The threshold GARCH (TGARCH) model, proposed by Glosten et al. [14] and Zakoian [30], can take the leverage effect into account. For the TGARCH(1,1) model, the specification[1] of the conditional variance is

$$\sigma_t^2 = \gamma + \alpha \sigma_{t-1}^2 + \beta u_{t-1}^2 + \delta u_{t-1}^2 d_{t-1} \qquad (9)$$

where d_t is the dummy variable taking value as follows:

$$d_{t-1} = \begin{cases} 1 & if \quad u_{t-1} < 0 \quad (bad\,news) \\ 0 & if \quad u_{t-1} > 0 \quad (good\,news). \end{cases}$$

From the model, we can see that the good news shocks have the impact of β while the bad news shocks have the impact of $\beta + \delta$. The asymmetry of the shock effects can be trivially checked. When $\delta = 0$, the model reduces to the simple GARCH(1,1) model and both the news will have symmetric effects, whereas $\delta > 0$ implies asymmetry in which the negative shocks have a greater effect on σ_t^2 than positive shocks and vice versa. In the general TGARCH(m,n) model, the conditional variance equation is given by

$$\sigma_t^2 = \gamma + \sum_{i=1}^{m} \alpha_i \sigma_{t-i}^2 \sum_{j=1}^{n} (\beta_j + \delta_j d_{t-j}) u_{t-j}^2, \qquad (10)$$

where γ, α_i, and β_j are non-negative parameters that satisfy similar conditions as the GARCH model.

---

[1]We use the model specification given by Asteriou and Hall [4], which is also known as the GJR model [14]. The original TGARCH model was introduced by Zakoian [30], and it considers conditional standard deviation instead of conditional variance.

3.5 PGARCH(1,1) Model

The power GARCH (PGARCH) model was introduced by Ding et al. [11]. Unlike previous GARCH models, the conditional variance is replaced by the conditional standard deviation to measure volatility. The PGARCH(1,1) model is given by:

$$\sigma_t^k = \gamma + \alpha \sigma_{t-1}^k + \beta(|\mu_{t-1}| - \delta \mu_{t-1})^k \tag{11}$$

For the general case, the equation of the PGARCH(m,n) model is given by

$$\sigma_t^k = \gamma + \sum_{i=1}^{m} \alpha_i \sigma_{t-i}^k + \sum_{j=1}^{n} \beta_j (|\mu_{t-j}| - \delta_j \mu_{t-j})^k \tag{12}$$

where α_i denotes the GARCH parameters; β_j denotes the ARCH parameters; δ_j denotes the leverage parameters; k is the power parameter with $k > 0$ and $|\delta_j| \leq 1$. It can be seen that if $\delta_j = 0$, the effects of the shocks are symmetric; otherwise, the effects of the shocks are asymmetric. Moreover, the PGARCH model is rather flexible compared to previous GARCH models. It can be seen that when $k = 2$, the PGARCH(m,n) model specification reduces to the conventional GARCH(m,n) model. But if $k = 1$, unlike the GARCH(m,n) model, the conditional standard deviation will be estimated instead of the conditional variance. Hence, it is possible to say that the GARCH model is a special case of PGARCH model.

4 Empirical Results

The results of the ADF unit root test from both the conventional approach and alternative approach using the AIC are shown in Table 2. It can be seen that they reach the same conclusion. Therefore, we could say that series for each of the currencies is stationary and is appropriate to be used for our investigation.

Table 3 shows the selected ARMA models for the conditional mean in the growth rate of the exchange rates for each currency according to the AIC from the set of ARMA(0,0), ARMA(1,0), ARMA(0,1) and ARMA(1,1) models. We can see that the best-fitting model for THB–CHY, THB–USD, and THB–MYR is ARMA(0,1); the best-fitting model for THB-JPY is ARMA(1,0) and the best-fitting model for THB–SGD is ARMA(1,1).

Given the selected mean equations, the different GARCH models are estimated. The results of the GARCH model selection are shown in Table 4. For the cases of THB–CHY and THB–USD, the symmetric GARCH(1,1) model with normal error distribution gives a better fit than the other GARCH models. In the cases of THB–JPY and THB–MYR, the asymmetric TGARCH(1,1) model with normal error distribution is chosen. Finally, the asymmetric TGARCH(1,1) with generalized error distribution (GED) is selected for the case of THB–SGD. After finding out which

Table 2 Unit root test results

Models	ADF test (original)			ADF test (AIC approach)		
	Lag length	ADF test statistic	Inference	AIC	Akaike weight	Inference
THB–CHY	0	−8.525***	Stationary	3.281	0.545	Stationary
THB–JPY	0	−9.592***	Stationary	4.415	0.553	Stationary
THB–USD	0	−8.598***	Stationary	3.340	0.545	Stationary
THB–MYR	0	−9.958***	Stationary	3.471	0.558	Stationary
THB–SGD	0	−9.308***	Stationary	3.007	0.552	Stationary

Note
1. Lag length selection is based on the Akaike information criterion (AIC)
2. If the Akaike weight is greater than 0.5, the stationary model is preferred
3. *** indicates 99% levels of confidence

Table 3 ARMA model selection for mean equation

Case	Model			
	ARMA(0,0)	ARMA(1,0)	ARMA(0,1)	ARMA(1,1)
THB–CHY	3.489	3.288	**3.285**	3.289
THB–JPY	4.500	**4.423**	4.442	4.427
THB–USD	3.490	3.348	**3.342**	3.349
THB–MYR	3.519	3.478	**3.474**	3.485
THB–SGD	3.104	3.014	2.993	**2.991**

models to be selected, we now discuss the ARCH, GARCH, and leverage effects for each case.

The empirical results of the selected models are given in Table 5. For the cases of THB–CHY and THB–USD, the GARCH(1,1) model is selected, and a similar conclusion is reached. The ARCH and the GARCH terms indicate that the lagged conditional variance and disturbance affect the conditional variance. The sum of the ARCH and the GARCH terms (persistence term) is less than one, which means that the shocks are not persistent to the conditional variance. However, the leverage effect could not be found as it is not considered in the GARCH(1,1) model.

In the case of THB–JPY, THB–MYR, and THB–SGD, the TGARCH(1,1) model is selected. For all of these cases, the sum of the ARCH and the GARCH terms is less than one, indicating that the variance process is mean reverting. The leverage effect exhibits a positive sign, implying the existence of asymmetric effect. This means there is a greater impact from negative shock than the one from positive shock on the volatility, provided that changes in negative and positive shocks are the same.

Table 4 GARCH model selection

Case (mean equation)	GARCH(1,1)			EGARCH(1,1)			TGARCH(1,1)			PGARCH(1,1)		
	Normal	Student's t	GED	Normal	Student's t	GED	Normal	Student's t	GED	Normal	Student's t	GED
THB–CHY [MA(1)]	**3.283**	3.293	3.287	3.305	3.309	3.302	3.293	3.301	3.296	3.297	3.298	3.292
THB–JPY [AR(1)]	4.386	4.376	4.382	4.348	4.354	4.354	**4.346**	4.355	4.354	4.350	4.357	4.357
THB–USD [MA(1)]	**3.351**	3.363	3.359	3.364	3.374	3.370	3.359	3.371	3.371	3.356	3.368	3.365
THB–MYR [MA(1)]	3.465	3.470	3.461	3.491	3.487	3.482	**3.384**	3.395	3.390	3.445	3.450	3.444
THB–SGD [ARMA(1,1)]	2.987	2.972	2.977	2.990	2.972	2.978	2.940	3.152	**2.922**	2.962	2.961	2.967

Table 5 Estimates of GARCH models for growth rate of monthly exchange rate

Model		THB–CHY GARCH(1,1)	THB–JPY TGARCH(1,1)	THB–USD GARCH(1,1)	THB–MYR TGARCH(1,1)	THB–SGD TGARCH(1,1)
Mean equation	AR(1)		0.165**			-0.503***
	MA(1)	0.386***		0.369***	0.239***	0.828***
Variance equation	Constant	0.164*	1.182***	0.255	0.124***	0.337*
	ARCH effect	-0.070**	0.181*	-0.066	-0.146***	-0.062***
	GARCH effect	0.964***	0.355***	0.909***	0.942***	0.591***
	Leverage effect		0.703*		0.270***	0.389*
	ARCH + GARCH effects	0.894	0.536	0.843	0.796	0.529
	AIC	3.283	4.346	3.351	3.384	2.922
	Error distribution	Normal	Normal	Normal	Normal	GED

Note
1. Using the maximum likelihood with BFGS approach for estimation
2. *, **, *** indicates statistical confidence at 90, 95, and 99% respectively

Table 6 Comparison of ARMA-GARCH models

Mean equation	Variance equation			
	GARCH(1,1)	EGARCH(1,1)	TGARCH(1,1)	PGARCH(1,1)
CASE: THB–CHY				
ARMA(0,0)	**3.452**	3.458	3.461	3.468
ARMA(1,0)	**3.275**	3.319	3.286	3.288
ARMA(0,1) (selected)	**3.283**	3.305	3.293	3.297
ARMA(1,1)	**3.248**	3.314	3.294	3.298
CASE: THB–JPY				
ARMA(0,0)	4.387	4.367	**4.362**	4.374
ARMA(1,0) (selected)	4.386	4.348	**4.346**	4.350
ARMA(0,1)	4.383	4.349	4.347	**4.345**
ARMA(1,1)	4.398	4.358	**4.357**	4.361
CASE: THB–USD				
ARMA(0,0)	**3.509**	3.518	3.521	3.531
ARMA(1,0)	**3.352**	3.380	3.359	3.355
ARMA(0,1) (selected)	**3.351**	3.364	3.359	3.356
ARMA(1,1)	**3.360**	3.375	3.367	3.362
CASE: THB–MYR				
ARMA(0,0)	3.534	3.541	**3.410**	3.500
ARMA(1,0)	3.481	3.500	**3.413**	3.467
ARMA(0,1) (selected)	3.465	3.491	**3.384**	3.445
ARMA(1,1)	3.485	3.507	**3.350**	3.462
CASE: THB–SGD				
ARMA(0,0)	3.119	3.147	**3.101**	3.114
ARMA(1,0)	3.024	3.029	3.031	**3.022**
ARMA(0,1)	2.984	3.002	2.990	**2.976**
ARMA(1,1) (selected)	2.977	2.978	**2.922**	2.967

Note
1. Residual distributions are considered as normal for all cases, except for THB–SGD which is a GED distribution
2. Numbers in **bold** indicate the most appropriate GARCH model for a given mean equation based on the AIC

5 Robustness: Sensitivity to Mean Equation

In this section, we investigate how the selection of the GARCH-family models responds to the change in the mean equation. As seen in many research studies in the volatility literature, the mean equations vary from the simple mean equation (i.e. regression with a constant term or ARMA(0,0), while some employ ARMA(p,q)). We believe that the mean equation specification plays an important role in the GARCH model selection. Asteriou and Hall [4] asserted that GARCH model may be sensitive to mean equation specification such that it could lead to undetermined or misleading results, and that researchers should exercise much caution when selecting the mean equation for a GARCH model. Therefore, by conducting all possible combinations of the ARMA-GARCH model, and together with Akaike Information Criteria, the robustness of the GARCH models under various mean equations could be found.

According to Table 6, it is evident that the optimal GARCH models are not stable for the cases of THB–JPY and THB–SGD. A shift from the selected mean equation leads to a different selected GARCH model. However, for the cases of THB–CHY and THB–USD the AIC suggests that GARCH(1,1) is the most suitable model for volatility modelling, regardless of the mean equation considered in this study. Additionally, TGARCH(1,1) is robust to mean equations misspecification in the case of THB–MYR. Nevertheless, considering all the cases, it is possible to conclude that GARCH model selection is still sensitive to the selection of mean equation; hence, all considered GARCH models in this study still suffer from the risk of misspecification in the mean equations.

6 Conclusion

In this paper, we investigated the GARCH models under different mean equations for the exchange rates of Thai baht. The monthly of the exchange rates of Thai baht over the period of January 2002 to March 2016 were employed. Five different currency exchanges were considered in the study based on the currencies of five top trade partners of Thailand, and these currencies are China yuan (CHY), Japanese yen (JPY), US dollar (USD), Malaysian ringgit (MYR), and Singapore dollar (SGD).

The results showed that the GARCH(1,1) model with normal error distribution is preferable in comparison with other models in the cases of THB–CHY and THB–USD. As for the cases of THB–JPY and THB–MYR, TGARCH(1,1) with normal error distribution is preferred, while TGARCH(1,1) with generalised error distribution fits the data better for THB–SGD.

In addition, we also investigated the mean equation sensitivity check to see how the GARCH model responds to changes in the mean equation which plays an important role in the selection of the GARCH model. We found that the optimal GARCH models are not stable for the cases of THB–JPY and THB–SGD. However, for the cases of THB–CHY and THB–USD, the GARCH(1,1) model is selected regardless of the mean equation considered in this study, whereas the TGARCH(1,1) model is found to be robust for THB–MYR.

Nevertheless, considering all the cases, it is possible to conclude that GARCH model selection is sensitive to the selection of the mean equation. Hence, the GARCH models in this study still suffer from the risk of having misspecification in the mean equations. The findings of this study, as we expected, confirm that researchers should pay attention to mean equation specification when it comes to volatility modelling. Moreover, further investigation in developing robust volatility modelling together with different diagnostic tools and alternative volatility models could enable us to have a closer look at volatility.

Acknowledgements The authors would like to thank the anonymous reviewer for suggestions which have improved the quality of this paper. This research is supported by Puay Ungpakoyn Centre of Excellence in Econometrics, Faculty of Economics, Chiang Mai University.

References

1. Ahmed AEM, Suliman SZ (2011) Modeling stock market volatility using GARCH models evidence from Sudan. Int J Bus Soc Sci 2(23):114–128
2. Ali G (2013) EGARCH, GJR-GARCH, TGARCH, AVGARCH, NGARCH, IGARCH and APARCH models for pathogens at marine recreational sites. J Stat Econom Methods 2:57–73
3. Anderson DR, Burnham KP, Thompson WL (2000) Null hypothesis testing: problems, prevalence, and an alternative. J Wildl Manag 64:912–923
4. Asteriou D, Hall SG (2016) Applied econometrics, 3rd edn. Palgrave, London
5. Auboin M, Ruta M (2013) The relationship between exchange rates and international trade: a literature review. World Trade Rev 12:577–605
6. Bollerslev T (1986) Generalized autoregressive conditional heteroskedasticity. J Econom 31:307–327
7. Brownlees CT, Gallo GM (2010) Comparison of volatility measures: a risk management perspective. J Finan Econom 8:29–56
8. Byers JD, Peel DA (1995) Bilinear quadratic ARCH and volatility spillovers in inter-war exchange rates. Appl Econ Lett 2:215–219
9. Department of Trade Negotiation (2015) Thailand's trade partner ranking, Nonthaburi. http://www.dtn.go.th/images/89/Trade/traderank1258.pdf. Accessed 7 Mar 2016
10. Dickey DA, Fuller WA (1981) Likelihood ratio statistics for autoregressive time series with a unit root. Econometrica 49:1057–1072
11. Ding Z, Granger CWJ, Engle RF (1993) A long memory property of stock market returns and a new model. J Empir Finan 1:83–106
12. Donaldson RG, Kamstra MJ (2005) Volatility forecasts, trading volume and the arch versus option-implied volatility trade-off. J Finan Res 28:519–538
13. Engle RF (1982) Autoregressive conditional heteroscedasticity with estimates of the variance of United Kingdom inflation. Econometrica 50:987–1007
14. Glosten LR, Jagannathan R, Runkle DE (1993) On the relation between the expected value and the volatility of the nominal excess return on stocks. J Financ 48:1779–1801
15. Hansen PR, Lunde A (2005) A forecast comparison of volatility models: does anything beat a GARCH(1,1)? J Appl Econom 20:873–889
16. Hooy C-W, Baharumshah AZ (2015) Impact of exchange rate volatility on trade: empirical evidence for the East Asian economics. Malays J Econ Stud 52:75

17. Lee KY (1991) Are the GARCH models best in out-of-sample performance? Econ Lett 37: 305–308
18. Nelson DB (1991) Conditional heteroskedasticity in asset returns: a new approach. Econometrica 59:347–370
19. Nimanussornkul K, Nimanussornkul C, Kanjanakaroon P, Punnarong S (2009) Modeling unemployment volatility in the US, Europe, and Asia. Europe, and Asia, 15 May 2009
20. Pilbeam K, Langeland KN (2014) Forecasting exchange rate volatility: GARCH models versus implied volatility forecasts. Int Econ Econ Policy 12:127–142
21. Praprom C, Sriboonchitta S (2014) Extreme value Copula analysis of dependences between exchange rates and exports of Thailand. In: Huynh V-N, Kreinovich V, Sriboonchitta S (eds.) Modeling dependence in econometrics: selected papers of the seventh international conference of the Thailand econometric society, faculty of economics, Chiang Mai University, Thailand, 8–10 January 2014. Springer International Publishing, pp 187–199
22. Prasad ES, Gable JA (1998) International evidence on the determinants of trade dynamics. Staff Pap 45:401–439
23. Rahmatsyah T, Rajaguru G, Siregar RY (2002) Exchange-rate volatility, trade and fixing for life in Thailand. Jpn World Econ 14:445–470
24. Sang WC, Sriboonchitta S, Huang WT, Wiboonpongse A (2013) Modeling the volatility of rubber price return using VARMA GARCH model. Taiwan Electron Period Serv 6:1–15
25. Simoes A, Landry D, Hidalgo C, Teng M (2016) The observatory of economic complexity, The MIT Media Lab. http://atlas.media.mit.edu/en/profile/country/tha/. Accessed 7 Mar 2016
26. Sriboonchitta S, Kreinovich V (2009) Asymmetric heteroskedasticity models: a new justification. Int J Intell Technol Appl Stat 2:1–12
27. Sriboonchitta S, Nguyen HT, Wiboonpongse A, Liu J (2013) Modeling volatility and dependency of agricultural price and production indices of Thailand: static versus time-varying Copulas. Int J Approx Reason 54:793–808
28. Todaro MP, Smith SC (2015) Economic development, 12th edn. Pearson, Boston
29. Tse YK, Tsui AKC (1997) Conditional volatility in foreign exchange rates: evidence from the Malaysian ringgit and Singapore dollar. Pac Basin Finan J 5:345–356
30. Zakoian J-M (1994) Threshold heteroskedastic models. J Econ Dyn Control 18:931–955

Joint Determinants of Foreign Direct Investment (FDI) Inflow in Cambodia: A Panel Co-integration Approach

Theara Chhorn, Jirakom Sirisrisakulchai, Chukiat Chaiboonsri
and Jianxu Liu

Abstract Globalization and modernization have generated the new opportunities for Multinational Enterprises (MNEs) to invest in foreign countries. Especially, many emerging and developing countries are making efforts actively to attract foreign direct investment (FDI) inflow in the purpose of boosting economic growth and development. This paper investigates the determinants of Cambodia's inward FDI within the time interval from 1995 to 2014. Panel co-integration approach, namely Full Modified Ordinary Least Square (FMOLS) and Dynamic Ordinary Least Square (DOLS) are proposed to estimate the long run coefficients. Our analysis shows that most of the variables are statistically significant except for population growth rate. Market size and financial development are, as expected, positively correlated whereas macroeconomic instability and cost of living are negatively associated but poor institution is, as unexpected, positively associated to inward FDI. The sign of ECT $(t-1)$ coefficient from panel causality analysis is significantly negative for GDP to FDI equation. It is indicated that economic growth and FDI is bidirectional causal relationship in the short run and the long run. The result from measurement predictive accuracy obtained from out of sample ex-post forecasting (2013–2014) confirmed that panel DOLS has a good predictive power to apply the long run ex-ante forecasting of Cambodia's inward FDI. Thus, our findings suggest that improving macroeconomic indicators, administrative barrier and financial instrument and development are the crucial policies to attract more inward FDI in the upcoming period.

Keywords Panel co-integration approach · Out of sample ex-post forecasting · Diebold and Mariano Test · Foreign direct investment · Cambodia

T. Chhorn (✉) · J. Sirisrisakulchai · C. Chaiboonsri · J. Liu
Faculty of Economics, Chiang Mai University, Chiang Mai, Thailand
e-mail: chhorntheara91@gmail.com

© Springer International Publishing AG 2017
V. Kreinovich et al. (eds.), *Robustness in Econometrics*,
Studies in Computational Intelligence 692, DOI 10.1007/978-3-319-50742-2_38

1   Introduction

Since the Eclectic Paradigm or OLI theory of [21] has developed to explain the phenomenon of foreign direct investment (FDI) theory through three approaches namely Ownership, Location and International Advantages, the assumption of FDI seems to be plainly comprehensive. Still, the context of FDI becomes an enlargeable concerning channel for researchers and as well as investors to study and conduct the research studies. Correspondingly, it is quietly important for developing countries and Least Developing Countries (LDCs) to enhance their economic performances, financial systems and policies to attract more inward FDI.

In the theoretical frameworks, FDI approach always comes up with the question why Multinational Enterprises (MNEs) have encouraged investing and establishing the entity facilitates in the host countries rather than staying in the domestic market. That's because they searched thereby the opportunities to benefit from innovation and/or technological development and the new market shares to operate their investments. More precisely, what is strongly attracted them is due to political and economic stability, policy certainty for investors and elevated geopolitical risks, natural resource abundant, tax burden or trade barrier, favorably of market and regional and international integration, open economy to the global market would be considered as the main factors to attract foreign entities and MNEs since generally foreign investors always want to invest in which uncertainty expectation is comparatively low, [17], [29] and [48]. On the other hand, in the empirical analysis, FDI approach accounts to investigate either in term of partial or general equilibrium frameworks. In the partial equilibrium framework, FDI was irritable with the firm's decision making and other specific variables while in the general equilibrium's one, [35] proposed Knowledge-Capital model to unify both horizontal and vertical motivation of FDI decision.

FDI was connected with employment creation, technological transfer and growth spillover, rising income and consumer expenditure, enhanced international trade performance and also a broadened tax base. Beside the realization of those advancements, many governments of the world economy, either developing or developed countries, are traversing in ever-increasing competitiveness of the country to attract and sustain FDI inwards as the manner unprecedented in the history of economic development (Lall 1998). In the context of Cambodia, after recovering from the deterioration and international isolation and due to social stability, political liberation, regional and global integration, favorable economic conditions and especially the presence of United Nations Transitional Authority in Cambodia (UNTAC) in 1992, the country has developed and improved remarkably to the new era of transitional and fresh development nation and become one of the most attracting destination for investment. Nearly the last two decades indicated that the average annual growth rate of Cambodia's gross demotic product (GDP) is approximately over 7.3% (World Bank Database 2015). Accordingly, Cambodia was ranked among the fast-growing regions in the world. The growth rate of GDP has achieved from the potential sectors such tourism, service-banking, agriculture, textiles, and especially FDI in small and medium enterprises (SMEs). FDI has come up to get widely concerning channel from the Royal Government of Cambodia (RGC) as a major potential contribution factor in boosting economic growth and increase household income nearly two decades up to present.

Cambodia's inward FDI has been augmented dramatically alongside with many political-economic reasons. For instant, in 1994 and 2005, the law on investment was created and Cambodia's Special Economic Zones (CSEZs) was constructed respectively. These were making FDI inflow raised considerably and approximately almost USD 1.7 billion (World Bank 2014) contributed to national economic. This among was received from China's investment as the premier, followed by Korea, ASEAN memberships, European Union (EU) and United Stated of American (USA). FDI inflow as percentage of GDP from 1993 to 2015 was averagely accounted 5.7% and approximated almost USD 10.9 billion (World Bank Database 2015).

The RGC has recognized many important factors to make out the facilitative condition for investment climate to encourage both domestic and foreign investment flow, to challenge with others nearby countries and as well as lead country into the international economy system. Beside this renovation, Cambodia has attracted more inward FDI, trade partnership that jointly assist to facilitate technological transformations, create new jobs, new business ideas and systems as an international standard and integrate her economy into the world. Accordingly, that would be an effective mechanism to robust economic growth, poverty reduction and sustainable development.

FDI based on approved investments in 2015 was approximately almost USD 1610 million (CDC 2015). Garment and textile industries are the main sources of FDI since investors are enjoyed with low cost and tax incentive, better investment climate, duty-quota free access to EU and US market since 2001 up to presence, the development of banking system and as well as macroeconomic stability (Chanthol 2015). Although, the growth rate of Cambodia's inward FDI has increased dramatically and contributed to the national economy in many sectors, the analysis of economic factors which may influence to this expansion is not yet well-understood empirically. Correspondingly, the main contribution of the study is to investigate the direction of foreign direct investment (FDI) and economic growth (GDP), the vital of economic factors influencing and driving the growth rate of FDI inflow in Cambodia from 1995 to 2014 and as well as to adopt an out of sample forecasting from 2013 to 2014.

As the result, the structural designed of the study is followed: Sect. 2 is to present some empirical literature reviews whereas Sect. 3 aims to explain the methodology using in the study and Sect. 4 is to interpret the empirical outcomes. Finally, the conclusion remarks will be stated combining together with some policy implications.

2 Data Collection and Sources

The selected variables using in the estimated regression are extracted annually from different sources. FDI inflow measuring in million dollars divided by Cambodia's gross domestic product (at 2000 prices) is extracted from UNCTAD and CEIC data

manager from by Chiang Mai university (CMU). The gross domestic product (GDP) growth rate and per capita, population growth rate, inflation rate, exchange rate and interest rate are extracted from World Development Indicator (WDI), Asia Development Bank (ADB). The degree of control of corruption index is imported from World Governance Indicator (WGI), World Bank.

3 Empirical Literature Reviews

In the recent literature reviews, market size adopting as the proxy of GDP growth rate and per capital, GNP growth rate and per capital and also population growth rate etc. was considered as an essential factor in analyzing the movement of FDI inflow. Simply, the large market size indicated the large size of customers and large potential demand in the country. It is also one of the most necessary condition to search for economic of scales and efficient utilization. In the theoretical point of views, investors who are market seeking-FDI is always aimed to exploit the possibilities of the new market particularly market growth [22]. Mall [34], Bilgili et al. [8], Cuyvers et al. [14] showed that market size is significantly impacted to FDI inflow. Macroeconomic instability, is another crucial factor, using as the proxy of exchange rate and inflation rate, is caused foreign investors considering whether to seek their market in the host country or stay domestically. In the empirical research indicated that, the magnitude of an influence of exchange rate on FDI is ambiguous and significantly positive impacted to FDI, Phillips [44], Solomon [47]. Narayanamurthy [49] indicated an influence of inflation rate of the host country may be delayed the foreign entities to consider investing. Financial development knowing as an emerging of banking system and instrument, using as the proxy of interest rate for the determined variable, is also a crucial factor among others. The development of banking sectors is indicated the rehabilitative or facilitative of capital movement and funds transferring from the host country to the home country or in the host country alone. Hira Aijaz Ahmed [46] indicated that interest rate is positively related to FDI while Cuyvers et al. [14] found there is insignificantly between interest rate and FDI. Another essential variable, institutional factor is one of the most concerning channel to explain the movement of FDI inflow. In the theoretical framework, [1] indicated that foreign investors are mostly paying attention by the better environment of politics in which corruption tends to be relatively low while in the empirical researches, Arbatli [5] and Cuyvers et al. [14] denoted that the country risk variable or socio-economic instability is highly significant to determine FDI inflow.

4 Methodology

This research study employs a panel co-integration approach, namely FMOLS and DOLS to estimate the long run coefficients of the determinants of Cambodia's FDI inflow from 1995 to 2014. The diagnostic tests such panel unit root tests, panel cross-sectional dependency tests and panel co-integration tests are primitively validated to detect the stationary, correlation dependency and long run equilibrium of the observations using in the model.

4.1 Specification of the Function: Cambodia's Foreign Direct Investment Inflow

Cambodia's FDI inflow has been attracted mostly from ASEAN region, developed east Asia and industry countries, the empirical study which seeks to identify the economic factors influencing to inward FDI, is not yet done in the recent period. Therefore, Cambodia's inward FDI would be modeled by a simple equation as follows:

$$FDI_{it} = \begin{pmatrix} Market & Size \\ Macroeconomic & Instability \\ Financial & Development \\ Institutional & Factor \end{pmatrix}_{it} \tag{1}$$

Market size refers to any economic factors that used ad the proxy of gross domestic product (GDP) either per capita and growth rate or population growth etc., macroeconomic instability considers to an uncertainty of exchange rate, inflation rate and price level whereas financial development adopts as a proxy of interest rate and lastly, institutional factor applies as a proxy of degree of control of corruption index which may affect to investment decision whether are willing to invest and/or start up the firms entities in the host country. The choice of selected variables are considered in reviewing many comparative theories and empirical research studies as presented in the preceding section. As the result, the Eq. (1) can be rewritten to the regression equation following the methodology of Cuyvers et al. [14] and Dauti [15] to adopt the pragmatic approach as follows:

$$\Delta FDI_{it} = \alpha_i + \beta_1 GDPGR_{it} + \beta_2 GDPPC_{it} + \beta_3 POPU_{it} + \beta_4 INTER_{it} + \beta_5 INFLA_{it} \\ + \beta_6 EXR_{it} + \beta_7 CORUP_{it} + \beta_8 CL_{it} + \varepsilon_{it}, \tag{2}$$

where

- α_i and ε_{it} are a constant term and an idiosyncratic error term respectively
- FDI_{it} is the foreign direct investment inflow by home countries j to the host country i at time t divided by real GDP at 2000 prices
- $GDPGR_{it}$ is the difference between the gross domestic product (GDP) growth rate of the host country i to home countries j at time t
- $GDPPC_{it}$ is the relative radio of gross domestic product (GDP) per capita of the host country i to home countries j at time t

- $POPU_{it}$ is the relative radio of population growth rate of the host country i to home countries j at time t
- $INTER_{it}$ is the relative radio of lending interest rate of the host country i to home countries j at time t
- $INFLA_{it}$ is the difference between inflation rate of the host country i to home countries j at time t
- EXR_{it} is the relative radio of official exchange rate of the host country i to home countries j at time t
- $CORUP_{it}$ is the proxy of the relative radio of level of corruption index of the host country i to home countries j at time t
- CL_{it} is the proxy of relative cost of living index, follows the methodology of Wong et al. [52] as follows: $CL_{it} = \ln[\frac{CPI_{it}/CPI_{jt}}{EXR_{it}/EXR_{jt}}]$.

4.2 Estimating Cambodia's Foreign Direct Investment Inflow via Panel Co-integration Model

Panel FMOLS, Pedroni et al. [39, 40] and DOLS, Kao and Chiang [28, 29] and Pedroni [40] are efficient techniques to eliminate serial correlation and endogeneity and were asymptotically normal distribution. Using these models are facilitated to establish the regression without requirement of taking the difference of integrated variables. The regression equation of panel data would be modeled as follows:

$$Y_{it} = \alpha_i + \beta_i X'_{it} + \varepsilon_{it}, \tag{3}$$

where

- i is the number of cross-sectional data with N and t is time dimension with T number
- β_i and ε_{it} are the coefficient and error term respectively
- α_i is an unknown intercept for each entity (n entity-specific intercepts) and $i = 1, 2, \ldots, n$
- Y_{it} is an explained variable where i is entity and t is time
- X'_{it} is the explanatory variables.

4.3 Fully Modified Ordinary Least Square (FMOLS) Estimator

FMOLS was firstly proposed by Pesaran et al. [42] and developed by Pedroni [40] as a non-parametric adjustment for endogeneity and serial correlation. It is consistency and asymptotically unbiased in heterogeneous panel model to allow an analyzing for the superior flexibility in the existence of heterogeneous in co-integration, Breitung and Pesaran [9, 42]. Therefore, the non-parametric coefficient of FMOLS was derived from OLS's to robust the result is:

$$\hat{\beta}_{OLS} = [\sum_{i=1}^{N} \sum_{t=1}^{T} (X_{it} - \bar{X}_1)^2]^{-1} \sum_{i=1}^{N} \sum_{t=1}^{T} (X_{it} - \bar{X}_1)(Y_{it} - \bar{Y}_1), \tag{4}$$

To correct for endogeneity and serial correlation, Pedroni [39, 40] has suggested the group-means FMOLS estimator that incorporates [43] semi-parametric correction to OLS estimator in adjusting for the heterogeneity that is presented in the dynamics underlying X and Y. Therefore, getting the non-parametric coefficient of FMOLS converged to normal distribution as follows:

$$\hat{\beta}_{FMOLS} = \hat{\beta}_{OLS} - \beta = [\sum_{i=1}^{N} \sum_{t=1}^{T} (X_{it} - \bar{X}_1)^2]^{-1} \sum_{i=1}^{N} \sum_{t=1}^{T} (X_{it} - \bar{X}_1)(Y_{it}^* - T\gamma_T^n),$$

$$\tag{5}$$

where

- X_{it} is the explanatory variables and X_1 is the mean of explanatory variables
- Y_{it}^* is regressand adjustment of the error term and explained variable
- $T\gamma_T^n$ is constant term and $\hat{\beta}_{FMOLS}$ is fully modified OLS estimator.

4.4 Dynamic Ordinary Least Square (DOLS) Estimator

Conversely, given the efficiency outcomes in both homogeneous and heterogeneous in panel co-integration model and opposing to panel FMOLS, panel DOLS is the fully parametric adjustment for endogeneity and serial correlation and it is the long run coefficient by taking into account the lead and lag values of variables [28, 29, 39]. The general regression equation of DOLS replaced from (3) is as follows:

$$Y_{it} = \alpha_i + \beta_i X_{it}' + \sum_{j=-q}^{q} C_{ij} \Delta X_{it+j} + V_{it}, \tag{6}$$

where

- α_i is a constant term and Y_{it} is an explained variable
- X_{it}' is the explanatory variables at time t

- q and $-q$ are lead and lag value respectively
- β_i is DOLS coefficient estimator obtained from i^{th} unit in panel
- X_{it+j} is the explanatory variables using lead and lag dynamic in DOLS
- $V_{it}^* = \sum_{j=-q}^{q} C_{ij} \Delta X_{it+j} + V_{it}$ and ΔX_{it+j} is differential term of X_{it}'

So, getting the coefficient of DOLS estimator:

$$\hat{\beta}_{DOLS} = N^{-1} [\sum_{i=1}^{N} (\sum_{t=1}^{T} z_{it} z_{it}')^{-1} (\sum_{t=1}^{T} z_{it} \bar{y}_{it})], \tag{7}$$

where

- $\hat{\beta}_{DOLS}$ is a dynamic OLS estimator
- z_{it} is the vector regressand in $2(p+1) \times 1$ dimension and
- $z_{it} = (x_{it} - \bar{x}_1, \Delta x_{it-p}, \dots, \Delta x_{it+p})$.

4.5 Panel Granger Causality in Bi-variate Based VEC Model

Panel Granger-Causality developed by [23] based Error Correction Terms (ECT) will be modeled to identify whether there exists short and long run bidirectional causality between FDI and economic growth in subject to the regression equation as follows:

$$\Delta(1-L) \begin{bmatrix} GDP_{it} \\ FDI_{it} \end{bmatrix} = (1-L) \begin{bmatrix} \alpha_{i,GDP} \\ \alpha_{i,FDI} \end{bmatrix} + \sum_{i=1}^{p} (1-L) \begin{bmatrix} \varphi_{11ip} & \varphi_{12ip} \\ \varphi_{21ip} & \varphi_{22ip} \end{bmatrix} \begin{bmatrix} GDP_{it-p} \\ FDI_{it-p} \end{bmatrix}$$
$$+ \begin{bmatrix} \beta_{GDP_i} \\ \beta_{FDI_i} \end{bmatrix} ECT_{t-1} + \begin{bmatrix} \varepsilon_{1t} \\ \varepsilon_{2t} \end{bmatrix}, \tag{8}$$

The Eq. (8) can be derived into Eqs. (9) and (10) as follows:

$$\Delta FDI_{it} = \alpha_0 + \sum_{i=1}^{m} \varphi_{11} \Delta FDI_{j,t-1} + \sum_{i=1}^{n} \varphi_{12} \Delta GDP_{j,t-1} + \beta_1 ECT_{t-1} + \varepsilon_{1t}, \tag{9}$$

$$\Delta GDP_{it} = \alpha_0 + \sum_{i=1}^{m} \varphi_{21} \Delta GDP_{j,t-1} + \sum_{i=1}^{n} \varphi_{22} \Delta FDI_{j,t-1} + \beta_2 ECT_{t-1} + \varepsilon_{2t}, \tag{10}$$

where ECT_{t-1} is the error correction term to determine the long run coefficient, p denote the lag and length, $(1-L)$ is the first difference operation, ε_{it} is the error term, φ_{ik} denote the parameter indicating the speed of adjustment to the equilibrium level after the shock and β_i, φ_{ik} and α_0 denoted the estimated parameter to be found. Thus, from Eqs. (9) and (10), the long run Granger causality was tested under the null hypothesis as follows:

- H_0: $\beta_1 = \beta_2 = 0$ for all i and k dimension

The strong granger causality will be tested by:

- H_0: $\varphi_{12} = \beta_1 = 0$ and $\varphi_{22} = \beta_2 = 0$ for all i and k dimension.

4.6 Out of Sample Ex-post Forecasting of Cambodia's Foreign Direct Investment Inflow

Adjacent to estimate the long run coefficient of the determinants of Cambodia's FDI inflow, the study aims continually to adopt in sample estimating from panel DOLS and panel Autoregressive-Distributed Lag (ARDL) so-called Pooled Mean Group PMG) estimation and out of sample ex-post forecasting to examine whether panel DOLS is the best model to adopt the long run ex-ante forecasting. Accordingly, the ex-post estimating uses in-sample observation from 1995 to 2012 and out of sample forecasting from 2013 to 2014 as an one step-ahead to check the measurement predictive accuracy. Still, Diebold and Mariano [19] test will be validated to test the equality of two predictive accuracy in order to compare two of out of sample ex-post forecasting equations.

To get the measurement predictive accuracy, let $\bar{y}_{it}$ be the forecast of variable i for period t, and let y_{it} be the actual value. $\bar{y}_{it}$ can be a prediction for one period ahead. Assuming that observations on $\bar{y}_{it}$ and y_{it} are available for $t = 1, 2, \ldots, T$. Correspondingly, the four most common measures of forecasting accuracy that have been used to evaluate ex-post forecasts are:

$$RMSE = \sqrt{\frac{1}{T} \sum_{t=1}^{T} (y_{it} - \bar{y}_{it})^2}, \tag{11}$$

$$MAE = \frac{1}{T} \sum_{t=1}^{T} |y_{it} - \bar{y}_{it}|, \tag{12}$$

$$MAPE = \frac{1}{n} \sum_{t=1}^{n} \frac{y_{it} - \bar{y}_{it}}{|y_{it}|} \times 100, \tag{13}$$

$$U = \frac{\sqrt{\frac{1}{T} \sum_{t=1}^{T} (y_{it} - \bar{y}_{it})^2}}{\sqrt{\frac{1}{T} \sum_{t=1}^{T} (\Delta y_{it})^2}}, \tag{14}$$

Diebold and Mariano Test

Diebold-Mariano (D-M) test is a statistical econometric method developed by Diebold and Mariano [19] which is asymptotically normal distribution, $N(0, 1)$.

It is used to compare between two predictive models and not proposed to compare the estimated models. D-M test is equated as follows:

$$S = \frac{\bar{d}}{\sqrt{\frac{\hat{V}}{\bar{d}} \times T}}, \quad \text{where} \tag{15}$$

$$\frac{\hat{V}}{\bar{d}} = \hat{\gamma}_0 + 2 \sum_{j=1}^{h-1} \hat{\gamma}_j, \tag{16}$$

$\hat{\gamma}_j = cov(d_t, d_{t-j})$ denoted a consistent estimate of the asymptotic variance of $\sqrt{\frac{T}{\bar{d}}}$ $\bar{d} = \frac{1}{n_2} \sum_{i=t_1}^{T} d_t$ and $t = t_1, \ldots T$ is the ex-post forecasting period with the total of n_2 forecasting and the lost differential between two forecasting accuracy is:

$$d_t = g(e_{1t}) - g(e_{2t}), \tag{17}$$

where e_{it} denoted the forecasting error and $g(e_{it})$ denoted the square (squared error loss) or the absolute value of e_{it}. The null hypothesis of the test is: $H_0 : E(d_t) = 0 \forall t$, meaning that two competing predictive accuracy is equaled, while the alternative one is, $H_a : E(d_t) \neq 0 \forall t$, meaning that two competing predictive accuracy is unequaled.

5 Results and Discussion

5.1 Panel Unit Root Tests

Panel unit root tests, [4, 24, 25, 32] and [33], were adopted to detect the stationary of the variables under the null hypothesis of having a common unit root across the country groups. Therefore, Table 1 reported that few tests could not reject the null hypothesis at level. Otherwise, the results are fairly enabling in conclusive that these series are non-stationary for few tests and they were become stationary after taking the first difference, I(1).

5.2 Panel Cross-Sectional Dependency Tests

Panel cross-sectional dependency tests developed by [10, 41] scaled LM and CD and [7] bias-corrected scaled LM, had introduced four statistics to test the null hypothesis of absence of cross-sectional dependency in panel data analysis will be used. Since the observation contains number of the cross-sectional, $N = 12$ and time, $T = 19$, that is suitable to apply the CD and LM statistic. Consequently, Table 2 identified that

Table 1 Panel unit root tests

Series	Individual unit roots			Common unit roots		Heteroscadastic	
	IPS	Fisher-ADF	Fisher-PP	LLC	Breitung	Hadri Z-stat.	Con. Z-stat.
FDI	−0.50*	117.83*	213.7*	−15.99*	−6.45*	5.36*	13.4*
GDPGR	−5.65*	74.09*	204.36*	−3.62*	−8.42*	4.75*	8.09*
GDPPC	−3.52*	50.63*	67.54*	−4.5*	−4.61*	2.14*	8.62*
POPU	−4.97*	83.7*	128.35*	−3.72*	1.45	35.4*	15.48*
INTER	−8.37*	106.32*	235.42*	−4.12*	−4.43*	−1.3	23.9*
INFLA	−12.24*	150.78*	245.61*	−13.1*	−10.87*	18.2*	21.6*
EXR	−4.03*	56.54*	58.39*	−7.22*	−6.17*	6.07*	4.74*
CORUP	−5.23*	71.3*	98.26*	−8.34*	0.55	35.9*	34.5*
CL	−8.22*	103.88*	228.49*	−8.78*	−6.28*	8.01*	20.27*

The sign * denote the rejection of null hypothesis at 1 and 5% level respectively
Note The optimal length and lag was selected based on the SIC criterion. All tests are assumed asymptotic normality and taken individual effects, individual linear trends

Table 2 Panel cross-sectional dependency tests

Stat. series	FDI stat.	GDPGR stat.	GDPPC stat.	POPU stat.	INTER stat.	INFLA stat.	EXR stat.	CORUP stat.
CD_{LM_1}	214.39*	459.48*	1247.27*	229.52*	876.73*	694.01*	787.45*	854.84*
CD_{LM_2}	11.87*	33.20*	101.77*	13.19*	69.52*	53.62*	61.75*	67.62*
CD_{LM_3}	11.554*	32.89*	101.46*	12.87*	69.21*	53.3*	61.43*	67.22*
CD	7.445*	19.66*	23.82*	4.56*	29.22*	25.51*	27.35*	28.71*

Note The sign * denote the rejection of null hypothesis at 1% level of significant
• Panel cross-sectional dependency tests take 66 degree of freedom for all observations
• The CD statistic is distributed as a two-tailed standard normal distribution and LM statistic as a $\chi^2_{N(N+1)}$ distribution

p-value can be rejected at 1% level for all statistics, meaning that there is no correlation within the cross-sectional group and all variables are dependency in difference shocked in which they are exposed to.

5.3 Panel Co-integration Tests

To identify co-integrated among the variables, [26, 27], Error Correction (EC) panel based co-integration [50] and Pedroni (1999) co-integration tests will be employed with the respect to heterogeneous panel techniques. Firstly, Pedroni co-integration test introduced seven panel group such panel v-Statistic, panel rho-Statistic, panel PP-Statistic and panel ADF-Statistic, rho-Statistic, group PP-Statistic and group ADF-

Statistic, assuming asymptotically normal distributed tests. Accordingly, Table 3 reported that most of the statistics have p-value less than 5%. By looking to model 1 and model 2 are indicated 5 statistics can be rejected the null hypothesis of absence of co-integration at 1% level and 1 statistic was rejected at 10% level. Thus, the result strongly implied that the long run equilibrium was presented in the subject to Pedroni co-integration test.

Secondly, Fisher-Johansen co-integration tests introduced trace and max-eigen statistic based Fisher method. Therefore, according to Table 4 indicated that both trace and max-eigen test have p-value less than 5% from none to at most 4 level. Those outcomes implied that the long run equilibrium is strongly existed among the selected variables. The null hypothesis that each series have unit root and no co-integration among them ($r = 0$) can be rejected at 1% level in conclusive to $\lambda_{trace(r)}$ and $\lambda_{max(r,r+1)}$ values for all models.

Panel co-integration test based Kao residual was assuming pooled and LLC technique. As the result, Table 5 reported that p-value is less than 5% for ADF; that's implied that it can be rejected the null hypothesis at 1% level. Thus, the long run equilibrium among the variables was strongly exited. Yet, taking into consideration another test, EC panel co-integration test [50] introduced four statistics to test the hypothesis of co-integration idea. Accordingly, Table 5 summarized that three tests among four can be rejected the null hypothesis at 1% level, meaning that there is also co-integrated based EC panel co-integration test.

Shortly, from four different panel co-integration tests, the study found that there strongly existed the long run equilibrium among the selected variables. Thus, in

Table 3 Pedroni based co-integration test (1999)

Within-dimension				
Panel group statistics	Model 1		Model 2	
	t-statistic	Weighted t-statistic	t-statistic	Weighted t-statistic
Panel v-Statistic	9.7427*	−3.0737	19.7*	−1.0962
Panel rho-Statistic	1.2424	0.9875	−1.2406*	−0.6947
Panel PP-Statistic	4.0757	−6.8366*	2.6939	−7.7654*
Panel ADF-Statistic	4.1246	−6.8825*	2.7006	−7.4239*
Between-dimension				
	Model 1		Model 2	
	t-statistic		t-statistic	
Group rho-Statistic	2.2478***		0.4573	
Group PP-Statistic	−6.664*		−13.0894*	
Group ADF-Statistic	−8.5054*		−14.31*	

Note The sign * and *** denoted the significant level of 1 and 10% respectively

• Model 1 and model 2 are referring to a regression without deterministic intercept or trend and with deterministic intercept and trend, respectively

Table 4 Johansen-Fisher based co-integration test

Co-integration (r)	$\lambda_{trace}(r)$			$\lambda_{max}(r, r+1)$		
Null hypothesis	Model 1	Model 2	Model 3	Model 1	Model 2	Model 3
$r = 0$	84.77*	13.86	135.9*	84.77*	540.6*	135.9*
$r \leq 1$	507.9*	204.0*	546.9*	361.1*	204.0*	367.6*
$r \leq 2$	283.5*	426.7*	377.1*	176.8*	376.0*	209.7*
$r \leq 3$	144.4*	215.1*	202.9*	115.6*	141.6*	140.4*
$r \leq 4$	56.00*	98.67*	88.95*	53.65*	76.81*	62.99*
$r \leq 5$	30.83	45.81*	55.27*	30.83	45.81*	55.27*

The sign * denoted the significant level of 1 and 5% respectively

Note Model 1, model 2 and model 3 referred to a regression which takes linear deterministic trend, linear deterministic trend (restricted) and no deterministic trend respectively

Table 5 Kao based residual and error correction panel co-integration tests

Kao based residual co-integration test		
Series	t-statistic	
ADF	−37.24*	
Residual	0.003	
HAC variance	0.002	
EC panel based co-integration test		
Series	Value	Z-value
G_t	−7.346	−17.527*
G_a	−45.903	−15.419*
P_t	−3.957	2.095
P_a	−33.409	−11.054*

Note The sign * denoted the significant level of 1 and 5% respectively

addition to those results, panel FMOLS and DOLS estimator shall be adopted to estimate the long run coefficients in the next section.

5.4   Long Run Coefficient from Panel Co-integration Models

Taking into account the long run coefficient obtaining from panel FMOLS estimator identified that most of explanatory variables are statistically significant except for EXR and POPU. Panel FMOLS estimator is adopted 95 and 99% of coefficient of confidential interval (CCI) to check the long run coefficients at difference level was indicated in Table 6 as follows:

Market size, adopted as the proxy of GDPGR, GDPPC and POPU, is positively associated to FDI inflow except for POPU is negatively impacted but insignificant. The result suggests that once economic growth rate and per capita in the host country is remain high, foreign investor as the market and resource seeking-FDI is likely to invest due to the fact of rising income per capita of domestic household. Yet, according to World Bank stated that Cambodia approved from the lower income country (LIC) status to the middle lower income, 2016. Cambodia's household income (GDP per capita) was increased regressively from 781.91$ to 1168.04$ per year from 2010 to 2015 respectively (Statista 2016). Noteworthy, the large market size also expressed the emerging of regionalization and economic of scale in the host country. By looking into POPU variable, is negatively influenced to FDI but insignificant. In facts, it is quite important to notify that is because most population in Cambodia are farmer and unskilled with lack of education and another reason is that so far Cambodia is an emerging and new market for foreign investors, population may not effect strongly to their investment decision. On the reflected to the study of Akin [2], in terms of cohort size, old and young age cohorts weaken FDI inflow.

Macroeconom instability applied as the proxy of inflation rate and exchange rate, INFLA and EXR, is negatively influenced to FDI inflow but EXR is insignificant. From the theoretical point of views, Dixit-Pindyck [20] notified that high level of inflation and exchange rate could signal macroeconomic instability resulted in decreasing inward FDI to the host country. Thus, foreign investors will probably change from market seeking-FDI to export-substituting FDI in addition to the existence of uncertainty with the regard to present net value (PNV) and future net value (FNV) of investment. Accordingly, they will build the plant in the home country instead. Furthermore, this finding is due to the facts that Cambodia's currency is highly dollarized and in the estimated regression, the study applied the radio of home country currency over US's per dollar against the host country. Hence, the depreciation of US's currency will delay market-seeking FDI into the host country. Yet, consistent to most of the study in developing countries by Arbatli [5] found that the effective exchange rate is negatively associated to FDI and the depreciation in host country currency is negatively influenced to inward FDI to that country.

Financial development, applied as the proxy of INTER, is negatively associated to FDI inflow. The long run coefficient indicated that it is −0.0375 in the overall

Table 6 Long run coefficient from panel FMOLS estimator

FDI is an explained variable		Coefficient	95% of CCI		99% of CCI	
			Low	High	Low	High
Market size	GDPGR	0.0200*	0.0164	0.0238	0.0152	0.0250
	GDPPC	12.0831*	14.0935	10.0728	14.7438	9.4225
	POPU	−0.0002	−0.0003	−0.0002	−0.0003	−0.0001
Financial development	INTER	−0.0375*	−0.0548	−0.0200	−0.0605	−0.0145
Macroeconomic instability	INFLA	−0.0064*	−0.0089	−0.0038	−0.0097	−0.0030
	EXR	−0.6192	−0.7409	−0.4975	−0.7802	−0.4581
Institutional factor	CORUP	0.0008*	0.0002	0.0014	0.0009	0.0016

Note The sign * denoted the significant level of 1 and 5% respectively
• Panel FMOLS applied pooled estimation using full sample from 1995–2014

of CCI. This is opposite to the outcomes from panel DOLS estimator, which is positively influenced to FDI inflow. The negative sign is also identified that the foreign investors whom come to invest in Cambodia mostly have enough capital to set up their business without requirement the pre-funds from government as pre-capital operation. Moreover, investors may consider saving interest rate rather than lending or loaning interest rate.

Another one, institutional factor, adopted as the proxy of degree of control of corruption index, CORUP, is positively influenced to inward FDI. This is likely due to the facts that corruption is affected directly and indirectly to some specific location firms in the host country. Accordingly, foreign investors may know rationally how to deal with the high level of corruption country resulted in why it may not affect strongly to investment decision. Still, this result is also consistent to most of the study of Azam [6] found that political risk has positively significant impacted to FDI in LDCs. Noteworthy, since China is the highest investor in Cambodia, this finding is reflected to the study of Buckley [11] and Kolstad [31] showed that poor institution, high degree of corruption and lack of rule of law mostly attracted China investors.

Meanwhile panel DOLS estimator uses to examine an out of sample ex-post forecasting (2013–2014), the study used in sample observation from 1995 to 2012 whereas dropping some variables such INFLA and EXR and adding CL instead. Similarly to panel FMOLS, panel DOLS estimator result was indicated in Table 7 as follows. Accordingly, with the respect to GDPGR, GDPPC, INTER and CL are statistically significant meanwhile POPU is showed the insignificant relationship. Market size is positively affected to FDI inflow excepts for POPU is negatively significant at 10% level.

Conversely to panel FMOLS, panel DOLS estimator indicated that financial development is positively affected to FDI inflow. This finding is likely due to the facts that, since the late 1997 after Asia Crisis was eliminated, banking sector started to be involved crucially and potentially in economic transaction and system in Cambodia. Presently, there are 36 commercial banks, 11 specialized banks, 40 licensed micro-finance institutions and 33 registered-micro finance operators have operated in the financial sectors in Cambodia. Similarly, Desbordes [18] and Choong [13] found

Table 7 Long run coefficient from panel DOLS estimator

FDI is an explained variable		Coefficient	95% of CCI		99% of CCI	
			Low	High	Low	High
Market size	GDPGR	0.0023*	0.0013	0.0034	0.0008	0.0038
	GDPPC	0.047*	0.0418	0.0527	0.0392	0.0549
	POPU	−0.00001	−0.0001	0.0001	−0.0001	0.0001
Financial development	INTER	0.0067*	0.0054	0.0081	0.0049	0.0087
Relative cost of living	CL	−0.0009*	−0.0010	−0.0006	−0.0014	−0.0004

Note The sign * denoted the significant level of 1 and 5% respectively

• Panel DOLS used lead and lag 1 of heterogeneous in the long run and adopted pooled estimation of in-sample observation from 1995 to 2012

that the development of the domestic financial system is an important prerequisite for FDI inflow.

The last one, relative cost of living index, CL, is negatively influenced to FDI inflow. It was strongly implied that if the relative living cost of the host country comparing to the home country is higher, foreign investor is differ or inhibit to invest since they will produce with high input prices. The result is likely due to the consideration that if the price in the host country is velocity rapidly, investors could not guarantee to adopt the price of their products in the market.

5.5 Panel Granger Causality in Bi-variate Based VEC Model

By looking into the residuals of the short and long run equilibrium from panel Granger causality based VEC model, employing to investigate the directional causality of FDI and economic growth. As is apparent in Table 8 indicated that the long run coefficient of ECT(t − 1) of FDI equation is negatively insignificant and conversely for GDP equation is negatively significant (ECT equaled to −0.49 for GDP), meaning that there existed the long run relationship from GDP to FDI equation or there is bidirectional causality between economic growth and FDI. The short run coefficient of GDP equation is 54.74 and significant, meaning that there is short run Granger causality from economic growth to FDI. Reflecting from FDI to GDP equation in the short run indicated positively coefficient, 0.0084 but insignificant. The result implied that there is no directional causality in the bi-variate model from FDI to economic growth equation.

5.6 In Sample Estimating and Out of Sample Ex-post Forecasting

The empirical result from ex-post forecasting is estimated based in-sample observation from 1995 to 2012 and performed out of sample forecasting from 2013 to

Table 8 Long and short run coefficient from panel Granger-causality analysis

Dependent Variables	Short run causality		Long run causality	Joint causality F-test
	DGDPGR(−1)	DFDI(−1)	ECT(t − 1)	
ΔFDI	0.0084	−0.1755	−0.1284	0.0048
	(0.0572)	(−0.0378)	(−0.0408)	(7.1442)
ΔGDPGR	−0.0791	54.7409**	−0.4909***	36.5568***
	(−1.0348)	(2.2777)	(−6.3481)	(3.2340)

Note The sign *** and ** denoted the significant level of 1, 5 and 10% respectively

Table 9 Out of sample Ex-post Forecasting from 2013 to 2014

Panel Estimator	RMSE	MAE	MAPE	U	DM Test
DOLS	1.6193	0.3411	1167	0.9915	1.3052
ARDL	1.6208	0.3382	2119	0.9860	(0.3005)

Note The value insights the parenthesis indicated p-value

2014 as an one step-ahead forecasting in order to obtain the measurement predictive accuracy such RMSE, MAE, MAPE and U. D-M test is calculated based on two predictive equations from panel DOLS and panel ARDL estimator. Table 9 reported that the measurement predictive accuracy such RMSE or MAE obtained from panel DOLS estimator is produced smaller error rather than panel ARDL. Still, D-M test cannot reject the null hypothesis at 5% level. Therefore, based on the result from the measurement error was indicated that panel DOLS is the best efficiency estimator to adopt the long run ex-ante forecasting.

6 Concluding Remarks

The panel econometric models of the heterogeneous had given the suitable and favorable benefits to examine the economic characteristic of Cambodia's inward FDI that leaded to investigate the determinants in the purpose of policy implications and approach within the international facilities. The study uses cross-sectional data from 12 home countries during the time interval of 1995 to 2014 and applies a panel co-integration approach to estimate the long run coefficients and several diagnostic tests are used to detect the stationary, correlation dependency and co-integrated among the variables.

The result from panel Granger causality based VEC model signified that there is bidirectional causality from economic growth to FDI equation in the short and long run and un-bidirectional causality from FDI to economic growth equation. In fact, thanks to economic growth which has played an essential role in encouraging inward FDI, remaining stable and approximating averagely 7% per year, foreign investors will consider in setting up business in terms of both realized and approved FDI in the short and the long run.

The outcomes from the measurement predictive accuracy such RMSE, MAE, MAPE and U obtaining from in sample estimating (1995–2012) and out of sample ex-post forecasting (2013–2014), indicated that panel DOLS is the efficiency model in applying the long run ex-ante forecasting of Cambodia's FDI inflow in the upcoming period.

With the respect to panel FMOLS and DOLS estimator indicated that market size, financial development and poor institution are positively associated to inward FDI. This is likely appreciated for Cambodia's government due to the facts that Cambodia has enjoyed nearly the last two decades from economic growth rate, it has been alerting the significant signal for both domestic and international investors, although most of Cambodia population, joining in the labor market are less educative or unskilled and some are decided to be immigration in other developed country such Thailand, Malaysia and Korea.

Although the global financial crisis during the period of 2008 and 2009 was impacted a little bit to financial sector particularly banking sector, NBC has tried hardly and technically to sustain financial institution and management meanwhile Cambodia's financial sector has being inserted into the era of specialization with the development of financial institutions and instrument, emerging of security markets in the early 2012 which will be continually prioritized to launch the government securities in the upcoming years. It could be identified the sensible signal to both domestic and foreign investors who are interested to invest not only direct investment but also an opportunity to gather the stock market.

Such the worthy notice, macroeconomic instability and cost of living index were discovered negatively influencing to inward FDI. Although, inflation and exchange rate are economically stable after the global financial crisis was eliminated and recovery in 2009, NBC has technically tried to adopt monetary policy either contractionary or expansionary to promote de-dollarization, maintain the domestic price and as well as sustain the financial development. Especially, since Cambodia's currency was pegged to USA's currency, they also irritated to encourage Khmer Riel currency being circulated strongly and confidentially in the domestic market and to keep exchange rate stable otherwise their intervention has decelerated the decline in 2012. Yet, exchange rate is likely to depreciate due to the reduction in exports and lower inward FDI. However, the high degree of dollarization that has been circulated almost 90% of the total transaction inside Cambodia's economy system, implying the signal of movement in which will be moderately impacted to economy and investment climate in both the short run and the long run.

With the regard to institutional factor, the study found that high degree of control of corruption is positively related to inward FDI. However, that's still not implied a pretty signal for Cambodia in the long run, especially to catch up ethical investors from those who are escaped investing where corruption or central government are exceptionally existed and hardly to be eliminated. In Cambodia, corruption was ranked 156th in the status of corruption perception index (TI 2014). As a helping-hand effect, it was considered undoubtedly one of the biggest challenges and frontiers harming socioeconomic development and investment climate. Domestically, where it has been occurred nearly the overall level of society affecting business efficiency resulted in raising the production cost and capital requirement to startup the business, the household incomes might probably not distribute equally to all levels of economics resulted in income inequality in the society. As the result of rising corruption, an implementation of various legislatures in curbing corruption and anti-corruption mechanism are taking hand in removing. Accordingly, in 2006, RGC established the Anti-Corruption Unit (ACU) to put in place. More notices, with this respect, those who are responsible in charge of One-stop-Service in facilitating an approved and realized investment should pay more attention to eliminate these dirty issues to assemble the technical and standard business from the ethic countries such Japan and EU etc. meanwhile they are not yet ready to invest highly capital into Cambodia.

References

1. Agarwal S, Ramaswami SN (1992) Choice of foreign market entry mode: impact of ownership, location and internationalization factors. J Int Bus stud, 23(1): 1–28
2. Akin MS (2009) How is the market size relevant as a determinant of FDI in developing countries? A research on population and the cohort size. Int Symposium on Sustain Dev 425–429
3. Ali S, Guo W (2005) Determinants of FDI in China. J Global Bus Technol 1(2):21–33
4. Anyanwu JC (2011) Determinants of foreign direct investment inflows to Africa, 1980–2007. Working paper series No 136, African Development Bank, Tunis, Tunisia
5. Arbatli E (2011) Economic policies and FDI inflows to emerging market economies. IMF working paper, Middle East and Central Asia Department
6. Azam M, Ahmad SA (2013) The effects of corruption on foreign direct investment inflows: some empirical evidence from less developed countries. J Appl Sci Res 9:3462–3467. ISSN: 1819–544X
7. Baltagi B, Feng Q, Chihwa K (2012) A Lagrange Multiplier Test for Cross-SectionalDependence in a Fixed Effects Panel Data Model. Center for Policy Research. Paper193
8. Bilgili F, Halci, NS, Dogan L (2012) The determinants of FDI in Turkey: a Markov regime-switching approach. Econ Model 29:1161–1169. Faculty of Economics and Administrative Sciences, Erciyes University, Turkey
9. Breitung J (2000) The local power of some unit root tests for panel data. In: Baltagi B (ed) Nonstationary panels, panel cointegration, and dynamic panels. Advances in econometrics, vol 15. JAI Press, Amsterdam, pp 161–178
10. Breusch T, Pagan A, (1980) The Lagrange multiplier test and its application to model specification in econometrics. Rev Econ Stud 47: 239–253
11. Buckley PJ, Clegg J, Wang C (2004) The relationship between inward foreign direct investment and the performance of domestically-owned Chinese manufacturing industry. Multinatl Bus Rev 12(3):23–40
12. Carr DL, Markusen JR, Maskus KE (1998) Estimating the knowledge-capital model of the multinational enterprise, National Bureau of Economic Research, Cambridge, October 1998
13. Choong C-K, Lam S-Y (2011) Foreign direct investment, financial development and economic growth: panel data analysis. IUP J Appl Econ 10(2):57–73
14. Cuyvers L, Soeng R, Plasmans J, Van Den Bulcke D (2011) Determinants of foreign direct investment in Cambodia. J Asian Econ 22(3):222–234
15. Dauti B (2015) Determinants of FDI inflow in South East European Countries. In: panel estimation, Tirana international conference, Economics Faculty, State University of Tetova, 11–13 December 2008
16. De Vita G (2008) Determinants of FDI and Portfolio flows to developing countries: a panel co-integration analysis. Eur J Econ Finan Adm Sci 13:161–168. Coventry University
17. Denisia V (2010) Foreign direct investment theories: An Overview of the main FDI theories, Vol.2(2). Academy of Economic Studies, Bucharest
18. Desbordes R, Wei S-J (2014) The effects of financial development on foreign direct investment. World bank policy research working paper no. 7065
19. Diebold FX, Mariano R (1995) Comparing predictive accuracy. J Bus Econ Stat 13:253–265. Economic Research, Cambridge, October 1998
20. Dixit A, Pindyck R (1994) Investment under uncertainty. Princeton University Press, Princeton
21. Dunning, John (1979) Toward an Eclectic Theory of International Production: Some Empirical Tests. J Int Bus Stud

22. Dunning, John H (2000–2004) The eclectic paradigm as an envelope for economic and business theories of MNE activity
23. Granger CWJ (1986) Developments in the study of cointegrated Economic Variables. Oxford B Econ Stat 48:213–228
24. Hadri K (2000) Testing for stationarity in heterogeneous panel data. J Economet 3:148–161
25. Im KS, Pesaran MH, Shin Y (2003). Testing for unit roots in heterogeneous panels. J Economet 115:53–74
26. Johansen S. (1998) Statistrcal analysis of cointegration vectors. J Econ Dynam and Control 12231–254. North-Holland
27. Kao C (1999) Spurious regression and residual-based tests for cointegration in panel data. J of Economet 90:1–44
28. Kao C (1999) Spurious regression and residual-based tests for cointegration in panel data. J Econ 90:1–44
29. Kao C, Chiang MH (2000) On the estimation and inference of a cointegrated regression in panel data. Adv Econ 15:179–222
30. Kariuki C (2015) The determinants of foreign direct investment in the African Union. J Econ Bus Manage 3(3):346–351
31. Kolstad I, Wiig A (2009) What determines Chinese outward FDI? CMI working paper WP 2009: 3, Chr. Michelsen Institute, Bergen, 19 pp
32. Levin A, Lin C-F, Chu C-SJ (2002) Unit root tests in panel data: Asymptotic and finite-sample properties. J Economet 108:1–24
33. Maddala, GS Wu S (1999) A comparative study of unit root tests with panel dataand a new simple test. Oxford B Econ St 61:631–652
34. Mall S (2013) Foreign direct investment inflows in Pakistan: a time series analysis with autoregressive distributive lag (ARDL) approach. Int J Comput Appl 78(5):7–16
35. Markusen JR, Anthony JV (1998) Multinational firms and the new trade theory. J Int Econ 46(2):183–203
36. Mottaleb KA, Kalirajan K (2013) Determinants of foreign direct investment in developing countries: a comparative analysis. ASARC working paper 2010/13
37. Mughal MM, Akram M (2011) Does market size affect FDI? The case of Pakistan. Interdisc J Contemp Res Bus 2:237–247. Hailey College of Commerce, University of the Punjab, Lahore
38. Pedroni P (1996) Fully modified OLS for heterogeneous cointegrated panels. In: Nonstationary panels, panel cointegration and dynamic panels, vol 15. Elsevier Science Inc., pp 93–130
39. Pedroni P (2004) Panel cointegration: asymptotic and finite sample properties of pooled time series tests with an application to the PPP hypothesis. Indiana University Mimeo, New Results
40. Pedroni P (2004) Panel cointegration: asymptotics and finite sample properties of pooled time series tests with an application to the PPP hypothesis. Manuscript, Department of Economics, Indiana University
41. Pesaran MH (2004) General diagnostic tests for cross section dependence in panels. University of Cambridge, Faculty of economics, Cambridge Working papers in economics No. 0435
42. Pesaran MH, Shin Y, Smith R (1999) Pooled mean group estimation of dynamic heterogeneous panels. J Am Stat Assoc 94:621–634
43. Phillips PCB, Hansen BE (1990) Statistical inference in instrumental variables regressions with I(1) Processes, Rev Econ Stud 57:99–125
44. Phillips S, Ahmadi-Esfahani FZ (2006) Exchange rates and foreign direct investment: theoretical models and empirical evidence. Aust J Agric Resourc Econ 52:505525
45. Ramirez MD (2006) A panel unit root and panel cointegration test of the complementarity hypothesis in the Mexican case, 1960–2001. Center discussion paper no 942, Economic Growth Center Yale University, New Haven
46. Siddiqui HAA, Aumeboonsuke V (2014), Role of Interest Rate in attracting the FDI: Study on ASEAN 5 Economy, Int J Tech Res Appl e-ISSN: 2320–8163
47. Solomon Charis M, Islam A, Bakar R (2015) Attracting foreign direct investment: the case of malaysia. Int Bus Manage 9:349–357

48. UNCTAD, World Investment Report 2015: Reforming international investment governance, ISBN: 978-92-1-112891-8
49. Vijayakumar N, Sridharan P, Rao KCS (2010) Determinants of FDI in BRICS countries: a panel analysis. Int J Bus 5:1–13
50. Westerlund J (2007) Testing for error correction in panel data. Oxford Bull Econ Stat 69:0305–9049
51. Westerlund J, Endgerton D (2006) A panel bootstrap cointegration test. Department of Economics, Lund University, Sweden
52. Wong KK et al (2006) Bayesian models for tourism demand forecasting. Tourism Manage 27:773–780
53. Yin F, Ye M, Xu L (2014) Location determinants of foreign direct investment in services: evidence from Chinese provincial-level data. Asia Research Centre working paper 64

The Visitors' Attitudes and Perceived Value Toward Rural Regeneration Community Development of Taiwan

Wan-Tran Huang, Chung-Te Ting, Yu-Sheng Huang and Cheng-Han Chuang

Abstract The purpose of the rural regeneration plan carried out for years is mainly for rural sustainable development, which makes communities change and indirectly attracts many tourists. Especially the rural experience tourism emerged recently drives the rural economy grow entirely, enriches the rural environment and style, and also increases many job opportunities and accelerates the prosperity of local communities. Although the booming tourism increases the number of travelers and facilitates the local development, it has the cognitive deficiency in the aspect of ecology maintenance. As a result, the conservation and the economic development fail to reach a balance.

In this study, we will take the Wu Mi Le community of Tainan as an example to analyze the cognitive elements of the rural regeneration, and use the cluster analysis to discuss the preference of difference groups to travel experience. In addition, we will further use the contingent valuation method (CVM) to measure the willingness to pay (WTP) of tourists to the rural maintenance and the tourist activities in this study.

The research results are summarized as below: 1. The environment conservations cognition is firstly considered for tourists to the rural regeneration communities; 2. The multi-existence group has a higher contribution in rural development; 3. Tourists think the maintained value is higher than the recreation value.

Keywords Rural regeneration · CVM · WTP

W.-T. Huang (✉)
Department of Business Administration, Asia University, Taichung, Taiwan
e-mail: wthuangwantran7589@gmail.com

C.-T. Ting · Y.-S. Huang
Department of Tourism, Food and Beverage Management, Chang Jung Christian University, Tainan City, Taiwan
e-mail: ctting@mail.cjcu.edu.tw

Y.-S. Huang
e-mail: yshuang@mail.cjcu.edu.tw

C.-H. Chuang
Department of Business Administration, Chang Jung Christian University, Tainan City, Taiwan
e-mail: hyusam@gmail.com

© Springer International Publishing AG 2017
V. Kreinovich et al. (eds.), *Robustness in Econometrics*,
Studies in Computational Intelligence 692, DOI 10.1007/978-3-319-50742-2_39

637

1   Introduction

1.1   Motivation

The rural regeneration plan carried out by the Taiwanese government is mainly for promoting the rural sustainable development, including activation and regeneration of the rural area, improvement of community infrastructure, beautification of ecological environment, and reinforcement of resident consensus. In recent years, the rural development of Taiwan has been mainly based on "Production, Life and Ecology", and expanded to recreation, and improvement of peasants' social welfare. Lots of researches indicate that the rural regeneration not only remarkably improves the rural planning, but also facilitates the rural tourism benefit and brings elements important to rural economic development [1–5]. Although the rural tourism facilitates the rural economic development, tourists know little about the rural ecology maintenance. As a result, the rural conservation and the economic development fail to reach a balance [1, 5, 6].

Therefore, this study will discuss cognitive elements of tourists about the rural regeneration plan, evaluate the difference of different cognitive groups to WTP of the rural regeneration, and then evaluates the maintained value and the recreation value for increasing the rural regeneration, which will be helpful to interested parties to understand the benefit of carrying out the rural regeneration plan.

1.2   Wu Mi Le Community, Houbi Township in Tainan

In the study, the Wu Mi Le Community, Houbi Township in Tainan is taken as the object of study. Houbi Township is an important granary production area. The Wu Mi Le is on the Northwest side of the Houbi Township. Wu Mi Le actually means "Let It Be" to describes how a group of Taiwan peasants continue rice farming in the case that the external environment becomes extremely difficult. In recent years, as the introduction of tourist activities, the Wu Mi Le community has become widely well known. Moreover, the Let It Be community is one of pilot areas carried out the rural regeneration plan by the Taiwanese government, and also is the important development area among the 4,232 rural communities in Taiwan.

In conclusion, this study will discuss tourists' cognition of the rural regeneration communities, analyze tourists' WTP of the rural regeneration communities, and evaluate the economic benefit from the rural regeneration plan. In the past, literatures about the rural regeneration mostly investigate the benefit to rural areas and communities brought by the rural regeneration through literature review and depth interview.

Researches focus on physical facilities (infrastructures, communities and transfer systems, cultural relics and farmhouse repair), environmental facilities (beautification of overall community environment, and improvement of soil, water resources and ecological environment), life improvement (residents' inhabiting quality, residents' additional income), and effects on improvement of health care, education and more budget subsidy. Therefore, this study will measure the benefit of tourists to rural tourism through questionnaire survey, analyze preferences of different tourists through cluster analysis, and evaluate WTP of the rural regeneration development through CVM.

2 Application of Methodology and Questionnaire

2.1 Methodology

The study uses the contingent valuation method (CVM) to evaluate the benefit of people after receiving food and farming education. CVM sets a hypothetical question through questionnaires, simulates various market situations, and changes the quantity or quality of people to environment into the willingness-to-pay (WTP) or the willingness-to-accept (WTA), and evaluates changes in consumers' utility level through model deduction.

The study evaluates WTP about promotion of food and farming education through CVM. After people receive food and farming education, i.e., improve people's cognition to food materials sources (e.g., when people know that restaurants' food materials are from local production, it can let people understand the cognitive level of food materials.), people are asked if they are willing to pay more money to support the promotion of the government in food and farming education. The expression is listed as below.

$$WTP_i = f(x_i) + \varepsilon_i$$

where WTP_i represents the cost that people are willing to pay, ε_i is the residual and meets the assumption of $N(0, {}^2)$, and x_i is the explanatory variable vector of the i_{th} respondents.

In addition, the study uses the double-bounded dichotomous choice to estimate people's WTP about promotion of food and farming education by the government through CVM inquiry. This not only can reduce respondents' pressure and avoid starting point bias or range deviation, but also can deliver more messages [7–9]. The inquiry of the double-bounded dichotomous choice uses the first minimum payment cost as criterion (e.g., T). When people said they are willing to pay for the first time, it shall inquire the WTP cost for the second time. The WTP cost of the second time is twice of that of the first time (2T). If people said they are not willing to pay, the WTP cost of the second time is half of that of the first time (0.5T) (Fig. 1).

Fig. 1 An example of
double-bounded
dichotomous

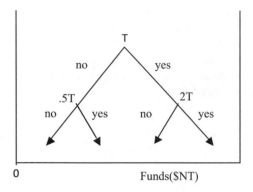

2.2 Questionnaire Design

The questionnaire is issued to tourists of Wu Mi Le community of Houbi Township in Tainan. The pretesting questionnaire are total of 50, and the formal questionnaire are total of 462. Total of 416 questionnaires are returned. The questionnaire content consists of 3 parts such as personal information, travelling experience, rural regeneration cognition and rural regeneration WTP price. There are totally 3 questions as below:

◇ Question 1: Did you participate in activities held in rural communities?
◇ Question 2: How much would you like to pay for the free experience activities provided by communities in order to maintain recreation resources?
◇ Question 3: How much would you like to pay in order to maintain these cultural heritages with historical significance?

3   Results

3.1   Analysis of Rural Regeneration Cognition

The questionnaire is based on questionnaires designed by [4, 10], and issued to tourists to Wu Mi Le communities. In the study, KMO and Bartlett of the rural regeneration cognition are tested. As shown in Table 1, KMO value is 0.711 > 0.6, indicating the common factor of the factor analysis and extraction is significant in effect. Bartlett test value is 1170.67 and $P = 0.000 < \alpha = 0.01$, so it is of significance. As for the factory analysis result of the rural regeneration cognition, it uses the principal component analysis to extract 3 factors and rename "environment conservation", "cultural heritages", "economic value".

Table 1 KMO and Bartlett test

Kaiser-Meyer-Olkin		0.711
Bartlett test	Similar to chi-square distribution	1170.67
	Df	136
	Significance	0.000***

Note *** indicates significant level of 1% respectively

3.2 Cognitive Differences of Rural Regeneration

The previous analyses show 3 factor dimensions, which are respectively "environment conservation", "cultural heritages", "economic value". This study further uses the cluster analysis to divide it into two clusters according to different characteristics. Details are as follows:

(1) Cluster 1: in this cluster, there are totally 242 tourists (66.7%). The level of agreement is the maximum in both "environment conservation" and "economic value". In the "cultural heritages", the level of agreement is higher than the 2nd cluster, showing this group of interviewees attaches great importance to every aspect. In addition to the economic benefit brought by policy, the main consideration is in the hope that the rural environment and the cultural assets can be carefully maintained. Therefore, the cluster is named "multi-existence".

(2) Cluster 2: in this cluster, there are totally 121 tourists (33.3%). The level of agreement is the maximum in the "cultural heritages", followed by the "environment conservation" and the "economic value", showing this group of interviewees hopes the rural regeneration should focus on the overall rural landscape and earnings brought. Therefore, this cluster is named "value recreation".

This study will investigate tourists economic backgrounds and travelling characteristics on the basis of value-recreation and multi-existence. It finds that the multi-existence cluster is mainly female, and the value-recreation is mainly male. As for the monthly income, the multi-existence cluster is higher than the value-recreation cluster in average monthly income. The multi-existence cluster accounts for more than 90% in the revisit willing and also is the highest in the personal spending and the participation of rural activities.

4 The WTP of Rural Regeneration

In this study, it will use questionnaire to further investigate tourists to Let It Be communities, and use the double-bounded dichotomous choice to establish a positivism model for evaluation of the rural regeneration through CVMs situational design.

CVM has long been one of the important methods used for a wide range of natural resource and environmental assessments. With the assumption of a virtual market, it reflects the measurement of natural resources and environment where market prices do not exist. In addition, an appropriate form of price inquiry method is used for a questionnaire survey to understand peoples preferences for natural resources and environment, and then estimate the amount the respondent is willing to pay or the compensation amount the respondent is willing to accept. The study uses a single-bound dichotomous choice model of the close-ended bidding method and a double-bound dichotomous choice model to estimate the value people are willing to pay for improvement of rural regeneration, with a reference to the theoretical model of [7, 11]. The structure is described as follows.

$$\ln L^s(\theta) = \sum_{i=1}^{N} \{d_i^Y \ln \pi^Y(B_i^s) + d_i^N \ln \pi^N(B_i^s)\}$$

$$= \sum_{i=1}^{N} \{d_i^Y \ln[1 - G(B_i; \theta)] + d_i^N \ln G(B_i, \theta)\}$$

The log-likelihood function of respondent (N) and the given price (B_i, B_i^U, B_i^D) is as follows:

$$\ln L^D(\theta) = \sum_{i=1}^{N} (d_i^{YY} \ln \pi^{YY}(B_i, B_i^U) + d_i^{NN} \ln \pi^{NN}(B_i, B_i^D)$$

$$+ d_i^{YN} \ln \pi^{YN}(B_i, B_i^U) + d_i^{NY} \ln \pi^{NY}(B_i, B_i^D))$$

4.1 Model Variable Setting

When the positivism model is set in this study, it respectively takes the tourism value and the maintenance value into consideration. Three variables such as income, times and stay are selected. The positivism model of the rural regeneration WTP price is as below.

$$\ln \text{WTP1} = \int (income, \, TIMES, \, STAY, \, d1, \, d2, \, fac1, \, fac2, \, fac3)$$

$$\ln \text{WTP2} = \int (income, \, TIMES, \, STAY, \, d1, \, d2, \, fac1, \, fac2, \, fac3)$$

where, InWTP1 is the WTP amount of the tourism value; InWTP2 is the WTP amount of the maintenance value; $(fac1)$ is the environment conservation factor; $(fac2)$ is the cultural heritage factor; $(fac3)$ is the economic value factor; $(d1)$ is the virtual variable of the community activity where 1 means participated, and 0 means not participated; ($d2$) is the cluster's virtual variable where 1 means the first group, and 2 means the second group.

4.2 Rural Regeneration Value Assessment

Under the hypothesis of Log-normal distribution, Weibull distribution and Gamma distribution, this study assesses the positivism model and deletes controversial samples, uses the survival regression analysis to obtain the estimated parameter of the explanatory variable and the functional parameter so as to investigate factors affecting the rural regeneration WTP price. Details are listed in Tables 2 and 3.

4.2.1 Estimated Result of Tourism Value

The estimated result of tourism value (Table 4) shows that in the three distributions of the evaluation function the income of the social economy is plus, and its value t is remarkable respectively under 1, 5 and 10% significant levels; the times is minus, and its value t is remarkable respectively under 1, 5 and 10% significant levels; the stay is plus, and its value t is remarkable for Weibull distribution under 10% significant level and for Gamma distribution under 5 and 10% except for Log-normal distribution. The variable (1) participating social activities is plus, and its value t is remarkable respectively under 1, 5 and 10% significant levels. The economic value (3) of the rural regeneration cognition is plus, and its value t remarkable for Log-normal distribution and Weibull distribution under 5 and 10% significant levels, and for Gamma distribution under 10% significant level.

4.2.2 Estimated Result of Maintenance Value

The estimated result of tourism value (Table 4) shows that in the three distributions of the evaluation function the income of the social economy is plus, and its value t is remarkable for Log-normal distribution and Weibull distribution respectively under

Table 2 One-way analysis of variance

Factor dimension	Multi-existence $n = 242$	Value-recreation $n = 121$	F-value
Environment conservation	4.56	3.97	301.992***
Cultural heritage	4.35	4.02	101.374***
Economic value	4.46	3.84	190.44***

Note *** indicates significant level of 1% respectively

Table 3 Estimated result of tourism values evaluation function

Variable	Evaluation function's probability distribution		
	Log-normal distribution	Weibull distribution	Gamma distribution
Intercept	0.062	−0.316	0.513
	−0.001	−0.042	−0.093
Social economy variables			
INCOME	0.308	0.364	0.258
	(12.83)***	(19.812)***	(7.759)***
TIMES	−0.176	−0.174	−0.178
	(8.609)***	(9.238)***	(8.563)***
STAY	0.167	0.145	0.174
	(−3.982)	(3.244)*	(4.292)**
Participate community activities or not			
d1	0.441	0.472	0.417
	(13.265)***	(16.327)***	(12.141)***
Cluster			
d2	0.024	0.078	−0.006
	(−0.011)	(−0.129)	(−0.001)
Rural regeneration cognition			
fac1	0.183	0.133	0.18
	(−0.604)	(−0.309)	(−0.638)
fac2	−0.162	−0.099	−0.179
	(−0.741)	(−0.315)	(−0.929)
fac3	0.247	0.277	0.236
	(4.027)**	(5.552)**	(3.689)*
Log-likelihood	−339.246	−349.401	−337.934
Restricted	−360.674	−372.316	−358.418
Log-likelihood			
Log-likelihood ratio	42.856	45.83	40.968
N	363		

Note *, ** and *** indicate significant levels of 10, 5 and 1% respectively

1, 5 and 10% significant levels, and for Gamma distribution respectively under 5 and 10% significant levels; times is minus; stay is plus. The variable (1) participating social activities is plus, and its value t is remarkable respectively under 1, 5 and 10% significant levels. The economic value (3) of the rural regeneration cognition is plus, and its value t is remarkable for Weibull distribution under 10% significant level.

Table 4 Estimated result of maintenance value's evaluation function

	Evaluation function's probability distribution		
	Log-normal distribution	Weibull distribution	Gamma distribution
Intercept	0.927	0.53	1.064
	−0.304	−0.119	−0.38
Social economy			
INCOME	0.238	0.291	0.224
	(7.03)***	(12.077)***	(5.397)**
TIMES	−0.104	−0.098	−0.105
	(−2.554)	(−2.556)	(−2.54)
STAY	0.111	0.095	0.112
	(−1.591)	(−1.363)	(−1.572)
Participate community activities or not			
d1	0.451	0.446	0.45
	(12.714)***	(14.215)***	(12.622)***
Cluster			
d2	0.034	0.113	0.171
	(−0.021)	(−0.253)	(−0.005)
Rural regeneration cognition			
fac1	0.206	0.122	0.213
	(−0.681)	(−0.246)	(−0.737)
fac2	−0.068	−0.001	−0.077
	(−0.122)	0	(−0.152)
fac3	0.148	0.207	0.14
	(−1.346)	(2.976)*	(−1.152)
Log-likelihood	−242.736	−247.533	242.642
Restricted	−255.492	−262.307	−255.014
Log-likelihood			
Log-likelihood ratio	25.512	29.548	24.744
N	363		

Note *, ** and *** respectively indicate they are remarkable under 1, 5 and 10% significant levels

4.2.3   Result of Rural Regeneration WTP

After both the tourism value and the maintenance value are estimated by the WTP value evaluation model, the result obtained is further used for investigating the money value of each rural regeneration function. Through [12] the use of median will not be easily affected by the extreme value when evaluating the confidence interval between the mean value and the median of the WTP price. Therefore, this study will estimate the result according to each rural regeneration WTP value.

The total of 363 questionnaires collected for interviewees in the WTP price show that tourists' WTP amount is higher than that of the tourism value compared to the maintenance value, indicating tourists are willing to pay a greater amount of money to maintain the overall environment and landscape of the rural regeneration communities and do not want these buildings and environments with long history to perish because of disrepair and not being carefully maintained.

Item	Mean (NTD/person)	95% confidence interval (NTD/person)
Recreation value	153	(111,211)
Maintained value	177	128,246)

5 Conclusion

(1) Rural regeneration cognition: it can investigate from three dimensions, including "environment conservation", "cultural heritage" and "economic value". Although our body and mind can be relaxed through rural tourism and landscape appreciation, we do not want the increased number of tourists to destroy the environment. Buildings and landscapes conserved in early stage can be presented to tourists through implementation of the rural regeneration plan. Besides activating rural economy and reducing urban-rural gap, the government endows rural areas with another value through the rural regeneration plan.

(2) Analysis of group cognitive differences: it mainly consists of the "multi-existence" attaching importance to all functions, and the "value recreation" attaching more importance to rural cultural heritage. Generally speaking, the "multi-existence" group prefers the average monthly income, stay, revisit willing, personal spending and participation of rural activities.

(3) Evaluation of rural regeneration WTP value: the evaluation function's esti-mated result shows that other results are similar to the estimated result of the tourism value except for the non-significant times and stay as for the maintenance values evaluation result. It is explained as below: the income of social economy is plus and significant, indicating the higher the income, and higher the WTP value will be; the times is minus and significant, indicating the higher the number of travelling to rural areas, the lower the WTP price will be, and the maintenance value is not significant; the stay is plus and significant, indicating the longer you stay in a community, the higher the WTP price will be, and the maintenance value is not significant; the par-ticipation of community activities is plus and significant, indicating the higher you are willing to participate in community activities, the higher the WTP price will be; the economic value of the rural regeneration cognition is plus and significant, indi-cating tourists mainly take the rural macroeconomic value into consideration as for the rural regeneration policy, and hope that they can directly be helpful to peasants.

In addition, the "maintained value" is higher than the "recreation value" for tourists, and it also can reflect tourists' cognition of rural regeneration functions. The cognition level of the "environment conservation" is higher, so we can estimate that the reason why its WTP price is higher.

References

1. Lee KY, Lan LC, Fang CL, Wang JH (2014) Impact of participation in rural regeneration program on satisfaction of rural residents with community development. Surv Res Method Appl 32:12–51
2. Liu CW (2009) Conservation benefits of public facilities at the areas of rural community consolidation. PhD. Thesis, National Chung Hsing University in Taiwan
3. Tung YC (2009) A study of rural roads establishment from the perspective of sustainable development. Master Thesis, National Chung Hsing University in Taiwan
4. Huang MY (2011) The guide of rural regeneration promotion, soil and water conservation bureau of Nantou
5. Haven Tang C, Jones E (2012) Local leadership for rural tourism development: a case study of adventa, Monmouthshire. Tourism Manag Perspect 4:28–35
6. Tsai CY, Wei L, Chang FH (2008) The study of residents' recreational impact and sense of community on the attitude of ecotourism development in Penghu Area. Sport Recreation Res J 2(4):14–29
7. Hoehn J, Randall A (1987) A satisfactory benefit cost indicator from contingent valuation. J Environ Econ Manag 14:226–247
8. Duffield JW, Patterson DA (1991) Inference and optimal design for a welfare measure in dichotomous choice contingent valuation. Land Econ 67(2):225–239
9. Cameron TA (1988) A new paradigm for valuing non-market goods using referendum data-maximum-likelihood estimation by censored logistic-regression. J Environ Econ Manag 15(3):355–379
10. Li MT, Huang YC (2006) The research of tourist's willingness to pay for interpretative service: a case of ecological reserve area in Kentin National Park. J Leisure Tourism Ind Res 1(1):19–33
11. Suzuki Y (2008) Food in schools programme healthier cookery clubs. J Jpn Assoc Home Econ Educ 51(1):11–18
12. Hanemann WM, Loomis J, Kanninen B (1991) Statistical efficiency of double-bounded dichotomous choice contingent valuation. Am J Agric Econ 73(4):1255–1263

Analyzing the Contribution of ASEAN Stock Markets to Systemic Risk

Roengchai Tansuchat, Woraphon Yamaka, Kritsana Khemawanit
and Songsak Sriboonchitta

Abstract In this paper, seven stock markets from six countries (Thailand, Malaysia, Indonesia, Vietnam, the Philippines, and Singapore) and their risk contribution to ASEAN stock system are investigated using the Component Expected Shortfall approach. Prior to computing this systemic risk measure, we need to compute a dynamic correlation, thus the study proposes a Markov Switching copula with time varying parameter to measure the dynamic correlation between each pair of stock market index and ASEAN stock system. The empirical results show that Philippines stock index contributed the highest risk to the ASEAN stock system.

Keywords Markov switching model copula · Time varying dependence · CES · ASEAN stock markets

1 Introduction

Although economic growth in ASEAN countries has been quite favorable in general, it can be disrupted or even reversed by various factors as we have witnessed from such situation as the financial crisis in 2008–2009 in Thailand or the political disorders elsewhere. These situations can be referred as a risk that might occur in the future.

After the establishment of the Association of Southeast Asian Nations (ASEAN), it is crucial to observe the roles and impacts of the seven leading ASEAN financial markets which consist of the Stock Exchange of Thailand, Bursa Malaysia, Ho Chi Minh Exchange, Hanoi Stock Exchange, the Philippine Stock Exchange, Singapore Stock Exchange, and Indonesia Stock Exchange. These stock markets can potentially

R. Tansuchat (✉) · W. Yamaka · K. Khemawanit · S. Sriboonchitta
Faculty of Economics, Chiang Mai University, Chiang Mai, Thailand
e-mail: roengchaitan@gmail.com

W. Yamaka
e-mail: woraphon.econ@gmail.com

© Springer International Publishing AG 2017
V. Kreinovich et al. (eds.), *Robustness in Econometrics*,
Studies in Computational Intelligence 692, DOI 10.1007/978-3-319-50742-2_40

stimulate the ASEAN economic growth for functioning as the large source of capital investment. After the formal establishment of the ASEAN Community in 2015, ASEAN countries become more integrated and thereby leading to fewer trade barriers and more collaboration among the various stock markets of ASEAN. Although cross-border collaboration of ASEAN countries can promote ASEAN stock markets and offer more opportunities to investors across the region, it can also bring a large financial risk to a country as well as across the ASEAN countries. Therefore, it will be a great benefit to the ASEAN if we can quantify the contribution of each stock market to the overall risk of the ASEAN stock system. To achieve our goal, this study considers Component Expected Shortfall (CES) concept proposed by Banulescu and Dumitrescu [1]. This new approach provides several advantages like that it can be used to assess the contribution of each stock market to the overall risk of the system at a precise date. In the real application, the study of Liu et al. [9] examined the volatility and dependence for systemic risk measurement using copula model with CES. Their work found that CES can explain the financial crisis risk in 2009 and that the risk contribution was lower in pre-crisis period when compared to the post crisis time. Hence, we expect that CES becomes a good candidate tool for policy makers to select which stock markets to monitor, with a view to discourage the accumulation of systemic risk.

Prior to measuring one-period-ahead, the time-varying correlations of ASEAN and individual stock market need to be computed. Banulescu and Dumitrescu [1] and Liu et al. [9] proposed a Dynamic conditional correlation (DDC) GJR-GARCH(1,1) model to compute conditional volatility, standardized residuals for the ASEAN and each country. However, the linear correlation and normality assumption of the model might not be appropriate and accurate for measuring the correlation between two financial markets. In reality, finance asset return has the presence of heavy tails and asymmetry correlation thus implementing DCC-GJR-GARCH may lead to inadequate CES estimation. To overcome these problems, the study proposed an alternative model, a Markov Switching dynamic copula as advanced by Silva Filho et al. [4] to compute the dynamic correlation of market pair. This model takes an advantage of the copula approach of Sklar theorem to construct the joint distribution of the different marginal distribution with different copula structure. Hence, the model becomes more flexible to capture both linear and nonlinear and both symmetric and asymmetric correlation between ASEAN and individual stock market. In addition, we also take into account the non-linearity and asymmetric dependence of the financial data since financial markets are likely to be more dependent in market downturn than in market upturn, see Chokethaworn et al. [3], Fei et al. [5], Filho and Ziegelmann [6], Pathairat et al. [12].

The rest of this paper is organized as follows: Sects. 1, 2 and 3 present the approaches that we employ in this study. In Sect. 4, we explain the data and the empirical results and Sect. 5 provides a conclusion of this study.

2 Methodology

2.1 ARMA-GARCH Model

The log-difference of each stock index (y_t) is modeled by univariate $ARMA(p, q)$ with $GARCH(1, 1)$. This study used $GARCH(1, 1)$ since it is able to reproduce the volatility dynamics of financial data, while leading to no autocorrelation in the ARMA process. In our case, the $ARMA(p, q) - GARCH(1, 1)$ (where p is the order of AR and q is the order of MA) is given by

$$y_t = \mu + \sum_{i=1}^{p} \phi_i y_{t-i} + \sum_{i=1}^{q} \psi_i \varepsilon_{t-i} + \varepsilon_t \tag{1}$$

$$\varepsilon_t = h_t z_t \tag{2}$$

$$h_t^2 = \varpi + \alpha_1 \varepsilon_{t-1}^2 + \beta_1 h_{t-1}^2 \tag{3}$$

where μ, ϕ_i, ψ_i, ϖ, α_1 and β_1 are the unknown parameters of the model, ε_t is the white noise process at time t, h_t^2 is the variance of error at time t, z_t is standardized residuals and it must satisfy the condition of being independently and identically distributed. We also assume that ε_t has a student-t distribution with mean 0, variance σ^2, and degree of freedom v, i.e., $\varepsilon_t \sim t(0, \sigma^2, v)$. Some standard restrictions on the GARCH parameters are given such that $\varpi > 0, \alpha_1 > 0, \beta_1 > 0$ and $\alpha_1 + \beta_1 < 1$.

2.2 Conditional Copula Model

Sklar theorem showed a way to construct a joint distribution function using copula approach. By the theorem, let H be the joint distribution of random variable $(x_1, x_2, ..., x_n)$ with marginals $F_1(x_1), F_2(x_2), ..., F_n(x_n)$, then the joint cumulative distribution function (cdf) can be represented according to

$$H(x_1, x_2, ..., x_n) = C(F(x_1), F_2(x_2), ..., F_n(x_n)) \tag{4}$$

when $F_i(x)$ are continuous functions, then Eq. (1) provides a unique representation of cdf for any random variables or Copula is unique. In this study, we aim to analyze the dynamic dependence of two dimension copula, therefore, according to Pattan (2006), we can rewrite (Eq. 4) in the form of conditional copula such that

$$H(x_1, x_2 | \omega) = C(F_1(x_1 | \omega), F_2(x_2 | \omega)) \tag{5}$$

where ω is a 1 dimension conditioning variable of x_1 or x_2 and F_1 and F_2 become the conditional distribution of $x_1 \,|\omega$ and $x_2 \,|\omega$, respectively. Thus, we can obtain the conditional density function by differentiating (Eq. 5) with respect to x_1 and x_2.

$$
\begin{aligned}
h(x_1, x_2 \,|\omega) &= \frac{\partial^2 H(x_1, x_2 \,|\omega)}{\partial x_1, x_2} \\
&= \frac{\partial F_1(x_1 \,|\omega)}{\partial x_1} \cdot \frac{\partial F_2(x_2 \,|\omega)}{\partial x_2} \cdot \frac{\partial^2 C(F_1(x_1 \,|\omega), F_2(x_2 \,|\omega) \,|\omega)}{\partial u_1 \partial u_2} \qquad (6) \\
&= f_1(x_1 \,|\omega) \cdot f_2(x_2 \,|\omega) \cdot c(u_1, u_2 \,|\omega)
\end{aligned}
$$

where $u_1 = F_1(x_1 \,|\omega)$ and $u_2 = F_2(x_2 \,|\omega)$ and these marginal distributions (u_1, u_2) are uniform in the [0, 1]. In this dynamic case, Patton (2006) suggested allowing the dependence parameter (θ_t) to vary over time in the ARMA (1,10) process, as follows:

$$
\theta_t = \Lambda(a + b\theta_{t-1} + \varphi \Gamma_t) \qquad (7)
$$

where $\Lambda(\cdot)$ is the logistic transformation for each copula function, a is the intercept term, b is the estimated coefficient of AR and Γ_t is the forcing variable which is defined as

$$
\Gamma_t = \begin{cases} \frac{1}{10} \sum_{j=1}^{10} F_1^{-1}(u_{1,t-j}) F_2^{-1}(u_{2,t-j}) & elliptical \\ \frac{1}{10} \sum_{j=1}^{10} |u_{1,t-j} - u_{2,t-j}| & Archimedean \end{cases}
$$

In the Copula model, there are two main classes of the copulas namely, Elliptical class and Archimedean class. Both classes contain copula families that are used to join the marginal distribution. In the case of Elliptical copula, there are two symmetric copula families consisting Gaussian and the Student-t copulas. Both families have a similar structure except for their tail dependence. The Student-t copula has shown to be generally superior to the Normal copula since it has tail dependence. As for the Archimedean case, it is an alternative class of copulas with asymmetric tail dependence, meaning that dependence in lower tail can be larger than dependence in upper tail and vice-versa.

In the most recent development, there are many copula functions being proposed to join the marginal distribution; see, e.g., [2, 8]. In this study, we consider 5 conditional copula families consisting Gaussian copula, Student-t copula, Gumbel copula, Clayton copula, and Symmetrized JoeClayton (SJC) copula to analyze the structure of dependence between each stock market and ASEAN market (see the copula functions in Tofoli et al. [15].

2.3   Regime-Switching Copula

There are many evidences regarding financial returns tending to exhibit different patterns of dependence such as those from the works of Silva Filho et al. [4], Tofoli et al. [15], and Pastpipatkul et al. [12]. These studies arrived at similar conclusion that stock markets exhibit the different degree of dependence over time and Tofoli et al. [15] specifically mentioned that stock returns tend to be more dependent during crisis period or high volatility period while likely to be less dependent in the market upturn or low volatility period. For these reasons, the dependence structure of the variables may be determined by a hidden Markov chain with two states (Tofoli et al. [15]). Hence, in this study, it is reasonable to extend the time varying copula of Patton [14] to the Markov Switching of Hamilton [7] and thus we have a Markov-switching copula with time-varying dependence (MS-Cop) to model dependence parameter (θ_t). The study allows the (θ_t) to vary across the economic regime, say the upturn market (regime 1) and downturn market (regime 2). Thus, θ_t is assumed to be governed by an unobserved variable (S_t).

$$\theta_t = \theta_{t(S_t=1)} + \theta_{t(S_t=2)} \tag{8}$$

where $\theta_{t(S_t=1)}$ and $\theta_{t(S_t=2)}$ are time varying dependence parameter for regime 1 and regime 2, respectively. Thus, when the regime switching is taken into account in dependence parameter, then we can rewrite the dynamic function with ARMA(1,10) process Eq. (7) for two regimes as

$$\theta_{(S_t=1),t} = \Lambda(a_{(S_t=1)} + b\theta_{(S_t=1),t-1} + \varphi\Gamma_t)$$
$$\theta_{(S_t=2),t} = \Lambda(a_{(S_t=2)} + b\theta_{(S_t=2),t-1} + \varphi\Gamma_t) \tag{9}$$

where there is only intercept term of time varying (Eq. 9) $a_{(S_t=i)}$, $i = 1, 2$, that is governed by state. In this study, the unobservable regime ($S_t = 2$) is governed by the first order Markov chain, meaning that the probability of this time t is governed by $t-1$, hence, we can write the following transition probabilities (P):

$$p_{ij} = Pr(S_t = j \,|S_{t-1} = i) \quad and \quad \sum_{j=1}^{2} p_{ij} = 1 \quad i, j = 1, 2 \tag{10}$$

where p_{ij} is the probability of switching from regime i to regime j, and these transition probabilities can be formed in a transition matrix P, as follows:

$$P = \begin{bmatrix} p_{11} & p_{12} = 1 - p_{11} \\ p_{21} = 1 - p_{22} & p_{22} \end{bmatrix} \tag{11}$$

2.4 Copula Likelihood Estimation

Since the computation of the ML estimate may be difficult to find the optimal solution for a large number of unknown parameters, the two-stage maximum likelihood (ML) approach, as proposed by Patton [14] and Tofoli et al. [15], is conducted in this study to estimate the MS-Cop model. In the first step, we estimate and select the parameters of the best fit marginal distributions for individual variables from ARMA(p,q)-GARCH process. In the second step, we estimate the dependence structure of the MS-Cop. According to (Eq. 6), let $\Theta = \{\omega_1, \omega_2, \theta_t\}$ we can derive the likelihood function of a single regime conditional copula as

$$L(\Theta \,|x_1, x_2) = f_1(x_1 \,|\omega_1) \cdot f_2(x_2 \,|\omega_2) \cdot c(u_1, u_1 \,|\theta_t)$$

where $f_1(x_1 \,|\omega_1)$ and $f_2(x_2 \,|\omega_2)$ are the density function of the marginal distribution which are assumed to be fixed obtaining from ARMA(p,q)-GARCH process in the first step. $c(u_1, u_2 \,|\theta_t)$ is the density function of the conditional copula. Note that the study is considering two-regime MS-Cop, thus we can rewrite the single regime conditional copula to be two-regime MS-Cop as:

$$
L(\Theta_{S_t} \,|x_1, x_2)
$$
$$
= \sum_{t=1}^{T} \log \left[\sum_{S_t=1}^{2} [f_1(x_1 \,|\omega_1) \cdot f_2(x_2 \,|\omega_2) \cdot c(u_1, u_2 \,|\theta_{(S_t=i),t})] \cdot \Pr(S_t = i \,|\xi_{t-1}) \right]
$$
(12)

where $Pr(S_t = i \,|\xi_{t-1})$ is the filtered probabilities and ξ_{t-1} is the all information up to time $t - 1$, $\Phi_{S_t, t-1}, x_{1,t-1}, x_{2,t-1}$. To compute the $Pr(S_t = i \,|\xi_{t-1})$, we employ a Kims filter as described in Kim and Nelson [11]. The estimation in this second step is performed by maximizing the copula log-likelihood Eq. (12).

3 Component Expected Shortfall

In this section, we introduce a Component Expected Shortfall (CES) which is proposed in Banulescu and Dumitrescu (2012). We apply the MS-Cop to CES in order to assess the contribution of an individual stock in ASEAN to the risk of the ASEAN stock system at a precise date. Let r_{it} denote the return of stock index i at time t and r_{mt} denote the aggregate return of the ASEAN stock index at time t.

$$r_{mt} = \sum_{i=1}^{n} w_{it} \cdot r_{it}$$
(13)

where w_{it} is an individual weight the value-weighted of stock index i, $i = 1, ..., n$, at each date under analysis. These weights are given by the relative of stock index i capitalization to ASEAN stock system. And CES is defined as the part of Expected Shortfall (ES) of the ASEAN stock index due to ith stock index

$$
CES_{it} = \frac{w_{it} \partial ES_{m,t-1}(C)}{\partial w_{it}} \tag{14}
$$
$$
= -w_{it} E_{t-1}(r_{it} | r_{mt} < C)
$$

where $E_{t-1}(r_{it} | r_{mt} < C) = \partial ES_{m,t-1}(C)/\partial w_{it}$ is the Marginal Expected Shortfall (MES) which measures the marginal contribution of individual stock index to the risk of the ASEAN stock index.

$$
MES_{mt} = \left[h_{it} \cdot \kappa_{it} \frac{\sum_{t=1}^{T} \Upsilon_{mt} \Phi(\frac{C-\Upsilon_{mt}}{h_{mt}})}{\sum_{t=1}^{T} \Phi(\frac{C-\Upsilon_{mt}}{h_{mt}})} \right]
$$
$$
+ \left[h_{it} \cdot \sqrt{1 - \kappa_{it}} \frac{\sum_{t=1}^{T} e_{it} \Phi(\frac{C-\Upsilon_{mt}}{h_{mt}})}{\sum_{t=1}^{T} \Phi(\frac{C-\Upsilon_{mt}}{h_{mt}})} \right] \tag{15}
$$

where $\Upsilon_{mt} = r_{mt}/h_{mt}$ and $e_{it} = (r_{it}/h_{it}) - \kappa_{it}$ are standardized ASEAN market return and stock index i, which h_{mt} and h_{it} are the variance of error at time t. $C = 1/h_{mt}$ is the threshold value which is assumed to depend on the distribution of the r_{mt}. Φ is the cumulative normal distribution function and κ_{it} is the time varying Kendall s tau which can be tranformed from the expected dependence parameter $(E\kappa_t)$,

$$
E\kappa_t = \sum_{j=1}^{2} [\kappa_{(S_t=j),t}] \cdot [Pr(S_t = j | \xi_{t-1}) \times P]
$$

However, our study aims to assess the contribution of risk of each stock market to the ASEAN stock system, thus it is better to measure the risk in terms of percentage by

$$
CES_{it}\% = (CES_{it}/\sum_{i=1}^{n} CES_{it}) \times 100
$$

4 Data and Empirical Results

In this study, we use the data set comprising the Stock Exchange of Thailand index (SET), Indonesia Stock Exchange index (IDX), the Philippine Stock Exchange (PSE), Bursa Saham Kuala Lumpur Stock Exchange (BURSA), Straits Times stock index (STI), Ho Chi Minh Stock Index (HOC) and Hanoi Stock Exchange index(HN). The data set consists of weekly frequency collected from the period of January 1, 2009 to June 8, 2016, covering 388 observations. All the series have been transformed into the difference of the logarithm. And the ASEAN market index is based

Table 1 Descriptive statistics on ASEAN index

	SUM ASEAN	SET	IDX	BURSA	STI	HOC	HN	PSEI
Mean	0.0029	0.0032	0.0035	0.0016	0.0015	0.0025	0.0026	0.0037
Med	0.004	0.0059	0.004	0.002	0.0024	0.0022	0.0034	0.0047
Max	0.0975	0.0994	0.099	0.0568	0.1639	0.1202	0.1066	0.0913
Min	−0.0738	−0.1	−0.108	−0.0694	−0.104	−0.1633	−0.1498	−0.1287
Std.	0.0189	0.025	0.0238	0.015	0.0228	0.0335	0.0322	0.0251
Skew	−0.0991	−0.3708	−0.2792	−0.3585	0.6942	−0.1913	−0.1327	−0.5563
Kurtosis	6.3125	4.7446	5.9074	5.3807	10.8985	5.1584	5.0532	6.0172
JB	174.360*	56.895*	138.774*	97.8811*	1018.303*	76.082*	67.861*	163.742*
ADF-test								
None	−17.561*	−19.097*	−19.446*	−19.042*	−18.391*	−17.837*	−17.780*	−20.475*
Intercept	−17.935*	−19.396*	−19.854*	−19.261*	−18.456*	−17.904*	−17.867*	−20.927*
Intercept and Trend	−18.331*	−19.742*	−20.403*	−19.738*	−18.782*	−17.943*	−17.924*	−21.135*

Source Calculation

Note: * is significant at 1% level

on the stocks in only the seven stock markets of our interest. The computation of this index is defined as the value-weighted average of all stock index returns.

4.1 Modeling Marginal Distributions

For the first state, we use each ASEAN indexes prices to calculate the natural log returns defined as $r_{i,t} = ln(P_{i,t}) - ln(P_{i,t-1})$ where $P_{i,t}$ is the ith index price at time t, and $r_{i,t}$ is the ith log return index price at time t. The descriptive statistics of ASEAN returns are shown in Table 1 which is clear that mean of each ASEAN variable is positive with the highest mean returns being PSEI (0.0037), the lowest mean return being STI (0.0015), and that the standard deviation in HOC is the highest (0.0335) and that in BURSA the lowest (0.0150). In terms of skewness and kurtosis, the values of skewness are small but the values of kurtosis are large. So these mean that the distributions of ASEAN returns have fatter tail instead of normal distribution and Jaque-Bera test rejected the null hypothesis, thus the return series has non-normal distribution.

Moreover, in order to check unit roots in the series, the Augmented Dickey-Fuller (ADF) tests are applied. The test results at 0.01 statistical significance level 0.01 indicated that all series of ASEAN returns are stationary. Table 2 presents the coefficient for the ARMA(p,q)-GARCH(1,1) with student-t distribution for each ASEAN return series. The optimum lag for ARMA(p,q)-GARCH(1,1) is selected by the minimum Akaike information criterion (AIC) and Bayesian information criterion (BIC) value. The estimated equations of SUM ASEAN, SET, and IDX are

Table 2 Estimate of ARMA(q,p)-GARCH(1,1) in student-t distribution

	SUM ASEAN	SET	IDX	BURSA	STI	HOC	HN	PSEI
μ	0.004***	0.001	0.002	0.002***	0.002*	0.001	0.001*	0.005
	(−0.001)	(−0.001)	(−0.001)	(−0.001)	(−0.001)	(−0.001)	(−0.001)	(−0.001)
AR(1)	0.391***	0.562***	0.603***	−1.353***	−0.909***	0.911***	0.905***	−1.066***
	(−0.033)	(−0.045)	(−0.059)	(−0.014)	(−0.023)	(−0.029)	(−0.024)	(−0.055)
AR(2)	0.328***	−0.256***	0.522***	−0.958***	−0.891***	–	–	−0.117**
	(−0.044)	(−0.049)	(−0.054)	(−0.014)	(−0.021)			(−0.049)
AR(3)	−0.904***	0.657***	−0.155***	–	–	–	–	–
	(−0.034)	(−0.044)	(−0.054)					
MA(1)	−0.398***	−0.575***	−0.708***	1.395***	0.974***	−0.951***	−0.950***	0.974***
	(−0.025)	(−0.016)	(−0.016)	(−0.011)	(−0.003)	(−0.027)	(−0.024)	(−0.018)
MA(2)	−0.336***	0.291***	−0.454***	0.982***	0.990***	–	–	–
	(−0.033)	(−0.014)	(−0.005)	(−0.011)	(−0.002)			
MA(3)	0.957***	−0.710***	0.173***	–	–	–	–	–
	(−0.024)	(−0.020)	(−0.004)					
ω	0.000*	0.000**	0.000*	0.000	0.000*	0.000	0.000	0.000
	(0.000)	(0.000)	(0.000)	(0.000)	(0.000)	(0.000)	(0.000)	(0.000)
α	0.154**	0.219***	0.138**	0.115**	0.098**	0.048**	0.026*	0.073**
	(−0.062)	(−0.082)	(−0.058)	(−0.056)	(−0.041)	(−0.023)	(−0.015)	(−0.036)
β	0.790***	0.643***	0.809**	0.802***	0.861***	0.936***	0.971***	0.879***
	(−0.073)	(−0.113)	(−0.067)	(−0.095)	(−0.044)	(−0.032)	(−0.016)	(−0.055)
ν	5.009***	9.892***	4.554***	4.360***	4.637***	7.205***	7.139***	5.191***
	(−1.305)	(−5.053)	(−1.208)	(−1.195)	(−1.021)	(−2.319)	(−2.480)	(−1.335)
AIC	−5.3193	−4.689	−4.843	−5.722	−5.076	−4.102	−4.246	−4.677
BIC	−5.2046	−4.575	−4.729	−5.628	−4.982	−4.029	−4.174	−4.594
$Q^2(10)$	0.4154	0.2548	0.2478	0.2448	0.5784	0.9854	0.8845	0.5946
KS-test	0.5757	0.6584	0.5447	0.9871	0.9844	0.4354	0.8841	0.6844

Source Calculation

Note: (1) *, **, and *** denote significant at 90, 95 and 99%, respectively
(2) In the bracket is standard error

ARMA(3,3)-GARCH(1,1), BURSA and STI are ARMA(2,2)-GARCH(1,1), VI and VAI are ARMA(1,1)-GARCH(1,1), and PSEI is ARMA(2,1)-GARCH(1,1). Furthermore, the coefficient of each equation is statistically significant at 1% in most cases which means that the t distribution assumption for ARMA-GARCH model is reasonable.

In addition, the autocorrelation test (LjungBox test) and the KolmogorovSmirnov test (KS-test) are also shown in this table. The p-value of the KS-test suggests that the probabilities of the integral transform of the standardized residuals are uniform in the [0, 1] interval. Additionally, the p-value of the LjungBox-test of autocorrelation on standardized residuals with 10 lags, $Q^2(10)$, confirms that we cannot reject the 5% significance level; thus, there is no autocorrelation in any of the series.

4.2 Model Fit

MS-Cop models are estimated by different copula functions, and selection of the most appropriate structure dependence between each pair in this section is based on the lowest Akaiki information criterion (AIC) and Bayesian information criterion (BIC). Table 3 presents various copula functions of MS-Cop model. It contains the AIC and BIC for each copula model. These are evaluated at the highest value of copula log likelihood. The result showed that Clayton copula yields the lowest AIC and BIC for IDX-ASEAN pair, STI pair, and PHI-pair; while for SET-ASEAN, BURSA-ASEAN, HOC-ASEAN, and HN-ASEAN pairs, Gumbel copula provides the best structural fit.

Table 3 Family selection of each pair copula

AIC BIC	SET ASEAN	IDX ASEAN	BURSA ASEAN	STI ASEAN	HOC ASEAN	HN ASEAN	PHI ASEAN
Gaussian	225.2815	374.7991	212.0398	313.8305	56.7444	53.3816	536.4979
	248.8749	398.3925	235.6333	337.424	80.3379	76.9751	560.0914
Student- t	236.8088	382.0257	218.9943	336.3531	56.6692	56.6692	542.4576
	268.2667	413.4837	250.4523	367.811	88.1271	88.1271	573.9156
Clayton	211.3228	305.5207	187.2049	282.9581	46.2591	45.8371	439.9532
	234.9163	329.1142	210.7983	306.5516	69.8526	69.4306	463.5466
Gumbel	192.52	333.1427	180.1803	297.4253	51.0105	45.0553	447.7723
	216.1134	356.7362	203.7738	321.0088	74.604	68.6487	471.3658
SJC	255.0907	381.9507	226.8514	342.2112	69.2851	63.4991	527.4651
	294.4132	421.2731	266.1739	381.5336	108.6076	102.8216	566.7876

Source Calculation

4.3 Results of Estimated Parameters

Table 4 reports the estimated parameters of the MS-Cop for seven pairs of market returns. The models present a dynamic copula equation and the result showed that all stock pairs provide an evidence of the lower value of intercept coefficient in regime 1, $a_{(S_t=1)}$, than the value of the regime intercept coefficient in regime 2 $a_{(S_t=2)}$. Thus, we can interpret regime 1 as the low dependence regime, while regime 2 as the high dependence regime. Moreover, many recent studies, such as the studies by Tofoli et al. [15] and Karimalis and Nimokis [10], suggested that the degree of dependence during market upturns is less than that during market downturns. Thus, we will indicate the high dependence regime as the market downturn regime and the low dependence regime as the market upturn regime. Furthermore, we take into consideration the estimated coefficient, b, which is related to the autoregressive parameter component in the dynamic equation. Different results have been obtained from these coefficients. We found that the autoregressive parameter components of $c_{set,Asean}$, $c_{bursa,Asean}$, $c_{hoc,Asean}$, $c_{hn,Asean}$, and $c_{psi,Asean}$ have a negative sign, indicating that those pair relations are persistent over time, while the autoregressive parameter components of $c_{idx,Asean}$ and $c_{sti,Asean}$ have a positive sign, indicating that those pair relations are not persistent over time. As for the distance from the perfect correlation in the dependence dynamics co-movement, φ, the results also provide a different sign for each pair return. We found that the φ of $c_{set,Asean}$, $c_{idx,Asean}$, $c_{bursa,Asean}$, $c_{sti,Asean}$, and $c_{psi,Asean}$ has a negative sign, indicating that the greater distance from the perfect correlation can decrease their dependence, while the φ of $c_{hoc,Asean}$ and $c_{hn,Asean}$ has a positive sign, indicating that the greater distance from the perfect correlation can increase their dependence.

In addition, the transition probabilities p_{11} and p_{22} of all pair dependences are also reported in Table 4. We denote the probabilities p_{11} and p_{22} as the probabilities of

Table 4 Estimated parameters from Markov-switching dynamic copula

	$c_{set,Asean}$	$c_{idx,Asean}$	$c_{bursa,Asean}$	$c_{sti,Asean}$	$c_{hoc,Asean}$	$c_{hn,Asean}$	$c_{psi,Asean}$
$a_{(S_t=1)}$	1.7228	1.5386	1.599	1.282	0.2374	0.4255	1.8777
	(−0.0673)	(−0.4746)	(−0.0698)	(−0.0737)	(−0.4384)	(−0.0539)	(−0.583)
$a_{(S_t=2)}$	4.53	2.6278	4.9678	3.3328	6.801	10.6893	6.0509
	(−0.0001)	(−0.9729)	(−0.0002)	(−2.769)	(−22.582)	(−0.0001)	(−6.7414)
b	−0.3592	0.047	−0.3904	0.0773	−0.1345	−0.0857	−0.0289
	(−0.0001)	(−0.1854)	(−0.0001)	(−0.0826)	(−0.4144)	(−0.0041)	(−0.1807)
φ	−1.2757	−1.8663	−0.5262	−0.7013	0.1472	0.5014	−1.4553
	(−0.0001)	(−0.4534)	(−0.0045)	(−0.8727)	(−0.9533)	(−0.4512)	(−0.1541)
Transition probabilities							
p_{11}	0.9992	0.9999	0.9799	0.9999	0.9999	0.9999	0.9999
p_{22}	0.9987	0.9999	0.9984	0.9999	0.9999	0.9999	0.9999

Source Calculation
Note: In the bracket is standard error

staying in their own regime. We can observe that both regimes are persistent because of the high values obtained for the probabilities p_{11} and p_{22}.

4.4 Risk Measure

To achieve our goal of study, in this section, we extend our results obtained from the MS-Cop to assess the contribution of individual stock market to systemic risk at time. The study employed CES approach as a tool to assess the percentage of each stock markets contribution to the risk of the ASEAN stock system. The analysis is performed for almost eight years of samples from 2009 to 2016, coinciding with the period of Hamburger crisis of the United States of America (USA) (2009) and European debt crisis (2002-present). The study of Pastpipatkul et al. [13] investigated and found the effect of these crises on some countries in the ASEAN. Therefore, it is reasonable to measure the contribution of risk under these periods in order to check whether CES can identify the systemic financial risk or not.

As we mentioned in the introduction, the main purpose of the study is to access to contribution of each stock index to the ASEAN stock system. We also aim to identify the riskiest of the seven stock markets in the ASEAN by directly ranking the markets according to their CES%. According to Figs. 1, 2, 3, 4, 5, 6 and 7, these figures display the expected dependence (measured by Kendall tau) between individual stock index and the ASEAN stock system (upper panel); and the percentage of each

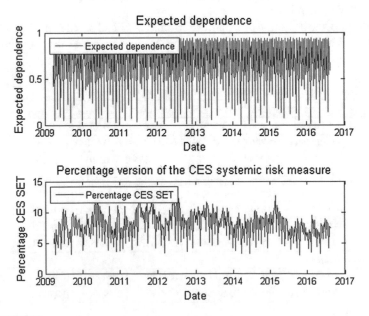

Fig. 1 CES SET

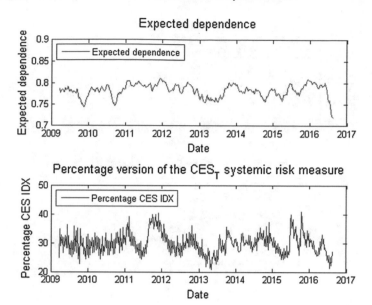

Fig. 2 CES IDX

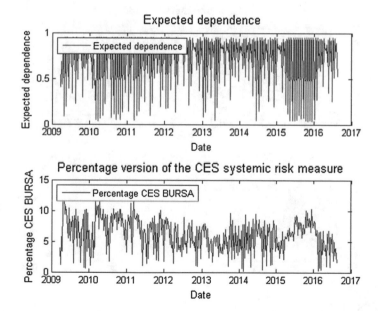

Fig. 3 CES BURSA

individual stock index in the risk of the ASEAN market system (measured by CES). Let consider the upper panel of all pairs, the results show that the expected dependencies are varying over time and provide an evidence of positive dependence. These

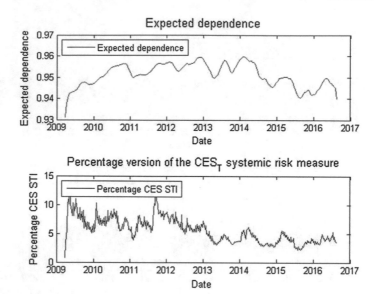

Fig. 4 CES STI

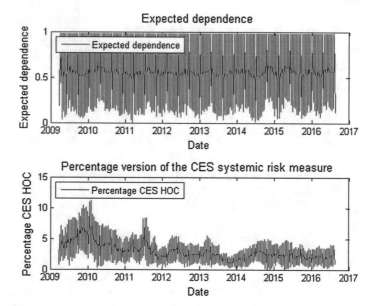

Fig. 5 CES HOC

indicate that ASEAN stock markets have the same movement direction through-out the sampling period. However, we can obviously notice that the time varying Kendalls tau, which was obtained from the estimated bivariate time varying dependence copula parameters, shows different results regarding correlation. The results of

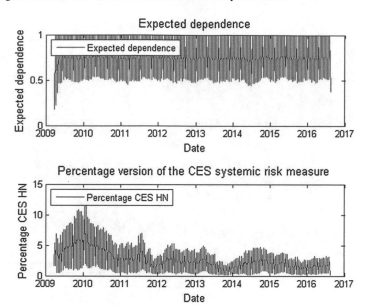

Fig. 6 CES HN

ASEAN-SET, ASEAN-BURSA, ASEAN-HOC, and ASEAN-HN pairs illustrate a highly fluctuating correlation over time where the values vary between 0.1–1, except for the ASEAN-HN where the value of time varying Kendalls tau varies between 0.5–1. Meanwhile ASEAN-IDX, ASEAN-STI, and ASEAN-PHI seem to have a lower fluctuating correlation where the values vary between 0.2–0.6, 0.93–0.96, and 0.2–0.5 for ASEAN-IDX, ASEAN-STI, and ASEAN-PHI, respectively. These evidences can be explained in various ways. Firstly, our results confirm that there exists a different degree of dependence between individual stock index and ASEAN stock system over time. Secondly, there is a positive co-movement between individual stock index and ASEAN stock system.

Then, let consider the lower panel of Figs. 1, 2, 3, 4, 5, 6 and 7 which presents the total loss of ASEAN stock system attributable to the seven stock markets for the period 2009–2016. There are several interesting findings that can be observed when we focus on the individual stock market results. We can observe that during 2008–2009 ASEAN-BURSA, ASEAN-STI, ASEAN-HOC, and ASEAN-HN seem to contribute a higher risk to ASEAN stock system when compared with their overall usual risk. This period coincides with the time of Hamburger financial crisis in USA. During 2009–2016, many emerging stock markets including ASEAN stock markets have experienced great growth after the crisis in 2008 since the Federal Reserve of USA introduced an unconventional Quantitative easing (QE) policy that led to a capital outflow from USA to the emerging markets. However, this large capital brought somewhat unwelcome pressure on stock price and a high volatility in the markets as well. In addition, we observe that CES% is also high in the periods of

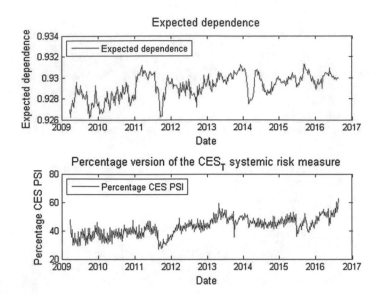

Fig. 7 CES PSI

2012 and 2014 in the cases of SET, IDX, PHI, and STI market indexes. We found
that those two sub-periods are corresponding to the European debt crisis in 2012 and
QE tapering in 2014. If we consider the amplitude of CES% in these two periods;
in the first sub-period, we can see that PSI and IDX contribute the highest risk to
the overall ASEAN stock system while HOC and HN contribute the lowest risk
to ASEAN. In the second sub-period, we also observe that PSI is the highest risk
contributor to the ASEAN stock system with the value of CES% more than 50%.
The further interesting results of HOC and HN are also obtained. The evolution of
CES% in these two markets perform similar level of contribution to the ASEAN
financial risk. This can indicate that Vietnam stock markets seemed not affected by
external factors or they had low interaction with global financial market as well as
the ASEAN. Moreover, we notice that the evolution of CES% of these two countries
took place very often and exhibited very high fluctuation. Consequently, decision
about Vietnams stock regulations has to be made very often.

5 Conclusion

This study aims to assess the risk contribution of seven ASEAN stock markets to the
aggregate ASEAN stock system. It is very important to analyze this issue because
it may have significant implications for the development of ASEAN stock market
and the regulation of the markets and their mechanisms. Thus, the study employed a

Component Expected Shortfall (CES) measure proposed by Banulescu and Dumitrescu [1] as a tool for assessing the contribution of each stock market to the overall risk of the ASEAN stock system. Instead of using the DCC-GARCH model to measure the dynamic correlation, the present study aims to relax the strong assumption of linear and normal correlation by using the copula approach. Thus, the study proposed to employ a Markov Switching copula with time varying parameter as a tool to measure the dependence between individual stock index and ASEAN stock system and the obtained best fit dependence parameters are used to compute the time varying correlation Kendall's tau.

Our findings on the degree of dependence are in line with previous findings in the literature. However, we clearly show that the degree of dependence can vary over time and the regime switching needs to be taken into account. In addition, the time varying risk contribution is considered here. We found that the Philippines stock index contributed the highest risk to the ASEAN stock system. Our results are very important to the policy makers or the regulators of each stock market since they can impose a specific policy to stabilize their stock markets when the financial risk is likely to occur. Moreover, our result will give a benefit to the investors by helping them to invest their money in the appropriate stock market.

Acknowledgements The authors are grateful to Puay Ungphakorn Centre of Excellence in Econometrics, Faculty of Economics. Chiang Mai University for the financial support.

References

1. Banulescu GD, Dumitrescu EI (2013) Which are the SIFI? A component expected shortfall (CES) approach to systemic risk. Preprint, University of Orlans
2. Candido Silva Filho O, Augusto Ziegelmann F (2014) Assessing some stylized facts about finacial market indexes: a Markov copula approach. J Econ Stud 41(2):253–271
3. Chokethaworn K et al (2013) The dependence structure and co-movement toward between thais currency and malaysians currency: markov switching model in dynamic copula approach (MSDC). Procedia Economics and Finance, pp 152–161
4. da Silva Filho OC, Ziegelmann FA, Dueker MJ (2012) Modeling dependence dynamics through copulas with regime switching. Insur Math Econ 50(3):346–356
5. Fei F et al (2013) modeling dependence in cds and equity markets: dynamic copula with markov-switching. Cass Business School
6. Filho OCS, Ziegelmann FA (2014) Assessing some stylized factsabout financial market indexes: a Markov copula approach. J Econ Stud 41:253–271
7. Hamilton JD (1989) A new approach to the economic analysis of nonstationary time series and the business cycle. Econometrica J Econ Soc 57:357–384
8. Lin J, Wu X (2013) Smooth Tests of Copula Specification under General Censorship
9. Liu J et al (2015) Volatility and dependence for systemic risk measurement of the international financial system. In: 4th international symposium on integrated uncertainty in knowledge modelling and decision making, pp 403–414
10. Karimalis EN, Nomikos N (2014) Measuring systemic risk in the European banking sector: A Copula CoVaR approach
11. Kim C-J, Nelson CR (1999) State-space models with regime switching: classical and Gibbs-sampling approaches with applications. MIT Press

12. Pastpipatkul P, Yamaka W, Sriboonchitta S (2016) Analyzing financial risk and co-movement of gold market, and indonesian, philippine, and thailand stock markets: dynamic copula with markov-switching. In: Causal Inference in Econometrics. Springer International Publishing, pp 565–586
13. Pastpipatkul P, Yamaka W, Wiboonpongse A, Sriboonchitta S (2015) Spillovers of quantitative easing on financial markets of Thailand, Indonesia, and the Philippines. In: Integrated Uncertainty in Knowledge Modelling and Decision Making. Springer International Publishing, pp 374–388
14. Patton AJ (2006) Modelling asymmetric exchange rate dependence. Int Econ Rev 47(2):527–556
15. Tofoli PV, Ziegelmann FA, Silva Filho OC (2013) A comparison Study of Copula Models for European Financial Index Return

Estimating Efficiency of Stock Return with Interval Data

**Phachongchit Tibprasorn, Chatchai Khiewngamdee,
Woraphon Yamaka and Songsak Sriboonchitta**

Abstract Existing studies on capital asset pricing model (CAPM) have basically focused on point data which may not concern about the variability and uncertainty in the data. Hence, this paper suggests the approach that gains more efficiency, that is, the interval data in CAPM analysis. The interval data is applied to the copula-based stochastic frontier model to obtain the return efficiency. This approach has proved its efficiency through application in three stock prices: Apple, Facebook and Google.

Keywords Capital asset pricing model · Stochastic frontier · Copula · Interval data

1 Introduction

"If there is only information of the quantities of input and output, and there is no information on input or output prices, then the type of efficiency that can be measured is technical efficiency (TE)" [1]. In the production analysis, TE is a measure of the effectiveness, that is, a given set of inputs is used to produce an output. A firm exists technically efficient if it produces the maximum output with the minimum quantity of inputs such as labour, capital and technology. To measure TE, stochastic frontier approach is considered. It is applied in many studies, in particular the area of the production of agriculture and industry as well as the macroeconomic fields. However, recently, there is a work of Hasan et al. [2] applying the stochastic frontier model (SFM) to the financial area, that is, a stock market. He investigated the TE of selected companies of Bangladesh stock market in the Dhaka Stock Exchange (DSE) market in which the technical inefficiency effects are defined by one of two error component, say Truncated normal or half-normal errors. However, the assumption of the multivariate normal distribution on the joint between the two error components, of the conventional model, is the strong assumption that need to relax. Therefore, the copula function is employed to join the two errors of the financial returns as presented in Tibprasorn et al. [3]. In their work, they employed a copula function to join two error components in the SFM and they found that the price of stocks

P. Tibprasorn (✉) · C. Khiewngamdee · W. Yamaka · S. Sriboonchitta
Faculty of Economics, Chiang Mai University, Chiang Mai, Thailand
e-mail: phachongchit_t@cmu.ac.th

© Springer International Publishing AG 2017   667
V. Kreinovich et al. (eds.), *Robustness in Econometrics*,
Studies in Computational Intelligence 692, DOI 10.1007/978-3-319-50742-2_41

is not over or underestimated in terms of pricing efficiency (measured by technical efficiency). Hence, in this study, we aim to measure the efficiency from the financial market through the capital asset pricing model (CAPM) of Sharpe [4]. The model becomes one of the fundamental tenants in financial theory, and then is applied in various studies to analyze the performance of mutual funds, individual stock returns and other portfolios.

The technique of CAPM is to compare the historical risk-adjusted stock returns (return of stock minus the return of risk-free) with the risk-adjusted market return, and then use a linear regression to fit a straight line. Each point in the line can be represented as the risk-adjusted return of the stock and that of the market return over one time period in the past. Typically, the data used in the CAPM is a single point data, mostly a closing price. However, Neto and Carvalho [5, 6] argued that the implementation of single point data is too restrictive to represent complex data in the real world since it does not take into account the variability and uncertainty inherent to the data. We found that the stock data varies over time and presents in high fluctuation in every single day. Then, there exists the highest and lowest values or the boundary values of an interval [7]. To overcome this problem, the interval data is considered for this study to capture all information set at each time point. The minimum and maximum recorded values will offer a more complete insight about the phenomenon than the point data, i.e. closing price.

In recent years, time-series models such as linear regression, quantile regression, and Markov switching model has been applied to the CAPM to quantify the relationship between the beta risk of stock return and market return. However, few studies have been reported concerning the CAPM using interval data. For instance, the study carried out by Piamsuwannakit et al. [8] applied the concept of the interval-valued data to the CAPM. They found that interval-valued data is more reasonable than the single point through the application of CAPM. Thus, we are going to apply the CAPM in copula based stochastic frontier model to quantify the beta risk and the efficiency of the stock. To the best of our knowledge, the estimation of copula based stochastic frontier model has not been considered yet. This fact becomes one of motivations for this paper.

This paper is structured as follows: the next section introduces the copula-based stochastic frontier model with interval data. Section 3 explains about the estimation of CAPM with copula-based stochastic frontier. The estimation results is presented in Sect. 4 while the conclusion is given in the last section.

2 The Copula-Based Stochastic Frontier Model with Interval Data

In this section, we establish the copula-based stochastic frontier model with interval data by employing the powerful method dealing with the interval data, namely the center method proposed by Billard and Diday [9, 10], to estimate this model.

2.1 Interval Data

In many studies in financial econometrics, they usually use the closing price of a stock one day which is denoted as X^c in order to take into account the estimated results. But in the stock market, stock price is moving vary within the High-low price during the day. Absolutely, the closing price is contained in the range of daily interval stock prices, $P_i = [X^L, \ldots, X^c, \ldots, X^H]$. The closing price could be either the lower or upper bound price. To find the more appropriate value than the closing price to present the best value in the range of P_i, we employ the center method which proposed by Billard and Diday [9, 10] to enhance our estimation and prediction.

According to Billard and Diday [9, 10], the interval-valued data is considered as Symbolic Data Analysis (SDA) which is the extension of classical exploratory data analysis and statistical methods. This type of data appears when the observed values are intervals, namely highest and lowest value, of the set of real number. Thus, this kind of data can be used to explain the uncertainty or variability presented in the data.

2.2 The Model with the Center Method

The stochastic frontier model (SFM) proposed by Aigner et al. [11] is given by

$$Y = f(X; \beta) + (V - U), \tag{1}$$

where Y is an output given set of inputs X; β is a vector of unknown parameters to be estimated; the stochastic noise, V, is the symmetric error and the technical inefficiency, U, is the non-negative error. We assume these two error components to be independent.

Following the center method proposed by Billard and Diday [9, 10], we can write the SFM with interval data as

$$Y^L = f(X^L; \beta^L) + (V^L - U^L) \tag{2}$$
$$Y^H = f(X^H; \beta^H) + (V^H - U^H), \tag{3}$$

where $Y = [Y^L, Y^H]$, $X = [X^L, X^H]$, $V = [V^L, V^H]$ and $U = [U^L, U^H]$. Thus, the vector of parameters based on the center method is following

$$Y^c = f(X^c; \beta^c) + (V^c - U^c), \tag{4}$$

where $Y^c = (Y^L + Y^H)/2$, $X^c = (X^L + X^H)/2$ and superscript c denotes the midpoint value. β^c and the two error components (V^c and U^c) are obtained from the center method.

Following Eq. 4, the technical efficiency based on the center method (TE^c) is estimated by

$$TE^c = \exp(-U^c). \tag{5}$$

However, there may exist the dependency between U^c and V^c. Thus, by Sklar's Theorem, the joint of cumulative distribution function (cdf) of U^c and V^c is

$$H(u^c, v^c) = Prob(U^c \leqslant u^c, V^c \leqslant v^c) \tag{6a}$$

$$= C_\theta(F_{U^c}(u^c), F_{V^c}(v^c)), \tag{6b}$$

where $C_\theta(\cdot, \cdot)$ is denoted as the bi-variate copula with unknown parameter θ. Following Smith [12], transforming (U^c, V^c) to (U^c, ξ^c) as the probability density function (pdf) of (U^c, ξ^c) then we get

$$h(u^c, \varepsilon^c) = f_{U^c}(u^c) f_{V^c}(u^c + \varepsilon^c) c_\theta(F_{U^c}(u^c), F_{V^c}(u^c + \varepsilon^c)), \tag{7}$$

where $f_{U^c}(u^c)$ and $f_{V^c}(v^c)$ are the marginal density of $H(u^c, v^c)$, the composite error $\xi^c = \varepsilon^c(-\infty < \varepsilon^c < \infty)$, $v^c = u^c + \varepsilon^c$ and $c_\theta(\cdot, \cdot)$ is the copula density of $C_\theta(\cdot, \cdot)$. Thus, the pdf of ε^c is obtained by

$$h_\theta(\varepsilon^c) = \int_0^\infty h(u^c, \varepsilon^c) du^c \tag{8a}$$

$$= \int_0^\infty f_{U^c}(u^c) f_{V^c}(u^c + \varepsilon^c) c_\theta(F_{U^c}(u^c), F_{V^c}(u^c + \varepsilon^c)) du^c. \tag{8b}$$

In this study, we employ the maximum simulated likelihood (MSL) technique mentioned in Wiboonpongse [13] to approximate integral of $h_\theta(\varepsilon^c)$. Thus, the pdf of ε^c is given by

$$h_\theta(\varepsilon^c) \approx \frac{1}{J} \sum_{j=1}^{J} \frac{h(u_j^c, \varepsilon_i^c)}{f(u_j^c)}, \tag{9}$$

where $f(u_j^c)$ is the pdf of u_j^c and $j = 1, \ldots, J$ is a sequence of random drawn from the distribution of u_j^c.

Therefore, the likelihood function for copula-based stochastic frontier model is defined by

$$L(\beta^c, \sigma_{v^c}, \sigma_{u^c}, \theta^c) = \prod_{i=1}^{N} h_\theta(y_i^c - x_i^{c'} \beta^c) = \prod_{i=1}^{N} h_\theta(\varepsilon_i^c), \tag{10}$$

where σ_{v^c} and σ_{u^c} are the scale parameters of marginal distribution of V^c and U^c, θ^c is the parameter of copula and $i = 1, \ldots, N$ is the number of observations.

Taking a natural logarithm, the log-likelihood function for copula-based stochastic frontier model becomes

$$\ln L(\beta^c, \sigma_{v^c}, \sigma_{u^c}, \theta^c) = \sum_{i=1}^{N} \ln h_\theta(\varepsilon_i^c) \approx \sum_{i=1}^{N} \ln \frac{1}{J} \sum_{j=1}^{J} \frac{h(u_j^c, \varepsilon_i^c)}{f(u_j^c)}. \tag{11}$$

Following Battese and Coelli [14], the technical efficiency of each copula (TE_θ) can be computed by

$$TE_\theta = E[\exp(-U^c)|\varepsilon^c] \tag{12a}$$

$$= \frac{1}{h_\theta(\varepsilon^c)} \int_0^\infty \exp(-u^c)h(u^c, \varepsilon^c)du^c \tag{12b}$$

$$= \frac{\int_0^\infty \exp(-u^c) f_{U^c}(u^c) f_{V^c}(u^c + \varepsilon^c)c_\theta(F_{U^c}(u^c), F_{V^c}(u^c + \varepsilon^c))du^c}{\int_0^\infty f_{U^c}(u^c) f_{V^c}(u^c + \varepsilon^c)c_\theta(F_{U^c}(u^c), F_{V^c}(u^c + \varepsilon^c))du^c}. \tag{12c}$$

and using Monte Carlo integration. Thus, we obtain

$$TE_\theta = \frac{\sum_{i=1}^{N} \exp(-u_i^c) f_{U^c}(u_i^c) f_{V^c}(u_i^c + \varepsilon_i^c)c_\theta(F_{U^c}(u_i^c), F_{V^c}(u_i^c + \varepsilon_i^c))}{\sum_{i=1}^{N} f_{U^c}(u_i^c) f_{V^c}(u_i^c + \varepsilon_i^c)c_\theta(F_{U^c}(u_i^c), F_{V^c}(u_i^c + \varepsilon_i^c))}. \tag{13}$$

Note that u_i^c follows the cumulative distribution of U^c which is assumed to be distributed as $HN(0, \sigma_{u^c})$. The density function follows

$$f_{U^c}(u^c; \sigma_{u^c}) = \frac{2}{\sqrt{2\pi\sigma_{u^c}^2}} \exp\{-\frac{(u^c)^2}{2\sigma_{u^c}^2}\}. \tag{14}$$

And v_i^c follows the distribution of V^c which is assumed to be distributed as $N(0, \sigma_{v^c}^2)$. The density can be written as following

$$f_{V^c}(v^c; \sigma_{v^c}) = \frac{1}{\sqrt{2\pi o_{v^c}^2}} \exp\{-\frac{(v^c)^2}{2\sigma_{v^c}^2}\}. \tag{15}$$

3 Estimation of CAPM with Copula-Based Stochastic Frontier

To compare the return efficiencies across individual stocks, we start with the CAPM suggested by Jensen [15] as a benchmark for presenting the relationship between the expected excess return (risk premium) and its systematic risk (market risk).

This relationship claims that the expected excess return on any stock is directly proportional to its systematic risk, β. For any stock s the β is calculated by

$$E(R_s) = \beta_s E(R_m), \tag{16}$$

where $E(R_s)$ is the expected excess return on stock s, β_s is its systematic risk and $E(R_m)$ is the expected excess return on market.

In this study, the copula-based stochastic frontier model is proposed to capture the deviation of interval-stock return from its efficient frontier. According to Eq. 4, the model can be represented as

$$E\left(R_{s,t}^c\right) = \beta_{0,s}^c + \beta_{1,s}^c E\left(R_{m,t}^c\right) + V_{s,t}^c - U_{s,t}^c, \tag{17}$$

where $E\left(R_{s,t}^c\right)$ is the midpoint value of the expected excess return on stock s at time t; $E\left(R_{m,t}^c\right)$ is the midpoint value of the expected return on market at time t. In this model, the estimation of intercept and beta risk, $\beta_{0,s}^c$ and $\beta_{1,s}^c$, are based only on the midpoint value of the interval-stock data; $V_{s,t}^c$ reflects the stochastic nature of the frontier itself at time t (e.g., measurement error of stock s) and $U_{s,t}^c$ reflects return inefficiency resulting in a non-negative error term at time t. This value of inefficiency can be defined as return inefficiency which shows the failure to achieve the efficient return as a result of uncontrollable events, such as an asymmetric information.

Ten copulas, consisting Gaussian, Student-t, Clayton, Gumbel, Frank, Joe, rotated Clayton, rotated Gumbel, rotated Joe, and Independence, are considered to capture a degree of the dependence between $V_{s,t}^c$ and $U_{s,t}^c$. We then, maximize the simulated likelihood function (Eq. 11) to obtain all parameters of the model. Therefore, the return efficiency (or RE) of stock s at time t for our model is given by

$$RE_{s,t}^c = E[\exp(-U_{s,t}^c)|\varepsilon^c]. \tag{18}$$

4 Empirical Results

The data contains the prices of Apple Inc. (AAPL), Facebook Inc. (FB) and Alphabet Inc. (GOOGL) in the NASDAQ Stock Market, and we provide the NASDAQ index as a benchmark. The interval-stock return and interval-market return are computed using the weekly high/low price of stock. United States Government 10-Year Bond Yield is used as a risk free rate. The data are weekly data taken from May 2012 to March 2016, and collected by Thomson Reuters. Figure 1 and Table 1 display the summary statistics for the variables.

By applying the model in Sect. 3, the return efficiency of each stock ($RE_{s,t}^c$) is estimated, and the empirical results are shown in Fig. 2 and Table 2. According to the lowest AIC, the best model is the one based on Joe copula for the case of AAPL,

Fig. 1 The weekly
high/low returns of AAPL,
FB, GOOGL and NASDAQ
index

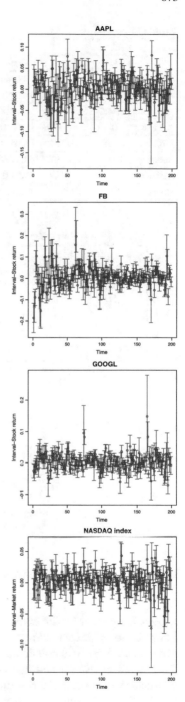

Table 1 Descriptive statistics

	AAPL		FB		GOOGL		NASDAQ	
	High	Low	High	Low	High	Low	High	Low
Mean	0.0485	−0.0319	0.0440	−0.0317	0.0341	−0.0310	0.0249	−0.0221
Median	0.0436	−0.0195	0.0351	−0.0213	0.0295	−0.0218	0.0215	−0.0135
Maximum	0.2483	0.0673	0.3318	0.1224	0.1648	0.0727	0.1160	0.0442
Minimum	−0.0908	−0.2809	−0.1166	−0.2545	−0.0668	−0.2298	−0.0745	−0.2313
Std. dev.	0.0482	0.0569	0.0569	0.0518	0.0397	0.0457	0.0261	0.0358
Skewness	0.4315	−1.5357	1.4497	−1.2913	0.5799	−1.2441	0.4824	−2.0841
Kurtosis	3.9144	6.3449	7.5281	6.7120	3.7086	5.5226	4.8992	10.1115
No. obs.	200	200	200	200	200	200	200	200

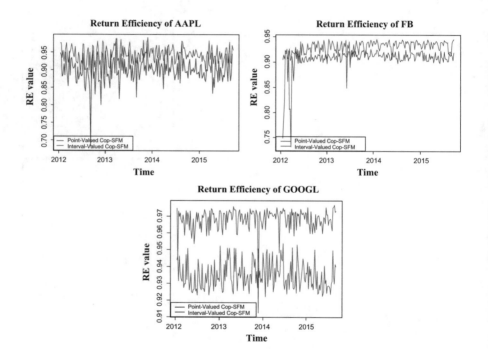

Fig. 2 Return efficiencies of each stock

rotated Joe copula for the case of FB and Clayton copula for the case of GOOGL. Figure 2 shows the estimation of return efficiencies of AAPL, FB and GOOGL stocks using interval data and point data. We attempt to compare the performances between these two models. We find that the movement of return efficiencies using interval data tends to have a similar pattern with the return efficiencies using point data based on this data set. However, the return efficiencies of these three stocks estimated

Table 2 Parameter estimations

Parameter	AAPL	FB	GOOGL
β_o^c	0.0575	0.0793	0.0336
	(0.0045)	(0.0043)	(0.0021)
β_1^c	1.0830	1.2266	0.9716
	(0.0856)	(0.1254)	(0.0756)
σ_v^c	0.0382	0.0634	0.0322
	(0.0855)	(0.0052)	(0.0029)
σ_u^c	0.0736	0.0980	0.0410
	(0.0051)	(0.0050)	(0.0021)
θ^c	5.1092	5.1587	5.1244
	(0.8590)	(0.9008)	(1.2398)
Min $RE_{s,t}^c$	0.8675	0.7320	0.9126
Max $RE_{s,t}^c$	0.9907	0.9452	0.9768
Mean $RE_{s,t}^c$	0.9432	0.9255	0.9683
Name of Copula	Joe	Rotated Joe	Clayton
Log L	474.2569	355.2151	519.6174
AIC	−938.5137	−700.4302	−1029.2350

from interval data are higher than point data, meaning that they can reflect more information in the market. Thus, these results lead to the conclusion that the return efficiencies estimated by the model based on the interval data are mostly higher than the one based on point data. We also conclude that interval data can be used as an alternative data to measure the return efficiency.

According to values as shown in Table 2, the interval-return efficiencies of each stock are 0.8675–0.9907 (average 0.9432), 0.7320–0.9452 (average 0.9255) and 0.9126–0.9768 (average 0.9683) for AAPL, FB and GOOGL, respectively. These suggest that all of three stock returns are quite efficient. We can imply that the stock price nearly reflect all relevant information in the market. However, we find that when the interval return is wider, the return efficiency becomes lower. In other words, the return efficiency tends to be low when the market exists high fluctuation return.

Following the CAPM approach, the stock whose beta less than one is called defensive stock and it is called aggressive stock when the beta is greater than one. Table 2 shows that only the beta of GOOGL is less than one, while the others are greater than one. Thus, only GOOGL is considered to be the defensive stock, but the aggressive stock for AAPL and FB. Clearly, these empirical results suggest that our model can be used as an alternative strategy for selecting stocks into the portfolio. However, we should consider both beta risk and return efficiency in order to select the appropriate stocks.

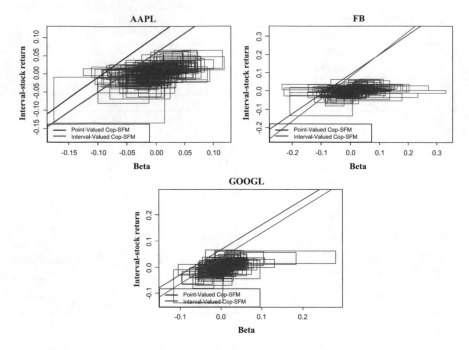

Fig. 3 Security market line obtained using both interval and point data

Additionally, we consider the security market line (SML) as in Fig. 3. The upper line is obtained from the estimation with point data and the lower line is obtained from the estimation with interval data. From these pictures, we observe that the SML from interval data is fitter with the samples than the SML from point data. Therefore, we can conclude that the beta risk (β_1^c) estimated using interval data reflects information in the market better than β_1^c estimated using point data.

5 Conclusions

In the production analysis, technical efficiency (TE) is the effectiveness which a given set of inputs is used to produce an output. The efficiency is obtained if a firm can produce the maximum quantity of output with the minimum of inputs. In this study, the concept of TE is applied to the financial analysis through the capital asset pricing model (CAPM). The study aims to quantify the return efficiency (RE) of the individual stock using TE value in order to assess the efficiency of stock price. To calculate the TE, this paper concerns the uses of interval data rather than the classical type of data, i.e. the point data. We employ the center method of Billard and Diday to find the best value in the range of interval data.

The contribution of this research is twofold. First, the estimation of frontier and efficiency is applied to the CAPM to obtain the beta risk and RE (measured by TE). Second, the interval data is conducted here to capture the variability and uncertainty in the data. Empirical study is also conducted in this paper. Our results show that AAPL, FB and GOOGL stocks are quite efficiency and almost reflect all relevant information in the market. In addition, we observe that when the interval return is wider, the RE becomes lower. In other words, the RE tends to be low when the market exists high fluctuation return. Thus, the investors should take into account the interval range of stock return prior to the selection of appropriate stock for their portfolios.

Finally, to assess the performance of interval data in our CAPM analysis, we conduct the interval data to measure the beta risk (β) and RE. Then, we compare it with the β and the RE obtained from the model based on point data. The results lead to the conclusion that the β obtained from interval data is more likely to represent information, in terms of maximum and minimum price at each time point. The results also illustrate that the RE obtained from the interval data is higher than the one that obtained from point data. Additionally, the security market line (SML) shows that the line from interval data is fitter with the samples than the line from point data, and hence, the RE obtained from point data is less accurate.

Acknowledgements We are grateful for financial support from Puey Ungpakorn Center of Excellence in Econometrics, Faculty of Economics, Chiang Mai University.

References

1. Kokkinou A (2009) Stochastic frontier analysis: empirical evidence on Greek productivity. In: 4th Hellenic observatory PhD symposium on contemporary Greece & Cyprus, LSE, London
2. Hasan MZ, Kamil AA, Mustafa A, Baten MA (2012) Stochastic frontier model approach for measuring stock market efficiency with different distributions. PloS One 7(5):e37047
3. Tibprasorn P, Autchariyapanitkul K, Chaniam S, Sriboonchitta S (2015) A Copula-based stochastic frontier model for financial pricing. In: International symposium on integrated uncertainty in knowledge modeling and decision making. Springer International Publishing, Cham, pp 151–162
4. Sharpe WF (1964) Capital asset prices: a theory of market equilibrium under conditions of risk. J Finan 19:425–442
5. Neto EDAL, de Carvalho FDA (2008) Centre and range method for fitting a linear regression model to symbolic interval data. Comput Stat Data Anal 52(3):1500–1515
6. Neto EDAL, de Carvalho FDA (2010) Constrained linear regression models for symbolic interval-valued variables. Comput Stat Data Anal 54(2):333–347
7. Rodrigues PM, Salish N (2015) Modeling and forecasting interval time series with threshold models. Adv Data Anal Classif 9(1):41–57
8. Piamsuwannakit S, Autchariyapanitkul K, Sriboonchitta S, Ouncharoen R (2015) Capital asset pricing model with interval data. In: International symposium on integrated uncertainty in knowledge modeling and decision making. Springer International Publishing, Cham, pp 163–170

9. Billard L, Diday, E (2000) Regression analysis for interval-valued data. Data analysis, clas-
    sification and related methods. In: Proceedings of the seventh conference of the international
    federation of classification societies (IFCS00). Springer, Belgium, pp 369–374
10. Billard L, Diday E (2002) Symbolic regression analysis. Classification, clustering and data
    analysis. In: Proceedings of the eighth conference of the international federation of classification
    societies (IFCS02), Springer, Poland, pp 281–288
11. Aigner D, Lovell K, Schmidt P (1977) Formulation and estimation of stochastic frontier func-
    tion models. J Econ 6(1):21–37
12. Smith MD (2004) Stochastic frontier models with correlated error components. The University
    of Sydney
13. Wiboonpongse A, Liu J, Sriboonchitta S, Denoeux T (2015) Modeling dependence between
    error components of the stochastic frontier model using Copula: application to inter crop coffee
    production in Northern Thailand frontier. Int J Approx Reason 65:33–34
14. Battese GE, Coelli TJ (1988) Prediction of firm-level technical efficiencies with a generalized
    frontier production function and panel data. J Econ 38(3):387–399
15. Jensen MC (1968) The performance of mutual funds in the period 1945–1964. J Finan
    23(2):389–416

The Impact of Extreme Events on Portfolio in Financial Risk Management

K. Chuangchid, K. Autchariyapanitkul and S. Sriboonchitta

Abstract We use the concept of copula and extreme value theory to evaluate the impact of extreme events such as flooding, nuclear disaster, etc. on the industry index portfolio. A t copulas based on GARCH model is applied to explain a portfolio risk management with high-dimensional asset allocation. Finally, we calculate the condition Value-at-Risk (CVaR) with the hypothesis of t joint distribution to construct the potential frontier of the portfolio during the times of crisis.

Keywords Extreme value theory · GARCH · Multivariate t copula · CVaR · Expected shortfall

1 Introduction

Large and unpredictable financial fluctuation associated risk for investment. The credit crisis and natural disasters cause massive losses for investors. For example, Insurance company have to reserve enough of money for their claims when the catastrophically large losses occur. The problem of understanding economic fluctuations is essential. The economic theories which treat "bubbles and crashes" as outliers to described variations which do not agree with existing theories. There are no outliers; there is a power law! The power law can model enormous and unpredictable changes in financial markets.

The main issue in portfolio optimization where the asset allocation is described by random variables. In this case, the choices of the potential portfolio rely on the underlying assumption on a behavior of the asset and the selection of the measure of

K. Chuangchid · K. Autchariyapanitkul (✉)
Faculty of Economics, Maejo University, Chiang Mai 50290, Thailand
e-mail: Kittawit_a@mju.ac.th

K. Chuangchid
e-mail: kaekanta@hotmail.com

S. Sriboonchitta
Faculty of Economics, Chiang Mai University, Chiang Mai 52000, Thailand
e-mail: songsakecon@gmail.com

© Springer International Publishing AG 2017   679
V. Kreinovich et al. (eds.), *Robustness in Econometrics*,
Studies in Computational Intelligence 692, DOI 10.1007/978-3-319-50742-2_42

risk. However, the correlation between asset returns is entirely explained by the linear equation, and the efficient portfolios are usually described by the conventional mean and variance model as we can found in Markowitz [1]. In particular, a correlation provides a knowledge of dependence structure of random variables in the linear situation, but it may be inappropriate for the financial analysis (see, Autchariyapanitkul et al. [2–4]). The dependency between the main factors in the portfolio has to be considered. An incorrect model may cause the loss on portfolio and miss-specification to evaluate the liability. There are many papers showed the superiority of copula to model dependence. The reason they are not comfortable to use correlation approach because of its failure to capture the tails dependency (see, Artzner et al. [9] and Szegö [10]) and extreme events (see, Longin and Solnik [11], Hartmann et al. [12]). Copulas can be easily show multivariate distributions and offer much more flexibility than the conventional one (see, Autchariyapanitkul et al. [2, 3], Kiatmanaroch [5], Kreinovich [6, 7], Sirisrisakulchai [8]).

Based on the studied from Harvey and Siddique [13], considered multivariate Generalized AutoRegressive Conditional Heteroskedasticity (GARCH) model with skewness. The same as in Chiang et al. [14], created a dynamic portfolio of crude oil, soybean, and corn by ARJI and GARCH models to calculate the value at risk(VaR). This model is used to explain the time-varying conditional correlation, but they can not perform asymmetry in asymptotic tail dependence. Thus, we introduce an optional method to formulate the relationship between a multivariate data, beyond assuming any restriction in marginal distributions by using a copula theory.

The need of extreme value theory (EVT) was introduced in this study to analyze the impact of the extreme events. (i.e., Global financial crisis in 2007–2008, Flooding in Thailand 2011, Fukushima Daiichi nuclear disaster in 2011, Quantitative Easing (QE) in 2013, etc.) On the portfolio, asset returns over the past years. Our primary concerns were the appearance of significant large values of X in data sets. These large values are called extremes (or extreme events) which appear "under of the tail" of the distribution F of X. Thus, we are concerned mainly with modeling the tail of F, and not the whole distribution F. For example, Wang et al. [15] applied the method of GRACH-EVT-Copula model to studied the risk of foreign exchange portfolio, the results suggested that t copula and Claton copula are well explained for the structural relations among the portfolios. The same way as in Singh et al. [16] and Allen et al. [17] used EVT to measure the market risk in the $S\&P500$ and ASX-All Ordinaries stock markets.

In this article, we are using a multivariate t copula which is applied to portfolio optimization in financial risk management. In general, multivariate t copula is the widely applied in the context of modeling multivariate financial analysis, and show the superior to the normal copula (see, Romano [18], Chan and Kroese [19]). Similarly, the works from Kole et al. [20] provided the test of fit to a selection of the right copula for an asset portfolio, the results clearly showed that t copulas are better than Gumbel and Gaussian copulas. Thus, t copulas may be considered for measuring the risk of portfolio investment.

This study focused on industry index of the Stock Exchange of Thailand (SET). We used the concept of EVT and copula to measure the risk of multi-dimensional industry index portfolios. Thus, the contributes to this study can be summarized in two folds. First, we emphasize that the multivariate t copula illustrates the asymmetric dependence structure and evaluates the complex nonlinear relations among financial portfolios. Second, we use the n-dimensional of industry index with the EVT to show the significant impact of shocks to the returns of the portfolio.

The remainder of this article is arranged according to the following topic: Sect. 2 gives the theoretical background of GARCH model and extreme value theory, while the empirical results in Sect. 3 and the final section gives concluding remarks.

2 Theoretical Background

2.1 GARCH

GARCH model was introduced by Bollerslev (1986), which can relax an assumption that volatility is a constant over time because GARCH can be captured the characteristics of financial time series data (heteroscedasticity and volatility). If the data has a skewness or heavy tail, We can choose an innovation that supports this information. Then, ARMA(p,q) and GARCH(k,l) are defined by

$$r_t = \mu + \sum_{i=1}^{p} \phi_i r_{t-i} + \sum_{i=1}^{q} \psi_i \varepsilon_{t-i} + \varepsilon_t$$

$$\varepsilon_t = \sigma_t \cdot v_t$$

$$\sigma_t^2 = \omega + \sum_{i=1}^{k} \alpha_i \varepsilon_{i-t}^2 + \sum_{i=1}^{l} \beta_i \sigma_{t-i}^2$$

where $\sum_{i=1}^{n} \phi_i < 1$, $\omega > 0$, $\alpha_i, \omega_i \geq 0$ and $\sum_{i=1}^{k} \alpha_i + \sum_{i=1}^{l} \beta_i \leq 1$, v_t is an standardized residual of a chosen innovation. In this case, we used t distribution because the data was considered as a heavy tail distribution, which well defined for the financial time series data.

2.2 Extreme Value Theory (EVT)

The methods of EVT including Block Maxima model and Peaks over Threshold model (POT). Suppose we use POT approach to modeling extremes, namely regarding as extremes those observations which are above some high threshold u. Then we will use the parametric form of Pareto distribution for tail estimation. We take those which are above u. Note that, while "high" threshold level u makes the approximation of F_u by a generalized Pareto distribution (GPD) more reasonable, it will make estimators more volatile since there will be fewer observations among the i.i.d. Sample $X_1, \ldots, X_n$ drawn from X to use.

Here we fit the Pareto model (with support $1 + \dfrac{\gamma x}{\beta} > 0$)

$$\mathscr{P}_{\gamma, \beta}(x) = 1 - (1 + \frac{\gamma x}{\beta})^{-\frac{1}{\gamma}}, \; for \; \gamma > 0 \tag{1}$$

using the sample $Y_j = X_{ij} - u$, where $X_{ij} > u$, $j = 1, 2, \ldots, k(u)$. The log-likelihood is

$$\log L(\gamma, \beta) = -k(u) \log \beta - (1 + \frac{1}{\gamma}) \sum_{i=1}^{k}(u) \log(1 + \frac{\gamma Y_i}{\beta}), \tag{2}$$

From which MLE (γ, β) of (γ, β) can be obtained.

Our distribution F is heave-tailed, i.e. $\bar{F}(x) = L(x) x^{-\frac{1}{\gamma}}$. Not only γ is estimated, we can also estimate the "tail" of F, i.e., $\bar{F}(x)$ for $x > u$ (considered as "where the tail begins"). Indeed, from

$$F_u(y) = (F(u + y) - F(u))/(1 - F(u)) \tag{3}$$

we set $x = u + y > u$, then

$$1 - F_u(x - u) = \bar{F}_u(x - u) = 1 - (F(x) - F(u))/(1 - F(u)) \tag{4a}$$
$$= (1 - F(x))/(1 - F(u)) = \bar{F}(x)/\bar{F}(u) \tag{4b}$$

so that

$$\bar{F}(x) = \bar{F}(u)(x - u) \tag{5}$$

and we estimate $\bar{F}(x)$ by the estimates of $\bar{F}(u)$ and $\bar{F}_u(x - u)$ as follows. Since u is taken larege enough, $F_u(x - u)$ is approximate by $1 - (1 + \dfrac{\gamma(x - u)}{\beta})^{-\frac{1}{\gamma}}$, so that an estimate of $\bar{F}_u(x - u)$ is $(1 + \frac{\hat{\gamma}(x-u)}{\hat{\beta}})^{-\frac{1}{\gamma}}$. On the other hand, $\bar{F}(u)$ is estimates by $\dfrac{1}{n} \sum_{j=1}^{n} 1_{(X_j > u)}$. Thus, for $x > u$, $\bar{F}(x)$ is estimated by

$$[\frac{1}{n} \sum_{j=1}^{n} 1_{(X_j > u)}][(1 + \frac{\hat{\gamma}(x - u)}{\hat{\beta}})^{-\frac{1}{\gamma}}] \tag{6}$$

or approximately, for $x > u$

$$F(x) = 1 - [\frac{1}{n} \sum_{j=1}^{n} 1_{(X_j > u)}][(1 + \frac{\hat{\gamma}(x - u)}{\hat{\beta}})^{-\frac{1}{\gamma}}] \tag{7}$$

from which we can extrapolate beyond the available data, i.e. estimate $P(X > x)$ for x greater than the maximum order statistic $x_{(n)}$ in the data set, namely using

$$\left[\frac{1}{n}\sum_{j=1}^{n}1_{(X_j>u)}\right]\left[\left(1+\frac{\hat{\gamma}(x-u)}{\hat{\beta}}\right)^{-\frac{1}{\hat{\gamma}}}\right] \tag{8}$$

Considering $x > x_{(n)}$ means that we would like to know whether $X > x$ in the future. Now we cannot estimate $P(X > x)$ correctly when $x > x_{(n)}$ (x is out of range of the data). Thus, we can use the empirical distribution F_n of F, according to the order statistics $X_{(1)\le X_{(2)}\cdots\le X_{(n)}}$, since $P(X > x)$ will be estimated by $1 - F_n(x) = 0$. Using the above estimation procedure for heavy-tailed distributions, we can, therefore, deal with potentially catastrophic events which have not yet occurred.

On the estimation of $VaR_p(X)$, here is what we could do. From the estimated distribution of F for $x > u$,

$$F(x) = 1 - \left[\frac{1}{n}\sum_{j-1}^{n}1_{(X_j>u)}\right]\left[\left(1+\frac{\hat{\gamma}(x-u)}{\hat{\beta}}\right)^{-\frac{1}{\hat{\gamma}}}\right] \tag{9}$$

We estimate is large quantiles p (i.e., for $p > F(u)$) by inverting it (solving $F(VaR_p(X)) = p$), yielding

$$VaR_p(X) = u + \frac{\hat{\beta}}{\hat{\gamma}}\left([\frac{n}{\sum_{j=1}^{n}1_{(X_j>u)}}(1-p)]^{-\gamma} - 1\right) \tag{10}$$

2.3 GARCH-EVT Model

To eliminate heteroskedasticity, first, we fit the historical index returns by using GRACH(1,1) and obtained the innovations of the process. Usually, these innovations are assumed to be a normal distribution. However, these errors tend to be a heavier tail rather than the normal one. Then, we used the GPD assumption to described the innovation behavior, let z_i represent residual in the upper or lower tail and the empirical distribution. Thus, the marginal distribution of error is

$$F(z_i) = \begin{cases} \dfrac{n_{u_i^L}}{n}\left(1+\gamma_i^L\dfrac{u_i^L - z_i}{\beta_i^L}\right)^{-\frac{1}{\gamma_i^L}} & : z_i < u_i^L \\ \varphi(z_i) & : u_i^L < z_i < u_i^R \\ 1 - \dfrac{n_{u_i^R}}{n}\left(1+\gamma_i^R\dfrac{z_i - u_i^R}{\beta_i^R}\right)^{-\frac{1}{\gamma_i^R}} & : z_i > u_i^R, \end{cases} \tag{11}$$

where β is scale parameter, γ is shape parameter and u_i^L, u_i^R are lower and upper threshold respectively. $\varphi(z_i)$ is the empirical distribution, n is the number of z_i. While, $n_{u_i^L}$ is the number of innovations with the value less than u_i^L and $n_{u_i^R}$ is the number of innovations with the value greater than u_i^R.

An appropriate choice for u could be selected by using the mean excess function $e(u) = E(X - u | X > u)$. indeed, for u sufficiently large, $F_u(x)$ is approximately the generalized Pareto distribution (GPD) $\mathscr{P}_{\gamma,\beta}(x) = 1 - (1 + \frac{\gamma x}{\beta})^{-\frac{1}{\gamma}}$. Now, if X is distributed as $\mathscr{P}_{\gamma,\beta}(x)$, then for $u < x_F$,

$$e(u) = \frac{\beta + \gamma u}{1 - \gamma}, \tag{12}$$

for $\beta + \gamma u > 0$. Note that we need $\gamma < 1$, i.e., the heavy-tailed distribution must have at least a finite mean.

The above linearity of the mean excess function forms the basis for deciding the threshold based on its empirical counter-part $e_n(u)$, where

$$e_n(u) = \sum_{i=1}^{n} (X_i - u) 1_{(X_i > u)} / \sum_{i=1}^{n} 1_{(X_i > u)} \tag{13}$$

Suppose the observations support a GPD model over a high threshold then the plot $u, e_n(u)$ should become increasingly linear for higher values of u. We plot u against $e_n(u)$ and pick out u where there is linearity.

3   Applications to the Real World Data

In this study, The data attain of 2421 daily logarithm returns during 2004–2013 are obtained from DataStream. There are Agro & Food Industry (AGRO), Consumer Products (CONS), Financials (FIN), Industrials (INDUS), Property & Construction (PROP), Resources (RESO), Services (SERV), and Technology (TECH). Table 1 gives a summary statistics.

Formulating the tails of a distribution with GPD needs the data to be i.i.d. For example, the sample auto-correlation function (ACF) of the industry index returns related with the AGRO index reveal some weak serial correlation. However, the ACF of the squared returns shows the degree of persistence in variance and suggests that GARCH modeling may vary significantly in the tail. We use AGRO index as a sample, the results are as the following (Fig. 1).

Table 1 Summary statistics

	AGRO	CONS	FIN	INDUS	PROP	RESO	SERV	TECH
Mean	0.0005	0.0001	0.0002	0.0001	0.0001	0.0002	0.0004	0.0002
Median	0.0008	0.0003	.0001	0.0003	0.0005	0.0002	0.0004	0.0004
Max.	0.063	0.068	0.107	0.082	0.100	0.126	0.081	0.129
Min.	−0.094	−0.066	−0.193	−0.128	−0.155	−0.172	−0.112	−0.208
SD.	0.0111	0.0069	0.0171	0.0165	0.0150	0.0179	0.0118	0.0168
Skew.	−0.596	−0.659	−0.760	−0.412	−0.702	−0.477	−0.957	−0.713
Kurt.	8.704	18.187	14.475	8.266	11.348	11.516	11.843	15.911
Obs.	2421	2421	2421	2421	2421	2421	2421	2421

All values are the log returns

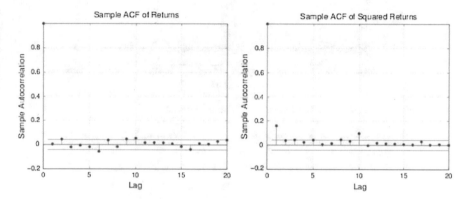

Fig. 1 ACF of returns and ACF of squared returns

Moreover, the standardized residuals of each industry index are modeled as a standardized t-distribution to satisfy for the fat tails often related to index returns. For AGRO index, the lower graph clearly exhibits the heteroskedasticity present in the filtered residuals. These figures show the underlying zero mean, unit variance and i.i.d (Fig. 2).

Comparing the auto-correlation function of standardized residuals (ACFs) to the squared ACFs of the index returns exhibits that the standardized residuals are approximately i.i.d as show in the lower graphs (Fig. 3).

We calculate the empirical CDF of each industry index by using Gaussian kernel. We obtain the distribution by applying EVT to residuals in each tail. Then, we use the maximum likelihood to estimate a parametric GDP for those extreme residuals which exceed the threshold. The Figures 4 and 5 show the empirical CDF and upper tail of standardized residuals, respectively, these are useful to investigate the GPD fit in more detail. So, the GPD model is a good fit to the data (Table 2).

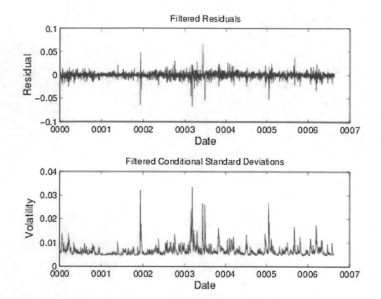

Fig. 2 Filtered residuals and filtered conditional standard deviations

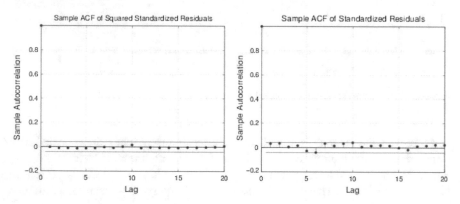

Fig. 3 ACFs of residuals and the squared ACFs

Table 3 shows the multivariate t copula parameters. We can construct efficient portfolio by using this matrix.

Resulting in the t copulas parameters, given the equally weight the Table 4 exhibits the value of VaR and CVaR at levels of 10, 5 and 1%. We emphasize that the estimated CVaR converges to, -0.1406, -0.0976 and -0.0788 at 1, 5 and 10% levels in period $t + 1$, respectively (Fig. 6).

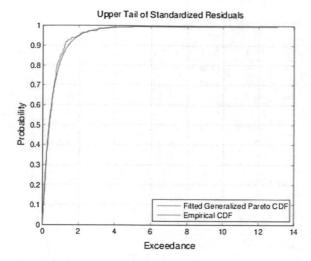

Fig. 4 The empirical CDF

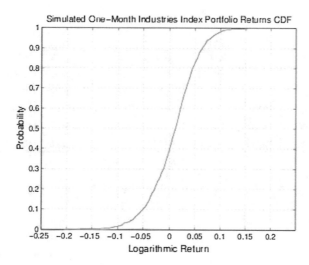

Fig. 5 Upper tail of standardized residuals

Table 2 Estimations of AR-GARCH parameters

	AGRO	CONS	FIN	INDUS	PROP	RESO	SERV	TECH
C	0.0007	0.0003	0.0003	0.0004	0.0006	0.0004	0.0010	0.0004
	(0.0002)	(0.0001)	(0.0003)	(0.0003)	(0.0003)	(0.0002)	(0.0001)	(0.0002)
AR(1)	0.0358	0.0001	0.0514	0.0664	0.0800	0.0271	0.0580	0.0037
	(0.0210)	(0.0209)	(0.0212)	(0.0210)	(0.0213)	(0.0211)	(0.0213)	(0.0211)
K	0.0001	0.0001	0.0001	0.0001	0.0001	0.0001	0.0010	0.0001
	(0.0001)	(0.0001)	(0.0001)	(0.0001)	(0.0001)	(0.0001)	(0.0001)	(0.0001)
GARCH(1)	0.8306	0.7008	0.8469	0.8458	0.8332	0.8455	0.8143	0.7035
	(0.0202)	(0.0380)	(0.0208)	(0.0173)	(0.0191)	(0.0194)	(0.0222)	(0.0463)
ARCH(1)	0.0982	0.1495	0.0457	0.0970	0.0707	0.0850	0.0848	0.0638
	(0.0220)	(0.0378)	(0.0151)	(0.0189)	(0.0186)	(0.0197)	(0.0232)	(0.0269)
Leverage(1)	0.0842	0.0768	0.1264	0.0732	0.1145	0.0836	0.0952	0.1535
	(0.0294)	(0.0454)	(0.0275)	(0.0247)	(0.0256)	(0.0275)	(0.0275)	(0.0411)
DoF	5.2104	3.8256	7.2344	6.6058	7.1788	7.0096	6.4982	6.1534
	(0.4876)	(0.2898)	(0.6636)	(0.6904)	(0.6791)	(0.6579)	(0.6387)	(0.5603)
LL	6725.9574	7814.3908	6758.9113	6840.9501	7115.0829	6676.7602	7723.2792	6725.9574

() standard error is in parenthesis, C and K are constant terms

Table 3 Correlation matrix of t copula

	AGRO	CONS	FIN	INDUS	PROP	RESO	SERV	TECH
AGRO	1	0.0258	0.0067	0.0395	0.0109	0.0036	0.0664	0.0174
CONS	0.0258	1	0.2569	0.1985	0.0335	0.2039	0.2397	0.1148
FIN	0.0067	0.2569	1	0.4483	0.0476	0.6603	0.4917	0.2075
INDUS	0.0395	0.1985	0.4483	1	0.0433	0.4652	0.4394	0.1760
PROP	0.0109	0.0335	0.0476	0.0433	1	0.0306	0.2483	0.2329
RESO	0.0036	0.2039	0.6603	0.4652	0.0306	1	0.4334	0.1602
SERV	0.0664	0.2397	0.4917	0.4394	0.2483	0.4334	1	0.1741
TECH	0.0174	0.1148	0.2075	0.1760	0.2329	0.1602	0.1741	1

DoF = 13.9664

Table 4 Expected shortfall of asset returns for an equal weight

Level of significant	Expected returns	VaR	CVaR
10%	0.2164	−0.0501	−0.0788
5%	0.2164	−0.0693	−0.0976
1%	0.2164	−0.1108	−0.1406

Fig. 6 A CDF of simulated 22 trading days portflolio returns

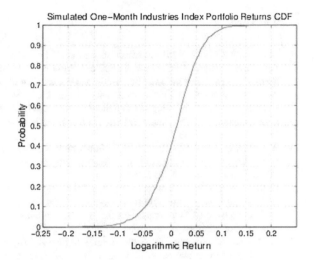

Simulated One–Month Industries Index Portfolio Returns CDF

Another scenario illustrates the Monte Carlo simulation to generate a set of 20,000 samples. Given the level of significant, at 5%, we are using the mean-CVaR to obtained the efficient frontier and the optimal weight of the assets in portfolio given the ES. Finally, we obtained the simulated returns of each industry index, report of 21.6436 maximum gain and 20.4451 maximum loss, as well with the VaR and CVaR at various confidence levels.

4 Concluding Remarks

In this article, we illustrate the complexity of extreme value theory and t copulas to construct the efficient portfolio in the situation of extreme events have occurred. We are using CVaR and mean-CVaR to attain the effect of extreme events on the portfolio management. We applied the multivariate t copula to formulated the relationship between industry index of Stock Exchange of Thailand. We can summarize our study in two folds. First, given the extreme value and t- copulas methods to modeling the dependence between tails of distributions and to estimating CVaR. Second, eight different industry index n-dimensional portfolio have been examined. The results show that filtered historical simulation, extreme value, and t-copulas approaches are the most appropriate for risk-management techniques in the times of crisis.

Acknowledgements We are grateful to Prof. Dr. Hung T. Nguyen for his constructive comments and suggestions.

References

1. Markowitz H (1952) Portfolio selection. J Finan 7(1):77–91
2. Autchariyapanitkul K, Chainam S, Sriboonchitta S (2014) Portfolio optimization of stock returns in high-dimensions: a copula-based approach. Thai J Math 11–23
3. Autchariyapanitkul K, Piamsuwannakit S, Chanaim S, Sriboonchitta S (2015) Optimizing stock returns portfolio using the dependence structure between capital asset pricing models: a vine copula-based approach. In: Causal inference in econometrics. Springer International Publishing, pp 319–331
4. Autchariyapanitkul K, Chanaim S, Sriboonchitta S (2015) Quantile regression under asymmetric Laplace distribution in capital asset pricing model. In: Econometrics of risk. Springer International Publishing, pp 219–231
5. Kiatmanaroch T, Puarattanaarunkorn O, Autchariyapanitkul K, Sriboonchitta S (2015) Volatility linkages between price returns of crude oil and crude palm oil in the ASEAN region: a copula based GARCH approach. In: Integrated uncertainty in knowledge modelling and decision making. Springer International Publishing, pp 428–439
6. Kreinovich V, Nguyen HT, Sriboonchitta S (2013) Why clayton and gumbel copulas: a symmetry-based explanation. In: Uncertainty analysis in econometrics with applications, vol 200. Springer, Heidelberg, pp 79–90
7. Kreinovich V, Nguyen HT, Sriboonchitta S (2014) How to detect linear dependence on the copula level? In: Modeling dependence in econometrics, vol 251. Springer, pp 63–79
8. Sirisrisakulchai J, Sriboonchitta S (2014) Modeling dependence of accident-related outcomes using pair copula constructions for discrete data. In: Modeling dependence in econometrics, vol 251. Springer International Publishing, pp 215–228
9. Artzner P, Delbaen F, Eber J-M, Heath D (1999) Coherent measures of risk. Math Finan 9:203–228
10. Szegö G (2005) Measures of risk. Eur J Oper Res 163(1):5–19
11. Longin F, Solnik B (2001) Extreme correlation of international equity markets. J Finan 56(2):649–676
12. Hartmann P, Straetmans S, De Vries CG (2004) Asset market linkages in crisis periods. Rev Econ Stat 86(1):313–326
13. Harvey CR, Siddique A (1999) Autoregressive conditional skewness. J Finan Quant Anal 34(04):465–487
14. Chiang TC, Zheng D (2010) An empirical analysis of herd behavior in global stock markets. J Banking Finan 34(8):1911–1921
15. Wang ZR, Chen XH, Jin YB, Zhou YJ (2010) Estimating risk of foreign exchange portfolio: using VaR and CVaR based on GARCHEVT-Copula model. Phy A Stat Mech Appl 389(21):4918–4928
16. Singh AK, Allen DE, Robert PJ (2013) Extreme market risk and extreme value theory. Math Comput Simul 94:310–328
17. Allen DE, Singh AK, Powell RJ (2013) EVT and tail-risk modelling: evidence from market indices and volatility series. North Am J Econ Finan 26:355–369
18. Romano C (2002) Applying copula function to risk management, Capitalia, Italy. http://www.icer.it/workshop/Romano.pdf
19. Chan JC, Kroese DP (2010) Efficient estimation of large portfolio loss probabilities in t-copula models. Eur. J. Oper. Res. 205(2):361–367
20. Kole E, Koedijk K, Verbeek M (2007) Selecting copulas for risk management. J Bank Finance 31(8):2405–2423

Foreign Direct Investment, Exports and Economic Growth in ASEAN Region: Empirical Analysis from Panel Data

Pheara Pheang, Jianxu Liu, Jirakom Sirisrisakulchai,
Chukiat Chaiboonsri and Songsak Sriboonchitta

Abstract The major purpose of this research study is twofold. Firstly, to examine the causal relationship among foreign direct investment (FDI), exports, and economic growth of ASEAN economy comprising Cambodia, Lao PDR, Malaysia, Philippines, Singapore, Thailand, and Vietnam, by using panel VECM covering from 2000 to 2014. Secondly, to estimate the impact of FDI and exports on ASEAN economy. The dummy variable representing the financial crisis in 2008 is used to see the real effect in this study. The empirical results indicate that bidirectional causal relation between economic growth and exports is found in ASEAN association while there are two unidirectional causal linkages between FDI-economic growth and FDI-exports as the causal direction running from FDI to economic growth and running from FDI to exports in ASEAN economy. Based on the findings from panel dynamic ordinary least square (DOLS) and fully modified ordinary least square (FMOLS) methods, the elasticity of GDP with respect to FDI is 0.048 and 0.044% and respect to exports is 0.547 and 0.578%. Therefore, it can be concluded that FDI and exports are significant aspects which positively impact on ASEAN economic development.

Keywords FDI · Exports · Economic growth · VECM · Panel data

1 Introduction

In recent decades of ASEAN, foreign direct investment (FDI) and international trade (importantly exports) seem to correspond progressively to the regional development since ASEAN was created and integration policies were formed. Possibly, this creation and policies make the ASEAN region become broader liberalization through tariff reduction policies, trade facilitation programs, and greater business opportunities, etc. However, these judgments were not proved evidently whether and how these three economic terms of ASEAN region relate with each other. In ASEAN region, each country has a distinctive history of policy formation appearing to have

P. Pheang · J. Liu (✉) · J. Sirisrisakulchai · C. Chaiboonsri · S. Sriboonchitta
Faculty of Economics, Chiang Mai University, Chiang Mai, Thailand
e-mail: liujianxu1984@163.com

© Springer International Publishing AG 2017

691

V. Kreinovich et al. (eds.), *Robustness in Econometrics*,
Studies in Computational Intelligence 692, DOI 10.1007/978-3-319-50742-2_43

different system in economic practice. Following the diversion, all members seem to involve in the liberalization structure, and this ASEAN cooperation creates the opportunities to lure the FDI with the good options for multinational companies to take advantages. The cooperation of country developmental diversity not only attract FDI, but it also enhances the integration of production network and trade within the region. Also, it facilitates technology transfer to face the challenge and integration with the global economy through trade liberalization. In 1997, ASEAN was effected by Asian crisis but had rebounded later. Unfortunately, the financial crisis took place again in 2008 but ASEAN can be recuperated and grew rapidly after this critical era.

Furthermore, the agreement at Bali ASEAN summit enhances the outward orientation for productivity growth and technology transfer, and this orientation is also appealed via the promotion of FDI. Related to the challenges, ASEAN Economic Community (AEC) Blueprint prescribes many implementations of trade facilitation programs and agreements to satisfy the standardize trade. Due to these implementations, ASEAN has liberalized the trade and alters the economic structure to adapt with ASEAN's aim of purposes. Most of investors in the greater economy countries of ASEAN may consider to locate new production entities in ASEAN nearby lower economy countries according to lower labor costs and opportunities to expand the new markets and production networks. These can be a factor of investment growth in ASEAN and also increasing in export flows to supply more to the world. In last decades, the rapid growth of foreign trade in ASEAN is noted as the exports has risen from \$750.9 billion in 2006 to \$1,292.4 billion in 2014 even facing of the financial crisis in 2008, and it seems less suffered of the crisis on ASEAN's exports [1]. Comparing to the exports of other regions such as United State, China, Europe and Japan, ASEAN ranked in the top among the regions while ASEAN reached \$55.15 billion in 2015.

Based on the above description, it is intuitively sure that FDI and exports relate strongly with the economic growth of ASEAN and there are many facilitation policies to stimulate these economic terms. With the developing momentum of ASEAN currently, it seems to be almost no confirmation concerning the relations among these major economic terms in the ASEAN. Hence, the purpose of this research is to examine the relationship among FDI, exports, and economic growth in ASEAN region over 2000–2014 with the panel data approach.

This research will contribute to two aspects for scholars and policy makers: First, as most of the previous research studies discovered only the bivariate relationship among FDI, exports, and economic growth. This study is vital to understand the channeling effects of the three variables via panel data approach. Second, this paper focuses on the ASEAN region – region of emerging economy. Also, no previous empirical analysis studied on the causal and impact relationship among FDI, exports, and economic growth in ASEAN region. Thus, this research can be used to fill the gaps. The results of this paper are useful for researchers and policy makers of ASEAN to identify the major economic matters for development effectively and efficiently, and design the correct principles to enlarge the economic size of ASEAN. Moreover, the findings can be used as the based figures for comparing the development change in the next 10 or 15 years after AEC launching at the end of 2015.

The rest of the paper is organized as follows. Section 2 provides a brief review of the empirical literature on the relationship among FDI, exports and economic growth. Section 3 presents the data source and description. Section 4 reports the methodology and empirical results. Section 5 gives the concluding remarks.

2   Literature Review

In the recent literature, many researchers chose to work on FDI, exports, and economic growth on their studies while the three variables are being the most interesting economic terms for development. Yet, those studies are mostly examined in bi-variate relationship. Few studies were conducted by using the multivariate relationship among the three variables, and fewer were done in panel analysis.

In the study of FDI and economic growth linkage, Abbes et al. [2] found that FDI had a unidirectional causality on economic growth for 65 countries during 1980–2010. Similarly, Pegka [3] publicized that the hypothesis of long run association between FDI and economic growth in Eurozone countries during 2002–2012 was supported. As well as a study of Gui-Diby [4] displayed that FDI inflow was definitely a significant impact of economic growth in the African during 1980–1994 with the addition that the impact of FDI on economic growth was negative in the period 1980–1994 but the impact was positive for 1995–2009.

In the exports and economic growth nexus, Shihab et al. [5] defined that the causal effect was obtained from GDP to exports in Jordan over 2000–2012. Mehdi and Zaroki [6] presented that the export increment delivered the positive effect on the growth of Irans economy for 1961–2006. Abbas [7] revealed only production growth was able to magnify the export size in Pakistan during 1975–2010.

In the FDI and exports nexus, the findings of Sultan [8] stipulated that the long run dynamic relation of FDI and exports was found with one-way granger causality direction running from exports to FDI in India covering 1980–2010. Rahmaddi and Ichihashi [9] elucidated that FDI inflow significantly completed the exports of manufacturing sectors in Indonesia during 1990–2008, but all manufacturing sectors were obtained FDI - exports complementary, the FDI effects on exports varied between sectors based on the effect of other related aspects. Dritsaki and Dritsaki [10] concluded that there was the presence of cointegration and dual directional granger causal relation among FDI and exports for EU members during 1995–2010.

Moreover, there are several papers performed the relation of the multi-variables but those studies only identified on one way direction or one endogenous variable while other papers proposed only time series approach. There was hardly ever analyzed papers using panel approach. Those papers of the trivariate relationship in time series method are raised in the review. Obviously, Szkorupov [11] illustrated that there were the findings of causal connection among GDP, exports, and FDI while the exports and FDI increment positively impacted the economic size in Slovakia over 2001–2010. Keho [12] presented the conclusion with the mixed results of the study as bidirectional and unidirectional causation was found in various countries in African area for the period of 1970–2013. Saba-Sabir [13] found dual directional causal linkage between FDI and exports, and unilateral causality direction running from GDP to FDI and exports without causal relation from FDI or exports to GDP in Pakistan over 1970–2012.

The few studies of multivariate relationship in panel data approach in recent literature are also described in the reviews. Likewise, Sothan [14] concluded that the hypothesis of bidirectional causality between FDI and economic growth and between exports and economic growth was accepted in 21 Asian countries during 1980–2013. Mehrara et al. [15] concluded the study on the developing country during 1980–2008 into three parts such as (1) the presence of bidirectional causation between FDI and GDP was found. (2) the presence of unidirectional causation between exports and GDP with the direction from exports to GDP. (3) no any granger causality running from FDI or GDP to exports in both short and long run. Moreover, Stamatiou and Dritsakis [16] confirmed the presence of the bilateral causality between exports and GDP and the presences of unilateral causality from GDP to employment and from exports to FDI for new EU members over 1995–2013.

As above description, it can be informed that least papers conducted researched studies on the tri-variate relationship among FDI, exports, and economic growth in panel data approach in the recent literature. Also, there is almost never a study of these selected variables in ASEAN region. Most of their analyses indicate that FDI and exports have positive effects on economic growth and the causality analyses reveal varied results: univariate, bivariate, or no causal relation. The different results may depend on the different methodology, different treatment of variables (real or nominal), different studied period, country, and presence of other related variables.

3   Data Source and Description

The annual data of FDI, exports, and economic growth of ASEAN-7 countries namely Cambodia, Lao PDR, Malaysia, Philippines, Singapore, Thailand and Vietnam during 2000–2014 are used in US million dollars and obtained from World Indicators (WDI) published online by the World Bank. The description for the three variables of ASEAN economy are summarized in the scatter graph of Fig. 1.

Through the scatter graph of ASEAN-7 countries, it reports that the trend of FDI and exports respected to economic growth are upward, then it means that both FDI and exports have positive relationships with economic growth of ASEAN. Moreover, the moving trend of exports looks higher than the trend of FDI, thus the gradual rising of economic growth via the larger exports is greater than the larger FDI.

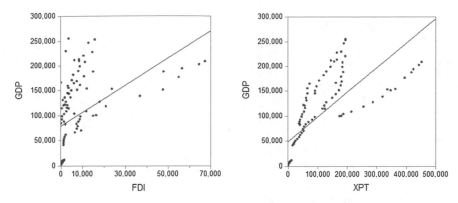

Fig. 1 Descriptive Scatter Graph of ASEAN-7 countries (in US $Million)

4 Methodology and Empirical Results

To reach the objective of the study on the relationship among FDI, exports, and economic growth in ASEAN, it is required to satisfy the four main stages of econometric
approach.

4.1 Panel Unit Root Test

First, panel unit root test is used to verify the property of the data and verify the
integration order to avoid spurious estimation. There are two main processes of
panel unit root test: Common unit root processes and Individual unit root processes.
To measure the stationarity of the series precisely, the panel unit root method is
going to be conducted by using one common unit root test method (Levin-Lin-Chu
(LLC)) and one individual unit root test (Im, Pesaran, Shin (IPS)) which both are the
latest developed methods among each process. LLC test is based on the ADF test
by assuming homogeneity in the dynamics of the autoregressive coefficient for all
the time series across the cross section units [17]. IPS test basically follows the ADF
test by allowing the heterogeneity in the dynamic of the autoregressive coefficient
for all the time series across the cross section units [18].

   Based on the Schwarz Info Criterion (SIC), the stationary test of LLC and IPS
with the existence of intercept and trend in the equation are presented in Table 1.
The results are shown in the test of both the level series and first difference series.
Both LLC and IPS display different diagnostic results among LGDP, LFDI, LXPT
as there are mixed results of LGDP in both test types, stationary in LLC test but
non-stationary in IPS test. Yet, both tests provide the same results of LFDI and
LXPT cases, as LFDI in both tests are stationary at the 1% level and LXPT are non-
stationary. Further, all series of both tests result all stationary at the 1% significant

Table 1 Panel unit root test results

Variables	Level series		First difference series	
	LLC	IPS	LLC	IPS
LGDP	−2.824***	−0.381	−6.358***	−3.225***
	(0.0024)	(0.3512)	(0.0000)	(0.0006)
LFDI	−5.108***	−4.148***	−10.863***	−7.494***
	(0.0000)	(0.0000)	(0.0000)	(0.0000)
LXPT	−0.562	0.182	−6.307***	−4.187***
	(0.286)	(0.572)	(0.0000)	(0.0000)

Notes:
- The null Hypothesis: Series has a unit root
- The values in parentheses refer to p. value
- (*, **, and ***) denote rejection of the null hypothesis at the 1%, 5%, and 10% level of significance, respectively.

level in the first difference. Hence, it indicates that the series are non-stationary and not cointegrated at I (0), but the series are stationary at I (1) in the panel ASEAN-7.

4.2 Panel Johansen Co-integration Test

While the variables are stationary at I (1), these series are probably cointegrated in order I (1) and the cointegration among FDI, exports, and economic growth are possibly found. Then, the panel cointegration test is also used to confirm the presence of the cointegration of the three variables. Johansen Fisher panel cointegration test proposed by Madala and Wu (1999) is applied to test the null hypothesis of no long run relationship among variables. This test is based on the unrestricted cointegration rank test (trace and maximum eigenvalue) [19]. Table 2 presents the results of panel cointegration test. The cointegrating relations are significant in both trace and maximum eigenvalue methods. The presence of long run cointegration is found at the 5% significant level even using with only intercept or with both intercept and trend in the equation. Therefore, this panel cointegration test concludes that there exists 3 cointegration vectors among the triple variables in the sample of ASEAN-7.

4.3 Panel Long Run Elasticity Test

As the cointegrated variables are given, the impact study of FDI and exports on economic growth is added to examine the long run elasticity in ASEAN-7 economy. The OLS is the most biased and inconsistent analysis for pertaining the panel cointegration method [3]. Then, panel dynamic ordinary least square methods (DOLS) by Kao and Chiang (1999) and fully modified ordinary least square methods (FMOLS) by Pedroni (2000) are the better methods to mitigate the faintness. The DOLS is a

Table 2 Panel Co-integration Test

Johansen-Fisher panel Co-integration test

Individual intercepts

Variables	Trace test	P.value	Max-Eigenvalue test	P.value
None	83.810***	(0.0000)	61.720***	(0.0000)
At most 1	37.690***	(0.0006)	27.600**	(0.0161)
At most 2	33.060***	(0.0028)	33.060***	(0.0028)

Individual intercepts

Variables	Trace test	P.value	Max-Eigenvalue test	P.value
None	115.500***	(0.0000)	98.660***	(0.0000)
At most 1	37.990***	(0.0005)	23.290*	(0.0558)
At most 2	26.900**	(0.0199)	26.900**	(0.0199)

Notes:
• The null Hypothesis: There is no co-integration among the variables
• The values in parentheses refer to p. value
• (***, **, and *) denote rejection of the null hypothesis at the 1%, 5%, and 10% level of significance, respectively.

Table 3 Panel DOLS and FMOLS estimates

Dependent variable: LGDP	DOLS	FMOLS
Independent variables	Coef.	Coef.
LFDI	0.048**	0.044**
	(0.0141)	(0.0151)
LXPT	0.547**	0.578**
	(0.0314)	(0.0434)
Adjusted R-squared	0.9987	0.9981
S.E of Regression	0.0607	0.0707

Notes:
• The null hypothesis: There is no long run equilibrium co-integration within the variables
• The values in parentheses refer to standard error
• (***, **, and *) denote rejection of the null hypothesis at the 1%, 5%, and 10% level of significance, respectively.

parametric normal distribution estimation augmented by particular lead and lag of explanatory term to remove the endogenous effects and serial correlation and the FMOLS is a non-parametric normal distribution process which is capable to deal with serial correlation problems [20].

From Table 3 presenting the panel DOLS and FMOLS estimates, it can be summarized that the coefficients of inward FDI and exports in the long run are positively and statistically significant at most 5% level.

In the DOLS, the long run elasticity of inward FDI and exports on GDP are found 0.048 and 0.547, respectively. This indicates that one percent increases in FDI will enhance economic development for panel of ASEAN-7 countries by about 0.048%, and one percent grows in exports will boost the economic growth for panel data of ASEAN-7 countries by about 0.547%. Similarly, through the FMOLS, the elasticity of FDI inflow and exports with respect to GDP are respectively resulted 0.044 and 0.578. It represents that one percent increases in FDI will foster the economic advancement for the panel of ASEAN-7 countries by about 0.044%, and one percent increases in exports will induce economic progress for the panel of ASEAN-7 countries by 0.578%. Hence, the findings illustrate that both FDI and exports are significantly crucial for the development of ASEAN economy, and the findings also confirm that the impact of FDI and exports on the economic growth via the DOLS and FMOLS are not far different in the studied period.

4.4 Panel Estimation

Also, panel estimation conducted with fixed effects and random effects is taken into account. This panel estimation will allow the readers to discern the impact comparison between the fixed effects and random effects method and which method is the most fitted estimation in this study. The equation of panel estimation with both effects methods can be written as below:

$$LGDP_{i,t} = \beta_0 + \beta_1 LFDI_{i,t} + \beta_2 LXPT_{i,t} + \gamma_i + \lambda_t + \varepsilon_{i,t}, \tag{1}$$

where β_0 is a constant intercept. β_1 and β_2 are constant slope coefficients of FDI and exports respectively, γ_i is the country specific effect. λ_t represents time specific effect, $\varepsilon_{i,t}$ is residual terms.

Fixed effects method assumes that the certain unobserved variables of each specific country (country specific effect) are fixed over time and may affect the economic growth by correlating with the exogenous variables in the equation [21]. with the existence of time specific effect, the country specific effect in fixed effects method may provide the coefficients unbiased and consistent estimations. Random effects method assumes that the country specific effect is random and not correlated with the exogenous variables in the equation. Then, all the country specific intercept terms are random across the cross section units. For finding the most fitted estimation, the correlated random effects method – Hausman test – is used to answer this enquiry. Two specifications are arranged in this study, specification 1 involves the estimations without including the dummy variable, and the specification 2 involves the estimation taken with the dummy variable. Both specifications and methods are estimated consistently as the estimation are corrected for both cross sectional heteroskedasticity and serial correlation (cross sectional clustered) by robust coefficient covariance method: White period [22]. Table 4 reports the results of panel estimation using fixed effects and random effects method.

Referring to the specification 1, the Hausman test statistic is not significant enough to reject the null hypothesis of no correlation between individual effects and other

Table 4 Panel estimation with robustness standard error using fixed and random effects

Dependent variable: LGDP	Specification 1		Specification 2	
	Fixed effects	Random effects	Fixed effects	Random effects
C	5.722***	5.563***	6.542***	6.295***
	(0.5200)	(0.6782)	(0.4503)	(0.6999)***
LFDI	0.031***	0.029***	0.029***	0.027***
	(0.0083)	(0.0074)	(0.0049)	(0.0045)
LXPT	0.463***	0.479***	0.378***	0.403***
	(0.0485)	(0.0511)	(0.0417)	(0.0472)
DUM			0.186***	0.174***
			(0.0368)	(0.0361)
Hausman test				
Chi-Sq.	0.0000	Random	57.2290	Fixed
P.value	1.0000	Effects	0.0000	Effects
Adjusted R-Squared	0.9983	0.7963	0.9991	0.9011
S.E of regression	0.0639	0.0584	0.0473	0.0501

Notes:
- The null hypothesis: There is no long run equilibrium co-integration with the variables
- The values in parentheses refer to standard error
- Robustness Coefficient Co-variance: White Period
- (***, **, and *) denote rejection of the null hypothesis at the 1%, 5%, and 10% level of significance, respectively.

regressors at the 10% level. Then, it defines that random effects method is more appropriate in this impact study, and the outcomes of the Hausman test indicate that the long run coefficients of all ASEAN-7 countries are not equaled or heterogeneous. Through the results, FDI and exports are significant at the 5% level in both methods and contain positive effects supporting to the theory. This means that both FDI and exports positively impact the economic growth of ASEAN. The impact coefficient of FDI and exports on economic growth are found 0.029 and 0.479, respectively. It represents that an increasing 1% in FDI makes the economic growth of ASEAN by about 0.029% while an increasing 1% in exports fosters the economic growth of ASEAN by about 0.479%. For specification 2 which includes the presence of dummy variable, fixed effects method seems to become more appropriate than random effects as the statistic of Hausman test is significantly rejected the null hypothesis at the 1% level, and the coefficient of FDI and exports of specification 2 found 0.028% and 0.378%, respectively are lower than specification 1, but the significant level looks more robust. Moreover, coefficient of dummy variable is oddly found positive number, this reveals that dummy variable seems to provide better condition for economic development of ASEAN. The coefficient is about 0.18 in both methods and significant at the 1% level. It shows that the world financial crisis in 2008 does not affect negatively to the ASEAN region, as some of ASEAN countries, especially the least developing countries may find more opportunities to improve their economic conditions. Then, the post world financial crisis results average increase of ASEAN economic growth by 0.18%. Probably, it seems that the destination for FDI is transferred to emerging regions and leads the growth of FDI magnitude in ASEAN while the economic situation in the European Union (EU) is not convenient for investors since the financial crisis. During the financial crisis, the decline in the share of FDI might result from the lack of confidence of EU investors in their own regional market and transfer from EU countries to the world outside EU (European commission, 2012). According to Dr. Sandra Seno-Alday, a researcher at the Sydney Southeast Asia Centre (SSEAC), the behavior of the Southeast Asian region following the 2008 crisis was opposite to what it was in Asian crisis in internet bubble in 1997 and 2000. After 2008, the rest of the world faced serious economic recession while Southeast Asia's regional economy grew slowly but notably did not decline. The reason of this positive effects of the crisis is revealed that Southeast Asia region has a strong export orientation. Therefore, fostering closer international trade involves the advance of flexible, internationally-oriented industries and greater robustness against the shock. Totally, the results provide a supporting to the theoretical enlightenment mentioning FDI and exports are crucial components of economic development, and respect to the previous research literature.

4.5   VECM Granger Causality Test

The above tests confirm the long run cointegration among the variables, but the existence of cointegration does not illustrate the direction of causality, then granger causality analysis is used to examine the causal direction of the triple variables. This granger causality may require employing panel vector error correction model (VECM)(Granger 1988) to recognize the cointegration equations and the estimated coefficients in the long run and short run causality. In VECM, the long run dynamic relations will be revised by short run relations which is obscured by the disturbances in long run equilibrium. The VECM equation of the study is specified as follows:

$$\Delta LGDP_{i,t} = \alpha_{1i} + \beta_{1i} ECT_{i,t-1} + \sum_{k=1}^{q} \delta_{1i,k} \Delta LGDP_{i,t-k} + \sum_{k=1}^{q} \gamma_{1i,k} \Delta LFDI_{i,t-k}$$

$$+ \sum_{k=1}^{q} \theta_{1i,k} \Delta LXPT_{i,t-k} + \varepsilon_{1i,t}, \tag{2}$$

$$\Delta LFDI_{i,t} = \alpha_{2i} + \beta_{2i} ECT_{i,t-1} + \sum_{k=1}^{q} \delta_{2i,k} \Delta LFDI_{i,t-k} + \sum_{k=1}^{q} \gamma_{2i,k} \Delta LGDP_{i,t-k}$$

$$+ \sum_{k=1}^{q} \theta_{2i,k} \Delta LXPT_{i,t-k} + \varepsilon_{2i,t}, \tag{3}$$

$$\Delta LXPT_{i,t} = \alpha_{3i} + \beta_{3i}ECT_{i,t-1} + \sum_{k=1}^{q} \delta_{3i,k}\Delta LXPT_{i,t-k} + \sum_{k=1}^{q} \gamma_{3i,k}\Delta LGDP_{i,t-k}$$

$$+ \sum_{k=1}^{q} \theta_{3i,k}\Delta LFDI_{i,t-k} + \varepsilon_{3i,t}, \tag{4}$$

where $\Delta LGDP_{i,t}$, $\Delta LFDI_{i,t}$, $\Delta LXPT_{i,t}$ are the first difference of natural logarithmic gross domestic product (GDP), foreign direct investment (FDI, and exports (XPT), respectively. $\alpha_{ni}(n = 1, 2, 3)$ is cross country effect. k ($k = 1, 2, ...q$)) is the optimal lag order determined by the Schwarz Information Criterion (SIC). β_{ni} ($n = 1, 2, 3$) is the coefficient of variables in one period lagged (speed of adjustment). $ECT_{i,t-1}$ is the estimated Error Correction Term (ECT) in one lag period generated from the long run cointegration form. $\varepsilon_{ni,t}$ ($n = 1, 2, 3$) is the error term with zero mean.

To proceed VECM, ECT stationarity and lag length order are needed to check [23]. The results of ECT stationary test illustrate that all ECTs are stationary at the 1% significant level in every method. Next, the lag order selection is conducted via VAR model to find the optimal lag length in the estimation. The selection is explored with the maximum lag 4. Based on the results of lag order selection in Table 5, the results confirm lag order 1 and 2 are the optimal lag lengths for granger causality of ASEAN. Nonetheless, three of six lag selection methods (LR, FPE, and AIC) confirm lag 2 while LogL method is unavailable within maximum lag 4, then the granger causality estimated in this study is determined by optimal lag length $p = 2$.

Subsequently, VECM for causality analysis is officially tested. In the estimating process of VECM (2), the dummy variable of the financial crisis effect during 2008 is involved by setting 0 for the period of 2000–2007 and 1 for 2008–2014. Table 6 reports the findings of both long run and short run granger causality. The long run granger causality is shown by the significant t-statistic of lagged ECT, and the short run granger causal direction is expressed by wald coefficient test. The results provide the similar causality relations among variables in short run and long run. For long

Table 5 Lag order selection

Lags	LogL	LR	FPE	AIC	SC	HQ
1	190.127	NA	1.82E-06	–4.705	–4.431*	–4.595*
2	201.141	20.311*	1.73E-06*	–4.757*	–4.209	–4.538
3	209.628	14.990	1.75E-06	–4.744	–3.922	–4.415
4	214.836	8.792	1.94E-06	–4.645	–3.549	–4.207

Notes:
* indicates lag order selected by the criterion
LR: sequential modified LR test statistic (each test at 5% level)
FPE: Final prediction error
AIC: Akaike information criterion
SC: Schwarz information criterion
HQ: Hannan-Quinn information criterion

Table 6 Panel granger causality - VECM (2, Dummy)

Dependent variables	Independent variables (Sources of causation)				
	Long run	Short run			Crisis
	ECT_{t-1}	$\triangle$LGDP	$\triangle$LFDI	$\triangle$LXPT	Dummy
$\triangle$LGDP	-0.002***		4.6987*	4.2972^w	-0.024***
	(0.0037)		(0.0954)	(0.1166)	(0.0002)
$\triangle$LFDI	0.024	0.5185		3.1127	-0.195
	(0.2791)	(0.7716)		(0.1948)	(0.3621)
$\triangle$LXPT	-0.007***	4.6235*	11.4426***		-0.078***
	(0.0009)	(0.0991)	(0.0033)		(0.0001)

Notes:
- Wald statistic reports with short run changes in the independent variables
- T-statistic of lagged ECT reports long run relations
- The values in parentheses refer to p-value
- (***, **, *, and ^{w}) denote rejection of the null hypothesis at the 1%, 5%, 10% and 15% level of significance, respectively.

run, there exists bilateral causality between exports and economic growth, and also two unilateral causality running in the direction from FDI to economic growth and exports. Evidently, these long run causal relationships are found at the 1% level of significance. For the short run relationship (see Fig. 2), exports and economic growth have an interacting causal relation with each other as the findings in the study confirm the presence of bidirectional causal relationship between exports and GDP which is significant at most 15% level. On the other hand, FDI has causal connections at the 5% and 10% level of significance with economic growth and exports respectively as only the unidirectional causality running from FDI to GDP and from FDI to exports are found. This evidence reveals that FDI and exports play an important role in economic booster of ASEAN. However, FDI is the major economic term to be more concentrated as FDI is not only an element of promoting ASEAN's economic development directly but also diffusely via the growth of export volumes. Additionally, the growth of domestic products and the larger amount of foreign capital are the two major factors in enhancing export volumes of ASEAN. Not to mention, the coefficients of dummy variable are shown in negative values and significant at the 1% level, excluding only FDI which is displayed in negative value but insignificant. These values prove that the financial crisis in 2008 may affect negatively on the causation of the economic growth and exports in ASEAN region.

Through the reinforcing effects of FDI inflow, the policy for economic development in open economy should put priority on FDI as FDI is the crucial source for economic and export growth. Then, ASEAN countries should improve the infrastructure, human capital and technology progress to attract investors and exist higher productions to enlarge export volume and use exports and FDI to induce the growth.

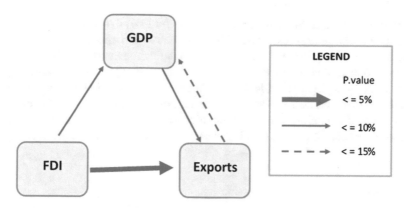

Fig. 2 Granger causality relation of ASEAN regions

5 Concluding Remarks

The objective of this paper is to discover the multivariate relationship between FDI, exports, and economic growth in ASEAN region. To reach the objective, the diagnostic check for variable stationarity is conducted via LLC and IPS tests. Then, panel cointegration test based on trace and maximum eigenvalue is used to examine the long run cointegrating relationship among the variables. Lately, the VECM granger causality test is applied to determine the causal direction in ASEAN region.

The empirical findings illustrate that all variable series are cointegrated at I (1) confirmed by LLC and IPS tests and lead the three variables to be cointegrated, then VECM condition is fulfilled to use for causality study through the results of panel Johansen cointegration test and error correction term. Following the existence of cointegration of FDI, exports, and economic growth, the DOLS and FMOLS estimation methods result positive impact of FDI and exports on economic growth by 0.048% and 0.547%, respectively in the DOLS and by 0.044% and 0.578%, correspondingly in the FMOLS. Moreover, the comparing between fixed and random effects for the best fitted method is performed with the presence of financial crisis, and fixed effects method is reported as the suitable method for this study.

For the empirical results of granger causality analysis, the causal relations of triple variables are displayed in both short run and long run. In long run, it is found that both FDI and exports generate the economic advancement while FDI and GDP also enhance the exports. Through the short run causality results, the reinforcing effect of FDI spillovers should be concentrated on. For our policy recommendation, we suggest a policy design for luring FDI which may induce the growth in economy and complement export orientation as the exports at least jointly promote economic growth at the weak 15% level. Exports can be an economic term to impact stronger on economic development and to prevent the ASEAN from the crises as seen in the previous history of the financial crisis in 2008. Probably, the technological progress, human capital, financial development, and structural change should also be considered, as these components are vital sources for FDI advancement and the development for developing countries of ASEAN. In conclusion, the causality relationship studied by combining data for the 7 countries illustrates an interesting and reasonable causality connections among FDI, exports, and economic growth of ASEAN region and can be used as a general rule of economic policy for ASEAN. Finally, the future research studies can be focused on the development change of ASEAN economy with these major economic terms or included other effecting factors in the next 10 or 15 years because there can be a big change after launching AEC process.

References

1. Yearbook AS (2015) ASEAN Statistical Yearbook 2014. ASEAN Publications, Jakarta
2. Abbes SM, Mostfa B, Seghir G, Zakarya GY (2015) Causal interactions between FDI, and economic growth: evidence from dynamic panel co-integration. Procedia Econ. Finan. 23:276–290
3. Pegkas P (2015) The impact of FDI on economic growth in Eurozone countries. J. Econ. Asymmetries 12(2):124–132
4. Gui-Diby SL (2014) Impact of foreign direct investments on economic growth in Africa: evidence from three decades of panel data analyses. Res. Econ. 68:248–256
5. Shihab RA, Soufan T, Abdul-Khaliq S (2014) The causal relationship between exports and economic growth in Jordan. Int. J. Bus. Soc. Sci. 5(3):147–156
6. Mehdi S, Zaroki S (2012) The study examining the effect of export growth on economic growth in Iran. Editorial Note Words Board Editor 2 Profile Authors Included Number 3 Inf. Contributors 5(1):21
7. Abbas S (2012) Causality between exports and economic growth: investigating suitable trade policy for Pakistan. Eurasian J. Bus. Econ. 5(10):91–98
8. Sultan ZA (2013) A causal relationship between FDI inflows and export: the case of India. J. Econ. Sustain. Dev. 4(2):1–9
9. Rahmaddi R, Ichihashi M (2012) The impact of foreign direct investment on host country exports: Sector based evidence from Indonesian manufacturing. Hiroshima University, Graduate School for International Development and Cooperation (IDEC)
10. Dritsaki C, Dritsaki M (2012) Exports and FDI: a Granger causality analysis in a heterogeneous panel. Econ. Bull. 32(4):3128–3139
11. Szkorupov Z (2014) A causal relationship between foreign direct investment, economic growth and export for Slovakia. Procedia Econ. Finan. 15:123–128
12. Keho Y (2015) Foreign direct investment, exports and economic growth: some African evidence. J. Appl. Econ. Bus. Res. 5(4):209–219
13. Saba-Sabir MS (2015) Causality relationship among foreign direct investments, gross domestic product and exports for Pakistan. Eur. J. Bus. Manag. 7(13):8
14. Sothan S (2015) Foreign direct investment, exports, and long-run economic growth in Asia: panel cointegration and causality analysis. Int. J. Econ. Finan. 8(1):26
15. Mehrara M, Haghnejad A, Dehnavi J, Meybodi FJ (2014) Dynamic causal relationships among GDP, exports, and foreign direct investment (FDI) in the developing countries. Int. Lett. Soc. Humanistic Sci. 3:1–19
16. Stamatiou P, Dritsakis N (2015) Granger Causality Relationship between Foreign Direct Investments, Exports, Unemployment and Economic Growth. A Panel Data Approach for the New EU Members
17. Levin A, Lin C-F, Chu C-SJ (2002) Unit root tests in panel data: asymptotic and finite-sample properties. J. Econ. 108(1):1–24

18. Im KS, Pesaran MH, Shin Y (2003) Testing for unit roots in heterogeneous panels. J. Econ. 115(1):53–74
19. Maddala GS, Wu S (1999) A comparative study of unit root tests with panel data and a new simple test. Oxf. Bull. Econ. Stat. 61(S1):631–652
20. Saikkonen P (1991) Asymptotically efficient estimation of cointegration regressions. Econometric Theor. 7(01):1–21
21. Hsiao FS, Hsiao M-CW (2006) FDI, exports, and GDP in East and Southeast Asia-panel data versus time-series causality analyses. J. Asian Econ. 17(6):1082–1106
22. IHS global Inc. "Eviews 9 Help" ISBN:978-1-880411-09-4
23. Hamit-Haggar M (2012) Greenhouse gas emissions, energy consumption and economic growth: a panel cointegration analysis from Canadian industrial sector perspective. Energy Econ. 34(1):358–364

Printed in the United States
By Bookmasters